# Extensions and Relaxations

Mathematics and Its Applications



Volume 542

# Extensions and Relaxations

*by*

A.G. Chentsov

and

S.I. Morina

*Institute of Mathematics and Mechanics,*
*Ural Branch of Russian Academy of Sciences,*
*Ekaterinburg, Russia*

KLUWER ACADEMIC PUBLISHERS
DORDRECHT / BOSTON / LONDON

A C.I.P. Catalogue record for this book is available from the Library of Congress.

ISBN 978-90-481-6001-3

---

Published by Kluwer Academic Publishers,
P.O. Box 17, 3300 AA Dordrecht, The Netherlands.

Sold and distributed in North, Central and South America
by Kluwer Academic Publishers,
101 Philip Drive, Norwell, MA 02061, U.S.A.

In all other countries, sold and distributed
by Kluwer Academic Publishers,
P.O. Box 322, 3300 AH Dordrecht, The Netherlands.

*Printed on acid-free paper*

# Contents

# Preface

Questions concerning the application of various extensions and relaxations in extremal problems and problems of attainability under constraints are considered in this monograph. It continues a large series of the authors' publications in which a quite general approach to the investigation of the whole class of problems of the above type has been formed. These problems have no stability property in the traditional sense. However, the specific character of perturbations of a system of constraints (relaxation of conditions) makes the instability arising under relaxation, to a certain extent, useful, because the real possibility of attaining that or another aim is essentially extended. But there arises the question of 'true' possibilities taking into account the effect of the relaxation of a system of constraints. This question is of interest not only from the theoretical point of view but it is important for practice. In particular, it is significant in engineering investigations connected with constructing concrete systems of control. It is known that in many control problems arising in practice, there is a necessity of realizing some hypothetical regime with a high degree of accuracy. At the same time, tools, which are used for this goal and are natural from the engineering point of view, are unfit for the setting of the corresponding mathematical control problem in 'closed' form.

For example, the so called relay controlled functions, which mathematicians should define as piecewise constant and continuous from the right or left depending on the particularity of problems, are widely used in practice. Such controls correspond to the nature of the engineering setting oriented towards solving concrete problems by means of accessible technics. At the same time, in nonlinear control problems with so called geometric constraints, which were first investigated systematically by Pontryagin, it is often required to realize the sliding mode along some surfaces (the known 'bang bang' principle). The natural question of interpreting the above sliding mode (which is not realized by means of relay controls) stays into the background if we know the way to construct the real trajectory closed to the required sliding regime. However, the latter is often unknown, and, at the same time, is ideal in its essence. We will be in a difficult situation if we restrict ourselves to only real trajectories and, hence, controls (or the others which are suitable in the sense of technical realization). Since in this case, to a considerable extent, we do not have enough perfect tools in present day mathematics. For example, if, in addition, it is required to optimize some quality index of a technical system, then it may be that the extremum is not attained. However, this does not reduce the importance of the problem, but we can lose the possibility of applying necessary conditions of optimality because there is no object to which they can be applied. At the same time, an effective enumeration of realized controls is impossible in practice. It is worth noting that we must admit errors in the sliding mode for relay controls that can not be foreseen beforehand. In a word, we must review the problem of setting itself. In this sense, we should recall the useful advice for practices in Chapter III of Warga's monograph [117]. It seems that the notion of control should be properly improved itself at the stage of mathematical investigation of the engineering problem. Warga's monograph (and a series of other works) gives a good example in this connection. As a matter of fact, it turns out well to model generalized 'control' for the above mentioned problems.

As a result, a harmonic mathematical theory was founded, elements of which came from the theory of optimal control. One of the most important and graceful points of the above theory are the constructions of metrizable compactifications of sets in functional spaces of the usual controls in the presence of some connections corresponding to substantive sense of the problem. We note that this constructions are useful for the theory of Pontryagin's maximum principle, in which formulation of the setting provided *a priori* with the existence theorem for the optimal control was of great importance. In this connection the investigations of Gamkrelidze should be mentioned, and, in particular, his monograph [78] in which the above extension construction was used in the optimal time problem, which is very important for theory and practice. In other control problems (with other constraints) there arise difficulties under realization of analogues of the above mentioned scheme.

For example, under the extension of control problems with impulse constraints there arises the effect connected with the product of a discontinuous function and a generalized function. Here we find great difficulties in constructing a rigorous theory. At the same time it turns out well to formalize generalized controls in many settings of problems of control by linear systems, and hence the realization of the compactification with connections is possible as a matter of principle; however, non-metrizable topologies can be required in this case. In particular, the topology of pointwise convergence may be more natural for equipping the space of trajectories. In connection with the use of generalized controls in linear systems with impulse constraints, we note the remarkable monograph of Krasovskii *The theory of motion control* which also contains other graceful constructions of contemporary control theory. This monograph and the related circle of Krasovskii's works gave an impulse to many subsequent investigations. Natural difficulties were also marked.

We will not discuss details but note that when solving control problems the basic constructions must be carried over to the level of so called generalized problems in which a rather intricate technique from other fields of mathematics would be applied; in particular, topology and measure theory. It is quite natural that these two disciplines are reflected in the details in Warga's monograph, and they form a great part of it. From our point of view this turned out useful for sections of 'pure' mathematics that define quality solution of the above mentioned generalized problems. In particular, there arise an additional stimulus for investigation of spaces of measures and some corresponding topological equipment. Some investigations of such a kind are presented in this work and in the previous monographs [32, 35] of one of the authors. We note that the abstract settings are motivated by concrete practical problems but are not reduced to them. Namely, extension constructions are useful for some problems arising in 'pure' mathematics and, in particular, in measure theory. In turn, the latter plays an important role in contemporary control theory, game theory, mathematical economics, and other disciplines of applied mathematics. In this connection, it makes sense to touch upon the problems of measure theory in detail.

The Lebesgue investigations and, in particular, his new construction of the integral gave a powerful impulse to the development of this important discipline. We refer the interested reader to the monographs on measure theory of Halmos [75], Dunford and Schwartz [66], and many other. We note only some circumstances connected with the investigation of spaces of measures, including finitely additive measures. The connection of measures and linear continuous functionals on some Banach spaces are well known (recall, for example, the famous Riesz theorem). This connection allows us to use the elaborate apparatus of functional analysis in measure theory. In particular, these representations define 'convenient' conditions of compactness in the form of Alaoglu's theorem following from the fundamental Tichonoff theorem. Of course, this scheme of embedding into the space generalized elements with the use of compactifications was employed by many authors for problems of classical control theory and the theory of differential games, the calculus of variations [120], game theory and some other fields of applied mathematics.

We focus our attention on a quite new element—the use of finitely additive measures as a 'material' for constructing extensions. Many sections of this monograph are concerned with this. In this connection it is worth noting, in the authors' opinion, that finitely additive measure theory is very interesting mathematical discipline which was founded in the works of many remarkable mathematicians. We would especially note the works of Maharam, Lipecki, Christensen, Hewitt and Yoside, Mahnard, Bhaskara Rao and Bhaskara Rao. The sufficiently complete exposition of the finitely additive measure theory is given in the monographs of Dunford and Schwartz [66], Semadeni [110], Bhaskara Rao and Bhaskara Rao [106].

We use various classes of finitely additive measures for constructing well posed extensions of different substantive problems. Here we continue to investigate the property of asymptotic non-sensitivity of extremal problems and problems of attainability under relaxation of part of the constraints. This property is important for practice and it was established (see Chentsov's monographs [32, 35]) by means of applying finitely additive measures. In the authors' opinion equipping substantive problems with peculiar finitely additive structures, including multi-topological constructions, can be of interest to specialists on measure theory. In this connection some questions which are concerned with universal integrability of bounded functions (see investigations of Leader and Maharam) are included in this book. In fact, ideas of compactifications are also used here. Besides, some 'unusual' extensions of the length function, which is primarily defined on the family of all intervals of the interval $[0, 1]$, are realized in the class of purely finitely additive measures. These extensions use elements of asymptotic constructions, which were earlier applied for determining attraction sets in problems of attainability under relaxation of constraints. Lastly, the authors see fit to study the problem of extension of so called multivalued quasi-strategies—control procedures, which are applied in the theory of differential games. Here multivalued mappings, whose values are attraction sets, are realized in quite different substantive problems, and conditions of non-anticipation of these multi-valued mappings are investigated. We note that the notion of quasi-strategy (mono-valued) ascend from works on the theory of differential games of Roxin, Elliott and Kalton. In investigations of one of the authors, multi-valued and 'compactified' versions of this important notion were used. This construction was applied for the determination of controlling procedures in differential games with geometric constraints on the choice of controls. The scheme considered in this monograph should be used in game problems with the simplest impulse constraints. Such use is natural because, in this version, an extension construction similar to that of problems of attainability in the presence of constraints is realized. In this sense we have a logical continuation of the whole circle of the authors' investigations on well posed extensions of unstable problems, including control problems with impulse and other constraints. On the other hand, constructions of multi-valued quasi-strategies is in reality connected with a known method, which is primarily elaborated for solving differential games and then is applied in other substantive problems. We mean the so called programmed iterations method, one of the variants of which conceptually connected with multi-valued quasi-strategies is considered in this monograph. This method was suggested by one of the authors in 1974 for solving nonlinear differential games in the Krasovskii formalization. On the one hand, this formalization admits a natural engineering realization based on step by step schemes of control by feedback, and on the other hand it admits receiving thoughtful results of qualitative character. One of them is the Krasovskii and Subbotin fundamental theorem on the alternative (see, for example, the monograph [88]), which determines the fact of existence and the structure of the solution of a positional differential game. The statements about the existence of a saddle point in different classes of positional strategies are a corollary of this theorem. In works of Krasovskii and his pupils auxiliary programmed constructions are used for constructing optimal strategies in the class of so called regular differential games. In fact, these constructions determined a direct passage from programmed controls to controls by the feedback principle. If the above conditions of regularity are not satisfied then such a passage is significantly complicated and is reduced to a recursive procedure, which uses an universal (for all iterations) problem of programmed control on every step. This is the point of early constructions of the programmed iterations method (see works of Chentsov and Chistyakov in 1975–1979 and the review of [112]). In a sense these procedures should be called 'undirected' if constructions of solving strategies are meant. As a matter of the fact, the pay off function and so called stable bridges are realized as limits of these iterative procedures. On this basis, controlling procedures can be constructed according to known rules. Latterly, other, 'direct' in a sense, versions of the iteration method were realized, which deliver, as a limit, a multi-valued quasi-strategy or its analogue. One of the chapters of the monograph is devoted to a systematic exposition of this method and, in this sense, it is consistent with the subsequent consideration of extension constructions for multi-valued quasi-strategies. On the other hand, this variant of the iteration method is the way to construct fixed points of some 'programmed' (in some sense) operator; here, topological constructions are used similarly to those which were applied for obtaining attraction sets in problems of 'asymptotic attainability'.

Thus in the monograph quite different problems are investigated which apply constructions connected, directly or indirectly, with the extension of spaces. It seems that these constructions conceptually form some approach common enough to the investigation of non-regular problems of various nature. We hope this approach will be of interest to the reader.

ALEXANDER CHENTSOV, SVETLANA MORINA

# Chapter 1

# PHASE CONSTRAINTS AND BOUNDARY CONDITIONS IN LINEAR CONTROL PROBLEMS

## 1.1 SOME CONTROL PROBLEMS: ATTAINABILITY DOMAINS AND THEIR APPLICATION

We touch upon some questions connected with control under constraints. Consider a dynamical system $\Sigma$ functioning on the finite time interval $[t_0, \vartheta_0]$, $t_0 < \vartheta_0$; an initial state of the system is given: $x(t_0) = x_0$. Moreover, some control program $f = (f(t), t_0 \le t < \vartheta_0)$ acts on the input of the system $\Sigma$.

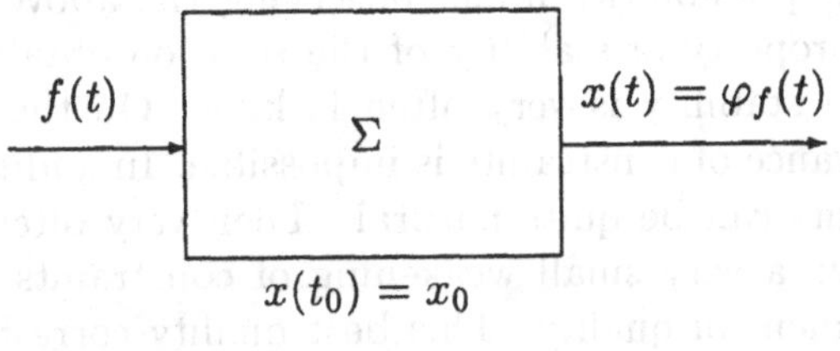


A concrete trajectory $x(t) = \varphi_f(t)$ of $\Sigma$ is then realized. This trajectory is defined by the natural 'input output' operator, which transforms the control function $f$. Of course, very different variants of the representation for this operator are possible (it should be added that, in many important cases, the employment of discrete time is advisable, but we do not consider such cases). Very often the system $\Sigma$ (and its input output operator) is defined by means of ordinary differential equations. We will consider only such cases. In addition, the concrete nature of $\Sigma$ is non-essential for us. Conversely, for engineers this nature is very important because the question of the corresponding realization plays a quite essential role. But we do not touch upon this question. We are oriented only towards mechanical systems (airplanes, spacecrafts, and others). This circumstance implies corresponding settings. So we consider the transformation $f \to \varphi_f$. In addition, the concrete choice of $f$ is restricted by some set $F$ of all possible control programs. Along with the constraint $f \in F$, others are possible. In particular, some of such 'new' constraints can be defined in terms of corresponding trajectories. So the restriction $\varphi_f \in \Phi$, where $\Phi$ is some fixed functional set, can be imposed. In concrete engineering problems this $\Phi$-constraint can be of different nature. For example, if we deal with the motion of an airplane, then the following requirement arises very often: the height of the airplane flight must be not less than a given positive number. Here we have a variant of the so called phase constraints. Moreover, for this airplane the time of the arrival at a given point is fixed. We get a boundary condition. Finally, the airplane has a reserve of fuel. Thus we get a resource constraint. It is possible to point out many other constraints typical for such problems of control by an airplane. In addition, the constraints are observed approximately. For example, the height of the flight is kept with some precision. But the above mentioned real constraints 'form' a corresponding mathematical setting (moreover, many other factors influence the setting). We use some quite concrete values of parameters in this setting. Fixing these values (i.e., fixing some constraints chosen approximately), we can solve our problem. For example, under these conditions we can optimize some criterion. In other cases we investigate our possibilities under observing the above constraints. In any case the property of stability of the solution obtained is worth considering. Such stability is very often lacking. On the other hand, the precise observance of constraints is impossible. In addition, a weakening of constraints can be quite natural. Then very often we get the 'useful' instability: a very small weakening of constraints leads to the essential improvement of quality. This best quality corresponds to the engineer's approach: all solutions have an approximate character (this question was discussed in [117, Ch. III]). So our goal corresponds to the engineer's look on constructing complete systems under conditions of different errors and elements of uncertainty. But the following mathematical problem arises: how to find new possibilities connected with the effect of the approximate observance of constraints? Indeed, rejecting the representation of the problem of attainability in terms of precise constraints, we lose the corresponding traditional mathematical apparatus. As a result, we obtain the problem of asymptotic attainability or (in other cases) of asymptotic optimization. In fact, we have a problem of asymptotic analysis; special mathematical constructions are required for solving this problem. This book is devoted to special constructions of such a kind which use the notion of well posed extension.

We consider the simplest class of control systems with constraints on the choice of control programs. Namely, in this chapter we discuss some effects connected with the observance of constraints and conditions on values of trajectories of linear control systems. Basic attention is paid to integral constraints of various types. But some effects arise in control systems with geometric constraints on the choice of controls. A systematic investigation of dynamical systems with such constraints was begun by Pontryagin.

## 1.2 CONTROL BY A MASS POINT UNDER INTEGRAL CONSTRAINTS

In this section we investigate (on the informative level) some problems of control by a traditional mechanical system. We consider the vector mass point

$$\ddot{y}(t) = m(t)f(t) \tag{1.2.1}$$

on the time interval $[0, \vartheta_0]$, where $\vartheta_0$ is a given positive number. In addition, $y(t) \in \mathbb{R}^r$, where $r$ is a natural number. Here $f$ is a control function on $[0, \vartheta_0[$. The values $m(t)$ can characterize the mass and the force direction at the current time. The function $m$ is supposed to be discontinuous. For simplicity we denote the sets $[0, \vartheta_0]$ and $[0, \vartheta_0[$ by $I_0$ and $I$, respectively.

Various stipulations with respect to the right hand side of (1.2.1) can be considered. We can suppose that $f$ is a vector function and $m$ is a given real-valued function simulating the respective variation of mass. Conversely, it is possible to consider the case when $f$ is a real-valued function and $m$ is a vector function. Now we agree on the second stipulation. For simplicity, suppose that $f$ is a piecewise constant (p.c.) and continuous from the right (c.f.r.) real-valued function on $I$. In addition, suppose that $m$ is the function acting from $I$ into $\mathbb{R}^r$, whose components $m_1, \ldots, m_r$ are uniform limits of corresponding sequences of p.c. and c.f.r. real-valued functions on $I$. Fix the initial conditions: $y(0) = y_0 \in \mathbb{R}^r$ and $\dot{y}(0) = \dot{y}_0 \in \mathbb{R}^r$. We postulate that the choice of $f$ must satisfy the following constraints:

$$\Big((y_f(t_i))_{i\in\overline{1,k}}, (\dot{y}_f(t'_j))_{j\in\overline{1,l}}\Big) \in \mathbf{Y}, \quad \int_0^{\vartheta_0} |f(t)|\, dt \le c. \tag{1.2.2}$$

In (1.2.2) $k$ and $l$ are natural numbers, $t_1 \in I_0, \ldots, t_k \in I_0$, $t'_1 \in I_0, \ldots, t'_l \in I_0$, the set $\mathbf{Y}$ is a subset of the product of $\mathbb{R}^k$ and $\mathbb{R}^l$, and

$c \in [0, \infty[$ is a recourse constant. In fact, $\mathbf{Y}$ is a subset of $\mathbb{R}^{k+l}$; in addition, suppose that $\mathbf{Y}$ is closed in $\mathbb{R}^{k+l}$ with the ordinary topology of coordinate-wise convergence. We consider the attainability domain at the time $\vartheta_0$ under the constraints (1.2.2). Namely, (1.2.2) defines a concrete subset $F_d$ of the set $F$ of all p.c. and c.f.r. real-valued functions on $I$: $f \in F_d$ iff (1.2.2) holds, where $y_f = (y_f(t) \in \mathbb{R}^r, t \in I)$ is the $f$-trajectory corresponding to the control $f$. Then the attainability domain can be defined in the form $G \triangleq \{y_f(\vartheta) : f \in F_d\}$. In the sequel we consider various relaxations of (1.2.2). In addition, we use perturbations of the $\mathbf{Y}$-constraint in (1.2.2) or the $c$-constraint. In the simplest case the replacement of $\mathbf{Y}$ by an (Euclidean) $\varepsilon$-neighborhood is used, where $\varepsilon \in ]0, \infty[$. This replacement generates the corresponding change of the attainability domain: $G \to G_\varepsilon$. In addition, the $\varepsilon$-weakening of the $\mathbf{Y}$-constraints implies a change of $F_d$: $F_d \to F_d^{(\varepsilon)}$. The set $F_d^{(\varepsilon)}$ consists of all $f \in F$ satisfying a condition similar to (1.2.2); the set $\mathbf{Y}$ is replaced by its $\varepsilon$-neighborhood in this new condition. In this case the set-valued dependence $(G_\varepsilon, \varepsilon > 0)$ converges to some limit ATT as $\varepsilon \downarrow 0$. This limit can be regarded as an attraction set. Note that another variant of weakening the constraints (1.2.2) can be realized. For example, it is possible to replace the set $\mathbf{Y}$ by its $\varepsilon$-neighborhood, and the number $c$ by $c+\varepsilon$ (in connection with weakening the resource constraint, see [32, Ch. 5], [35, Ch. 3]). In this case we obtain a new attainability domain $G^{(\varepsilon)}$. Moreover, we obtain a new attraction set in the form of a corresponding limit of the dependence $(G^{(\varepsilon)}, \varepsilon > 0)$ as $\varepsilon \downarrow 0$. However, in reality the last attraction set coincides with ATT. Thus we have the simplest statement about an asymptotic non-sensitivity under the weakening of a part of the constraints. Simultaneously the initial problem of constructing the attainability domain can be unstable. In the following we systematically investigate the properties of attraction sets under the weakening of constraints like (1.2.2). We consider the given system of constraints both as basic and as model for other very important problems. In this chapter we first discuss the last possibility. Namely, we use a particular case of (1.2.2) as an auxiliary construction for solving the problem of control with phase constraints. Let us consider this problem on the informative level.

**The control problem with phase constraints.**

Let $(N_t)_{t \in I_0}$ be a mapping from $I_0$ into the family $\mathcal{K}_r$ of all nonempty compact subsets of $\mathbb{R}^r$ (i.e., the family of all bounded and closed sets). Suppose that this mapping is continuous in the sense of the Hausdorff metric of $\mathcal{K}_r$. Of course, we equip $\mathbb{R}^r$ with the Euclidean norm $\|\cdot\|$.

Below we consider corresponding precise definitions. We denote by $N_t^{[\varepsilon]}$ the closed Euclidean $\varepsilon$-neighborhood of the set $N_t$ for $t \in I_0$.

Consider (as basic) the following constraints on the choice of $f \in F$:

$$\left(\forall t \in I_0 : y_f(t) \in N_t\right) \& \left(\int_0^{\vartheta_0} |f(t)|\, dt \leq c\right). \tag{1.2.3}$$

Let $F^{(d)}$ be the set of all $f \in F$ satisfying the constraints (1.2.3). Of course, $y_f$ corresponds to the previous definitions under fixed initial conditions $(y_0, \dot{y}_0)$. As earlier, we will consider the attainability domain. But we allow various variants of this domain. In particular, the following sets can be taken:

$$\left\{y_f(\vartheta_0) : f \in F^{(d)}\right\}, \left\{\dot{y}_f(\vartheta_0) : f \in F^{(d)}\right\},$$
$$\text{and } \left\{(y_f(\vartheta_0), \dot{y}_f(\vartheta_0)) : f \in F^{(d)}\right\}.$$

In applied problems one of these variants can be of interest. We denote by $G^{(d)}$ the attainability domain corresponding to the requiered variant. Along with (1.2.3) consider the following weakened constraints:

$$\left(\forall t \in I_0 : y_f(t) \in N_t^{[\varepsilon]}\right) \& \left(\int_0^{\vartheta_0} |f(t)|\, dt \leq c\right), \tag{1.2.4}$$

where $\varepsilon \in ]0, \infty[$. Moreover, we can admit the perturbation $c \to c + \varepsilon$. But now we restrict ourselves to the case (1.2.4). For each $\varepsilon \in ]0, \infty[$ we denote by $F_\varepsilon^{(d)}$ the set of all $f \in F$ for which (1.2.4) is fulfilled; moreover, we denote by $G_\varepsilon^{(d)}$ the attainability domain under the replacement $F^{(d)} \to F_\varepsilon^{(d)}$. Of course, we postulate that $G^{(d)}$ and $G_\varepsilon^{(d)}$ correspond to the common variant of the choice of the space of attainable elements. Namely, we suppose that

$$\begin{aligned}
&\left(G^{(d)} = \{y_f(\vartheta_0) : f \in F^{(d)}\}\right)\\
&\& \left(\forall \varepsilon \in ]0, \infty[: G_\varepsilon^{(d)} = \{y_f(\vartheta_0) : f \in F_\varepsilon^{(d)}\}\right)\\
&\vee \left(G^{(d)} = \{\dot{y}_f(\vartheta_0) : f \in F^{(d)}\}\right)\\
&\& \left(\forall \varepsilon \in ]0, \infty[: G_\varepsilon^{(d)} = \{\dot{y}_f(\vartheta_0) : f \in F_\varepsilon^{(d)}\}\right)\\
&\vee \left(G^{(d)} = \{(y_f(\vartheta_0), \dot{y}_f(\vartheta_0)) : f \in F^{(d)}\}\right)\\
&\& \left(\forall \varepsilon \in ]0, \infty[: G_\varepsilon^{(d)} = \{(y_f(\vartheta_0), \dot{y}_f(\vartheta_0)) : f \in F_\varepsilon^{(d)}\}\right).
\end{aligned} \tag{1.2.5}$$

We choose one of these variants. In addition, we obtain the usual domain of attainability $G^{(d)}$ and the attraction set ATT defined as the intersection of all sets $\overline{G_\varepsilon^{(d)}}$, $\varepsilon \in ]0,\infty[$, where the overline denotes the closure in the respective finite-dimensional space with the topology of coordinate-wise convergence. Always, $\overline{G^{(d)}} \subset \text{ATT}$; very often, $\overline{G^{(d)}} \neq \text{ATT}$.

**Example.** Let $\vartheta_0 = 1$, $r = 1$, $y_0 = \dot{y}_0 = 0$, $N_t \equiv \{0\}$ (a singleton). Suppose that $m(t) \equiv 1$. We choose the second variant in (1.2.5). Thus we consider the attainability domain relative to the velocity coordinate under phase constraints relative to the geometric coordinate. Note that $\forall f \in F\ \forall t \in I_0$:

$$y_f(t) = \int_0^t (t-\tau)f(\tau)\, d\tau.$$

Therefore $y_f(t) \not\equiv 0$ under $f(t) \not\equiv 0$. As a consequence the set $F^{(d)}$ consists of one element corresponding to the control $f_0 \in F$ such that $f_0(t) \equiv 0$. Hence $G^{(d)} = \{0\}$, i.e., $G^{(d)}$ is a singleton.

Let us consider a natural variant of the $\varepsilon$-weakening of the phase constraints. We fix $\varepsilon \in ]0,\infty[$. In this concrete case $N_t^{[\varepsilon]} \equiv [-\varepsilon, \varepsilon]$, and (1.2.4) has the following form:

$$\left(\forall t \in I_0 : |y_f(t)| \leq \varepsilon\right) \,\&\, \left(\int_0^1 |f(t)|\, dt \leq c\right). \tag{1.2.6}$$

As a consequence we obtain a very essential expansion of the admissible set. Namely, for any $a \in [-c, c]$ the set $F_\varepsilon^{(d)}$ contains some control $f_a \in F$ for which $\dot{y}_{f_a}(1) = a$. Indeed, choose

$$\delta_\varepsilon \triangleq \inf\left(\left\{1, \frac{\varepsilon}{c+1}\right\}\right).$$

Then $\delta_\varepsilon \in ]0,1]$. Let $f_a \in F$ be the following function:

$$f_a(t) = \begin{cases} 0, & t \in [0, 1-\delta_\varepsilon[, \\ a/\delta_\varepsilon, & t \in [1-\delta_\varepsilon, 1[. \end{cases} \tag{1.2.7}$$

From (1.2.7) we have the equality

$$\dot{y}_{f_a}(1) = \int_0^1 f_a(t)\, dt = \int_{1-\delta_\varepsilon}^1 f_a(t)\, dt = a. \tag{1.2.8}$$

Let us verify the relations (1.2.6). We have

$$\int_0^1 |f_a(t)|\, dt = \int_{1-\delta_\varepsilon}^1 |f_a(t)|\, dt = |a| \leq c.$$

On the other hand, $y_f(t) = 0$ for $t \in [0, 1-\delta_\varepsilon]$. Moreover, for $t \in [1-\delta_\varepsilon, 1]$ we have

$$y_{f_a}(t) = \int_0^t (t-\tau) f_a(\tau)\, d\tau = \int_{1-\delta_\varepsilon}^t (t-\tau) f_a(\tau)\, d\tau$$
$$= \frac{a}{\delta_\varepsilon} \int_{1-\delta_\varepsilon}^t (t-\tau)\, d\tau.$$

Hence $\forall t \in [1-\delta_\varepsilon, 1]$ we obtain

$$|y_{f_a}(t)| \leq \frac{|a|}{\delta_\varepsilon}(t-1+\delta_\varepsilon)^2 \leq |a|\delta_\varepsilon \leq c\delta_\varepsilon < \varepsilon.$$

Thus $f_a \in F_\varepsilon^{(d)}$. By (1.2.5) and (1.2.8) we have $a \in G_\varepsilon^{(d)}$, and hence $[-c, c] \subset G_\varepsilon^{(d)}$. But from the second condition of (1.2.4) we have $\forall f \in F_\varepsilon^{(d)}$:

$$|\dot{y}_{f_a}(1)| = \left| \int_0^1 f_a(t)\, dt \right| \leq \int_0^1 |f_a(t)|\, dt \leq c.$$

As a consequence $G_\varepsilon^{(d)} \subset [-c, c]$. Hence $G_\varepsilon^{(d)} = [-c, c]$ for $\varepsilon \in ]0, \infty[$. Of course, we obtain the set ATT coinciding with $[-c, c]$. For $c > 0$ the closure of $G(d)$ defined as $\{0\}$ and the set ATT are very different. Thus we obtain an unstable problem. However, the effect of such instability can be regarded as useful: we obtain an essential expansion of our possibilities. Of course, this example should be considered only as a model one.

Returning to the general setting, we consider the weakening of constraints (1.2.4). The Cauchy formula should be used for representation of the trajectory $y_f(\cdot)$. In this case, for $f \in F$ and $t \in I_0$ we have

$$y_f(t) = y_0 + \dot{y}_0 t + \int_0^t (t-\tau) m(\tau) f(\tau)\, d\tau;$$

here and below the Riemann integral of a vector function is defined as the respective vector of integrals of scalar components of the vector function.

Note a useful estimate. If $t_1 \in I_0$ and $t_2 \in [t_1, \vartheta_0]$, then

$$y_f(t_2) = y_f(t_1) + (t_2 - t_1)\dot{y}_f(t_1) + \int_{t_1}^{t_2} (t_2 - \tau) m(\tau) f(\tau) d\tau. \quad (1.2.9)$$

We use the known semigroup property. From (1.2.9) we have $\forall f \in F$ $\forall t_1 \in I_0$ $\forall t_2 \in [t_1, \vartheta_0]$:

$$\| y_f(t_2) - y_f(t_1) \| \leq (t_2 - t_1) \| \dot{y}_f(t_1) \| + \int_{t_1}^{t_2} (t_2 - \tau) \| m(\tau) \| \cdot | f(\tau) |\, d\tau. \quad (1.2.10)$$

Introduce the set $\mathbb{F}[c]$ of all functions $f \in F$ for which the second condition in (1.2.3) holds. Let $f \in \mathbb{F}[c]$. Then $\forall t \in I_0$:

$$\| \dot{y}_f(t) \| \leq \| \dot{y}_0 \| + \int_0^t \| m(\tau) \| \cdot | f(\tau) | \, d\tau. \tag{1.2.11}$$

In (1.2.11) the vector function $m(\cdot)$ is bounded. Therefore (see (1.2.4)) one can choose $c_1 \in [0, \infty[$ such that $\forall f \in \mathbb{F}[c]$ $\forall t \in I_0$: $\|\dot{y}_f(t)\| \leq c_1$. Furthermore, in (1.2.10) one can choose $c_2 \in [0, \infty[$ such that $\forall f \in \mathbb{F}[c]$ $\forall t_1 \in I_0$ $\forall t_2 \in [t_1, \vartheta_0]$:

$$\int_{t_1}^{t_2} (t_2 - \tau) \| m(\tau) \| \cdot | f(\tau) | \, d\tau \leq c_2(t_2 - t_1).$$

As a consequence there exists $c_3 \in [0, \infty[$ such that $\forall f \in \mathbb{F}[c]$ $\forall t_1 \in I_0$, and $\forall t_2 \in [t_1, \vartheta_0]$:

$$\| y_f(t_2) - y_f(t_1) \| \leq c_3(t_2 - t_1). \tag{1.2.12}$$

In reality, the second condition in (1.2.4) implies (1.2.12). Here only the fact that $f \in F$ possesses the property

$$\int_0^{\vartheta_0} | f(t) | \, dt \leq c$$

is of importance. Thus if $f \in F$ and the last inequality is satisfied, then the relation (1.2.12) is valid for $t_1 \in I_0$ and $t_2 \in [t_1, \vartheta_0]$. Hence we can state the following. If constraints in the control problem are fulfilled approximately, then it is sufficient to analyze some finite collections of conditions on the mass point trajectory. Namely, the following requirement can be imposed:

$$y_f(t_1) \in N_{t_1}, \ldots, y_f(t_m) \in N_{t_m}, \tag{1.2.13}$$

where $m \geq 2$, $0 = t_1 < \ldots < t_m = \vartheta_0$, and the greatest of numbers $t_{i+1} - t_i$, $i \in \overline{1, m-1}$ is small enough. As a result we obtain some particular case of (1.2.2). Since the dependence $t \mapsto N_t$ is continuous in the Hausdorff metric, we guarantee that constraints of the type (1.2.4) will be valid under the satisfaction of (1.2.13). The step of the partition of $I_0$ in (1.2.13) should be chosen in dependence on the parameter $\varepsilon$, $\varepsilon > 0$. In addition, weakened versions of (1.2.13) (i.e., approximate satisfaction of the phase constraints) can be used

$$y_f(t_1) \in N_{t_1}^{[\alpha]}, \ldots, y_f(t_m) \in N_{t_m}^{[\alpha]}. \tag{1.2.14}$$

Here $m$ and $(t_1, \ldots, t_m)$ correspond to (1.2.13), and $\alpha \in ]0, \infty[$. In addition, the last (energetic) requirement in (1.2.4) is assumed. If the number $\varepsilon \in ]0, \infty[$ in (1.2.4) is given, then one can choose $m$, $(t_1, \ldots, t_m)$, and $\alpha \in ]0, \infty[$ such that for $f \in F$ satisfying the energetic constraint and (1.2.14) the first requirement in (1.2.4) is satisfied. As a result the following version of constraints on the choice of $f \in F$ can be used when investigating the relaxations on the basis of (1.2.4):

$$\left(\forall t \in K : y_f(t) \in N_t^{[\alpha]}\right) \& \left(\int_0^{\vartheta_0} |f(t)|\, dt \leq c\right); \tag{1.2.15}$$

here $K$ is a finite subset of $I_0$ and $\alpha > 0$. The pair $(K, \alpha)$ is regarded as a parameter. In addition, we make the set $K$ larger and larger, and the number $\alpha \in ]0, \infty[$ smaller and smaller. Now we consider some equivalent transformation of the first requirement in (1.2.15). Let us look at the first condition in (1.2.15) from another point of view.

Suppose that $\forall t \in I_0 : N_t \neq \varnothing$. This natural condition makes the problem under consideration substantial. Denote by $\mathfrak{N}$ the set of all mappings $y$ from $I_0$ into $\mathbb{R}^r$ such that $\forall t \in I_0 : y(t) \in N_t$. Thus $\mathfrak{N}$ is the product of all sets $N_t$, $t \in I_0$. By the axiom of choice $\mathfrak{N} \neq \varnothing$. Then the first relation in (1.2.15) is equivalent to the following condition on the choice of $f \in F$:

$$\exists \mathbf{y} \in \mathfrak{N}\ \forall t \in K : \| y_f(t) - \mathbf{y}(t) \| \leq \alpha. \tag{1.2.16}$$

Indeed, by the definition of $\mathfrak{N}$ the first relation in (1.2.15) immediately follows from (1.2.16). Let now the first condition in (1.2.15) be valid. Choose

$$(z_t)_{t \in K} \in \prod_{t \in K} N_t \tag{1.2.17}$$

such that $\forall t \in K : \|y_f(t) - z_t\| \leq \alpha$. The possibility of such choice follows from the first condition in (1.2.15). Choose $z^0 \in \mathfrak{N}$ and define $\mathbf{y} \in \mathfrak{N}$ as follows:

$$(\forall t \in K : \mathbf{y}(t) \triangleq z_t) \& (\forall t \in I_0 \setminus K : \mathbf{y}(t) = z^0(t)).$$

Then (1.2.16) is true. Thus (1.2.15) implies (1.2.16). This property shows that conditions of the type (1.2.15) can be used to represent an approximate observance of the phase constraints. In turn, conditions of the type (1.2.15) with the use of Cauchy's formula can be reduced to conditions of the belonging of values of a vector integral to a given set.

Consider the set $\mathcal{R}$ of all $r$-vector functions on $I_0$. In this set we pick out in a natural way the set $\widetilde{\mathbb{N}}$ of all functions $y \in \mathcal{R}$ such that $\forall t \in I_0$:

$$y_0 + t\dot{y}_0 + y(t) \in N_t.$$

It is convenient to describe this set as a Cartesian product, setting $\forall t \in I_0$:

$$\tilde{N}_t \triangleq \{z - (y_0 + t\dot{y}_0) : z \in N_t\}. \qquad (1.2.18)$$

Then $\tilde{\mathbb{N}}$ is the Cartesian product of all sets $\tilde{N}_t$, $t \in I_0$. Each set $\tilde{N}_t$ is compact since each set $N_t$ possesses this property. Hence the requirement for $f \in F$ is equivalent to the statement $\forall t \in I_0$:

$$\int_0^t (t-\tau)m(\tau)f(\tau)\,d\tau \in \tilde{N}_t.$$

In turn, the latter means that

$$\left(\int_0^t (t-\tau)m(\tau)f(\tau)\,d\tau\right)_{t\in I_0} \in \tilde{\mathbb{N}}.$$

By the Tichonoff theorem the set $\tilde{\mathbb{N}}$ is compact in the topology of pointwise convergence of the space $\mathcal{R}$; in particular, it is closed. If $K \in \mathrm{Fin}(I_0)$ (here and below the symbol $\mathrm{Fin}(I_0)$ stands for the family of all finite subsets of $I_0$) and $\alpha \in ]0,\infty[$, we denote by $F(K,\alpha)$ the set of all $f \in F$ satisfying the second condition in (1.2.4) and such that

$$\exists \mathbf{n} \in \tilde{\mathbb{N}}\ \forall t \in K: \left\| \left(\int_0^t (t-\tau)m(\tau)f(\tau)\,d\tau\right) - \mathbf{n}(t)\right\| \le \alpha, \qquad (1.2.19)$$

and by $G(K,\alpha)$ the corresponding attainability domain, determined like (1.2.5) under the replacement of $F_\varepsilon^{(d)}$ by $F(K,\alpha)$. Let $s = |K|$, where $|K|$ is the number of elements of $K$. Introduce the bijective numeration

$$(t_i)_{i\in\overline{1,s}} : \overline{1,s} \to K$$

of the set $K$ (in fact, a permutation of elements of $K$ is introduced). Then the condition (1.2.19) is equivalent to the following one:

$$\left(\int_0^{t_i} (t_i-\tau)m(\tau)f(\tau)\,d\tau\right)_{i\in\overline{1,s}} \in \prod_{i=1}^{s} \tilde{N}_{t_i}^{[\alpha]}, \qquad (1.2.20)$$

which, by (1.2.18), is equivalent to (1.2.14). Thus the equivalence of (1.2.14) and (1.2.19) is established under the condition when time parameters of (1.2.14) correspond to the bijective numeration of $K$. This allows us to connect conditions of the form (1.2.19) and those of the first part in (1.2.4).

For $\xi \in I_0$ and $\alpha \in ]0,\infty[$, let $F_\xi(\alpha)$ be the set of all $f \in F$ satisfying the constraints

$$\left(y_f(\xi) \in N_\xi^{[\alpha]}\right) \,\&\, \left(\int_0^{\vartheta_0} |f(t)|\,dt \le c\right), \qquad (1.2.21)$$

and $G_\xi(\alpha)$ be the corresponding attainability domain. Thus in (1.2.4), (1.2.19), and (1.2.21) we have introduced different variants of relaxations of the basic condition (1.2.3). Our goal is to study limits of the attainability domains $G_\varepsilon^{(d)}$ and $G(K,\alpha)$ as $\varepsilon \downarrow 0$, $\alpha \downarrow 0$ and expanding $K$. Moreover, interrelations between this limits is the subject to be investigated in the next section.

## 1.3 EQUIVALENCE OF DIFFERENT VARIANTS OF RELAXATIONS

From the definition of $F_\varepsilon^{(d)}$ and $F(K,\alpha)$, it is clear that

$$\forall K \in \mathrm{Fin}(I_0)\ \forall \varepsilon \in ]0,\infty[: F_\varepsilon^{(d)} \subset F(K,\varepsilon). \tag{1.3.1}$$

At the same time,

$$\forall \varepsilon \in ]0,\infty[\ \exists K \in \mathrm{Fin}(I_0)\ \exists \alpha \in ]0,\infty[: F(K,\alpha) \subset F_\varepsilon^{(d)}. \tag{1.3.2}$$

Indeed, let $\varepsilon \in ]0,\infty[$. Since the dependence of $N_t$ on $t \in I_0$ is a continuous function with values in the space of compacta in $\mathbb{R}^r$ with the Hausdorff metric, it is uniformly continuous. This means that for the number $\varepsilon/3$ one can point out $\delta \in ]0,\infty[$ such that $\forall t' \in I_0\ \forall t'' \in I_0$:

$$(|t'-t''| < \delta) \Rightarrow (\rho_{\mathcal{K}}(N_{t'}, N_{t''}) < \varepsilon/3), \tag{1.3.3}$$

where $\rho_{\mathcal{K}}$ is the Hausdorff metric of the family $\mathcal{K}_r$ of all nonempty compact sets in $\mathbb{R}^r$. Choose a natural number $p \in \mathcal{N}$ such that

$$\left(\vartheta_0 p^{-1} < \frac{\varepsilon}{3c_3}\right) \& \left(\vartheta_0 p^{-1} < \delta\right),$$

where $\delta$ satisfies condition (1.3.3) and $c_3$ corresponds to (1.2.12). We define the set $K$ by means of the uniform grid with the step $\vartheta_0 p^{-1}$. Let

$$t_l^{(p)} \triangleq \frac{l}{p}\vartheta_0, \quad l \in \overline{0,p}.$$

We set $K \triangleq \{t_j^{(p)} : j \in \overline{0,p}\}$ and assume that $\alpha \triangleq \varepsilon/3$. Let $f' \in F(K,\alpha)$ for these $K$ and $\alpha$. Choose arbitrarily $t \in I_0$, and take $q \in \overline{1,p}$ such that $t \in [t_{q-1}^{(p)}, t_q^{(p)}]$. By the choice of $p \in \mathcal{N}$ and relation (1.2.12) we have

$$\left\| y_{f'}(t) - y_{f'}(t_q^{(p)}) \right\| \le c_3 \vartheta_0 p^{-1} < \varepsilon/3. \tag{1.3.4}$$

In addition, by the choice of $K$ and $\varepsilon$ and relations (1.2.18), (1.2.19) we have

$$y_{f'}(t_q^{(p)}) \in N_{t_q^{(p)}}^{[\varepsilon/3]}. \tag{1.3.5}$$

From (1.3.4) and (1.3.5) it follows that

$$y_{f'}(t) \in N^{[2\varepsilon/3]}_{t_q^{(p)}}. \tag{1.3.6}$$

With taking (1.3.3) into account we obtain the inclusion

$$N^{[2\varepsilon/3]}_{t_q^{(p)}} \subset N^{[\varepsilon]}_t \tag{1.3.7}$$

which in combination with (1.3.6) gives the inclusion $y_{f'}(t) \in N^{[\varepsilon]}_t$. In turn, the latter means that $f' \in F^{(d)}_\varepsilon$ and consequently (1.3.2) is proved.

Recall that for relaxation (1.2.4) the corresponding attraction set was denoted by ATT. Let $\mathrm{Att}_1$ be the attraction set corresponding to variant (1.2.19) of relaxation of constraints. Namely,

$$\mathrm{Att}_1 = \bigcap_{\substack{K \in \mathrm{Fin}(I_0)\\ \alpha \in ]0,\infty[}} \overline{G(K,\alpha)}. \tag{1.3.8}$$

From the definitions of $G^{(d)}_\varepsilon$, $G(K,\alpha)$ and relations (1.3.1), (1.3.2) it follows that $\mathrm{ATT} = \mathrm{Att}_1$. Indeed, let $x \in \mathrm{ATT}$. Fix any $K_* \in \mathrm{Fin}(I_0)$ and $\alpha_* \in ]0,\infty[$. Let us prove that $x \in \overline{G(K_*,\alpha_*)}$. Since $x \in \mathrm{ATT}$, we have $\forall \varepsilon \in ]0,\infty[: x \in \overline{G^{(d)}_\varepsilon}$. In particular, $x \in \overline{G^{(d)}_{\alpha_*}}$. From (1.3.1) it follows that $F^{(d)}_{\alpha_*} \subset F(K_*,\alpha_*)$ and hence $G^{(d)}_{\alpha_*} \subset G(K_*,\alpha_*)$. From this we get $x \in \overline{G(K_*,\alpha_*)}$. Since $K_*$ and $\alpha_*$ was chosen arbitrarily, we have $x \in \mathrm{Att}_1$. Let now $x \in \mathrm{Att}_1$. Taking into account (1.3.2) and reasoning as above, we get the inclusion $x \in \mathrm{ATT}$.

In addition, note that $\forall \alpha \in ]0,\infty[\ \forall \xi \in I_0 : F_\xi(\alpha) = F(K,\alpha)$ for $K = \{\xi\}$. On the other hand,

$$\forall \alpha \in ]0,\infty[\ \forall K \in \mathrm{Fin}(I_0) : F(K,\alpha) = \bigcap_{\xi \in K} F_\xi(\alpha).$$

We get a useful representation of the sets $F(K,\alpha)$ in terms of weakened versions (1.2.21) of the basic system of constraints. We will return to this representation in Section 4.13, where the representation of the set of admissible generalized programmed controls for this problem will be considered in detail. These constructions require a formalization developed enough and use elements of measure theory and topology.

We shall give one more representation of attraction sets. Such a representation allows us to describe attraction sets by means of sequences, in contrast to the general case when nets are used for this purpose. Let

$\forall n \in \mathcal{N} : F_n = F(K_n, 1/n)$, where

$$K_n = \left\{ \frac{i\vartheta_0}{n} : i \in \overline{0,n} \right\}. \quad (1.3.9)$$

Denote by $G_n$ the corresponding attainability domain. Let $\mathrm{Att}_{\mathcal{N}}$ be the attraction set, i.e.,

$$\mathrm{Att}_{\mathcal{N}} = \bigcap_{n \in \mathcal{N}} \overline{G_n}.$$

Introduce the family $\mathcal{Z} = \{F(K,\alpha) : K \in \mathrm{Fin}(I_0), \alpha \in ]0,\infty[\}$.

PROPOSITION 1.3.1 *The sequence $(F_n)_{n\in\mathcal{N}}$ is a fundamental sequence for the family $\mathcal{Z}$, i.e.,*

$$\forall K \in \mathrm{Fin}(I_0)\ \forall \alpha \in ]0,\infty[\ \exists s \in \mathcal{N} : F_s \subset F(K,\alpha). \quad (1.3.10)$$

**Scheme of the proof.** Fix any $K' \in \mathrm{Fin}(I_0)$ and $\alpha' \in ]0,\infty[$. From (1.3.1) we have $F^{(d)}_{\alpha'} \subset F(K',\alpha')$. Analogously to (1.3.3) we get that for this $\alpha'$ one can point out $\delta \in ]0,\infty[$ such that $\forall t' \in I_0\ \forall t'' \in I_0$:

$$(|t' - t''| < \delta) \Rightarrow (\rho_K(N_{t'}, N_{t''}) < \alpha'/3).$$

Choose $n \in \mathcal{N}$ such that

$$n > \max\left\{ \frac{3\vartheta_0}{\alpha'}, \frac{3c_3\vartheta_0}{\alpha'}, \frac{\vartheta_0}{\delta} \right\}.$$

We further argue as in the proof of the relation (1.3.2).

From the definition of $F_n$ and (1.3.10) we get the equality

$$\mathrm{Att}_1 = \mathrm{Att}_{\mathcal{N}}. \quad (1.3.11)$$

## 1.4 CONCLUSION

We have touched upon only one control problem with integral constraints of impulse type. Such constraints mean that the system has a certain reserve of fuel, which is natural for engineering problems. This constraint determines the possibility of functioning for the technical system. At the same time, this condition is important in corresponding theoretical constructions, since it endows the space of controls with the property of pre-compactness. Such a combination is of interest for us: restrictions that are actual for practice lead to a 'good' mathematical setting. However, it is worth noting that the class of usual controls—functions of time—is often not sufficient for the well posed mathematical setting of problem without loss of its specialities, which is of interest in engineering practice; let us again note the remark of [117] (that the application of generalized controls is useful for practice), which is addressed not only to mathematicians (and perhaps mainly not to them) but to engineers. In this chapter we are concerned with other conditions that are typical for engineering problems. In particular, the possibility of the constructing of an (asymptotic) analogue of the attainability domain is analyzed under the condition when phase constraints with respect to a part of coordinates are approximately observed. These constraints correspond to the geometrical coordinates of the mass point. The investigation of asymptotic regimes of control under constraints of such a type is of apparent interest for practice.

We see that different approaches to constructing approximative schemes are possible. However, these approaches interlock at the level of limit representations. We will study this phenomenon, which was considered earlier in [32, 35] and in a series of other works. Following the approach suggested in these monographs, we will use finitely additive measures (FAM) as generalized controls. These FAM satisfy some restrictions, which are abstract analogues of the above substantive conditions imposed on the energy and the realization of trajectories. Thus, using the notion of total variation we introduce the generalized impulse constraint that is an analogue of the second condition in (1.2.2). Evidently, our abstract condition, which is of inequality type, can not be regarded as the restriction on the fuel reserve. Alike, FAM used as analogues of controls are objects of other nature in comparison with usual control functions. All of these abstractions can not be introduced into engineering settings in their natural form. However, in these settings they will correspond to effects connected with limits of either 'results' of the substantive problem (extremum, regularization of attainability domains and 'true' trajectory bundles, which are possible under the fulfillment of all requirements to the system). In other words, our rather abstract generalized problems supply us with 'proper' asymptotics of controls and trajectories realized. Of course, a formalization developed enough is required for this purpose. We will construct it in the next chapter.

# Chapter 2

# GENERAL STRUCTURES

## 2.1 INTRODUCTION

The problem of constructing extensions and relaxations is connected with obtaining representations with the use of some specific transformations of topological spaces. These transformations are very often compactifications of the space of solutions. But it is possible to construct examples for which the corresponding compactifications are impossible (see, in particular, [35, p. 156]). Therefore other (more general) constructions should be found. These constructions mean a topological improvement of the space of solutions of the initial problem too. One of approaches is connected with different localizations of compactifications and is realized by employing perfect or almost perfect [71] mappings. Moreover, we use some analogue of 'usual' compactness. For example, we use the notions of countably compact sets in topological space (TS). Finally, below we consider topological constructions in interrelation with measurable structures. Therefore previously it was important to consider set-theoretic notions and some introductory notions of general topology. In this chapter we recall a series of the well known properties. The most of constructions of this chapter are used in the sequel. But some of them are given for completeness of the presentation. When considering auxiliary constructions, it is useful to discuss some interpretations and connections of them with other notions. For example, when introducing filters and ultrafilters of measurable spaces, it is advisable to give the corresponding Stone representations (the space of Stone representation is considered together with a measurable structure). In Section 2.5 we give a common point of view with respect to constructions of continuity and measurability (the last notion is regarded as main). In the sequel we use it in the basic constructions connected with continuity. But the point of view connected with measurability seems to be highly useful for us.

## 2.2 SETS AND FAMILIES OF SETS

We use the standard set-theoretic symbolics including quantifiers ($\forall$, $\exists$), sheafs (&, $\vee$, $\Rightarrow$, $\Leftrightarrow$, and other). We use the following special symbols: def (by definition), $\triangleq$ (equals by definition), $\exists!$ (there exists and unique). This is consistent with [35, p. 37]. Namely, expressions $\mathbf{S}[X]$ and $\mathbf{S}[X \neq \varnothing]$ are exploited for the abbreviated notation of propositions '$X$ is a set' and '$X$ is a nonempty set' respectively. Moreover, the expression

$$\exists_X \mathbf{S}[X] \ (\exists_X \mathbf{S}[X \neq \varnothing])$$

replaces the proposition: there exists a set (a nonempty set) $X$. Finally, the expression

$$\forall_X \mathbf{S}[X] \ (\forall_X \mathbf{S}[X \neq \varnothing])$$

replaces the proposition: for each (for each nonempty) set $X$. In connection with the two last stipulations, we recall that the 'notion' of the set of all sets is contradictory.

We consider families of sets. In addition, in the following we suppose that a family (of sets) is the set each element of which is a set. Sometimes, a family stands for the term 'class of sets'. We use following designations compatible with [35, Section 3.3]. Namely, if $\mathbf{S}[H]$, then $\mathcal{P}(H)$ is def the family of all subsets of $H$, and $2^H \triangleq \mathcal{P}(H) \setminus \{\varnothing\}$; moreover, by $\mathrm{Fin}(H)$ we denote the family of all nonempty finite subsets of $H$ and suppose $\forall P \in \mathrm{Fin}(H)$:

$$(\mathrm{Fin})[H|P] \triangleq \{Q \in \mathrm{Fin}(H) | P \subset Q\} \in 2^{\mathrm{Fin}(H)}.$$

In particular, it is possible to use these designations when $H$ is a family (of course, we suppose that the reader is acquainted with basic set-theoretic operations: we use union, intersection, difference, product of sets and other standard operations without additional clarification; we exploit traditional definitions for the one-element set (singleton) $\{x\}$ corresponding to the object $x$ and for the ordered pair $(u, v)$ of the objects $u$ and $v$). If $\mathbf{S}[A]$ and $\mathbf{S}[B]$, then the set $A \times B \triangleq \{(a, b) : a \in A, b \in B\}$ generates the family $\mathcal{P}(A \times B)$ of all relations in $A \times B$; in particular, it is possible to consider the case $A = B$. Suppose that $\forall_T \mathbf{S}[T] \ \forall \zeta \in \mathcal{P}(T \times T) \ \forall a \in T \ \forall b \in T$ def:

$$(a \zeta b) \Leftrightarrow ((a, b) \in \zeta). \tag{2.2.1}$$

We use binary relations (2.2.1) with specific properties. Among all such relations we select pre-orders, orders and directions. We introduce the sets of pre-orders, orders, and directions respectively. Then $\forall_T \mathbf{S}[T]$:

$$(\mathrm{Ord})[T] \triangleq \{\zeta \in \mathcal{P}(T \times T)|\ (\forall t \in T:\ t\,\zeta\,t)\&(\forall x \in T\ \forall y \in T\ \forall z \in T: ((x\,\zeta\,y)\&(y\,\zeta\,z)) \Rightarrow (x\zeta z))\} \in \mathcal{P}(\mathcal{P}(T \times T)); \tag{2.2.2}$$

elements of the set (2.2.2) are exactly pre-orders on $T$. Let $\forall_T \mathbf{S}[T]$:

$$(\mathrm{Ord})_0[T] \triangleq \{\zeta \in (\mathrm{Ord})[T]|\ \forall x \in T\ \forall y \in T: ((x\,\zeta\,y)\&(y\,\zeta\,x)) \Rightarrow (x = y)\}; \tag{2.2.3}$$

of course, (2.2.3) is the set of all orders on $T$ (a binary relation $\zeta$ (2.2.1) on $T$ is an order on $T$ iff $\zeta$ is reflexive, transitive, and antisymmetric). We introduce $\forall_T \mathbf{S}[T]$:

$$(\mathrm{DIR})[T] \triangleq \{\zeta \in (\mathrm{Ord})[T]|\ \forall x \in T\ \forall y \in T\ \exists z \in T:\ (x\,\zeta\,z)\&(y\,\zeta\,z)\}; \tag{2.2.4}$$

elements of the set (2.2.4) are exactly directions on $T$. In the following, very often elements of (2.2.2)–(2.2.4) are designated by specific symbols: $\preceq$, $\ll$, and other. If $\mathbf{S}[A \neq \varnothing]$ and $\preceq \in (\mathrm{DIR})[A]$, then $(A, \preceq)$ is called the directed set (we consider only nonempty directed sets). Suppose that $\forall_A \mathbf{S}[A]\ \forall_B \mathbf{S}[B]$:

$$B^A \triangleq \{f \in \mathcal{P}(A \times B)|\ \forall a \in A\ \exists! b \in B:\ (a, b) \in f\}. \tag{2.2.5}$$

In (2.2.5) the set of all mappings (operators, functions) from $A$ into $B$ is introduced. So, functions are specific relations in $A \times B$. If $\mathbf{S}[A]$, $\mathbf{S}[B]$, $f \in B^A$ and $a \in A$, then $f(a) \in B$ has the property $(a, f(a)) \in f$. We distinguish $f \in B^A$ and $f(a) \in B$; the last object is the image of $f$ at the point $a \in A$. We distinguish the image of a point and the image of a set. In this connection we recall that $\forall_A \mathbf{S}[A]\ \forall_B \mathbf{S}[B]\ \forall f \in B^A$:

$$(\forall U \in \mathcal{P}(A):\ f^1(U) \triangleq \{f(x):\ x \in U\} \in \mathcal{P}(B)) \&(\forall V \in \mathcal{P}(B):\ f^{-1}(V) \triangleq \{x \in A|\ f(x) \in B\} \in \mathcal{P}(A)). \tag{2.2.6}$$

In (2.2.6) the image and inverse image of a set under a mapping are introduced. If $\mathbf{S}[A]$, $\mathbf{S}[B]$, $f \in B^A$ and $C \in \mathcal{P}(A)$ are given, then $(f|C) \triangleq f \cap (C \times B) \in B^C$ is the restriction of $f$ to the set $C$; if $c \in C$, then $(f|C)(c) = f(c)$.

If $\mathbf{S}[A \neq \varnothing]$, $\preceq \in (\mathrm{DIR})[A]$, $\mathbf{S}[B]$ and $h \in B^A$ are given, then the triplet $(A, \preceq, h)$ is called a net in $B$. So, nets are defined by operators on directed sets. Moreover, $\forall_{\mathbf{D}} \mathbf{S}[\mathbf{D} \neq \varnothing]\ \forall \preceq \in (\mathrm{DIR})[\mathbf{D}]$:

$$(\preceq\text{–cof})[\mathbf{D}] \triangleq \{H \in \mathcal{P}(\mathbf{D})|\ \forall d \in \mathbf{D}\ \exists \delta \in H:\ d \preceq \delta\} \in \mathcal{P}(2^{\mathbf{D}}). \tag{2.2.7}$$

Elements of the family (2.2.7) are cofinal (with respect to $(\mathbf{D}, \preceq)$) subsets of $\mathbf{D}$. We have $\forall_{\mathbf{D}}\, \mathbf{S}[\mathbf{D} \neq \varnothing]\ \forall \zeta \in (\mathrm{DIR})[\mathbf{D}]\ \forall H \in (\preceq\!-\mathrm{cof})[\mathbf{D}]$:

$$\zeta \cap (H \times H) \in (\mathrm{DIR})[H]. \tag{2.2.8}$$

In (2.2.8) the natural restriction of the direction $\zeta$ is introduced; in designations of (2.2.8) the pair $(H, \zeta \cap (H \times H))$ is a directed set. Let $\forall_A\, \mathbf{S}[A \neq \varnothing]\ \forall_B\, \mathbf{S}[B \neq \varnothing]\ \forall \preceq \in (\mathrm{DIR})[A]\ \forall \angle \in (\mathrm{DIR})[B]$:

$$\begin{aligned}(\mathrm{Isot})[A; \preceq; B, \angle] \triangleq \{g \in B^A |\ (g^1(A) \in (\angle\!-\mathrm{cof})[B])\&(\forall \alpha_1 \in A\ \forall \alpha_2 \in A :\\ (\alpha_1 \preceq \alpha_2) \Rightarrow (g(\alpha_1) \angle g(\alpha_2)))\}.\end{aligned} \tag{2.2.9}$$

Elements of (2.2.9) are isotone mappings from $(A, \preceq)$ into $(B, \angle)$, whose images are cofinal. If $(B, \angle, h)$ is a net in the set $H$, $(A, \preceq)$ is a directed set (nonempty), and $g$ is an element of the set (2.2.9), then $(A, \preceq, h \circ g)$ is a new net in $H$ called the isotone subnet of $(B, \angle, h)$.

Consider some specific types of families. Let $\forall_X\, \mathbf{S}[X]$:

$$\begin{aligned}\pi[X] \triangleq \{\mathcal{X} \in \mathcal{P}(\mathcal{P}(X))\ |\ (\varnothing \in \mathcal{X})\&(X \in \mathcal{X})\\ \&(\forall A \in \mathcal{X}\ \forall B \in \mathcal{X} :\ A \cap B \in \mathcal{X})\}.\end{aligned} \tag{2.2.10}$$

Elements of the set (2.2.10) are multiplicative families of subsets of $X$ with 'zero' and 'unit'. In the following we call elements of (2.2.10) simply multiplicative families of subsets of $X$. Of course,

$$\forall_X\, \mathbf{S}[X] :\ \pi[X] \subset 2^{\mathcal{P}(X)}.$$

We use (2.2.10) as the basis for development of set-theoretic constructions in two following directions: constructing topological structures and constructing measurable structures. Suppose that $\forall_T\, \mathbf{S}[T]$:

$$\begin{aligned}(\mathrm{top})[T] &\triangleq \left\{\tau \in \pi[T]\ \middle|\ \forall \xi \in \mathcal{P}(\tau) :\ \bigcup_{G \in \xi} G \in \tau\right\}\\ &= \left\{\tau \in \pi[T]\ \middle|\ \forall \xi \in 2^\tau :\ \bigcup_{G \in \xi} G \in \tau\right\}.\end{aligned} \tag{2.2.11}$$

Elements of (2.2.11) are exactly topologies of the set T. If $\mathbf{S}[T]$ and $\tau \in (\mathrm{top})[T]$, then $(T, \tau)$ is a topological space (TS). Suppose that $\forall_T\, \mathbf{S}[T]$ $\forall \tau \in (\mathrm{top})[T]\ \forall M \in \mathcal{P}(T)$:

$$\mathbb{N}^0_\tau[M] \triangleq \{G \in \tau |\ M \subset G\}. \tag{2.2.12}$$

Elements of (2.2.12) are exactly open neighborhoods of $M$ in the TS $(T, \tau)$. By (2.2.12) we introduce 'arbitrary' neighborhoods, following the definition of [4]. Namely, $\forall_T \mathbf{S}[T]\ \forall \tau \in (\text{top})[T]\ \forall M \in \mathcal{P}(T)$:

$$\mathbb{N}_\tau[M] \triangleq \{H \in \mathcal{P}(T) |\ \exists G \in \mathbb{N}^0_\tau[M] :\ G \subset H\}. \qquad (2.2.13)$$

For (2.2.12) and (2.2.13) it is useful to consider the particular case when $M$ is an one-element set. Namely, $\forall_T \mathbf{S}[T]\ \forall \tau \in (\text{top})[T]\ \forall x \in T$:

$$(N^0_\tau(x) \triangleq \mathbb{N}^0_\tau[\{x\}])\ \&\ (N_\tau(x) \triangleq \mathbb{N}_\tau[\{x\}]). \qquad (2.2.14)$$

We obtain the family of neighborhoods of an arbitrary point in a TS (in the following we will see that the second family in (2.2.14) is a filter of subsets of the given TS). From (2.2.12)–(2.2.14) we obtain the obvious property: if $(T, \tau)$ is a TS and $x \in T$, then $N^0_\tau(x) = \{G \in \tau |\ x \in G\}$ and

$$\begin{aligned} N_\tau(x) &= \{H \in \mathcal{P}(T)\ |\ \exists G \in N^0_\tau(x) :\ G \subset H\} \\ &= \{H \in 2^T\ |\ \exists G \in N^0_\tau(x) :\ G \subset H\}; \end{aligned}$$

in addition, $N^0_\tau(x) \subset N_\tau(x)$ and $\forall H \in N_\tau(x) :\ x \in H$. We introduce the notions of $T_1$-space and $T_2$-space (in the following we do not use $T_0$-spaces); see [71, Ch. I]. Suppose that $\forall_T \mathbf{S}[T]$:

$$\begin{aligned} (\mathcal{D}-\text{top})[T] \triangleq &\{\tau \in (\text{top})[T] |\ \forall x \in T\ \forall y \in T \setminus \{x\}\ \exists H \in N_\tau(x) : y \notin H\} \\ = &\{\tau \in (\text{top})[T] |\ \forall x \in T\ \forall y \in T \setminus \{x\}\ \exists G \in N^0_\tau(x) :\ y \notin G\}. \end{aligned} \qquad (2.2.15)$$

If $\mathbf{S}[T]$ and $\tau \in (\mathcal{D} - \text{top})[T]$, then $(T, \tau)$ is called the $T_1$-space or attainable space [1]. Finally, we introduce Hausdorff spaces as basic TS. Suppose that $\forall_T \mathbf{S}[T]$:

$$\begin{aligned} (\text{top})_0[T] \triangleq &\{\tau \in (\text{top})[T] |\ \forall x \in T\ \forall y \in T \setminus \{x\}\ \exists H_1 \in N_\tau(x) \\ &\exists H_2 \in N_\tau(y) :\ H_1 \cap H_2 = \varnothing\} \\ = &\{\tau \in (\text{top})[T] |\ \forall x \in T\ \forall y \in T \setminus \{x\} \\ &\exists G_1 \in N^0_\tau(x)\ \exists G_2 \in N^0_\tau(y)\ : G_1 \cap G_2 = \varnothing\}. \end{aligned} \qquad (2.2.16)$$

If $\mathbf{S}[T]$ and $\tau \in (\text{top})_0[T]$, then the pair $(T, \tau)$ is called the Hausdorff space. Of course,

$$(\text{top})_0[T] \subset (\mathcal{D} - \text{top})[T].$$

Several separation axioms were earlier introduced. These axioms are connected with separation of points. We shall return to axioms of such a type. We now introduce the notion of closed set. We have $\forall_T \mathbf{S}[T]$ $\forall \tau \in (\text{top})[T]$:

$$\mathcal{F}_\tau \triangleq \{T \setminus G :\ G \in \tau\} = \{F \in \mathcal{P}(T) |\ T \setminus F \in \tau\} \in \pi[T]. \qquad (2.2.17)$$

Each element of the family $\mathcal{F}_\tau$ is a closed set (in the sense of $(T,\tau)$). Properties of the family (2.2.17) follow from (2.2.11). We have $\forall_T\, \mathbf{S}[T]$:

$$\begin{aligned}(\mathrm{top})^0[T] &\triangleq \{\tau \in (\mathrm{top})[T] \mid \forall F \in \mathcal{F}_\tau\ \forall x \in T \setminus F\ \exists H_1 \in \mathbb{N}_\tau[F] \\ &\qquad \exists H_2 \in N_\tau(x) :\ H_1 \cap H_2 = \varnothing\} \\ &= \{\tau \in (\mathrm{top})[T] \mid \forall F \in \mathcal{F}_\tau \forall x \in T \setminus F\ \exists G_1 \in \mathbb{N}_\tau^0[F] \\ &\qquad \exists G_2 \in N_\tau^0(x) : G_1 \cap G_2 = \varnothing\};\end{aligned} \tag{2.2.18}$$

if $\tau \in (\mathrm{top})^0[T]$, then $(T,\tau)$ is called a regular TS [71, Ch. 1]. If $\mathbf{S}[T]$ and $\tau \in (\mathcal{D}-\mathrm{top})[T] \cap (\mathrm{top})^0[T]$, then $(T,\tau)$ is called a $T_3$-space [71, Ch. 1]. Obviously, $\forall_T\, \mathbf{S}[T]$:

$$\begin{aligned}(\mathbf{n}-\mathrm{top})[T] &\triangleq \{\tau \in (\mathrm{top})[T] \mid \forall F_1 \in \mathcal{F}_\tau\ \forall F_2 \in \mathcal{F}_\tau :\ (F_1 \cap F_2 = \varnothing) \\ &\qquad \Rightarrow (\exists H_1 \in \mathbb{N}_\tau[F_1]\ \exists H_2 \in \mathbb{N}_\tau[F_2] :\ H_1 \cap H_2 = \varnothing)\} \\ &= \{\tau \in (\mathrm{top})[T] \mid \forall F_1 \in \mathcal{F}_\tau\ \forall F_2 \in \mathcal{F}_\tau :\ (F_1 \cap F_2 = \varnothing) \\ &\qquad \Rightarrow (\exists G_1 \in \mathbb{N}_\tau^0[F_1]\ \exists G_2 \in \mathbb{N}_\tau^0[F_2] :\ G_1 \cap G_2 = \varnothing)\};\end{aligned} \tag{2.2.19}$$

if $\tau \in (\mathbf{n}-\mathrm{top})[T]$, then $(T,\tau)$ is called a normal space. If $\mathbf{S}[T]$ and $\tau \in (\mathcal{D}-\mathrm{top})[T] \cap (\mathbf{n}-\mathrm{top})[T]$, then $(T,\tau)$ is called a $T_4$-space. As usual we introduce the notion of a centered family. Namely, if $\mathcal{H}$ is a nonempty family of sets, then

$$\mathbf{Z}[\mathcal{H}] \triangleq \left\{\mathcal{U} \in 2^{\mathcal{H}} \mid \forall \mathcal{K} \in \mathrm{Fin}(\mathcal{U}) :\ \bigcap_{K \in \mathcal{K}} K \neq \varnothing\right\}. \tag{2.2.20}$$

In (2.2.20) one can use the case $\mathcal{H} = \mathcal{F}_\tau$, where $\tau$ is a topology. We have (see [1, 4, 71]) $\forall_T\, \mathbf{S}[T]$

$$\begin{aligned}(\mathbf{c}-\mathrm{top})[T] &\triangleq \left\{\tau \in (\mathrm{top})[T] \mid \forall \xi \in 2^\tau :\ (T = \bigcup_{G \in \xi} G) \right. \\ &\qquad \left. \Rightarrow (\exists \mathcal{K} \in \mathrm{Fin}(\xi) : T = \bigcup_{G \in \mathcal{K}} G)\right\} \\ &= \left\{\tau \in (\mathrm{top})[T] \mid \forall \mathcal{H} \in \mathbf{Z}[\mathcal{F}_\tau] :\ \bigcap_{H \in \mathcal{H}} H \neq \varnothing\right\};\end{aligned} \tag{2.2.21}$$

in the case $\tilde{\tau} \in (\mathbf{c}-\mathrm{top})[T]$ we have, in the form of $(T,\tilde{\tau})$, a compact space. Suppose that $\forall_T\, \mathbf{S}[T]$:

$$(\mathbf{c}-\mathrm{top})_0[T] = (\mathbf{c}-\mathrm{top})[T] \cap (\mathrm{top})_0[T]. \tag{2.2.22}$$

If $\mathbf{S}[T]$ and $\tau \in (\mathbf{c} - \text{top})_0[T]$, then $(T, \tau)$ is called a compactum. The definitions (2.2.21) and (2.2.22) are very important; we use topologies of the sets (2.2.21) and (in particular) (2.2.22) in constructions of compactifications.

In the following we denote by $\mathbb{R}$ the set of all real numbers with the usual order $\leq$ and suppose that $\mathcal{N} \triangleq \{1; 2; ...\}$, $\mathcal{N} \subset \mathbb{R}$. Let $\forall k \in \mathcal{N}$:

$$(\overline{1,k} \triangleq \{i \in \mathcal{N} \mid i \leq k\}) \& (\overrightarrow{k,\infty} \triangleq \{i \in \mathcal{N} \mid k \leq i\}).$$

By virtue of $(\mathbb{R}, \leq)$ we construct the natural pointwise order in spaces of functionals. Namely, suppose that $\forall_X \mathbf{S}[X \neq \varnothing]$ $\forall f \in \mathbb{R}^X$ $\forall g \in \mathbb{R}^X$ def:

$$(f \leqq g) \Leftrightarrow (\forall x \in X : \ f(x) \leq g(x)).$$

Moreover, if $\mathbf{S}[X \neq \varnothing]$, then $\mathbb{O}_X \in \mathbb{R}^X$ is def the function for which $\forall x \in X : \ \mathbb{O}_X(x) \triangleq 0$. In the following, linear operations, product and order in spaces of functionals are defined only pointwise. Suppose that $\forall_X \mathbf{S}[X \neq \varnothing]$:

$$\mathbb{B}(X) \triangleq \{f \in \mathbb{R}^X \mid \exists c \in [0, \infty[\ \forall x \in X : \ |f(x)| \leq c\}.$$

Suppose that elements of $\mathcal{N}$ (or positive integers) are not sets. In this connection we make the natural stipulation (here and below we use the index form of notation of functions): if $\mathbf{S}[A]$ and $n \in \mathcal{N}$, then $A^n$ is used instead of $A^{\overline{1,n}}$ (so, $A^n$ is the set of all collections

$$(a_i)_{i \in \overline{1,n}} : \ \overline{1,n} \to A;$$

in other words, $A^n$ is $A \times ... \times A$; here we keep in mind the $n$-multiple product). The above stipulation about $\mathcal{N}$ excludes an ambiguity in designations of sets of mappings. Recall that for each set $A$ the set of all sequences

$$(a_i)_{i \in \mathcal{N}} : \mathcal{N} \to A$$

(in $A$) is denoted by $A^{\mathcal{N}}$. A family of sets can be used instead of $A$. Suppose that $\forall_T \mathbf{S}[T]$:

$$(\mathbf{c}_{\mathcal{N}} - \text{top})[T] \triangleq \Big\{ \tau \in (\text{top})[T] \mid \forall (G_i)_{i \in \mathcal{N}} \in \tau^{\mathcal{N}} : \ (T = \bigcup_{i \in \mathcal{N}} G_i) \Rightarrow (\exists n \in \mathcal{N} : \ T = \bigcup_{i=1}^{n} G_i) \Big\}; \tag{2.2.23}$$

if $\tilde{\tau} \in (\mathbf{c}_{\mathcal{N}} - \mathrm{top})[T]$, then $(T, \tilde{\tau})$ is called a countably compact space. Of course, $\forall_T\, \mathbf{S}[T]$:

$$(\mathbf{c} - \mathrm{top})[T] \subset (\mathbf{c}_{\mathcal{N}} - \mathrm{top})[T]. \tag{2.2.24}$$

We note an important representation of countably compact spaces. Namely, $\forall_X\, \mathbf{S}[X]$:

$$(\mathbf{c}_{\mathcal{N}} - \mathrm{top})[X] = \Big\{ \tau \in (\mathrm{top})[X] \mid \forall (H_i)_{i \in \mathcal{N}} \in \mathcal{F}_{\tau}^{\mathcal{N}} : (\forall m \in \mathcal{N} : \bigcap_{i=1}^{m} H_i \neq \varnothing) \Rightarrow (\bigcap_{i \in \mathcal{N}} H_i \neq \varnothing) \Big\}.$$

This representation is analogous to (2.2.21). We use this property in the next chapter.

We now end the topological introduction and consider some notions of measure theory. In addition, we consider constructions of topological and measurable spaces as the whole. In this connection we give many attention to questions of interrelation of the above mentioned types of structures. We return to (2.2.10). Suppose that $\forall_X\, \mathbf{S}[X]$ $\forall \mathcal{L} \in \pi[X]$ $\forall A \in \mathcal{P}(X)$ $\forall m \in \mathcal{N}$:

$$\Delta_m(A, \mathcal{L}) \triangleq \{(L_i)_{i \in \overline{1,m}} \in \mathcal{L}^m \mid (A = \bigcup_{i=1}^{m} L_i)\ \&\ (\forall p \in \overline{1,m}\ \forall q \in \overline{1,m} \setminus \{p\} : L_p \cap L_q = \varnothing)\}. \tag{2.2.25}$$

We call elements of the set (2.2.25) the ordered $\mathcal{L}$-partitions of $A$. Suppose that $\forall_X\, \mathbf{S}[X]$:

$$\Pi[X] \triangleq \{\mathcal{L} \in \pi[X] \mid \forall L \in \mathcal{L}\ \exists n \in \mathcal{N} :\ \Delta_n(X \setminus L, \mathcal{L}) \neq \varnothing\}. \tag{2.2.26}$$

If $\mathbf{S}[X]$ and $\mathcal{L} \in \Pi[X]$, then $(X, \mathcal{L})$ is called a measurable space with semi-algebra of sets; elements of (2.2.26) are exactly semi-algebras of subsets of $X$. We have $\forall_X\, \mathbf{S}[X]$:

$$\begin{aligned} (\mathrm{alg})[X] &\triangleq \{\mathcal{L} \in \Pi[X] \mid \forall L \in \mathcal{L} :\ X \setminus L \in \mathcal{L}\} \\ &= \{\mathcal{L} \in \pi[X] \mid \forall L \in \mathcal{L} :\ X \setminus L \in \mathcal{L}\} \\ &= \{\mathcal{L} \in \mathcal{P}(\mathcal{P}(X)) \mid (\varnothing \in \mathcal{L})\ \&\ (\forall L \in \mathcal{L} :\ X \setminus L \in \mathcal{L}) \\ &\qquad \&\ (\forall A \in \mathcal{L}\ \forall B \in \mathcal{L} : A \cup B \in \mathcal{L})\}. \end{aligned} \tag{2.2.27}$$

Of course, elements of the set (2.2.27) are exactly algebras of subsets of $X$. Finally, $\forall_X\ \mathbf{S}[X]$:

$$(\sigma-\text{alg})[X] \triangleq \left\{\mathcal{L} \in (\text{alg})[X] \mid \forall (L_i)_{i\in\mathcal{N}} \in \mathcal{L}^{\mathcal{N}} : \bigcup_{i\in\mathcal{N}} L_i \in \mathcal{L}\right\}. \quad (2.2.28)$$

Elements of (2.2.28) are exactly $\sigma$-algebras of subsets of $X$. We have $\forall_X\ \mathbf{S}[X]\ \forall \mathcal{X} \in \mathcal{P}(\mathcal{P}(X))$:

$$(\sigma-\text{alg})[X|\mathcal{X}] \triangleq \{\mathcal{L} \in (\sigma-\text{alg})[X] \mid \mathcal{X} \subset \mathcal{L}\} \in 2^{(\sigma-\text{alg})[X]} \quad (2.2.29)$$

(in (2.2.29) we use the following obvious property: $\mathcal{P}(X) \in (\sigma-\text{alg})[X]$). As usual we obtain $\forall_X\ \mathbf{S}[X]\ \forall \mathcal{X} \in \mathcal{P}(\mathcal{P}(X))$:

$$\sigma_X^0(\mathcal{X}) \triangleq \left(\bigcap_{\mathcal{L}\in(\sigma-\text{alg})[X|\mathcal{X}]} \mathcal{L}\right) \in (\sigma-\text{alg})[X]. \quad (2.2.30)$$

In addition, (2.2.30) is the $\sigma$-algebra of subsets of $X$ generated by the family $\mathcal{X}$. Of course, $\forall_X\ \mathbf{S}[X]\ \forall \mathcal{X} \in \mathcal{P}(\mathcal{P}(X))$:

$$(\sigma_X^0(\mathcal{X}) \in (\sigma-\text{alg})[X|\mathcal{X}])\&(\forall \mathcal{L} \in (\sigma-\text{alg})[X|\mathcal{X}] : \sigma_X^0(\mathcal{X}) \subset \mathcal{L}). \quad (2.2.31)$$

In (2.2.31) we have the well known property: (2.2.30) is the smallest element of the family (2.2.29). Note that the construction realized in terms of monotone classes [103, Ch. I] can be used for representation of (2.2.31). The natural union of topological and measurable structures is realized in terms of Borel structures. In this connection we note that (2.2.11) and (2.2.17) imply the following property: if $\mathbf{S}[X]$ and $\tau \in (\text{top})[X]$, then

$$(\sigma-\text{alg})[X|\tau] = (\sigma-\text{alg})[X|\mathcal{F}_\tau].$$

As a corollary we have $\forall_X\ \mathbf{S}[X]\ \forall \tau \in (\text{top})[X]$:

$$\mathcal{B}_\sigma(\tau) \triangleq \sigma_X^0(\tau) = \sigma_X^0(\mathcal{F}_\tau) \in (\sigma-\text{alg})[X]. \quad (2.2.32)$$

Of course, the $\sigma$-algebra (2.2.32) is called the $\sigma$-algebra of Borel subsets of the TS $(X,\tau)$; we call $(X,\mathcal{B}_\sigma(\tau))$ the Borel space generated by the initial TS $(X,\tau)$. The construction on the basis of (2.2.32) is traditional in classical measure theory. We note an analogue of (2.2.32) connected with the above mentioned structures (see [66, p. 70–74]).

Suppose that $\forall_X\ \mathbf{S}[X]\ \forall \mathcal{U} \in \mathcal{P}(\mathcal{P}(X))\ \forall \mathcal{V} \in \mathcal{P}(\mathcal{P}(X))$:

$$\mathcal{U}\{\cap\}\mathcal{V} \triangleq \{U \cap V : (U,V) \in \mathcal{U} \times \mathcal{V}\}. \quad (2.2.33)$$

Note that (2.2.33) does not depend on $X$ (if $\mathbf{S}[X_1]$, $\mathcal{U} \in \mathcal{P}(\mathcal{P}(X_1))$ and $\mathcal{V} \in \mathcal{P}(\mathcal{P}(X_1))$, then the following equality holds:

$$\begin{aligned}&\{H \in \mathcal{P}(X)|\ \exists U \in \mathcal{U}\ \exists V \in \mathcal{V}:\ H = U \cap V\}\\ =&\{H \in \mathcal{P}(X_1)|\ \exists U \in \mathcal{U}\ \exists V \in \mathcal{V}:\ H = U \cap V\}).\end{aligned}$$

Then (see [21, p. 74]), $\forall_X\ \mathbf{S}[X]\ \forall \tau \in (\mathrm{top})[X]$:

$$(\mathbf{s}-\mathbf{a})[\tau] \triangleq \tau\{\cap\}\mathcal{F}_\tau \in \Pi[X]. \tag{2.2.34}$$

In addition, (2.2.34) has an extremal property. Namely, $\forall_X\ \mathbf{S}[X]\ \forall \tau \in (\mathrm{top})[X]$:

$$(\tau \cup \mathcal{F}_\tau \subset (\mathbf{s}-\mathbf{a})[\tau])\&(\forall \mathcal{L} \in \Pi[X]:\ (\tau \cup \mathcal{F}_\tau \subset \mathcal{L}) \Rightarrow ((\mathbf{s}-\mathbf{a})[\tau] \subset \mathcal{L})). \tag{2.2.35}$$

Thus (2.2.34) is the semi-algebra of subsets of X 'generated' (see (2.2.35)) by the family $\tau \cup \mathcal{F}_\tau$. And what is more, $\forall_X\ \mathbf{S}[X]\ \forall \tau \in (\mathrm{top})[X]\ \forall \mathcal{L} \in \pi[X]$:

$$(\tau \cup \mathcal{F}_\tau \subset \mathcal{L}) \Rightarrow ((\mathbf{s}-\mathbf{a})[\tau] \subset \mathcal{L}).$$

Here we use (2.2.10). Note that algebras of sets are used most often in FAM theory. In this connection we recall that the construction similar to (2.2.29)–(2.2.32) is realized for obtaining the algebra of sets generated by an arbitrary family of subsets of 'unit'. Note that $\forall_X\ \mathbf{S}[X]\ \forall \mathcal{X} \in \mathcal{P}(\mathcal{P}(X))$:

$$(\mathrm{alg})[X|\mathcal{X}] \triangleq \{\mathcal{L} \in (\mathrm{alg})[X]|\ \mathcal{X} \subset \mathcal{L}\} \in 2^{(\mathrm{alg})[X]}.$$

Following the construction similar to (2.2.30), we introduce $\forall_X\ \mathbf{S}[X]\ \forall \mathcal{X} \in \mathcal{P}(\mathcal{P}(X))$:

$$a_X^0(\mathcal{X}) \triangleq \left(\bigcap_{\mathcal{L} \in (\mathrm{alg})[X|\mathcal{X}]} \mathcal{L}\right) \in (\mathrm{alg})[X]. \tag{2.2.36}$$

Of course, we call (2.2.36) the algebra generated by $\mathcal{X}$. In addition, $\forall_X\ \mathbf{S}[X]\ \forall \mathcal{X} \in \mathcal{P}(\mathcal{P}(X))$:

$$(a_X^0(\mathcal{X}) \in (\mathrm{alg})[X|\mathcal{X}])\&(\forall \mathcal{L} \in (\mathrm{alg})[X|\mathcal{X}]:\ a_X^0(\mathcal{X}) \subset \mathcal{L}). \tag{2.2.37}$$

Recall that in [103, Ch. I] some concrete method of determining the algebra (2.2.36) is given (see Proposition 1.2.2 of [103]). We restrict ourselves to the case when the initial family is a semi-algebra of sets. Then we have $\forall_X\ \mathbf{S}[X]\ \forall \mathcal{X} \in \Pi[X]$:

$$a_X^0(\mathcal{X}) = \{H \in \mathcal{P}(X)|\ \exists n \in \mathcal{N}:\ \Delta_n(H, \mathcal{X}) \neq \varnothing\}. \tag{2.2.38}$$

As for the representation (2.2.38) for the algebra (2.2.37) in the considered particular case, we note that it is possible to obtain very simple constructions for (2.2.37) in terms of (2.2.34) and (2.2.38) in the case $\mathcal{X} = \tau \in (\text{top})[X]$. Namely, $\forall_X \ \mathbf{S}[X] \ \forall \tau \in (\text{top})[X]$:

$$a_X^0(\tau) = a_X^0((\mathbf{s}-\mathbf{a})[\tau]) = \{H \in \mathcal{P}(X) | \ \exists n \in \mathcal{N}: \ \Delta_n(H, (\mathbf{s}-\mathbf{a})[\tau]) \neq \varnothing\}. \tag{2.2.39}$$

Note that elements of the semi-algebra (2.2.34) are locally closed sets in the corresponding TS (see [71, Ch. 2]). It is useful to take into account that $\forall_X \ \mathbf{S}[X] \ \forall \mathcal{X} \in \mathcal{P}(\mathcal{P}(X))$:

$$\sigma_X^0(\mathcal{X}) = \sigma_X^0(a_X^0(\mathcal{X})). \tag{2.2.40}$$

From (2.2.32), (2.2.39) and (2.2.40) we have $\forall_X \ \mathbf{S}[X] \ \forall \tau \in (\text{top})[X]$:

$$\mathcal{B}_\sigma(\tau) = \sigma_X^0(a_X^0(\tau)) = \sigma_X^0(a_X^0((\mathbf{s}-\mathbf{a})[\tau])). \tag{2.2.41}$$

In addition, in constructions (2.2.40) and (2.2.41) it is useful to exploit the representation of the $\sigma$-algebra generated by the algebra of sets in terms of monotone class (see, for example [103, Ch. I]). Suppose $\forall_X \ \mathbf{S}[X]$ $\forall \mathcal{L} \in \pi[X]$:

$$(\mathbf{SA})[\mathcal{L}] \triangleq \{L \in \mathcal{L} | \ \exists n \in \mathcal{N}: \ \Delta_n(X \setminus L, \mathcal{L}) \neq \varnothing\}. \tag{2.2.42}$$

**Proposition** 2.2.1 $\forall_X \ \mathbf{S}[X] \ \forall \mathcal{L} \in \pi[X]$: $(\mathbf{SA})[\mathcal{L}] \in \Pi[X]$.

**Proof.** Fix $\mathbf{S}[X]$ and $\mathcal{L} \in \pi[X]$. Let $\Omega \triangleq (\mathbf{SA})[\mathcal{L}]$. It is obvious that $\varnothing \in \Omega$ and $X \in \Omega$. Let $A \in \Omega$ and $B \in \Omega$. Then $A \cap B \in \mathcal{L}$. Choose $m \in \mathcal{N}$, $n \in \mathcal{N}$, $(A_i)_{i \in \overline{1,m}} \in \Delta_m(X \setminus A, \mathcal{L})$ and $(B_j)_{j \in \overline{1,n}} \in \Delta_n(X \setminus B, \mathcal{L})$. Such choice is possible by (2.2.42). Note that

$$\begin{aligned} X \setminus (A \cap B) &= (X \setminus A) \cup (A \setminus B) = (X \setminus A) \cup (A \cap (X \setminus B)) \\ &= \Big(\bigcup_{i=1}^{m} A_i\Big) \cup \Big(A \cap \Big(\bigcup_{j=1}^{n} B_j\Big)\Big) = \Big(\bigcup_{i=1}^{m} A_i\Big) \cup \Big(\bigcup_{j=1}^{n} (A \cap B_j)\Big). \end{aligned} \tag{2.2.43}$$

In addition, $A \cap B_j \in \mathcal{L}$ for $j \in \overline{1,n}$. Introduce

$$(C_k)_{k \in \overline{1,m+n}}: \ \overline{1, m+n} \to \mathcal{L},$$

setting

$$(\forall i \in \overline{1,m}: \ C_i \triangleq A_i) \& (\forall i \in \overline{m+1, m+n}: \ C_i \triangleq A \cap B_{i-m}).$$

In addition, $A_i \cap A = \varnothing$ for $i \in \overline{1,m}$. By (2.2.43) we have the property

$$(C_k)_{k \in \overline{1,m+n}} \in \Delta_{m+n}(X \setminus (A \cap B), \mathcal{L}).$$

As a corollary we obtain $A \cap B \in \Omega$. Since the choice of $A$ and $B$ was arbitrary, we have $\Omega \in \pi[X]$. Fix $\Lambda \in \Omega$. Choose (see (2.2.42)) $p \in \mathcal{N}$ and $(\Lambda_i)_{i \in \overline{1,p}} \in \Delta_p(X \setminus \Lambda, \mathcal{L})$. Let $q \in \overline{1,p}$. Then $\Lambda_q \in \mathcal{L}$. In addition, $\Lambda_q \subset X \setminus \Lambda$. Consider the collection

$$(M_i)_{i \in \overline{1,p}} : \overline{1,p} \to \mathcal{L} \tag{2.2.44}$$

for which $(\forall i \in \overline{1,p} \setminus \{q\} : M_i \triangleq \Lambda_i) \& (M_q \triangleq \Lambda)$. In addition, the union of all sets of collection (2.2.44) is a subset of $X \setminus \Lambda_q$. Let $x_0 \in X \setminus \Lambda_q$. Then $(x_0 \in \Lambda) \vee (\exists i \in \overline{1,p} : x_0 \in \Lambda_i)$. We have

$$(x_0 \in \Lambda) \Rightarrow \left( x_0 \in \bigcup_{i=1}^{p} M_i \right).$$

Let $\exists i \in \overline{1,p} : x_0 \in \Lambda_i$. Fix such $i \in \overline{1,p}$. Since $x_0 \notin \Lambda_q$, we have $i \neq q$. Therefore $i \in \overline{1,p} \setminus \{q\}$ and $M_i = \Lambda_i$. In this case $x_0$ is an element of the union of all sets $M_i$, $i \in \overline{1,p}$. Therefore we always have the property

$$x_0 \in \bigcup_{i=1}^{p} M_i.$$

We obtain the inclusion $X \setminus \Lambda_q \subset \bigcup_{i=1}^{p} M_i$. So $X \setminus \Lambda_q$ coincides with the union of all sets $M_i$, $i \in \overline{1,p}$. By the choice (2.2.44) we have $(M_i)_{i \in \overline{1,p}} \in \Delta_p(X \setminus \Lambda_q, \mathcal{L})$. From (2.2.42) we obtain the property $\Lambda_q \in \Omega$. Since the choice of $q$ was arbitrary, we get

$$(\Lambda_i)_{i \in \overline{1,p}} : \overline{1,p} \to \Omega.$$

Therefore $(\Lambda_i)_{i \in \overline{1,p}} \in \Delta_p(X \setminus \Lambda, \Omega)$. Since the choice of $\Lambda$ was arbitrary too, we have $\forall L \in \Omega \ \exists n \in \mathcal{N}$: $\Delta_n(X \setminus L, \Omega) \neq \varnothing$. As a corollary $\Omega \in \Pi[X]$. $\square$.

Consider the very important notion of the filter of subsets of a given set. In addition, we operate with filters of measurable structures. Let $\forall_X \ \mathbf{S}[X] \ \forall \mathcal{X} \in \pi[X]$:

$$\mathbb{F}^*(\mathcal{X}) \triangleq \{\mathcal{H} \in 2^{\mathcal{X}} | \ (\varnothing \notin \mathcal{H}) \& (\forall A \in \mathcal{H} \ \forall B \in \mathcal{H} : \ A \cap B \in \mathcal{H}) \\ \& \ (\forall U \in \mathcal{H} : \ \{V \in \mathcal{X} | \ U \subset V\} \subset \mathcal{H})\}; \tag{2.2.45}$$

we call elements of the family (2.2.45) $\mathcal{X}$-filters of X. Recall that under conditions realizing (2.2.45), $X$ is the union of all sets of $\mathcal{X}$; so the set $X$ is defined by $\mathcal{X}$ in a unique way. In connection with (2.2.45) we note that $\forall_X \ \mathbf{S}[X] \ \forall \tau \in (\text{top})[X] \ \forall x \in X$:

$$(N_\tau^0(x) \in \mathbb{F}^*(\tau)) \& (N_\tau(x) \in \mathbb{F}^*(\mathcal{P}(X))). \tag{2.2.46}$$

In (2.2.46) we have two variants of filters. Note that in topology one usually consider filters in the space of all subsets of a given TS. In this connection the second variant of (2.2.46) is especially important. If $\mathbf{S}[X]$, then elements of the set $\mathbb{F}^*(\mathcal{P}(X))$ are called filters of the set $X$. Note that $\forall_X \mathbf{S}[X] \ \forall_{\mathbf{D}} \mathbf{S}[\mathbf{D} \neq \varnothing] \ \forall \preceq \in (\mathrm{DIR})[\mathbf{D}] \ \forall h \in X^{\mathbf{D}}$:

$$(X - \mathrm{ass})[\mathbf{D}; \preceq; h] \triangleq \{M \in \mathcal{P}(X) \mid \exists d \in \mathbf{D} \ \forall \delta \in \mathbf{D} : (d \preceq \delta) \Rightarrow (h(\delta) \in M)\} \in \mathbb{F}^*(\mathcal{P}(X)). \tag{2.2.47}$$

We call (2.2.47) the filter associated with the net $(\mathbf{D}, \preceq, h)$. In terms of the filters (2.2.47) the Moor–Smith convergence is defined. Namely, def $\forall_X \mathbf{S}[X] \ \forall_{\mathbf{D}} \mathbf{S}[\mathbf{D} \neq \varnothing] \ \forall \tau \in (\mathrm{top})[X] \ \forall \preceq \in (\mathrm{DIR})[\mathbf{D}] \ \forall f \in X^{\mathbf{D}} \ \forall x \in X$:

$$((\mathbf{D}, \preceq, f) \xrightarrow{\tau} x) \Leftrightarrow (N_\tau(x) \subset (X - \mathrm{ass})[\mathbf{D}; \preceq; f]). \tag{2.2.48}$$

The relation (2.2.48) permits us to introduce the natural convergence of filters of the set $X$, using the expression on the right hand side of (2.2.48) as a basis. In general, nets and filters generate equivalent theories in topological aspect. This circumstance is noted (for example) in [81]. We consider sequences as a particular case of nets. In addition, $\mathcal{N}$ is equipped with the natural order $\leq_{\mathcal{N}}$ induced from $(\mathbb{R}, \leq)$ by the rule analogous to that used in the left hand side of (2.2.8). As a result, $(\mathcal{N}, \leq_{\mathcal{N}})$ is a directed set. If $\mathbf{S}[A]$ and $(a_i)_{i \in \mathcal{N}} \in A^{\mathcal{N}}$, then we consider the sequence $\mathbf{a} \triangleq (a_i)_{i \in \mathcal{N}}$ as the net $(\mathcal{N}, \leq_{\mathcal{N}}, \mathbf{a})$ in $A$. Of course, we use (2.2.48) in this particular case. But we exploit a more traditional designation. Namely, if $\mathbf{S}[X]$, $\tau \in (\mathrm{top})[X]$, $(x_i)_{i \in \mathcal{N}} \in X^{\mathcal{N}}$, and $x \in X$, then we use the expression $(x_i)_{i \in \mathcal{N}} \xrightarrow{\tau} x$ instead of the expression

$$(\mathcal{N}, \leq_{\mathcal{N}}, (x_i)_{i \in \mathcal{N}}) \xrightarrow{\tau} x.$$

As a result we have $\forall_X \mathbf{S}[X] \ \forall \tau \in (\mathrm{top})[X] \ \forall (x_i)_{i \in \mathcal{N}} \in X^{\mathcal{N}} \ \forall x \in X$:

$$((x_i)_{i \in \mathcal{N}} \xrightarrow{\tau} x) \Leftrightarrow (\forall H \in N_\tau(x) \ \exists n \in \mathcal{N} \ \forall k \in \overrightarrow{n, \infty} : x_k \in H). \tag{2.2.49}$$

The definition (2.2.49) following from (2.2.48) is usually used in the case of metrizable spaces. It is useful also to note the following circumstance. In metrizable compact spaces the procedures of the passage to subsequences of a given sequence of elements in the corresponding space are often used. In TS (on the basis of (2.2.9)) an analogous procedure of the passage to a subnet of a given net is used. If the latter is a sequence of elements of the initial space, then employing subnets is essential for the above procedure. Namely, we realize subnets of sequences in TS of a general kind. In this connection we introduce the corresponding variant of (2.2.9), setting $\forall_{\mathbf{D}}\ \mathbf{S}[\mathbf{D} \neq \varnothing]\ \forall \preceq \in (\mathrm{DIR})[\mathbf{D}]$:

$$\begin{aligned}(\mathrm{isot})[\mathbf{D};\preceq] &\triangleq (\mathrm{Isot})[\mathbf{D};\preceq;\mathcal{N};\leq_{\mathcal{N}}]\\ &= \{\rho \in \mathcal{N}^{\mathbf{D}} \mid (\forall n \in \mathcal{N}\ \exists d \in \mathbf{D} : n \leq \rho(d))\\ &\qquad \&\ (\forall d_1 \in \mathbf{D}\ \forall d_2 \in \mathbf{D} : (d_1 \preceq d_2) \Rightarrow (\rho(d_1) \leq (\rho(d_2)))\}.\end{aligned} \tag{2.2.50}$$

If $\mathbf{S}[A]$, $\mathbf{a} \triangleq (a_i)_{i\in\mathcal{N}}$ is a sequence in $A$, $(\mathbf{D},\preceq)$, $\mathbf{D} \neq \varnothing$, is a directed set, and $\rho \in (\mathrm{isot})[\mathbf{D};\preceq]$ (see (2.2.50)), then the triplet

$$(\mathbf{D},\preceq,\mathbf{a}\circ\rho) = (\mathbf{D},\preceq,(a_{\rho(d)})_{d\in\mathbf{D}})$$

is the (isotone) subnet of the sequence $\mathbf{a}$.

Finally, we introduce sequentially compact spaces. We use the following designation: $\mathbf{N} \triangleq \{(k_i)_{i\in\mathcal{N}} \in \mathcal{N}^{\mathcal{N}} \mid \forall s \in \mathcal{N} : \ k_s < k_{s+1}\}$. In these terms it is possible to introduce the set of all sequentially compact topologies of a given set. Namely, $\forall_X\ \mathbf{S}[X]$:

$$\begin{aligned}(\mathbf{c}_{\mathrm{seq}} - \mathrm{top})[X] \triangleq \{\tau \in (\mathrm{top})[X] \mid\ &\forall (x_i)_{i\in\mathcal{N}} \in X^{\mathcal{N}}\ \exists \rho \in \mathbf{N}\\ &\exists x \in X : (x_{\rho(s)})_{s\in\mathcal{N}} \xrightarrow{\tau} x\}.\end{aligned}$$

We supplement (2.2.24) by the following known property [71]. Namely, $\forall_T\ \mathbf{S}[T]$:

$$(\mathbf{c}_{\mathrm{seq}} - \mathrm{top})[T] \subset (\mathbf{c}_{\mathcal{N}} - \mathrm{top})[T].$$

Then (2.2.24) and the last relation characterize countably compact spaces as the most general space respectively to properties of the compactness type. Later, we add several useful representations in terms of subspaces to the given brief information. Moreover, it is possible to find the above mentioned notions in specific literature (see, in particular, [1, 71, 81, 90]).

Consider several well known properties of filters (2.2.45). We note that $\forall_X\ \mathbf{S}[X]\ \forall \mathcal{X} \in \pi[X]\ \forall K \in 2^X$:

$$\{H \in \mathcal{X} \mid K \subset H\} \in \mathbb{F}^*(\mathcal{X}).$$

In particular, in the last relation the case when $K$ is an one-element set is possible. Suppose that $\forall_X\ \mathbf{S}[X]\ \forall \mathcal{X} \in \pi[X]$:

$$\mathbb{F}_0^*(\mathcal{X}) \triangleq \{\mathcal{G} \in \mathbb{F}^*(\mathcal{X}) \mid \forall \mathcal{H} \in \mathbb{F}^*(\mathcal{X}) : \ (\mathcal{G} \subset \mathcal{H}) \Rightarrow (\mathcal{G} = \mathcal{H})\}. \tag{2.2.51}$$

Elements of the family (2.2.51) are called ultrafilters of the space $(X,\mathcal{X})$; this space is generated by equipping the set $X$ with a multiplicative structure defined in terms of $\mathcal{X} \in \pi[X]$.

PROPOSITION 2.2.2 $\forall_X\, \mathbf{S}[X]\ \forall \mathcal{X} \in \Pi[X]\, \forall x \in X$:

$$\{L \in \mathcal{X} |\ x \in L\} \in \mathbb{F}_0^*(\mathcal{X}). \tag{2.2.52}$$

PROOF. The proof is practically obvious. Indeed, fix $\mathbf{S}[X]$, $\mathcal{X} \in \Pi[X]$ and $x \in X$. Denote by $\mathcal{G}$ the family on the left hand side of (2.2.52). We know that $\mathcal{G} \in \mathbb{F}^*(\mathcal{X})$. Let $\mathcal{H} \in \mathbb{F}^*(\mathcal{X})$ and $\mathcal{G} \subset \mathcal{H}$. Then $\mathcal{G} = \mathcal{H}$. Indeed, suppose the contrary: $\mathcal{G} \neq \mathcal{H}$. Then $\mathcal{H} \setminus \mathcal{G} \neq \varnothing$. Fix $H \in \mathcal{H} \setminus \mathcal{G}$. Then $\mathcal{H} \subset \mathcal{X}$ and

$$(x \in H) \Rightarrow (H \in \mathcal{G}).$$

So, $x \notin H$. Then $x \in X \setminus H$. But from (2.2.26) for some $n \in \mathcal{N}$ we have

$$\Delta_n(X \setminus H, \mathcal{X}) \neq \varnothing.$$

Choose $(H_i)_{i \in \overline{1,n}} \in \Delta_n(X \setminus H, \mathcal{X})$. Then $X \setminus H$ is the union of all sets $H_i \in \mathcal{X}$, $i \in \overline{1,n}$. It is possible to choose $r \in \overline{1,n}$ for which $x \in H_r$. Then $H_r \in \mathcal{X}$ has the last property and hence $H_r \in \mathcal{G}$ (see (2.2.52)). Therefore $H \in \mathcal{H}$ and $H_r \in \mathcal{H}$. By (2.2.45) we have $H \cap H_r \in \mathcal{H}$. From (2.2.45) we obtain $H \cap H_r \neq \varnothing$. But $H_r \subset X \setminus H$. Therefore $H \cap H_r = \varnothing$. The obtained contradiction means that $\mathcal{G} = \mathcal{H}$. Since the choice of $\mathcal{H}$ was arbitrary, by (2.2.51) we have $\mathcal{G} \in \mathbb{F}_0^*(\mathcal{X})$. $\square$

PROPOSITION 2.2.3 $\forall_X\, \mathbf{S}[X]\ \forall \mathcal{X} \in \pi[X]\ \forall \mathcal{G} \in \mathbb{F}^*(X)\ \exists \mathcal{H} \in \mathbb{F}_0^*(\mathcal{X})$: $\mathcal{G} \subset \mathcal{H}$.

PROOF. Fix the set $X$, the family $\mathcal{X} \in \pi[X]$ and $\mathcal{G} \in \mathbb{F}^*(\mathcal{X})$. Consider the nonempty set

$$\mathbb{M} \triangleq \{\Omega \in \mathbb{F}^*(\mathcal{X}) |\ \mathcal{G} \subset \Omega\}.$$

Introduce the natural order $\preceq$ on $\mathbb{M}$, setting def $\forall \Omega_1 \in \mathbb{M}\ \forall \Omega_2 \in \mathbb{M}$:

$$(\Omega_1 \preceq \Omega_2) \Leftrightarrow (\Omega_1 \subset \Omega_2).$$

We use the natural ordering by inclusion. Consider $(\mathbb{M}, \preceq)$. In this space it is possible to introduce maximal elements (see [66, Ch. I]). Let

$$\begin{aligned}(\preceq -MAX)[\mathbb{M}] &\triangleq \{\mathcal{E} \in \mathbb{M} |\ \forall \mathcal{H} \in \mathbb{M}:\ (\mathcal{E} \preceq \mathcal{H}) \Rightarrow (\mathcal{E} = \mathcal{H})\} \\ &= \{\mathcal{E} \in \mathbb{M} |\ \forall \mathcal{H} \in \mathbb{M}:\ (\mathcal{E} \subset \mathcal{H}) \Rightarrow (\mathcal{E} = \mathcal{H})\}.\end{aligned} \tag{2.2.53}$$

In (2.2.53) we have the set of all maximal elements of $\mathbb{M}$ in the sense of the order $\preceq$. Below we use familiar Zorn's lemma [66, Ch. I]. Consider an arbitrary nonempty linearly ordered set $\mathbb{H}$, $\mathbb{H} \subset \mathbb{M}$. Then $\forall \mathcal{A} \in \mathbb{H}$ $\forall \mathcal{B} \in \mathbb{H}$:

$$(\mathcal{A} \subset \mathcal{B}) \vee (\mathcal{B} \subset \mathcal{A}). \tag{2.2.54}$$

Introduce the following nonempty subfamily of $\mathcal{X}$:

$$\mathfrak{G} \triangleq \bigcup_{\mathcal{H} \in \mathbb{H}} \mathcal{H} \in 2^{\mathcal{X}}. \tag{2.2.55}$$

Since $\mathbb{H} \subset \mathbb{F}^*(\mathcal{X})$, we have $\varnothing \notin \mathfrak{G}$. Moreover, from (2.2.45), (2.2.54) and (2.2.55) we obtain $\forall A \in \mathfrak{G}\ \forall B \in \mathfrak{G}$: $A \cap B \in \mathfrak{G}$. Finally, from (2.2.45) and (2.2.55) we have $\forall H \in \mathfrak{G}$: $\{S \in \mathcal{X} |\ H \subset S\} \subset \mathfrak{G}$. Of course, here we use the property that each element of $\mathbb{H}$ is an $\mathcal{X}$-filter. By (2.2.45) we obtain $\mathfrak{G} \in \mathbb{F}^*(\mathcal{X})$. In addition, $\forall \mathcal{H} \in \mathbb{H}$: $\mathcal{G} \subset \mathcal{H}$. As a corollary $\mathcal{G} \subset \mathfrak{G}$. We obtain $\mathfrak{G} \in \mathbb{M}$. But $\forall \mathcal{H} \in \mathbb{H}$: $\mathcal{H} \preceq \mathfrak{G}$. Therefore $\mathfrak{G}$ is a majorant of $\mathbb{H}$ in $(\mathbb{M}, \preceq)$. We have established that each nonempty linearly ordered set $\tilde{\mathbb{H}}$, $\tilde{\mathbb{H}} \subset \mathbb{M}$, has a majorant in $(\mathbb{M}, \preceq)$. The set (2.2.53) is not empty by Zorn's lemma. Choose $\mathcal{Y} \in (\preceq -MAX)[\mathbb{M}]$. Then $\mathcal{Y} \in \mathbb{M}$. As a corollary $\mathcal{Y} \in \mathbb{F}^*(\mathcal{X})$ and $\mathcal{G} \subset \mathcal{Y}$. Let $\mathcal{S} \in \mathbb{F}^*(\mathcal{X})$. Suppose that $\mathcal{Y} \subset \mathcal{S}$. Then $\mathcal{G} \subset \mathcal{S}$, since $\mathcal{G} \subset \mathcal{Y}$. Therefore $\mathcal{S} \in \mathbb{M}$ and $\mathcal{Y} \subset \mathcal{S}$. By properties of $\mathcal{Y}$ we obtain (see (2.2.53)) $\mathcal{Y} = \mathcal{S}$. Since the choice of $\mathcal{S}$ was arbitrary, from (2.2.51) it follows that $\mathcal{Y} \in \mathbb{F}_0^*(\mathcal{X})$ has the property $\mathcal{G} \subset \mathcal{Y}$. □

Note that statements of the type of Proposition 2.2.3 are usually considered in the case of filters of an algebra sets. In this case it is possible to establish some new useful facts. Namely, the following known [103, Ch. I] statement holds.

PROPOSITION 2.2.4 $\forall_X\ \mathbf{S}[X]\ \forall \mathcal{X} \in (\text{alg})[X]\ \forall \mathcal{H} \in \mathbb{F}_0^*(\mathcal{X})\ \forall H_1 \in \mathcal{X}$ $\forall H_2 \in \mathcal{X}$:

$$(H_1 \cup H_2 \in \mathcal{H}) \Rightarrow ((H_1 \in \mathcal{H}) \vee (H_2 \in \mathcal{H})).$$

PROOF. Fix $\mathbf{S}[X]$, $\mathcal{X} \in (\text{alg})[X]$, $\mathcal{H} \in \mathbb{F}_0^*(\mathcal{X})$, $U \in \mathcal{X}$ and $V \in \mathcal{X}$. Let $U \cup V \in \mathcal{H}$. Then $(U \in \mathcal{H}) \vee (V \in \mathcal{H})$. Indeed, suppose the contrary:

$$(U \in \mathcal{X} \setminus \mathcal{H}) \& (V \in \mathcal{X} \setminus \mathcal{H}). \tag{2.2.56}$$

Consider the family $\mathcal{C} \triangleq \{L \in \mathcal{X} |\ U \cup L \in \mathcal{H}\}$. Since $V \in \mathcal{C}$, we get $\mathcal{C} \in 2^{\mathcal{X}}$. Let $N \in \mathcal{C}$. Consider the family $\mathfrak{N} \triangleq \{H \in \mathcal{X} |\ N \subset H\}$. Since $N \in \mathcal{X}$, we have $N \in \mathfrak{N}$, and hence $\mathfrak{N} \in 2^{\mathcal{X}}$. Choose arbitrarily $\Gamma \in \mathfrak{N}$. Then $\Gamma \in \mathcal{X}$ and $N \subset \Gamma$. As a corollary we have the inclusion $U \cup N \subset U \cup \Gamma$. But $U \cup N \in \mathcal{H}$. Therefore $U \cup \Gamma \in \mathcal{H}$ and hence $\Gamma \in \mathcal{C}$. So $\mathfrak{N} \subset \mathcal{C}$. Since the choice of $N$ was arbitrary, we have

$$\forall G \in \mathcal{C}: \{H \in \mathcal{X} |\ G \subset H\} \subset \mathcal{C}. \tag{2.2.57}$$

Note that $U \cup \varnothing = U \notin \mathcal{H}$ (see (2.2.56)). Therefore $\varnothing \notin \mathcal{C}$. As a corollary $\mathcal{C} \subset \mathcal{X} \setminus \{\varnothing\} \subset 2^X$. This property supplements the relation (2.2.57). Choose two arbitrary sets $T \in \mathcal{C}$ and $\Theta \in \mathcal{C}$. Then $T \in \mathcal{X}$ and

$\Theta \in \mathcal{X}$. As a corollary $T \cap \Theta \in \mathcal{X}$. Moreover, $U \cup T \in \mathcal{H}$ and $U \cup \Theta \in \mathcal{H}$. By properties of each $\mathcal{X}$-filter we have

$$(U \cup T) \cap (U \cup \Theta) \in \mathcal{H}.$$

But $U \cup (T \cap \Theta)) = (U \cup T) \cap (U \cup \Theta)$; in this connection see [90, Ch. I, Section 4]. So $U \cup (T \cap \Theta) \in \mathcal{H}$ and $T \cap \Theta \in \mathcal{C}$. We have established that

$$\forall A \in \mathcal{C}\ \forall B \in \mathcal{C}:\ A \cap B \in \mathcal{C}. \tag{2.2.58}$$

From (2.2.45), (2.2.57) and (2.2.58) we have the property $\mathcal{C} \in \mathbb{F}^*(\mathcal{X})$. In addition, $V \in \mathcal{C}$ (see the basic supposition). Therefore $V \in \mathcal{C} \setminus \mathcal{H}$. Let $M \in \mathcal{H}$. As a corollary $\mathcal{W} \triangleq \{G \in \mathcal{X} | \ M \subset G\} \subset \mathcal{H}$. In addition, $U \cup M \in \mathcal{X}$ and consequently $U \cup M \in \mathcal{W}$. We obtain the property $U \cup M \in \mathcal{H}$. Therefore $M \in \mathcal{C}$. We have the inclusion $\mathcal{H} \subset \mathcal{C}$. So $\mathcal{C} \in \mathbb{F}^*(\mathcal{X})$: $\mathcal{H} \subset \mathcal{C}$. Since $\mathcal{H}$ is an ultrafilter, we have the equality $\mathcal{H} = \mathcal{C}$ which contradicts to the above mentioned property of $V$: $V \in \mathcal{C}$ and $V \notin \mathcal{H}$. So (2.2.56) is impossible, and $(U \in \mathcal{H}) \vee (V \in \mathcal{H})$. The proposition is proved. □

Recall that by (2.2.45) we have $\forall_X\ \mathbf{S}[X]\ \forall \mathcal{X} \in \pi[X]\ \forall \mathcal{H} \in \mathbb{F}^*(\mathcal{X})$: $X \in \mathcal{H}$. Therefore from Proposition 2.2.4 we have the following basic property. Namely, $\forall_X\ \mathbf{S}[X]\ \forall \mathcal{X} \in (\mathrm{alg})[X]\ \forall A \in \mathcal{X}\ \forall \mathcal{H} \in \mathbb{F}_0^*(\mathcal{X})$:

$$(A \in \mathcal{H}) \vee (X \setminus A \in \mathcal{H}). \tag{2.2.59}$$

From Proposition 2.2.3 we have the following obvious property. Namely, $\forall_X\ \mathbf{S}[X]\ \forall \mathcal{X} \in \pi[X]\ \forall \mathcal{G} \in \mathbb{F}^*(\mathcal{X})$:

$$\mathbb{F}_*^0(\mathcal{X}|\mathcal{G}) \triangleq \{\mathcal{H} \in \mathbb{F}_0^*(\mathcal{X}) |\ \mathcal{G} \subset \mathcal{H}\} \in 2^{\mathbb{F}_0^*(\mathcal{X})}. \tag{2.2.60}$$

By (2.2.60) we obtain that $\forall_X\ \mathbf{S}[X]\ \forall \mathcal{X} \in \pi[X]\ \forall \mathcal{G} \in \mathbb{F}^*(\mathcal{X})$:

$$\bigcap_{\mathcal{H} \in \mathbb{F}_*^0(\mathcal{X}|\mathcal{G})} \mathcal{H} \in \mathcal{P}(\mathcal{X}).$$

PROPOSITION 2.2.5 $\forall_X\ \mathbf{S}[X]\ \forall \mathcal{X} \in (\mathrm{alg})[X]\ \forall \mathcal{G} \in \mathbb{F}^*(\mathcal{X})$:

$$\mathcal{G} = \bigcap_{\mathcal{H} \in \mathbb{F}_*^0(\mathcal{X}|\mathcal{G})} \mathcal{H}.$$

PROOF. We fix $\mathbf{S}[X]$, $\mathcal{X} \in (\mathrm{alg})[X]$ and $\mathcal{G} \in \mathbb{F}^*(\mathcal{X})$. Denote by $\mathcal{K}$ the intersection of all families $\mathcal{H} \in \mathbb{F}_*^0(\mathcal{X}|\mathcal{G})$. Then $\mathcal{K} \in \mathcal{P}(\mathcal{X})$. In addition, $\mathcal{G} \subset \mathcal{K}$. Therefore by (2.2.45) we have the property $\mathcal{K} \in 2^{\mathcal{X}}$. Let $A \in \mathcal{X} \setminus \mathcal{G}$. Then by the implication (see (2.2.45))

$$(\mathcal{P}(A) \cap \mathcal{G} \neq \varnothing) \Rightarrow (A \in \mathcal{G}) \tag{2.2.61}$$

we have the following property: $\mathcal{P}(A) \cap \mathcal{G} = \varnothing$. Therefore $\forall B \in \mathcal{X} \cap \mathcal{P}(A) : B \notin \mathcal{G}$. Recall that $X \backslash A \in \mathcal{X}$; in this connection see (2.2.27). We note that by (2.2.60) the statement (2.2.59) is true for each $\mathcal{H} \in \mathbb{F}_*^0(\mathcal{X}|\mathcal{G})$. If $L \in \mathcal{G}$, then the sets $L \cap A \in \mathcal{X}$ and $L \cap (X \setminus A) \in \mathcal{X}$ possess the property

$$L = (L \cap A) \cup (L \cap (X \setminus A)),$$

from which the relation $L \cap (X \setminus A) \neq \varnothing$ follows (see (2.2.61)). We have established that $X \setminus A \in \mathcal{X}$ has the property

$$\forall G \in \mathcal{G} : G \cap (X \setminus A) \neq \varnothing. \tag{2.2.62}$$

Using (2.2.62) we introduce the following family:

$$\mathcal{C} \triangleq \{C \in \mathcal{X} | \; \exists G \in \mathcal{G} : G \cap (X \setminus A) \subset C\}.$$

We have $\mathcal{G} \subset \mathcal{C}$. Then $\mathcal{C} \in 2^{\mathcal{X}}$ (see (2.2.45)). In addition, by (2.2.62) we have $\varnothing \notin \mathcal{C}$. Let $Y \in \mathcal{C}$ and $Z \in \mathcal{C}$. Then $Y \cap Z \in \mathcal{X}$ (see (2.2.27)) and for some $\widetilde{Y} \in \mathcal{G}$ and $\widetilde{Z} \in \mathcal{G}$ the following relations hold:

$$(\widetilde{Y} \cap (X \setminus A) \subset Y) \& (\widetilde{Z} \cap (X \setminus A) \subset Z).$$

But $\widetilde{Y} \cap \widetilde{Z} \in \mathcal{G}$ (see (2.2.45)) and

$$(\widetilde{Y} \cap \widetilde{Z}) \cap (X \setminus A) \subset Y \cap Z.$$

So $Y \cap Z \in \mathcal{C}$. We have established that

$$\forall A' \in \mathcal{C} \; \forall B' \in \mathcal{C} : A' \cap B' \in \mathcal{C}.$$

By the definition of $\mathcal{C}$ we have the following obvious property: if $C \in \mathcal{C}$, then $\{L \in \mathcal{X} | \; C \subset L\} \subset \mathcal{C}$. By (2.2.45) we get $\mathcal{C} \in \mathbb{F}^*(\mathcal{X})$. Moreover, $(\mathcal{G} \subset \mathcal{C}) \& (X \setminus A \in \mathcal{C})$. By Proposition 2.2.3 one can choose $\mathcal{B} \in \mathbb{F}_0^*(\mathcal{X}) : \mathcal{C} \subset \mathcal{B}$. As a result $(\mathcal{G} \subset \mathcal{B}) \& (X \setminus A \in \mathcal{B})$. Then from (2.2.60) it follows that $\mathcal{B} \in \mathbb{F}_*^0(\mathcal{X}|\mathcal{G})$. It is possible to verify that $A \notin \mathcal{B}$ (see (2.2.45)). As a corollary $A \notin \mathcal{K}$. Of course, then $A \in \mathcal{X} \setminus \mathcal{K}$. We have established the inclusion

$$\mathcal{X} \setminus \mathcal{G} \subset \mathcal{X} \setminus \mathcal{K}.$$

Since $\mathcal{G} \subset \mathcal{X}$ and $\mathcal{K} \subset \mathcal{X}$, we have $\mathcal{K} \subset \mathcal{G}$. So $\mathcal{K} = \mathcal{G}$. □

PROPOSITION 2.2.6 $\forall_X \mathbf{S}[X] \; \forall \mathcal{X} \in (\mathrm{alg})[X]$:

$$\mathbb{F}_0^*(\mathcal{X}) = \{\mathcal{H} \in \mathbb{F}^*(\mathcal{X}) | \; \forall A \in \mathcal{X} : (A \in \mathcal{H}) \vee (X \setminus A \in \mathcal{H})\}. \tag{2.2.63}$$

PROOF. Fix $\mathbf{S}[X]$ and $\mathcal{X} \in (\mathrm{alg})[X]$. Denote by $\mathbf{F}$ the set on the right hand side of (2.2.63). From Proposition 2.2.4 we have the inclusion

$\mathbb{F}_0^*(\mathcal{X}) \subset \mathbf{F}$. Choose arbitrarily $\mathcal{H} \in \mathbf{F}$. So $\mathcal{H} \in \mathbb{F}^*(\mathcal{X})$ and $\forall A \in \mathcal{X}$ : $(A \in \mathcal{H}) \vee (X \setminus A \in \mathcal{H})$. Let $\mathfrak{B} \in \mathbb{F}^*(\mathcal{X})$ and $\mathcal{H} \subset \mathfrak{B}$. Then $\mathcal{H} = \mathfrak{B}$. Indeed, suppose the contrary; then $\mathfrak{B} \setminus \mathcal{H} \neq \varnothing$. Let $M \in \mathfrak{B} \setminus \mathcal{H}$. As a corollary (see (2.2.63)) $X \setminus M \in \mathcal{H}$. By the choice of $\mathfrak{B}$ we obtain that $X \setminus M \in \mathfrak{B}$. Therefore $\varnothing = M \cap (X \setminus M) \in \mathfrak{B}$. But the last statement is impossible. The obtained contradiction means that $\mathcal{H} = \mathfrak{B}$. We have established that $\mathcal{H} \in \mathbb{F}_0^*(\mathcal{X})$; in this connection see (2.2.51). So $\mathbf{F} \subset \mathbb{F}_0^*(\mathcal{X})$. □

We note that properties similar to Propositions 2.2.3–2.2.6 are used in the theory of Boolean algebras (see, for example, [111]). We recall these properties in connection with the following consideration of (0, 1)-measures.

We note a useful construction (see, for example, [103, Ch. I]) connected with imbedding of an algebra of sets in the space of ultrafilters of this algebra. Namely, $\forall_X\, \mathbf{S}[X]\ \forall \mathcal{X} \in (\mathrm{alg})[X]\ \forall L \in \mathcal{X}$:

$$(\mathcal{X} - ST)[L] \triangleq \{\mathcal{H} \in \mathbb{F}_0^*(\mathcal{X}) |\ L \in \mathcal{H}\}. \qquad (2.2.64)$$

The sets (2.2.64) are connected with the space of Stone representation [103, Ch. I]. Namely, $\forall_X\, \mathbf{S}[X]\ \forall \mathcal{X} \in (\mathrm{alg})[X]$:

$$(\mathbf{ST})[\mathcal{X}] \triangleq \{(\mathcal{X} - ST)[L] :\ L \in \mathcal{X}\}. \qquad (2.2.65)$$

Consider some properties of families of the kind (2.2.65).

PROPOSITION 2.2.7 *$\forall_X\, \mathbf{S}[X]\ \forall \mathcal{X} \in (\mathrm{alg})[X]$:*

$$(\mathbf{ST})[\mathcal{X}] \in (\mathrm{alg})[\mathbb{F}_0^*(\mathcal{X})].$$

PROOF. Fix $\mathbf{S}[X]$ and $\mathcal{X} \in (\mathrm{alg})[X]$. From (2.2.45), (2.2.51), (2.2.64) and (2.2.65) we have $(\mathcal{X} - ST)[\varnothing] = \varnothing$. Therefore $\varnothing \in (\mathbf{ST})[\mathcal{X}]$, since $\varnothing \in \mathcal{X}$. Let $M \in \mathcal{X}$ and $N \in \mathcal{X}$. Choose arbitrarily $\mathfrak{X} \in (\mathcal{X} - ST)[M \cup N]$. Then (see (2.2.64)) $\mathfrak{X} \in \mathbb{F}_0^*(\mathcal{X})$ and $M \cup N \in \mathfrak{X}$. By Proposition 2.2.4 we obtain that $M \in \mathfrak{X}$ or $N \in \mathfrak{X}$. Therefore by (2.2.64) we get

$$(\mathfrak{X} \in (\mathcal{X} - ST)[M]) \vee (\mathfrak{X} \in (\mathcal{X} - ST)[N]).$$

We have established the inclusion $(\mathcal{X} - ST)[M \cup N] \subset (\mathcal{X} - ST)[M] \cup (\mathcal{X} - ST)[N]$. Let $\mathfrak{M} \in (\mathcal{X} - ST)[M]$. Then $\mathfrak{M} \in \mathbb{F}_0^*(\mathcal{X})$ and $M \in \mathfrak{M}$. In addition, the set $M \cup N \in \mathcal{X}$ has the property $M \cup N \in \mathfrak{M}$. From (2.2.64) we have $\mathfrak{M} \in (\mathcal{X} - ST)[M \cup N]$. So $(\mathcal{X} - ST)[M] \subset (\mathcal{X} - ST)[M \cup N]$. Let $\mathfrak{N} \in (\mathcal{X} - ST)[N]$. From (2.2.64) it follows that $\mathfrak{N} \in \mathbb{F}_0^*(\mathcal{X})$ and $N \in \mathfrak{N}$. By (2.2.45) we obtain (for the set $M \cup N \in \mathcal{X}$) the property $M \cup N \in \mathfrak{N}$. Therefore by (2.2.64) we have $\mathfrak{N} \in (\mathcal{X} - ST)[M \cup N]$. So $(\mathcal{X} - ST)[N] \subset (\mathcal{X} - ST)[M \cup N]$. We have established the inclusion

$$(\mathcal{X} - ST)[M] \cup (\mathcal{X} - ST)[N] \subset (\mathcal{X} - ST)[M \cup N].$$

As a corollary $(\mathcal{X}-ST)[M]\cup(\mathcal{X}-ST)[N] = (\mathcal{X}-ST)[M\cup N] \in (\mathbf{ST})[\mathcal{X}]$ (see (2.2.65)). Since the choice of $M$ and $N$ was arbitrary, it follows that

$$\forall \mathbf{A} \in (\mathbf{ST})[\mathcal{X}]\ \forall \mathbf{B} \in (\mathbf{ST})[\mathcal{X}] :\ \mathbf{A} \cup \mathbf{B} \in (\mathbf{ST})[\mathcal{X}]. \tag{2.2.66}$$

Recall that $\forall \mathcal{H} \in \mathbb{F}_0^*(\mathcal{X}) :\ X \in \mathcal{H}$. We use (2.2.45) and (2.2.51). By (2.2.64) we have $(\mathcal{X}-ST)[X] = \mathbb{F}_0^*(\mathcal{X})$. As a corollary $\mathbb{F}_0^*(\mathcal{X}) \in (\mathbf{ST})[\mathcal{X}]$. Fix $\mathbf{R}_* \in (\mathbf{ST})[\mathcal{X}]$ and choose $R_* \in \mathcal{X}$ such that the representation $\mathbf{R}_* = (\mathcal{X} - ST)[R_*]$ holds. Consider the set $\mathbb{F}_0^*(\mathcal{X}) \setminus \mathbf{R}_*$. Note that $X \setminus R_* \in \mathcal{X}$. In addition, by (2.2.65) we have the property

$$(\mathcal{X} - ST)[X \setminus R_*] \in (\mathbf{ST})[\mathcal{X}]. \tag{2.2.67}$$

By Proposition 2.2.6 we obtain that $\forall \mathcal{H} \in \mathbb{F}_0^*(\mathcal{X}) :\ (R_* \in \mathcal{H}) \vee (X \setminus R_* \in \mathcal{H})$. Therefore by (2.2.64),

$$\mathbb{F}_0^*(\mathcal{X}) = (\mathcal{X} - ST)[R_*] \cup (\mathcal{X} - ST)[X \setminus R_*]. \tag{2.2.68}$$

Moreover, the sets on the right hand side of (2.2.68) have empty intersection. Indeed, suppose the contrary:

$$(\mathcal{X} - ST)[R_*] \cap (\mathcal{X} - ST)[X \setminus R_*] \neq \varnothing. \tag{2.2.69}$$

Choose an element $\mathcal{U}$ of the set on the left hand side of (2.2.69). Then by (2.2.64), $R_* \in \mathcal{U}$ and $X \setminus R_* \in \mathcal{U}$. Hence $\varnothing = R_* \cap (X \setminus R_*) \in \mathcal{U}$, because $\mathcal{U} \in \mathbb{F}^*(\mathcal{X})$ and (therefore) $\mathcal{U}$ is a multiplicative family. But the relation $\varnothing \in \mathcal{U}$ is impossible (see (2.2.45)). The contradiction implies the required property

$$(\mathcal{X} - ST)[R_*] \cap (\mathcal{X} - ST)[X \setminus R_*] = \varnothing. \tag{2.2.70}$$

From (2.2.68) and (2.2.70) we have the relation

$$\mathbb{F}_0^*(\mathcal{X}) \setminus \mathbf{R}_* = \mathbb{F}_0^*(\mathcal{X}) \setminus (\mathcal{X} - ST)[R_*] = (\mathcal{X} - ST)[X \setminus R_*].$$

From the last relation and (2.2.67) we obtain the property

$$\mathbb{F}_0^*(\mathcal{X}) \setminus \mathbf{R}_* \in (\mathbf{ST})[\mathcal{X}].$$

But the choice of $\mathbf{R}_*$ was arbitrary. It follows that

$$\forall \mathbf{R} \in (\mathbf{ST})[\mathcal{X}] :\ \mathbb{F}_0^*(\mathcal{X}) \setminus \mathbf{R} \in (\mathbf{ST})[\mathcal{X}]. \tag{2.2.71}$$

We obtain that (2.2.65) is the nonempty family of subsets of $\mathbb{F}_0^*(\mathcal{X})$ with the following properties (see (2.2.66), (2.2.71)):

$$\begin{aligned} &(\varnothing \in (\mathbf{ST})[\mathcal{X}])\&(\forall \mathbf{L} \in (\mathbf{ST})[\mathcal{X}] : \mathbb{F}_0^*(\mathcal{X}) \setminus \mathbf{L} \in (\mathbf{ST})[\mathcal{X}]) \\ &\&(\forall \mathbf{A} \in (\mathbf{ST})[\mathcal{X}]\ \forall \mathbf{B} \in (\mathbf{ST})[\mathcal{X}] : \mathbf{A} \cup \mathbf{B} \in (\mathbf{ST})[\mathcal{X}]). \end{aligned} \tag{2.2.72}$$

From (2.2.27) and (2.2.72) we have $(\mathbf{ST})[\mathcal{X}] \in (\text{alg})[\mathbb{F}_0^*(\mathcal{X})]$. □

We note that by Proposition 2.2.2, $\forall_X\ \mathbf{S}[X \neq \varnothing]\ \forall \mathcal{X} \in \Pi[X] : \mathbb{F}_0^*(\mathcal{X}) \neq \varnothing$. In particular, we have the following property: if $\mathbf{S}[X \neq \varnothing]$ and $\mathcal{X} \in (\text{alg})[X]$, then $\mathbf{S}[\mathbb{F}_0^*(\mathcal{X}) \neq \varnothing]$ and (2.2.65) is an algebra of subsets of $\mathbb{F}_0^*(\mathcal{X})$. Hence

$$(\mathbb{F}_0^*(\mathcal{X}), (\mathbf{ST})[\mathcal{X}]) \tag{2.2.73}$$

is a measurable space with an algebra of sets. We call (2.2.73) the space of Stone representation.

## 2.3 SUBSPACES

In this section we deal with some constructions connected with the well known notion of subspace. In addition, here it is possible to consider subspaces of TS or measurable spaces. And what is more, two different cases have much in common. The natural problem of restriction of ultrafilters adjoins to the two above variants. Therefore we use a highly general approach. We exploit (2.2.33) in the case when one of the families used in (2.2.33) is an one-element set. So $\forall_X\ \mathbf{S}[X]\ \forall \mathcal{X} \in \mathcal{P}(\mathcal{P}(X))$ $\forall Y \in \mathcal{P}(X)$:

$$\mathcal{X}|_Y \triangleq \{U \cap Y : U \in \mathcal{X}\} = \mathcal{X}\{\cap\}\mathcal{Y}|_{\mathcal{Y}=\{Y\}}. \tag{2.3.1}$$

We do not use the symbol $X$ (see the remark after (2.2.33)) in this designation. One can use an element of $\pi[X]$ in place of $\mathcal{X}$. In particular, one can take a topology, a semi-algebra, an algebra or a $\sigma$-algebra of $X$ for $\mathcal{X}$ (see (2.3.1)). In this connection we note that $\forall_X\ \mathbf{S}[X]\ \forall \mathcal{X} \in \pi[X]$ $\forall Y \in \mathcal{X}$:

$$\mathcal{X}|_Y = \{Z \in \mathcal{X} |\ Z \subset Y\} = \mathcal{X} \cap \mathcal{P}(Y). \tag{2.3.2}$$

In this representation we take into account (2.2.10). Note the following simple property. Namely, $\forall_X\ \mathbf{S}[X]\ \forall \mathcal{X} \in \pi[X]\ \forall Y \in \mathcal{P}(X)$:

$$\mathcal{X}|_Y = \{U \cap Y : U \in \mathcal{X}\} \in \pi[Y]. \tag{2.3.3}$$

The relations (2.3.2) and (2.3.3) have many useful concrete variants. We begin with (2.3.2). In particular, $\forall_X\ \mathbf{S}[X]\ \forall \tau \in (\text{top})[X]$:

$$(\forall G \in \tau : \tau|_G = \tau \cap \mathcal{P}(G))\&(\forall F \in \mathcal{F}_\tau : \mathcal{F}_\tau|_F = \mathcal{F}_\tau \cap \mathcal{P}(F)). \tag{2.3.4}$$

Here we use (2.2.11) and (2.2.17). In terms of (2.3.3) open and closed subspaces of a TS are defined. In the general case we have $\forall_X\ \mathbf{S}[X]$ $\forall \tau \in (\text{top})[X]\ \forall Y \in \mathcal{P}(X)$:

$$\tau|_Y = \{G \cap Y : G \in \tau\} \in (\text{top})[Y]. \tag{2.3.5}$$

If $(X, \tau)$ is a TS, $Y \in \mathcal{P}(X)$ and $\theta = \tau|_Y$, then $(Y, \theta)$ is called a subspace of $(X, \tau)$. Note that the operation of the passage to a subspace of a TS has the transitivity property. Namely, if $(U, \tau_1)$, $(V, \tau_2)$ and $(W, \tau_3)$ are TS for which $(V, \tau_2)$ is a subspace of $(U, \tau_1)$ and $(W, \tau_3)$ is a subspace of $(V, \tau_2)$, then $(W, \tau_3)$ is a subspace of $(U, \tau_1)$. Many important constructions of general topology are defined in terms of subspaces. First we introduce $\forall_X \mathbf{S}[X] \ \forall \tau \in (\text{top})[X]$:

$$(\tau - \text{comp})[X] \triangleq \{K \in \mathcal{P}(X) | \ \tau|_K \in (\mathbf{c} - \text{top})[K]\}$$
$$= \{K \in \mathcal{P}(X) | \ \forall \mathcal{G} \in 2^\tau : \ (K \subset \bigcup_{G \in \mathcal{G}} G) \Rightarrow (\exists \mathcal{K} \in \text{Fin}(\mathcal{G}) :$$
$$K \subset \bigcup_{G \in \mathcal{K}} G)\}. \tag{2.3.6}$$

It is very convenient to use the natural combination of (2.2.21) and (2.3.6). Of course, exactly elements of the family (2.3.6) are compact in TS $(X, \tau)$. In addition, $\forall_X \mathbf{S}[X] \ \forall \tau \in (\mathbf{c} - \text{top})[X]$:

$$\mathcal{F}_\tau \subset (\tau - \text{comp})[X].$$

Moreover, it is possible to establish the natural connection of the filters of neighborhoods of a point in a TS and in the corresponding subspace of the initial TS. Namely, from (2.3.1) we have $\forall_X \mathbf{S}[X] \ \forall \tau \in (\text{top})[X]$ $\forall Y \in \mathcal{P}(X) \ \forall y \in Y$:

$$N^0_{\tau|_Y}(y) = N^0_\tau(y)|_Y. \tag{2.3.7}$$

As a corollary we obtain $\forall_X \mathbf{S}[X] \ \forall \tau \in (\text{top})[X] \ \forall Y \in \mathcal{P}(X) \ \forall y \in Y$:

$$N_{\tau|_Y}(y) = N_\tau(y)|_Y. \tag{2.3.8}$$

In connection with (2.2.47), (2.2.48) and (2.3.8) we note the following useful property: if $\mathbf{S}[X]$, $\mathbf{S}[\mathbf{D} \neq \varnothing]$, $\tau \in (\text{top})[X]$, $\preceq \in (\text{DIR})[\mathbf{D}]$, $Y \in \mathcal{P}(X)$, $f \in Y^{\mathbf{D}}$ and $y \in Y$, then

$$((\mathbf{D}, \preceq, f) \xrightarrow{\tau} y) \Leftrightarrow ((\mathbf{D}, \preceq, f) \xrightarrow{\tau|_Y} y). \tag{2.3.9}$$

In (2.3.9) we consider the net $(\mathbf{D}, \preceq, f)$ in two ways. Namely, on the left hand side of (2.3.9) this net is regarded as a net in the TS $(X, \tau)$; moreover, the point $y$ is treated as an element of $X$. We interpret $(\mathbf{D}, \preceq, f)$ on the right hand side of (2.3.9) as a net in the TS $(Y, \vartheta)$, where $\vartheta \triangleq \tau|_Y$. As a useful particular case we note the corresponding sequential analogue of (2.3.9). Namely, $\forall_X \mathbf{S}[X] \ \forall \tau \in (\text{top})[X] \ \forall Y \in \mathcal{P}(X) \ \forall (y_i)_{i \in \mathcal{N}} \in Y^{\mathcal{N}}$ $\forall y \in Y$:

$$((y_i)_{i \in \mathcal{N}} \xrightarrow{\tau} y) \Leftrightarrow ((y_i)_{i \in \mathcal{N}} \xrightarrow{\tau|_Y} y).$$

In connection with the employment of subspaces it is advisable to compare the closures of a set in a TS and in a subspace of this TS. But first we recall important notion of the closure operator itself.

As usual [1, 4, 71, 66, 81, 90] we suppose that $\forall_X\ \mathbf{S}[X]\ \forall \tau \in (\text{top})[X]\ \forall A \in \mathcal{P}(X)$:

$$\begin{aligned}\text{cl}(A,\tau) &\triangleq \{x \in X \mid \forall H \in N_\tau(x) :\ H \cap A \neq \varnothing\}\\ &= \{x \in X \mid \forall G \in N_\tau^0(x) :\ G \cap A \neq \varnothing\}.\end{aligned} \tag{2.3.10}$$

Of course, along with (2.3.10) it is possible to note several equivalent representations of the closure operator. We recall only two representations. If $\mathbf{S}[X]$, $\tau \in (\text{top})[X]$ and $A \in \mathcal{P}(\mathcal{X})$, then $\text{cl}(A,\tau)$ (2.3.10) is the smallest (in the sense of inclusion) element of the nonempty family $\{F \in \mathcal{F}_\tau \mid A \subset F\}$. Moreover, $\forall_X\ \mathbf{S}[X]\ \forall \tau \in (\text{top})[X]\ \forall A \in \mathcal{P}(X)$:

$$\begin{aligned}\text{cl}(A,\tau) = \{x \in X \mid \exists_{\mathbf{D}}\ \mathbf{S}[\mathbf{D} \neq \varnothing]\ \exists \preceq \in (\text{DIR})[\mathbf{D}]\ \exists f \in A^{\mathbf{D}} :\\ (\mathbf{D}, \preceq, f) \overset{\tau}{\to} x\}.\end{aligned} \tag{2.3.11}$$

In (2.3.11) we have the statement of the well known Birkhoff theorem. We note that $\forall_X\ \mathbf{S}[X]\ \forall \tau \in (\text{top})[X]\ \forall Y \in \mathcal{P}(X)$:

$$\mathcal{F}_{\tau|_Y} = \mathcal{F}_\tau|_Y. \tag{2.3.12}$$

From (2.3.12) and the representation of the closure of a set in the form of the smallest element in the class of all closed supersets, we have $\forall_X\ \mathbf{S}[X]$ $\forall \tau \in (\text{top})[X]\ \forall Y \in \mathcal{P}(X)\ \forall A \in \mathcal{P}(Y)$:

$$\text{cl}(A, \tau|_Y) = \text{cl}(A,\tau) \cap Y. \tag{2.3.13}$$

We use the notion of subspace of a TS in definitions of countably compact sets in an arbitrary TS. Suppose that $\forall_X\ \mathbf{S}[X]\ \forall \tau \in (\text{top})[X]$:

$$(\mathcal{N} - \tau - \text{comp})[X] \triangleq \{C \in \mathcal{P}(X) \mid \tau|_C \in (\mathbf{c}_{\mathcal{N}} - \text{top})[C]\}. \tag{2.3.14}$$

Elements of (2.3.14) are countably compact sets in a given topological space. Of course, $\forall_X\ \mathbf{S}[X]\ \forall \tau \in (\text{top})[X]$:

$$\begin{aligned}(\mathcal{N} - \tau - \text{comp})[X] = \{K \in \mathcal{P}(X) \mid \forall (G_i)_{i\in\mathcal{N}} \in \tau^{\mathcal{N}} : (K \subset \bigcup_{i\in\mathcal{N}} G_i)\\ \Rightarrow (\exists m \in \mathcal{N} :\ K \subset \bigcup_{i=1}^{m} G_i)\}.\end{aligned}$$

It is useful to note that $\forall_X\ \mathbf{S}[X]\ \forall \tau \in (\mathbf{c}_{\mathcal{N}} - \text{top})[X]$:

$$\mathcal{F}_\tau \subset (\mathcal{N} - \tau - \text{comp})[X].$$

Let us introduce sequentially compact sets in an arbitrary TS. Namely, $\forall_X \mathbf{S}[X] \ \forall \tau \in (\text{top})[X]$:

$$(\tau - \text{comp}_{\text{seq}})[X] \triangleq \{K \in \mathcal{P}(X) | \ \tau|_K \in (\mathbf{c}_{\text{seq}} - \text{top})[X]\}. \quad (2.3.15)$$

It is possible to easily establish that $\forall_X \mathbf{S}[X] \ \forall \tau \in (\mathbf{c}_{\text{seq}} - \text{top})[X]$:

$$\mathcal{F}_\tau \subset (\tau - \text{comp}_{\text{seq}})[X].$$

Using (2.3.15) and the corresponding definition of the previous section, we can verify that $\forall_X \mathbf{S}[X] \ \forall \tau \in (\text{top})[X]$:

$$(\tau - \text{comp}_{\text{seq}})[X] = \{K \in \mathcal{P}(X) | \ \forall (x_i)_{i \in \mathcal{N}} \in K^{\mathcal{N}} \exists l \in \mathbf{N}$$
$$\exists x \in K : \ (x_{l(s)})_{s \in \mathcal{N}} \xrightarrow{\tau} x\}. \quad (2.3.16)$$

Note that representations conceptually similar to (2.3.16) can be constructed for (2.3.6) and (2.2.14). We consider these representations, omitting corresponding proofs. First we note some simple properties. Suppose $\forall_X \mathbf{S}[X]$:

$$(\text{top})_I[X] \triangleq \{\tau \in (\text{top})[X] | \ \forall x \in X \ \exists (H_i)_{i \in \mathcal{N}} \in N_\tau(x)^{\mathcal{N}}$$
$$\forall H \in N_\tau(x) \ \exists k \in \mathcal{N} : \ H_k \subset H\}. \quad (2.3.17)$$

If $\mathbf{S}[X]$ and $\tau \in (\text{top})_I[X]$, then (see (2.3.17)) we call $(X, \tau)$ a TS with the first axiom of countability. Note that $\forall_X \mathbf{S}[X] \ \forall \tau \in (\text{top})_I[X]$:

$$(\tau \in (\mathbf{c}_{\text{seq}} - \text{top})[X]) \Leftrightarrow (\tau \in (\mathbf{c}_{\mathcal{N}} - \text{top})[X]). \quad (2.3.18)$$

As for (2.3.18) see, for example, the natural combination of Theorems 3.10.3 and 3.10.30 of [71]. Moreover, see [81, Theorem 5.5]. We introduce the known notion of sequentially closed sets. Suppose that $\forall_X \mathbf{S}[X] \ \forall \tau \in (\text{top})[X]$:

$$\mathcal{F}_{\text{seq}}[\tau] \triangleq \{H \in \mathcal{P}(X) \ | \forall (h_i)_{i \in \mathcal{N}} \in H^{\mathcal{N}} \ \forall x \in X :$$
$$((h_i)_{i \in \mathcal{N}} \xrightarrow{\tau} x) \Rightarrow (x \in H)\}. \quad (2.3.19)$$

Elements of (2.3.19) are exactly sequentially closed subsets of $X$. Note that $\forall_X \mathbf{S}[X] \ \forall \tau \in (\text{top})_0[X]$:

$$((\tau - \text{comp})[X] \subset \mathcal{F}_\tau) \& ((\tau - \text{comp}_{\text{seq}})[X] \subset \mathcal{F}_{\text{seq}}[\tau]). \quad (2.3.20)$$

We use (2.3.16) in the second statement of (2.3.20). In (2.3.20) we have very natural properties connecting the variants of compactness and closedness. Note that from properties of the previous section we have $\forall_X \mathbf{S}[X] \ \forall \tau \in (\text{top})[X]$:

$$((\tau - \text{comp})[X] \subset (\mathcal{N} - \tau - \text{comp})[X])$$
$$\&((\tau - \text{comp}_{\text{seq}})[X] \subset (\mathcal{N} - \tau - \text{comp})[X]). \tag{2.3.21}$$

In (2.3.21) we use the relations (2.2.24), (2.3.6), (2.3.14) and (2.3.15); moreover, we use the natural connection of the properties of compactness and sequential compactness. In connection with (2.3.21) it is useful to give representations of these properties in terms of convergence, which is typical for applications. Note that $\forall_X \mathbf{S}[X] \ \forall \tau \in (\text{top})[X]$:

$$(\mathcal{N} - \tau - \text{comp})[X] = \{K \in \mathcal{P}(X) | \ \forall (x_i)_{i \in \mathcal{N}} \in K^{\mathcal{N}} \ \exists_{\mathbf{D}} \mathbf{S}[\mathbf{D} \neq \varnothing]$$
$$\exists \preceq \in (\text{DIR})[\mathbf{D}] \ \exists h \in (\text{isot})[\mathbf{D}; \preceq]$$
$$\exists x \in K : \ (\mathbf{D}, \preceq, (x_{h(d)})_{d \in \mathbf{D}}) \xrightarrow{\tau} x\}. \tag{2.3.22}$$

In connection with (2.3.22) see (2.2.50). We omit a highly simple proof of (2.3.22), using the natural combination of statements similar to Theorem 3.10.3 of [71] and the reasoning of the kind [71, Ch. 2]. In connection with (2.3.20) and (2.3.22) we note that $\forall_X \mathbf{S}[X] \ \forall \tau \in (\text{top})_0[X]$:

$$(\mathcal{N} - \tau - \text{comp})[X] \subset \mathcal{F}_{\text{seq}}[\tau].$$

As for the property of compactness, we have an analogue of (2.3.22) (see [81, Ch. 2]). Namely, $\forall_X \mathbf{S}[X] \ \forall \tau \in (\text{top})[X]$:

$$(\tau - \text{comp})[X] = \{K \in \mathcal{P}(X) | \ \forall_{\mathbf{D}} \mathbf{S}[\mathbf{D} \neq \varnothing] \ \forall \preceq \in (\text{DIR})[\mathbf{D}] \forall f \in K^{\mathbf{D}}$$
$$\exists_{\mathbf{T}} \mathbf{S}[\mathbf{T} \neq \varnothing] \ \exists \ll \in (\text{DIR})[\mathbf{T}] \exists l \in (\text{Isot})[\mathbf{T}; \ll; \mathbf{D}; \preceq]$$
$$\exists x \in K : \ (\mathbf{T}, \ll, f \circ l) \xrightarrow{\tau} x\}. \tag{2.3.23}$$

Of course, (2.3.22) and (2.3.23) are supplemented by (2.3.16). The triplet of relations (2.3.16), (2.3.22) and (2.3.23) is connected with the construction using subspaces of TS. We will return to properties of the compactness kind when considering problems of extensions.

Now we return to the general definition (2.3.1) in connection with measurable spaces. We consider (2.3.1) in the case $\mathcal{X} \in \pi[X]$. In addition, we usually suppose that $\mathcal{X} \in \Pi[X]$, $\mathcal{X} \in (\text{alg})[X]$ or $\mathcal{X} \in (\sigma - \text{alg})[X]$. Moreover, later we consider some constructions for filters of measurable spaces for which replacement of the measurable structure is assumed. In addition, (2.3.2) and (2.3.3) define the very important particular case of 'measurable' subspaces. In (2.3.3) the case of a 'non-measurable' subspace is possible. Note that a semi-algebra or an $\sigma$-algebra of sets can be used in place of $\mathcal{X}$ in (2.3.3). In the last case we have a subspace of the standard measurable space.

PROPOSITION 2.3.1 $\forall_X \mathbf{S}[X] \; \forall \mathcal{X} \in \Pi[X] \; \forall Y \in \mathcal{P}(X)$: $\mathcal{X}|_Y \in \Pi[Y]$.

PROOF. Let $\mathbf{S}[X]$, $\mathcal{X} \in \Pi[X]$, and $Y \in \mathcal{P}(X)$. Then by (2.3.3), $\mathcal{Y} \triangleq \mathcal{X}|_Y \in \pi[Y]$. Fix $\Lambda \in \mathcal{Y}$. Choose $L \in \mathcal{X}$ such that $\Lambda = L \cap Y$. By (2.2.26) the set $X \setminus L$ admits a finite partition by sets of $\mathcal{X}$. Let $n \in \mathcal{N}$ be a number for which $\Delta_n(X \setminus L, \mathcal{X}) \neq \varnothing$. Choose $(L_i)_{i \in \overline{1,n}} \in \Delta_n(X \setminus L, \mathcal{X})$. Then $(L_i)_{i \in \overline{1,n}} \in \mathcal{X}^n$ has the following properties:

$$(X \setminus L = \bigcup_{i=1}^{n} L_i) \& (\forall p \in \overline{1,n} \; \forall q \in \overline{1,n} \setminus \{p\} : \; L_p \cap L_q = \varnothing). \quad (2.3.24)$$

Note that $\forall i \in \overline{1,n}$: $\Lambda_i \triangleq Y \cap L_i \in \mathcal{Y}$. From (2.3.24) we have the property $\Lambda_p \cap \Lambda_q = \varnothing$ for $p \in \overline{1,n}$ and $q \in \overline{1,n} \setminus \{p\}$. Consider the set $Y \setminus \Lambda$. We have

$$\begin{aligned} Y \setminus \Lambda &= Y \setminus (L \cap Y) = Y \cap (X \setminus (L \cap Y)) \\ &= Y \cap ((X \setminus L) \cup (X \setminus Y)) \\ &= (Y \cap (X \setminus L)) \cup (Y \cap (X \setminus Y)) \\ &= Y \cap (X \setminus L) = Y \cap (\bigcup_{i=1}^{n} L_i) = \bigcup_{i=1}^{n} \Lambda_i. \end{aligned}$$

By (2.2.25) we obtain

$$(\Lambda_i)_{i \in \overline{1,n}} \in \Delta_n(Y \setminus \Lambda, \mathcal{Y}).$$

Since the choice of $\Lambda$ was arbitrary, by (2.2.26) we have $\mathcal{Y} \in \Pi[Y]$.

PROPOSITION 2.3.2 $\forall_X \mathbf{S}[X] \quad \forall \mathcal{X} \in (\text{alg})[X] \quad \forall Y \in \mathcal{P}(X)$: $\mathcal{X}|_Y \in (\text{alg})[Y]$.

PROOF. Fix $\mathbf{S}[X]$, $\mathcal{X} \in (\text{alg})[X]$ and $Y \in \mathcal{P}(X)$. Then $\mathcal{Y} \triangleq \mathcal{X}|_Y \in \pi[Y]$. Let $A \in \mathcal{Y}$. Choose a set $B \in \mathcal{X}$ for which $A = Y \cap B$. Consider $Y \setminus A \in \mathcal{P}(Y)$. By (2.2.27) we obtain that $X \setminus B \in \mathcal{X}$. Therefore

$$\begin{aligned} Y \setminus A &= Y \setminus (Y \cap B) = Y \cap (X \setminus (Y \cap B)) \\ &= Y \cap ((X \setminus Y) \cup (X \setminus B)) = (Y \cap (X \setminus Y)) \cup (Y \cap (X \setminus B)) \\ &= Y \cap (X \setminus B) \in \mathcal{Y}. \end{aligned}$$

Since the choice of $A$ was arbitrary, we obtain that $\mathcal{Y} \in (\text{alg})[Y]$ (see (2.2.27)).

PROPOSITION 2.3.3 $\forall_X \mathbf{S}[X]\ \forall \mathcal{X} \in (\sigma-\text{alg})[X]\ \forall Y \in \mathcal{P}(X):\ \mathcal{X}|_Y \in (\sigma-\text{alg})[Y]$.

PROOF. Let $\mathbf{S}[X]$, $\mathcal{X} \in (\sigma-\text{alg})[X]$ and $Y \in \mathcal{P}(X)$. Then $\mathcal{Y} \triangleq \mathcal{X}|_Y \in (\text{alg})[Y]$ (see Proposition 2.3.2). Let

$$(\Lambda_i)_{i\in\mathcal{N}} : \mathcal{N} \to \mathcal{Y}.$$

Hence $\forall j \in \mathcal{N}:\ \mathcal{X}_j \triangleq \{L \in \mathcal{X} |\ Y \cap L = \Lambda_j\} \in 2^{\mathcal{X}}$. From the axiom of choice it follows that the product of all sets $\mathcal{X}_i$, $i \in \mathcal{N}$, is a nonempty set. Choose some element $(L_i)_{i\in\mathcal{N}}$ of the above mentioned product. Then $(L_i)_{i\in\mathcal{N}}$ is a sequence in $\mathcal{X}$ and the union of all sets $(L_i)_{i\in\mathcal{N}}$ is an element of $\mathcal{X}$. Consequently

$$\bigcup_{i\in\mathcal{N}} \Lambda_i = \bigcup_{i\in\mathcal{N}} (Y \cap L_i) = Y \cap (\bigcup_{i\in\mathcal{N}} L_i) \in \mathcal{Y}.$$

Since the choice of $(\Lambda_i)_{i\in\mathcal{N}}$ was arbitrary, we obtain the property $\mathcal{Y} \in (\sigma-\text{alg})[Y]$.

We recall that (see (2.3.2)) $\forall_X \mathbf{S}[X]\ \forall \mathcal{X} \in \Pi[X]\ \forall Y \in \mathcal{X}$:

$$\mathcal{X}|_Y = \{Z \in \mathcal{X} |\ Z \subset Y\} = \mathcal{X} \cap \mathcal{P}(Y) \in \Pi[Y] \cap 2^{\mathcal{X}}. \tag{2.3.25}$$

From (2.3.25) and Proposition 2.3.2 we have $\forall_X \mathbf{S}[X]\ \forall \mathcal{X} \in (\text{alg})[X]\ \forall Y \in \mathcal{X}$:

$$\mathcal{X}|_Y = \{Z \in \mathcal{X} |\ Z \subset Y\} = \mathcal{X} \cap \mathcal{P}(Y) \in (\text{alg})[Y] \cap 2^{\mathcal{X}}. \tag{2.3.26}$$

From (2.3.26) and Proposition 2.3.3 we obtain that $\forall_X \mathbf{S}[X]\ \forall \mathcal{X} \in (\sigma-\text{alg})[X]\ \forall Y \in \mathcal{X}$:

$$\mathcal{X}|_Y = \{Z \in \mathcal{X} |\ Z \subset Y\} = \mathcal{X} \cap \mathcal{P}(Y) \in (\sigma-\text{alg})[Y] \cap 2^{\mathcal{X}}. \tag{2.3.27}$$

We have two versions of the common object called subspace. These versions are typical for topological constructions of measure theory. We use the two above mentioned types of subspaces for constructing extensions.

We should note several statements about constructing algebras and $\sigma$-algebras of sets generated by 'arbitrary' families in subspaces.

PROPOSITION 2.3.4 $\forall_X \mathbf{S}[X]\ \forall \mathcal{H} \in \mathcal{P}(\mathcal{P}(X))\ \forall A \in \mathcal{P}(X)$:

$$a_A^0(\mathcal{H}|_A) = a_X^0(\mathcal{H})\ |_A\ .$$

The corresponding proof is very simple. Fix $\mathbf{S}[X]$, $\mathcal{H} \in \mathcal{P}(\mathcal{P}(X))$ and $A \in \mathcal{P}(X)$. Then $a_X^0(\mathcal{H}) \in (\text{alg})[X]$ and $\mathcal{H}|_A \in \mathcal{P}(\mathcal{P}(A))$. In addition, for $a_X^0(\mathcal{H})|_A \in (\text{alg})[A]$ (see Proposition 2.3.2) we have

$$\mathcal{H}\ |_A \subset a_X^0(\mathcal{H})|_A.$$

As a corollary, from (2.2.37) we obtain the following inclusion:

$$a_A^0(\mathcal{H}|_A) \subset a_X^0(\mathcal{H})|_A. \tag{2.3.28}$$

Consider the family

$$\mathcal{D} \triangleq \{M \in \mathcal{P}(X) |\ A \cap M \in a_A^0(\mathcal{H}|_A)\} \in (\text{alg})[X]. \tag{2.3.29}$$

To prove (2.3.29), we note that for $D \in \mathcal{D}$ we have

$$A \cap (X \setminus D) = A \setminus D = A \setminus (A \cap D) \in a_A^0(\mathcal{H}|_A).$$

Fix $\Lambda \in \mathcal{H}$. Then $A \cap \Lambda \in \mathcal{H}|_A$ and (as a consequence) $A \cap \Lambda \in a_A^0(\mathcal{H}|_A)$. The last property means that $\Lambda \in \mathcal{D}$ (see (2.3.29)). We have $\mathcal{H} \subset \mathcal{D}$. By (2.3.29) we obtain the inclusion $a_X^0(\mathcal{H}) \subset \mathcal{D}$ and hence, by (2.3.1),

$$a_X^0(\mathcal{H})|_A \subset a_A^0(\mathcal{H}|_A). \tag{2.3.30}$$

From (2.3.28) and (2.3.30) we have the required statement.

It is easily possible to obtain the following property.

LEMMA 2.3.1 *$\forall_X\, \mathbf{S}[X]\ \forall \mathcal{X} \in \mathcal{P}(\mathcal{P}(X))\ \forall A \in \mathcal{P}(X)$:*

$$\sigma_A^0(\mathcal{X}|_A) \subset \sigma_X^0(\mathcal{X})|_A.$$

The proof follows from Proposition 2.3.3.

LEMMA 2.3.2 *$\forall_X\, \mathbf{S}[X]\ \forall A \in \mathcal{P}(X)\ \forall \mathcal{L} \in (\sigma - \text{alg})[A]$: $\{H \in \mathcal{P}(X) |\ A \cap H \in \mathcal{L}\} \in (\sigma - \text{alg})[X]$.*

The proof is obvious and we omit it. A useful statement follows from Lemmas 2.3.1 and 2.3.2.

PROPOSITION 2.3.5 *$\forall_X\, \mathbf{S}[X]\ \forall \mathcal{X} \in \mathcal{P}(\mathcal{P}(X))\ \forall A \in \mathcal{P}(X)$:*

$$\sigma_A^0(\mathcal{X}|_A) = \sigma_X^0(\mathcal{X})|_A.$$

Note two obvious properties connected with the last two propositions. Namely, $\forall_X\, \mathbf{S}[X]\ \forall \mathcal{X} \in \mathcal{P}(\mathcal{P}(X))\ \forall A \in \mathcal{P}(X)$:

$$\begin{aligned} &((a_A^0(\mathcal{X}|_A) \subset a_X^0(\mathcal{X})) \Leftrightarrow (A \in a_X^0(\mathcal{X}))) \\ &\& \left((\sigma_A^0(\mathcal{X}|_A) \subset \sigma_X^0(\mathcal{X})) \Leftrightarrow (A \in \sigma_X^0(\mathcal{X}))\right). \end{aligned} \tag{2.3.31}$$

In (2.3.31) we have the natural conditions of measurability of the corresponding subspaces of the basic measurable space.

## 2.4 EXTENSION OF FILTERS

In this section we deal with a problem logically connected with constructions of subspaces. We consider different measurable structures on the same set. Therefore in this section we fix a nonempty set $X$. Namely, $\mathbf{S}[X \neq \varnothing]$ ( the case $X = \varnothing$ is not considered because then $\mathbb{F}^*(\mathcal{X}) = \varnothing$ for $\mathcal{X} \in \pi[X]$). We operate with $\mathcal{X} \in \pi[X]$ for constructing 'measurable' spaces $(X, \mathcal{X})$. In such spaces, we consider filters and, in particular, ultrafilters. Suppose that $\forall \mathcal{M} \in \pi[X]$:

$$\pi_0[X;\mathcal{M}] \triangleq \pi[X] \cap \mathcal{P}(\mathcal{M}) = \{\mathcal{L} \in \pi[X] | \ \mathcal{L} \subset \mathcal{M}\}. \qquad (2.4.1)$$

For $\mathcal{M} \in \pi[X]$ and $\mathcal{L} \in \pi_0[X;\mathcal{M}]$ we have (in the form of $(X, \mathcal{L})$) an analogue of a subspace of $(X, \mathcal{M})$. If $\mathcal{M} \in \pi[X]$ and $\mathfrak{G} \in 2^{\mathcal{M}}$, then we set

$$\psi[\mathcal{M};\mathfrak{G}] \triangleq \{M \in \mathcal{M} | \ \exists G \in \mathfrak{G}: \ G \subset M\}. \qquad (2.4.2)$$

The operation (2.4.2) is typical under constructing filters in terms of bases. In particular, in place of $\mathfrak{G}$ one can use an $\mathcal{L}$-filter for $\mathcal{L}$ of the set (2.4.1). In addition,

$$\forall \mathcal{M} \in \pi[X] \ \forall \mathcal{C} \in \pi_0[X;\mathcal{M}] \ \forall \mathcal{X} \in \mathbb{F}^*(\mathcal{C}): \ \psi[\mathcal{M};\mathcal{X}] \in \mathbb{F}^*(\mathcal{M}). \quad (2.4.3)$$

Moreover, it is useful to note that $\forall \mathcal{M} \in \pi[X] \ \forall \mathcal{C} \in \pi_0[X;\mathcal{M}] \ \forall \mathcal{X} \in \mathbb{F}^*(\mathcal{C})$:

$$\psi[\mathcal{M};\mathcal{X}] \cap \mathcal{C} = \mathcal{X}. \qquad (2.4.4)$$

So (2.4.3) and (2.4.4) define some variant of the extension of filters. As a corollary we obtain $\forall \mathcal{M} \in \pi[X] \ \forall \mathcal{C} \in \pi_0[X;\mathcal{M}]$:

$$\mathbb{F}^*(\mathcal{C}) = \{\mathcal{X} \cap \mathcal{C}: \ \mathcal{X} \in \mathbb{F}^*(\mathcal{M})\}. \qquad (2.4.5)$$

In (2.4.3)–(2.4.5) we have a very general variant of extensions of filters.

PROPOSITION 2.4.1 *$\forall \mathcal{D} \in \pi[X] \ \forall \mathcal{U} \in \pi_0[X;\mathcal{D}] \ \forall \mathcal{V} \in \mathbb{F}_0^*(\mathcal{U}) \ \exists \mathcal{H} \in \mathbb{F}_0^*(\mathcal{D})$: $\mathcal{H} \cap \mathcal{U} = \mathcal{V}$.*

PROOF. Fix $\mathcal{D} \in \pi[X]$, $\mathcal{U} \in \pi_0[X;\mathcal{D}]$ and $\mathcal{V} \in \mathbb{F}_0^*(\mathcal{U})$. Then $\mathcal{U} \in \pi[X]$ and $\mathcal{U} \subset \mathcal{D}$. Since $\mathcal{V} \in \mathbb{F}^*(\mathcal{U})$, by (2.4.5) we have $\mathcal{W} \in \mathbb{F}^*(\mathcal{D})$ such that $\mathcal{V} = \mathcal{W} \cap \mathcal{U}$. By Proposition 2.2.3 we can choose $\mathcal{H} \in \mathbb{F}_0^*(\mathcal{D})$ for which $\mathcal{W} \subset \mathcal{H}$. Then by (2.4.5) we obtain the property

$$\mathcal{H} \cap \mathcal{U} \in \mathbb{F}^*(\mathcal{U}): \ \mathcal{V} \subset \mathcal{H} \cap \mathcal{U}.$$

But $\mathcal{V}$ is an ultrafilter of $\mathcal{U}$. Therefore by (2.2.51) we have $\mathcal{V} = \mathcal{H} \cap \mathcal{U}$.

PROPOSITION 2.4.2 *$\forall \mathcal{M} \in \pi[X] \ \forall \mathfrak{G} \in 2^{\mathcal{M}} \ \forall \mathcal{W} \in \mathbb{F}^*(\mathcal{M})$: $(\mathfrak{G} \subset \mathcal{W}) \Rightarrow (\psi[\mathcal{M};\mathfrak{G}] \subset \mathcal{W})$.*

PROOF. Fix $\mathcal{M} \in \pi[X]$, $\mathfrak{G} \in 2^{\mathcal{U}}$ and $\mathcal{W} \in \mathbb{F}^*(\mathcal{M})$. Let $\mathfrak{G} \subset \mathcal{W}$. Choose arbitrarily $H \in \psi[\mathcal{M}; \mathfrak{G}]$. Then $H \in \mathcal{M}$ and $G \subset H$, where $G \in \mathfrak{G}$. Fix such set $G$. In addition, $G \in \mathcal{W}$, and by (2.2.45)

$$\{V \in \mathcal{M} |\ G \subset V\} \subset \mathcal{W}.$$

Therefore $H \in \mathcal{W}$. We obtain the inclusion $\psi[\mathcal{M}; \mathfrak{G}] \subset \mathcal{W}$. □

We note the following property. For $\mathcal{M} \in \pi[X]$, $\mathcal{C} \in \pi_0[X; \mathcal{M}]$ and $\mathcal{V} \in \mathbb{F}_0^*(\mathcal{C})$, the statement $\psi[\mathcal{M}; \mathcal{V}] \notin \mathbb{F}_0^*(\mathcal{M})$ can be true. Consider the following

**Example.** Let the set $X$ have the property that $X \setminus \{x\} \neq \varnothing$ for each $x \in X$. Suppose that $\mathcal{M} \triangleq \mathcal{P}(X)$, $\mathcal{C} \triangleq \{\varnothing; X\}$, and $\mathcal{V} = \{X\}$. Then by (2.4.2) we obtain $\psi[\mathcal{M}; \mathcal{V}] = \{X\} = \mathcal{V}$. Fix $x_0 \in X$. Then $\mathcal{P}(X) \in \Pi[X]$ and by Proposition 2.2.2, $\mathcal{W} \triangleq \{L \in \mathcal{P}(X) |\ x_0 \in L\} \in \mathbb{F}_0^*(\mathcal{P}(X))$. Then $\{x_0\} \neq X$ and hence $\{x_0\} \in \mathcal{W} \setminus \psi[\mathcal{M}; \mathcal{V}]$. Of course, $X \in \mathcal{W}$ and $\psi[\mathcal{M}; \mathcal{V}] \subset \mathcal{W}$. But $\psi[\mathcal{M}; \mathcal{V}] \neq \mathcal{W}$ and by (2.2.51) $\psi[\mathcal{M}; \mathcal{V}] \notin \mathbb{F}_0^*(\mathcal{P}(X))$.

Note that $\forall \mathcal{L} \in \Pi[X]$:

$$\mathcal{L} \in \pi_0[X; a_X^0(\mathcal{L})].$$

Therefore by (2.4.3) we have $\forall \mathcal{L} \in \Pi[X]\ \forall \mathcal{X} \in \mathbb{F}^*(\mathcal{L})$:

$$\psi[a_X^0(\mathcal{L}); \mathcal{X}] \in \mathbb{F}^*(a_X^0(\mathcal{L})). \tag{2.4.6}$$

In the last statement we can suppose that $\mathcal{X} \in \mathbb{F}_0^*(\mathcal{L})$. We will establish that in this case (2.4.6) is an ultrafilter.

LEMMA 2.4.1 *If $\mathcal{X} \in (\mathrm{alg})[X]$, $\mathcal{H} \in \mathbb{F}_0^*(\mathcal{X})$, $m \in \mathcal{N}$, and $(L_i)_{i \in \overline{1,m}} \in \mathcal{X}^m$, then*

$$\left( \bigcup_{i=1}^{m} L_i \in \mathcal{H} \right) \Rightarrow (\exists j \in \overline{1,m} :\ L_j \in \mathcal{H}).$$

The lemma is proved by induction on the basis of Proposition 2.2.4.

PROPOSITION 2.4.3 *If $\mathcal{L} \in \Pi[X]$ and $\mathcal{U} \in \mathbb{F}_0^*(\mathcal{L})$, then $\psi[a_X^0(\mathcal{L}); \mathcal{U}] \in \mathbb{F}_0^*(a_X^0(\mathcal{L}))$.*

PROOF. Fix $\mathcal{L} \in \Pi[X]$ and $\mathcal{U} \in \mathbb{F}_0^*(\mathcal{L})$. In particular, $\mathcal{U} \in \mathbb{F}^*(\mathcal{L})$. In addition, $\forall \mathcal{H} \in \mathbb{F}^*(\mathcal{L})$:

$$(\mathcal{U} \subset \mathcal{H}) \Rightarrow (\mathcal{U} = \mathcal{H}). \tag{2.4.7}$$

The property (2.4.7) follows from (2.2.51). Let $\mathcal{A} \triangleq a_X^0(\mathcal{L}) \in (\mathrm{alg})[X]$. Then $(X, \mathcal{A})$ is a measurable space with an algebra of sets. Note that by (2.4.2)

$$\mathcal{V} \triangleq \psi[\mathcal{A}; \mathcal{U}] = \{A \in \mathcal{A} |\ \exists G \in \mathcal{U} :\ G \subset A\}. \tag{2.4.8}$$

On the other hand, by Proposition 2.4.1 we can choose $\mathcal{W} \in \mathbb{F}_0^*(\mathcal{A})$ for which $\mathcal{W} \cap \mathcal{L} = \mathcal{U}$. Let $W \in \mathcal{W}$. By (2.2.38) we can choose $q \in \mathcal{N}$ such that $\Delta_q(W, \mathcal{L}) \neq \varnothing$. Let $(D_i)_{i \in \overline{1,q}} \in \Delta_q(W, \mathcal{L})$. Since $W$ is the union of the sets $D_i$, $i \in \overline{1,q}$, by Lemma 2.4.1 we can choose $\nu \in \overline{1,q}$ such that $D_\nu \in \mathcal{W}$. Since $D_\nu \in \mathcal{L}$, we obtain the property $D_\nu \in \mathcal{U}$. But $D_\nu \subset W$. From (2.4.8) it follows that $W \in \mathcal{V}$. We have established the inclusion $\mathcal{W} \subset \mathcal{V}$. In addition, $\mathcal{V} \in \mathbb{F}^*(\mathcal{A})$; in this connection see (2.4.6). Then by (2.2.51) we have $\mathcal{W} = \mathcal{V}$, and hence $\mathcal{V} \in \mathbb{F}_0^*(\mathcal{A})$. By (2.4.8) we obtain $\psi[\mathcal{A}; \mathcal{U}] \in \mathbb{F}_0^*(\mathcal{A})$. □

COROLLARY 2.4.1 $\forall \mathcal{L} \in \Pi[X]\ \forall \mathcal{U} \in \mathbb{F}_0^*(\mathcal{L})\ \exists \mathcal{V} \in \mathbb{F}_0^*(a_X^0(\mathcal{L}))$: $\mathcal{V} \cap \mathcal{L} = \mathcal{U}$.

The proof follows from (2.4.4) and Proposition 2.4.3. See also Proposition 2.4.1.

If $\mathcal{L} \in \Pi[X]$, then by (2.4.6) and Proposition 2.4.3 we have the mapping

$$\mathcal{U} \mapsto \psi[a_X^0(\mathcal{L}); \mathcal{U}] : \mathbb{F}^*(\mathcal{L}) \to \mathbb{F}^*(a_X^0(\mathcal{L})), \tag{2.4.9}$$

the image of $\mathbb{F}_0^*(\mathcal{L})$ under which is contained in $\mathbb{F}_0^*(a_X^0(\mathcal{L}))$. We consider below the mapping (2.4.9) and its restriction to $\mathbb{F}_0^*(\mathcal{L})$. The last case is very important in the following. Since $X \neq \varnothing$, from Proposition 2.2.2 it follows that $\mathbb{F}_0^*(\mathcal{L}) \neq \varnothing$. By Proposition 2.4.3, for $\mathcal{L} \in \Pi[X]$ the restriction of the operator (2.4.9) to the set $\mathbb{F}_0^*(\mathcal{L})$ is the mapping

$$\mathcal{U} \mapsto \psi[a_X^0(\mathcal{L}); \mathcal{U}] : \mathbb{F}_0^*(\mathcal{L}) \to \mathbb{F}_0^*(a_X^0(\mathcal{L})). \tag{2.4.10}$$

We recall that by (2.4.4) $\forall \mathcal{L} \in \Pi[X]\ \forall \mathcal{X} \in \mathbb{F}_0^*(\mathcal{L})$:

$$\psi[a_X^0(\mathcal{L}); \mathcal{X}] \cap \mathcal{L} = \mathcal{X}. \tag{2.4.11}$$

Consider a natural combination of (2.4.10) and (2.4.11). We obtain the following

PROPOSITION 2.4.4 *If $\mathcal{L} \in \Pi[X]$, then the mapping (2.4.10) is an injection from $\mathbb{F}_0^*(\mathcal{L})$ into $\mathbb{F}_0^*(a_X^0(\mathcal{L}))$; namely, $\forall \mathcal{U}_1 \in \mathbb{F}_0^*(\mathcal{L})\ \forall \mathcal{U}_2 \in \mathbb{F}_0^*(\mathcal{L})$:*

$$(\psi[a_X^0(\mathcal{L}); \mathcal{U}_1] = \psi[a_X^0(\mathcal{L}); \mathcal{U}_2]) \Rightarrow (\mathcal{U}_1 = \mathcal{U}_2). \tag{2.4.12}$$

PROOF. Fix $\mathcal{L} \in \Pi[X]$. Then we have the mapping (2.4.10). Let $\mathcal{U}_1 \in \mathbb{F}_0^*(\mathcal{L})$ and $\mathcal{U}_2 \in \mathbb{F}_0^*(\mathcal{L})$ have the property $\psi[a_X^0(\mathcal{L}); \mathcal{U}_1] = \psi[a_X^0(\mathcal{L}); \mathcal{U}_2]$. Then by (2.4.11) we obtain

$$\mathcal{U}_1 = \psi[a_X^0(\mathcal{L}); \mathcal{U}_1] \cap \mathcal{L} = \psi[a_X^0(\mathcal{L}); \mathcal{U}_2] \cap \mathcal{L} = \mathcal{U}_2.$$

So we have established (2.4.12).

We note that by (2.4.4) it is possible to consider (2.4.3) as extension of $\mathcal{C}$-filters 'on $\mathcal{M} \in \pi[X]$'. Analogously, by (2.4.11) we have an extension of $\mathcal{L}$-ultrafilters 'on $a_X^0(\mathcal{L})$' in the form of (2.4.10). We will return to this procedure of the filter extension after the corresponding consideration of questions of finitely additive measure (FAM) theory. In particular, we will supplement Proposition 2.4.4 by statements using the very useful connection of the ultrafilters theory and constructions of two-valued FAM (see, in particular, [63]; moreover, these questions are considered in [35, Ch. 7]).

## 2.5 MEASURABILITY AND CONTINUITY

In this section we consider (very briefly) some notions connected with two classes of mappings. And what is more, we will unite these classes in some natural sense. Suppose that $\forall_X\, \mathbf{S}[X \neq \varnothing]\ \forall_Y\, \mathbf{S}[Y \neq \varnothing]\ \forall \mathcal{X} \in \mathcal{P}(\mathcal{P}(X))\ \forall \mathcal{Y} \in \mathcal{P}(\mathcal{P}(X))$:

$$(\mathrm{Meas})[X; \mathcal{X}; Y; \mathcal{Y}] \triangleq \{f \in Y^X \mid \forall A \in \mathcal{Y}:\ f^{-1}(A) \in \mathcal{X}\}. \tag{2.5.1}$$

We can assume two possible concrete variants for the spaces $(X, \mathcal{X})$ and $(Y, \mathcal{Y})$ in (2.5.1). Namely, we consider the case when $(X, \mathcal{X})$ and $(Y, \mathcal{Y})$ are TS. Then (2.5.1) is reduced to the conception of continuous mappings. On the other hand, we use (in (2.5.1)) standard measurable spaces for $(X, \mathcal{X})$ and $(Y, \mathcal{Y})$. In this case it is usually presupposed that $\mathcal{X}$ is a $\sigma$-algebra of subsets of $X$ and $\mathcal{Y}$ is a $\sigma$-algebra of subsets of $Y$. We consider the two above mentioned types of structures in parallel. So $\forall_X\, \mathbf{S}[X \neq \varnothing]\ \forall_Y\, \mathbf{S}[Y \neq \varnothing]\ \forall \tau \in (\mathrm{top})[X]\ \forall \vartheta \in (\mathrm{top})[Y]$:

$$\begin{aligned} C(X, \tau, Y, \vartheta) &\triangleq (\mathrm{Meas})[X; \tau; Y; \vartheta] \\ &= \{f \in Y^X \mid \forall F \in \mathcal{F}_\vartheta:\ f^{-1}(F) \in \mathcal{F}_\tau\} \\ &= (\mathrm{Meas})[X; \mathcal{F}_\vartheta; Y; \mathcal{F}_\tau]. \end{aligned} \tag{2.5.2}$$

We use the relation (2.2.17) in (2.5.2). It is possible to give several other useful representations of the spaces of continuous mappings. We note some of them without proof (see [1, 4, 66, 71, 81]). Of course, it is possible to find these representations in [1, 4, 66, 71, 81, 90]. We give a brief summary. Suppose that $\forall_X\, \mathbf{S}[X \neq \varnothing]\ \forall_Y\, \mathbf{S}[Y \neq \varnothing]\ \forall \tau \in (\mathrm{top})[X]$ $\forall \vartheta \in (\mathrm{top})[Y]\ \forall x \in X$:

$$\begin{aligned} C_{\mathrm{loc}}(X; \tau; Y; \vartheta | x) &\triangleq \{f \in Y^X \mid \forall U \in N_\vartheta(f(x))\ \exists V \in N_\tau(x):\ f^1(V) \subset U\} \\ &= \{f \in Y^X \mid \forall \tilde{U} \in N_\vartheta^0(f(x))\ \exists \tilde{V} \in N_\tau^0(x):\ f^1(\tilde{V}) \subset \tilde{U}\}. \end{aligned} \tag{2.5.3}$$

In (2.5.3) the set of all operators $f$ from $(X, \tau)$ into $(Y, \vartheta)$ with the property of local continuity at $x$ is considered. Below we give a few of very useful equivalent representations (see [1, 4, 66, 71, 81]). Namely, in addition to (2.5.2) we note that $\forall_X \mathbf{S}[X \neq \varnothing]\ \forall_Y \mathbf{S}[Y \neq \varnothing]\ \forall \tau \in (\text{top})[X]$ $\forall \vartheta \in (\text{top})[Y]$:

$$\begin{aligned} C(X,\tau,Y,\vartheta) &= \bigcap_{x \in X} C_{\text{loc}}(X,\tau,Y,\vartheta|x) \\ &= \{f \in Y^X |\ \forall A \in \mathcal{P}(Y) : \text{cl}(f^{-1}(A),\tau) \subset f^{-1}(\text{cl}(A,\vartheta))\} \\ &= \{f \in Y^X |\ \forall \tilde{A} \in \mathcal{P}(X) : f^1(\text{cl}(\tilde{A},\tau)) \subset \text{cl}(f^1(\tilde{A}),\vartheta)\} \\ &= \{f \in Y^X | \forall_{\mathbf{D}}\, \mathbf{S}[\mathbf{D} \neq \varnothing]\ \forall \preceq \in (\text{DIR})[\mathbf{D}] \forall g \in X^{\mathbf{D}}\ \forall x \in X : \\ &\qquad ((\mathbf{D},\preceq,g) \xrightarrow{\tau} x) \Rightarrow ((\mathbf{D},\preceq,f \circ g) \xrightarrow{\vartheta} f(x))\}. \end{aligned} \tag{2.5.4}$$

We use (2.5.2)–(2.5.4) in topological constructions of extensions. Of course, (2.5.4) is supplemented by relations connected with representations in terms of the fundamental system of neighborhoods. Moreover, the representations in terms of sequential continuity can be used in particular cases. Now we do not consider these possibilities but discuss versions of (2.5.1) connected with mappings on measurable spaces. Here we have measurability in the usual sense. Moreover, arbitrary sets can be equipped with measurable structures generated by a given mapping. From (2.5.1) we obtain that $\forall_X \mathbf{S}[X \neq \varnothing]\ \forall_Y \mathbf{S}[Y \neq \varnothing]\ \forall \mathcal{X} \in (\sigma\text{-alg})[X]$ $\forall \mathcal{Y} \in (\sigma - \text{alg})[Y]$:

$$(\text{Meas})[X;\mathcal{X};Y;\mathcal{Y}] = \{f \in Y^X |\ \forall A \in \mathcal{Y} :\ f^{-1}(A) \in \mathcal{X}\} \in 2^{(Y^X)}; \tag{2.5.5}$$

note that all constant functions from $X$ into $Y$ are elements of (2.5.5). Of course, other families of sets can be used in (2.5.5) instead of $\sigma$-algebras. And what is more, the employment of more simple families (in place of $\mathcal{Y}$) is often very useful. But the supposition that the family $\mathcal{X}$ in (2.5.5) is a $\sigma$-algebra of sets is essential.

It is advisable to introduce the natural notion of the image and the inverse image of a family. We suppose that $\forall_X \mathbf{S}[X \neq \varnothing]\ \forall_Y \mathbf{S}[Y \neq \varnothing]$ $\forall f \in Y^X\ \forall \mathcal{X} \in \mathcal{P}(\mathcal{P}(X))$:

$$f^1[\mathcal{X}] \triangleq \{f^1(A) :\ A \in \mathcal{X}\} = \{B \in \mathcal{P}(Y) |\ \exists A \in \mathcal{X} :\ B = f^1(A)\}.$$

Note the following useful circumstance. Namely, if $\mathbf{S}[X \neq \varnothing]$, $\mathbf{S}[Y \neq \varnothing]$, $\mathbf{S}[Z \neq \varnothing]$, $f \in Y^X \cap Z^X$ and $\mathcal{X} \in \mathcal{P}(\mathcal{P}(X))$, then

$$\begin{aligned} &\{B \in \mathcal{P}(Y) |\ \exists A \in \mathcal{X} :\ B = f^1(A)\} \\ =&\{B \in \mathcal{P}(Z) |\ \exists A \in \mathcal{X} :\ B = f^1(A)\}. \end{aligned} \tag{2.5.6}$$

We use (2.2.6) in the last relation. With regard to (2.5.6) we do not use the corresponding symbol $Y$ in the first relation of $f^1[\mathcal{X}]$ , since here $Y$ is a fictitious variable.

But one more notion of the family image for which $Y$ is not fictitious can be introduced. Namely, $\forall_X \mathbf{S}[X \neq \varnothing]\ \forall_Y \mathbf{S}[Y \neq \varnothing]\ \forall f \in Y^X\ \forall \mathcal{X} \in \mathcal{P}(\mathcal{P}(X))$:

$$f_Y[\mathcal{X}] \triangleq \{H \in \mathcal{P}(Y) |\ f^{-1}(H) \in \mathcal{X}\}. \tag{2.5.7}$$

The family (2.5.7) is regarded as an image of the family $\mathcal{X}$ too. Finally, let us introduce the inverse image of a family. Namely, $\forall_X \mathbf{S}[X \neq \varnothing]$ $\forall_Y \mathbf{S}[Y \neq \varnothing]\ \forall f \in Y^X\ \forall \mathcal{Y} \in \mathcal{P}(\mathcal{P}(Y))$:

$$f^{-1}[\mathcal{Y}] \triangleq \{f^{-1}(B):\ B \in \mathcal{Y}\} = \{A \in \mathcal{P}(X) |\ \exists B \in \mathcal{Y}:\ A = f^{-1}(B)\}. \tag{2.5.8}$$

It is possible to reduce (2.5.1) to these terms: if $\mathbf{S}[X \neq \varnothing]$, $\mathbf{S}[Y \neq \varnothing]$, $\mathcal{X} \in \mathcal{P}(\mathcal{P}(X))$, $\mathcal{Y} \in \mathcal{P}(\mathcal{P}(Y))$, then

$$(\text{Meas})[X; \mathcal{X}; Y; \mathcal{Y}] = \{f \in Y^X |\ f^{-1}[\mathcal{Y}] \subset \mathcal{X}\}. \tag{2.5.9}$$

From (2.5.2) we have $\forall_X \mathbf{S}[X \neq \varnothing]\ \forall_Y \mathbf{S}[Y \neq \varnothing]\ \forall \tau \in (\text{top})[X]\ \forall \vartheta \in (\text{top})[Y]$:

$$C(X, \tau, Y, \vartheta) = \{f \in Y^X |\ f^{-1}[\vartheta] \subset \tau\} = \{f \in Y^X |\ f^{-1}[\mathcal{F}_\vartheta] \subset \mathcal{F}_\tau\}.$$

Note several obvious properties connected with linearly ordered families of measurable structures. Namely, $\forall_E \mathbf{S}[E]\ \forall \mathbb{H} \in 2^{\pi[E]}$:

$$(\forall \mathcal{H}_1 \in \mathbb{H}\ \forall \mathcal{H}_2 \in \mathbb{H}:\ (\mathcal{H}_1 \subset \mathcal{H}_2) \vee (\mathcal{H}_2 \subset \mathcal{H}_1)) \Rightarrow \left(\bigcup_{\mathcal{H} \in \mathbb{H}} \mathcal{H} \in \pi[E]\right). \tag{2.5.10}$$

By (2.5.10) we can verify that $\forall_E \mathbf{S}[E]\ \forall \mathcal{H} \in 2^{\Pi[E]}$:

$$(\forall \mathcal{H}_1 \in \mathbb{H}\ \forall \mathcal{H}_2 \in \mathbb{H}:\ (\mathcal{H}_1 \subset \mathcal{H}_2) \vee (\mathcal{H}_2 \subset \mathcal{H}_1)) \Rightarrow \left(\bigcup_{\mathcal{H} \in \mathbb{H}} \mathcal{H} \in \Pi[E]\right). \tag{2.5.11}$$

Finally, analogously to (2.5.10) and (2.5.11), we have the property of measurable spaces with algebras of sets. Namely, $\forall_E \mathbf{S}[E]\ \forall \mathbb{H} \in 2^{(\text{alg})[E]}$:

$$(\forall \mathcal{H}_1 \in \mathbb{H}\ \forall \mathcal{H}_2 \in \mathbb{H}:\ (\mathcal{H}_1 \subset \mathcal{H}_2) \vee (\mathcal{H}_2 \subset \mathcal{H}_1)) \Rightarrow \left(\bigcup_{\mathcal{H} \in \mathbb{H}} \mathcal{H} \in (\text{alg})[E]\right). \tag{2.5.12}$$

In (2.5.10)–(2.5.12) we have the characteristic property of 'simple' measurable structures. Note several obvious properties connected with the inverse images of families. So $\forall_X \mathbf{S}[X \neq \varnothing]$ $\forall_Y \mathbf{S}[Y \neq \varnothing]$ $\forall f \in Y^X$:

$$\begin{aligned}&(\forall \mathcal{Y} \in \pi[Y]: f^{-1}[\mathcal{Y}] \in \pi[X])\\ \&&(\forall \mathcal{Y} \in \Pi[Y]: f^{-1}[\mathcal{Y}] \in \Pi[X])\\ \&&(\forall \mathcal{Y} \in (\mathrm{alg})[Y]: f^{-1}[\mathcal{Y}] \in (\mathrm{alg})[X])\\ \&&(\forall \mathcal{Y} \in (\sigma - \mathrm{alg})[Y]: f^{-1}[\mathcal{Y}] \in (\sigma - \mathrm{alg})[X]).\end{aligned} \tag{2.5.13}$$

The property (2.5.13) means that measurable structures are preserved under taking inverse images. Analogously, similar properties are preserved under taking the 'image' of the type (2.5.7). Namely, $\forall_X \mathbf{S}[X \neq \varnothing]$ $\forall_Y \mathbf{S}[Y \neq \varnothing]$ $\forall f \in Y^X$:

$$\begin{aligned}&(\forall \mathcal{X} \in \pi[X]: f_Y[\mathcal{X}] \in \pi[Y])\&(\forall \mathcal{X} \in (\mathrm{alg})[X]: f_Y[\mathcal{X}] \in (\mathrm{alg})[Y])\\ \&&(\forall \mathcal{X} \in (\sigma - \mathrm{alg})[X]: f_Y[X] \in (\sigma - \mathrm{alg})[Y]).\end{aligned} \tag{2.5.14}$$

It is possible to connect the following simple properties with (2.5.13) and (2.5.14). Namely, $\forall_X \mathbf{S}[X \neq \varnothing]$ $\forall_Y \mathbf{S}[Y \neq \varnothing]$ $\forall \mathcal{Y} \in \mathcal{P}(\mathcal{P}(Y))$ $\forall f \in Y^X$:

$$(a_X^0(f^{-1}[\mathcal{Y}]) = f^{-1}[a_Y^0(\mathcal{Y})])\&(\sigma_X^0(f^{-1}[\mathcal{Y}]) = f^{-1}[\sigma_Y^0(\mathcal{Y})]). \tag{2.5.15}$$

**Remark.** Note the natural connection of Propositions 2.3.4, 2.3.5 and (2.5.15). Fix a set $U$, a family $\mathcal{H} \in \mathcal{P}(\mathcal{P}(U))$ and a set $V \in 2^U$. Consider $a_V^0(\mathcal{H}|_V)$ and $\sigma_V^0(\mathcal{H}|_V)$. Suppose that $X \triangleq V$, $Y \triangleq U$, $\mathcal{Y} \triangleq \mathcal{H}$. Moreover, suppose that $f \in Y^X$ is defined by the rule: $f(x) \triangleq x$ for $x \in X$. So we have the natural embedding $V$ into $U$. Then $f^{-1}[\mathcal{Y}] = f^{-1}[\mathcal{H}]$ is the family of all sets $f^{-1}(H)$, $H \in \mathcal{H}$. But $\forall H \in \mathcal{P}(U)$: $f^{-1}(H) = \{x \in V \mid f(x) \in H\} = V \cap H$. Hence (see (2.3.1)) in the given case

$$f^{-1}[\mathcal{Y}] = f^{-1}[\mathcal{H}] = \{V \cap H : H \in \mathcal{H}\} = \mathcal{H}|_V \in \mathcal{P}(\mathcal{P}(V)). \tag{2.5.16}$$

Therefore we have

$$(a_V^0(\mathcal{H}|_V) = a_X^0(f^{-1}[\mathcal{Y}]))\&(\sigma_V^0(\mathcal{H}|_V) = \sigma_X^0(f^{-1}[\mathcal{Y}])). \tag{2.5.17}$$

By analogy with (2.5.16) we obtain the equalities

$$\begin{aligned}&(f^{-1}[a_Y^0(\mathcal{Y})] = \{f^{-1}(S) : S \in a_Y^0(\mathcal{Y})\}\\ &\qquad = \{S \cap V : S \in a_Y^0(\mathcal{Y})\} = a_Y^0(\mathcal{Y})|_X = a_U^0(\mathcal{H})|_V)\\ \&&(f^{-1}[\sigma_Y^0(\mathcal{Y})] = \{f^{-1}(S) : S \in \sigma_Y^0(\mathcal{Y})\}\\ &\qquad = \{S \cap V : S \in \sigma_Y^0(\mathcal{Y})\} = \sigma_Y^0(\mathcal{Y})|_X = \sigma_U^0(\mathcal{H})|_V).\end{aligned} \tag{2.5.18}$$

Of course, we use (2.3.1) and the above representation of the sets $f^{-1}(H)$, $H \in \mathcal{P}(Y)$. From (2.5.15)–(2.5.17) we obtain

$$\begin{aligned}&\left(a_V^0(\mathcal{H}|_V) = a_X^0(f^{-1}[\mathcal{Y}]) = f^{-1}[a_Y^0(\mathcal{Y})] = a_U^0(\mathcal{H})|_V\right)\\ \& &\left(\sigma_V^0(\mathcal{H}|_V) = \sigma_X^0(f^{-1}[\mathcal{Y}]) = f^{-1}[\sigma_Y^0(\mathcal{Y})] = \sigma_U^0(\mathcal{H})|_V\right).\end{aligned} \tag{2.5.19}$$

Hence we obtain the statements of Propositions 2.3.4 and 2.3.5 using (2.5.15). The considered approach employing the natural embedding of a subset of the given set into this set permits us to obtain many useful results concerning constructions connected with the corresponding subspaces.

It is useful to note some representations of measurable functions. The following property is connected with (2.5.15). Namely, $\forall_X\, \mathbf{S}[X \neq \varnothing]$ $\forall_Y\, \mathbf{S}[Y \neq \varnothing]\ \forall \mathcal{X} \in (\sigma - \text{alg})[X]\ \forall \mathcal{Y} \in (\sigma - \text{alg})[Y]\ \forall f \in Y^X$:

$$\begin{aligned}(f \in (\text{Meas})[X;\mathcal{X};Y;\mathcal{Y}]) \Leftrightarrow &(\exists \mathcal{H} \in \mathcal{P}(\mathcal{P}(Y)) : (\mathcal{Y} = \sigma_Y^0(\mathcal{H}))\\ &\&(f^{-1}[\mathcal{H}] \subset \mathcal{X})).\end{aligned} \tag{2.5.20}$$

Moreover, we note the following well known property: a continuous function is a Borel measurable function. We use (2.2.32) and (2.5.2). Then $\forall_X\, \mathbf{S}[X \neq \varnothing]\ \forall_Y\, \mathbf{S}[Y \neq \varnothing]\ \forall \tau \in (\text{top})[X]\ \forall \vartheta \in (\text{top})[Y]$:

$$C(X,\tau,Y,\vartheta) \subset (\text{Meas})[X;\mathcal{B}_\sigma(\tau);Y;\mathcal{B}_\sigma(\vartheta)]. \tag{2.5.21}$$

Of course, it is possible to connect (2.5.20) and (2.5.21). Elements of the set on the right hand side of (2.5.21) are called Borel functions. It is important to recall some known representations connected with employment of (2.5.20) in the case when the real line with the Borel $\sigma$-algebra is used for $(Y,\mathcal{Y})$.

In the following we use

$$\begin{aligned}&(\tau_{\mathbb{R}} \triangleq \{G \in \mathcal{P}(\mathbb{R}) \mid \forall x \in G\ \exists \varepsilon \in ]0,\infty[ : ]x - \varepsilon, x + \varepsilon[ \subset G\} \in (\text{top})[\mathbb{R}])\\ &\qquad \& \ (\tau_\partial \triangleq \mathcal{P}(\mathbb{R}) \in (\text{top})[\mathbb{R}]).\end{aligned} \tag{2.5.22}$$

The first topology in (2.5.22) is the ordinary topology of $\mathbb{R}$. The second topology (in (2.5.22)) is the discrete topology of $\mathbb{R}$. Then $\mathcal{B}_\sigma(\tau_{\mathbb{R}})$ is the $\sigma$-algebra of Borel subsets of $\mathbb{R}$. We call the pair $(\mathbb{R}, \mathcal{B}_\sigma(\tau_{\mathbb{R}}))$ the Borel real line. It is useful to note several representations of $(\text{Meas})[X;\mathcal{X};\mathbb{R};\mathcal{B}_\sigma(\tau_{\mathbb{R}})]$, where $(X,\mathcal{X})$ is a measurable space. We recall that $(\mathbb{R},\tau_{\mathbb{R}})$ is a separable space. Namely,

$$\exists (c_i)_{i \in \mathcal{N}} \in \mathbb{R}^{\mathcal{N}}\ \forall t \in \mathbb{R}\ \forall \varepsilon \in ]0,\infty[\ \exists k \in \mathcal{N} : \ |t - c_k| < \varepsilon. \tag{2.5.23}$$

The rational numbers can be used in (2.5.23). In addition, $\forall (c_i)_{i\in\mathcal{N}} \in \mathbb{R}^{\mathcal{N}}$:

$$\begin{aligned}&(\forall t \in \mathbb{R}\ \forall \varepsilon \in ]0,\infty[\ \exists k \in \mathcal{N} : |t - c_k| < \varepsilon)\\ \Rightarrow&(\forall G \in \tau_{\mathbb{R}}\ \exists (\varepsilon_i)_{i\in\mathcal{N}} \in \mathbb{R}^{\mathcal{N}} : G = \bigcup_{i\in\mathcal{N}} ]c_i - \varepsilon_i, c_i + \varepsilon_i[).\end{aligned} \tag{2.5.24}$$

In (2.5.24) we assume that in the notation $]a, b[$ the case $b \leq a$ is possible; of course, $]a, b[= \varnothing$ for $b \leq a$ . Therefore $\forall_X\ \mathbf{S}[X \neq \varnothing]\ \forall \mathcal{X} \in (\sigma-\text{alg})[X]$ $\forall f \in \mathbb{R}^X$:

$$(f \in (\text{Meas})[X;\mathcal{X};\mathbb{R};\tau_{\mathbb{R}}]) \Leftrightarrow (\forall c \in \mathbb{R}\ \forall \varepsilon \in ]0,\infty[: f^{-1}(]c-\varepsilon, c+\varepsilon[) \in \mathcal{X}).$$

As a corollary (see (2.5.20)) we obtain that $\forall_X\ \mathbf{S}[X \neq \varnothing]\ \forall \mathcal{X} \in (\sigma-\text{alg})[X]$:

$$\begin{aligned}(\text{Meas})[X;\mathcal{X};\mathbb{R};\mathcal{B}_\sigma(\tau_{\mathbb{R}})] &= (\text{Meas})[X;\mathcal{X};\mathbb{R};\tau_{\mathbb{R}}])\\ &=\{f \in \mathbb{R}^X \mid \forall A \in \mathcal{B}_\sigma(\tau_{\mathbb{R}}) : f^{-1}(A) \in \mathcal{X}\}\\ &=\{f \in \mathbb{R}^X \mid \forall c \in \mathbb{R} : f^{-1}(]-\infty, c[) \in \mathcal{X}\}\\ &=\{f \in \mathbb{R}^X \mid \forall c \in \mathbb{R} : f^{-1}(]-\infty, c]) \in \mathcal{X}\}\\ &=\{f \in \mathbb{R}^X \mid \forall c \in \mathbb{R} : f^{-1}(]c, \infty[) \in \mathcal{X}\}\\ &=\{f \in \mathbb{R}^X \mid \forall c \in \mathbb{R} : f^{-1}([c, \infty[) \in \mathcal{X}\}.\end{aligned} \tag{2.5.25}$$

In the following the real line $\mathbb{R}$ with the ordinary topology $\tau_{\mathbb{R}}$ is used very often. We suppose that $\forall_X\ \mathbf{S}[X \neq \varnothing]\ \forall \tau \in (\text{top})[X]$:

$$\mathbb{C}(X,\tau) \triangleq C(X,\tau,\mathbb{R},\tau_{\mathbb{R}}).$$

From (2.5.21) it follows that if $\mathbf{S}[X \neq \varnothing]$ and $\tau \in (\text{top})[X]$, then

$$\mathbb{C}(X,\tau) \subset (\text{Meas})[X;\mathcal{B}_\sigma(\tau);\mathbb{R};\mathcal{B}_\sigma(\tau_{\mathbb{R}})]. \tag{2.5.26}$$

From (2.5.26) we have the well known statement that each continuous functional has the property of Borel measurability. Recall a general property connected with (2.5.1). Namely, $\forall_X\ \mathbf{S}[X \neq \varnothing]\ \forall_Y\ \mathbf{S}[Y \neq \varnothing]$ $\forall_Z\ \mathbf{S}[Z \neq \varnothing]\ \forall \alpha \in 2^{\mathcal{P}(X)}\ \forall \beta \in 2^{\mathcal{P}(Y)}\ \forall \gamma \in 2^{\mathcal{P}(Z)}\ \forall f \in (\text{Meas})[X;\alpha;Y;\beta]$ $\forall g \in (\text{Meas})[Y;\beta;Z;\gamma]$:

$$g \circ f \in (\text{Meas})[X;\alpha;Z;\gamma]. \tag{2.5.27}$$

We supplement (2.5.27) by the following obvious property concerning the restriction of a function. Namely, $\forall_X\ \mathbf{S}[X \neq \varnothing]\ \forall_Y\ \mathbf{S}[Y \neq \varnothing]\ \forall \mathcal{X} \in \mathcal{P}(\mathcal{P}(X))\ \forall \mathcal{Y} \in \mathcal{P}(\mathcal{P}(Y))\ \forall f \in (\text{Meas})[X;\mathcal{X};Y;\mathcal{Y}]\ \forall A \in 2^X$:

$$(f|A) \in (\text{Meas})[A;\mathcal{X}|_A;Y;\mathcal{Y}]. \tag{2.5.28}$$

We use (2.3.1) in (2.5.28). Of course, it is possible to use (2.5.27) and (2.5.28) in the following two particular cases: 1) $\mathcal{X}$ and $\mathcal{Y}$ are topologies; 2) $\mathcal{X}$ and $\mathcal{Y}$ are $\sigma$-algebras of subsets of $X$ and $Y$ respectively. So $\forall_X \mathbf{S}[X \neq \varnothing]$ $\forall_Y \mathbf{S}[Y \neq \varnothing]$ $\forall_Z \mathbf{S}[Z \neq \varnothing]$ $\forall \tau \in (\mathrm{top})[X]$ $\forall \vartheta \in (\mathrm{top})[Y]$ $\forall \zeta \in (\mathrm{top})[Z]$ $\forall f \in C(X, \tau, Y, \vartheta)$ $\forall g \in C(Y, \vartheta, Z, \zeta)$:

$$g \circ f \in C(X, \tau, Z, \zeta). \tag{2.5.29}$$

In addition, (2.5.29) follows from (2.5.2) and (2.5.27). Similarly, from (2.3.5), (2.5.2) and (2.5.28) we have $\forall_X \mathbf{S}[X \neq \varnothing]$ $\forall_Y \mathbf{S}[Y \neq \varnothing]$ $\forall \tau \in (\mathrm{top})[X]$ $\forall \vartheta \in (\mathrm{top})[Y]$ $\forall f \in C(X, \tau, Y, \vartheta)$ $\forall A \in 2^X$:

$$(f|A) \in C(A, \tau|_A, Y, \vartheta). \tag{2.5.30}$$

By (2.5.29) and (2.5.30) the case 1) is realized. The case 2) is obtained analogously: in (2.5.27) and (2.5.28) one should set $\mathcal{X} \in (\sigma - \mathrm{alg})[X]$ and $\mathcal{Y} \in (\sigma - \mathrm{alg})[Y]$. From comparison of 1) and 2) and from other constructions of this section we have the profound analogy of the properties of continuity and measurability. Note that from (2.5.2), (2.5.21) and (2.5.27) one can extract the following property: if $\mathbf{S}[X \neq \varnothing]$, $\mathbf{S}[Y \neq \varnothing]$, $\mathbf{S}[Z \neq \varnothing]$, $\mathcal{X} \in (\sigma - \mathrm{alg})[X]$, $\tau \in (\mathrm{top})[Y]$ $\vartheta \in (\mathrm{top})[Z]$, $f \in (\mathrm{Meas})[X; \mathcal{X}; Y; \mathcal{B}_\sigma(\tau)]$ and $g \in C(Y, \tau, Z, \vartheta)$, then

$$g \circ f \in (\mathrm{Meas})[X; \mathcal{X}; Z; \mathcal{B}_\sigma(\vartheta)]. \tag{2.5.31}$$

The property (2.5.31) is often used in the theory of topological measures. From (2.5.25) and (2.5.31) we have $\forall_X \mathbf{S}[X \neq \varnothing]$ $\forall_Y \mathbf{S}[Y \neq \varnothing]$ $\forall \mathcal{X} \in (\sigma - \mathrm{alg})[X]$ $\forall \tau \in (\mathrm{top})[Y]$ $\forall f \in (\mathrm{Meas})[X; \mathcal{X}; Y; \mathcal{B}_\sigma(\tau)]$ $\forall g \in \mathbb{C}(Y, \tau)$:

$$g \circ f \in (\mathrm{Meas})[X; \mathcal{X}; \mathbb{R}; \mathcal{B}_\sigma(\tau_{\mathbb{R}})]. \tag{2.5.32}$$

## 2.6 TOPOLOGICAL BASES AND SUBBASES

In this section we return to procedures of constructing topologies by more simple families. If $\mathcal{H}$ is a family of sets, then it is possible to determine the union of all sets of $\mathcal{H}$. In case $\mathcal{H} = \varnothing$ we obtain $\varnothing$. We have $\forall_X \mathbf{S}[X]$ $\forall \mathcal{X} \in \mathcal{P}(\mathcal{P}(X))$:

$$\{\cup\}(\mathcal{X}) \triangleq \{\bigcup_{B \in \mathcal{H}} B : \mathcal{H} \in \mathcal{P}(\mathcal{X})\} \in \mathcal{P}(\mathcal{P}(X)). \tag{2.6.1}$$

It is useful to note that $\forall_X \mathbf{S}[X]$ $\forall \mathcal{H} \in \mathcal{P}(\mathcal{P}(X))$:

$$\mathcal{H} \subset \{\cup\}(\mathcal{H}).$$

It is possible to construct topologies on the basis of (2.6.1). In this connection we note that $\forall_X \mathbf{S}[X]$:

$$\begin{aligned}(\mathrm{BAS})[X] \triangleq \{\beta \in \mathcal{P}(\mathcal{P}(X)) | \, (X = \bigcup_{H \in \beta} H) \& (\forall H_1 \in \beta \, \forall H_2 \in \beta \\ \forall x \in H_1 \cap H_2 \, \exists H_3 \in \beta : \, (x \in H_3) \\ \& \, (H_3 \subset H_1 \cap H_2))\} \\ = \{\beta \in \mathcal{P}(\mathcal{P}(X)) | \, \{\cup\}(\beta) \in (\mathrm{top})[X]\}.\end{aligned} \quad (2.6.2)$$

Note the obvious property: if $X$ is a set, then $(\mathrm{top})[X] \subset (\mathrm{BAS})[X]$. We call elements of (2.6.2) topological bases of $X$. In addition, $\forall_X \mathbf{S}[X]$ $\forall \beta \in \mathcal{P}(\mathcal{P}(X))$:

$$\{\cup\}(\beta) = \{H \in \mathcal{P}(X) | \, \forall h \in H \, \exists B \in \beta : \, (h \in B) \& (B \subset H)\}. \quad (2.6.3)$$

From (2.6.2) and (2.6.3) we have the very useful representation of the topologies generated by topological bases. Of course, $\forall_X \mathbf{S}[X \neq \varnothing]$:

$$(\mathrm{BAS})[X] \subset 2^{\mathcal{P}(X)}.$$

We consider the case of nonempty bases as basic. Note that $\forall_X \mathbf{S}[X]$ $\forall \mathcal{X} \in \mathcal{P}(\mathcal{P}(X))$:

$$\{\cap - \mathrm{Fin}\}(\mathcal{X}) \triangleq \{\bigcap_{H \in \mathcal{K}} H : \, \mathcal{K} \in \mathrm{Fin}(\mathcal{X})\} \in \mathcal{P}(\mathcal{P}(X)). \quad (2.6.4)$$

The following obvious property is realized by (2.6.4): under conditions providing (2.6.4), the inclusion $\mathcal{X} \subset \{\cap - \mathrm{Fin}\}(\mathcal{X})$ holds. Let $\forall_X \mathbf{S}[X]$:

$$(\mathbf{s} - \mathrm{BAS})[X] \triangleq \{\mathcal{H} \in \mathcal{P}(\mathcal{P}(X)) | \, \{\cap - \mathrm{Fin}\}(\mathcal{H}) \in (\mathrm{BAS})[X]\}. \quad (2.6.5)$$

Elements of (2.6.5) are called topological subbases. From (2.6.5) we obtain that $\forall_X \mathbf{S}[X \neq \varnothing]$:

$$(\mathbf{s} - \mathrm{BAS})[X] = \left\{\mathcal{H} \in 2^{\mathcal{P}(X)} \;\Big|\; \bigcup_{H \in \mathcal{H}} H = X\right\}.$$

Each (topological) base or subbase generates some topology. But very often it is required to realize the inverse operation. Namely, we have a topology and it is required to determine the bases (or subbases) generating this topology. In addition, $\forall_X \mathbf{S}[X] \, \forall \tau \in (\mathrm{top})[X]$:

$$(\tau - \mathrm{BAS})_0[X] \triangleq \{\beta \in (\mathrm{BAS})[X] | \, \tau = \{\cup\}(\beta)\} \in 2^{\mathcal{P}(\tau)}. \quad (2.6.6)$$

In (2.6.6) bases of the topology $\tau$ are considered. Suppose that $\forall_X\ \mathbf{S}[X]$:

$$(0-\text{top})[X] \triangleq \{\tau \in (\text{top})[X] |\ \exists \beta \in (\tau-\text{BAS})_0[X]:\ \beta \subset \mathcal{F}_\tau\}. \quad (2.6.7)$$

If $X$ is a nonempty set and $\tau \in (0-\text{top})[X] \cap (\mathcal{D}-\text{top})[X]$, then $(X,\tau)$ is called the zero-dimensional TS [71]. Zero-dimensional topologies (see (2.6.7)) are highly useful in constructions of extension [32, 35, 46]. Local bases are useful in the following. Suppose that $\forall_X\ \mathbf{S}[X]\ \forall\tau \in (\text{top})[X]$ $\forall x \in X$:

$$(x-\text{bas})[\tau] \triangleq \{\mathcal{G} \in 2^{N_\tau(x)} |\ \forall H \in N_\tau(x)\ \exists G \in \mathcal{G}:\ G \subset H\}. \quad (2.6.8)$$

By definitions of Section 2.2 and (2.6.8) we have $\forall_X\ \mathbf{S}[X]\ \forall\tau \in (\text{top})[X]$ $\forall x \in X$:

$$N_\tau^0(x) \in (x-\text{bas})[\tau].$$

It is convenient to characterize the property of the local continuity of functions in the terms of (2.6.8). From (2.5.3) and (2.6.8) we have the following well known property. Namely, $\forall_X\ \mathbf{S}[X \neq \varnothing]\ \forall_Y\ \mathbf{S}[Y \neq \varnothing]\ \forall\tau \in (\text{top})[X]\ \forall\vartheta \in (\text{top})[Y]\ \forall x \in X\ \forall f \in Y^X$:

$$(f \in C_{\text{loc}}(X,\tau,Y,\vartheta|x)) \Leftrightarrow (\exists \mathcal{U} \in (x-\text{bas})[\tau]\ \exists \mathcal{V} \in (f(x)-\text{bas})[\vartheta]$$
$$\forall V \in \mathcal{V}\ \exists U \in \mathcal{U}:\ f^1(U) \subset V).$$

We recall (see [1, 4, 66, 71, 81]) that $\forall \xi \in \mathbb{R}$: $\{]\xi-\varepsilon,\xi+\varepsilon[:\ \varepsilon \in ]0,\infty[\} \in (\xi-\text{bas})[\tau_{\mathbb{R}}]$. Here (2.5.22) is used. In specific monographs on topology one can find many other useful representations in terms of bases (global or local); in particular, see [1, 4, 66, 71, 81].

We now consider two questions connected with the axioms of countability. First we consider these notions in terms of the corresponding sequences. We recall (2.3.17) and the notion of TS with the first axiom of countability. In this connection we introduce the natural notion of the sequential closure. Namely, $\forall_X\ \mathbf{S}[X]\ \forall\tau \in (\text{top})[X]\ \forall A \in \mathcal{P}(X)$:

$$(\text{seqcl})[A;\tau] \triangleq \{x \in X |\ \exists (a_i)_{i\in\mathcal{N}} \in A^{\mathcal{N}}:\ (a_i)_{i\in\mathcal{N}} \xrightarrow{\tau} x\}. \quad (2.6.9)$$

We call the set (2.6.9) the sequential closure of $A$ in the TS $(X,\tau)$. Then by (2.3.19) we have $\forall_X\ \mathbf{S}[X]\ \forall\tau \in (\text{top})[X]$:

$$\mathcal{F}_{\text{seq}}[\tau] = \{H \in \mathcal{P}(X) |\ H = (\text{seqcl})[H;\tau]\}.$$

Sequentially closed sets are exactly fixed points of the operator of sequential closure. Note that (2.6.9) is used very often in applied mathematics under the corresponding supposition relative to the TS. Here it is possible to recall the operator in metrizable and in pseudo-metrizable TS.

But in reality the productive application of this notion can be realized in more general cases of TS. In particular, note that in TS with the first axiom of countability the operations of closure and sequential closure coincide. Namely, $\forall_X \mathbf{S}[X]\ \forall \tau \in (\text{top})_I[X]\ \forall A \in \mathcal{P}(X)$:

$$\text{cl}(A,\tau) = \{x \in X \mid \exists (a_i)_{i\in\mathcal{N}} \in A^{\mathcal{N}} : (a_i)_{i\in\mathcal{N}} \xrightarrow{\tau} x\} = (\text{seqcl})[A;\tau]. \tag{2.6.10}$$

From the properties of (2.3.10) it follows that $\forall_X \mathbf{S}[X]\ \forall \tau \in (\text{top})[X]$:

$$\mathcal{F}_\tau = \{F \in \mathcal{P}(X) \mid F = \text{cl}(F,\tau)\}.$$

As a consequence, from (2.6.10) we have $\forall_X \mathbf{S}[X]\ \forall \tau \in (\text{top})_I[X]$:

$$\mathcal{F}_\tau = \mathcal{F}_{\text{seq}}[\tau].$$

It is possible to supplement (2.6.10) by several other useful properties. We note only the following well known property.

PROPOSITION 2.6.1 $\forall_X \mathbf{S}[X \neq \varnothing]\ \ \forall_Y \mathbf{S}[Y \neq \varnothing]\ \ \forall \tau \in (\text{top})_I[X]$ $\forall \vartheta \in (\text{top})[Y]$:

$$C(X,\tau,Y,\vartheta) = \{f \in Y^X \mid \forall (x_i)_{i\in\mathcal{N}} \in X^{\mathcal{N}}\ \forall x \in X : ((x_i)_{i\in\mathcal{N}} \xrightarrow{\tau} x) \Rightarrow ((f(x_i))_{i\in\mathcal{N}} \xrightarrow{\vartheta} f(x))\}. \tag{2.6.11}$$

PROOF. We fix two TS $(X,\tau)$, $X \neq \varnothing$, and $(Y,\vartheta)$, $Y \neq \varnothing$. Let $\tau \in (\text{top})_I[X]$. Denote by $\mathbf{C}$ the set on the right hand side of (2.6.11). Then by (2.5.4) and definitions of Section 2.2 (see (2.2.14)) we have

$$C(X,\tau,Y,\vartheta) \subset \mathbf{C}. \tag{2.6.12}$$

In reality, (2.6.12) is true in the general case of the TS $(X,\tau)$ and $(Y,\vartheta)$. Let $\varphi \in \mathbf{C}$. So $\varphi \in Y^X$. Moreover, $\forall (x_i)_{i\in\mathcal{N}} \in X^{\mathcal{N}}\ \forall x \in X$:

$$((x_i)_{i\in\mathcal{N}} \xrightarrow{\tau} x) \Rightarrow ((\varphi(x_i))_{i\in\mathcal{N}} \xrightarrow{\vartheta} \varphi(x)). \tag{2.6.13}$$

Then (2.6.10) defines the set $\text{cl}(A,\tau)$. Fix $A \in \mathcal{P}(X)$. Let $y_* \in \varphi^1(\text{cl}(A,\tau))$. Choose $x_* \in \text{cl}(A,\tau)$ for which $y_* = \varphi(x_*)$. Using (2.6.10), we choose a sequence $(a_i)_{i\in\mathcal{N}} \in A^{\mathcal{N}}$ for which

$$(a_i)_{i\in\mathcal{N}} \xrightarrow{\tau} x_*. \tag{2.6.14}$$

Then $(\varphi(a_i))_{i\in\mathcal{N}} : \mathcal{N} \to \varphi^1(A)$. In addition, from (2.6.13) and (2.6.14) we get the convergence

$$(\varphi(a_i))_{i\in\mathcal{N}} \xrightarrow{\vartheta} y_*.$$

From this we obtain the property $y_* \in \mathrm{cl}(\varphi^1(A), \vartheta)$ (see (2.2.14)). Since the choice of $y_*$ was arbitrary, the inclusion $\varphi^1(\mathrm{cl}(A, \tau)) \subset \mathrm{cl}(\varphi^1(A), \vartheta)$ is established. But the choice of $A$ was arbitrary too. Therefore by (2.5.4) we have the property $\varphi \in C(X, \tau, Y, \vartheta)$. So the inclusion inverse to (2.6.12) is established. □

In conclusion of the section we consider the very important notion of the Tichonoff product of TS. We recall the well known notion of the Cartesian product of sets. Namely, $\forall_X\, \mathbf{S}[X \neq \varnothing]\ \forall_Y\, \mathbf{S}[Y]\ \forall (A_x)_{x \in X} \in \mathcal{P}(Y)^X$:

$$\prod_{x \in X} A_x \triangleq \{f \in Y^X \mid \forall s \in X : f(s) \in A_s\}. \tag{2.6.15}$$

By the axiom of choice (we admit this axiom) we have the following property: if $\mathbf{S}[X \neq \varnothing]$, $\mathbf{S}[Y \neq \varnothing]$ and $(A_x)_{x \in X} \in (2^Y)^X$, then

$$\prod_{x \in X} A_x \neq \varnothing. \tag{2.6.16}$$

In the construction of the Tichonoff product an equipment of all sets $A_x$, $x \in X$, with the corresponding topologies is supposed. But in the following it is sufficient to consider the case when $A_x \equiv A$ in (2.6.15) and (2.6.16). So we restrict ourselves to consideration of a topological equipment of sets $Y^X$, where $X$ and $Y$ are nonempty sets. The corresponding versions of the Tichonoff products have the sense of distinctive Tichonoff degrees. The given construction is connected in a natural way with some specific topological base. But at first it is possible to consider some scheme of determining the weakest topology of a functional space for which all functions of a given set are continuous.

If $\mathbf{S}[U \neq \varnothing]$, $\mathbf{S}[V \neq \varnothing]$, $\vartheta \in (\mathrm{top})[V]$, and $\mathbb{H} \in \mathcal{P}(V^U)$, then

$$\mathbf{T}_{\mathrm{top}}(U|\vartheta, \mathbb{H}) \triangleq \{\tau \in (\mathrm{top})[U] \mid \mathbb{H} \subset C(U, \tau, V, \vartheta)\} \in 2^{(\mathrm{top})[U]}; \tag{2.6.17}$$

really, $\mathcal{P}(U) \in \mathbf{T}_{\mathrm{top}}(U|\vartheta, \mathbb{H})$. In addition, it is possible to introduce the smallest element of the set (2.6.17) in the sense of inclusion. Namely, $\forall_U\, \mathbf{S}[U \neq \varnothing]\ \forall_V\, \mathbf{S}[V \neq \varnothing]\ \forall \vartheta \in (\mathrm{top})[V]\ \forall \mathbb{H} \in \mathcal{P}(V^U)$:

$$\tau^0(U|\vartheta, \mathbb{H}) \triangleq \left( \bigcap_{\tau \in \mathbf{T}_{\mathrm{top}}(U|\vartheta, \mathbb{H})} \tau \right) \in \mathbf{T}_{\mathrm{top}}(U|\vartheta, \mathbb{H}). \tag{2.6.18}$$

The property (2.6.18) follows from relations of Section 2.5. If $\mathbf{S}[U \neq \varnothing]$, $\mathbf{S}[V \neq \varnothing]$, $\vartheta \in (\mathrm{top})[V]$, and $h \in V^U$, then

$$\tau^0(U|\vartheta, \{h\}) = h^{-1}[\vartheta].$$

In a more general case the corresponding topology (2.6.18) is realized in a more complicated manner. But it is possible to construct the corresponding topological base: if $\mathbf{S}[U \neq \varnothing]$, $\mathbf{S}[V \neq \varnothing]$, $\vartheta \in (\text{top})[V]$, and $\mathbb{H} \in 2^{(V^U)}$, then the family $\beta^0(U|\vartheta, \mathbb{H})$ of all sets $T \in \mathcal{P}(U)$ for which

$$\exists m \in \mathcal{N}\ \exists (h_i)_{i\in\overline{1,m}} \in \mathbb{H}^m\ \exists (G_i)_{i\in\overline{1,m}} \in \vartheta^m : T = \bigcap_{i=1}^{m} h_i^{-1}(G_i) \quad (2.6.19)$$

is a topological base of $U$, i.e., $\beta^0(U|\vartheta, \mathbb{H}) \in (\text{BAS})[U]$; and what is more, $\tau^0(U|\vartheta, \mathbb{H}) = \{\cup\}(\beta^0(U|\vartheta, \mathbb{H}))$ and as a consequence we have

$$\beta^0(U|\vartheta, \mathbb{H}) \in (\tau^0(U|\vartheta, \mathbb{H}) - \text{BAS})_0[U]. \quad (2.6.20)$$

So by (2.6.3), (2.6.6) and (2.6.20) the topology (2.6.18) is constructed. Of course, we consider sets of the type (2.6.19) as simplest.

For realization of the corresponding Tichonoff degree we discuss an obvious concrete variant of (2.6.18). We consider the case when, for $\mathbf{S}[X \neq \varnothing]$, $\mathbf{S}[Y \neq \varnothing]$ and $\tau \in (\text{top})[Y]$ (in (2.6.17)), the following model is used : $U \triangleq Y^X$, $V = Y$, $\vartheta = \tau$, and $\mathbb{H}$ is some specific subset of $V^U = Y^{(Y^X)}$. We introduce the well known projection operation. Namely, $\forall_X\ \mathbf{S}[X \neq \varnothing]\ \forall_Y\ \mathbf{S}[Y \neq \varnothing]\ \forall x \in X$:

$$(x - \text{proj})[X; Y] \triangleq (f(x))_{f\in Y^X} \in Y^{(Y^X)}. \quad (2.6.21)$$

In (2.6.21) the projection operator is introduced. If the point $x$ in (2.6.21) is changed, then we obtain the required variant of the set $\mathbb{H}$ in (2.6.17) and (2.6.18). If $\mathbf{S}[X \neq \varnothing]$ and $\mathbf{S}[Y \neq \varnothing]$, then we denote by $(\text{Proj})[X; Y]$ the set of all mappings $(x - \text{proj})[X; Y]$, $x \in X$; i.e.,

$$(\text{Proj})[X; Y] \triangleq \{(x - \text{proj})[X; Y] : x \in X\}$$

and hence, by (2.6.17), for $\tau \in (\text{top})[Y]$ we obtain

$$\begin{aligned} &\mathbf{T}_{\text{top}}(Y^X|\tau, (\text{Proj})[X; Y]) \\ &= \{\mathbf{t} \in (\text{top})[Y^X] |\ \forall x \in X : (x - \text{proj})[X; Y] \in C(Y^X, \mathbf{t}, Y, \tau)\} \\ &= 2^{(\text{top})[Y^X]}. \end{aligned}$$

The weakest topology of the last set is the required Tichonoff degree of the TS $(Y, \tau)$. Namely, $\forall_X\ \mathbf{S}[X \neq \varnothing]\ \forall_Y\ \mathbf{S}[Y \neq \varnothing]\ \forall \tau \in (\text{top})[Y]$:

$$\otimes^X(\tau) \triangleq \tau^0(Y^X|\tau, (\text{Proj})[X; Y]) \in \mathbf{T}_{\text{top}}(Y^X|\tau, (\text{Proj})[X; Y]). \quad (2.6.22)$$

In addition, it is possible to define the corresponding (to (2.6.22)) topological base in terms of the property (2.6.19). Namely, by (2.6.20) and (2.6.22) we obtain $\forall_X\ \mathbf{S}[X \neq \varnothing]\ \forall_Y\ \mathbf{S}[Y \neq \varnothing]\ \forall \tau \in (\text{top})[Y]$:

$$\beta^0_{\otimes}(X, Y, \tau) \triangleq \beta^0(Y^X|\tau, (\text{Proj})[X; Y]) \in (\otimes^X(\tau) - \text{BAS})_0[Y^X]. \quad (2.6.23)$$

In (2.6.23) we have a natural base of the Tichonoff degree of the TS $(Y,\tau)$. Note that $\forall_X\, \mathbf{S}[X \neq \varnothing]\ \forall_Y\, \mathbf{S}[Y \neq \varnothing]\ \forall \tau \in (\mathrm{top})[Y]\ \forall (G_x)_{x\in X} \in \tau^X$:

$$(\exists K \in \mathrm{Fin}(X)\ \forall u \in X \setminus K:\ G_u = Y) \Rightarrow \left(\prod_{x\in X} G_x \in \beta^0_{\otimes}(X,Y,\tau)\right).$$

Recall some simple properties of the Tichonoff degree of a given TS. If $\mathbf{S}[X \neq \varnothing]$, $\mathbf{S}[Y \neq \varnothing]$ and $\tau \in (\mathrm{top})_0[Y]$, then

$$\otimes^X(\tau) \in (\mathrm{top})_0[Y^X]. \tag{2.6.24}$$

By (2.6.24) the property of separability of the initial TS is preserved for the Tichonoff degree. Note that $\forall_X\, \mathbf{S}[X \neq \varnothing]\ \forall_Y\, \mathbf{S}[Y \neq \varnothing]\ \forall \tau \in (\mathrm{top})[Y]\ \forall E \in \mathcal{F}_{\otimes^X(\tau)}$:

$$\begin{aligned}&(\forall x \in X:\ \mathrm{cl}(\{e(x):\ e \in E\},\tau) \in (\tau-\mathrm{comp})[Y])\\ &\qquad \Rightarrow (E \in (\otimes^X(\tau)-\mathrm{comp})[Y^X]).\end{aligned} \tag{2.6.25}$$

The relation (2.6.25) is a natural corollary of the well known Tichonoff theorem; see, for example, [81, Ch. 7]. Using the statements of [81, Ch. 7], we obtain that $\forall_X\, \mathbf{S}[X \neq \varnothing]\ \forall_Y\, \mathbf{S}[Y \neq \varnothing]\ \forall \tau \in (\mathrm{top})_0[Y]$:

$$\begin{aligned}&(\otimes^X(\tau)-\mathrm{comp})[Y^X]\\ =&\{F \in \mathcal{F}_{\otimes^X(\tau)} \mid \forall x \in X: \mathrm{cl}(\{f(x):\ f \in F\},\tau) \in (\tau-\mathrm{comp})[Y]\}\\ =&\{F \in \mathcal{F}_{\otimes^X(\tau)} \mid \forall x \in X:\ \{f(x):\ f \in F\} \in (\tau-\mathrm{comp})[Y]\};\end{aligned} \tag{2.6.26}$$

in (2.6.26) we have the very useful necessary and sufficient conditions of compactness for sets in the above mentioned variant of the Tichonoff product. We note that it is possible to assume that $E = Y^X$ in (2.6.25) (for the case when $\mathbf{S}[X \neq \varnothing]$, $\mathbf{S}[Y \neq \varnothing]$, and $\tau \in (\mathrm{top})[Y]$); then $\{e(x):\ e \in E\} = Y$. In this case we have $\otimes^X(\tau) \in (\mathbf{c}-\mathrm{top})[Y^X]$ for $\tau \in (\mathbf{c}-\mathrm{top})[Y]$. This statement is a particular case of the Tichonoff theorem. We note that this theorem is the basis of constructions used in the known Alaoglu theorem [66, Ch. V]. The latter plays the very important role in extension constructions for spaces. Of course, we have $\forall_X\, \mathbf{S}[X \neq \varnothing]$ $\forall_Y\, \mathbf{S}[Y \neq \varnothing]\ \forall \tau \in (\mathrm{top})[Y]\ \forall (K_x)_{x\in X} \in \mathrm{Fin}(Y)^X$:

$$\prod_{x\in X} K_x \in (\otimes^X(\tau)-\mathrm{comp})[Y^X];$$

in this example it is useful to employ a more general variant of the Tichonoff theorem.

In the following it is more important to consider the cases $(Y,\tau) = (\mathbb{R},\tau_{\mathbb{R}})$ or $(Y,\tau) = (\mathbb{R},\tau_{\partial})$ in (2.6.24). Let us discuss two given cases in detail. First consider the case $(Y,\tau) = (\mathbb{R},\tau_{\mathbb{R}})$. Of course, we have a variant of (2.6.22). But it is advisable to use specific properties of $(\mathbb{R},\tau_{\mathbb{R}})$. Suppose $\forall_X\, \mathbf{S}[X \neq \varnothing]\ \forall f \in \mathbb{R}^X\ \forall K \in \mathrm{Fin}(X)\ \forall \varepsilon \in ]0,\infty[$:

$$\mathbb{N}_X(f,K,\varepsilon) \triangleq \{s \in \mathbb{R}^X \mid \forall x \in K : |f(x) - s(x)| < \varepsilon\}. \tag{2.6.27}$$

On the basis of (2.6.27) we construct the corresponding family of sets. If $\mathbf{S}[X \neq \varnothing]$ and $f \in \mathbb{R}^X$ then denote by $\mathfrak{N}_X^0(f)$ the family of all sets

$$\mathbb{N}_X(f,K,\varepsilon),\ (K,\varepsilon) \in \mathrm{Fin}(X)\times]0,\infty[.$$

Finally, define $\forall_X\, \mathbf{S}[X \neq \varnothing]$:

$$\mathfrak{N}_X \triangleq \bigcup_{f\in\mathbb{R}^X} \mathfrak{N}_X^0(f). \tag{2.6.28}$$

From the definition of the topology $\tau_{\mathbb{R}}$ it follows that if $\mathbf{S}[X \neq \varnothing]$, $G \in \otimes^X(\tau_{\mathbb{R}})$ and $g \in G$, then for some $K \in \mathrm{Fin}(X)$ and $\varepsilon \in ]0,\infty[$ the inclusion $\mathbb{N}_X(g,K,\varepsilon) \subset G$ holds. In this connection see (2.6.19) and (2.6.23). From (2.6.2), (2.6.27) and (2.6.28) we obtain that $\forall_X\, \mathbf{S}[X \neq \varnothing]$: $\mathfrak{N}_X \in (\mathrm{BAS})[\mathbb{R}^X]$. Then it is possible to use (2.6.3). If $\mathbf{S}[X \neq \varnothing]$, then $\otimes^X(\tau_{\mathbb{R}}) = \{\cup\}(\mathfrak{N}_X)$ and by (2.6.6) we have

$$\mathfrak{N}_X \in (\otimes^X(\tau_{\mathbb{R}}) - \mathrm{BAS})_0[\mathbb{R}^X]. \tag{2.6.29}$$

In (2.6.29) we have a more simple base of the Tichonoff degree of $(\mathbb{R},\tau_{\mathbb{R}})$. Note that by (2.6.3), (2.6.27)–(2.6.29) we obtain $\forall_X\, \mathbf{S}[X \neq \varnothing]\ \forall G \in \mathcal{P}(\mathbb{R}^X)$:

$$(\forall g \in G\ \exists K \in \mathrm{Fin}(X)\ \exists \varepsilon \in ]0,\infty[\colon \mathbb{N}_X(g,K,\varepsilon) \subset G) \Rightarrow (G \in \otimes^X(\tau_{\mathbb{R}})).$$

As a consequence we obtain the following very suitable representation: if $\mathbf{S}[X \neq \varnothing]$, then

$$\begin{aligned}\otimes^X(\tau_{\mathbb{R}}) = \{G \in \mathcal{P}(\mathbb{R}^X) \mid \forall f \in G\ \exists K \in \mathrm{Fin}(X)\ \exists \varepsilon \in ]0,\infty[\colon \\ \mathbb{N}_X(f,K,\varepsilon) \subset G\} \in (\mathrm{top})_0[\mathbb{R}^X];\end{aligned} \tag{2.6.30}$$

so $(\mathbb{R}^X, \otimes^X(\tau_{\mathbb{R}}))$ is defined by the rule (2.6.30). For this TS we consider a useful variant of the local topological base. For this we note that by (2.6.27) and (2.6.30), $\forall_X\, \mathbf{S}[X \neq \varnothing]\ \forall f \in \mathbb{R}^X\ \forall G \in N^0_{\otimes^X(\tau_{\mathbb{R}})}(f)\ \exists K \in \mathrm{Fin}(X)\ \exists \varepsilon \in ]0,\infty[$:

$$\mathbb{N}_X(f,K,\varepsilon) \subset G.$$

As a corollary we obtain the important property of local bases. Namely, if $\mathbf{S}[X \neq \varnothing]$ and $f \in \mathbb{R}^X$, then

$$\mathfrak{N}_X^0(f) \in (f - \text{bas})[\otimes^X(\tau_{\mathbb{R}})]. \tag{2.6.31}$$

In (2.6.31) we have a very 'suitable' local base. Recall that elements of this base are exactly the sets (2.6.27). Note that by (2.6.31) we obtain the property that if $\mathbf{S}[X \neq \varnothing]$, $(a_x)_{x \in X} \in \mathbb{R}^X$ and $(b_x)_{x \in X} \in \mathbb{R}^X$, then

$$\prod_{x \in X} [a_x, b_x] = \{f \in \mathbb{R}^X | \ \forall x \in X : (a_x \leq f(x)) \& (f(x) \leq b_x)\} \in \mathcal{F}_{\otimes^X(\tau_{\mathbb{R}})},$$

and for $u \in X$ the following property holds:

$$(\{e(u) : e \in \prod_{x \in X} [a_x, b_x]\} = \varnothing) \vee (\{e(u) : e \in \prod_{x \in X} [a_x, b_x]\} = [a_u, b_u]);$$

of course, always $\{e(u) : e \in \prod_{x \in X} [a_x, b_x]\} \in (\tau_{\mathbb{R}} - \text{comp})[\mathbb{R}]$. As a consequence of (2.6.26) we have $\forall_X \ \mathbf{S}[X \neq \varnothing] \ \forall (a_x)_{x \in X} \in \mathbb{R}^X \ \forall (b_x)_{x \in X} \in \mathbb{R}^X$:

$$\prod_{x \in X} [a_x, b_x] \in (\otimes^X(\tau_{\mathbb{R}}) - \text{comp})[\mathbb{R}^X]. \tag{2.6.32}$$

The relation (2.6.32) is very useful when proving the Alaoglu theorem (see [66, Ch. V]).

Consider other important variant of the topological equipment of $\mathbb{R}^X$, where $X$ is a nonempty set. Namely, we equip $\mathbb{R}$ with the discrete topology. If $\mathbf{S}[X \neq \varnothing]$, $f \in \mathbb{R}^X$ and $K \in \text{Fin}(X)$, then we denote by $\mathrm{N}_X^{(\partial)}(f, K)$ the set of all $g \in \mathbb{R}^X$ such that $\forall x \in K : \ f(x) = g(x)$. Of course, $\forall_X \ \mathbf{S}[X \neq \varnothing] \ \forall f \in \mathbb{R}^X \ \forall K \in \text{Fin}(X) \ \forall g \in \mathrm{N}_X^{(\partial)}(f, K)$:

$$\mathrm{N}_X^{(\partial)}(f, K) = \mathrm{N}_X^{(\partial)}(g, K). \tag{2.6.33}$$

For $\mathbf{S}[X \neq \varnothing]$ and $f \in \mathbb{R}^X$ we introduce the family $\mathfrak{N}_X^{(\partial)}(f) \triangleq \{\mathrm{N}_X^{(\partial)}(f, K) : \ K \in \text{Fin}(X)\}$. Finally, by analogy with (2.6.28) we consider the union of all such families. Namely, $\forall_X \ \mathbf{S}[X \neq \varnothing]$:

$$\mathfrak{N}^{(\partial)}[X] \triangleq \bigcup_{f \in \mathbb{R}^X} \mathfrak{N}_X^{(\partial)}(f) = \{E \in \mathcal{P}(\mathbb{R}^X) | \ \exists f \in \mathbb{R}^X \ \exists K \in \text{Fin}(X) : E = \mathrm{N}_X^{(\partial)}(f, K)\} \in (\text{BAS})[\mathbb{R}^X]. \tag{2.6.34}$$

By (2.6.34) some topology of $\mathbb{R}^X$ is defined. It is possible easily to verify that $\otimes^X(\tau_\partial) = \{\cup\}(\mathfrak{N}^{(\partial)}[X])$ for each nonempty set $X$, i.e.,

$$\mathfrak{N}^{(\partial)}[X] \in (\otimes^X(\tau_\partial) - \text{BAS})_0[\mathbb{R}^X]. \tag{2.6.35}$$

As a corollary of (2.6.35) we have $\forall_X\, \mathbf{S}[X \neq \varnothing]$:

$$\otimes^X(\tau_\partial) = \{G \in \mathcal{P}(\mathbb{R}^X) | \ \forall g \in G\ \exists K \in \mathrm{Fin}(X) : \mathbb{N}_X^{(\partial)}(g, K) \subset G\} \in (\mathrm{top})_0[\mathbb{R}^X]. \tag{2.6.36}$$

In (2.6.36) we use (2.6.24). Consider some properties of the TS

$$(\mathbb{R}^X, \otimes^X(\tau_\partial)), \tag{2.6.37}$$

where $\mathbf{S}[X \neq \varnothing]$. We note that from (2.6.8) and (2.6.36) the useful property follows (see (2.2.14)): if $\mathbf{S}[X \neq \varnothing]$ and $f \in \mathbb{R}^X$, then

$$\mathfrak{N}_X^{(\partial)}(f) \in (f - \mathrm{bas})[\otimes^X(\tau_\partial)].$$

Therefore the local bases of the TS (2.6.37) are defined (we recall that by (2.6.33) and (2.6.36) we have $\mathfrak{N}_X^{(\partial)}(f) \subset N^0_{\otimes^X(\tau_\partial)}(f)$). Moreover, we note that $\forall_X\, \mathbf{S}[X \neq \varnothing]\ \forall f \in \mathbb{R}^X\ \forall K \in \mathrm{Fin}(X)\ \forall g \in \mathbb{R}^X \setminus \mathbb{N}_X^{(\partial)}(f, K)$ $\exists K' \in 2^K$:

$$\mathbb{N}_X^{(\partial)}(g, K') \subset \mathbb{R}^X \setminus \mathbb{N}_X^{(\partial)}(f, K).$$

From this statement and (2.6.36) we obtain the property that if $\mathbf{S}[X \neq \varnothing]$, then

$$\mathfrak{N}^{(\partial)}[X] \subset \mathcal{F}_{\otimes^X(\tau_\partial)}. \tag{2.6.38}$$

From (2.6.7), (2.6.35) and (2.6.38) we have $\forall_X\, \mathbf{S}[X \neq \varnothing]$:

$$\otimes^X(\tau_\partial) \in (0 - \mathrm{top})[\mathbb{R}^X]. \tag{2.6.39}$$

Note that by (2.6.38) and (2.6.39) all sets of the family (2.6.34) are open-closed.

Compare the topologies (2.6.30) and (2.6.36). From (2.5.22) we have the obvious property $\tau_{\mathbb{R}} \subset \tau_\partial$. Therefore from (2.5.1) and (2.5.2) it follows that $\forall_E\, \mathbf{S}[E \neq \varnothing]\ \forall \tau \in (\mathrm{top})[E]$:

$$C(E, \tau, \mathbb{R}, \tau_\partial) \subset C(E, \tau, \mathbb{R}, \tau_{\mathbb{R}}).$$

Therefore by (2.6.17) we have $\forall_U\, \mathbf{S}[U \neq \varnothing]\ \forall \mathbb{H} \in \mathcal{P}(\mathbb{R}^U)$:

$$\mathbf{T}_{\mathrm{top}}(U|\tau_\partial, \mathbb{H}) \subset \mathbf{T}_{\mathrm{top}}(U|\tau_{\mathbb{R}}, \mathbb{H}).$$

As a corollary, from (2.6.18) it follows that $\forall_U\, \mathbf{S}[U \neq \varnothing]\ \forall \mathbb{H} \in \mathcal{P}(\mathbb{R}^U)$:

$$\tau^0(U|\tau_\partial, \mathbb{H}) \in \mathbf{T}_{\mathrm{top}}(U|\tau_{\mathbb{R}}, \mathbb{H}).$$

We have the natural estimate: if $\mathbf{S}[U \neq \varnothing]$ and $\mathbb{H} \in \mathcal{P}(\mathbb{R}^U)$, then

$$\tau^0(U|\tau_{\mathbb{R}}, \mathbb{H}) \subset \tau^0(U|\tau_\partial, \mathbb{H}). \tag{2.6.40}$$

From (2.6.22) and (2.6.40) we obtain that $\forall_X \mathbf{S}[X \neq \varnothing]$:

$$\otimes^X(\tau_{\mathbb{R}}) \subset \otimes^X(\tau_\partial). \tag{2.6.41}$$

From (2.6.41) we get corresponding estimates for subspaces. Namely, by (2.3.1) and (2.6.41) we have $\forall_X \mathbf{S}[X \neq \varnothing]\ \forall L \in \mathcal{P}(\mathbb{R}^X)$:

$$\otimes^X(\tau_{\mathbb{R}})|_L \subset \otimes^X(\tau_\partial)|_L. \tag{2.6.42}$$

We use (2.6.42) in Chapter 4. Namely, we will consider a connection of two different topologies of the space of finitely additive measures (FAM) with a bounded total variation. Note that the TS (2.6.37) and subspaces of spaces of the type (2.6.37) were used in [35, 32] and in many works devoted to extensions and relaxations.

In connection with (2.6.42) we make a remark connected with the passage to a subspace. Namely, by (2.3.1) and (2.6.2) we have $\forall_E \mathbf{S}[E]$ $\forall \beta \in (\mathrm{BAS})[E]\ \forall H \in \mathcal{P}(E)$:

$$\beta|_H = \{U \cap H :\ U \in \beta\} \in (\mathrm{BAS})[H]. \tag{2.6.43}$$

And what is more, for (2.6.43) we have a useful property connected with generated topologies. Namely, $\forall_E \mathbf{S}[E]\ \forall \beta \in (\mathrm{BAS})[E]\ \forall H \in \mathcal{P}(E)$:

$$\{\cup\}(\beta)|_H = \{\cup\}(\beta|_H). \tag{2.6.44}$$

From (2.6.6), (2.6.43) and (2.6.44) we have $\forall_E \mathbf{S}[E]\ \forall \tau \in (\mathrm{top})[E]\ \forall \beta \in (\tau - \mathrm{BAS})_0[E]$:

$$\beta|_H \in (\tau|_H - \mathrm{BAS})_0[H]. \tag{2.6.45}$$

It is possible to realize (2.6.45) for the spaces used in (2.6.42). We consider these constructions in Chapter 4 only for some cases of the spaces of measures. In particular, we use (2.6.37) in these cases. In this connection recall (2.6.39). The given property characterizes the spaces of the type (2.6.37) as very nonstandard TS. These TS can be used for constructing different 'pathologies' of topological character. It is useful to note that the closure and the sequential closure of a set are different notions. In this connection consider a simple example connected with topology used in Chapter 4.

**Example.** Let $X = \mathbb{R}$. So elements of $\mathbb{R}^X$ are functions from $\mathbb{R}$ into $\mathbb{R}$, i.e., $\mathbb{R}^{\mathbb{R}} = \{\mathbb{R} \to \mathbb{R}\}$. In this case we use the TS (2.6.37). Let

$$A \triangleq \{f \in \mathbb{R}^X \mid \exists K \in \mathrm{Fin}(\mathbb{R})\ \forall x \in \mathbb{R} \setminus K :\ f(x) \neq 0\}. \tag{2.6.46}$$

Consider the function $\mathbb{O} = \mathbb{O}_{\mathbb{R}}$ for which $\forall x \in \mathbb{R} :\ \mathbb{O}(x) = 0$. Compare $\mathrm{cl}(A, \otimes^{\mathbb{R}}(\tau_\partial))$ and $(\mathrm{seqcl})[A; \otimes^{\mathbb{R}}(\tau_\partial)]$ defined in the form of the right hand side of (2.6.9) (see (2.6.10)). We note that

$$\mathbb{O} \in \mathrm{cl}(A, \otimes^{\mathbb{R}}(\tau_\partial)) \setminus (\mathrm{seqcl})[A; \otimes^{\mathbb{R}}(\tau_\partial)]. \tag{2.6.47}$$

By (2.6.46) we will establish that $\mathbb{O} \in \mathrm{cl}(A; \otimes^{\mathbb{R}}(\tau_\partial))$. We use (2.6.38). Let

$$H^* \in N_\tau(\mathbb{O})|_{\tau=\otimes^{\mathbb{R}}(\tau_\partial)}.$$

Then by (2.6.8) and (2.6.38) we can choose a set $G^* \in \mathfrak{N}_{\mathbb{R}}^{(\partial)}(\mathbb{O})$ for which $G^* \subset H^*$. For some $K^* \in \mathrm{Fin}(\mathbb{R})$ we have the representation $G^* = \mathrm{N}_{\mathbb{R}}^{(\partial)}(\mathbb{O}; K^*)$. Fix $K^*$ and construct the function $f^* \in \mathbb{R}^{\mathbb{R}}$ by the following rule:

$$(\forall x \in K^* : \ f^*(x) \triangleq 0)\&(\forall x \in \mathbb{R} \setminus K^* : \ f^*(x) \triangleq 1).$$

From (2.6.46) we obtain $f^* \in A$. On the other hand, $f^* \in \mathrm{N}_{\mathbb{R}}^{(\partial)}(\mathbb{O}; K^*)$, i.e., $f^* \in G^*$. So $f^* \in A \cap H^*$. Since the choice of $H^*$ was arbitrary, by (2.3.10) we have $\mathbb{O} \in \mathrm{cl}(A, \otimes^{\mathbb{R}}(\tau_\partial))$. In addition, $\mathbb{O} \notin (\mathrm{seqcl})[A; \otimes^{\mathbb{R}}(\tau_\partial)]$. Suppose the contrary: $\mathbb{O} \in (\mathrm{seqcl})[A; \otimes^{\mathbb{R}}(\tau_\partial)]$. By (2.6.9) we can choose

$$(a_i)_{i\in\mathcal{N}} : \mathcal{N} \to A$$

for which $(a_i)_{i\in\mathcal{N}} \overset{\otimes^{\mathbb{R}}(\tau_\partial)}{\to} \mathbb{O}$. Using the axiom of choice and (2.6.46) we choose a sequence

$$(K_i)_{i\in\mathcal{N}} : \mathcal{N} \to \mathrm{Fin}(\mathbb{R})$$

for which $\forall j \in \mathcal{N}\ \forall x \in \mathbb{R} \setminus K_j : \ a_j(x) \neq 0$. Then the set

$$\mathbf{K} \triangleq \bigcup_{i\in\mathcal{N}} K_i \in 2^{\mathbb{R}}$$

is not more than countable. Since the cardinality of $\mathbb{R}$ is continuum, we have $\mathbb{R} \setminus \mathbf{K} \neq \varnothing$. Let $u \in \mathbb{R} \setminus \mathbf{K}$. Then $\{u\} \in \mathrm{Fin}(\mathbb{R})$ and

$$\mathrm{N}_{\mathbb{R}}^{(\partial)}(\mathbb{O}, \{u\}) = \{f \in \mathbb{R}^{\mathbb{R}} |\ f(u) = 0\} \in \mathfrak{N}_{\mathbb{R}}^{(\partial)}(\mathbb{O}).$$

Therefore by (2.6.8) and (2.6.38) we have

$$\mathrm{N}_{\mathbb{R}}^{(\partial)}(\mathbb{O}, \{u\}) \in N_{\otimes^{\mathbb{R}}(\tau_\partial)}(\mathbb{O}).$$

From the convergence of $(a_i)_{i\in\mathcal{N}}$ we obtain that $\exists k \in \mathcal{N}\ \forall i \in \overrightarrow{k,\infty}$: $a_i \in \mathrm{N}_{\mathbb{R}}^{(\partial)}(\mathbb{O}, \{u\})$. Here we use (2.2.49). Fix $j \in \mathcal{N}$ for which $a_j \in \mathrm{N}_{\mathbb{R}}^{(\partial)}(\mathbb{O}, \{u\})$. Then $a_j(u) = 0$. But

$$\mathbb{R} \setminus \mathbf{K} = \bigcap_{i\in\mathcal{N}} (\mathbb{R} \setminus K_i)$$

and hence $\mathbb{R} \setminus \mathbf{K} \subset \mathbb{R} \setminus K_j$ and $u \in \mathbb{R} \setminus K_j$. So, $a_j(u) \neq 0$. The obtained contradiction proves the property $\mathbb{O} \notin (\mathrm{seqcl})[A; \otimes^{\mathbb{R}}(\tau_\partial)]$. So (2.6.47) is established.

## 2.7 METRIZABLE AND PSEUDO-METRIZABLE SPACES

In this section we consider a very brief summary of notions connected with metric and pseudo-metric spaces. If $\mathbf{S}[X \neq \emptyset]$, then we denote by $(\mathbf{p} - \mathrm{Dist})[X]$ the set of all mappings

$$\rho: X \times X \to [0, \infty[ \tag{2.7.1}$$

for each of which

$$\begin{aligned} &(\forall x \in X: \ \rho(x, x) = 0) \\ &\&(\forall x_1 \in X \ \forall x_2 \in X: \ \rho(x_1, x_2) = \rho(x_2, x_1)) \\ &\&(\forall x_1 \in X \ \forall x_2 \in X \ \forall x_3 \in X: \ \rho(x_1, x_3) \leq \rho(x_1, x_2) + \rho(x_2, x_3)). \end{aligned} \tag{2.7.2}$$

The mappings (2.7.1) with the properties (2.7.2) are called pseudo-metrics of the set $X$. If $\mathbf{S}[X \neq \emptyset]$ and $\rho \in (\mathbf{p} - \mathrm{Dist})[X]$, then the pair $(X, \rho)$ is called the pseudo-metric space. Let $\forall_X \ \mathbf{S}[X \neq \varnothing]$:

$$\begin{aligned} (\mathrm{Dist})[X] \triangleq \{\rho \in (\mathbf{p} - \mathrm{Dist})[X] | \ \forall x_1 \in X \ \forall x_2 \in X: \\ (\rho(x_1, x_2) = 0) \Rightarrow (x_1 = x_2)\}. \end{aligned} \tag{2.7.3}$$

Elements of the set (2.7.3) are called metrics or distances of the set $X$. If $\mathbf{S}[X \neq \emptyset]$ and $\rho \in (\mathrm{Dist})[X]$, then the pair $(X, \rho)$ is called the metric space.

**Example 2.7.1.** Let $\mathbf{S}[X \neq \emptyset]$ and $f \in \mathbb{R}^X$. Moreover, assume that it is possible to choose $x^{(1)} \in X$ and $x^{(2)} \in X \setminus \{x^{(1)}\}$ for which $f(x^{(1)}) = f(x^{(2)})$. Introduce the function (2.7.1) by the following condition: if $x_1 \in X$ and $x_2 \in X$, then $\rho(x_1, x_2) = |f(x_1) - f(x_2)|$. Then

$$\rho \in (\mathbf{p} - \mathrm{Dist})[X] \setminus (\mathrm{Dist})[X].$$

So the given mapping $\rho$ is a pseudo-metric but not metric of $X$.

**Example 2.7.2.** Let $\mathbf{S}[X \neq \emptyset]$ and $\rho$ (2.7.1) is defined by the following stipulation: if $x_1 \in X$ and $x_2 \in X$, then

$$((x_1 = x_2) \Rightarrow (\rho(x_1, x_2) \triangleq 0))\&((x_1 \neq x_2) \Rightarrow (\rho(x_1, x_2) \triangleq 1)).$$

Then $\rho \in (\mathrm{Dist})[X]$. This metric $\rho$ is called discrete (the discrete topology of $X$ is realized by this metric). The given pair $(X, \rho)$ is a metric space.

Returning to the general case, we introduce a natural notion of a ball in a pseudo-metric space. Namely, if $\mathbf{S}[X \neq \emptyset]$, $\rho \in (\mathbf{p} - \mathrm{Dist})[X]$, $x \in X$, and $\varepsilon \in \mathbb{R}$, then we set

$$(\mathbf{B}_\rho(x, \varepsilon) \triangleq \{\tilde{x} \in X | \ \rho(x, \tilde{x}) \leq \varepsilon\})\&(\mathbf{B}^0_\rho(x, \varepsilon) \triangleq \{\tilde{x} \in X | \ \rho(x, \tilde{x}) < \varepsilon\}). \tag{2.7.4}$$

Of course, in (2.7.4) we have the empty set for $\varepsilon < 0$. Moreover, the second set in (2.7.4) is also the empty set for $\varepsilon = 0$. If $\rho \in (\mathrm{Dist})[X]$, then the first set in (2.7.4) is the one-element set $\{x\}$ for $\varepsilon = 0$. The case $\varepsilon > 0$ is basic (in (2.7.4)). The first set in (2.7.4) is called a closed ball in $(X, \rho)$. The second set in (2.7.4) is called an open ball in $(X, \rho)$. In addition, $\forall_X\, \mathbf{S}[X \neq \varnothing]\ \forall \rho \in (\mathbf{p} - \mathrm{Dist})[X]$:

$$\begin{aligned}\tau_\rho^\natural(X) &\triangleq \{G \in \mathcal{P}(X) |\ \forall x \in G\ \exists \varepsilon \in ]0, \infty[ :\ \mathbf{B}_\rho^0(x, \varepsilon) \subset G\} \\ &= \{G \in \mathcal{P}(X) |\ \forall x \in G\ \exists \varepsilon \in ]0, \infty[ :\ \mathbf{B}_\rho(x, \varepsilon) \subset G\} \in (\mathrm{top})[X].\end{aligned} \tag{2.7.5}$$

By (2.7.5) the pseudo-metric space $(X, \rho)$ generates the TS $(X, \tau_\rho^\natural(X))$. Recall (see [81]) that a TS $(X, \tau)$, $X \neq \emptyset$, is called pseudo-metrizable in the case when there exists a pseudo-metric $\rho \in (\mathbf{p} - \mathrm{Dist})[X]$ for which $\tau = \tau_\rho^\natural(X)$. Analogously, a TS $(X, \tau)$, $X \neq \emptyset$, is called metrizable in the case when there exists a metric $\rho \in (\mathrm{Dist})[X]$ for which $\tau = \tau_\rho^\natural(X)$. We note that $\forall_X\, \mathbf{S}[X \neq \varnothing]\ \forall \rho \in (\mathbf{p} - \mathrm{Dist})[X]\ \forall x \in X\ \forall \varepsilon \in [0, \infty[$:

$$(\mathbf{B}_\rho^0(x, \varepsilon) \in \tau_\rho^\natural(X)) \& (\mathbf{B}_\rho(x, \varepsilon) \in \mathcal{F}_{\tau_\rho^\natural(X)}). \tag{2.7.6}$$

Moreover, note that $\forall_X\, \mathbf{S}[X \neq \varnothing]$:

$$(\mathrm{Dist})[X] = \{\rho \in (\mathbf{p} - \mathrm{Dist})[X] |\ \tau_\rho^\natural(X) \in (\mathrm{top})_0[X]\}.$$

The proof of (2.7.6) is obvious; we omit it. In connection with (2.7.6) the following useful circumstance should be noted. Namely, $\forall_X\, \mathbf{S}[X \neq \varnothing]$ $\forall \rho \in (\mathbf{p} - \mathrm{Dist})[X]\ \forall x \in X$:

$$\begin{aligned}&(\forall \varepsilon \in ]0, \infty[ :\ \mathbf{B}_\rho^0(x, \varepsilon) \in N^0_{\tau_\rho^\natural(X)}(x)) \\ &\& (N_{\tau_\rho^\natural(X)}(x) = \{H \in \mathcal{P}(X) |\ \exists \varepsilon \in ]0, \infty[ :\ \mathbf{B}_\rho^0(x, \varepsilon) \subset H\}).\end{aligned} \tag{2.7.7}$$

The relation (2.7.7) characterizes the system of open balls with a given center as a local topological base or as a fundamental system of neighborhoods of a point. In reality, the family of open balls is the topological base. This conclusion follows from (2.7.5) and (2.7.6). Indeed, we have (see (2.6.2), (2.6.3), (2.6.6)) $\forall_X\, \mathbf{S}[X \neq \varnothing]\ \forall \rho \in (\mathbf{p} - \mathrm{Dist})[X]$:

$$\{\mathbf{B}_\rho^0(x, \varepsilon) :\ x \in X,\ \varepsilon \in ]0, \infty[\} \in (\tau_\rho^\natural(X) - \mathrm{BAS})_0[X].$$

Therefore sets of the topology (2.7.5) are exactly unions of open balls. Returning to (2.7.7), we note that $\forall_X\, \mathbf{S}[X \neq \varnothing]\ \forall \rho \in (\mathbf{p} - \mathrm{Dist})[X]$ $\forall x \in X$:

$$\{\mathbf{B}_\rho^0(x, \varepsilon) :\ \varepsilon \in ]0, \infty[\} \in (x - \mathrm{bas})[\tau_\rho^\natural(X)]. \tag{2.7.8}$$

We use (in (2.7.8)) (2.6.8) and (2.7.7). Of course, the statement (2.7.8) will be true if the family $\{\mathbf{B}^0_\rho(x, n^{-1}) : n \in \mathcal{N}\}$ is used in its left hand side. As a corollary, $\forall_X\ \mathbf{S}[X \neq \varnothing]\ \forall \rho \in (\mathbf{p} - \mathrm{Dist})[X]$:

$$\tau^\natural_\rho(X) \in (\mathrm{top})_I[X]. \tag{2.7.9}$$

Hence (2.6.10) holds in metrizable space. Here it is sufficient to compare (2.6.10) and (2.7.9). Note that by (2.2.49), (2.7.4) and (2.7.7) we have $\forall_X\ \mathbf{S}[X \neq \varnothing]\ \forall \rho \in (\mathbf{p} - \mathrm{Dist})[X]\ \forall (x_i)_{i \in \mathcal{N}} \in X^{\mathcal{N}}\ \forall x \in X$:

$$((x_i)_{i\in\mathcal{N}} \xrightarrow{\tau^\natural_\rho(X)} x) \Leftrightarrow ((\rho(x_i, x))_{i\in\mathcal{N}} \xrightarrow{\tau_{\mathbb{R}}} 0). \tag{2.7.10}$$

In connection with the designation of the last convergence it is advisable to take into account the traditional stipulation: the corresponding index $\tau_{\mathbb{R}}$ is omitted. Namely, if $(\xi_i)_{i\in\mathcal{N}} \in \mathbb{R}^{\mathcal{N}}$ and $\xi \in \mathbb{R}$, then, as usual, we suppose by definition that

$$((\xi_i)_{i\in\mathcal{N}} \to \xi) \Leftrightarrow (\forall \varepsilon \in ]0, \infty[\ \exists m \in \mathcal{N}\ \forall k \in \overrightarrow{m, \infty} :\ |\xi_k - \xi| < \varepsilon);$$

of course, $(\xi_i)_{i\in\mathcal{N}}$ converges to $\xi$ in this traditional sense if and only if $(\xi_i)_{i\in\mathcal{N}} \xrightarrow{\tau_{\mathbb{R}}} \xi$. Therefore from (2.7.10) we obtain $\forall_X\ \mathbf{S}[X \neq \varnothing]\ \forall \rho \in (\mathbf{p} - \mathrm{Dist})[X]\ \forall (x_i)_{i\in\mathcal{N}} \in X^{\mathcal{N}}\ \forall x \in X$:

$$((x_i)_{i\in\mathcal{N}} \xrightarrow{\tau^\natural_\rho(X)} x) \Leftrightarrow ((\rho(x_i, x))_{i\in\mathcal{N}} \to 0). \tag{2.7.11}$$

Of course, from (2.6.10), (2.7.9), and (2.7.11) we have the useful 'sequential" representation of the closure operation: if $\mathbf{S}[X \neq \emptyset]$, $\rho \in (\mathbf{p} - \mathrm{Dist})[X]$ and $A \in \mathcal{P}(X)$, then

$$\begin{aligned} \mathrm{cl}(A, \tau^\natural_\rho(X)) &= (\mathrm{seqcl})[A; \tau^\natural_\rho(X)] \\ &= \{x \in X \mid \exists (a_i)_{i\in\mathcal{N}} \in A^{\mathcal{N}} :\ (\rho(a_i, x))_{i\in\mathcal{N}} \to 0\}. \end{aligned} \tag{2.7.12}$$

Moreover, from (2.3.10) and (2.7.7) we obtain $\forall_X\ \mathbf{S}[X \neq \varnothing]\ \forall \rho \in (\mathbf{p} - \mathrm{Dist})[X]\ \forall A \in \mathcal{P}(X)$:

$$\mathrm{cl}(A, \tau) = \{x \in X \mid \forall \varepsilon \in ]0, \infty[:\ \mathbf{B}^0_\rho(x, \varepsilon) \cap A \neq \emptyset\}.$$

Introduce the natural notion of the distance from point to set. Namely, $\forall_X\ \mathbf{S}[X \neq \varnothing]\ \forall \rho \in (\mathbf{p} - \mathrm{Dist})[X]\ \forall A \in 2^X\ \forall x \in X$:

$$(\rho - \inf)[x; A] \triangleq \inf(\{\rho(x, y) :\ y \in A\}) \in [0, \infty[. \tag{2.7.13}$$

By (2.7.13) we introduce a useful function. In addition, $\forall \rho \in (\mathbf{p} - \mathrm{Dist})[X]\ \forall A \in 2^X\ \forall y \in X\ \forall z \in X$:

$$|(\rho - \inf)[y; A] - (\rho - \inf)[z; A]| \leq \rho(y, z). \tag{2.7.14}$$

We obtain the property of uniform continuity of $(\rho-\inf)[\cdot;A]$. It is possible to add a representation of the closure operator in terms of (2.7.13). Namely, $\forall_X\,\mathbf{S}[X\neq\varnothing]\;\forall\rho\in(\mathbf{p}-\mathrm{Dist})[X]\;\forall A\in 2^X$:

$$\mathrm{cl}(A,\tau_\rho^\natural(X))=\{x\in X|\ (\rho-\inf)[x;A]=0\}. \tag{2.7.15}$$

In (2.7.15) it is sufficient to use the relation (2.7.8). By analogy with (2.7.4) it is possible to introduce $\varepsilon$-neighborhoods of nonempty sets in a pseudo-metric space. We suppose that $\forall_X\,\mathbf{S}[X\neq\varnothing]\;\forall\rho\in(\mathbf{p}-\mathrm{Dist})[X]$ $\forall A\in 2^X\;\forall\varepsilon\in\mathbb{R}$:

$$\begin{aligned}&(\mathbf{B}_\rho[A;\varepsilon]\triangleq\{x\in X\mid(\rho-\inf)[x;A]\leq\varepsilon\})\\ &\&(\mathbf{B}^0_\rho[A;\varepsilon]\triangleq\{x\in X\mid(\rho-\inf)[x;A]<\varepsilon\}).\end{aligned} \tag{2.7.16}$$

From (2.7.13), (2.7.15), and (2.7.16), under conditions defining (2.7.15) we obtain $\mathrm{cl}(A,\tau_\rho^\natural(X))=\mathbf{B}_\rho[A;0]$. In connection with the natural construction of neighborhoods we note a representation following from (2.7.5), (2.7.7), and (2.7.14). Namely, $\forall_X\,\mathbf{S}[X\neq\varnothing]\;\forall\rho\in(\mathbf{p}-\mathrm{Dist})[X]$ $\forall A\in 2^X\;\forall\varepsilon\in]0,\infty[$:

$$\mathbf{B}^0_\rho[A;\varepsilon]\in\mathbb{N}^0_{\tau_\rho^\natural(X)}[A]. \tag{2.7.17}$$

Neighborhoods of the type (2.7.17) are used under relaxation of constraints very often. One can easily verify that $\forall_X\,\mathbf{S}[X\neq\varnothing]\;\forall_Y\,\mathbf{S}[Y\neq\varnothing]$ $\forall\rho\in(\mathbf{p}-\mathrm{Dist})[X]\;\forall\vartheta\in(\mathrm{top})[Y]$:

$$\begin{aligned}C(X,\tau_\rho^\natural(X),Y,\vartheta)=\{f\in Y^X\mid\forall(x_i)_{i\in\mathcal{N}}\in X^{\mathcal{N}}\;\forall x\in X:\\ (\rho(x_i,x))_{i\in\mathcal{N}}\to 0)\Rightarrow((f(x_i))_{i\in\mathcal{N}}\overset{\vartheta}{\to}f(x))\}.\end{aligned} \tag{2.7.18}$$

To prove (2.7.18), it is sufficient to use Proposition 2.6.1, (2.7.9), and (2.7.11). So the property of continuity is equivalent to the sequential continuity for mappings from a pseudo-metrizable space in an arbitrary TS. Of course (see (2.7.9) and Section 2.6), $\forall_X\,\mathbf{S}[X\neq\varnothing]\;\forall\rho\in(\mathbf{p}-\mathrm{Dist})[X]$:

$$\mathcal{F}_{\tau_\rho^\natural(X)}=\mathcal{F}_{\mathrm{seq}}[\tau_\rho^\natural(X)]. \tag{2.7.19}$$

As for (2.7.12),(2.7.18), and (2.7.19), constructions based on the sequential approach in topological representations are used in pseudo-metrizable and (in particular) metrizable TS. In this connection recall (2.3.18) and (2.7.9). Note the following obvious property concerning questions of separability. Namely, $\forall_X\,\mathbf{S}[X\neq\varnothing]\;\forall\rho\in(\mathbf{p}-\mathrm{Dist})[X]$ $\forall M\in\mathcal{F}_{\tau_\rho^\natural(X)}\;\forall N\in\mathcal{F}_{\tau_\rho^\natural(X)}$

$$(M\cap N=\varnothing)\Rightarrow(\exists G_1\in\mathbb{N}^0_{\tau_\rho^\natural(X)}[M]\;\exists G_2\in\mathbb{N}^0_{\tau_\rho^\natural(X)}[N]:G_1\cap G_2=\varnothing). \tag{2.7.20}$$

From (2.2.19) and (2.7.20) we have $\forall_X \mathbf{S}[X \neq \varnothing]\ \forall \rho \in (\mathbf{p} - \mathrm{Dist})[X]$:

$$\tau_\rho^\natural(X) \in (\mathbf{n} - \mathrm{top})[X]. \tag{2.7.21}$$

In connection with (2.7.21) we note [81, Ch. 4]. From (2.7.3), (2.7.7), and (2.7.21) we obtain $\forall_X \mathbf{S}[X \neq \varnothing]\ \forall \rho \in (\mathrm{Dist})[X]$:

$$\tau_\rho^\natural(X) \in (\mathbf{n} - \mathrm{top})[X] \cap (\mathrm{top})_0[X]. \tag{2.7.22}$$

Therefore $(X, \tau_\rho^\natural(X))$ is a $T_4$-space.

In conclusion of the section we note some properties connected with the compactness of metrizable spaces. Sequential constructions are usually used here. In particular, variants of sequential compactness are considered. First we discuss these variants for metrizable spaces. Then we investigate general questions of comparison of compactness, countable compactness and sequential compactness of sets in TS.

We note that $\forall_X \mathbf{S}[X \neq \varnothing]\ \forall \rho \in (\mathbf{p}{-}\mathrm{Dist})[X]\ \forall (x_i)_{i\in\mathcal{N}} \in X^{\mathcal{N}}\ \forall x \in X$:

$$(\exists \alpha \in \mathbf{N} : (\rho(x_{\alpha(s)}, x))_{s\in\mathcal{N}} \to 0)$$

$$\Leftrightarrow (\forall \varepsilon \in ]0, \infty[\ \forall m \in \mathcal{N}\ \exists j \in \overrightarrow{m, \infty} : x_j \in \mathbf{B}_\rho^0(x, \varepsilon)). \tag{2.7.23}$$

As a consequence we obtain the following obvious property. Namely $\forall_X \mathbf{S}[X \neq \varnothing]\ \forall \rho \in (\mathbf{p} - \mathrm{Dist})[X]$:

$$\begin{aligned}(\tau_\rho^\natural(X) \in (\mathbf{c} - \mathrm{top})[X]) \Rightarrow &(\forall (x_i)_{i\in\mathcal{N}} \in X^{\mathcal{N}}\ \exists \alpha \in \mathbf{N}\ \exists x \in X : \\ &(\rho(x_{\alpha(s)}, x))_{s\in\mathcal{N}} \to 0).\end{aligned} \tag{2.7.24}$$

We recall (2.7.11) in connection with (2.7.24). By the corresponding definition from Section 2.2 we have $\forall_X \mathbf{S}[X \neq \varnothing]\ \forall \rho \in (\mathbf{p} - \mathrm{Dist})[X]$:

$$(\tau_\rho^\natural(X) \in (\mathbf{c} - \mathrm{top})[X]) \Rightarrow (\tau_\rho^\natural(X) \in (\mathbf{c}_{\mathrm{seq}} - \mathrm{top})[X]). \tag{2.7.25}$$

The relations (2.2.21) and (2.7.23) are used when proving the known property (2.7.24). An analogous reasoning can be found in [104]. Special arguments are given in [1, 71, 81].

Suppose that $\forall_X \mathbf{S}[X \neq \varnothing]\ \forall \rho \in (\mathbf{p} - \mathrm{Dist})[X]$:

$$\begin{aligned}((\mathrm{comp})_0[X;\rho] \triangleq \{H \in 2^X\ |&\forall (x_i)_{i\in\mathcal{N}} \in H^{\mathcal{N}}\ \exists \alpha \in \mathbf{N}\ \exists x \in X : \\ &(\rho(x_{\alpha(s)}, x))_{s\in\mathcal{N}} \to 0\}) \\ \&((\mathrm{comp})[X;\rho] \triangleq \{K \in 2^X\ |&\forall (x_i)_{i\in\mathcal{N}} \in K^{\mathcal{N}}\ \exists \alpha \in \mathbf{N}\ \exists x \in K : \\ &(\rho(x_{\alpha(s)}, x))_{s\in\mathcal{N}} \to 0\}).\end{aligned} \tag{2.7.26}$$

In (2.7.26) we consider the compactness in pseudo-metrizable spaces. Sometimes, sets of the first family in (2.7.26) are called relatively compact in $(X, \rho)$. And what is more, sometimes these sets (elements of $(\text{comp})_0[X;\rho]$) are called compact sets in $(X, \rho)$. Respectively, sets of the second family in (2.7.26) often are called compact in themselves in the sense of $(X, \rho)$. In reality these notions assume a very natural interpretation in terms of TS. We will consider this later. We now use the terminology typical for applied fields. As for (2.7.26), it is useful to keep in mind (2.3.16) and (2.7.10). Namely, from these relations we have the following representation: if $\mathbf{S}[X \neq \varnothing]$ and $\rho \in (\mathbf{p} - \text{Dist})[X]$, then

$$\begin{aligned}(\text{comp})[X;\rho] &= \{K \in 2^X \mid \forall (x_i)_{i\in\mathcal{N}} \in K^{\mathcal{N}}\ \exists \alpha \in \mathbf{N}\ \exists x \in K : \\ &\qquad (x_{\alpha(s)})_{s\in\mathcal{N}} \overset{\tau_\rho^\natural(X)}{\longrightarrow} x\} \\ &= (\tau_\rho^\natural(X) - \text{comp}_{\text{seq}})[X] \setminus \{\varnothing\}.\end{aligned} \tag{2.7.27}$$

By (2.7.27) we realize the representation of the 'metric" compactness in topological terms. In addition, we use here the sequential compactness. From (2.7.24) and (2.7.26) we obtain the obvious property. Namely, $\forall_X\ \mathbf{S}[X \neq \varnothing]\ \forall \rho \in (\mathbf{p} - \text{Dist})[X]$:

$$(\tau_\rho^\natural(X) - \text{comp})[X] \setminus \{\varnothing\} \subset (\text{comp})[X;\rho]. \tag{2.7.28}$$

There is a simple connection of families in (2.7.26). Indeed, $\forall_X\ \mathbf{S}[X \neq \varnothing]$ $\forall \rho \in (\mathbf{p} - \text{Dist})[X]\ \forall H \in (\text{comp})_0[X;\rho]$:

$$\text{cl}(H, \tau_\rho^\natural(X)) \in (\text{comp})[X;\rho].$$

We note also a useful property of continuous functionals on relatively compact sets in pseudo-metric spaces. Namely, from (2.5.25) and (2.7.18) we have $\forall_X\ \mathbf{S}[X \neq \varnothing]\ \forall \rho \in (\mathbf{p} - \text{Dist})[X]$:

$$\begin{aligned}\mathbb{C}(X, \tau_\rho^\natural(X)) = \{f \in \mathbb{R}^X \mid \forall (x_i)_{i\in\mathcal{N}} \in X^{\mathcal{N}}\ \forall x \in X : \\ (\rho(x_i, x))_{i\in\mathcal{N}} \to 0) \Rightarrow ((f(x_i))_{i\in\mathcal{N}} \to f(x))\}.\end{aligned}$$

We obtain that if $\mathbf{S}[X \neq \varnothing]$, $\rho \in (\mathbf{p} - \text{Dist})[X]$, $f \in \mathbb{C}(X, \tau_\rho^\natural(X))$ and $H \in (\text{comp})_0[X;\rho]$, then

$$\exists c \in [0, \infty[\ \forall h \in H : |f(h)| \leq c. \tag{2.7.29}$$

Of course, (2.7.29) implies many generalizations connected with properties of continuous and semi-continuous functionals on compact sets. We do not consider these generalizations now.

Let us return to (2.7.25) and (2.7.28). It is advisable to introduce 'compact' pseudo-metrics. Suppose that $\forall_X \mathbf{S}[X \neq \varnothing]$:

$$((\mathbf{c},\mathbf{p}) - \mathrm{Dist})[X] \triangleq \{\rho \in (\mathbf{p} - \mathrm{Dist})[X] \mid \forall (x_i)_{i\in\mathcal{N}} \in X^{\mathcal{N}}\ \exists \alpha \in \mathbf{N}\ \exists x \in X : (\rho(x_{\alpha(s)}, x))_{s\in\mathcal{N}} \to 0\}. \tag{2.7.30}$$

It is possible to supplement (2.7.30) by a statement about restrictions of a pseudo-metric. Namely, $\forall_X \mathbf{S}[X \neq \varnothing]\ \forall \rho \in (\mathbf{p} - \mathrm{Dist})[X]\ \forall A \in 2^X$:

$$(\rho \mid A \times A) = (\rho(z))_{z\in A\times A} \in (\mathbf{p} - \mathrm{Dist})[A]. \tag{2.7.31}$$

Using (2.3.5) and (2.7.31), we obtain $\forall_X \mathbf{S}[X \neq \varnothing]\ \forall \rho \in (\mathbf{p} - \mathrm{Dist})[X]$ $\forall A \in 2^X$:

$$\tau_\rho^\natural(X) \mid_A = \tau^\natural_{(\rho\mid A\times A)}(A). \tag{2.7.32}$$

By (2.7.32) we characterize a subspace of pseudo-metrizable space. In addition to (2.7.29) we note that $\forall_X \mathbf{S}[X \neq \varnothing]\ \forall \rho \in ((\mathbf{c},\mathbf{p}) - \mathrm{Dist})[X]$ $\forall A \in 2^X\ \exists x_A \in X\ \forall x \in X$:

$$(\rho - \inf)[x; A] \leq (\rho - \inf)[x_A; A]. \tag{2.7.33}$$

To prove (2.7.33) we use the obvious property that if $\mathbf{S}[X \neq \varnothing]$ and $\rho \in ((\mathbf{c},\mathbf{p}) - \mathrm{Dist})[X]$, then $X \in (\mathrm{comp})[X;\rho]$. Moreover, (2.7.14) can be used here. Note that by (2.7.30) and (2.7.32), $\forall_X \mathbf{S}[X \neq \varnothing]\ \forall \rho \in$ $(\mathbf{p} - \mathrm{Dist})[X]$:

$$(\mathrm{comp})[X;\rho] = \{K \in 2^X \mid (\rho \mid K \times K) \in ((\mathbf{c},\mathbf{p}) - \mathrm{Dist})[K]\}. \tag{2.7.34}$$

In (2.7.34) we characterize elements of the family (2.7.27) in terms of subspaces of an initial pseudo-metric space: if $\mathbf{S}[X \neq \varnothing]$, $\rho \in (\mathbf{p} - \mathrm{Dist})[X]$, $A \in 2^X$, and $\rho_A$ is the pseudo-metric (2.7.31), then $(A, \rho_A)$ can be regarded as a subspace of the pseudo-metric space $(X, \rho)$. We call elements of the set (2.7.30) the compact pseudo-metric. In particular, compact metrics can be considered on this basis. Employing (2.7.33), we obtain the property that $\forall_X \mathbf{S}[X \neq \varnothing]\ \forall \rho \in ((\mathbf{c},\mathbf{p}) - \mathrm{Dist})[X]\ \forall \varepsilon \in ]0,\infty[$ $\exists m \in \mathcal{N}\ \exists (x_i)_{i\in\overline{1,m}} \in X^m$:

$$X = \bigcup_{i=1}^{m} \mathbf{B}^0_\rho(x_i, \varepsilon); \tag{2.7.35}$$

this is the property of complete boundedness of $(X, \rho)$. As a consequence (see (2.7.26)) we have $\forall_X \mathbf{S}[X \neq \varnothing]\ \forall \rho \in (\mathbf{p} - \mathrm{Dist})[X]\ \forall H \in$

$(\text{comp})_0[X;\rho]$ $\forall \varepsilon \in ]0,\infty[$ $\exists m \in \mathcal{N}$ $\exists (x_i)_{i\in\overline{1,m}} \in X^m$:

$$H \subset \bigcup_{i=1}^{m} \mathbf{B}^0_\rho(x_i,\varepsilon). \tag{2.7.36}$$

In the sequel the following obvious corollary of (2.7.36) is useful. If $\mathbf{S}[X \neq \varnothing]$, $\rho \in (\mathbf{p}-\text{Dist})[X]$, $K \in (\text{comp})[X;\rho]$ and $\varepsilon \in ]0,\infty[$, then one can choose $m \in \mathcal{N}$ and $(x_i)_{i\in\overline{1,m}} \in K^m$ for which (2.7.36) is correct for $H = K$. Moreover, if $\mathbf{S}[X \neq \varnothing]$ and $\rho \in ((\mathbf{c},\mathbf{p})-\text{Dist})[X]$, then $(X,\rho)$ is complete; i.e., $\forall (x_i)_{i\in\mathcal{N}} \in X^{\mathcal{N}}$:

$$(\forall \varepsilon \in ]0,\infty[\ \exists m \in \mathcal{N}\ \forall p \in \overrightarrow{m,\infty}\ \forall q \in \overrightarrow{m,\infty} : \rho(x_q,x_p) < \varepsilon)$$
$$\Rightarrow (\exists x \in X : (\rho(x_i,x))_{i\in\mathcal{N}} \to 0). \tag{2.7.37}$$

In this connection we introduce complete pseudo-metrics. If $\mathbf{S}[X \neq \varnothing]$, then we denote by $(\mathbf{p}^0-\text{Dist})[X]$ the set of all $\rho \in (\mathbf{p}-\text{Dist})[X]$ for which (2.7.37) is valid; note that by (2.7.37) we have

$$((\mathbf{c},\mathbf{p})-\text{Dist})[X] \subset (\mathbf{p}^0-\text{Dist})[X].$$

In reality we have $\forall_X\ \mathbf{S}[X \neq \varnothing]$ $\forall \rho \in (\mathbf{p}^0-\text{Dist})[X]$ :

$$(\forall \varepsilon \in ]0,\infty[\ \exists m \in \mathcal{N}\ \exists (x_i)_{i\in\overline{1,m}} \in X^m : X = \bigcup_{i=1}^{m} \mathbf{B}^0_\rho(x_i,\varepsilon))$$
$$\Rightarrow (\rho \in ((\mathbf{c},\mathbf{p})-\text{Dist})[X]). \tag{2.7.38}$$

It is useful to note that $\forall_X\ \mathbf{S}[X \neq \varnothing]$ $\forall \rho \in (\mathbf{p}^0-\text{Dist})[X]$ $\forall F \in \mathcal{F}_{\tau_\rho^\natural(X)} \setminus \{\varnothing\}$:

$$(\rho \mid F \times F) \in (\mathbf{p}^0-\text{Dist})[F]. \tag{2.7.39}$$

So the completness is inherited by closed subspaces. In the last property it is advisable to use a natural combination of (2.7.31), (2.7.32), and (2.7.38). Namely, if $X$, $\rho$, and $F$ satisfy conditions defining (2.7.39), then

$$(\forall \varepsilon \in ]0,\infty[\ \exists m \in \mathcal{N}\ \exists (x_i)_{i\in\overline{1,m}} \in X^m : F \subset \bigcup_{i=1}^{m} \mathbf{B}^0_\rho(x_i,\varepsilon))$$
$$\Rightarrow (F \in (\text{comp})[X;\rho]).$$

Moreover, it is possible to verify that $\forall_X\ \mathbf{S}[X \neq \varnothing]$ $\forall \rho \in (\mathbf{p}^0-\text{Dist})[X]$:

$$\{H \in 2^X \mid \forall \varepsilon \in ]0,\infty[\ \exists m \in \mathcal{N}\ \exists (x_i)_{i\in\overline{1,m}} \in X^m : H \subset \bigcup_{i=1}^{m} \mathbf{B}^0_\rho(x_i,\varepsilon)\}$$
$$=(\text{comp})_0[X;\rho]. \tag{2.7.40}$$

In (2.7.40) we use (2.7.36). Finally, $\forall_X\, \mathbf{S}[X \neq \varnothing]\ \forall \rho \in ((\mathbf{c},\mathbf{p})-\mathrm{Dist})[X]$:

$$\tau_\rho^\natural(X) \in (\mathbf{c}-\mathrm{top})[X]. \tag{2.7.41}$$

The property (2.7.41) supplements (2.7.25). As a consequence we have $\forall_X\, \mathbf{S}[X \neq \varnothing]$:

$$((\mathbf{c},\mathbf{p})-\mathrm{Dist})[X] = \{\rho \in (\mathbf{p}-\mathrm{Dist})[X] \mid \tau_\rho^\natural(X) \in (\mathbf{c}-\mathrm{top})[X]\}. \tag{2.7.42}$$

From (2.7.34) and (2.7.42) we obtain the property that $\forall_X\, \mathbf{S}[X \neq \varnothing]$ $\forall \rho \in (\mathbf{p}-\mathrm{Dist})[X]$:

$$(\mathrm{comp})[X;\rho] = (\tau_\rho^\natural(X) - \mathrm{comp})[X] \setminus \{\varnothing\}. \tag{2.7.43}$$

The relations (2.7.42) and (2.7.43) define a natural connection of the compactness properties considered in terms of TS and in terms of characteristic notions of metric spaces. The relation (2.7.27) characterizes the obtained representations in the natural terms of sequential compactness. A (closed) segment of the real line is the simplest case of the compact metric space. In this connection we introduce the metric

$$\mathbf{d}_{|\cdot|} \in (\mathrm{Dist})[\mathbb{R}]$$

for which $\forall x \in \mathbb{R}\ \forall y \in \mathbb{R}$: $\mathbf{d}_{|\cdot|}(x,y) \triangleq |x-y|$. Of course,

$$\tau_{\mathbb{R}} = \tau_\rho^\natural(\mathbb{R})\,\big|_{\rho=\mathbf{d}_{|\cdot|}}.$$

In this case we have $\forall a \in \mathbb{R}\ \forall b \in [a,\infty[$:

$$(2^{[a,b]} \subset (\mathrm{comp})_0[\mathbb{R};\mathbf{d}_{|\cdot|}])\ \&\ ([a,b] \in (\mathrm{comp})[\mathbb{R};\mathbf{d}_{|\cdot|}]).$$

From (2.7.43) we obtain $\forall a \in \mathbb{R}\ \forall b \in [a,\infty[$:

$$[a,b] \in (\tau_{\mathbb{R}} - \mathrm{comp})[\mathbb{R}] \setminus \{\varnothing\}. \tag{2.7.44}$$

The relation (2.7.44) generates many analogues for finite-dimensional spaces. We do not consider these statements. We only discuss a very useful compactness property for sets in the space of continuous functions. This property is connected with familiar Arzelà's theorem. First, using (2.7.36), we obtain $\forall_X\, \mathbf{S}[X \neq \varnothing]\ \forall \rho \in ((\mathbf{c},\mathbf{p})-\mathrm{Dist})[X]\ \exists x \in X\ \exists \varepsilon \in\ ]0,\infty[$:

$$X = \mathbb{B}_\rho^0(x,\varepsilon).$$

This means that the nonempty number set $\{\rho(x_1,x_2) : (x_1,x_2) \in X \times X\}$ is bounded under the conditions above. Suppose that $\forall_X\, \mathbf{S}[X \neq \varnothing]\ \forall \rho \in ((\mathbf{c},\mathbf{p})-\mathrm{Dist})[X]$:

$$(\rho-\mathrm{diam})[X] \triangleq \sup(\{\rho(x_1,x_2) : (x_1,x_2) \in X \times X\}). \tag{2.7.45}$$

In (2.7.45) we have a nonnegative number. Note that $\forall_X \mathbf{S}[X \neq \varnothing]$ $\forall_Y \mathbf{S}[Y \neq \varnothing]$ $\forall \rho \in ((\mathbf{c},\mathbf{p}) - \mathrm{Dist})[Y]$ $\forall g \in Y^X$ $\forall h \in Y^X$ $\forall x \in X$:

$$0 \leq \rho(g(x), h(x)) \leq (\rho - \mathrm{diam})[Y].$$

Therefore $\forall_X \mathbf{S}[X \neq \varnothing]$ $\forall_Y \mathbf{S}[Y \neq \varnothing]$ $\forall \rho \in ((\mathbf{c},\mathbf{p}) - \mathrm{Dist})[Y]$ $\forall g \in Y^X$ $\forall h \in Y^X$:

$$(\rho - \sup)[g; h] \triangleq \sup(\{\rho(g(x), h(x)) : x \in X\}) \in [0, (\rho - \mathrm{diam})[Y]\,]. \tag{2.7.46}$$

PROPOSITION 2.7.1 *Let* $\mathbf{S}[X \neq \varnothing]$, $\mathbf{S}[Y \neq \varnothing]$ *and* $\rho \in ((\mathbf{c},\mathbf{p}) - \mathrm{Dist})[Y]$. *Then the mapping* $\rho^*$ *defined as*

$$(g, h) \mapsto (\rho - \sup)[g; h] : Y^X \times Y^X \to [0, \infty[$$

*is a pseudo-metric of* $Y^X$: $\rho^* \in (\mathbf{p} - \mathrm{Dist})[Y^X]$.

This proposition is well known (see [1, 4, 71]). The corresponding proof is obvious. Note the following useful property. Namely, $\forall_X \mathbf{S}[X \neq \varnothing]$ $\forall_Y \mathbf{S}[Y \neq \varnothing]$ $\forall \rho_1 \in (\mathbf{p} - \mathrm{Dist})[X]$ $\forall \rho_2 \in ((\mathbf{c},\mathbf{p}) - \mathrm{Dist})[Y]$ $\forall (g_i)_{i \in \mathcal{N}} \in C(X, \tau^{\natural}_{\rho_1}(X), Y, \tau^{\natural}_{\rho_2}(Y))^{\mathcal{N}}$ $\forall g \in Y^X$:

$$(((\rho_2 - \sup)[g_i; g])_{i \in \mathcal{N}} \to 0) \Rightarrow (g \in C(X, \tau^{\natural}_{\rho_1}(X), Y, \tau^{\natural}_{\rho_2}(Y))). \tag{2.7.47}$$

In (2.7.47) we have (in essence) the closedness property for the set of all continuous mappings from a pseudo-metrizable space into a compact pseudo-metrizable space. Indeed, consider (2.7.12) and Proposition 2.7.1. If $\mathbf{S}[X \neq \varnothing]$, $\mathbf{S}[Y \neq \varnothing]$, $r \in (\mathbf{p} - \mathrm{Dist})[X]$, $\rho \in ((\mathbf{c},\mathbf{p}) - \mathrm{Dist})[Y]$, and $\rho^* \in (\mathbf{p} - \mathrm{Dist})[Y^X]$ correspond to Proposition 2.7.1, then from (2.7.12) and (2.7.47) it follows that

$$C(X, \tau^{\natural}_r(X), Y, \tau^{\natural}_\rho(Y)) = \mathrm{cl}(C(X, \tau^{\natural}_r(X), Y, \tau^{\natural}_\rho(Y)), \tau^{\natural}_{\rho^*}(Y^X)) \in \mathcal{F}_{\tau^{\natural}_{\rho^*}(Y^X)}.$$

Note the following obvious property. If $p \in \mathcal{N}$ and $q \in \mathcal{N}$, then $\exists m \in \mathcal{N}$ $\exists g \in (\overline{1,p^q})^m : \overline{1,p^q} = \{g(i) : i \in \overline{1,m}\}$. Here the corresponding stipulation of Section 2.2 is kept in mind. Using this simple property, one can verify that $\forall_X \mathbf{S}[X \neq \varnothing]$ $\forall_Y \mathbf{S}[Y \neq \varnothing]$ $\forall r \in ((\mathbf{c},\mathbf{p}) - \mathrm{Dist})[X]$

$\forall \rho \in ((\mathbf{c},\mathbf{p}) - \mathrm{Dist})[Y]\ \forall H \in 2^{C(X,\tau_r^\natural(X),Y,\tau_\rho^\natural(Y))}$:

$$\Bigl(\forall \varepsilon \in ]0,\infty[\ \exists \delta \in ]0,\infty[\ \forall h \in H\ \forall x_1 \in X\ \forall x_2 \in X: (r(x_1,x_2) < \delta) \Rightarrow (\rho(h(x_1),h(x_2)) < \varepsilon)\Bigr) \Rightarrow \Bigl(\forall \tilde{\varepsilon} \in ]0,\infty[\ \exists m \in \mathcal{N}\ \exists (g_i)_{i \in \overline{1,m}} \in (Y^X)^m : H \subset \bigcup_{i=1}^{m} \{s \in Y^X \mid (\rho - \mathrm{sup})[g_i; s] < \tilde{\varepsilon}\}\Bigr). \tag{2.7.48}$$

The relation (2.7.48) is supplemented up to the following obvious property. If $X$, $Y$, $\rho$ and $\rho^*$ correspond to the conditions of Proposition 2.7.1, then $\rho^* \in (\mathbf{p}^0 - \mathrm{Dist})[Y^X]$. Thus the natural combination of (2.7.37) and (2.7.48) can be considered. But it is advisable to regard the nonempty set of all continuous functions as a pseudo-metric space. Here we use (2.7.31). If $\mathbf{S}[X \neq \varnothing]$, $\mathbf{S}[Y \neq \varnothing]$, $r \in (\mathbf{p}-\mathrm{Dist})[X]$, $\rho \in ((\mathbf{c},\mathbf{p})-\mathrm{Dist})[Y]$, then we set that

$$(\rho - \mathrm{Sup})[r] : C(X,\tau_r^\natural(X),Y,\tau_\rho^\natural(Y)) \times C(X,\tau_r^\natural(X),Y,\tau_\rho^\natural(Y)) \to [0,\infty[ \tag{2.7.49}$$

is by definition the mapping for which $\forall g \in C(X,\tau_r^\natural(X),Y,\tau_\rho^\natural(Y))\ \forall h \in C(X,\tau_r^\natural(X),Y,\tau_\rho^\natural(Y))$:

$$(\rho - \mathrm{Sup})[r](g,h) \triangleq (\rho - \mathrm{sup})[g;h]; \tag{2.7.50}$$

for (2.7.49) and (2.7.50) with taking (2.7.31) into account we obtain that

$$(\rho - \mathrm{Sup})[r] \in (\mathbf{p} - \mathrm{Dist})[C(X,\tau_r^\natural(X),Y,\tau_\rho^\natural(Y))]. \tag{2.7.51}$$

From (2.7.40), (2.7.48) and (2.7.51) we have $\forall_X\ \mathbf{S}[X \neq \varnothing]\ \forall_Y\ \mathbf{S}[Y \neq \varnothing]$ $\forall r \in ((\mathbf{c},\mathbf{p}) - \mathrm{Dist})[X]\ \forall \rho \in ((\mathbf{c},\mathbf{p}) - \mathrm{Dist})[Y]\ \forall H \in 2^{C(X,\tau_r^\natural(X),Y,\tau_\rho^\natural(Y))}$:

$$\Bigl(\forall \varepsilon \in ]0,\infty[\ \exists \delta \in ]0,\infty[\ \forall x_1 \in X\ \forall x_2 \in X: (r(x_1,x_2) < \delta) \Rightarrow (\forall h \in H : \rho(h(x_1),h(x_2)) < \varepsilon)\Bigr) \Rightarrow \Bigl(H \in (\mathrm{comp})_0[C(X,\tau_r^\natural(X),Y,\tau_\rho^\natural(Y)); (\rho - \mathrm{Sup})[r]]\Bigr). \tag{2.7.52}$$

In essence, (2.7.52) determines the generalized Arzelà theorem. Note the well known property of uniform continuity for an arbitrary continuous function from a compact pseudo-metrizable space into a pseudo-metrizable space. Namely, $\forall_X \mathbf{S}[X \neq \varnothing]\ \forall_Y \mathbf{S}[Y \neq \varnothing]\ \forall r \in ((\mathbf{c}, \mathbf{p}) - \mathbf{Dist})[X]\ \forall \rho \in (\mathbf{p} - \mathbf{Dist})[Y]\ \forall g \in C(X, \tau_r^\natural(X), Y, \tau_\rho^\natural(Y))\ \forall \varepsilon \in ]0, \infty[\ \exists \delta \in ]0, \infty[\ \forall x_1 \in X\ \forall x_2 \in X:$

$$(r(x_1, x_2) < \delta) \Rightarrow (\rho(g(x_1), g(x_2)) < \varepsilon); \tag{2.7.53}$$

(2.7.53) is often used in traditional constructions of extension. With taking (2.7.53) into account we obtain $\forall_X \mathbf{S}[X \neq \varnothing]\ \forall_Y \mathbf{S}[Y \neq \varnothing]\ \forall r \in ((\mathbf{c}, \mathbf{p}) - \mathbf{Dist})[X]\ \forall \rho \in ((\mathbf{c}, \mathbf{p}) - \mathbf{Dist})[Y]$:

$$\begin{aligned}&(\mathrm{comp})_0[C(X, \tau_r^\natural(X), Y, \tau_\rho^\natural(Y)); (\rho - \mathrm{Sup})[r]]\\ &= \{H \in 2^{C(X, \tau_r^\natural(X), Y, \tau_\rho^\natural(Y))} \mid \forall \varepsilon \in ]0, \infty[\ \exists \delta \in ]0, \infty[\ \forall x_1 \in X\ \forall x_2 \in X:\\ &\qquad (r(x_1, x_2) < \delta) \Rightarrow (\forall h \in H: \rho(h(x_1), h(x_2)) < \varepsilon)\}.\end{aligned} \tag{2.7.54}$$

In connection with (2.7.54) it is useful to keep in mind the above relation for two families in (2.7.26) in terms of the closer operator. Note that from (2.3.20), (2.7.7), and (2.7.43) it follows that $\forall_X \mathbf{S}[X \neq \varnothing]\ \forall \rho \in (\mathbf{Dist})[X]$:

$$(\mathrm{comp})[X; \rho] \subset \mathcal{F}_{\tau_\rho^\natural(X)}. \tag{2.7.55}$$

Return to (2.7.30). Let $\forall_X \mathbf{S}[X \neq \varnothing]$:

$$\begin{aligned}(\mathbf{c} - \mathbf{Dist})[X] &\triangleq ((\mathbf{c}, \mathbf{p}) - \mathbf{Dist})[X] \cap (\mathbf{Dist})[X]\\ &= \{\rho \in (\mathbf{Dist})[X] \mid \forall (x_i)_{i \in \mathcal{N}} \in X^{\mathcal{N}}\ \exists \alpha \in \mathbf{N}\ \exists x \in X:\\ &\qquad (\rho(x_{\alpha(s)}, x))_{s \in \mathcal{N}} \to 0\}.\end{aligned} \tag{2.7.56}$$

From (2.7.56) and (2.7.46) with $\rho \in (\mathbf{c} - \mathbf{Dist})[X]$ in Proposition 2.7.1, we obtain $\rho^* \in (\mathbf{Dist})[Y^X]$. As a consequence of (2.7.49)–(2.7.50) we have $\forall_X \mathbf{S}[X \neq \varnothing]\ \forall_Y \mathbf{S}[Y \neq \varnothing]\ \forall r \in ((\mathbf{c}, \mathbf{p}) - \mathbf{Dist})[X]\ \forall \rho \in (\mathbf{c} - \mathbf{Dist})[Y]$:

$$(\rho - \mathrm{Sup})[r] \in (\mathbf{Dist})[C(X, \tau_r^\natural(X), Y, \tau_\rho^\natural(Y))]. \tag{2.7.57}$$

We now can use (2.7.57) in (2.7.55). From (2.7.43), (2.7.53), (2.7.55), and (2.7.57) we obtain the following

PROPOSITION 2.7.2 *Let* $\mathbf{S}[X \neq \varnothing]$, $\mathbf{S}[Y \neq \varnothing]$, $r \in ((\mathbf{c}, \mathbf{p}) - \mathbf{Dist})[X]$, $\rho \in (\mathbf{c} - \mathbf{Dist})[Y]$, *and*

$$\theta \triangleq \tau^\natural_{(\rho - \mathrm{Sup})[r]}(C(X, \tau_r^\natural(X), Y, \tau_\rho^\natural(Y))).$$

*Then*

$$\begin{aligned}&(\mathrm{comp})[C(X,\tau_r^{\natural}(X),Y,\tau_\rho^{\natural}(Y));(\rho-\mathrm{Sup})[r]]\\&=(\theta-\mathrm{comp})[C(X,\tau_r^{\natural}(X),Y,\tau_\rho^{\natural}(Y))]\setminus\{\varnothing\}\\&=\{K\in\mathcal{F}_\theta\setminus\{\varnothing\}\mid\forall\varepsilon\in]0,\infty[\ \exists\delta\in]0,\infty[\ \forall x_1\in X\ \forall x_2\in X:\\&\qquad(r(x_1,x_2)<\delta)\Rightarrow(\forall h\in K:\rho(h(x_1),h(x_2))<\varepsilon)\}.\end{aligned}$$

In the proof we use the property that the closure of a relatively compact set in a metric space is compact in this space.

## 2.8 CLOSED AND PERFECT MAPPINGS

In this section we very briefly discuss two important types of functional spaces. They are required in the next chapter. Thus we consider some types of continuous functions acting in TS. We use the well known notions from general topology (see [71, 81]). We follow definitions in the form of [35, p.36]. Suppose that $\forall_X \mathbf{S}[X \neq \varnothing]$ $\forall_Y \mathbf{S}[Y \neq \varnothing]$ $\forall \tau \in (\mathrm{top})[X]$ $\forall \vartheta \in (\mathrm{top})[Y]$:

$$C_{\mathrm{cl}}(X,\tau,Y,\vartheta) \triangleq \{f \in C(X,\tau,Y,\vartheta) \mid \forall F \in \mathcal{F}_\tau : f^1(F) \in \mathcal{F}_\vartheta\}. \quad (2.8.1)$$

We call elements of the set (2.8.1) closed mappings in the sense of the TS $(X,\tau)$ and $(Y,\vartheta)$. Of course, we consider only continuous mappings preserving closedness of the image. Let $\forall_X \mathbf{S}[X \neq \varnothing]$ $\forall_Y \mathbf{S}[Y \neq \varnothing]$ $\forall \tau \in (\mathrm{top})[X]$ $\forall \vartheta \in (\mathrm{top})[Y]$:

$$\begin{aligned}C_{\mathbf{ap}}(X,\tau,Y,\vartheta) \triangleq \{f \in C_{\mathrm{cl}}(X,\tau,Y,\vartheta) \mid \forall y \in Y :&\\ f^{-1}(\{y\}) \in (\tau-\mathrm{comp})[X]\}.&\quad (2.8.2)\end{aligned}$$

Following [71, p.244], we call elements of the set (2.8.2) almost perfect mappings. Finally, $\forall_X \mathbf{S}[X \neq \varnothing]$ $\forall_Y \mathbf{S}[Y \neq \varnothing]$ $\forall \tau \in (\mathrm{top})[X]$ $\forall \vartheta \in (\mathrm{top})[Y]$:

$$\begin{aligned}C_{\mathbf{qp}}(X,\tau,Y,\vartheta) \triangleq \{f \in C_{\mathrm{cl}}(X,\tau,Y,\vartheta) \mid \forall y \in Y :&\\ f^{-1}(\{y\}) \in (\mathcal{N}-\tau-\mathrm{comp})[X]\}.&\quad (2.8.3)\end{aligned}$$

We call elements of the set (2.8.3) quasi-perfect mappings. Note that an obvious property is extracted from (2.8.1). Namely, $\forall_X \mathbf{S}[X \neq \varnothing]$ $\forall_Y \mathbf{S}[Y \neq \varnothing]$ $\forall \tau \in (\mathrm{top})[X]$ $\forall \vartheta \in (\mathrm{top})[Y]$:

$$C_{\mathrm{cl}}(X,\tau,Y,\vartheta) = \{f \in Y^X \mid \forall A \in \mathcal{P}(X) : f^1(\mathrm{cl}(A,\tau)) = \mathrm{cl}(f^1(A),\vartheta)\}. \quad (2.8.4)$$

To prove this it is sufficient to compare (2.5.4) and (2.8.1). Closed mappings are mappings preserving the closure operation. Properties of almost perfect mappings are considered in [71]. It should be noted that an almost perfect mapping defined on a Hausdorff space is called the perfect mapping (see [71, Ch. 3]).

Let us note some obvious properties. Firstly, from (2.3.6) and (2.5.2) the well known property follows. Namely, $\forall_X \mathbf{S}[X \neq \varnothing]$ $\forall_Y \mathbf{S}[Y \neq \varnothing]$ $\forall \tau \in (\text{top})[X]$ $\forall \vartheta \in (\text{top})[Y]$ $\forall f \in C(X,\tau,Y,\vartheta)$ $\forall K \in (\tau - \text{comp})[X]$:

$$f^1(K) \in (\vartheta - \text{comp})[Y]. \tag{2.8.5}$$

From (2.3.20) and (2.8.5) we obtain $\forall_X \mathbf{S}[X \neq \varnothing]$ $\forall_Y \mathbf{S}[Y \neq \varnothing]$ $\forall \tau \in (\text{top})[X]$ $\forall \vartheta \in (\text{top})_0[Y]$ $\forall f \in C(X,\tau,Y,\vartheta)$ $\forall K \in (\tau - \text{comp})[X]$:

$$f^1(K) \in \mathcal{F}_\vartheta. \tag{2.8.6}$$

As a consequence we have the following known property (see [71, 81]).If $\mathbf{S}[X \neq \varnothing]$, $\mathbf{S}[Y \neq \varnothing]$, $\tau \in (\mathbf{c} - \text{top})[X]$, and $\vartheta \in (\text{top})_0[Y]$, then

$$C_{\mathbf{ap}}(X,\tau,Y,\vartheta) = C(X,\tau,Y,\vartheta). \tag{2.8.7}$$

To prove (2.8.7), owing to (2.8.1) and (2.8.2) it is sufficient to establish the inclusion

$$C(X,\tau,Y,\vartheta) \subset C_{\mathbf{ap}}(X,\tau,Y,\vartheta).$$

Fix $f \in C(X,\tau,Y,\vartheta)$. Let $F \in \mathcal{F}_\tau$. By properties of Section 2.3 we have $F \in (\tau - \text{comp})[X]$ and by (2.8.6) we obtain $f^1(F) \in \mathcal{F}_\vartheta$. By (2.8.1) we have $f \in C_{\text{cl}}(X,\tau,Y,\vartheta)$. Let $y \in Y$. From (2.2.15) and (2.2.16) it follows that $\{y\} \in \mathcal{F}_\vartheta$. But $f$ is a continuous function. Therefore by (2.5.2) $f^{-1}(\{y\}) \in \mathcal{F}_\tau$. Since $(X,\tau)$ is a compact space, we obtain $f^{-1}(\{y\}) \in (\tau - \text{comp})[X]$. But the choice of $y$ was arbitrary. Therefore from (2.8.2) we have $f \in C_{\mathbf{ap}}(X,\tau,Y,\vartheta)$. The equality (2.8.7) is established.

## 2.9 LOCALLY COMPACT SPACES

In this section we consider a very brief summary of notions connected with local TS and some their analogues. As a rule, the classical notion of local compactness is sufficient in concrete settings. But in some logical constructions it is advisable to consider variants of properties of such a kind. In this connection we recall the notion of locally quasi-compact TS (see [4]). Let $\forall_X \mathbf{S}[X]$:

$$(\mathbf{c} - \text{top})_{\text{loc}}[X] \triangleq \{\tau \in (\text{top})[X] \mid \forall x \in X : N_\tau(x) \cap (\tau - \text{comp})[X] \neq \varnothing\}. \tag{2.9.1}$$

In (2.9.1) we have introduced a TS whose each point has a compact neighborhood. It is useful to note that $\forall_X \mathbf{S}[X]$ $\forall \tau \in (\text{top})[X]$:

$$(\forall x \in X \exists G \in N_\tau^0(x) : \text{cl}(G,\tau) \in (\tau - \text{comp})[X]) \Rightarrow (\tau \in (\mathbf{c} - \text{top})_{\text{loc}}[X]). \tag{2.9.2}$$

The proof of (2.9.2) is obvious. Moreover, from (2.3.12), (2.3.13) and (2.9.1) we obtain $\forall_X\ \mathbf{S}[X]\ \forall \vartheta \in (\mathbf{c}-\mathrm{top})_{\mathrm{loc}}[X]\ \forall x \in X\ \exists G \in N_\vartheta^0(x)$ $\exists H \in N_\vartheta(x)$:

$$(G \subset H)\ \&\ (\mathrm{cl}(G,\vartheta) \cap H \in (\vartheta\,|_H - \mathrm{comp})[H]). \tag{2.9.3}$$

The relation (2.9.3) defines the natural connection of (2.9.1) and the traditional notion of local compactness. In addition, $\forall_X\ \mathbf{S}[X]$:

$$\begin{aligned}&(\mathbf{c}-\mathrm{top})^0_{\mathrm{loc}}[X]\\ &\triangleq\{\tau \in (\mathrm{top})_0[X] \mid \forall x \in X\ \exists G \in N_\tau^0(x) : \mathrm{cl}(G,\tau) \in (\tau - \mathrm{comp})[X]\}\\ &=(\mathbf{c}-\mathrm{top})_{\mathrm{loc}}[X] \cap (\mathrm{top})_0[X].\end{aligned} \tag{2.9.4}$$

In (2.9.4) we have introduced the set of all Hausdorff topologies converting $X$ into a locally compact space in the traditional sense. In this connection, by (2.9.1) we have $\forall_X\ \mathbf{S}[X]$:

$$\begin{aligned}(\mathbf{c}-\mathrm{top})^0_{\mathrm{loc}}[X] &= \{\tau \in (\mathrm{top})_0[X] \mid \forall x \in X : N_\tau(x) \cap (\tau - \mathrm{comp})[X] \neq \varnothing\}\\ &= (\mathbf{c}-\mathrm{top})_{\mathrm{loc}}[X] \cap (\mathrm{top})_0[X].\end{aligned} \tag{2.9.5}$$

Suppose that $\forall_X\ \mathbf{S}[X]$:

$$\begin{aligned}(\mathbf{c}-\mathrm{top})^{\mathrm{loc}}[X] \triangleq \{\tau \in (\mathrm{top})[X] \mid \forall x \in X\ \exists G \in N_\tau^0(x) :\\ \mathrm{cl}(G,\tau) \in (\tau - \mathrm{comp})[X]\}\end{aligned} \tag{2.9.6}$$

The connection of (2.9.1) and (2.9.6) is obvious: if $\mathbf{S}[X]$, then (see (2.9.2), (2.9.6))

$$(\mathbf{c}-\mathrm{top})[X] \subset (\mathbf{c}-\mathrm{top})^{\mathrm{loc}}[X] \subset (\mathbf{c}-\mathrm{top})_{\mathrm{loc}}[X]. \tag{2.9.7}$$

From (2.9.4) and (2.9.6) we have the following property: if $\mathbf{S}[X]$, then

$$(\mathbf{c}-\mathrm{top})^0_{\mathrm{loc}}[X] = (\mathbf{c}-\mathrm{top})^{\mathrm{loc}}[X] \cap (\mathrm{top})_0[X]. \tag{2.9.8}$$

In (2.9.5) and (2.9.8) we have two representations of Hausdorff locally compact space. In (2.9.6) the traditional notion of locally compact space is given.

Let us note some other properties. If $\mathbf{S}[X]$, $\tau \in (\mathbf{c}-\mathrm{top})^{\mathrm{loc}}[X]$ and $F \in \mathcal{F}_\tau$, then

$$\tau\,|_F \in (\mathbf{c}-\mathrm{top})^{\mathrm{loc}}[F].$$

Thus a closed subset of a locally compact TS is a locally compact space. As a consequence $\forall_X\ \mathbf{S}[X]\ \forall \tau \in (\mathbf{c}-\mathrm{top})^0_{\mathrm{loc}}[X]\ \forall F \in \mathcal{F}_\tau$:

$$\tau\,|_F \in (\mathbf{c}-\mathrm{top})^0_{\mathrm{loc}}[F].$$

Finally, note that $\forall_X\ \mathbf{S}[X]\ \forall \tau \in (\mathbf{c}-\mathrm{top})^0_{\mathrm{loc}}[X]\ \forall x \in X\ \forall H_1 \in N_\tau(x)$ $\exists H_2 \in N_\tau(x) \cap (\tau - \mathrm{comp})[X]$:

$$H_2 \subset H_1. \tag{2.9.9}$$

From (2.6.8) and (2.9.9) we have the important property that if $\mathbf{S}[X]$, $\tau \in (\mathbf{c}-\mathrm{top})^0_{\mathrm{loc}}[X]$ and $x \in X$, then

$$N_\tau(x) \cap (\tau - \mathrm{comp})[X] \in (x - \mathbf{bas})[\tau]. \tag{2.9.10}$$

By (2.9.10) each point in a locally compact Hausdorff TS has the local topological base in the form of the family of all compact neighborhoods.

Finally, we give a simple example of a locally compact Hausdorff space. Namely, $\forall_X\ \mathbf{S}[X]\ \forall \tau \in (\mathbf{c}-\mathrm{top})_0[X]\ \forall G \in \tau$:

$$\tau\ \big|_G \in (\mathbf{c}-\mathrm{top})^0_{\mathrm{loc}}[G]. \tag{2.9.11}$$

So in (2.9.11) we have the following statement: each open subspace of a compactum is a locally compact Hausdorff space.

## 2.10 CONCLUSION

The matter of this chapter is connected with topology and, on the whole, corresponds to [4, 71, 81, 89]. Quite general topological and, in many cases, non-metrizable spaces are used in the main part when equipping the spaces of generalized elements. The corresponding spaces of usual solutions are embedded in the above 'generalized' spaces as everywhere dense subsets. In this connection we note the monographs [117, 120] and other. We should in detail analyze the compactifications that preserve some 'connections' stipulated by a concrete problem. Quite general knowledge on this account are given in the present chapter; moreover, notions connected with measurable spaces and mappings are introduced. In addition, the concept of the Tichonoff product is also examined. In particular, the Tichonoff product of samples of the real line with the discrete topology is considered; this is essential for the subsequent exposition.

At the same time, we give some constructions which are not used in the next chapters but are natural from the point of view intrinsic relations arising when considering general mathematical introductory notions. In particular, this is concerned with the Stone representation space, which is realized in the class of ultrafilters equipped with an appropriate measurable structure and, which will be more natural, with the uniformity connected with this structure. Such an approach, which is only outlined in this chapter, can be used for constructing extensions that are regarded in a broad sense; however, we do not use this approach in the sequel. We exploit, to a considerable extent, (0, 1)-measures, which are interpreted as ultrafilters but are usually studied in the framework of constructions of the FAM theory. In the sequel we, in a small way, use constructions of metrizable compactifications, which are applied for extension of problems of calculus of variations and optimal control with geometric constraints (see [78, 88, 112, 117, 120] and other). Nevertheless, the compactness in metric terms is given here and this seems to be advisable both in connection with the importance of compactification in the general theory of extension and with the possible development of respective analogues in the spirit of [78, 117, 120] (see also [33, 105]).

# Chapter 3

# TOPOLOGICAL CONSTRUCTIONS OF EXTENSIONS AND RELAXATIONS

## 3.1 INTRODUCTION

In this part we deal with the quite general property of the regularity type, which is intrinsic to solutions of many problems of investigating the attainability under constraints. In this connection let us consider the following natural scheme arising in many applied problems (for example, see constructions of Chapter 1; moreover, see [32, 35, 40, 47]).

Let **F**, **X** and **H** be nonempty sets. Moreover, let **Y** be a subset of **X**. Finally, fix two mappings:

$$s : \mathbf{F} \to \mathbf{X}; \quad h : \mathbf{F} \to \mathbf{H}. \tag{3.1.1}$$

Consider the condition $s(f) \in \mathbf{Y}$ on the choice of $f \in \mathbf{F}$. Then $s^{-1}(\mathbf{Y}) = \{f \in \mathbf{F} \mid s(f) \in \mathbf{Y}\}$ is the corresponding set of admissible elements. We investigate the possibility to attain points of **H** on values of the mapping $h$ (3.1.1). Then the set

$$h^1(s^{-1}(\mathbf{Y})) = \{h(f) : f \in s^{-1}(\mathbf{Y})\} \tag{3.1.2}$$

has the sense of the attainability domain in control theory. In many applied problems perturbations of **Y** should be considered. Namely, if **Y** is replaced by a set $\tilde{\mathbf{Y}}$, $\tilde{\mathbf{Y}} \subset \mathbf{X}$, then (3.1.2) is replaced by $h^1(s^{-1}(\tilde{\mathbf{Y}}))$. Very often we obtain a saltus of the attainable set (3.1.2) although **Y** and $\tilde{\mathbf{Y}}$ are 'near'. In the following (within this chapter) we investigate only the case $\mathbf{Y} \subset \tilde{\mathbf{Y}}$. We consider only the weakening of the **Y**-constraint. The corresponding examples arise, in particular, in control theory. Recall the problems considered in Chapter 1; moreover, see [32, 35, 40, 47].

Returning to the general case, we note that when employing the above mentioned weakening of the **Y**-constraint a system of new requirements of $\widetilde{\mathbf{Y}}$-constraints type is usually used. Namely, we often have some family of subsets of $\mathbf{X}$. Sets of this family are used instead of $\mathbf{Y}$. As a result we obtain a family of sets like $h^1(s^{-1}(\widetilde{\mathbf{Y}}))$. In many cases the initial family of 'weakened' versions of $\mathbf{Y}$ permits us to realize some passage to a limit. This limit passage can be connected with the 'process' of more precise definition of approximate variants of the $\mathbf{Y}$-constraint. Very often this process is characterized by narrowing of a neighborhood of $\mathbf{Y}$. Of course, this stipulation means the corresponding topological equipment of $\mathbf{X}$. This equipment is often realized via a metrization of $\mathbf{X}$. We equip $\mathbf{X}$ with a topology $\tau^{(1)} \in (\text{top})[\mathbf{X}]$. So in this variant of the weakening of the $\mathbf{Y}$-constraint we use neighborhoods in the capacity of $\widetilde{\mathbf{Y}}$ and interpret the respective limit passage as the natural narrowing of a neighborhood of $\mathbf{Y}$. In addition, $\mathbb{N}_{\tau^{(1)}}[\mathbf{Y}]$ can be realized as the family of sets having the sense of weakened versions of $\mathbf{Y}$. But sometimes the employment of all neighborhoods is not natural. For example, in the case of the metrizable space $\mathbf{X}$ it is natural to use only $\varepsilon$-neighborhoods of $\mathbf{Y}$, $\varepsilon > 0$. In this connection we consider very different families of 'admissible' neighborhoods of the set $\mathbf{Y}$. In addition, we strive to universality of the result with respect to the above mentioned family.

Thus we can assume that the corresponding family $\mathcal{H} \triangleq \{h^1(s^{-1}(G)) : G \in \mathfrak{G}\}$ of attainable sets arises for the given family $\mathfrak{G}$ of admissible neighborhoods. Of course, the given admissibility of $\mathfrak{G}$ corresponds to the considered concrete problem. In particular, the case $\mathfrak{G} = \mathbb{N}_{\tau^{(1)}}[\mathbf{Y}]$ is possible. The role of $\tau^{(1)}$ are reduced to the determining of the family $\mathfrak{G}$. In addition, the set $\mathbf{H}$ should be equipped with some topology. The basic aim of this equipment is constructing the limit of the family $\mathcal{H}$. Namely, realizing the transformation

$$\mathfrak{G} \to \mathcal{H} \tag{3.1.3}$$

of weakened versions of the considered problem, we form (for the family $\mathcal{H}$) the corresponding attraction set. In this case (3.1.3) is analogous to the transformation of $\mathbf{Y}$ into the set (3.1.2).

So we equip the set $\mathbf{H}$ with a topology $\tau^{(2)} \in (\text{top})[\mathbf{H}]$. Then it is possible (in correspondence with (3.1.3)) to replace (3.1.2) by the following set

$$\begin{aligned} \text{Att} &\triangleq \bigcap_{H \in \mathcal{H}} \text{cl}(H, \tau^{(2)}) = \bigcap_{G \in \mathfrak{G}} \text{cl}(h^1(s^{-1}(G)), \tau^{(2)}) \\ &= \bigcap_{U \in s^{-1}[\mathfrak{G}]} \text{cl}(h^1(U), \tau^{(2)}). \end{aligned} \tag{3.1.4}$$

In (3.1.4) we use (2.5.8). But other variants of the limit representation can be considered instead of (3.1.4). For simplicity we now restrict ourselves to the construction (3.1.4). We do the replacement

$$h^1(s^{-1}(\mathbf{Y})) \to \mathrm{Att}. \tag{3.1.5}$$

Of course, in (3.1.4) and (3.1.5), Att $=$ Att$(\mathfrak{G}, \tau^{(2)})$. An indirect dependence of Att on $\tau^{(1)}$ arises here, since $\mathfrak{G} \subset \mathrm{N}_{\tau^{(1)}}[\mathbf{Y}]$. On the contrary, the set on the left hand side of (3.1.5) does not depend on the above mentioned parameters from the asymptotic setting. Therefore obtaining the conditions for which the set (3.1.4) or its corresponding analogue has a 'perceptible' independence with respect to $\mathfrak{G}$ is of interest for us (the choice of $\tau^{(2)}$ is more definite in concrete problems). Under these conditions we obtain some independence with respect to $\tau^{(1)}$ (the independence within some bounds). We call the corresponding attraction set 'universal' in the given bounds. These bounds $\mathfrak{G}_1$, $\mathfrak{G}_2$ determine the possible variation of $\mathfrak{G}$, $\mathfrak{G}_1 \subset \mathfrak{G} \subset \mathfrak{G}_2$; under this variation the attraction set coincides with (3.1.4). Of course, in this chapter many questions of such a type are considered. Now we restrict ourselves to the discussion of the method of investigation. The employment of a model is laid in its basis. Namely, we introduce a TS $(\mathbf{K}, \tau^{(3)})$, $\mathbf{K} \neq \varnothing$, and a triplet of mappings

$$m : \mathbf{F} \to \mathbf{K}; \quad g : \mathbf{K} \to \mathbf{X}; \quad \omega : \mathbf{K} \to \mathbf{H}. \tag{3.1.6}$$

The mapping $m$ in (3.1.6) realizes the imbedding of $\mathbf{F}$ into TS $(\mathbf{K}, \tau^{(3)})$ called the space of generalized elements (GE). Very often we can suppose that $m^1(\mathbf{F}) = \{m(f) : f \in \mathbf{F}\}$ is the set everywhere dense in $(\mathbf{K}, \tau^{(3)})$: $\mathrm{cl}(m^1(\mathbf{F}), \tau^{(3)}) = \mathbf{K}$. In this case each GE $k \in \mathbf{K}$ assumes an approximate realization in $\mathbf{F}$: one can choose a net $(D, \preceq, \varphi)$ in $\mathbf{F}$ for which

$$(D, \preceq, m \circ \varphi) \overset{\tau^{(3)}}{\to} k. \tag{3.1.7}$$

Then the choice of an approximate solution in $\mathbf{F}$ is replaced by the corresponding choice of $k \in \mathbf{K}$. Of course, such conclusion is true under some additional supposition with respect to $g$ and $\omega$ in (3.1.6). We suppose that

$$g : (\mathbf{K}, \tau^{(3)}) \to (\mathbf{X}, \tau^{(1)}), \quad \omega : (\mathbf{K}, \tau^{(3)}) \to (\mathbf{H}, \tau^{(2)}) \tag{3.1.8}$$

are continuous mappings, and

$$(s = g \circ m) \& (h = \omega \circ m). \tag{3.1.9}$$

By (3.1.8) and (3.1.9) we extend the convergence (3.1.7) to the spaces $(\mathbf{X}, \tau^{(1)})$ and $(\mathbf{H}, \tau^{(2)})$. Of course, the model

$$(\mathbf{K}, \tau^{(3)}, m, g, \omega) \tag{3.1.10}$$

can be chosen in various ways.

Very often concrete realizations of (3.1.10) are connected with the case when $(\mathbf{K}, \tau^{(3)})$ is a subspace of the space of measures equipped with the $*$-weak topology. The following standard problem can be examined in terms of the model (3.1.10). Namely, consider the triplet $(\mathbf{K}, \mathbf{X}, \mathbf{H})$ of sets and the pair $(g, \omega)$ of the mappings (3.1.8). We will investigate possibilities of the realization of the values $\omega(k) \in \mathbf{H}$ under the condition

$$g(k) \in \mathbf{Y}. \tag{3.1.11}$$

This constraint forms the new admissible set $g^{-1}(\mathbf{Y})$. Then the solution of the new problem is determined as the set

$$\omega^1(g^{-1}(\mathbf{Y})). \tag{3.1.12}$$

Of course, (3.1.12) is analogous to the usual admissible set (3.1.2). But the new problem on the standard attainability is connected with better parameters. Namely, (3.1.8) determines two continuous mappings. The set $\mathbf{K}$ is equipped with a topology. It is required to form conditions to the model (3.1.10) under which the replacement (3.1.5) is characterized in terms of (3.1.10). One of variants of these conditions is the realization of (3.1.10) as a compactification. The last requirement consists in the following: $(\mathbf{K}, \tau^{(3)})$ is a compact space (i.e., $\tau^{(3)} \in (\mathbf{c}-\mathrm{top})[\mathbf{K}]$); moreover, $(\mathbf{H}, \tau^{(2)})$ must be a Hausdorff space (i.e., $\tau^{(2)} \in (\mathrm{top})_0[\mathbf{H}]$). This compactification is similar to those in general topology. But the conditions (3.1.9) (i.e., the connections defined in terms of continuous mappings (3.1.8)) are singularities with respect to standard topological compactifications. Note that the above compactification in terms of (3.1.8), (3.1.9) and (3.1.10) may be impossible. In this case new approaches are required. We begin the investigation of such new approaches in terms of more general topological constructions connected with notions of Chapter 2. In particular, we use designations and definitions introduced in this chapter.

## 3.2 TOPOLOGICAL PROPERTIES OF ATTRACTION SETS

In this section we discuss some very general topological constructions which, in the following, are applied to the problem considered in Section 3.1. Without additional clarification we use designations and notions of Chapter 2. Moreover, some new notions are introduced. Suppose $\forall_X \, \mathbf{S}[X]$:

$$\mathcal{B}[X] \triangleq \{\mathcal{X} \in 2^{\mathcal{P}(X)} \mid \forall A \in \mathcal{X} \, \forall B \in \mathcal{X} \, \exists C \in \mathcal{X} : C \subset A \cap B\}. \tag{3.2.1}$$

Elements of (3.2.1) are families with the property of distinctive proximability. The most typical case of such a family is a filter base. This case is characterized by the following additional condition on the choice of $\mathcal{X} \in \mathcal{B}[X]$: $\varnothing \notin \mathcal{X}$. Fixing the set $X$, we introduce the set of all bases of filters. Suppose $\forall_X\ \mathbf{S}[X]$:

$$\mathcal{B}_0[X] \triangleq \{\mathcal{X} \in \mathcal{B}[X] \mid \varnothing \notin \mathcal{X}\}. \tag{3.2.2}$$

From (2.2.45) and (3.2.2) we obtain the natural property: if $\mathbf{S}[X]$ and $\mathcal{X} \in \mathcal{B}_0[X]$, then

$$\{U \in \mathcal{P}(X) \mid \exists V \in \mathcal{X} : V \subset U\} \in \mathbb{F}^*(\mathcal{P}(X)). \tag{3.2.3}$$

Thus each base of filter generates a quite defined filter of the corresponding set. In problems considered below the case $\mathcal{X} \in \mathcal{B}_0[X]$ (where $X$ is a nonempty set) is most often used. But in general constructions we assume the employment of $\mathcal{X} \in \mathcal{B}[X]$.

Let us consider one more particular case of a family of the set (3.2.1). Namely, we introduce families with countable bases. Suppose $\forall_X\ \mathbf{S}[X]$:

$$\mathcal{B}_{\mathcal{N}}(X) \triangleq \{\mathcal{H} \in \mathcal{B}[X] \mid \exists (H_i)_{i\in\mathcal{N}} \in \mathcal{H}^{\mathcal{N}}\ \forall H \in \mathcal{H}\ \exists k \in \mathcal{N} : H_k \subset H\}. \tag{3.2.4}$$

We call elements of the set (3.2.4) the semi-multiplicative families with a countable base. It is useful to note the following property [35, p. 37]: if $\mathbf{S}[X]$, $\mathcal{H} \in \mathcal{B}[X]$, $m \in \mathcal{N}$, and $(H_i)_{i\in\overline{1,m}} \in \mathcal{H}^m$, then $\exists H \in \mathcal{H}$: $H \subset \bigcap_{i=1}^m H_i$. Therefore (see (3.2.4)) $\forall_X\ \mathbf{S}[X]\ \forall \mathcal{X} \in \mathcal{B}_{\mathcal{N}}(X)\ \exists (H^{(i)})_{i\in\mathcal{N}} \in \mathcal{X}^{\mathcal{N}}\ \forall H \in \mathcal{X}\ \exists p \in \mathcal{N}\ \forall q \in \overrightarrow{p,\infty}$:

$$H^{(q)} \subset H. \tag{3.2.5}$$

To prove (3.2.5) we use the following reasoning. Suppose that $\mathbf{S}[X]$ and $\mathcal{X} \in \mathcal{B}_{\mathcal{N}}(X)$. Then $\mathcal{X}$ is a nonempty family of subsets of the set $X$. Using (3.2.4) we choose $(H_i)_{i\in\mathcal{N}} \in \mathcal{X}^{\mathcal{N}}$ for which

$$\forall H \in \mathcal{X}\ \exists k \in \mathcal{N} : H_k \subset H. \tag{3.2.6}$$

Then $\forall m \in \mathcal{N}$: $\mathcal{X}_m \triangleq \{H \in \mathcal{X} \mid H \subset \bigcap_{i=1}^m H_i\} \in 2^{\mathcal{X}}$. Consider the sequence

$$m \mapsto \mathcal{X}_m : \quad \mathcal{N} \to 2^{\mathcal{X}}. \tag{3.2.7}$$

By the axiom of choice the set-valued mapping (3.2.7) has a selector. Choose

$$(H^{(k)})_{k\in\mathcal{N}} \in \prod_{k\in\mathcal{N}} \mathcal{X}_k.$$

Thus we have the property: if $m \in \mathcal{N}$, then $H^{(m)}$ is a subset of the intersection of all sets $H_i$, $i \in \overline{1,m}$. Fix $H \in \mathcal{X}$ and by (3.2.6) let $p \in \mathcal{N}$

be a number for which $H_p \subset H$. If $q \in \overrightarrow{p, \infty}$, then $p \in \overline{1, q}$ and hence the chain of inclusions $H^{(q)} \subset H_p \subset H$ takes place. So (3.2.5) is proved.

We use (2.2.47) and (2.2.48). In addition (see [32, pp. 39,40] and [35, p. 35]), $\forall_U\ \mathbf{S}[U \neq \varnothing]\ \forall_V\ \mathbf{S}[V \neq \varnothing]\ \forall \mathcal{U} \in \mathcal{B}[U]\ \forall \tau \in (\mathrm{top})[V]\ \forall f \in V^U$:

$$\begin{aligned}(\tau - \mathrm{LIM})[\mathcal{U} \mid f] &\triangleq \bigcap_{U \in \mathcal{U}} \mathrm{cl}(f^1(U), \tau) \\ &= \{y \in V \mid \exists_{\mathbf{D}}\ \mathbf{S}[\mathbf{D} \neq \varnothing]\ \exists \preceq \in (\mathrm{DIR})[\mathbf{D}]\ \exists g \in U^{\mathbf{D}} : \\ &\qquad \big(\mathcal{U} \subset (U - \mathrm{ass})[\mathbf{D}; \preceq; g]\big)\ \&\ \big((\mathbf{D}, \preceq, f \circ g) \xrightarrow{\tau} y\big)\}.\end{aligned} \tag{3.2.8}$$

We omit the very obvious proof of (3.2.8). Note that in this proof it is advisable to use the well known construction of the directed product. Of course, (3.2.8) is an attraction set (AS). This fact is used in other versions of AS. In addition, we keep in mind that the last expression of (3.2.8) is connected with the important notion of approximate solution which is defined as a net. Other variants of AS are possible too. Indeed, some modifications of the last set in (3.2.8) can be applied for constructing AS. Consider one of such possibilities.

We introduce approximate solutions admitting some compactification. The latter is connected with an appropriate mapping realizing the imbedding into a compact set of a TS. So if $\mathbf{S}[D \neq \varnothing]$, $\mathbf{S}[X \neq \varnothing]$, $(\mathbf{T}, \theta)$ is a TS, and $r \in \mathbf{T}^X$, then we set

$$\mathcal{M}_{\mathbf{c}}(D, X, \mathbf{T}, \theta, r) \triangleq \{\varphi \in X^D \mid \exists K \in (\theta - \mathrm{comp})[\mathbf{T}] : (r \circ \varphi)^1(D) \subset K\}. \tag{3.2.9}$$

In (3.2.9) we have compactified mappings from $D$ into $X$; in addition, the triplet $(\mathbf{T}, \theta, r)$ determines the concrete variant of the above compactification. Consider the corresponding modification of AS. If $\mathbf{S}[U \neq \varnothing]$, $\mathcal{U} \in \mathcal{B}[U]$, $(V, \tau)$ and $(\mathbf{T}, \theta)$ are TS, $f \in V^U$, and $r \in \mathbf{T}^U$, then

$$\begin{aligned}&((\mathbf{c}\theta, \tau) - \mathrm{LIM})[\mathcal{U} \mid r; f] \\ &\triangleq \{v \in V \mid \exists_{\mathbf{D}}\ \mathbf{S}[\mathbf{D} \neq \varnothing]\ \exists \preceq \in (\mathrm{DIR})[\mathbf{D}]\ \exists \varphi \in \mathcal{M}_{\mathbf{c}}(\mathbf{D}, U, \mathbf{T}, \theta, r) : \\ &\qquad (\mathcal{U} \subset (U - \mathrm{ass})[\mathbf{D}; \preceq; \varphi])\ \&\ ((\mathbf{D}, \preceq, f \circ \varphi) \xrightarrow{\tau} v)\};\end{aligned} \tag{3.2.10}$$

recall that in (3.2.10) $\tau \in (\mathrm{top})[V]$ and $\theta \in (\mathrm{top})[\mathbf{T}]$. In addition, $\forall_U\ \mathbf{S}[U \neq \varnothing]\ \forall_V\ \mathbf{S}[V \neq \varnothing]\ \forall_{\mathbf{T}}\ \mathbf{S}[\mathbf{T} \neq \varnothing]\ \forall \mathcal{U} \in \mathcal{B}[U]\ \forall \tau \in (\mathrm{top})[V]\ \forall \theta \in (\mathrm{top})[\mathbf{T}]\ \forall f \in V^U\ \forall r \in \mathbf{T}^U$:

$$((\mathbf{c}\theta, \tau) - \mathrm{LIM})[\mathcal{U} \mid r; f] \subset (\tau - \mathrm{LIM})[\mathcal{U} \mid f]. \tag{3.2.11}$$

By (3.2.11) the triplet $(\mathbf{T}, \theta, r)$ defines some reduction of the broad AS (3.2.8). Note that concrete versions of (3.2.10) were used in [32, 35, 40,

47] in the form of a distinctive attractor of bounded convergence. We can treat (3.2.11) as the effect of some stabilizer.

Consider some variants of the family $\mathcal{U}$ in (3.2.11). We keep in mind the constructions of Section 3.1 ( in particular, see (3.1.4)). In this connection we note that by (2.5.7) and (3.2.1), $\forall_U \mathbf{S}[U]\ \forall_V \mathbf{S}[V]\ \forall \mathcal{V} \in \mathcal{B}[V]\ \forall t \in V^U$:

$$t^{-1}[\mathcal{V}] = \{t^{-1}(H) : H \in \mathcal{V}\} \in \mathcal{B}[U]. \tag{3.2.12}$$

In particular, the family of all neighborhoods of a set in TS can be used in (3.2.12) (we mean the corresponding concrete definition of $\mathcal{V}$ in (3.2.12)). Namely, $\forall_U \mathbf{S}[U]\ \forall_V \mathbf{S}[V]\ \forall \tau \in (\text{top})[V]\ \forall t \in V^U\ \forall M \in \mathcal{P}(V)$:

$$t^{-1}[\mathbb{N}_\tau[M]\,] \in \mathcal{B}[U].$$

But in some settings the employment of the family of all neighborhoods of a given set (in the capacity of $\mathcal{V}$ in (3.2.12)) is not natural. For example, in metric space the use of $\varepsilon$-neighborhoods of a set (see (2.7.16)) is more reasonable. Of course, the given variant is natural in the more general case of pseudometric space. Therefore we consider a specific construction (see [41, 47]) of uniformization of neighborhoods.

Let us introduce the rather general notion of some (partially) neighborhood-valued mappings. First consider an intermediate notion. Suppose that $\forall_X \mathbf{S}[X \neq \varnothing]\ \forall_Q \mathbf{S}[Q \neq \varnothing]\ \forall \tau \in (\text{top})[X]\ \forall M \in \mathcal{P}(X)$:

$$\mathcal{Q}_M(Q,\tau) \triangleq \{\Lambda \in \mathcal{P}(X)^{X\times Q} \mid \forall x \in M\ \forall q \in Q : \Lambda(x,q) \in N_\tau(x)\}. \tag{3.2.13}$$

Note that arbitrary sets can be used in the capacity of $Q$ in (3.2.13). The simplest case is defined by the stipulation $Q = ]0,\infty[$; this case is very suitable in pseudometric space. We note that $\forall_X \mathbf{S}[X \neq \varnothing]\ \forall_Q \mathbf{S}[Q \neq \varnothing]$ $\forall \tau \in (\text{top})[X]\ \forall M \in \mathcal{P}(X)\ \forall \Lambda \in \mathcal{Q}_M(Q,\tau)\ \forall q \in Q$:

$$\bigcup_{\mathbf{m}\in M} \Lambda(\mathbf{m},q) \in \mathbb{N}_\tau[M]. \tag{3.2.14}$$

From (3.2.13) we obtain $\forall_X \mathbf{S}[X \neq \varnothing]\ \forall_Q \mathbf{S}[Q \neq \varnothing]\ \forall M \in \mathcal{P}(X)$:

$$\mathcal{Q}_X(Q,\tau) \subset \mathcal{Q}_M(Q,\tau). \tag{3.2.15}$$

We call elements of the set on the right hand side of (3.2.15) neighborhood-valued mappings.

**Example.** Let $\mathbf{S}[X \neq \varnothing]$ and $\rho \in (\mathbf{p}-\text{Dist})[X]$. Consider $Q \triangleq ]0,\infty[$ and $\tau \triangleq \boldsymbol{\tau}_\rho^\natural(X)$ (see (2.7.5)). Of course, this stipulation is valid only in the given example. Introduce

$$\Lambda : X \times Q \to \mathcal{P}(X),$$

setting $\Lambda(x,q) \triangleq \mathbf{B}^0_\rho(x,q)$ for $x \in X$ and $q \in Q$. Using (2.7.7) and (3.2.13) we obtain

$$\Lambda \in \mathcal{Q}_X(Q,\tau). \tag{3.2.16}$$

We have the concrete neighborhood-valued mapping (3.2.16). We use (3.2.16) in (3.2.14) for $M \in 2^X$. Then by (2.7.13) and (2.7.16) we obtain $\forall q \in Q$:

$$\bigcup_{\mathbf{m}\in M} \Lambda(\mathbf{m},q) = \bigcup_{x\in M} \mathbf{B}^0_\rho(x,q) = \mathbf{B}^0_\rho[M;q]. \tag{3.2.17}$$

Indeed, let $z \in \mathbf{B}^0_\rho[M;q]$. Then $z \in X$ and $(\rho - \inf)[z;M] < q$. In addition, $\kappa \triangleq q - (\rho - \inf)[z;M] \in ]0,\infty[$ and by (2.7.13) we can choose $y \in M$ for which

$$\rho(z,y) < (\rho - \inf)[z;M] + \kappa.$$

By the choice of $\kappa$ we have $\rho(z,y) < q$, i.e., $z \in \mathbf{B}^0_\rho(y,q)$. Therefore the third set of (3.2.17) is a subset of the union of all sets $\mathbf{B}^0_\rho(m,q)$, $m \in M$. The opposite inclusion is obvious. In (3.2.17) we have a highly prevalent variant of (3.2.14), which is very often used in applied problems. □

Returning to the general part, we suppose $\forall_X\, \mathbf{S}[X \neq \varnothing]\ \forall_Q\, \mathbf{S}[Q \neq \varnothing]$ $\forall \tau \in (\mathrm{top})[X]\ \forall M \in \mathcal{P}(X)$:

$$(\mathrm{UNIF})[Q;\tau \mid M] \triangleq \{\Lambda \in \mathcal{Q}_X(Q,\tau) \mid \forall q_1 \in Q\ \forall q_2 \in Q\ \exists q_3 \in Q\ \forall \mathbf{m} \in M : \Lambda(\mathbf{m},q_3) \subset \Lambda(\mathbf{m},q_1) \cap \Lambda(\mathbf{m},q_2)\}. \tag{3.2.18}$$

In the example considered above the relation (3.2.16) defines an element of the set (3.2.18). It is possible to point out a lot of other examples of elements of (3.2.18). Note that $\forall_X\, \mathbf{S}[X \neq \varnothing]\ \forall_Q\, \mathbf{S}[Q \neq \varnothing]\ \forall \tau \in (\mathrm{top})[X]$ $\forall M \in \mathcal{P}(X)\ \forall \Lambda \in (\mathrm{UNIF})[Q;\tau \mid M]$:

$$\mathcal{U}(Q,\tau,M \mid \Lambda) \triangleq \{\bigcup_{\mathbf{m}\in M} \Lambda(\mathbf{m},q) : q \in Q\} \in \mathcal{B}[X] \cap \mathcal{P}(\mathrm{N}_\tau[M]). \tag{3.2.19}$$

The example mentioned above defines the natural concrete variant of (3.2.19): the set $\mathcal{U}(Q,\tau,M \mid \Lambda)$ is the family of all sets $\mathbf{B}^0_\rho[M;\varepsilon]$, $\varepsilon \in ]0,\infty[$. We omit the consideration of other concrete examples and formulate a number of very general statements about some universality with respect to variants of the weakening of constraints.

PROPOSITION 3.2.1 *Let* $\mathbf{S}[X \neq \varnothing]$, $\mathbf{S}[Q \neq \varnothing]$, $\mathbf{S}[U \neq \varnothing]$,$\mathbf{S}[V \neq \varnothing]$, $\tau_1 \in (\mathrm{top})[U]$, $\tau_2 \in (\mathrm{top})[V]$, $\mathbf{M} \in \mathcal{P}(U)$, $\Lambda \in (\mathrm{UNIF})[Q;\tau_1 \mid \mathbf{M}]$, $\alpha \in U^X$ *and* $\beta \in V^X$. *Moreover, let*

$$(\exists \mathbf{g} \in C(V,\tau_2,U,\tau_1) : \alpha = \mathbf{g} \circ \beta)$$
$$\&(\forall u \in U \setminus \mathbf{M}\ \exists q \in Q : \Lambda(u,q) \cap (\bigcup_{\nu\in\mathbf{M}} \Lambda(\nu,q)) = \varnothing). \tag{3.2.20}$$

*Then* $(\tau_2 - \text{LIM})[\alpha^{-1}[\mathbb{N}_{\tau_1}[\mathbf{M}]\,] \mid \beta] = (\tau_2 - \text{LIM})[\alpha^{-1}[\mathcal{U}(Q, \tau_1, \mathbf{M} \mid \Lambda)] \mid \beta]$.

PROOF. Using (3.2.20) we fix $\mathbf{g} \in C(V, \tau_2, U, \tau_1)$ with the property $\alpha = \mathbf{g} \circ \beta$. By (3.2.8) and (3.2.19) we have the inclusion

$$(\tau_2 - \text{LIM})[\alpha^{-1}[\mathbb{N}_{\tau_1}[\mathbf{M}]\,] \mid \beta] \subset (\tau_2 - \text{LIM})[\alpha^{-1}[\mathcal{U}(Q, \tau_1, \mathbf{M} \mid \Lambda)] \mid \beta]. \tag{3.2.21}$$

Let $\mu$ be an element of the set on the right hand side of (3.2.21). Using (3.2.8) we obtain that $\mu \in V$, and for some net $(D, \preceq, \rho)$ in $X$ the following relations are true:

$$(\alpha^{-1}[\mathcal{U}(Q, \tau_1, \mathbf{M} \mid \Lambda)] \subset (X - \text{ass})[D, \preceq, \rho])\ \&((D, \preceq, \beta \circ \rho) \xrightarrow{\tau_2} \mu). \tag{3.2.22}$$

In addition, $\mathbf{g}(\mu) \in \mathbf{M}$. Indeed, suppose the contrary: $\mathbf{g}(\mu) \in U \setminus \mathbf{M}$. We use the second statement of (3.2.20). Namely, choose $q^* \in Q$ for which

$$\Lambda(\mathbf{g}(\mu), q^*) \cap \left( \bigcup_{x \in \mathbf{M}} \Lambda(x, q^*) \right) = \varnothing. \tag{3.2.23}$$

Denote by $\Lambda^*$ the union of all sets $\Lambda(x, q^*)$, $x \in \mathbf{M}$. Then from (3.2.23) it follows that $\Lambda(\mathbf{g}(\mu), q^*) \cap \Lambda^* = \varnothing$. By (3.2.19) we have $\Lambda^* \in \mathcal{U}(Q, \tau_1, \mathbf{M} \mid \Lambda)$. Then by (3.2.10) we obtain $\alpha^{-1}(\Lambda^*) \in \alpha^{-1}[\mathcal{U}(Q, \tau_1, \mathbf{M} \mid \Lambda)\,]$. From (3.2.22) it follows that

$$\alpha^{-1}(\Lambda^*) \in (X - \text{ass})[D, \preceq, \rho]). \tag{3.2.24}$$

On the other hand, $\Lambda \in \mathcal{O}_U(Q, \tau_1)$ by (3.2.18), and as a corollary

$$\Lambda(\mathbf{g}(\mu), q^*) \in N_{\tau_1}(\mathbf{g}(\mu)) \tag{3.2.25}$$

(see (3.2.11)). From (2.5.4) and (3.2.22) and by the choice of $\mathbf{g}$ we have the convergence

$$(D, \preceq, \mathbf{g} \circ \beta \circ \rho) \xrightarrow{\tau_1} \mathbf{g}(\mu).$$

Using the representation of $\alpha$, we obtain the statement

$$(D, \preceq, \alpha \circ \rho) \xrightarrow{\tau_1} \mathbf{g}(\mu). \tag{3.2.26}$$

Then by (2.2.48) and (3.2.25) we have (see (3.2.26))

$$\Lambda(\mathbf{g}(\mu), q^*) \in (U - \text{ass})[D; \preceq; \alpha \circ \rho].$$

From this and (2.2.47) we obtain

$$\alpha^{-1}(\Lambda(\mathbf{g}(\mu), q^*)) \in (X - \text{ass})[D; \preceq; \rho]. \tag{3.2.27}$$

From (2.2.45), (2.2.47), (3.2.24) and (3.2.27) it follows that

$$\alpha^{-1}(\Lambda(\mathbf{g}(\mu), q^*) \cap \Lambda^*) = \alpha^{-1}(\Lambda(\mathbf{g}(\mu), q^*)) \cap \alpha^{-1}(\Lambda^*) \in (X - \mathrm{ass})[D; \preceq; \rho]. \tag{3.2.28}$$

But by (3.2.23) and the definition of $\Lambda^*$ we have

$$\alpha^{-1}(\Lambda(\mathbf{g}(\mu), q^*) \cap \Lambda^*) = \varnothing$$

and hence by (2.2.45) and (2.2.47) the relation (3.2.28) is impossible. The obtained contradiction shows that $\mathbf{g}(\mu) \in \mathbf{M}$. Therefore by (2.2.13) and (2.2.14) we have the inclusion

$$\mathbb{N}_{\tau_1}[\mathbf{M}] \subset N_{\tau_1}(\mathbf{g}(\mu)). \tag{3.2.29}$$

From (3.2.29) we obtain the inclusion

$$\alpha^{-1}[\mathbb{N}_{\tau_1}[\mathbf{M}]\,] \subset (X - \mathrm{ass})[D; \preceq; \rho]. \tag{3.2.30}$$

Indeed, let $\tilde{P}$ be an arbitrary set of the family on the left hand side of (3.2.30). Choose $P \in \mathbb{N}_{\tau_1}[\mathbf{M}]$ for which $\tilde{P} = \alpha^{-1}(P)$; we use (2.5.8). From (3.2.29) we have $P \in N_{\tau_1}(\mathbf{g}(\mu))$. By (2.2.48) and (3.2.26) we obtain the property:

$$P \in (U - \mathrm{ass})[D; \preceq; \alpha \circ \rho].$$

As a corollary, the following statement holds:

$$\tilde{P} = \alpha^{-1}(P) \in (X - \mathrm{ass})[D; \preceq; \rho].$$

But the choice of $\tilde{P}$ was arbitrary. Therefore (3.2.30) is established. Then, by (3.2.22), for the net $(D, \preceq, \rho)$ in $X$ we have

$$(\alpha^{-1}[\mathbb{N}_{\tau_1}[\mathbf{M}]\,] \subset (X - \mathrm{ass})[D; \preceq; \rho])\ \&\ ((D, \preceq, \beta \circ \rho) \overset{\tau_2}{\to} \mu). \tag{3.2.31}$$

From (3.2.8) and (3.2.31) we obtain

$$\mu \in (\tau_2 - \mathrm{LIM})[\alpha^{-1}[\mathbb{N}_{\tau_1}[\mathbf{M}]\,] \mid \beta].$$

Since the choice of $\mu$ was arbitrary, the inclusion opposite to (3.2.21) is established.

PROPOSITION 3.2.2 *Let* $\mathbf{S}[X \neq \varnothing]$, $\mathbf{S}[Q \neq \varnothing]$, $\mathbf{S}[U \neq \varnothing]$, $\mathbf{S}[V \neq \varnothing]$, $\mathbf{S}[\mathbf{T} \neq \varnothing]$, $\tau_1 \in (\mathrm{top})[U]$, $\tau_2 \in (\mathrm{top})[V]$, $\theta \in (\mathrm{top})[\mathbf{T}]$, $\mathbf{M} \in \mathcal{P}(U)$, $\Lambda \in (\mathrm{UNIF})[Q; \tau_1 \mid \mathbf{M}]$, $\alpha \in U^X$, $\beta \in V^X$, $r \in \mathbf{T}^X$, *and the statement (3.2.20) be valid. Then* $((\mathbf{c}\theta, \tau_2) - \mathrm{LIM})[\alpha^{-1}[\mathbb{N}_{\tau_1}[\mathbf{M}]\,] \mid r, \beta] = ((\mathbf{c}\theta, \tau_2) - \mathrm{LIM})[\alpha^{-1}[\mathcal{U}(Q, \tau_1, \mathbf{M} \mid \Lambda)] \mid r, \beta]$.

The proof is analogous to that of Proposition 3.2.1. Indeed, in the latter proof the representation in the form of the third set of (3.2.8) is used. Moreover, in (3.2.10) we have an analogous representation.

**Remark.** In Propositions 3.2.1 and 3.2.2 the property of asymptotic equivalence of different variants of the weakening of constraints is established. Namely, the $Y$-constraint is replaced by a series of weakened conditions. This series is defined in different ways. The first variant employs all neighborhoods. The second one is connected with the employment of a specific scheme of constructing neighborhoods. Some uniform equipment is used in this scheme. In the general case these variants of the weakening of the $Y$-constraint are not equivalent. Consider the following obvious example. Let $(U, \tau_1)$ be the plane $\mathbb{R}^2$ with the natural topology of coordinate-wise convergence. Suppose that $\mathbf{M} \in \mathcal{F}_{\tau_1}$ is the set of all vectors $x \triangleq (x_i)_{i\in\overline{1,2}} \in \mathbb{R}^2$ such that $(x_1 \leq 0) \lor (x_2 \leq 0)$. Consider the set $H_1$ of all $(x_i)_{i\in\overline{1,2}} \in U\setminus\mathbf{M}$ for which $x_2 < (x_1)^{-1}$. Then

$$G_1 \triangleq \mathbf{M} \cup H_1 \in \mathbb{N}_{\tau_1}[\mathbf{M}].$$

Introduce the family $\mathfrak{G}$ of all open Euclidean $\varepsilon$-neighborhoods of $\mathbf{M}$. Namely, recall the example above under the following condition: the pseudo-metric $\rho$ of the set $X = U$ is the metric generated by Euclidean norm of $U$. Then (3.2.17) defines the required variant of neighborhoods of $\mathfrak{G}$: $\mathfrak{G} = \{\mathbb{B}^0_\rho[M;q] : q \in ]0,\infty[\}$ (see (2.7.16)). We know that $\mathfrak{G}$ is realized in the form of (3.2.19) under some $\Lambda \in (\mathrm{UNIF})[Q;\tau_1 \mid \mathbf{M}]$, where $Q = ]0,\infty[$. The required remark is given after (3.2.18). It is possible to easily verify that

$$\forall G \in \mathfrak{G} : G \setminus G_1 \neq \varnothing.$$

We have established that $\mathfrak{G}$ is not a fundamental system of neighborhoods.

It is useful to note the case when (in an analogous situation) the equivalence of two variants of families of neighborhoods holds. So let $\mathbf{S}[X \neq \varnothing]$, $\rho \in (\mathbf{p}-\mathrm{Dist})[X]$, $Y \in (\mathrm{comp})[X;\rho]$, and $\mathfrak{G} \in 2^{\tau_\rho^\natural(X)}$; see Section 2.7. Recall that by (2.7.43) we have $Y \in (\tau_\rho^\natural(X)-\mathrm{comp})[X]$ and $Y \neq \varnothing$. Bearing in mind this property, we note that in the given case

$$\left(Y \subset \bigcup_{G\in\mathfrak{G}} G\right) \Rightarrow \left(\exists \varepsilon \in ]0,\infty[\ \forall y \in Y\ \exists G \in \mathfrak{G} : \mathbb{B}^0_\rho(y,\varepsilon) \subset G\right). \tag{3.2.32}$$

As for the proof of (3.2.32), see the well known Lebesgue lemma on covering (we use a very obvious modification of this lemma). In particular, it is possible to consider the case $\mathfrak{G} = \{\mathbb{G}\}$, where $\mathbb{G} \in \mathbb{N}^0_{\tau^\natural_\rho(X)}[Y]$. Then (see (3.2.17))

$$\exists \varepsilon \in ]0, \infty[: \ \mathbb{B}^0_\rho[Y; \varepsilon] \subset \mathbb{G}. \tag{3.2.33}$$

From (3.2.33) we obtain the natural equivalence of $\mathbb{N}_{\tau^\natural_\rho(X)}[Y]$ and the family $\{\mathbb{B}^0_\rho[Y; \varepsilon] : \varepsilon \in ]0, \infty[\}$. The last family is a variant of (3.2.19). We have established that under some additional suppositions of compactness type, statements similar to the basic equality of Propositions 3.2.1 and 3.2.2 hold 'automatically'. But the given question is nontrivial in the general case, and (3.2.20) can be regarded as respective sufficient conditions. The first part of (3.2.20) has the sense of representation in the form of a continuous superposition. The given and other subsequent constructions are considered in [40, 44, 47]. The very important aspect of these constructions is connected with imbedding of the conventional space of solutions in the space of generalized elements (GE). This imbedding is usually realized under the fulfillment of the density condition. In this connection we note the following general statement.

PROPOSITION 3.2.3 *Let* $\mathbf{S}[X \neq \varnothing]$, $\mathbf{S}[U \neq \varnothing]$, $\mathbf{S}[V \neq \varnothing]$, $\tau_1 \in (\mathrm{top})[U]$, $\tau_2 \in (\mathrm{top})[V]$, $p \in V^X$, $q \in C(V, \tau_2, U, \tau_1)$, $\mathbf{M} \in \mathcal{P}(U)$, *and*

$$V = \mathrm{cl}(p^1(X), \tau_2). \tag{3.2.34}$$

*Then the following estimate of the attraction set holds:*

$$q^{-1}(\mathbf{M}) \subset (\tau_2 - \mathrm{LIM})[(q \circ p)^{-1}[\mathbb{N}_{\tau_1}[\mathbf{M}]\,] \mid p]. \tag{3.2.35}$$

**Remark.** Relation (3.2.34) means the dense imbedding. By (3.2.34) each point of $V$ is approximately realizable in $X$. As a rule, extensions being considered in the sequel use dense imbeddings in the space of GE.

**Proof** of Proposition 3.2.3. Fix $v_0 \in q^{-1}(\mathbf{M})$. Then $v_0 \in V$ and $q(v_0) \in \mathbf{M}$. Fix $H^* \in (q \circ p)^{-1}[\mathbb{N}_{\tau_1}[\mathbf{M}]\,]$. Then (see Section 2.4) it is possible to choose $H_* \in \mathbb{N}_{\tau_1}[\mathbf{M}]$ for which $H^* = (q \circ p)^{-1}(H_*)$. Here it is convenient to use open neighborhoods of $\mathbf{M}$. Let $G_* \in \mathbb{N}^0_{\tau_1}[\mathbf{M}]$ possess the property $G_* \subset H_*$. In particular, we have $G_* \in N^0_{\tau_1}(q(v_0))$. Therefore $G_* \in \tau_1$ and $q(v_0) \in G_*$ (see Section 2.2). By the continuity of $q$ we have $q^{-1}(G_*) \in \tau_2$. Moreover, $v_0 \in V$ has the property $q(v_0) \in G_*$. Therefore $v_0 \in q^{-1}(G_*)$. As a corollary we obtain $q^{-1}(G_*) \in N^0_{\tau_2}(v_0)$. Let $G \in N^0_{\tau_2}(v_0)$. It is obvious that

$$q^{-1}(G_*) \cap G \in N^0_{\tau_2}(v_0). \tag{3.2.36}$$

Moreover, by (3.2.34) and the choice of $v_0$, we have $v_0 \in \mathrm{cl}(p^1(X), \tau_2)$. From (2.3.10) and (3.2.36) we obtain

$$q^{-1}(G_*) \cap G \cap p^1(X) \neq \varnothing. \tag{3.2.37}$$

Since the choice of $G$ was arbitrary, from (3.2.37) we have the useful property: if $\tilde{G} \in N^0_{\tau_2}(v_0)$, then

$$\tilde{G} \cap (q^{-1}(G_*) \cap p^1(X)) \neq \varnothing.$$

As a corollary we obtain

$$v_0 \in \mathrm{cl}(q^{-1}(G_*) \cap p^1(X), \tau_2). \tag{3.2.38}$$

Let $v_* \in q^{-1}(G_*) \cap p^1(X)$. Then, in particular, $\exists x \in X : v_* = p(x)$. Let $x_* \in X$ be the point of $X$ for which $v_* = p(x_*)$. Then $p(x_*) \in q^{-1}(G_*)$. Therefore we have $x_* \in p^{-1}(q^{-1}(G_*))$. Then $(q \circ p)(x_*) = q(p(x_*)) \in G_*$. Since $q \circ p \in U^X$, it follows that $x_* \in (q \circ p)^{-1}(G_*)$. Therefore $v_* \in p^1((q \circ p)^{-1}(G_*))$. Since the choice of $v_*$ was arbitrary, we obtain the inclusion

$$q^{-1}(G_*) \cap p^1(X) \subset p^1((q \circ p)^{-1}(G_*)).$$

Consequently, by (2.3.10) we have the inclusion

$$\mathrm{cl}\left(q^{-1}(G_*) \cap p^1(X), \tau_2\right) \subset \mathrm{cl}\left(p^1((q \circ p)^{-1}(G_*)), \tau_2\right).$$

By the choice of $G_*$ we have

$$v_0 \in \mathrm{cl}(p^1((q \circ p)^{-1}(H_*)), \tau_2). \tag{3.2.39}$$

Of course, in (3.2.39) we use the monotonicity property of the operations of image and inverse image. Moreover, we use (2.3.10). By the choice of $H_*$ we have (see (3.2.39)) the inclusion

$$v_0 \in \mathrm{cl}(p^1(H^*), \tau_2).$$

But the choice of $H^*$ was arbitrary too. Therefore

$$v_0 \in \bigcap_{H \in (q \circ p)^{-1}[\mathbb{N}_{\tau_1}[\mathbf{M}]]} \mathrm{cl}(p^1(H), \tau_2). \tag{3.2.40}$$

By (3.2.8) and (3.2.40) we have

$$v_0 \in (\tau_2 - \mathrm{LIM})[(q \circ p)^{-1}[\mathbb{N}_{\tau_1}[\mathbf{M}]] \mid p].$$

So (3.2.35) is established. □

PROPOSITION 3.2.4 *Let $X, U, V, \tau_1, \tau_2, p, q,$ and $\mathbf{M}$ correspond to the conditions of Proposition 3.2.3. Moreover, let (3.2.34) be valid and*

$$\forall u \in U \setminus \mathbf{M}\ \exists H_1 \in N_{\tau_1}(u)\ \exists H_2 \in \mathbb{N}_{\tau_1}[\mathbf{M}] : H_1 \cap H_2 = \varnothing. \tag{3.2.41}$$

*Then (3.2.35) turns into the equality:*

$$(\tau_2 - \text{LIM})[(q \circ p)^{-1}[\mathbb{N}_{\tau_1}[\mathbf{M}]\,] \mid p] = q^{-1}(\mathbf{M}). \tag{3.2.42}$$

PROOF. We recall that $p \in V^X$ and $q \in U^V$. Therefore $q \circ p \in U^X$. As a corollary, for $\tilde{H} \in \mathcal{P}(U)$ we have

$$(q \circ p)^{-1}(\tilde{H}) = p^{-1}(q^{-1}(\tilde{H})).$$

Moreover, for $\hat{H} \in \mathcal{P}(V)$ we have the inclusion $p^1(p^{-1}(\hat{H})) \subset \hat{H}$. It is natural to use the case $\hat{H} = q^{-1}(\tilde{H})$. Namely, for $\tilde{H} \in \mathcal{P}(U)$ we obtain

$$p^1((q \circ p)^{-1}(\tilde{H})) = p^1(p^{-1}(q^{-1}(\tilde{H}))) \subset q^{-1}(\tilde{H}). \tag{3.2.43}$$

From (3.2.43) we obtain

$$\begin{aligned} \Omega &\triangleq (\tau_2 - \text{LIM})[(q \circ p)^{-1}[\mathbb{N}_{\tau_1}[\mathbf{M}]\,] \mid p] \\ &= \bigcap_{E \in (q \circ p)^{-1}[\mathbb{N}_{\tau_1}[\mathbf{M}]\,]} \text{cl}(p^1(E), \tau_2) = \bigcap_{H \in \mathbb{N}_{\tau_1}[\mathbf{M}]} \text{cl}(p^1((q \circ p)^{-1}(H)), \tau_2) \\ &\subset \bigcap_{H \in \mathbb{N}_{\tau_1}[\mathbf{M}]} \text{cl}(q^{-1}(H), \tau_2) \subset \bigcap_{H \in \mathbb{N}_{\tau_1}[\mathbf{M}]} q^{-1}(\text{cl}(H, \tau_1)). \end{aligned} \tag{3.2.44}$$

In (3.2.44) we use the well known property of continuous functions. Namely, if $H \in \mathbb{N}_{\tau_1}[\mathbf{M}]$, then $\text{cl}(H, \tau_1) \in \mathcal{F}_{\tau_1}$ and $q^{-1}(\text{cl}(H, \tau_1)) \in \mathcal{F}_{\tau_2}$ has the property

$$q^{-1}(H) \subset q^{-1}(\text{cl}(H, \tau_1)).$$

The subsequent reasoning is obvious; as a result we obtain (3.2.44). Note that (2.5.4) can be used here. Choose arbitrarily

$$v \in V \setminus q^{-1}(\mathbf{M}). \tag{3.2.45}$$

Then $q(v) \in U$, but $q(v) \notin \mathbf{M}$. So $q(v) \in U \setminus \mathbf{M}$. By (3.2.41) for some neighborhoods

$$(H_1 \in N_{\tau_1}(q(v))) \ \&\ (H_2 \in \mathbb{N}_{\tau_1}[\mathbf{M}])$$

we have the following statement:

$$H_1 \cap H_2 = \varnothing.$$

Of course, $q(v) \notin \text{cl}(H_2, \tau_1)$ and $v \notin q^{-1}(\text{cl}(H_2, \tau_1))$. From (3.2.44) it follows that $v \notin \Omega$. So $v \in V \setminus \Omega$. We obtain the inclusion

$$V \setminus q^{-1}(\mathbf{M}) \subset V \setminus \Omega$$

because the choice of $v$ in (3.2.45) was arbitrary. As a corollary we have the inclusion $\Omega \subset q^{-1}(\mathbf{M})$. Using Proposition 3.2.3 (see (3.2.35)) and the definition of $\Omega$ (see (3.2.44)), we obtain (3.2.42). □

It is useful to consider a natural combination of Propositions 3.2.1 and 3.2.4.

PROPOSITION 3.2.5 *Let* $\mathbf{S}[X \neq \varnothing]$, $\mathbf{S}[U \neq \varnothing]$, $\mathbf{S}[V \neq \varnothing]$, $\tau_1 \in (\mathrm{top})[U]$, $\tau_2 \in (\mathrm{top})[V]$, $p \in V^X$, $q \in C(V, \tau_2, U, \tau_1)$, *and* $\mathbf{M} \in \mathcal{P}(U)$. *Let (3.2.34) hold. Moreover, let* $\mathbf{S}[Q \neq \varnothing]$ *and* $\Lambda \in (\mathrm{UNIF})[Q; \tau_1 \mid M]$. *Finally, let the second statement of (3.2.20) be valid. Then*

$$\begin{aligned} q^{-1}(\mathbf{M}) &= (\tau_2 - \mathrm{LIM})\left[(q \circ p)^{-1}[\mathbb{N}_{\tau_1}[\mathbf{M}]\,] \mid p\right] \\ &= (\tau_2 - \mathrm{LIM})\left[(q \circ p)^{-1}[\mathcal{U}(Q, \tau_1, \mathbf{M} \mid \Lambda)] \mid p\right]. \end{aligned}$$

The proof is an obvious corollary of (3.2.15), (3.2.19), Propositions 3.2.1 and 3.2.4.

**Remark.** In Propositions 3.2.1, 3.2.4, and 3.2.5 the separability property for a point and a set is used. The variant used in (3.2.20) is realized in the case when $\mathbf{M}$ is a nonempty set in a pseudo-metrizable TS. Namely, let $\rho \in (\mathbf{p} - \mathrm{Dist})[U]$, $\tau_1 \triangleq \tau_\rho^\natural(U)$ and $\mathbf{M} \in \mathcal{F}_{\tau_1} \setminus \{\varnothing\}$. If $u \in U \setminus \mathbf{M}$, then from (2.7.15) it follows that

$$0 < (\rho - \inf)[u; \mathbf{M}].$$

Let

$$\varepsilon \triangleq \frac{1}{3}(\rho - \inf)[u; \mathbf{M}].$$

Then $\mathbf{B}_\rho^0(u, \varepsilon) \cap \mathbf{B}_\rho^0[\mathbf{M}; \varepsilon] = \varnothing$. Indeed, if $v \in \mathbf{B}_\rho^0(u, \varepsilon) \cap \mathbf{B}_\rho^0[\mathbf{M}; \varepsilon]$, then $\forall x \in \mathbf{M}$:

$$\rho(u, x) < \varepsilon + \rho(v, x).$$

Therefore for $x \in \mathbf{M}$ we have the inequality $(\rho - \inf)[u, \mathbf{M}] - \varepsilon < \rho(v, x)$. Consequently

$$(\rho - \inf)[u, \mathbf{M}] - \varepsilon \leq (\rho - \inf)[v, \mathbf{M}] < \varepsilon.$$

We use (2.7.13) in the last relation. Then $(\rho - \inf)[u, \mathbf{M}] \leq 2\varepsilon < 3\varepsilon$, which is impossible. So the required property is established: $\mathbf{B}_\rho^0(u, \varepsilon)$ and $\mathbf{B}_\rho^0[\mathbf{M}; \varepsilon]$ are non-intersecting sets.

Now we use the construction of the example above (see (3.2.16) and (3.2.17)). In this example the neighborhood-valued mapping $\Lambda$ with the property $\Lambda(x, q) = \mathbb{B}_\rho^0(x, q)$, $x \in U$, $q \in Q$, is constructed under the replacement $X \to U$ and $Q = ]0, \infty[$. So we have an informative example of the second condition in (3.2.20). Let us consider the condition

(3.2.41). It is satisfied in the case when $\mathbf{M}$ is a closed set in a regular TS. Namely,

$$\forall_U \mathbf{S}[U \neq \varnothing]\, \forall \tau \in (\text{top})^0[U]\, \forall \mathbf{M} \in \mathcal{F}_\tau\, \forall u \in U \setminus \mathbf{M}\, \exists H_1 \in N_\tau(u)$$

$$\exists H_2 \in \mathbb{N}_\tau[\mathbf{M}] : H_1 \cap H_2 = \varnothing.$$

This property follows from (2.2.18). In this connection it is useful to keep in mind (2.2.15),(2.2.19) and (2.7.21). Of course, the second condition in (3.2.20) can be regarded as a variant of (3.2.41). To prove this property it is advisable to use (3.2.13) and (3.2.14).

**Remark.** Let us return to Proposition 3.2.4 and consider the question of the structure of approximate solution-nets generating points of the attraction set (3.2.42). So we investigate a concrete realization of $q^{-1}(\mathbf{M})$. Namely, if $v \in q^{-1}(\mathbf{M})$, then it is possible to choose a very simple variant of the approximate realization of $v$. In particular, this $v$ possesses the property $v \in V$. From (3.2.34) we obtain that $v \in \text{cl}(p^1(X), \tau_2)$. Therefore we can use the Birkhoff theorem. By relation (2.3.11) it is possible to choose some net $(\mathbf{D}, \preceq, \tilde{f})$ in the set $p^1(X)$ with the convergence

$$(\mathbf{D}, \preceq, \tilde{f}) \overset{\tau_2}{\to} v. \tag{3.2.46}$$

In addition,

$$\tilde{f} : \mathbf{D} \to p^1(X).$$

Therefore $\forall d \in \mathbf{D} : \tilde{X}_d \triangleq p^{-1}(\{\tilde{f}(d)\}) \neq \varnothing$. Thus

$$(\tilde{X}_d)_{d \in \mathbf{D}} : \mathbf{D} \to 2^X.$$

Using the axiom of choice we can choose a function

$$f \in \prod_{d \in \mathbf{D}} \tilde{X}_d.$$

Of course, $(\mathbf{D}, \preceq, f)$ is a net in $X$; in addition, $f \in X^{\mathbf{D}}$. Moreover, for $\delta \in \mathbf{D}$ we have

$$(p \circ f)(\delta) = p(f(\delta)) = \tilde{f}(\delta),$$

because $f(\delta) \in \tilde{X}_\delta$. Thus $\tilde{f} = p \circ f$. From (3.2.46) we obtain the convergence

$$(\mathbf{D}, \preceq, p \circ f) \overset{\tau_2}{\to} v. \tag{3.2.47}$$

We have established that one can choose some net $(\mathbf{D}, \preceq, f)$ in $X$ for which the convergence (3.2.47) holds. We will call each net $(\mathbf{D}, \preceq, f)$ in the set $X$ possessing the property (3.2.47) the $v$-approximate net. Such a net exists. And what is more, each $v$-approximate net in $X$ has the very important property: it satisfies constraints of asymptotic character generated by the family $(q \circ p)^{-1}[\mathbb{N}_{\tau_1}[\mathbf{M}]]$. Indeed, let $(\mathbf{D}, \preceq, f)$ be an arbitrary $v$-approximate net in the set $X$: $(\mathbf{D}, \preceq, f)$ is the net in $X$ for which (3.2.47) holds. Choose any set

$$A \in (q \circ p)^{-1}[\mathbb{N}_{\tau_1}[\mathbf{M}]]. \qquad (3.2.48)$$

Using (2.5.8) and (3.2.48) we choose the set $B \in \mathbb{N}_{\tau_1}[\mathbf{M}]$ for which $A = (q \circ p)^{-1}(B)$. Since $q(v) \in \mathbf{M}$, we have $B \in N_{\tau_1}(q(v))$. But $q \in C(V, \tau_2, U, \tau_1)$. From (2.5.4) and (3.2.47) we obtain the convergence

$$(\mathbf{D}, \preceq, q \circ p \circ f) \xrightarrow{\tau_1} q(v). \qquad (3.2.49)$$

From (2.2.48) and (3.2.49) we have the inclusion

$$N_{\tau_1}(q(v)) \subset (U - \mathrm{ass})[\mathbf{D}; \preceq; q \circ p \circ f].$$

From (2.2.47) it follows that $\exists d \in \mathbf{D}\ \forall \delta \in \mathbf{D}$:

$$(d \preceq \delta) \Rightarrow ((q \circ p)(f(\delta)) \in B).$$

By the choice of $B$ we obtain that $\exists d \in \mathbf{D}\ \forall \delta \in \mathbf{D}$:

$$(d \preceq \delta) \Rightarrow (f(\delta) \in A). \qquad (3.2.50)$$

From (2.2.47) and (3.2.50) it follows that

$$A \in (X - \mathrm{ass})[\mathbf{D}; \preceq; f].$$

But the choice (3.2.48) was arbitrary. Therefore the inclusion

$$(q \circ p)^{-1}[\mathbb{N}_{\tau_1}[\mathbf{M}]] \subset (X - \mathrm{ass})[\mathbf{D}; \preceq; f] \qquad (3.2.51)$$

is established. Recall that (3.2.47) holds. From (3.2.47) and (3.2.51) we obtain that the point $v \in V$ is realized by the net $(\mathbf{D}, \preceq, f)$ under the validity of (3.2.47) and (3.2.51). So we have a variant of the asymptotic realization of $v$ as an element of the attraction set

$$(\tau_2 - \mathrm{LIM})\left[(q \circ p)^{-1}[\mathbb{N}_{\tau_1}[\mathbf{M}]] \mid p\right]. \qquad (3.2.52)$$

Since the choice of $v$ and $(\mathbf{D}, \preceq, f)$ was arbitrary, we have a variant of asymptotic realization of the whole set (3.2.52) in the class of $v$-approximate nets under enumeration of all $v \in q^{-1}(\mathbf{M})$. The following scheme can be used for the concrete realization of the whole set (3.2.52). Namely, it is advisable to choose an arbitrary $v$-approximate net for each $v \in q^{-1}(\mathbf{M})$. As a result we obtain an approximate solution with respect to $(q \circ p)^{-1}[\mathbb{N}_{\tau_1}[\mathbf{M}]]$; this solution generates $v$ in terms of '$p$-image' of the given net. We see that the approximate realization of the set (3.2.42), (3.2.52) is connected only with the equality (3.2.34), which has the sense of a dense imbedding of $X$ into $(V, \tau_2)$. In this construction realized for points from $q^{-1}(\mathbf{M})$ we forget, in fact, neighborhoods of the set $\mathbf{M}$ in the TS $(U, \tau_1)$. We will return to this circumstance in the sequel.

## 3.3 REGULARIZING CONSTRUCTIONS FOR ATTAINABLE SETS

Returning to the scheme of Section 3.1, we recall that the problem of regularization of the set (3.1.2) was considered as a basic one. A typical situation arises in control theory (see Chapter I) when investigating the question of constructing attainability domains and bundles of trajectories . In the previous section we considered regularizing constructions for the admissible set $s^{-1}(\mathbf{Y})$. Now we combine this problem with the problem of extension of the goal operator $h$. We follow methods of [32, 35, 40, 41, 44, 47]. The most known construction is connected with compactifications (for example, see [32, p. 41]). These compactifications are similar to those used in general topology, but they have the following seculiarity: compactifications of the type [32, p. 41] (see also [117, Chapters III,IV]) are complicated by connections of the type (3.1.9).

For many problems considered in [35, 40, 41, 47] the above mentioned compactification is impossible. This circumstance is connected with the following property: 'unbounded' admissible sets arise in these problems. The 'unboundedness' is usually connected with the strong norm of the corresponding space conjugate to a Banach space. For such 'unbounded' problems we use more general approaches; these approaches are realized in a very concrete form in [35, 40, 41, 47]. Namely, the investigations [32, 35, 40, 41, 47] are connected with the specific space of generalized elements. In [32, 35, 40, 41, 47] and many other investigations finitely additive measures (FAM) are used in the capacity of these generalized elements. In [44] the corresponding attention was paid to more general constructions of TS (moreover, see [29, 30, 47]). The given approach permits us to embrace many different settings in a common scheme of investigation of asymptotic effects. In this respect, we follow this approach. We begin the study with a series of auxiliary statements.

PROPOSITION 3.3.1 *Let* $\mathbf{S}[X \neq \varnothing]$, $\mathbf{S}[V \neq \varnothing]$, $\mathbf{S}[W \neq \varnothing]$, $\mathcal{X} \in \mathcal{B}[X]$, $\tilde{\tau}_1 \in (\mathrm{top})[V]$, $\tilde{\tau}_2 \in (\mathrm{top})[W]$, $\alpha \in V^X$, *and* $\beta \in C(V, \tilde{\tau}_1, W, \tilde{\tau}_2)$. *Then*

$$\beta^1((\tilde{\tau}_1 - \mathrm{LIM})[\mathcal{X} \mid \alpha]) \subset (\tilde{\tau}_2 - \mathrm{LIM})[\mathcal{X} \mid \beta \circ \alpha]. \tag{3.3.1}$$

*If* $\beta \in C_{\mathbf{ap}}(V, \tilde{\tau}_1, W, \tilde{\tau}_2)$, *then (3.3.1) turns into the equality.*

PROOF. By the continuity of $\beta$ we have the following chain of inclusions (see (2.5.4), (3.2.8)):

$$\beta^1((\tilde{\tau}_1 - \mathrm{LIM})[\mathcal{X} \mid \alpha]) = \beta^1\left(\bigcap_{U \in \mathcal{X}} \mathrm{cl}(\alpha^1(U), \tilde{\tau}_1)\right)$$

$$\subset \bigcap_{U \in \mathcal{X}} \beta^1(\mathrm{cl}(\alpha^1(U), \tilde{\tau}_1)) \subset \bigcap_{U \in \mathcal{X}} \mathrm{cl}(\beta^1(\alpha^1(U)), \tilde{\tau}_2)$$

$$= \bigcap_{U \in \mathcal{X}} \mathrm{cl}((\beta \circ \alpha)^1(U), \tilde{\tau}_2) = (\tilde{\tau}_2 - \mathrm{LIM})[\mathcal{X} \mid \beta \circ \alpha]. \tag{3.3.2}$$

In (3.3.2) we use the following property: if $A \in \mathcal{P}(X)$, then

$$(\beta \circ \alpha)^1(A) = \{\beta(\alpha(x)) : x \in A\} = \{\beta(y) : y \in \alpha^1(A)\} = \beta^1(\alpha^1(A)).$$

Let $\beta \in C_{\mathbf{ap}}(V, \tilde{\tau}_1, W, \tilde{\tau}_2)$. Then by (2.8.2) $\beta \in C_{\mathrm{cl}}(V, \tilde{\tau}_1, W, \tilde{\tau}_2)$ and consequently we have (see (2.8.4)) $\forall A \in \mathcal{P}(V)$:

$$\beta^1(\mathrm{cl}(A, \tilde{\tau}_1)) = \mathrm{cl}(\beta^1(A), \tilde{\tau}_2). \tag{3.3.3}$$

Moreover, by (2.8.2) we have $\forall y \in W$:

$$\beta^{-1}(\{y\}) \in (\tilde{\tau}_1 - \mathrm{comp})[V]. \tag{3.3.4}$$

Choose an arbitrary point $w \in (\tilde{\tau}_2 - \mathrm{LIM})[\mathcal{X} \mid \beta \circ \alpha]$. Then $w \in W$. In addition, by (3.2.8) we have $w \in \mathrm{cl}((\beta \circ \alpha)^1(U), \tilde{\tau}_2)$ for $U \in \mathcal{X}$. From (3.3.3) we obtain $w \in \beta^1(\mathrm{cl}(\alpha^1(U), \tilde{\tau}_1))$ for $U \in \mathcal{X}$. Therefore the point $w$ is such that $\forall U \in \mathcal{X}$:

$$\mathrm{cl}(\alpha^1(U), \tilde{\tau}_1) \cap \beta^{-1}(\{w\}) \neq \varnothing. \tag{3.3.5}$$

Recall that $\mathcal{X} \neq \varnothing$ and for $U \in \mathcal{X}$ we have $w = \beta(z)$ for some $z \in \mathrm{cl}(\alpha^1(U), \tilde{\tau}_1)$; in addition, $z \in V$. Hence $\mathbf{S}[\beta^{-1}(\{w\}) \neq \varnothing]$. From (3.3.4) we obtain

$$\beta^{-1}(\{w\}) \in (\tilde{\tau}_1 - \mathrm{comp})[V]. \tag{3.3.6}$$

From (3.3.6) we have the property

$$\theta \triangleq \tilde{\tau}_1 \big|_{\beta^{-1}(\{w\})} \in (\mathrm{c} - \mathrm{top})[\beta^{-1}(\{w\})]. \tag{3.3.7}$$

So $(\beta^{-1}(\{w\}), \theta)$ is a nonempty compact TS. By (2.3.1) and (2.3.12) we have $\forall F \in \mathcal{F}_{\tilde{\tau}_1}$:

$$F \cap \beta^{-1}(\{w\}) \in \mathcal{F}_\theta.$$

In particular, from (3.3.5) we obtain $\forall U \in \mathcal{X}$:

$$\mathrm{cl}(\alpha^1(U), \tilde{\tau}_1) \cap \beta^{-1}(\{w\}) \in \mathcal{F}_\theta \setminus \{\varnothing\}. \tag{3.3.8}$$

We introduce the family

$$\mathfrak{X} \triangleq \{\mathrm{cl}(\alpha^1(U), \tilde{\tau}_1) \cap \beta^{-1}(\{w\}) : U \in \mathcal{X}\} \in 2^{\mathcal{F}_\theta \setminus \{\varnothing\}}. \tag{3.3.9}$$

Since $\mathcal{X} \in \mathcal{B}[X]$, we obtain (see Section 3.2) $\forall m \in \mathcal{N}\ \forall (U_i)_{i\in\overline{1,m}} \in \mathcal{X}^m$ $\exists U \in \mathcal{X}$: $U \subset \bigcap_{i=1}^{m} U_i$. By monotonicity of the closure and image operations, from (3.3.9) we obtain

$$\forall m \in \mathcal{N}\ \forall (\tilde{M}_i)_{i\in\overline{1,m}} \in \mathfrak{X}^m\ \exists \tilde{M} \in \mathfrak{X} : \tilde{M} \subset \bigcap_{i=1}^{m} \tilde{M}_i. \tag{3.3.10}$$

Therefore by (2.2.20), (3.3.9) and (3.3.10) we have $\mathfrak{X} \in \mathbf{Z}[\mathcal{F}_\theta]$. From (2.2.21) and (3.3.7) we obtain

$$\begin{aligned}\bigcap_{H\in\mathfrak{X}} H &= \bigcap_{U\in\mathcal{X}} (\mathrm{cl}(\alpha^1(U), \tilde{\tau}_1) \cap \beta^{-1}(\{w\})) \\ &= \left(\bigcap_{U\in\mathcal{X}} \mathrm{cl}(\alpha^1(U), \tilde{\tau}_1)\right) \cap \beta^{-1}(\{w\}) \\ &= (\tilde{\tau}_1 - \mathrm{LIM})[\mathcal{X} \mid \alpha] \cap \beta^{-1}(\{w\}) \neq \varnothing.\end{aligned}$$

But in this case $w \in \beta^1((\tilde{\tau}_1 - \mathrm{LIM})[\mathcal{X} \mid \alpha])$. Since the choice of $w$ was arbitrary, the following inclusion holds:

$$(\tilde{\tau}_2 - \mathrm{LIM})[\mathcal{X} \mid \beta \circ \alpha] \subset \beta^1((\tilde{\tau}_1 - \mathrm{LIM})[\mathcal{X} \mid \alpha]). \tag{3.3.11}$$

From (3.3.2) and (3.3.11) we obtain the coincidence of $(\tilde{\tau}_2 - \mathrm{LIM})[\mathcal{X} \mid \beta \circ \alpha]$ and $\beta^1((\tilde{\tau}_1 - \mathrm{LIM})[\mathcal{X} \mid \alpha])$. □

From Proposition 3.3.1 it follows that $\forall_X\, \mathbf{S}[X \neq \varnothing]$ $\forall_V\, \mathbf{S}[V \neq \varnothing]$ $\forall_W\, \mathbf{S}[W \neq \varnothing]$ $\forall \mathcal{X} \in \mathcal{B}[X]$ $\forall \tilde{\tau}_1 \in (\mathrm{top})[V]$ $\forall \tilde{\tau}_2 \in (\mathrm{top})[W]$ $\forall \alpha \in V^X$ $\forall \beta \in C_{\mathbf{ap}}(V, \tilde{\tau}_1, W, \tilde{\tau}_2)$:

$$(\tilde{\tau}_2 - \mathrm{LIM})[\mathcal{X} \mid \beta \circ \alpha] = \beta^1((\tilde{\tau}_1 - \mathrm{LIM})[\mathcal{X} \mid \alpha]). \tag{3.3.12}$$

In (3.3.12) we have a very important representation. The basic attraction set is represented in terms of a continuous image of some auxiliary attraction set. We strive to find an appropriate representation in terms of a space of GE for the last set. Proposition 3.2.4 realizes an example of such representation. Below we give some combinations of representations of such a type. Now we consider other conditions sufficient for the validity of (3.3.12). But before we note that constructions of [32, p. 41] can be considered as a particular case of Proposition 3.3.1 (we keep in mind the second part and (3.3.12)).

**Remark.** Let $\mathbf{S}[X \neq \varnothing]$, $\mathbf{S}[V \neq \varnothing]$, $\mathbf{S}[W \neq \varnothing]$, $\mathcal{X} \in \mathcal{B}[X]$, $\tilde{\tau}_1 \in (\mathbf{c} - \mathrm{top})[V]$, $\tilde{\tau}_2 \in (\mathrm{top})_0[W]$, $\alpha \in V^X$ and $\beta \in C(V, \tilde{\tau}_1, W, \tilde{\tau}_2)$. Then by (2.8.7), the equality (3.3.12) holds. Indeed, by (2.8.7) we have $\beta \in C_{\mathbf{ap}}(V, \tilde{\tau}_1, W, \tilde{\tau}_2)$. We now return to conditions guaranteeing the validity of (3.3.12). Thus we have (3.3.12) in the considered case characterized in terms of continuous mapping from a compact space in a Hausdorff TS.

We obtain the following

PROPOSITION 3.3.2 $\forall_X \mathbf{S}[X \neq \varnothing] \quad \forall_V \mathbf{S}[V \neq \varnothing] \quad \forall_W \mathbf{S}[W \neq \varnothing]$ $\forall \mathcal{X} \in \mathcal{B}[X]$ $\forall \tilde{\tau}_1 \in (\mathbf{c} - \text{top})[V]$ $\forall \tilde{\tau}_2 \in (\text{top})_0[W]$ $\forall \alpha \in V^X$ $\forall \beta \in C(V, \tilde{\tau}_1, W, \tilde{\tau}_2)$:

$$(\tilde{\tau}_2 - \text{LIM})[\mathcal{X} \mid \beta \circ \alpha] = \beta^1((\tilde{\tau}_1 - \text{LIM})[\mathcal{X} \mid \alpha]).$$

The approach used in Propositions 3.3.1 and 3.3.2 admits natural development to the case when model countably compact spaces are used. We begin with a natural analogue of the last proposition.

PROPOSITION 3.3.3 (SEE [44, 47]) $\forall_X \mathbf{S}[X \neq \varnothing] \quad \forall_V \mathbf{S}[V \neq \varnothing]$ $\forall_W \mathbf{S}[W \neq \varnothing]$ $\forall \mathcal{X} \in \mathcal{B}_{\mathcal{N}}(X)$ $\forall \tilde{\tau}_1 \in (\mathbf{c}_{\mathcal{N}} - \text{top})[V]$ $\forall \tilde{\tau}_2 \in (\mathcal{D} - \text{top})[W]$ $\forall \alpha \in V^X$ $\forall \beta \in C_{\text{cl}}(V, \tilde{\tau}_1, W, \tilde{\tau}_2)$:

$$(\tilde{\tau}_2 - \text{LIM})[\mathcal{X} \mid \beta \circ \alpha] = \beta^1((\tilde{\tau}_1 - \text{LIM})[\mathcal{X} \mid \alpha]).$$

PROOF. Fix $X, V, W, \mathcal{X}, \tilde{\tau}_1, \tilde{\tau}_2, \alpha, \beta$ in correspondence with the conditions. We have (3.3.1). Using (3.2.5) we choose $(H^{(i)})_{i \in \mathcal{N}} \in \mathcal{X}^{\mathcal{N}}$ with the property

$$\forall H \in \mathcal{X}\ \exists p \in \mathcal{N}\ \forall q \in \overrightarrow{p, \infty} : H^{(q)} \subset H. \tag{3.3.13}$$

Then by (3.2.8) and (3.3.13) we have the equalities

$$\begin{aligned} \Big( (\tilde{\tau}_2 - \text{LIM})[\mathcal{X} \mid \beta \circ \alpha] &= \bigcap_{i \in \mathcal{N}} \text{cl}((\beta \circ \alpha)^1(H^{(i)}), \tilde{\tau}_2) \\ &= \bigcap_{i \in \mathcal{N}} \text{cl}(\beta^1(\alpha^1(H^{(i)})), \tilde{\tau}_2) \Big) \\ \& \Big( (\tilde{\tau}_1 - \text{LIM})[\mathcal{X} \mid \alpha] &= \bigcap_{i \in \mathcal{N}} \text{cl}(\alpha^1(H^{(i)}), \tilde{\tau}_1) \Big). \end{aligned} \tag{3.3.14}$$

Using (2.8.4) and (3.3.14) we obtain

$$(\tilde{\tau}_2 - \text{LIM})[\mathcal{X} \mid \beta \circ \alpha] = \bigcap_{i \in \mathcal{N}} \beta^1(\text{cl}(\alpha^1(H^{(i)}), \tilde{\tau}_1)). \tag{3.3.15}$$

Choose an arbitrary point $w$ of the set (3.3.15). Then for $i \in \mathcal{N}$ we have $w \in \beta^1(\text{cl}(\alpha^1(H^{(i)}), \tilde{\tau}_1))$ and consequently

$$F_i \triangleq \text{cl}(\alpha^1(H^{(i)}), \tilde{\tau}_1) \cap \beta^{-1}(\{w\}) \neq \varnothing. \tag{3.3.16}$$

In addition, $\{w\} \in \mathcal{F}_{\tilde{\tau}_2}$ (see (2.2.15)). Therefore by (2.5.2) we have $\beta^{-1}(\{w\}) \in \mathcal{F}_{\tilde{\tau}_1}$. As a corollary, from (3.3.16) we obtain

$$\forall i \in \mathcal{N} : F_i \in \mathcal{F}_{\tilde{\tau}_1} \setminus \{\varnothing\}.$$

In the form of $(V, \tilde{\tau}_1)$ we have a countably compact space, and

$$(F_i)_{i\in\mathcal{N}} : \mathcal{N} \to \mathcal{F}_{\tilde{\tau}_1} \setminus \{\varnothing\}. \tag{3.3.17}$$

Let $m \in \mathcal{N}$. Consider the intersection of all sets $F_i, i \in \overline{1,m}$. Since $(H^{(i)})_{i\in\overline{1,m}} \in \mathcal{X}^m$, we obtain that (see Section 3.2) $\exists H \in \mathcal{X}$:

$$H \subset \bigcap_{i=1}^{m} H^{(i)};$$

hence by (3.3.13) we can choose $k \in \mathcal{N}$ such that

$$\forall j \in \overrightarrow{k,\infty} : H^{(j)} \subset \bigcap_{i=1}^{m} H^{(i)}.$$

In particular, we can consider the case $j = k$:

$$H^{(k)} \subset \bigcap_{i=1}^{m} H^{(i)}.$$

Then by (3.3.16) we have the inclusion

$$F_k \subset \bigcap_{i=1}^{m} F_i.$$

From (3.3.16) it follows that the intersection of all sets $F_i, i \in \overline{1,m}$, is a nonempty set. Recall that the choice of $m$ was arbitrary. We obtain $\forall p \in \mathcal{N}$:

$$\bigcap_{i=1}^{p} F_i \neq \varnothing.$$

Return to the relation (3.3.17). Due to the countable compactness of $(V, \tilde{\tau}_1)$ we obtain (see Section 2.2) the property

$$\bigcap_{i\in\mathcal{N}} F_i \neq \varnothing. \tag{3.3.18}$$

On the other hand, by (3.3.14) and (3.3.16) we have the equality

$$\begin{aligned}\bigcap_{i\in\mathcal{N}} F_i &= \left(\bigcap_{i\in\mathcal{N}} \mathrm{cl}(\alpha^1(H^{(i)}), \tilde{\tau}_1)\right) \cap \beta^{-1}(\{w\}) \\ &= (\tilde{\tau}_1 - \mathrm{LIM})[\mathcal{X} \mid \alpha] \cap \beta^{-1}(\{w\}).\end{aligned} \tag{3.3.19}$$

From (3.3.18) and (3.3.19) we obtain

$$w \in \beta^1((\tilde{\tau}_1 - \mathrm{LIM})[\mathcal{X} \mid \alpha]).$$

Since the choice of $w$ was arbitrary, the inclusion

$$(\tilde{\tau}_2 - \mathrm{LIM})[\mathcal{X} \mid \beta \circ \alpha] \subset \beta^1((\tilde{\tau}_1 - \mathrm{LIM})[\mathcal{X} \mid \alpha]) \tag{3.3.20}$$

is established. From Proposition 3.3.1 (see (3.3.1)) we obtain the required equality

$$(\tilde{\tau}_2 - \mathrm{LIM})[\mathcal{X} \mid \beta \circ \alpha] = \beta^1((\tilde{\tau}_1 - \mathrm{LIM})[\mathcal{X} \mid \alpha]).$$

PROPOSITION 3.3.4 (SEE [44, 47]) $\forall_X\, \mathbf{S}[X \neq \varnothing]$ $\forall_V\, \mathbf{S}[V \neq \varnothing]$ $\forall_W\, \mathbf{S}[W \neq \varnothing]$ $\forall \mathcal{X} \in \mathcal{B}_{\mathcal{N}}(X)$ $\forall \tilde{\tau}_1 \in (\mathrm{top})[V]$ $\forall \tilde{\tau}_2 \in (\mathrm{top})[W]$ $\forall \alpha \in V^X$ $\forall \beta \in C_{\mathrm{qp}}(V, \tilde{\tau}_1, W, \tilde{\tau}_2)$:

$$(\tilde{\tau}_2 - \mathrm{LIM})[\mathcal{X} \mid \beta \circ \alpha] = \beta^1((\tilde{\tau}_1 - \mathrm{LIM})[\mathcal{X} \mid \alpha]).$$

PROOF. Fix $(X, \mathcal{X}), (V, \tilde{\tau}_1), (W, \tilde{\tau}_2)$, $\alpha$ and $\beta$ in correspondence with the conditions of the given proposition. We have (3.3.1) again. Choosing $(H^{(i)})_{i \in \mathcal{N}}$ with the property (3.3.13), we obtain (3.3.14) again. Since by (2.8.3) $\beta \in C_{\mathrm{cl}}(V, \tilde{\tau}_1, W, \tilde{\tau}_2)$, we have (3.3.15); here (2.8.4) is used. Choose $w \in (\tilde{\tau}_2 - \mathrm{LIM})[\mathcal{X} \mid \beta \circ \alpha]$. As a consequence we again obtain (3.3.16). Of course, we preserve the notation $F_i$ used in the proof of Proposition 3.3.3. By properties of $\beta$ we have (from (2.8.3))

$$\beta^{-1}(\{w\}) \in (\mathcal{N} - \tilde{\tau}_1 - \mathrm{comp})[V]. \tag{3.3.21}$$

From the property similar to (3.3.16) we obtain $\mathbf{S}[\beta^{-1}(\{w\}) \neq \varnothing]$. By (2.3.14) and (3.3.21) we have the countably compact space

$$\left(\beta^{-1}(\{w\}), \tilde{\tau}_1 \mid_{\beta^{-1}(\{w\})}\right). \tag{3.3.22}$$

For brevity we denote by $\tilde{\theta}$ the topology of the space (3.3.22), i.e., $\tilde{\theta} \triangleq \tilde{\tau}_1 \mid_{\beta^{-1}(\{w\})}$. So (3.3.22) is $(\beta^{-1}(\{w\}), \tilde{\theta})$. Return to (3.3.16). It is obvious that $F_i \in \mathcal{F}_{\tilde{\theta}}$ for $i \in \mathcal{N}$; in this connection see (2.3.12) and (2.3.13). The property (3.3.18) is established by analogy with the proof of Proposition 3.3.3. Of course, here we use the countably compact space (3.3.22) and the property that the intersection of sets $F_1, \ldots, F_k$, where $k \in \mathcal{N}$, is not empty. The basic fact is that

$$\tilde{\theta} \in (\mathbf{c}_{\mathcal{N}} - \mathrm{top})[\beta^{-1}(\{w\})]$$

(see (2.3.14)). As a result we have (3.3.18). The sequel arguments are analogous to the proof of (3.3.20).

Let us note that, to some extent, the proving method for the second part of Proposition 3.3.1 (of Proposition 3.3.4) can be regarded as a local version of Proposition 3.3.2 (of Proposition 3.3.3).

Now we shall consider another construction of the attraction set. Namely, we shall give an appropriate representation of a set of the kind (3.2.10). We strive to restrict ourselves only to the property of continuity of the mapping $\beta$ in Proposition 3.3.1 and to transform (3.3.1) into the equality. The employment of attraction sets of the kind (3.2.10) provides new possibilities in the questions of asymptotic representation of analogues of the attainability property for elements of functional spaces.

PROPOSITION 3.3.5 $\forall_X \mathbf{S}[X \neq \varnothing]\forall_V \mathbf{S}[V \neq \varnothing]\forall_W \mathbf{S}[W \neq \varnothing]\forall \mathcal{X} \in \mathcal{B}(X)$ $\forall \tilde{\tau}_1 \in (\mathrm{top})[V]\ \forall \tilde{\tau}_2 \in (\mathrm{top})_0[W]\ \forall \alpha \in V^X\ \forall \beta \in C(V, \tilde{\tau}_1, W, \tilde{\tau}_2)$:

$$((\mathbf{c}\tilde{\tau}_1, \tilde{\tau}_2) - \mathrm{LIM})[\mathcal{X} \mid \alpha; \beta \circ \alpha] \subset \beta^1((\tilde{\tau}_1 - \mathrm{LIM})[\mathcal{X} \mid \alpha]). \qquad (3.3.23)$$

PROOF. Fix $X, V, W, \mathcal{X}, \tilde{\tau}_1, \tilde{\tau}_2, \alpha$ and $\beta$ in correspondence with the conditions of the given proposition. Fix an arbitrary point $w$ of the set on the left hand side of (3.3.23). Then by (3.2.10) $w \in W$. Let $\mathbf{S}[\mathbf{D} \neq \varnothing]$, $\preceq \in (\mathrm{DIR})[\mathbf{D}]$ and $\varphi \in \mathcal{M}_{\mathbf{c}}(\mathbf{D}, X, V, \tilde{\tau}_1, \alpha)$ be such that

$$\left(\mathcal{X} \subset (X - \mathrm{ass})[\mathbf{D}; \preceq; \varphi]\right) \ \& \ \left((\mathbf{D}, \preceq, \beta \circ \alpha \circ \varphi) \xrightarrow{\tilde{\tau}_2} w\right). \qquad (3.3.24)$$

Using (3.2.9) we obtain, in particular, that $\varphi \in X^{\mathbf{D}}$. Choose (see (3.2.9)) the set

$$\mathbb{K} \in (\tilde{\tau}_1 - \mathrm{comp})[V] \qquad (3.3.25)$$

for which $(\alpha \circ \varphi)^1(\mathbf{D}) \subset \mathbb{K}$. Of course, the set (3.3.25) is not empty. In addition, $\varphi^1(\mathbf{D}) \subset \alpha^{-1}(\mathbb{K})$ and $\mathbf{U} \triangleq \alpha^{-1}(\mathbb{K}) \in 2^X$. So $\mathbf{S}[\mathbf{U} \neq \varnothing]$ and $\varphi \in \mathbf{U}^{\mathbf{D}}$. Moreover, by the definition of inverse image

$$(\alpha \mid \mathbf{U}) = (\alpha(x))_{x \in \mathbf{U}} \in \mathbb{K}^{\mathbf{U}}.$$

We note that by (2.3.6) the relation

$$\theta_1 \triangleq \tilde{\tau}_1 \mid_{\mathbb{K}} \in (\mathbf{c} - \mathrm{top})[\mathbb{K}]$$

holds. Thus $(\mathbb{K}, \theta_1)$ is a compact TS. Introduce the nonempty set $\mathbb{W} \triangleq \beta^1(\mathbb{K})$. Then by (2.8.5) and (3.3.25) we have the property $\mathbb{W} \in (\tilde{\tau}_2 - \mathrm{comp})[W] \setminus \{\varnothing\}$. As a consequence

$$\theta_2 \triangleq \tilde{\tau}_2 \mid_{\mathbb{W}} \in (\mathbf{c} - \mathrm{top})[\mathbb{W}]. \qquad (3.3.26)$$

So $(\mathbb{W}, \theta_2)$, $\mathbb{W} \neq \varnothing$, is a compact TS. In addition,

$$\tilde{\beta} \triangleq (\beta \mid \mathbb{K}) \in C(\mathbb{K}, \theta_1, \mathbb{W}, \theta_2). \qquad (3.3.27)$$

Indeed, let $G \in \theta_2$. Then by (2.3.1) and (3.3.26) we can choose $\tilde{G} \in \tilde{\tau}_2$ for which $G = \tilde{G} \cap \mathbb{W}$. By the continuity of $\beta$ we have $\beta^{-1}(\tilde{G}) \in \tilde{\tau}_1$ and as a consequence

$$\tilde{\beta}^{-1}(\tilde{G}) = (\beta \mid \mathbb{K})^{-1}(\tilde{G}) = \beta^{-1}(\tilde{G}) \cap \mathbb{K} \in \theta_1. \qquad (3.3.28)$$

In (3.3.28) we use (2.3.1) and the definition of $\theta_1$. Note that by the definition of $\mathbb{W}$ we obtain $\forall v \in \mathbb{K}$: $\tilde{\beta}(v) = \beta(v) \in \mathbb{W}$. Therefore

$$\tilde{\beta}^{-1}(\tilde{G}) = \{v \in \mathbb{K} \mid \tilde{\beta}(v) \in \tilde{G}\} = \{v \in \mathbb{K} \mid \tilde{\beta}(v) \in G\} = \tilde{\beta}^{-1}(G).$$

From (3.3.28) we obtain $\tilde{\beta}^{-1}(G) \in \theta_1$. Since the choice of $G$ was arbitrary, we have the property (3.3.27); here we use (2.5.2). So (3.3.27) is established.

We have got the following auxiliary procession: $\mathbf{S}[\mathbf{U} \neq \varnothing]$, $\mathbf{S}[\mathbb{K} \neq \varnothing]$, $\mathbf{S}[\mathbb{W} \neq \varnothing]$, $\mathcal{X}_1 \triangleq \mathcal{X} \mid_{\mathbf{U}} \in \mathcal{B}[\mathbf{U}]$, $\theta_1 \in (\mathbf{c}-\mathrm{top})[\mathbb{K}]$, $\theta_2 \in (\mathbf{c}-\mathrm{top})_0[\mathbb{W}]$, $(\alpha \mid \mathbf{U}) \in \mathbb{K}^U$ and $\tilde{\beta} \in C(\mathbb{K}, \theta_1, \mathbb{W}, \theta_2)$.

**Remark.** Of course, we use (2.2.16), (2.2.22), (2.3.7) and (2.3.8). From these relations and (3.3.26) we obtain $\theta_2 \in (\mathrm{top})_0[\mathbb{W}]$.

So all conditions of Proposition 3.3.2 are valid. From this proposition we have

$$(\theta_2 - \mathrm{LIM})[\mathcal{X}_1 \mid \tilde{\beta} \circ (\alpha \mid \mathbf{U})] = \tilde{\beta}^1((\theta_1 - \mathrm{LIM})[\mathcal{X}_1 \mid (\alpha \mid \mathbf{U})]). \quad (3.3.29)$$

Let us consider $(\theta_1 - \mathrm{LIM})[\mathcal{X}_1 \mid (\alpha \mid \mathbf{U})]$. From (2.3.13) and (3.2.8) it follows that

$$\begin{aligned}(\theta_1 - \mathrm{LIM})[\mathcal{X}_1 \mid (\alpha \mid \mathbf{U})] &= \bigcap_{H \in \mathcal{X}_1} \mathrm{cl}((\alpha \mid \mathbf{U})^1(H), \theta_1) \\ &= \bigcap_{H \in \mathcal{X}_1} (\mathrm{cl}((\alpha \mid \mathbf{U})^1(H), \tilde{\tau}_1) \cap \mathbb{K}) \subset \bigcap_{H \in \mathcal{X}_1} \mathrm{cl}((\alpha \mid \mathbf{U})^1(H), \tilde{\tau}_1) \\ &= \bigcap_{H \in \mathcal{X}_1} \mathrm{cl}(\alpha^1(H), \tilde{\tau}_1) = \bigcap_{T \in \mathcal{X}} \mathrm{cl}(\alpha^1(T \cap \mathbf{U}), \tilde{\tau}_1) \\ &\subset \bigcap_{T \in \mathcal{X}} \mathrm{cl}(\alpha^1(T), \tilde{\tau}_1) = (\tilde{\tau}_1 - \mathrm{LIM})[\mathcal{X} \mid \alpha]. \end{aligned} \qquad (3.3.30)$$

In (3.3.30) we use (2.3.1) and (3.2.8). From (3.3.29) and (3.3.30) we obtain the inclusion

$$(\theta_2 - \mathrm{LIM})[\mathcal{X}_1 \mid \tilde{\beta} \circ (\alpha \mid \mathbf{U})] \subset \beta^1((\tilde{\tau}_1 - \mathrm{LIM})[\mathcal{X} \mid \alpha]). \qquad (3.3.31)$$

Note that $w$ is an element of the set (3.3.29). Indeed, $\beta \circ \alpha \circ \varphi \in W^{\mathbf{D}}$ is the function for which $(\beta \circ \alpha \circ \varphi)(d) = \beta(\alpha(\varphi(d)))$ for $d \in \mathbf{D}$. On the other hand,

$$\left(\varphi \in \mathbf{U}^{\mathbf{D}}\right) \& \left((\alpha \mid \mathbf{U}) \in \mathbb{K}^{\mathbf{U}}\right) \& \left(\tilde{\beta} \in \mathbb{W}^{\mathbb{K}}\right), \tag{3.3.32}$$

where $\mathbf{U} \subset X$, $\mathbb{K} \subset V$ and $\mathbb{W} \subset W$. From (3.3.32) we have

$$\tilde{\beta} \circ (\alpha \mid \mathbf{U}) \circ \varphi \in \mathbb{W}^{\mathbf{D}} \tag{3.3.33}$$

and, in particular, $\tilde{\beta} \circ (\alpha \mid \mathbf{U}) \circ \varphi \in W^{\mathbf{D}}$. Let us compare values of the functions $\beta \circ \alpha \circ \varphi$ and $\tilde{\beta} \circ (\alpha \mid \mathbf{U}) \circ \varphi$. For $d \in \mathbf{D}$ we have

$$\begin{aligned}(\tilde{\beta} \circ (\alpha \mid \mathbf{U}) \circ \varphi)(d) &= \tilde{\beta}((\alpha \mid \mathbf{U})(\varphi(d))) = \tilde{\beta}(\alpha(\varphi(d))) \\ &= \beta(\alpha(\varphi(d))) = (\beta \circ \alpha \circ \varphi)(d).\end{aligned}$$

Indeed, $\varphi(d) \in \mathbf{U}$ and as a consequence $(\alpha \mid \mathbf{U})(\varphi(d)) = \alpha(\varphi(d)) \in \mathbb{K}$; hence $\tilde{\beta}((\alpha \mid \mathbf{U})(\varphi(d))) = \tilde{\beta}(\alpha(\varphi(d))) = \beta(\alpha(\varphi(d)))$. We have established that

$$\tilde{\beta} \circ (\alpha \mid \mathbf{U}) \circ \varphi = \beta \circ \alpha \circ \varphi. \tag{3.3.34}$$

From (3.3.24) and (3.3.34) we obtain the convergence

$$(\mathbf{D}, \preceq, \tilde{\beta} \circ (\alpha \mid \mathbf{U}) \circ \varphi) \overset{\tilde{\tau}_2}{\to} w. \tag{3.3.35}$$

Note that $w \in \mathbb{W}$. Indeed, from (2.3.20) and the compactness of $\mathbb{W}$ in the Hausdorff space $(W, \tilde{\tau}_2)$ it follows that $\mathbb{W} \in \mathcal{F}_{\tilde{\tau}_2}$ and as a consequence $\mathbb{W} = \mathrm{cl}(\mathbb{W}, \tilde{\tau}_2)$. From (2.3.11), (3.3.33) and (3.3.35) we obtain $w \in \mathbb{W}$. Therefore from (2.3.9), (3.3.26), (3.3.33) and (3.3.35) it follows that

$$(\mathbf{D}, \preceq, \tilde{\beta} \circ (\alpha \mid \mathbf{U}) \circ \varphi) \overset{\theta_2}{\to} w. \tag{3.3.36}$$

Let us return to the first statement of (3.3.24). We use (2.2.47). Namely, $\forall H \in \mathcal{X}\ \exists d \in \mathbf{D}\ \forall \delta \in \mathbf{D}$:

$$(d \preceq \delta) \Rightarrow (\varphi(\delta) \in H).$$

By (2.3.1) and the definition of $\mathcal{X}_1$ we have $\forall \tilde{H} \in \mathcal{X}_1\ \exists d \in \mathbf{D}\ \forall \delta \in \mathbf{D}$:

$$(d \preceq \delta) \Rightarrow (\varphi(\delta) \in \tilde{H}). \tag{3.3.37}$$

In (3.3.37) we use the property that the inclusion $\varphi(d) \in \mathbf{U}$ holds for $\delta \in \mathbf{D}$. From (3.3.37) we obtain

$$\mathcal{X}_1 \subset (\mathbf{U} - \mathrm{ass})[\mathbf{D}; \preceq; \varphi]. \tag{3.3.38}$$

In particular, from (3.3.36) and (3.3.38) we obtain that $w \in \mathbb{W}$ is a point for which we can choose some net $(D, \angle, f)$ in $\mathbf{U}$ with the following properties:

$$\Big(\mathcal{X}_1 \subset (\mathbf{U} - \mathrm{ass})[D; \angle; f]\Big) \,\&\, \Big((D, \angle, \tilde{\beta} \circ (\alpha \mid \mathbf{U}) \circ f) \xrightarrow{\theta_2} w\Big). \quad (3.3.39)$$

From (3.2.8) and (3.3.39) we obtain

$$w \in (\theta_2 - \mathrm{LIM})[\mathcal{X}_1 \mid \tilde{\beta} \circ (\alpha \mid \mathbf{U})].$$

From (3.3.31) we have the required inclusion

$$w \in \beta^1((\tilde{\tau}_1 - \mathrm{LIM})[\mathcal{X} \mid \alpha]). \quad (3.3.40)$$

Since the choice of $w$ was arbitrary, from (3.3.40) we obtain the inclusion (3.3.23). □

Of course, we have the most interesting case when (3.3.23) is the equality. This fact is connected with the property similar to the local compactness of model space. We use (2.9.1).

PROPOSITION 3.3.6 $\forall_X \mathbf{S}[X \neq \varnothing] \forall_V \mathbf{S}[V \neq \varnothing] \forall_W \mathbf{S}[W \neq \varnothing]\, \forall \mathcal{X} \in \mathcal{B}(X)$ $\forall \tilde{\tau}_1 \in (\mathbf{c} - \mathrm{top})_{\mathrm{loc}}[V]\ \forall \tilde{\tau}_2 \in (\mathrm{top})_0[W]\ \forall \alpha \in V^X\ \forall \beta \in C(V, \tilde{\tau}_1, W, \tilde{\tau}_2)$:

$$((\mathbf{c}\tilde{\tau}_1, \tilde{\tau}_2) - \mathrm{LIM})[\mathcal{X} \mid \alpha; \beta \circ \alpha] = \beta^1((\tilde{\tau}_1 - \mathrm{LIM})[\mathcal{X} \mid \alpha]).$$

PROOF. Fix $X, V, W, \mathcal{X}, \tilde{\tau}_1, \tilde{\tau}_2, \alpha$ and $\beta$ in correspondence with the conditions of the given proposition. We use Proposition 3.3.5 (see (3.3.23)). Choose arbitrarily

$$w \in \beta^1((\tilde{\tau}_1 - \mathrm{LIM})[\mathcal{X} \mid \alpha]). \quad (3.3.41)$$

Using (3.3.9) we choose also $v \in (\tilde{\tau}_1 - \mathrm{LIM})[\mathcal{X} \mid \alpha]$ with the property $w = \beta(v)$. Then $v \in V$ and one can choose some net $(\mathbf{D}, \preceq, \mathbf{g})$ in the set $X$ for which

$$\Big(\mathcal{X} \subset (X - \mathrm{ass})[\mathbf{D}; \preceq; \mathbf{g}]\Big) \,\&\, \Big((\mathbf{D}, \preceq, \alpha \circ \mathbf{g}) \xrightarrow{\tilde{\tau}_1} v\Big). \quad (3.3.42)$$

We use (3.2.8) in this representation. Here $\mathbf{g} \in X^{\mathbf{D}}$ and $\alpha \circ \mathbf{g} \in V^{\mathbf{D}}$. Using the property $\tilde{\tau}_1 \in (\mathbf{c} - \mathrm{top})_{\mathrm{loc}}[V]$, we choose (see (2.9.1)) a set

$$\mathbf{C} \in N_{\tilde{\tau}_1}(v) \cap (\tilde{\tau}_1 - \mathrm{comp})[V]. \quad (3.3.43)$$

By (3.3.42) and (3.3.43) we obtain that $\mathbf{C} \in (V - \mathrm{ass})[\mathbf{D}; \preceq; \alpha \circ \mathbf{g}]$. We choose $\mathbf{d} \in \mathbf{D}$ for which $\forall d \in \mathbf{D} : (\mathbf{d} \preceq d) \Rightarrow ((\alpha \circ \mathbf{g})(d) \in \mathbf{C})$. Then we introduce $\mathbf{D}_0 \triangleq \{d \in \mathbf{D} \mid \mathbf{d} \preceq d\} \in 2^{\mathbf{D}}$. So $\mathbf{S}[\mathbf{D}_0 \neq \varnothing]$. Let $\ll \in (\mathrm{DIR})[\mathbf{D}_0]$ be the relation for which $\forall d_1 \in \mathbf{D}_0\ \forall d_2 \in \mathbf{D}_0$ def:

$$(d_1 \ll d_2) \Leftrightarrow (d_1 \preceq d_2). \tag{3.3.44}$$

Of course, (3.3.44) determines a concrete binary relation in $\mathbf{D}_0$. But the set $\mathbf{D}_0$ possesses the following property: $\mathbf{D}_0 \in (\preceq -\mathrm{cof})[\mathbf{D}]$ (see (2.2.7)). Therefore it is possible to use (2.2.8):

$$\preceq \cap (\mathbf{D}_0 \times \mathbf{D}_0) \in (\mathrm{DIR})[\mathbf{D}_0].$$

Suppose that by definition $\ll$ is the direction servicing (3.3.44): $\ll \triangleq \preceq \cap(\mathbf{D}_0 \times \mathbf{D}_0)$. Then (3.3.44) holds. In the form of $(\mathbf{D}_0, \ll)$ we have a nonempty directed set (we note that $\mathbf{d} \in \mathbf{D}_0$). In addition, $\tilde{\mathbf{g}} \triangleq (\mathbf{g} \mid \mathbf{D}_0) \in X^{\mathbf{D}_0}$; $\tilde{\mathbf{g}}(\delta) = \mathbf{g}(\delta)$ for $\delta \in \mathbf{D}_0$. Moreover, by the definition of $\mathbf{D}_0$ we have $\forall d \in \mathbf{D}_0 : (\alpha \circ \mathbf{g})(d) \in \mathbf{C}$. In other words, $\alpha \circ \tilde{\mathbf{g}} \in \mathbf{C}^{\mathbf{D}_0}$. So $(\mathbf{D}_0, \ll, \alpha \circ \tilde{\mathbf{g}})$ is a net in $\mathbf{C}$. From (3.3.42) we have the property

$$(\mathbf{D}_0, \ll, \alpha \circ \tilde{\mathbf{g}}) \xrightarrow{\tilde{\tau}_1} v. \tag{3.3.45}$$

Indeed, fix $H_* \in N_{\tilde{\tau}_1}(v)$. Then by (3.3.42) we can choose $d_* \in \mathbf{D}$ such that $\forall d \in \mathbf{D}$:

$$(d_* \preceq d) \Rightarrow (\alpha(\mathbf{g}(d)) \in H_*).$$

Using (2.2.4) we choose $d^* \in \mathbf{D}$ for which $(\mathbf{d} \preceq d^*)\ \&\ (d_* \preceq d^*)$. Then $d^* \in \mathbf{D}_0$. If $\tilde{\delta} \in \mathbf{D}_0$ satisfies the condition $d^* \ll \tilde{\delta}$, then $d_* \preceq \tilde{\delta}$ and as a consequence

$$(\alpha \circ \tilde{\mathbf{g}})(\tilde{\delta}) = \alpha(\tilde{\mathbf{g}}(\tilde{\delta})) = \alpha(\mathbf{g}(\tilde{\delta})) \in H_*.$$

Since the choice of $\tilde{\delta}$ was arbitrary, we have $H_* \in (V - \mathrm{ass})[\mathbf{D}_0; \ll; \alpha \circ \tilde{\mathbf{g}}]$. We use (2.2.47) in this statement. But the choice of $H_*$ was arbitrary. Hence $N_{\tilde{\tau}_1}(v) \subset (V - \mathrm{ass})[\mathbf{D}_0; \ll; \alpha \circ \tilde{\mathbf{g}}]$. By (2.2.48) we obtain the convergence (3.3.45). Recall that $\alpha \circ \tilde{\mathbf{g}} \in V^{\mathbf{D}_0}$. From (2.5.4) we obtain the convergence

$$(\mathbf{D}_0, \ll, \beta \circ \alpha \circ \tilde{\mathbf{g}}) \xrightarrow{\tilde{\tau}_2} \beta(v). \tag{3.3.46}$$

In an analogous way, from (3.3.42) we have the inclusion

$$\mathcal{X} \subset (X - \mathrm{ass})[\mathbf{D}_0; \ll; \tilde{\mathbf{g}}]. \tag{3.3.47}$$

Indeed, let $A_* \in \mathcal{X}$. By (3.3.42) we have $A_* \in (X - \mathrm{ass})[\mathbf{D}; \preceq; \mathbf{g}]$. Then by (2.2.47) we obtain that $A_* \in \mathcal{P}(X)$ possesses the property: for some $\delta_* \in \mathbf{D}$ the implication

$$(\delta_* \preceq \delta) \Rightarrow (\mathbf{g}(\delta) \in A_*) \tag{3.3.48}$$

holds for $\delta \in \mathbf{D}$. Using (2.2.4) we choose $\delta^* \in \mathbf{D}$ such that $(\mathbf{d} \preceq \delta^*)$ & $(\delta_* \preceq \delta^*)$. Then $\delta^* \in \mathbf{D}_0$. Let $\delta' \in \mathbf{D}_0$ possesses the property $\delta^* \ll \delta'$. By (3.3.44) we have $\delta^* \preceq \delta'$ and as a consequence $\delta_* \preceq \delta'$. Then by (3.3.48) we have $\tilde{\mathbf{g}}(\delta') = \mathbf{g}(\delta') \in A_*$. Since the choice of $\delta'$ was arbitrary, we obtain that

$$\exists \delta_1 \in \mathbf{D}_0 \; \forall \delta_2 \in \mathbf{D}_0 : (\delta_1 \ll \delta_2) \Rightarrow (\tilde{\mathbf{g}}(\delta_2) \in A_*).$$

Hence $A_* \in (X - \text{ass})[\mathbf{D}_0; \ll; \tilde{\mathbf{g}}]$. Since the choice of $A_*$ was arbitrary, we have (3.3.47). Thus $(\mathbf{D}, \ll, \tilde{\mathbf{g}})$ is a net in $X$ for which the relations (3.3.46) and (3.3.47) are valid. Moreover, we note that $(\alpha \circ \tilde{\mathbf{g}})^1(\mathbf{D}_0) \subset \mathbf{C}$, and by (3.2.9)

$$\tilde{\mathbf{g}} \in \mathcal{M}_{\mathbf{c}}(\mathbf{D}_0, X, V, \tilde{\tau}_1, \alpha). \tag{3.3.49}$$

By (3.2.10), (3.3.46), (3.3.47) and (3.3.49) we obtain

$$w = \beta(v) \in ((\mathbf{c}\tilde{\tau}_1, \tilde{\tau}_2) - \text{LIM})[\mathcal{X} \mid \alpha; \beta \circ \alpha].$$

Since the choice of $w$ was arbitrary, the inclusion

$$\beta^1((\tilde{\tau}_1 - \text{LIM})[\mathcal{X} \mid \alpha]) \subset ((\mathbf{c}\tilde{\tau}_1, \tilde{\tau}_2) - \text{LIM})[\mathcal{X} \mid \alpha; \beta \circ \alpha]$$

is established. From this inclusion and Proposition 3.3.5 we have the required equality. □

We have obtained several variants of the representation of AS in terms of some auxiliary AS in a model TS. But in Section 3.2 we have established a number of representations of these auxiliary AS. In the next section we connect two above types of representations. In addition, we return to the informative setting of Section 3.1.

## 3.4 STANDARD REALIZATION OF ATTRACTION SETS. I

We follow the notation of Section 3.1 (see, in particular, (3.1.1)). Let

$$(\mathbf{S}[\mathbf{F} \neq \varnothing]) \; \& \; (\mathbf{S}[\mathbf{X} \neq \varnothing]) \; \& \; (\mathbf{S}[\mathbf{H} \neq \varnothing]). \tag{3.4.1}$$

Moreover, let $\mathbf{Y} \in \mathcal{P}(\mathbf{X})$, $s \in \mathbf{X}^{\mathbf{F}}$ and $h \in \mathbf{H}^{\mathbf{F}}$. The regularization of the problem of constructing the set (3.1.2) is our main goal. For this we equip the sets $\mathbf{X}$ and $\mathbf{H}$ with topologies $\tau^{(1)}$ and $\tau^{(2)}$ respectively. So

$$(\tau^{(1)} \in (\text{top})[\mathbf{X}]) \; \& \; (\tau^{(2)} \in (\text{top})[\mathbf{H}]). \tag{3.4.2}$$

Questions concerning the concrete choice of the topologies (3.4.2) are discussed in Section 3.1. As a result we obtain two TS: $(\mathbf{X}, \tau^{(1)})$ and $(\mathbf{H}, \tau^{(2)})$. For brevity suppose that

$$\mathcal{Y} \triangleq \mathbb{N}_{\tau^{(1)}}[\mathbf{Y}]. \tag{3.4.3}$$

Then in terms of $\mathcal{Y}$ (3.4.3) it is possible to introduce the set

$$(\tau^{(2)} - \text{LIM})\left[s^{-1}[\mathcal{Y}] \mid h\right] \in \mathcal{F}_{\tau^{(2)}} \tag{3.4.4}$$

as some natural regularization of (3.1.2). But constructing (3.4.4) can be sufficiently complicated. On the other hand, the replacement of $\mathbf{Y}$ by the family $\mathcal{Y}$ of all neighborhoods of $\mathbf{Y}$ is natural far not always. In this connection see Section 3.2.

Let us introduce a model oriented to employing (3.1.12) as an AS. This model corresponds to (3.1.6), (3.1.8), and (3.1.10). Let $\mathbf{S}[\mathbf{K} \neq \varnothing]$ and $\tau^{(3)} \in (\text{top})[\mathbf{K}]$. Moreover, let

$$\left(m \in \mathbf{K}^{\mathbf{F}}\right) \;\&\; \left(g \in C(\mathbf{K}, \tau^{(3)}, \mathbf{X}, \tau^{(1)})\right) \;\&\; \left(\omega \in C(\mathbf{K}, \tau^{(3)}, \mathbf{H}, \tau^{(2)})\right). \tag{3.4.5}$$

Following (3.1.8) and (3.1.9) we postulate that the model (3.1.10) possesses the properties

$$(s = g \circ m) \;\&\; (h = \omega \circ m) \;\&\; (\mathbf{K} = \text{cl}(m^1(\mathbf{F}), \tau^{(3)})). \tag{3.4.6}$$

In this section we consider the case when (3.1.10) is given and the relations (3.4.5) and (3.4.6) are valid (the problem of the search of such model (3.1.10) is considered in other chapters).

**Remark.** Note that, in fact, the last requirement in (3.4.6) is not a constraint. Indeed, let the procession (3.1.10) satisfy only (3.4.5) and the two first conditions in (3.4.6). We admit that the last equality in (3.4.6) can be violated. We introduce $\mathbf{K}_1 \triangleq \text{cl}(m^1(\mathbf{F}), \tau^{(3)})$. Of course, $\mathbf{S}[\mathbf{K}_1 \neq \varnothing]$. Hence $\mathbf{K}_1 \in 2^{\mathbf{K}}$. Moreover, let

$$\tau_1^{(3)} \triangleq \tau^{(3)} \mid_{\mathbf{K}_1} \in (\text{top})[\mathbf{K}_1].$$

So $(\mathbf{K}_1, \tau_1^{(3)})$ is a natural subspace of $(\mathbf{K}, \tau^{(3)})$. Further, by (2.5.30) we have the properties

$$\left(g_1 \triangleq (g \mid \mathbf{K}_1) \in C(\mathbf{K}_1, \tau_1^{(3)}, \mathbf{X}, \tau^{(1)})\right)$$

$$\&\; \left(\omega_1 \triangleq (\omega \mid \mathbf{K}_1) \in C(\mathbf{K}_1, \tau_1^{(3)}, \mathbf{H}, \tau^{(2)})\right). \tag{3.4.7}$$

Of course, we also use (3.4.5). Note that $m^1(\mathbf{F}) \subset \mathbf{K}_1$. Hence $m : \mathbf{F} \to \mathbf{K}_1$. We now replace (3.1.10) by the new model

$$(\mathbf{K}_1, \tau_1^{(3)}, m, g_1, \omega_1). \tag{3.4.8}$$

In addition, $s = g_1 \circ m$ and $h = \omega_1 \circ m$. Indeed, by (3.4.7) $g_1 \circ m \in \mathbf{X}^{\mathbf{F}}$ and $(g_1 \circ m)(f) = g_1(m(f)) = g(m(f)) = (g \circ m)(f) = s(f)$ for $f \in \mathbf{F}$. Analogously, the equality $h = \omega_1 \circ m$ holds. Finally, by (2.3.13) we have the equality

$$\mathrm{cl}(m^1(\mathbf{F}), \tau_1^{(3)}) = \mathrm{cl}(m^1(\mathbf{F}), \tau^{(3)}) \cap \mathbf{K}_1 = \mathbf{K}_1.$$

Thus we obtain

$$(s = g_1 \circ m) \ \& \ (h = \omega_1 \circ m) \ \& \ (\mathbf{K}_1 = \mathrm{cl}(m^1(\mathbf{F}), \tau_1^{(3)})). \qquad (3.4.9)$$

These properties permit us to consider (3.4.8) instead of (3.1.10), preserving the corresponding analogue of (3.4.6) in the form of (3.4.9).

Let us return to the general case of the model (3.1.10) with the properties (3.4.5) and (3.4.6).

PROPOSITION 3.4.1 *The following chain of inclusions holds:*

$$\begin{aligned} h^1(s^{-1}(\mathbf{Y})) \subset \omega^1(g^{-1}(\mathbf{Y})) \subset \omega^1((\tau^{(3)} - \mathrm{LIM})[s^{-1}[\mathcal{Y}] \mid m]) \\ \subset (\tau^{(2)} - \mathrm{LIM})[s^{-1}[\mathcal{Y}] \mid h]. \end{aligned} \qquad (3.4.10)$$

PROOF. Let $z \in h^1(s^{-1}(\mathbf{Y}))$. Choose $f \in s^{-1}(\mathbf{Y})$ with the property $z = h(f)$. By (3.4.5) $m(f) \in \mathbf{K}$. In addition, by (3.4.6) $s(f) = g(m(f)) \in \mathbf{Y}$ and as a consequence $m(f) \in g^{-1}(\mathbf{Y})$. But from (3.4.6) it follows that $z = \omega(m(f)) \in \omega^1(g^{-1}(\mathbf{Y}))$. Since the choice of $z$ was arbitrary, we obtain the inclusion

$$h^1(s^{-1}(\mathbf{Y})) \subset \omega^1(g^{-1}(\mathbf{Y})). \qquad (3.4.11)$$

From (3.4.11) we have the first inclusion in (3.4.10). To prove the second one we use Proposition 3.2.3. In addition, in the conditions of this proposition we set $X = \mathbf{F}$, $(U, \tau_1) = (\mathbf{X}, \tau^{(1)})$, $(V, \tau_2) = (\mathbf{K}, \tau^{(3)})$, $p = m$, $q = g$, and $\mathbf{M} = \mathbf{Y}$. Employing (3.4.5) and (3.4.6) we obtain that all the conditions of Proposition 3.2.3 are valid. Therefore from (3.2.35) we have $g^{-1}(\mathbf{Y}) \subset (\tau^{(3)} - \mathrm{LIM})[s^{-1}[\mathcal{Y}] \mid m]$ (of course, we use (3.4.3) and (3.4.6)). From this inclusion we obtain

$$\omega^1(g^{-1}(\mathbf{Y})) \subset \omega^1((\tau^{(3)} - \mathrm{LIM})[s^{-1}[\mathcal{Y}] \mid m]). \qquad (3.4.12)$$

To prove the last inclusion in (3.4.10) we use Proposition 3.3.1, setting in its conditions $X = \mathbf{F}$, $\mathcal{X} = s^{-1}[\mathcal{Y}]$, $(V, \tilde{\tau}_1) = (\mathbf{K}, \tau^{(3)})$, $(W, \tilde{\tau}_2) = (\mathbf{H}, \tau^{(2)})$, $\alpha = m$, $\beta = \omega$. Therefore from (3.3.1) we obtain the inclusion

$$\omega^1((\tau^{(3)} - \mathrm{LIM})[s^{-1}[\mathcal{Y}] \mid m]) \subset (\tau^{(2)} - \mathrm{LIM})[s^{-1}[\mathcal{Y}] \mid h]. \qquad (3.4.13)$$

The chain (3.4.10) follows from (3.4.11)–(3.4.13).

PROPOSITION 3.4.2 *Suppose that* $(\mathbf{H}, \tau^{(2)})$ *is a Hausdorff TS, i.e.* $\tau^{(2)} \in (\mathrm{top})_0[\mathbf{H}]$. *Then*

$$\begin{aligned} h^1(s^{-1}(\mathbf{Y})) &\subset ((\mathbf{c}\tau^{(3)}, \tau^{(2)}) - \mathrm{LIM})[s^{-1}[\mathcal{Y}] \mid m; h] \\ &\subset \omega^1((\tau^{(3)} - \mathrm{LIM})[s^{-1}[\mathcal{Y}] \mid m]). \end{aligned} \tag{3.4.14}$$

PROOF. Let $z \in h^1(s^{-1}(\mathbf{Y}))$ and $f \in s^{-1}(\mathbf{Y})$ have the property $z = h(f)$. Introduce the sequence $\mathbf{f} \triangleq (f_i)_{i \in \mathcal{N}} \in \mathbf{F}^{\mathcal{N}}$ by the following rule: $f_j \triangleq f$ for $j \in \mathcal{N}$. Thus $\mathbf{f}$ is a stationary sequence in $\mathbf{F}$. Consider the net $(\mathcal{N}, \leq_{\mathcal{N}}, \mathbf{f})$ in $\mathbf{F}$ and the singleton $\{m(f)\} \in (\tau^{(3)} - \mathrm{comp})[\mathbf{K}]$. Then $(m \circ \mathbf{f})^1(\mathcal{N}) = \{m(f)\}$ and $\mathbf{f} \in \mathcal{M}_{\mathbf{c}}(\mathcal{N}, \mathbf{F}, \mathbf{K}, \tau^{(3)}, m)$. We note that by (3.4.3) the inclusion $\mathbf{Y} \subset H$ holds for $H \in \mathcal{Y}$. But $s(f) \in \mathbf{Y}$. Hence $\forall H \in \mathcal{Y} : s(f) \in H$. We obtain the obvious property: if $H \in \mathcal{Y}$, then $\mathbf{f}$ is a sequence in $s^{-1}(H)$. As a consequence $(\mathcal{N}, \leq_{\mathcal{N}}, \mathbf{f})$ is a net in $s^{-1}(H)$. Of course, $\forall H \in \mathcal{Y} : s^{-1}(H) \in (\mathbf{F} - \mathrm{ass})[\mathcal{N}; \leq_{\mathcal{N}}; \mathbf{f}]$. We obtain the following inclusion

$$s^{-1}[\mathcal{Y}] \subset (\mathbf{F} - \mathrm{ass})[\mathcal{N}; \leq_{\mathcal{N}}; \mathbf{f}]. \tag{3.4.15}$$

We note that $h \circ \mathbf{f}$ is a stationary sequence in $\mathbf{H}$: $(h \circ \mathbf{f})(i) = h(f)$ for $i \in \mathcal{N}$. Of course,

$$(\mathcal{N}, \leq_{\mathcal{N}}, h \circ \mathbf{f}) \stackrel{\tau^{(2)}}{\to} z. \tag{3.4.16}$$

From (3.2.10), (3.4.15) and (3.4.16) it follows that $z \in ((\mathbf{c}\tau^{(3)}, \tau^{(2)}) - \mathrm{LIM})[s^{-1}[\mathcal{Y}] \mid m; h]$. Since the choice of $z$ was arbitrary, the inclusion

$$h^1(s^{-1}(\mathbf{Y})) \subset ((\mathbf{c}\tau^{(3)}, \tau^{(2)}) - \mathrm{LIM})[s^{-1}[\mathcal{Y}] \mid m; h] \tag{3.4.17}$$

is established.

Let us now consider the following version of Proposition 3.3.5: $X = \mathbf{F}$, $\mathcal{X} = s^{-1}[\mathcal{Y}]$, $(V, \tilde{\tau}_1) = (\mathbf{K}, \tau^{(3)})$, $(W, \tilde{\tau}_2) = (\mathbf{H}, \tau^{(2)})$, $\alpha = m$, $\beta = \omega$. Then by Proposition 3.3.5 we obtain the inclusion

$$((\mathbf{c}\tau^{(3)}, \tau^{(2)}) - \mathrm{LIM})[s^{-1}[\mathcal{Y}] \mid m; h] \subset \omega^1((\tau^{(3)} - \mathrm{LIM})[s^{-1}[\mathcal{Y}] \mid m]). \tag{3.4.18}$$

From (3.4.17) and (3.4.18) we obtain (3.4.14).

PROPOSITION 3.4.3 *Let* $\tau^{(2)} \in (\mathrm{top})_0[\mathbf{H}]$ *and* $\tau^{(3)} \in (\mathbf{c} - \mathrm{top})_{\mathrm{loc}}[\mathbf{K}]$. *Then* $\forall \mathcal{X} \in \mathcal{B}[\mathbf{F}]$:

$$((\mathbf{c}\tau^{(3)}, \tau^{(2)}) - \mathrm{LIM})[\mathcal{X} \mid m; h] = \omega^1((\tau^{(3)} - \mathrm{LIM})[\mathcal{X} \mid m]).$$

The proof follows from Proposition 3.3.6 with $X = \mathbf{F}$, $(V, \tilde{\tau}_1) = (\mathbf{K}, \tau^{(3)})$, $(W, \tilde{\tau}_2) = (\mathbf{H}, \tau^{(2)})$, $\alpha = m$, $\beta = \omega$.

COROLLARY 3.4.1 *Let* $\tau^{(2)} \in (\mathrm{top})_0[\mathbf{H}]$ *and* $\tau^{(3)} \in (\mathbf{c}-\mathrm{top})_{\mathrm{loc}}[\mathbf{K}]$. *Then*

$$((\mathbf{c}\tau^{(3)}, \tau^{(2)}) - \mathrm{LIM})[s^{-1}[\mathcal{Y}] \mid m; h] = \omega^1((\tau^{(3)} - \mathrm{LIM})[s^{-1}[\mathcal{Y}] \mid m]).$$

The proof is obvious (see (3.2.10) and (3.4.3)). From the two last propositions one can see that the auxiliary AS $(\tau^{(3)} - \mathrm{LIM})[s^{-1}[\mathcal{Y}] \mid m]$ plays the very important role. To investigate the structure of this AS we introduce the following

CONDITION 3.4.1 $\forall \mathbf{x} \in \mathbf{X} \setminus \mathbf{Y}\ \exists H_1 \in N_{\tau^{(1)}}(\mathbf{x})\ \exists H_2 \in \mathcal{Y} : H_1 \cap H_2 = \varnothing.$

As for examples of problems for which this condition is valid, see the remark in the previous section.

PROPOSITION 3.4.4 *Let Condition 3.4.1 be valid. Then*

$$g^{-1}(\mathbf{Y}) = (\tau^{(3)} - \mathrm{LIM})[s^{-1}[\mathcal{Y}] \mid m]. \tag{3.4.19}$$

The proof follows from Proposition 3.2.4 with $X = \mathbf{F}$, $(U, \tau_1) = (\mathbf{X}, \tau^{(1)})$, $(V, \tau_2) = (\mathbf{K}, \tau^{(3)})$, $p = m$, $q = g$, and $M = \mathbf{Y}$. From (3.4.5), (3.4.6) and Condition 3.4.1 we obtain (3.2.34), (3.2.41) and as a consequence the required equality (3.4.19). Of course, we use (3.4.3) and (3.4.6).

THEOREM 3.4.1 *Let Condition 3.4.1 be valid. Moreover, let* $\omega \in C_{\mathbf{ap}}(\mathbf{K}, \tau^{(3)}, \mathbf{H}, \tau^{(2)})$. *Then*

$$\omega^1(g^{-1}(\mathbf{Y})) = \omega^1((\tau^{(3)} - \mathrm{LIM})[s^{-1}[\mathcal{Y}] \mid m]) = (\tau^{(2)} - \mathrm{LIM})[s^{-1}[\mathcal{Y}] \mid h]. \tag{3.4.20}$$

PROOF. We use (3.3.12) and Proposition 3.4.4. Indeed, in the conditions defining (3.3.12) we set: $X = \mathbf{F}$, $\mathcal{X} = s^{-1}[\mathcal{Y}]$, $(V, \tilde{\tau}_1) = (\mathbf{K}, \tau^{(3)})$, $(W, \tilde{\tau}_2) = (\mathbf{H}, \tau^{(2)})$, $\alpha = m$, $\beta = \omega$. By (3.3.12) and (3.4.6) we obtain the equality

$$(\tau^{(2)} - \mathrm{LIM})[s^{-1}[\mathcal{Y}] \mid h] = \omega^1((\tau^{(3)} - \mathrm{LIM})[s^{-1}[\mathcal{Y}] \mid m]).$$

Moreover, from Proposition 3.4.4 we have the equality

$$\omega^1((\tau^{(3)} - \mathrm{LIM})[s^{-1}[\mathcal{Y}] \mid m]) = \omega^1(g^{-1}(\mathbf{Y})).$$

From the two last equalities we obtain the statement of the theorem. □

**Remark.** In Theorem 3.4.1 we obtain the first important representation of the basic AS in terms of $\omega^1(g^{-1}(\mathbf{Y}))$ (3.1.12). The last set plays the role of an attainable set under the constraint (3.1.11). In fact, we have the new standard problem: it is required to define the possibilities to realize points $\omega(k)$ by enumerating the GE $k$ under the constraint (3.1.11).

PROPOSITION 3.4.5 *Let* $\tau^{(2)} \in (\text{top})_0[\mathbf{H}]$ *and Condition 3.4.1 be valid. Then*

$$h^1(s^{-1}(\mathbf{Y})) \subset ((\mathbf{c}\tau^{(3)}, \tau^{(2)}) - \text{LIM})[s^{-1}[\mathcal{Y}] \mid m; h] \subset \omega^1(g^{-1}(\mathbf{Y})).$$

The proof follows from Propositions 3.4.2 (see (3.4.14)) and 3.4.4.

THEOREM 3.4.2 *Let* $\tau^{(2)} \in (\text{top})_0[\mathbf{H}]$ *and* $\tau^{(3)} \in (\mathbf{c} - \text{top})_{\text{loc}}[\mathbf{K}]$. *Moreover, let Condition 3.4.1 be valid. Then*

$$((\mathbf{c}\tau^{(3)}, \tau^{(2)}) - \text{LIM})[s^{-1}[\mathcal{Y}] \mid m; h] = \omega^1(g^{-1}(\mathbf{Y})).$$

PROOF. Consider Proposition 3.3.6, setting in its conditions $X = \mathbf{F}$, $\mathcal{X} = s^{-1}[\mathcal{Y}]$, $(V, \tilde{\tau}_1) = (\mathbf{K}, \tau^{(3)})$, $(W, \tilde{\tau}_2) = (\mathbf{H}, \tau^{(2)})$, $\alpha = m$, $\beta = \omega$. By the statement of this proposition we have

$$((\mathbf{c}\tau^{(3)}, \tau^{(2)}) - \text{LIM})[s^{-1}[\mathcal{Y}] \mid m; h] = \omega^1((\tau^{(3)} - \text{LIM})[s^{-1}[\mathcal{Y}] \mid m]).$$

From Proposition 3.4.4 we obtain the required statement. □

Theorems 3.4.1 and 3.4.2 characterize the regularizing procedure as the replacement

$$h^1(s^{-1}(\mathbf{Y})) \to \omega^1(g^{-1}(\mathbf{Y})). \tag{3.4.21}$$

In the sequel we consider additional arguments in (3.4.21) favour. Now consider some modification of the two last theorems. In addition, we use neighborhood mappings.

Fix a nonempty set $\mathbf{Q}$ and a neighborhood mapping $\Lambda_0 \in (\text{UNIF})[\mathbf{Q}; \tau^{(1)} \mid \mathbf{Y}]$. In the TS $(\mathbf{X}, \tau^{(1)})$ introduce the family

$$\mathfrak{N} \triangleq \mathcal{U}(\mathbf{Q}, \tau^{(1)}, \mathbf{Y} \mid \Lambda_0) \in \mathcal{B}[\mathbf{X}] \tag{3.4.22}$$

with the property $\mathfrak{N} \subset \mathcal{Y}$. It is possible to consider $\mathfrak{N}$ (3.4.22) as an incomplete family of neighborhoods of $\mathbf{Y}$. Consider the variant of the weakening of the $\mathbf{Y}$-constraint, for which only neighborhoods from $\mathfrak{N}$ may be used. Note that by (3.2.13), (3.2.15) and (3.2.18) we have $\forall x \in \mathbf{X}$ $\forall q \in \mathbf{Q}$: $\Lambda_0(x, q) \in N_{\tau^{(1)}}(x)$. Of course, we use the obvious property

$$\Lambda_0 \in \mathcal{O}_{\mathbf{X}}(\mathbf{Q}, \tau^{(1)})$$

(see (3.4.2)). Thus we have established that

$$(\forall x \in \mathbf{X}\ \forall q \in \mathbf{Q} : \Lambda_0(x, q) \in N_{\tau^{(1)}}(x)) \ \&\ (\mathfrak{N} \subset \mathcal{Y}). \tag{3.4.23}$$

The property (3.4.23) is essential in connection with Condition 3.4.1. We introduce the following

CONDITION 3.4.2 $\forall \mathbf{x} \in \mathbf{X} \setminus \mathbf{Y}\ \exists q \in \mathbf{Q} : \Lambda_0(\mathbf{x}, q) \cap (\bigcup_{y \in \mathbf{Y}} \Lambda_0(y, q)) = \varnothing.$

**Remark.** Condition 3.4.2 is natural for many applications. Several variants of conditions of such a type were considered in Section 3.2. There the basic attention is given to pseudo-metrizable spaces (for example, see (3.2.17)). Moreover, it is possible to note some analogous constructions for the canonical bases of the Tichonoff product.

PROPOSITION 3.4.6 *Condition 3.4.2 implies Condition 3.4.1.*

The proof follows from (3.2.19) and (3.4.23). We use Proposition 3.4.6 in combination with Propositions 3.4.4, 3.4.5 and Theorems 3.4.1, 3.4.2.

PROPOSITION 3.4.7 *Let Condition 3.4.2 be valid. Then*

$$g^{-1}(\mathbf{Y}) = (\tau^{(3)} - \mathrm{LIM})[s^{-1}[\mathcal{Y}] \mid m] = (\tau^{(3)} - \mathrm{LIM})[s^{-1}[\mathfrak{N}] \mid m].$$

PROOF. From Propositions 3.4.4 and 3.4.6 we have (3.4.19). Further we use Proposition 3.2.1 under the following supposition: $X = \mathbf{F}$, $Q = \mathbf{Q}$, $(U, \tau_1) = (\mathbf{X}, \tau^{(1)})$, $(V, \tau_2) = (\mathbf{K}, \tau^{(3)})$, $\mathbf{M} = \mathbf{Y}$, $\Lambda = \Lambda_0$ $\alpha = s$, and $\beta = m$. Then by (3.4.5), (3.4.6) and Condition 3.4.2 we have (3.2.20). Therefore by (3.4.22) and Proposition 3.2.1 we obtain the equality

$$(\tau^{(3)} - \mathrm{LIM})[s^{-1}[\mathcal{Y}] \mid m] = (\tau^{(3)} - \mathrm{LIM})[s^{-1}[\mathfrak{N}] \mid m].$$

The proof is completed.

COROLLARY 3.4.2 *Let Condition 3.4.2 be valid. Then* $\forall \mathcal{X} \in \mathcal{B}[\mathbf{F}]$:

$$(s^{-1}[\mathfrak{N}] \subset \mathcal{X} \subset s^{-1}[\mathcal{Y}]) \Rightarrow \Big(g^{-1}(\mathbf{Y}) = (\tau^{(3)} - \mathrm{LIM})[\mathcal{X} \mid m]\Big).$$

The proof follows from (3.2.8) and Proposition 3.4.7.

The last property means an universality with respect to the choice of a concrete variant of the weakening of the $\mathbf{Y}$-constraint. The family $\mathcal{X}$ is directly connected with the choice of $f \in \mathbf{F}$. Note that in Proposition 3.4.7 and Corollary 3.4.2 some regularization of the problem of constructing the admissible set $s^{-1}(\mathbf{Y})$ is considered. We now begin the investigation of the problem of constructing the regularization of the attainable set $h^1(s^{-1}(\mathbf{Y}))$.

PROPOSITION 3.4.8 *Suppose that Condition 3.4.2 is valid and* $\mathfrak{N} \in \mathcal{B}_{\mathcal{N}}(\mathbf{X})$. *Moreover, let* $\tau^{(3)} \in (\mathbf{c}_{\mathcal{N}} - \mathrm{top})[\mathbf{K}]$ *and* $\tau^{(2)} \in (\mathcal{D} - \mathrm{top})[\mathbf{H}]$. *Finally, let* $\omega \in C_{\mathrm{cl}}(\mathbf{K}, \tau^{(3)}, \mathbf{H}, \tau^{(2)})$. *Then*

$$\omega^1(g^{-1}(\mathbf{Y})) = (\tau^{(2)} - \mathrm{LIM})[s^{-1}[\mathfrak{N}] \mid h]. \qquad (3.4.24)$$

PROOF. We use Proposition 3.3.3 under the following parameters: $X = \mathbf{F}$, $\mathcal{X} = s^{-1}[\mathfrak{N}]$, $(V, \tilde{\tau}_1) = (\mathbf{K}, \tau^{(3)})$, $(W, \tilde{\tau}_2) = (\mathbf{H}, \tau^{(2)})$, $\alpha = m$, and $\beta = \omega$. Of course, $s^{-1}[\mathfrak{N}] \in \mathcal{B}_{\mathcal{N}}(\mathbf{F})$. By Proposition 3.3.3 we obtain the equality

$$(\tau^{(2)} - \mathrm{LIM})[s^{-1}[\mathfrak{N}] \mid h] = \omega^1((\tau^{(3)} - \mathrm{LIM})[s^{-1}[\mathfrak{N}] \mid m]).$$

From Proposition 3.4.7 and the last equality we obtain the required representation (3.4.24).

PROPOSITION 3.4.9 *Let Condition 3.4.2 be valid. Moreover, let* $\mathfrak{N} \in \mathcal{B}_{\mathcal{N}}(\mathbf{X})$. *Finally, let* $\omega \in C_{\mathbf{qp}}(\mathbf{K}, \tau^{(3)}, \mathbf{H}, \tau^{(2)})$. *Then (3.4.24) holds.*

PROOF. We have $s^{-1}[\mathfrak{N}] \in \mathcal{B}_{\mathcal{N}}(\mathbf{F})$. From Proposition 3.4.7 it follows that $g^{-1}(\mathbf{Y}) = (\tau^{(3)} - \mathrm{LIM})[s^{-1}[\mathfrak{N}] \mid m]$. Hence we obtain the equality

$$\omega^1(g^{-1}(\mathbf{Y})) = \omega^1((\tau^{(3)} - \mathrm{LIM})[s^{-1}[\mathfrak{N}] \mid m]). \tag{3.4.25}$$

We use Proposition 3.3.4 under the following parameters: $X = \mathbf{F}$, $\mathcal{X} = s^{-1}[\mathfrak{N}]$, $(V, \tilde{\tau}_1) = (\mathbf{K}, \tau^{(3)})$, $(W, \tilde{\tau}_2) = (\mathbf{H}, \tau^{(2)})$, $\alpha = m$, and $\beta = \omega$. From Proposition 3.3.4 it follows that

$$(\tau^{(2)} - \mathrm{LIM})[s^{-1}[\mathfrak{N}] \mid h] = \omega^1((\tau^{(3)} - \mathrm{LIM})[s^{-1}[\mathfrak{N}] \mid m]). \tag{3.4.26}$$

From (3.4.25) and (3.4.26) we obtain (3.4.24).

**Remark.** Propositions 3.4.8 and 3.4.9 are oriented towards investigation of the very widespread case of the pseudo-metrizable and, in particular, metrizable TS $(\mathbf{X}, \tau^{(1)})$. Hence the family $\mathfrak{N}$ is very often defined as the family of all $\varepsilon$-neighborhoods of the set $\mathbf{Y}$. See a corresponding example in Section 3.2 (in particular, see (3.2.17)). In the particular case above the condition $\mathfrak{N} \in \mathcal{B}_{\mathcal{N}}(\mathbf{X})$ is valid (of course, we keep in mind that $\mathbf{Q}$ and $\Lambda_0$ correspond to this example; see (3.2.16)–(3.2.17)). In Propositions 3.4.8 and 3.4.9 we strive to use a countably compact space for $(\mathbf{K}, \tau^{(3)})$. We know that the requirement of compactness of TS is stronger (see (2.2.24) and examples of [71, Ch. 3]). As for the possibility of replacing $\mathfrak{N}$ by $\mathcal{Y}$ in the two last propositions, we note only the following

**Particular case.** Let $(\mathbf{X}, \tau^{(1)})$ be a pseudo-metrizable TS and $\rho \in (\mathbf{p}-\mathrm{Dist})[\mathbf{X}])$. Thus $\tau^{(1)} \triangleq \tau_\rho^\natural(\mathbf{X})$; see (2.7.5). In this remark we suppose that $\mathbf{Q} = ]0, \infty[$ and $\Lambda_0$ is such that $\forall x \in \mathbf{X}\ \forall q \in ]0, \infty[\colon\ \Lambda_0(x, q) = \mathbf{B}_\rho^0(x, q)$. Thus we follow the example of Section 3.2. We know that the basic requirement on the choice of $\Lambda_0$ is valid. Note that $\mathfrak{N}$ (3.4.22) is the family of all sets $\mathbf{B}_\rho^0[\mathbf{Y}; q]$, $q \in \mathbf{Q}$. But here we consider $\mathfrak{N}$ as an auxiliary family. Let $\mathbf{Y} \in (\mathrm{comp})[\mathbf{X}; \rho]$. Then $\forall G \in \mathbb{N}^0_{\tau^{(1)}}[\mathbf{Y}]\ \exists r \in ]0, \infty[$

$\forall y \in \mathbf{Y}$: $\mathbf{B}^0_\rho(y,r) \subset G$. This statement is a very particular case of the well known Lebesgue lemma on covering (see [81, Ch. 5]). By (2.7.16) we have $\forall H \in \mathcal{Y}\ \exists r \in ]0,\infty[$: $\mathbf{B}^0_\rho[\mathbf{Y};r] \subset H$. This property means a natural strengthening of (2.7.17). It is useful to note a connection with $\Lambda_0$ and $\mathfrak{N}$. Namely, in the case under consideration $\mathfrak{N}$ is a fundamental subfamily of $\mathcal{Y}$ (3.4.3): for $H_1 \in \mathcal{Y}$ one can choose a set $H_2 \in \mathfrak{N}$ such that $H_2 \subset H_1$. But $\mathfrak{N} \in \mathcal{B}_\mathcal{N}(\mathbf{X})$. Indeed, it is possible to set $\forall k \in \mathcal{N}: \tilde{H}_k \triangleq \mathbf{B}^0_\rho[\mathbf{Y};k^{-1}]$. Then $\forall H \in \mathcal{Y}\ \exists k \in \mathcal{N}: \tilde{H}_k \subset H$. Thus $\mathcal{Y} \in \mathcal{B}_\mathcal{N}(\mathbf{X})$ by (3.2.4). But we consider $s^{-1}[\mathcal{Y}]$. Then it is advisable to introduce the sequence

$$\left(s^{-1}(\tilde{H}_k)\right)_{k\in\mathcal{N}} : \mathcal{N} \to s^{-1}[\mathcal{Y}]$$

with the property of fundamentality: for any $U \in s^{-1}[\mathcal{Y}]$ one can choose $j \in \mathcal{N}$ for which $s^{-1}(\tilde{H}_j) \subset U$. We have established that in the considered case

$$s^{-1}[\mathcal{Y}] \in \mathcal{B}_\mathcal{N}(\mathbf{F}). \tag{3.4.27}$$

Using Propositions 3.3.3 and 3.3.4 (in the case above $\mathbf{Y} \in (\text{comp})[\mathbf{X};\rho]$), we obtain the following statement.

PROPOSITION 3.4.10 (*) *Let* $\tau^{(3)} \in (\mathbf{c}_\mathcal{N} - \text{top})[\mathbf{K}]$, $\tau^{(2)} \in (\mathcal{D} - \text{top})[\mathbf{H}]$ *and* $\omega \in C_{\text{cl}}(\mathbf{K},\tau^{(3)},\mathbf{H},\tau^{(2)})$. *Then*

$$(\tau^{(2)} - \text{LIM})[s^{-1}[\mathcal{Y}] \mid h] = \omega^1((\tau^{(3)} - \text{LIM})[s^{-1}[\mathcal{Y}] \mid m]). \tag{3.4.28}$$

PROOF. We use (3.4.27) and the following version of Proposition 3.3.3: $X = \mathbf{F}$, $\mathcal{X} = s^{-1}[\mathfrak{N}]$, $(V,\tilde{\tau}_1) = (\mathbf{K},\tau^{(3)})$, $(W,\tilde{\tau}_2) = (\mathbf{H},\tau^{(2)})$, $\alpha = m$, and $\beta = \omega$. From (3.4.6), (3.4.7) and Proposition 3.3.3 we obtain that $(\tau^{(2)} - \text{LIM})[s^{-1}[\mathcal{Y}] \mid h]$ and $\omega^1((\tau^{(3)} - \text{LIM})[s^{-1}[\mathcal{Y}] \mid m])$ coincide. Thus (3.4.28) is established.

PROPOSITION 3.4.11 (*) *Let* $\omega \in C_{\mathbf{qp}}(\mathbf{K},\tau^{(3)},\mathbf{H},\tau^{(2)})$. *Then (3.4.28) is valid.*

PROOF. We use (3.4.27) and Proposition 3.3.4. Indeed, consider the following version of Proposition 3.3.4: $X = \mathbf{F}$, $\mathcal{X} = s^{-1}[\mathcal{Y}]$, $(V,\tilde{\tau}_1) = (\mathbf{K},\tau^{(3)})$, $(W,\tilde{\tau}_2) = (\mathbf{H},\tau^{(2)})$, $\alpha = m$, and $\beta = \omega$. Then by (3.4.6), (3.4.27) and Proposition 3.3.4 we obtain (3.4.28). □

Of course, if we supplement the two last propositions by the requirement for Condition 3.4.1 to be valid, we obtain (in Propositions 3.4.10* and 3.4.11*) the chain of equalities (3.4.20). Here Proposition 3.4.4 is used. We note that in the considered particular case, Conditions 3.4.1 and 3.4.2 are equivalent. Recall that Propositions 3.4.10* and 3.4.11* are established under the following additional supposition with respect to

$(\mathbf{X}, \tau^{(1)})$ and $\mathbf{Y}$: $\tau^{(1)}$ is generated by the pseudo-metric $\rho \in (\mathbf{p}-\mathrm{Dist})[\mathbf{X}]$ (i.e., $\tau^{(1)} = \tau_\rho^\natural(\mathbf{X})$) for which $\mathbf{Y} \in (\mathrm{comp})[\mathbf{X}; \rho]$. Preserving this supposition we add an additional condition. Namely, until the end of this section we suppose that $\rho \in (\mathrm{Dist})[\mathbf{X}]$. Thus we now consider the case of metrizable space $(\mathbf{X}, \tau^{(1)})$. Of course, $\tau^{(1)} \in (\mathrm{top})_0[\mathbf{X}]$ in this case; see (2.7.7). Then by (2.3.20) $\mathbf{Y} \in \mathcal{F}_{\tau^{(1)}}$. From (2.2.15) and the relation $\tau^{(1)} = \tau_\rho^\natural(\mathbf{X}) \in (\mathrm{top})_0[\mathbf{X}]$ we have $\forall x \in \mathbf{X} : \{x\} \in \mathcal{F}_{\tau^{(1)}}$. Hence from (2.7.20) we obtain the validity of Condition 3.4.1 (see (2.2.14) and (3.4.3)). As a corollary, under conditions defining Propositions 3.4.10* and 3.4.11* we obtain the chain (3.4.20). Thus we have shown that the two following propositions hold.

STATEMENT 3.4.1 (*) *Let* $\rho \in (\mathrm{Dist})[\mathbf{X}]$, $\tau^{(1)} = \tau_\rho^\natural(\mathbf{X})$, *and* $\mathbf{Y} \in (\mathrm{comp})[\mathbf{X}; \rho]$. *Moreover, suppose that* $\tau^{(3)} \in (\mathbf{c}_{\mathcal{N}} - \mathrm{top})[\mathbf{K}]$, $\tau^{(2)} \in (\mathcal{D} - \mathrm{top})[\mathbf{H}]$, *and* $\omega \in C_{\mathrm{cl}}(\mathbf{K}, \tau^{(3)}, \mathbf{H}, \tau^{(2)})$. *Then (3.4.20) is valid.*

STATEMENT 3.4.2 (*) *Let* $\rho \in (\mathrm{Dist})[\mathbf{X}]$, $\tau^{(1)} = \tau_\rho^\natural(\mathbf{X})$, $\mathbf{Y} \in (\mathrm{comp})[\mathbf{X}; \rho]$, *and* $\omega \in C_{\mathbf{qp}}(\mathbf{K}, \tau^{(3)}, \mathbf{H}, \tau^{(2)})$. *Then (3.4.20) is valid.*

## 3.5 STANDARD REALIZATION OF ATTRACTION SETS. II

In this section we continue to investigate representations of the basic AS (3.4.4). We preserve the stipulations (3.4.1)–(3.4.3), (3.4.5), and (3.4.6). But here we consider a more traditional representation of AS in terms of compactness and some 'local compactifications'. We supplement Theorems 3.4.1 and 3.4.2. In addition, we use the set $\mathbf{Q}$, $\mathbf{Q} \neq \varnothing$, and the mapping $\Lambda_0$ from Section 3.4. We follow (3.4.22).

THEOREM 3.5.1 *Let Condition 3.4.2 be valid and* $\omega \in C_{\mathbf{ap}}(\mathbf{K}, \tau^{(3)}, \mathbf{H}, \tau^{(2)})$. *Then*

$$\omega^1(g^{-1}(\mathbf{Y})) = (\tau^{(2)} - \mathrm{LIM})[s^{-1}[\mathcal{Y}] \mid h] = (\tau^{(2)} - \mathrm{LIM})[s^{-1}[\mathfrak{M}] \mid h]. \tag{3.5.1}$$

PROOF. Since Condition 3.4.1 is valid (see Proposition 3.4.6), from Theorem 3.4.1 we have (3.4.20). From Proposition 3.4.7 we have the coincidence of the set (3.4.20) and $\omega^1((\tau^{(3)} - \mathrm{LIM})[s^{-1}[\mathfrak{M}] \mid m])$. Now we use the following concrete variant of (3.3.12): $X = \mathbf{F}$, $\mathcal{X} = s^{-1}[\mathfrak{M}]$, $(V, \tilde{\tau}_1) = (\mathbf{K}, \tau^{(3)})$, $(W, \tilde{\tau}_2) = (\mathbf{H}, \tau^{(2)})$, $\alpha = m$, and $\beta = \omega$. Then

$$(\tau^{(2)} - \mathrm{LIM})[s^{-1}[\mathfrak{M}] \mid h] = \omega^1((\tau^{(3)} - \mathrm{LIM})[s^{-1}[\mathfrak{M}] \mid m]). \tag{3.5.2}$$

From (3.4.20) and (3.5.2) we obtain (3.5.1). □

COROLLARY 3.5.1 *Let* $\tau^{(3)} \in (\mathbf{c} - \mathrm{top})[\mathbf{K}]$ *and* $\tau^{(2)} \in (\mathrm{top})_0[\mathbf{H}]$, *and Condition 3.4.2 be valid. Then (3.5.1) holds.*

To prove this it is sufficient to use (2.8.7). Corollary 3.5.1 is a typical example of employing a compactification for obtaining the representation of AS. Theorem 3.5.1 is 'less typical'.

PROPOSITION 3.5.1 *Let* $\tau^{(2)} \in (\mathrm{top})_0[\mathbf{H}]$ *and Condition 3.4.2 be valid. Then*

$$h^1(s^{-1}(\mathbf{Y})) \subset ((\mathbf{c}\tau^{(3)}, \tau^{(2)}) - \mathrm{LIM})[s^{-1}[\mathfrak{N}] \mid m; h] \subset \omega^1(g^{-1}(\mathbf{Y})). \tag{3.5.3}$$

PROOF. The first inclusion in (3.5.3) is established similarly to that in (3.4.14); see Proposition 3.4.2. We recall this reasoning very briefly, fixing $z \in h^1(s^{-1}(\mathbf{Y}))$ and choosing $f \in s^{-1}(\mathbf{Y})$ with the property $z = h(f)$. We introduce the stationary sequence $\mathbf{f} \in \mathbf{F}^{\mathcal{N}}$ for which $\forall i \in \mathcal{N} : \mathbf{f}(i) = f$. As a result we obtain the net $(\mathcal{N}, \leq_{\mathcal{N}}, \mathbf{f})$ in $\mathbf{F}$ for which $\mathbf{f} \in \mathcal{M}_{\mathrm{c}}(\mathcal{N}, \mathbf{F}, \mathbf{K}, \tau^{(3)}, m)$. Indeed, $\{m(f)\} \in (\tau^{(3)} - \mathrm{comp})[\mathbf{K}]$ possesses the property: $(m \circ \mathbf{f})^1(\mathcal{N}) = \{m(f)\}$. Since $\mathfrak{N} \subset \mathcal{Y}$, we have $\mathbf{Y} \subset H$ for $H \in \mathfrak{N}$ (we use (3.4.3) and (3.4.23)). But $s(\mathbf{f}(i)) = (s \circ \mathbf{f})(i) = s(f) \in \mathbf{Y}$ for $i \in \mathcal{N}$. Hence $\mathbf{f}$ is a sequence in $s^{-1}(\mathbf{Y})$. As a consequence, in the form of $\mathbf{f}$ we have a net in $s^{-1}(H)$ for $H \in \mathfrak{N}$. Therefore $s^{-1}[\mathfrak{N}] \subset (\mathbf{F} - \mathrm{ass})[\mathcal{N}; \leq_{\mathcal{N}}; \mathbf{f}]$. Moreover, $(h \circ \mathbf{f})(i) = h(f) = z$. Hence we obtain (3.4.16). From (3.2.10) it follows that $z \in ((\mathbf{c}\tau^{(3)}, \tau^{(2)}) - \mathrm{LIM})[s^{-1}[\mathfrak{N}] \mid m; h]$. The first inclusion of (3.5.3) is established. Consider Proposition 3.3.5 under the following conditions: $X = \mathbf{F}$, $\mathcal{X} = s^{-1}[\mathfrak{N}]$, $(V, \tilde{\tau}_1) = (\mathbf{K}, \tau^{(3)})$, $(W, \tilde{\tau}_2) = (\mathbf{H}, \tau^{(2)})$, $\alpha = m$, and $\beta = \omega$. Then by (3.4.6) we obtain the inclusion

$$((\mathbf{c}\tau^{(3)}, \tau^{(2)}) - \mathrm{LIM})[s^{-1}[\mathfrak{N}] \mid m; h] \subset \omega^1((\tau^{(3)} - \mathrm{LIM})[s^{-1}[\mathfrak{N}] \mid m]). \tag{3.5.4}$$

From Proposition 3.4.7 and (3.5.4) we obtain the second inclusion in (3.5.3). □

THEOREM 3.5.2 *Suppose that* $\tau^{(3)} \in (\mathbf{c} - \mathrm{top})_{\mathrm{loc}}[\mathbf{K}]$, $\tau^{(2)} \in (\mathrm{top})_0[\mathbf{H}]$, *and Condition 3.4.2 is valid. Then*

$$\begin{aligned} \omega^1(g^{-1}(\mathbf{Y})) &= ((\mathbf{c}\tau^{(3)}, \tau^{(2)}) - \mathrm{LIM})[s^{-1}[\mathcal{Y}] \mid m; h] \\ &= ((\mathbf{c}\tau^{(3)}, \tau^{(2)}) - \mathrm{LIM})[s^{-1}[\mathfrak{N}] \mid m; h]. \end{aligned}$$

PROOF. From Proposition 3.4.7 we have

$$\begin{aligned} \omega^1(g^{-1}(\mathbf{Y})) &= \omega^1((\tau^{(3)} - \mathrm{LIM})[s^{-1}[\mathcal{Y}] \mid m]) \\ &= \omega^1((\tau^{(3)} - \mathrm{LIM})[s^{-1}[\mathfrak{N}] \mid m]). \end{aligned} \tag{3.5.5}$$

Now we use the two following versions of Proposition 3.3.6: $X = \mathbf{F}$, $(\mathcal{X} = s^{-1}[\mathcal{Y}]) \vee (\mathcal{X} = s^{-1}[\mathfrak{N}])$, $(V, \tilde{\tau}_1) = (\mathbf{K}, \tau^{(3)})$, $(W, \tilde{\tau}_2) = (\mathbf{H}, \tau^{(2)})$, $\alpha = m$, and $\beta = \omega$. Then by this proposition and (3.4.6) we have

$$\Big(((\mathbf{c}\tau^{(3)}, \tau^{(2)}) - \mathrm{LIM})[s^{-1}[\mathcal{Y}] \mid m; h] = \omega^1((\tau^{(3)} - \mathrm{LIM})[s^{-1}[\mathcal{Y}] \mid m])\Big)$$

$$\& \left(((\mathbf{c}\tau^{(3)}, \tau^{(2)}) - \mathrm{LIM})[s^{-1}[\mathfrak{N}] \mid m; h] = \omega^1((\tau^{(3)} - \mathrm{LIM})[s^{-1}[\mathfrak{N}] \mid m])\right). \tag{3.5.6}$$

From (3.5.5) and (3.5.6) we obtain the required statement. □

Theorems 3.4.1, 3.4.2, 3.5.1, and 3.5.2 characterize (3.4.21) as a very universal regularizing approach (moreover, see other statements of the two last sections). It is possible to generalize these statements by using Corollary 3.4.2. Namely, from Theorem 3.5.1 we have

PROPOSITION 3.5.2 *Suppose that Condition 3.4.2 is valid and* $\omega \in C_{\mathbf{ap}}(\mathbf{K}, \tau^{(3)}, \mathbf{H}, \tau^{(2)})$. *Then* $\forall \mathcal{X} \in \mathcal{B}[\mathbf{F}]$:

$$(s^{-1}[\mathfrak{N}] \subset \mathcal{X} \subset s^{-1}[\mathcal{Y}]) \Rightarrow \Big(\omega^1(g^{-1}(\mathbf{Y})) = (\tau^{(2)} - \mathrm{LIM})[\mathcal{X} \mid h]\Big).$$

Furthermore, from Theorem 3.5.2 we have the following

PROPOSITION 3.5.3 *Let Condition 3.4.2 be valid. Moreover, let* $\tau^{(3)} \in (\mathbf{c} - \mathrm{top})_{\mathrm{loc}}[\mathbf{K}]$ *and* $\tau^{(2)} \in (\mathrm{top})_0[\mathbf{H}]$. *Then* $\forall \mathcal{X} \in \mathcal{B}[\mathbf{F}]$:

$$(s^{-1}[\mathfrak{N}] \subset \mathcal{X} \subset s^{-1}[\mathcal{Y}])$$

$$\Rightarrow \Big(\omega^1(g^{-1}(\mathbf{Y})) = ((\mathbf{c}\tau^{(3)}, \tau^{(2)} - \mathrm{LIM})[\mathcal{X} \mid m; h]\Big).$$

To prove the two last statements it is sufficient to use (3.2.8) and (3.2.10).

**Remark.** Consider the question of the asymptotic realization of elements of $\omega^1(g^{-1}(\mathbf{Y}))$, following the construction of Remark of Section 3.2. In addition, this construction is exhausting in the sense of the asymptotic realization of $\omega^1(g^{-1}(\mathbf{Y}))$. We shall very brief consider a corresponding modification of the scheme on the basis of (3.2.46), (3.2.47), (3.2.49), and (3.2.51). Thus by Theorems 3.4.1, 3.4.2, 3.5.1, and 3.5.2 the set $\omega^1(g^{-1}(\mathbf{Y}))$ exhausts (under natural conditions) all possibilities of asymptotic attainability under the weakening of the $\mathbf{Y}$-constraint. Therefore the question of realization of points of $\omega^1(g^{-1}(\mathbf{Y}))$ is very important.

Choose and fix an arbitrary point $z \in \omega^1(g^{-1}(\mathbf{Y}))$. Let $v \in g^{-1}(\mathbf{Y})$ possess the property $z = \omega(v)$. Of course, by (3.4.6) we have $v \in$

$\mathrm{cl}(m^1(\mathbf{F}), \tau^{(3)})$. Using the Birkhoff theorem we choose an arbitrary net $(D, \preceq, f)$ in $\mathbf{F}$ with the property

$$(D, \preceq, m \circ f) \overset{\tau^{(3)}}{\to} v. \tag{3.5.7}$$

In (3.5.7) we use arguments similar to the passage from (3.2.46) to (3.2.47). Such net $(D, \preceq, f)$ certainly exists. By (3.4.5) we have $\omega \circ m \circ f = h \circ f$. Using (2.5.4), (3.4.5), and (3.5.7) we obtain the convergence

$$(D, \preceq, h \circ f) \overset{\tau^{(2)}}{\to} z. \tag{3.5.8}$$

On the other hand, by (3.4.6) $g \circ m \circ f = s \circ f$. From (2.5.4),(3.4.5), and (3.5.7) we obtain the convergence

$$(D, \preceq, s \circ f) \overset{\tau^{(1)}}{\to} g(v). \tag{3.5.9}$$

But $g(v) \in \mathbf{Y}$ and as a consequence $\mathfrak{N} \subset \mathcal{Y} \subset N_{\tau^{(1)}}(g(v))$. From (2.2.48) and (3.5.9) we have $\mathfrak{N} \subset \mathcal{Y} \subset (\mathbf{X} - \mathrm{ass})[D; \preceq; s \circ f]$ and hence

$$s^{-1}[\mathfrak{N}] \subset s^{-1}[\mathcal{Y}] \subset (\mathbf{F} - \mathrm{ass})[D; \preceq; f]. \tag{3.5.10}$$

From (3.2.8), (3.5.8), and (3.5.10) we have a concrete approximate solution-net realizing the point $z$ in the form of $(D, \preceq, f)$. Thus on the basis of the Birkhoff theorem it is possible to realize the whole set $\omega^1(g^{-1}(\mathbf{Y}))$ in the sense of (3.2.8). We do not require any additional conditions concerning (3.4.5) and (3.4.6); of course, we realize (3.4.10) in the most accessible form. In addition, for many concrete settings it is possible to construct approximate solutions in a simpler way (see, in particular, [29, 30, 32, 35, 40, 41, 47, 117]), using the corresponding specific character.

## 3.6 REALIZATION OF ATTRACTION SETS BY MEANS OF NEIGHBORHOODS

Constructions on the basis of (3.2.8) and (3.2.10) permit us to realize attraction sets as a limit. Thus these relations generate limit representations of asymptotic attainability. It is very interesting to consider the question of the approximate realization of AS by 'usual' attainable sets under some concrete weakening of the $\mathbf{Y}$-constraint. Thus, following (3.4.3), we admit an arbitrary choice of $\tilde{\mathbf{Y}} \in \mathcal{Y}$ determining the constraint $s(f) \in \tilde{\mathbf{Y}}$ on the choice of $f \in \mathbf{F}$ (see Section 3.1). As a result we have $h^1(s^{-1}(\tilde{\mathbf{Y}}))$ and $\mathrm{cl}(h^1(s^{-1}(\tilde{\mathbf{Y}})), \tau^{(2)})$ in the family $\mathcal{P}(\mathbf{H})$. We are oriented towards the case when $\tilde{\mathbf{Y}}$ is 'near' to $\mathbf{Y}$. Are the two last sets 'near' to AS (3.4.4)? It is natural to connect the notion of 'nearness' with realization of the second set-closure in the form of a subset of the preassigned neighborhood of the set (3.4.4) in the TS $(\mathbf{H}, \tau^{(2)})$. Indeed, by (3.2.8) we have the inclusion

$$(\tau^{(2)} - \mathrm{LIM})[s^{-1}[\mathcal{Y}] \mid h] \subset \mathrm{cl}(h^1(s^{-1}(\tilde{\mathbf{Y}})), \tau^{(2)}) \tag{3.6.1}$$

(recall that $\tilde{\mathbf{Y}}$ is a neighborhood of $\mathbf{Y}$ in the TS $(\mathbf{X}, \tau^{(1)})$). With due account of (3.6.1) we obtain the following natural notion of nearness. Namely, we call the realization of AS (3.4.4) in the form of (3.2.8) the neighborhood one, if for each $\Lambda^* \in \mathbb{N}_{\tau^{(2)}}\left[(\tau^{(2)} - \mathrm{LIM})[s^{-1}[\mathcal{Y}] \mid h]\right]$ one can choose $\Lambda_* \in \mathcal{Y}$ such that $\forall \tilde{\mathbf{Y}} \in \mathcal{Y}$:

$$\left(\tilde{\mathbf{Y}} \subset \Lambda_*\right) \Rightarrow \left(\mathrm{cl}(h^1(s^{-1}(\tilde{\mathbf{Y}})), \tau^{(2)}) \subset \Lambda^*\right) \tag{3.6.2}$$

In (3.6.1), (3.6.2) we admit the employment of all $\tilde{\mathbf{Y}} \in \mathcal{Y}$ for weakening of the $\mathbf{Y}$-constraint. However, an analogous construction is introduced in the case when only neighborhoods $\tilde{\mathbf{Y}} \in \mathfrak{N}$ may be used for weakening of the $\mathbf{Y}$-constraint. A corresponding natural scheme is similar to (3.6.1) and (3.6.2). We do not now consider this variant. Returning to (3.6.1) and (3.6.2), we note that in problems of asymptotic realization an unessential difference between the set $h^1(s^{-1}(\tilde{\mathbf{Y}}))$ and its closure can be admitted. We note also that the AS coincides with the set (3.1.12) very often (see the corresponding sufficient conditions in Sections 3.4 and 3.5). Therefore the question of realization of the set (3.1.12) by means of neighborhoods seems quite natural. Generally speaking, this question is not reduced to the question of coincidence of the sets (3.1.12) and (3.4.4). Moreover, in general, these questions are different even for problems possessing some properties of the stability type.

**Example.** Let $\mathbf{F} \triangleq [-\pi/2, \pi/2] \times [0, \infty[$, $\mathbf{X} \triangleq [-\pi/2, \pi/2]$, $\mathbf{H} \triangleq \mathbb{R} \times \mathbb{R}$, and $\mathbf{Y} \triangleq [-\pi/2, 0]$. Moreover, let $s \in \mathbf{X}^{\mathbf{F}}$ be a function for which $s(\alpha, l) \triangleq \alpha$ for $\alpha \in [-\pi/2, \pi/2]$ and $l \in [0, \infty[$. Finally, suppose that $h \in \mathbf{H}^{\mathbf{F}}$ is defined by the rule: the equality $h(f) \triangleq (l \cos \alpha, l \sin \alpha)$ holds for $f \triangleq (\alpha, l) \in \mathbf{F}$. We set that $\tau^{(1)} \triangleq \tau_{\mathbb{R}} \mid_{\mathbf{X}}$ and $\tau^{(2)}$ is the usual topology of coordinate-wise convergence. It is easily established that (3.4.4) coincides with the set

$$A_0 \triangleq [0, \infty[\times] - \infty, 0] = h^1(s^{-1}(\mathbf{Y})). \tag{3.6.3}$$

Indeed, (3.6.3) is a subset of the set on the right hand side of (3.6.1) for each $\tilde{\mathbf{Y}} \in \mathcal{Y}$. Therefore $A_0$ is a subset of the set (3.4.4). We use (3.2.8). If $f \in \mathbf{F}$, $f = (\alpha, l)$, where $\alpha \in \mathbf{X}$ and $l \in [0, \infty[$, then $h(f) \in [0, \infty[\times \mathbb{R}$. The set $[0, \infty[\times \mathbb{R}$ is closed in $(\mathbf{H}, \tau^{(2)})$. Then by (3.2.8) the set (3.4.4) is a subset of $[0, \infty[\times \mathbb{R}$. Let $z \triangleq (z_1, z_2) \in ([0, \infty[\times \mathbb{R}) \setminus A_0$. Hence $z_2 > 0$. If $z_1 = 0$, then

$$z \notin \mathrm{cl}(h^1(s^{-1}([-\pi/2, \pi/4[)), \tau^{(2)}).$$

We use the fact that the inequality $w_2 \le w_1$ holds for $(w_1, w_2) \in h^1(s^{-1}([-\pi/2, \pi/4[))$. Thus for $z_1 = 0$ we have the property: $z$ is not element of (3.4.4). Let $z_1 > 0$. Then $\kappa \triangleq z_2/2z_1 \in ]0, \infty[$. Choose $\alpha_0 \in ]0, \pi/2[$ such that $\forall \alpha \in ]-\alpha_0, \alpha_0[: \mid \tan\alpha \mid < \kappa$. We use the continuity of the function

$$x \mapsto \tan x : \quad ]-\pi/2, \pi/2[ \to \mathbb{R}$$

at the point 0. Then for $\tilde{z} \triangleq (\tilde{z}_1, \tilde{z}_2) \in h^1(s^{-1}([-\pi/2, \alpha_0[))$ we have the property $\tilde{z}_2 < \kappa\tilde{z}_1$ under the condition $\tilde{z}_1 > 0$ (recall that $\kappa > 0$ and $z_2 = 2\kappa z_1$). If $\tilde{z}_1 = 0$, then $\mid \tilde{z}_1 - z_1 \mid = z_1 > 0$. If $\left(\tilde{z}^{(j)}\right)_{j\in\mathcal{N}}$ is some sequence in $h^1(s^{-1}([-\pi/2, \alpha_0[))$ for which

$$\left(\tilde{z}^{(j)}\right)_{j\in\mathcal{N}} \overset{\tau^{(2)}}{\to} z,$$

and $\left(\tilde{z}_1^{(j)}\right)_{j\in\mathcal{N}}, \left(\tilde{z}_2^{(j)}\right)_{j\in\mathcal{N}}$ are the components of this sequence (i.e., $\tilde{z}^{(k)} = (\tilde{z}_1^{(k)}, \tilde{z}_2^{(k)})$ for $k \in \mathcal{N}$), then one can choose $n \in \mathcal{N}$ for which $\tilde{z}_1^{(k)} > 0$ for $k \in \overrightarrow{n, \infty}$. Then $\tilde{z}_2^{(k)} < \kappa\tilde{z}_1^{(k)}$ for $k \in \overrightarrow{n, \infty}$. As a limit we obtain $z_2 = 2\kappa z_1 \le \kappa z_1$ and consequently $\kappa z_1 \le 0$. The last inequality is impossible since $z_1 > 0$ and $\kappa > 0$. The contradiction shows that

$$z \notin \mathrm{cl}(h^1(s^{-1}([-\pi/2, \alpha_0[)), \tau^{(2)}).$$

Since $[-\pi/2, \alpha_0[ \in \mathcal{Y}$, we obtain that $z$ is not element of (3.4.4). We see that each point of $([0, \infty[ \times \mathbb{R}) \setminus A_0$ is not element of (3.4.4). Then (3.4.4) is a subset of $A_0$. We obtain the coincidence of (3.4.4) and $A_0$. Thus $A_0$ is the required AS. Take $\zeta \in ]0, \infty[$ and introduce the (open) set

$$G_0 \triangleq \{\hat{z} \triangleq (\hat{z}_1, \hat{z}_2) \in \mathbb{R} \times \mathbb{R} \mid \hat{z}_2 < \zeta\} \in \tau^{(2)}.$$

Of course, $G_0 \in \mathbb{N}^0_{\tau^{(2)}}[A_0]$. Suppose that $\Lambda^* \triangleq G_0$. Let $\Lambda \in \mathcal{Y}$. Choose $\Lambda^0 \in \mathbb{N}^0_{\tau^{(1)}}[\mathbf{Y}]$ for which $\Lambda^0 \subset \Lambda$. We use (2.2.13). By (2.2.12) $\Lambda^0 \in \tau^{(1)}$ and $\mathbf{Y} \subset \Lambda^0$. Then $0 \in \Lambda^0$. Therefore one can choose $\xi \in ]0, \pi/2[$ for which $]-2\xi, 2\xi[ \subset \Lambda^0 \subset \Lambda$. Of course, $\tan\xi \in ]0, \infty[$. Hence for some $\tilde{q} \in ]0, \infty[$ we have $\tilde{q}\tan\xi > \zeta$. Suppose that $q_1 \triangleq \tilde{q}$ and $q_2 \triangleq \tilde{q}\tan\xi = q_1\tan\xi > \zeta$. Moreover, let $q \triangleq (q_1, q_2)$. Then $q \notin \Lambda^*$. Suppose that $\mathbf{l} \triangleq (q_1^2 + q_2^2)^{1/2}$; $\mathbf{l} > 0$. In addition, $\mathbf{f} \triangleq (\xi, \mathbf{l}) \in \mathbf{F}$; $h(\mathbf{f}) = (\mathbf{l}\cos\xi, \mathbf{l}\sin\xi)$. But

$$\mathbf{l}\cos\xi = q_1(1 + (\tan\xi)^2)^{1/2}\cos\xi = q_1\frac{1}{\cos\xi}\cos\xi = q_1;$$

$$\mathbf{l}\sin\xi = q_1(1 + (\tan\xi)^2)^{1/2}\sin\xi = q_1\tan\xi = q_2.$$

Thus $h(\mathbf{f}) = (q_1, q_2) = q \notin \Lambda^*$. On the other hand, $\xi = s(\mathbf{f}) \in \Lambda$. So, $\mathbf{f} \in s^{-1}(\Lambda)$. Therefore

$$q \in h^1(s^{-1}(\Lambda)) \setminus \Lambda^*.$$

As a consequence we have

$$q \in \mathrm{cl}(h^1(s^{-1}(\Lambda)), \tau^{(2)}) \setminus \Lambda^*.$$

Since the choice of $\Lambda$ was arbitrary, the following effect takes place: the realization of the set (3.4.4) by means of neighborhoods is impossible. Note that this example is connected with the procedure of the passage from the polar coordinate system to the Descartes one. In addition, the $\mathbf{Y}$-constraint is given. We admit transformation errors under calculation of the angle tangent.

Returning to the general case, we note that under conditions of the compactness type it is possible to obtain the required neighborhood realization of AS. In particular, we obtain the following assertion as an immediate corollary of statements of [71, Section 3.1].

PROPOSITION 3.6.1 *Let* $\mathbf{S}[U \neq \varnothing]$, $\mathbf{S}[V \neq \varnothing]$, $\mathcal{U} \in \mathcal{B}[U]$, $\tau \in (\mathrm{top})[V]$, $\varphi \in V^U$, *and* $\{H \in \mathcal{U} \mid \mathrm{cl}(\varphi^1(H), \tau) \in (\tau - \mathrm{comp})[V]\} \neq \varnothing$. *Then*

$$\begin{aligned} &\left(\mathbb{N}^0_\tau[(\tau - \mathrm{LIM})[\mathcal{U} \mid \varphi]\,] \subset \bigcup_{H \in \mathcal{U}} \mathbb{N}^0_\tau[\mathrm{cl}(\varphi^1(H), \tau)]\right) \\ \& &\left(\mathbb{N}_\tau[(\tau - \mathrm{LIM})[\mathcal{U} \mid \varphi]\,] \subset \bigcup_{H \in \mathcal{U}} \mathbb{N}_\tau[\mathrm{cl}(\varphi^1(H), \tau)]\right). \end{aligned} \tag{3.6.4}$$

To prove (3.6.4) it is sufficient to establish the first inclusion. But this inclusion follows from Corollary 3.1.5 of [71]. Certainly, from Proposition 3.6.1 we obtain the following

COROLLARY 3.6.1 $\forall_U\, \mathbf{S}[U \neq \varnothing]$ $\forall_V\, \mathbf{S}[V \neq \varnothing]$ $\forall_W\, \mathbf{S}[W \neq \varnothing]$ $\forall \mathcal{U} \in \mathcal{B}[U]$ $\forall \tau_1 \in (\mathbf{c}-\mathrm{top})[V]$ $\forall \tau_2 \in (\mathrm{top})_0[W]$ $\forall \alpha \in V^U$ $\forall \beta \in C(V, \tau_1, W, \tau_2)$ $\forall S \in \mathbb{N}_{\tau_2}[(\tau_2 - \mathrm{LIM})[\mathcal{U} \mid \beta \circ \alpha]\,]$ $\exists P \in \mathcal{U}$ $\forall Q \in \mathcal{U}$:

$$(Q \subset P) \Rightarrow (S \in \mathbb{N}_{\tau_2}[\mathrm{cl}((\beta \circ \alpha)^1(Q), \tau_2)]).$$

PROOF. Fix $U$, $V$, $W$, $\mathcal{U}$, $\tau_1$, $\tau_2$, $\alpha$, $\beta$, and $S$ in correspondence with the conditions of the corollary. Let $\varphi \triangleq \beta \circ \alpha$ and $H \in \mathcal{U}$. Then $\varphi^1(H) = \beta^1(\alpha^1(H))$, and by (2.8.2), (2.8.4) and (2.8.7) we have $\mathrm{cl}(\varphi^1(H), \tau_2) = \beta^1(\mathrm{cl}(\alpha^1(H), \tau_1))$, where $\mathrm{cl}(\alpha^1(H), \tau_1) \in \mathcal{F}_{\tau_1}$ and hence $\mathrm{cl}(\alpha^1(H), \tau_1) \in (\tau_1 - \mathrm{comp})[V]$ (see Section 3.3); by (2.8.5), we obtain $\mathrm{cl}(\varphi^1(H), \tau_2) \in (\tau_2 - \mathrm{comp})[W]$. The basic condition of Proposition 3.6.1 is established.

The last corollary says about the neighborhood realization of an abstract AS. Of course, we consider an auxiliary problem. Such consideration has a sense since we discuss some possible approaches. Namely, we continue to investigate constructions using countably compact model TS. We first consider some properties similar to Corollary 3.1.5 of [71].

PROPOSITION 3.6.2 *Let* $\mathbf{S}[X]$, $\tau \in (\mathbf{c}_{\mathcal{N}} - \mathrm{top})[X]$,

$$(F_i)_{i \in \mathcal{N}} : \mathcal{N} \to \mathcal{F}_\tau,$$

*and* $G \in \mathbb{N}_\tau^0 \left[\bigcap_{i \in \mathcal{N}} F_i\right]$. *Then*

$$\exists k \in \mathcal{N} : G \in \mathbb{N}_\tau^0 \left[\bigcap_{i=1}^{k} F_i\right]. \tag{3.6.5}$$

PROOF. Consider $\forall i \in \mathcal{N} : G_i \triangleq X \setminus F_i \in \tau$. Suppose that $\left(G^{(i)}\right)_{i \in \mathcal{N}} \in \tau^{\mathcal{N}}$ is defined by the rule $(G^{(1)} \triangleq G)$ & $(\forall i \in \overrightarrow{2, \infty} : G^{(i)} \triangleq G_{i-1})$. Then the union of all sets $G^{(i)}$, $i \in \mathcal{N}$, is a subset of $X$. But

$$\bigcap_{i \in \mathcal{N}} F_i \subset G,$$

and as a consequence we have the inclusion

$$X \setminus G \subset X \setminus (\bigcap_{i \in \mathcal{N}} F_i) = \bigcup_{i \in \mathcal{N}} (X \setminus F_i) = \bigcup_{i \in \mathcal{N}} G_i \subset \bigcup_{i \in \mathcal{N}} G^{(i)}.$$

Thus $X$ is the union of all sets $G^{(i)}$, $i \in \mathcal{N}$. Using (2.2.3) we obtain for some $n \in \mathcal{N}$ the equality

$$X = \bigcup_{i=1}^{n} G^{(i)}.$$

As a consequence we get the inclusion

$$X \setminus G \subset \bigcup_{i=1}^{n} G_i.$$

It follows that

$$\bigcap_{i=1}^{n} F_i = \bigcap_{i=1}^{n} (X \setminus G_i) = X \setminus \left( \bigcup_{i=1}^{n} G_i \right) \subset G.$$

COROLLARY 3.6.2 *Let* $\mathbf{S}[X]$, $\tau \in (\mathrm{top})[X]$, $(F_i)_{i \in \mathcal{N}} : \mathcal{N} \to \mathcal{F}_\tau$, *and* $G \in \mathbb{N}_\tau^0 \left[\bigcap_{i \in \mathcal{N}} F_i\right]$. *Moreover, suppose that* $\exists p \in \mathcal{N} : F_p \in (\mathcal{N} - \tau - \mathrm{comp})[X]$. *Then (3.6.5) holds.*

PROOF. Fix $p \in \mathcal{N}$ for which $F_p \in (\mathcal{N} - \tau - \text{comp})[X]$. Then (see (2.3.14))

$$\theta \triangleq \tau\mid_{F_p} \in (\mathbf{c}_{\mathcal{N}} - \text{top})[F_p] \tag{3.6.6}$$

Consider the countably compact space $(F_p, \theta)$. By (2.3.12) we have

$$\mathcal{F}_\theta = \mathcal{F}_\tau\mid_{F_p} = \{F_p \cap H : H \in \mathcal{F}_\tau\}.$$

Therefore $\forall i \in \mathcal{N} : F_i \cap F_p \in \mathcal{F}_\theta$ and we get

$$(F_i \cap F_p)_{i \in \mathcal{N}} : \mathcal{N} \to \mathcal{F}_\theta.$$

In addition, by the choice of $G$ we have $G \cap F_p \in \theta$ and $\bigcap_{i \in \mathcal{N}}(F_i \cap F_p) \subset G \cap F_p$. By Proposition 3.6.2 we obtain the property

$$\bigcap_{i=1}^{n}(F_i \cap F_p) \subset G \cap F_p$$

for some $n \in \mathcal{N}$. Let $q \triangleq \sup(\{n; p\})$; then $q \in \mathcal{N}$ possesses the property $\overline{1,n} \subset \overline{1,q}$ and $p \in \overline{1,q}$. Therefore

$$\bigcap_{i=1}^{q} F_i \subset G.$$

The relation (3.6.5) is established.

PROPOSITION 3.6.3 *Let* $\mathbf{S}[X]$, $\tau \in (\text{top})[X]$, $\mathcal{X} \in \mathcal{B}_{\mathcal{N}}(X) \cap \mathcal{P}(\mathcal{F}_\tau)$, *and* $G \in \mathbb{N}^0_\tau\left[\bigcap_{U \in \mathcal{X}} U\right]$. *Moreover, let* $(\mathcal{N} - \tau - \text{comp})[X] \cap \mathcal{X} \neq \varnothing$. *Then* $\exists H \in \mathcal{X} : G \in \mathbb{N}^0_\tau[H]$.

PROOF. We use (3.2.4). Fix $(H_i)_{i \in \mathcal{N}} \in \mathcal{X}^{\mathcal{N}}$ with the property $\forall \tilde{H} \in \mathcal{X}\ \exists k \in \mathcal{N} : H_k \subset \tilde{H}$. Then

$$\bigcap_{U \in \mathcal{X}} U = \bigcap_{i \in \mathcal{N}} H_i. \tag{3.6.7}$$

Choose an arbitrary $F \in (\mathcal{N} - \tau - \text{comp})[X] \cap \mathcal{X}$ and introduce the sequence

$$(\tilde{H}_i)_{i \in \mathcal{N}} : \mathcal{N} \to \mathcal{X},$$

setting $(\tilde{H}_1 \triangleq F)\&(\forall i \in \overrightarrow{2,\infty} : \tilde{H}_i \triangleq H_{i-1})$. From (3.6.7) we have

$$\bigcap_{U \in \mathcal{X}} U \subset \bigcap_{i \in \mathcal{N}} \tilde{H}_i \subset \bigcap_{i \in \mathcal{N}} H_i = \bigcap_{U \in \mathcal{X}} U.$$

As a consequence we obtain

$$\bigcap_{U \in \mathcal{X}} U = \bigcap_{i \in \mathcal{N}} \tilde{H}_i;$$

in addition, $\tilde{H}_1 \in (\mathcal{N} - \tau - \text{comp})[X]$. By the choice of $G$ and (3.6.8) we have $G \in \tau$ and $\bigcap_{i \in \mathcal{N}} \tilde{H}_i \subset G$. Thus

$$G \in \mathbb{N}_\tau^0 \left[ \bigcap_{i \in \mathcal{N}} \tilde{H}_i \right].$$

From Corollary 3.6.2 it follows that $\exists k \in \mathcal{N}$:

$$G \in \mathbb{N}_\tau^0 \left[ \bigcap_{i=1}^{k} \tilde{H}_i \right]. \tag{3.6.8}$$

Fix $k \in \mathcal{N}$ with the property (3.6.8). Recall that $\tilde{H}_1 \in \mathcal{X}, \ldots, \tilde{H}_k \in \mathcal{X}$. Using (3.2.1) and arguments of [35, p. 37], we obtain for some $H^* \in \mathcal{X}$ the inclusion

$$H^* \subset \bigcap_{i=1}^{k} \tilde{H}_i.$$

By (3.6.8) we have $G \in \mathbb{N}_\tau^0[H^*]$. □

**Remark.** If $\mathbf{S}[U \neq \varnothing]$, $\mathbf{S}[V]$, $\tau \in (\text{top})[V]$, $\mathcal{U} \in \mathcal{B}_\mathcal{N}(U)$, and $f \in V^U$, then

$$\{\text{cl}(f^1(H), \tau) : H \in \mathcal{U}\} \in \mathcal{B}_\mathcal{N}(V) \cap \mathcal{F}_\tau. \tag{3.6.9}$$

The proof of (3.6.9) uses (3.2.1), (3.2.4), and the monotonicity of the image and closure operations.

Using Proposition 3.6.3 and the last remark, we obtain the following

PROPOSITION 3.6.4 *Let* $\mathbf{S}[U \neq \varnothing]$, $\mathbf{S}[V \neq \varnothing]$, $\mathcal{U} \in \mathcal{B}_\mathcal{N}(U)$, $\tau \in (\text{top})[V]$, *and* $\varphi \in V^U$. *Moreover, suppose that* $\exists H \in \mathcal{U} : \text{cl}(\varphi^1(H), \tau) \in (\mathcal{N} - \tau - \text{comp})[V]$. *Then* $\forall G \in \mathbb{N}_\tau^0[(\tau - \text{LIM})[\mathcal{U} \mid \varphi]]\ \exists \tilde{M} \in \mathcal{U}$ :

$$G \in \mathbb{N}_\tau^0[\text{cl}(\varphi^1(\tilde{M}), \tau)].$$

PROOF. Consider the family $\mathcal{V} \triangleq \{\text{cl}(\varphi^1(H), \tau) : H \in \mathcal{U}\}$. Then, by the previous remark, $\mathcal{V} \in \mathcal{B}_\mathcal{N}(V) \cap \mathcal{F}_\tau$. In addition, by (3.2.8) we have the property

$$(\tau - \text{LIM})[\mathcal{U} \mid \varphi] = \bigcap_{H \in \mathcal{U}} \text{cl}(\varphi^1(H), \tau) = \bigcap_{P \in \mathcal{V}} P. \tag{3.6.10}$$

Fix $G \in \mathbb{N}^0_\tau[(\tau - \mathrm{LIM})[\mathcal{U} \mid \varphi]]$. With due account of (3.6.10) and Proposition 3.6.3 we obtain that $\exists H_* \in \mathcal{V} : G \in \mathbb{N}^0_\tau[H_*]$ (since $(\mathcal{N} - \tau - \mathrm{comp})[V] \cap \mathcal{V} \neq \varnothing$). By the definition of $\mathcal{V}$ we get that $\exists H^* \in \mathcal{U} : G \in \mathbb{N}^0_\tau[\mathrm{cl}(\varphi^1(H^*), \tau)]$. □

COROLLARY 3.6.3 *Let* $U, V, \mathcal{U}, \tau$, *and* $\varphi$ *satisfy all conditions of Proposition 3.6.4 (in particular,* $\mathrm{cl}(\varphi^1(H), \tau) \in (\mathcal{N} - \tau - \mathrm{comp})[V]$ *for some* $H \in \mathcal{U}$*). Then*

$$\begin{aligned}&\left(\mathbb{N}^0_\tau[(\tau - \mathrm{LIM})[\mathcal{U} \mid \varphi]] \subset \bigcup_{E \in \mathcal{U}} \mathbb{N}^0_\tau[\mathrm{cl}(\varphi^1(E), \tau)]\right)\\ \& &\left(\mathbb{N}_\tau[(\tau - \mathrm{LIM})[\mathcal{U} \mid \varphi]] \subset \bigcup_{E \in \mathcal{U}} \mathbb{N}_\tau[\mathrm{cl}(\varphi^1(E), \tau)]\right).\end{aligned} \tag{3.6.11}$$

The proof is obvious. Indeed, the first statement of (3.6.11) immediately follows from Proposition 3.6.4. The second one does from (2.2.13).

PROPOSITION 3.6.5 *Let* $\mathbf{S}[U \neq \varnothing]$, $\mathbf{S}[V \neq \varnothing]$, $\mathbf{S}[W \neq \varnothing]$, $\mathcal{U} \in \mathcal{B}_{\mathcal{N}}[U]$, $\tau_1 \in (\mathbf{c}_{\mathcal{N}} - \mathrm{top})[V]$, $\tau_2 \in (\mathrm{top})[W]$, $\alpha \in V^U$, $\beta \in C_{\mathrm{cl}}(V, \tau_1, W, \tau_2)$, *and* $\mathbf{S} \in \mathbb{N}_{\tau_2}[(\tau_2 - \mathrm{LIM})[\mathcal{U} \mid \beta \circ \alpha]\,]$. *Then* $\exists P \in \mathcal{U}\ \forall Q \in \mathcal{U}$:

$$(Q \subset P) \Rightarrow (\mathbf{S} \in \mathbb{N}_{\tau_2}[\mathrm{cl}((\beta \circ \alpha)^1(Q), \tau_2)]).$$

PROOF. By (2.8.4) we have $\forall E \in \mathcal{U}$:

$$\mathrm{cl}((\beta \circ \alpha)^1(E), \tau_2) = \mathrm{cl}((\beta^1(\alpha^1(E)), \tau_2) = \beta^1(\mathrm{cl}(\alpha^1(E), \tau_1)). \tag{3.6.12}$$

Fix $E \in \mathcal{U}$. Then $\mathrm{cl}(\alpha^1(E), \tau_1) \in \mathcal{F}_{\tau_1}$ and consequently we obtain $\mathrm{cl}(\alpha^1(E), \tau_1) \in (\mathcal{N} - \tau_1 - \mathrm{comp})[V]$ (see Section 2.3). Since, in particular, $\beta \in C(V, \tau_1, W, \tau_2)$, by (3.6.12) we have

$$\mathrm{cl}((\beta \circ \alpha)^1(E), \tau_2) \in (\mathcal{N} - \tau_2 - \mathrm{comp})[W]. \tag{3.6.13}$$

The property (3.6.13) is analogous to (2.8.5); see, for example, [71, Section 3.10]. Since $\mathbf{S}[\mathcal{U} \neq \varnothing]$, in particular, we obtain that (see (3.6.13))

$$\exists H \in \mathcal{U} : \mathrm{cl}((\beta \circ \alpha)^1(H), \tau_2) \in (\mathcal{N} - \tau_2 - \mathrm{comp})[W]. \tag{3.6.14}$$

We use Corollary 3.6.3 under the following conditions: $U$ and $\mathcal{U}$ are the same as in the corollary; further we take $(W, \tau_2)$ instead of $(V, \tau)$, and $\beta \circ \alpha$ instead of $\varphi$. Then from (3.6.14) and the corollary above we get

$$\mathbf{S} \in \bigcup_{E \in \mathcal{U}} \mathbb{N}_{\tau_2}[\mathrm{cl}((\beta \circ \alpha)^1(E), \tau_2)].$$

Let $P \in \mathcal{U}$ be a set for which $\mathbf{S} \in \mathbb{N}_{\tau_2}[\mathrm{cl}((\beta \circ \alpha)^1(P), \tau_2)]$. If $Q \in \mathcal{U}$ possesses the property $Q \subset P$, then $\mathrm{cl}((\beta \circ \alpha)^1(Q), \tau_2) \subset \mathrm{cl}((\beta \circ \alpha)^1(P), \tau_2)$. Therefore $\mathbf{S} \in \mathbb{N}_{\tau_2}[\mathrm{cl}((\beta \circ \alpha)^1(Q), \tau_2)]$ (see (2.2.13)). □

COROLLARY 3.6.4 *Let* $\mathbf{S}[U \neq \varnothing]$, $\mathbf{S}[V \neq \varnothing]$, $\mathbf{S}[W \neq \varnothing]$, $\mathcal{U} \in \mathcal{B}_{\mathcal{N}}[U]$, $\tau_1 \in (\mathbf{c}_{\mathrm{seq}} - \mathrm{top})[V]$, $\tau_2 \in (\mathrm{top})[W]$, $\alpha \in V^U$, $\beta \in C_{\mathrm{cl}}(V, \tau_1, W, \tau_2)$, *and* $\mathbf{S} \in \mathbb{N}_{\tau_2}[(\tau_2 - \mathrm{LIM})[\mathcal{U} \mid \beta \circ \alpha]\,]$. *Then* $\exists P \in \mathcal{U}\ \forall Q \in \mathcal{U}$:

$$(Q \subset P) \Rightarrow (\mathbf{S} \in \mathbb{N}_{\tau_2}[\mathrm{cl}((\beta \circ \alpha)^1(Q), \tau_2)]).$$

The proof follows from the inclusion $(\mathbf{c}_{\mathrm{seq}} - \mathrm{top})[V] \subset (\mathbf{c}_{\mathcal{N}} - \mathrm{top})[V]$ (see Section 2.2).

Thus we have a series of auxiliary statements characterizing the possibility of a neighborhood realization of AS. These statements are used in the main problem considered in Section 3.4. Firstly, we are interested in questions of a neighborhood realization of the set (3.1.12). This approach is connected with the reduction (3.4.21). An appropriate motivation is given in Sections 3.1 and 3.4; moreover, the example at the beginning of this section should be noted.

THEOREM 3.6.1 *Let Condition 3.4.1 be valid. Moreover, suppose that* $\tau^{(3)} \in (\mathbf{c} - \mathrm{top})[\mathbf{K}]$ *and* $\tau^{(2)} \in (\mathrm{top})_0[\mathbf{H}]$. *Then* $\forall \mathbf{S} \in \mathbb{N}_{\tau^{(2)}}[\omega^1(g^{-1}(\mathbf{Y}))]$ $\exists P \in \mathcal{Y}\ \forall Q \in \mathcal{Y}$:

$$(Q \subset P) \Rightarrow (\mathbf{S} \in \mathbb{N}_{\tau^{(2)}}[\mathrm{cl}(h^1(s^{-1}(Q)), \tau^{(2)})]). \tag{3.6.15}$$

**Discussion.** Under conditions of this theorem, by Theorem 3.4.1 and (2.8.7) we have the equality (3.4.20). Hence one can use (3.6.1) realizing an arbitrary neighborhood $\widetilde{\mathbf{Y}}$ of the set $\mathbf{Y}$. Namely, in (3.6.1) each $\widetilde{\mathbf{Y}}$ can be used to realize the inclusion

$$\omega^1(g^{-1}(\mathbf{Y})) \subset \mathrm{cl}(h^1(s^{-1}(\widetilde{\mathbf{Y}})), \tau^{(2)}). \tag{3.6.16}$$

If Condition 3.4.1 is valid, $\tau^{(2)} \in (\mathrm{top})_0[\mathbf{H}]$, and $\tau^{(3)} \in (\mathbf{c} - \mathrm{top})[\mathbf{H}]$, then by (3.6.16) we have

$$\forall \mathbb{Q} \in \mathcal{Y} : \omega^1(g^{-1}(\mathbf{Y})) \subset \mathrm{cl}(h^1(s^{-1}(\mathbb{Q})), \tau^{(2)}). \tag{3.6.17}$$

Compare (3.6.15) and (3.6.17). Note that (3.6.15) gives the natural neighborhood realization of the set (3.1.12). Therefore properties similar to (3.6.15) are very interesting from the practical point of view. Indeed, in this case the set $\mathrm{cl}(h^1(s^{-1}(Q)), \tau^{(2)})$ can be regarded (from a certain time) as a very precise approximation of $\omega^1(g^{-1}(\mathbf{Y}))$. But $\mathrm{cl}(h^1(s^{-1}(Q)), \tau^{(2)})$ corresponds to a concrete weakening of the $\mathbf{Y}$-constraint.

**Proof of Theorem 3.6.1.** Recall that by (2.8.7), (3.4.5), and Theorem 3.4.1 (see (3.4.20)) we have the equality $\omega^1(g^{-1}(\mathbf{Y})) = (\tau_2 - \text{LIM})[s^{-1}[\mathcal{Y}] \mid h]$. Hence, in order to establish (3.6.15) we can use Proposition 3.6.1 and its corollary, setting in their conditions $U = \mathbf{F}$, $(V, \tau_1) = (\mathbf{K}, \tau^{(3)})$, $(W, \tau_2) = (\mathbf{H}, \tau^{(2)})$, $\alpha = m$, and $\beta = \omega$. Of course, it is advisable to use (3.4.6): $h = \omega \circ m = \beta \circ \alpha$. Then, by Corollary 3.6.1 in the version above, we have $\forall \mathbf{S} \in \mathrm{N}_{\tau^{(2)}}[\omega^1(g^{-1}(\mathbf{Y}))]\ \exists \tilde{P} \in s^{-1}[\mathcal{Y}]\ \forall \tilde{Q} \in s^{-1}[\mathcal{Y}]$:

$$(\tilde{Q} \subset \tilde{P}) \Rightarrow (\mathbf{S} \in \mathrm{N}_{\tau^{(2)}}[\mathrm{cl}(h^1(\tilde{Q}), \tau^{(2)}]). \tag{3.6.18}$$

Fix $\mathbf{S} \in \mathrm{N}_{\tau^{(2)}}[\omega^1(g^{-1}(\mathbf{Y}))]$. Using (3.6.18) we choose $\hat{P} \in s^{-1}[\mathcal{Y}]$ for which the implication (3.6.18) is true for $\tilde{Q} \in s^{-1}[\mathcal{Y}]$ and $\tilde{P} = \hat{P}$. Employing (2.5.8) we obtain the equality $\hat{P} = s^{-1}(P)$ for some $P \in \mathcal{Y}$. Let $Q \in \mathcal{Y}$ be a set for which $Q \subset P$. Then we obtain the inclusion $\hat{Q} \subset \hat{P}$ for $\hat{Q} \triangleq s^{-1}(Q)$. Therefore by the above corollary of (3.6.18) we have $\mathbf{S} \in \mathrm{N}_{\tau^{(2)}}[\mathrm{cl}(h^1(\hat{Q}), \tau^{(2)})]$. Using the definition of $\hat{Q}$ we obtain the validity of the corollary of the implication (3.6.15).

Let us consider the question of the employment of countably compact model TS for investigation of the neighborhood realization of AS. We are oriented towards the case when weakening of the $\mathbf{Y}$-constraint is realized by means of the employment of $H \in \mathfrak{N}$. The last circumstance is connected with the fact that the supposition $\mathfrak{N} \in \mathcal{B}_{\mathcal{N}}(\mathbf{X})$ is quite natural in many concrete settings important in practice. Constructions considered in the particular case of Section 3.4 and concerning Propositions 3.4.10* and 3.4.11* also should be noted.

THEOREM 3.6.2 *Let Condition 3.4.2 be valid. Moreover, suppose that* $\mathfrak{N} \in \mathcal{B}_{\mathcal{N}}(\mathbf{X})$, $\tau^{(2)} \in (\mathcal{D} - \text{top})[\mathbf{H}]$, *and* $\tau^{(3)} \in (\mathbf{c}_{\mathcal{N}} - \text{top})[\mathbf{K}]$. *Finally, let* $\omega \in C_{\mathrm{cl}}(\mathbf{K}, \tau^{(3)}, \mathbf{H}, \tau^{(2)})$. *Then* $\forall \mathbf{S} \in \mathrm{N}_{\tau^{(2)}}[\omega^1(g^{-1}(\mathbf{Y}))]\ \exists P \in \mathfrak{N}\ \forall Q \in \mathfrak{N}$:

$$(Q \subset P) \Rightarrow (\mathbf{S} \in \mathrm{N}_{\tau^{(2)}}[\mathrm{cl}(h^1(s^{-1}(Q)), \tau^{(2)})]). \tag{3.6.19}$$

**Remark.** In connection with (3.6.19) it is advisable to recall the obvious corollary of (3.6.1). Namely, by (3.2.8), (3.4.22), and (3.4.23) we have $\forall H \in \mathfrak{N}$:

$$(\tau^{(2)} - \text{LIM})[s^{-1}[\mathfrak{N}] \mid h] \subset \mathrm{cl}(h^1(s^{-1}(H)), \tau^{(2)}). \tag{3.6.20}$$

In addition, the AS on the right hand side of (3.6.20) coincides (under natural sufficient conditions) with the set (3.1.12). In this connection see, in particular, Proposition 3.4.8. Therefore the validity of

(3.6.19) means (see (3.6.20)) a natural neighborhood realization of the set (3.1.12): $\mathrm{cl}(h^1(s^{-1}(Q)), \tau^{(2)})$ is 'near' to (3.1.12).

**Proof of Theorem 3.6.2.** First recall that by Proposition 3.4.8 the equality (3.4.24) holds. Consider the following concrete version of Proposition 3.6.5: $U = \mathbf{F}$, $\mathcal{U} = s^{-1}[\mathfrak{N}]$, $(V, \tau_1) = (\mathbf{K}, \tau^{(3)})$, $(W, \tau_2) = (\mathbf{H}, \tau^{(2)})$, $\alpha = m$, and $\beta = \omega$. We use (3.4.6). Fix $\mathbf{S} \in \mathbb{N}_{\tau^{(2)}}[\omega^1(g^{-1}(\mathbf{Y}))]$. Then, by Proposition 3.6.5, for some $\tilde{P} \in s^{-1}[\mathfrak{N}]$ it follows that $\forall \tilde{Q} \in s^{-1}[\mathfrak{N}]$:

$$(\tilde{Q} \subset \tilde{P}) \Rightarrow (\mathbf{S} \in \mathbb{N}_{\tau^{(2)}}[\mathrm{cl}(h^1(\tilde{Q}), \tau^{(2)})]). \tag{3.6.21}$$

Choose $P \in \mathfrak{N}$ for which $\tilde{P} = s^{-1}(P)$. If $Q \in \mathfrak{N}$ has the property $Q \subset P$, then $s^{-1}(Q) \in s^{-1}[\mathfrak{N}]$ and $s^{-1}(Q) \subset s^{-1}(P) = \tilde{P}$. By (3.6.21) we have $\mathbf{S} \in \mathbb{N}_{\tau^{(2)}}[\mathrm{cl}(h^1(s^{-1}(Q)), \tau^{(2)})]$. □

In connection with the neighborhood realization of AS when using all sets of $\mathcal{Y}$ to weak the $\mathbf{Y}$-constraint, we recall (along with Theorem 3.6.1) the particular case connected with (3.4.27).

**Remark.** Until the end of this section suppose that $\rho \in (\mathrm{Dist})[\mathbf{X}]$, $\tau^{(1)} \triangleq \tau_\rho^\natural(\mathbf{X})$, and $\mathbf{Y} \in (\mathrm{comp})[\mathbf{X}; \rho]$. Moreover, suppose that $\tau^{(3)} \in (\mathbf{c}_{\mathcal{N}} - \mathrm{top})[\mathbf{K}]$, $\tau^{(2)} \in (\mathcal{D} - \mathrm{top})[\mathbf{H}]$, and $\omega \in C_{\mathrm{cl}}(\mathbf{K}, \tau^{(3)}, \mathbf{H}, \tau^{(2)})$. Recall that (3.4.27) holds. As a result we can use Proposition 3.6.5 under the following parameters: $U = \mathbf{F}$, $\mathcal{U} = s^{-1}[\mathcal{Y}]$, $(V, \tau_1) = (\mathbf{K}, \tau^{(3)})$, $(W, \tau_2) = (\mathbf{H}, \tau^{(2)})$, $\alpha = m$, and $\beta = \omega$. Then by Proposition 3.6.5 and (3.4.6) we have the following statement for $\mathbf{S} \in \mathbb{N}_{\tau^{(2)}}[(\tau^{(2)} - \mathrm{LIM})[s^{-1}[\mathcal{Y}] \mid h]\,]$. Namely, $\exists \tilde{P} \in s^{-1}[\mathcal{Y}]\ \forall \tilde{Q} \in s^{-1}[\mathcal{Y}]$:

$$(\tilde{Q} \subset \tilde{P}) \Rightarrow (\mathbf{S} \in \mathbb{N}_{\tau^{(2)}}[\mathrm{cl}(h^1(\tilde{Q}), \tau^{(2)}]). \tag{3.6.22}$$

But by Statement 3.4.1* we have the coincidence of $\omega^1(g^{-1}(\mathbf{Y}))$ and $(\tau^{(2)} - \mathrm{LIM})[s^{-1}[\mathcal{Y}] \mid h]$. Therefore we can set $\mathbf{S} \in \mathbb{N}_{\tau_2}[\omega^1(g^{-1}(\mathbf{Y}))]$ in (3.6.22). Thus we obtain $\forall \mathbf{S} \in \mathbb{N}_{\tau^{(2)}}[\omega^1(g^{-1}(\mathbf{Y}))]\ \exists P \in \mathcal{Y}\ \forall Q \in \mathcal{Y}$:

$$(Q \subset P) \Rightarrow (\mathbf{S} \in \mathbb{N}_{\tau_2}[\mathrm{cl}(h^1(s^{-1}(Q)), \tau_2)]). \tag{3.6.23}$$

In (3.6.23) we have the statement about the neighborhood realization of the set (3.1.12). It is similar to Theorem 3.6.1. But in the case (3.6.23) we use the countably compact model TS $(\mathbf{K}, \tau^{(3)})$. In this connection we recall (2.2.24).

## 3.7 SOME QUESTIONS OF TOPOLOGICAL EQUIPMENT OF THE SPACE IN WHICH CONSTRAINTS ARE DEFINED

Let us return to the informative discussion of constructions defining the weakening of the $\mathbf{Y}$-constraint in Section 3.1. Recall that the initial setting of the non-perturbed problem of Section 3.1 contains (as parameters) only $\mathbf{F}, \mathbf{X}, \mathbf{H}, \mathbf{Y}, s$ and $h$. Under these parameters (3.1.2) is a natural solution of the problem of attainability under the $\mathbf{Y}$-constraint. The possible instability of the initial problem requires the employment of two new parameters: $\tau^{(1)}$ and $\tau^{(2)}$. The last topology is used in (3.1.4) and in other similar notions for the corresponding formalization of a new version of the attainability set. Very often (but not far always) the choice of $\tau^{(2)}$ is 'indisputable'. We follow this stipulation. Recall only that in many traditional concrete settings of control theory, a finite-dimensional arithmetical space with the usual topology of coordinate-wise convergence is used in place of $(\mathbf{H}, \tau^{(2)})$. Conversely, $\tau^{(1)}$ is a new very essential parameter of the 'asymptotic' setting, since the basic family $\mathcal{Y}$ is defined by $\tau^{(1)}$. At the same time, the concrete choice of $\tau^{(1)}$ is not so obvious.

**Example.** Let the $\mathbf{Y}$-constraint be generated by an infinite system of inequalities. In the informative setting this constraint is defined by the following stipulation: the choice of $f \in \mathbf{F}$ must satisfy the condition

$$\mathbf{s}_\gamma(f) \leq 0 \quad (\gamma \in \Gamma). \tag{3.7.1}$$

In (3.7.1) $\mathbf{s}_\gamma \in \mathbb{R}^{\mathbf{F}}$ for each $\gamma \in \Gamma$, where $\mathbf{S}[\Gamma \neq \varnothing]$. So we have the set $\{\mathbf{s}_\gamma : \gamma \in \Gamma\}$ of functionals on $\mathbf{F}$. Suppose that $\Gamma$ is an infinite set. The stipulation of Section 3.1 with respect to $\mathbf{F}, \mathbf{H}$ and $h$ is supposed to be fulfilled. It is possible to define $\mathbf{X}$ as $\mathbb{R}^\Gamma$. Moreover, we suppose (under $\mathbf{X} = \mathbb{R}^\Gamma$) that $\mathbf{Y}$ is the set of all non-positive functionals on $\Gamma$. Finally, we define $s \in \mathbf{X}^{\mathbf{F}}$ by the following rule:

$$s(f) \triangleq (\mathbf{s}_\gamma(f))_{\gamma \in \Gamma} \in \mathbb{R}^\Gamma \tag{3.7.2}$$

for $f \in \mathbf{F}$. In (3.7.2) we use the so called index form of definition of functions. As a result we obtain a concrete version of the non-perturbed setting of Section 3.1. Let us consider some variants of the weakening of the system (3.7.1). The first variant is defined by the following requirement:

$$\mathbf{s}_\gamma(f) \leq \varepsilon \quad (\gamma \in \Gamma), \tag{3.7.3}$$

where $\varepsilon > 0$. Of course, the parameter $\varepsilon$ can be changed. As a result we obtain a system of admissible sets corresponding to (3.7.3) under the enumeration of $\varepsilon$, $\varepsilon > 0$. Another variant of the weakening of (3.7.1) is defined by conditions of the type

$$\mathbf{s}_\gamma(f) \leq 0 \quad (\gamma \in K), \tag{3.7.4}$$

where $K \in \operatorname{Fin}(\Gamma)$. Varying $K$, we obtain the asymptotics of finite subsystems of the basic system (3.7.1). Finally, we can consider a combination of (3.7.3) and (3.7.4), introducing the requirement

$$\mathbf{s}_\gamma(f) \le \varepsilon \quad (\gamma \in K), \tag{3.7.5}$$

where $K \in \mathrm{Fin}(\Gamma)$ and $\varepsilon \in ]0,\infty[$. Of course, $K$ and $\varepsilon$ are changed. Thus we obtain a system admissible sets. For all the three variants (see (3.7.3)-(3.7.5)) we can choose some natural representations in terms of neighborhoods of $\mathbf{Y}$ (here it is advisable to use constructions of the type (3.4.22)). In the case (3.7.3) it is logical to use the well known topology of uniform convergence (see, for example, [71, Section 2.6]). For the neighborhood representation of (3.7.4) and (3.7.5) it is advisable to use constructions on the basis of the Tichonoff product of samples of $\mathbb{R}$ under its different equipment. Namely, the usual topology of pointwise convergence of $\mathbf{X} = \mathbb{R}^\Gamma$ is exploited in the variant on the basis of (3.7.5). In this variant of the Tichonoff product, $\mathbb{R}$ is equipped with the topology $\tau_{\mathbb{R}}$. The Tichonoff product of samples of $(\mathbb{R}, \tau_\partial)$ is used for the corresponding representation of (3.7.4). Now we omit a detailed discussion, referring the reader to more concrete statements of [29, 30, 32, 35, 40, 45, 47, 117]. Note only that for realization of these topological constructions it is useful to apply the known topological notions connected with the representation of topologies in terms of bases and subbases (see [81, 71] and Section 2.6). We see that various topological equipment of $\mathbf{X}$ is required for solving practical problems.

Returning to the general case we introduce two comparable topologies of $\mathbf{X}$. Such situation arises in a natural way in some abstract problems of control with integral constraints (see [40, 41, 45, 47]). In fact, we consider some range of topologies of $\mathbf{X}$.

So we introduce some alternations in the setting of Section 3.4. We preserve the supposition (3.4.1) and the stipulation with respect to $\mathbf{Y}$, $s$, $h$, and $\tau^{(2)}$. Thus $\mathbf{F}$,$\mathbf{X}$, $\mathbf{H}$ $\mathbf{Y}$, $s$, $h$ and $\tau^{(2)}$ correspond to all stipulations of Section 3.4. Let us fix two topologies:

$$(\boldsymbol{\tau}_\mathrm{l} \in (\mathrm{top})[\mathbf{X}]) \ \& \ (\boldsymbol{\tau}_\mathbf{u} \in (\mathrm{top})[\mathbf{X}]). \tag{3.7.6}$$

We postulate the inclusion

$$\boldsymbol{\tau}_\mathrm{l} \subset \boldsymbol{\tau}_\mathbf{u}. \tag{3.7.7}$$

By (3.3.7) the condition of compatibility of the topologies (3.7.6) is introduced. Namely, the topology $\boldsymbol{\tau}_\mathrm{l}$ is weaker than $\boldsymbol{\tau}_\mathbf{u}$. Thus $(\mathbf{X}, \boldsymbol{\tau}_\mathrm{l})$ and $(\mathbf{X}, \boldsymbol{\tau}_\mathbf{u})$ are two TS for which (3.7.7) holds. Note that by (2.2.17) and (3.7.7) we have

$$\mathcal{F}_{\boldsymbol{\tau}_\mathrm{l}} \subset \mathcal{F}_{\boldsymbol{\tau}_\mathbf{u}}.$$

From (3.7.7) and the representations in Section 2.2 (see (2.2.12)) it follows that $\forall T \in \mathcal{P}(\mathbf{X}) : \mathbb{N}^0_{\boldsymbol{\tau}_\mathrm{l}}[T] \subset \mathbb{N}^0_{\boldsymbol{\tau}_\mathbf{u}}[T]$. As a consequence, from

(2.2.13) we have $\forall E \in \mathcal{P}(\mathbf{X})$:

$$\mathbb{N}_{\boldsymbol{\tau}_\mathrm{l}}[E] \subset \mathbb{N}_{\boldsymbol{\tau}_\mathrm{u}}[E]. \tag{3.7.8}$$

From (2.2.14) and (3.7.8) we obtain $\forall x \in \mathbf{X}$:

$$N_{\boldsymbol{\tau}_\mathrm{l}}(x) \subset N_{\boldsymbol{\tau}_\mathrm{u}}(x). \tag{3.7.9}$$

The properties (3.7.8) and (3.7.9) are very useful in conditions similar to Condition 3.4.1. Returning to the constructions in Section 3.4, we make some correction of (3.4.5), replacing the relations (3.4.5) by the following supposition:

$$(m \in \mathbf{K}^{\mathbf{F}})\& \ (g \in C(\mathbf{K}, \tau^{(3)}, \mathbf{X}, \boldsymbol{\tau}_\mathrm{u})) \ \&(\omega \in C(\mathbf{K}, \tau^{(3)}, \mathbf{H}, \tau^{(2)})). \tag{3.7.10}$$

Finally, we preserve the suppositions (3.4.6). We note that the second condition in (3.7.10) can be interpreted as a universal continuity of $g$. Indeed, from (2.5.1), (2.5.2), and (3.7.7) we obtain that

$$\begin{aligned} C(\mathbf{K}, \tau^{(3)}, \mathbf{X}, \boldsymbol{\tau}_\mathrm{u}) &= \{u \in \mathbf{X}^{\mathbf{K}} \mid \forall G \in \boldsymbol{\tau}_\mathrm{u} : u^{-1}(G) \in \tau^{(3)}\} \\ &\subset \{u \in \mathbf{X}^{\mathbf{K}} \mid \forall G \in \boldsymbol{\tau}_\mathrm{l} : u^{-1}(G) \in \tau^{(3)}\} \\ &= C(\mathbf{K}, \tau^{(3)}, \mathbf{X}, \boldsymbol{\tau}_\mathrm{l}). \end{aligned}$$

As a consequence, from (3.7.10) we obtain the property:

$$g \in C(\mathbf{K}, \tau^{(3)}, \mathbf{X}, \boldsymbol{\tau}_\mathrm{l}). \tag{3.7.11}$$

From (3.7.10) and (3.7.11) it follows that the operator $g$ is continuous both in the sense of $(\mathbf{X}, \boldsymbol{\tau}_\mathrm{l})$ and $(\mathbf{X}, \boldsymbol{\tau}_\mathrm{u})$. In reality, we have the universal continuity in the range of topologies $\tau \in (\mathrm{top})[\mathbf{X}]$, $\boldsymbol{\tau}_\mathrm{l} \subset \tau \subset \boldsymbol{\tau}_\mathrm{u}$. However, we consider only the two topologies (3.7.6). Suppose that

$$(\mathcal{Y}_\mathrm{l} \triangleq \mathbb{N}_{\boldsymbol{\tau}_\mathrm{l}}[\mathbf{Y}])\& \ (\mathcal{Y}_\mathrm{u} \triangleq \mathbb{N}_{\boldsymbol{\tau}_\mathrm{u}}[\mathbf{Y}]). \tag{3.7.12}$$

In (3.7.8) we have introduced two natural variants of the family (3.4.3). Later we consider the possibility of restriction of $\mathcal{Y}_\mathrm{l}$. From (3.7.8) and (3.7.12) we have

$$\mathcal{Y}_\mathrm{l} \subset \mathcal{Y}_\mathrm{u}. \tag{3.7.13}$$

In the sequel we use a natural combination of (3.7.9) and (3.7.13). This combination is supplemented by the above universal continuity. Let us consider some estimates of AS for different variants of the topological equipment of $\mathbf{X}$. From Proposition 3.4.1 and (3.7.13) the obvious chain of inclusions follows:

$$h^1(s^{-1}(\mathbf{Y})) \subset \omega^1(g^{-1}(\mathbf{Y})) \subset \omega^1((\tau^{(3)} - \mathrm{LIM})[s^{-1}[\mathcal{Y}_\mathrm{u}] \mid m])$$

$$\subset (\tau^{(2)} - \mathrm{LIM})[s^{-1}[\mathcal{Y}_{\mathbf{u}}] \mid h] \subset (\tau^{(2)} - \mathrm{LIM})[s^{-1}[\mathcal{Y}_{\mathbf{l}}] \mid h]. \quad (3.7.14)$$

Moreover, from Proposition 3.4.1 and (3.7.13) we always have the chain

$$\omega^1((\tau^{(3)} - \mathrm{LIM})[s^{-1}[\mathcal{Y}_{\mathbf{u}}] \mid m]) \subset \omega^1((\tau^{(3)} - \mathrm{LIM})[s^{-1}[\mathcal{Y}_{\mathbf{l}}] \mid m])$$
$$\subset (\tau^{(2)} - \mathrm{LIM})[s^{-1}[\mathcal{Y}_{\mathbf{l}}] \mid h]. \quad (3.7.15)$$

Finally, from Proposition 3.4.2 we can obtain the chain of inclusions for AS of the (3.2.10) type. Namely, we have the following

PROPOSITION 3.7.1 *Let* $\tau^{(2)} \in (\mathrm{top})_0[\mathbf{H}]$. *Then*

$$\begin{aligned} h^1(s^{-1}(\mathbf{Y})) &\subset ((\mathbf{c}\tau^{(3)}, \tau^{(2)}) - \mathrm{LIM})[s^{-1}[\mathcal{Y}_{\mathbf{u}}] \mid m; h] \\ &\subset ((\mathbf{c}\tau^{(3)}, \tau^{(2)}) - \mathrm{LIM})[s^{-1}[\mathcal{Y}_{\mathbf{l}}] \mid m; h] \\ &\subset \omega^1((\tau^{(3)} - \mathrm{LIM})[s^{-1}[\mathcal{Y}_{\mathbf{l}}] \mid m]) \\ &\subset (\tau^{(2)} - \mathrm{LIM})[s^{-1}[\mathcal{Y}_{\mathbf{l}}] \mid h]. \end{aligned}$$

The corresponding proof is obtained by an direct combination of Proposition 3.4.2 and (3.7.15).

Using (3.2.10) we can easily establish

PROPOSITION 3.7.2 *Let* $\tau^{(2)} \in (\mathrm{top})_0[\mathbf{H}]$. *Then*

$$\begin{aligned} h^1(s^{-1}(\mathbf{Y})) &\subset ((\mathbf{c}\tau^{(3)}, \tau^{(2)}) - \mathrm{LIM})[s^{-1}[\mathcal{Y}_{\mathbf{u}}] \mid m; h] \\ &\subset \omega^1((\tau^{(3)} - \mathrm{LIM})[s^{-1}[\mathcal{Y}_{\mathbf{u}}] \mid m]) \\ &\subset (\tau^{(2)} - \mathrm{LIM})[s^{-1}[\mathcal{Y}_{\mathbf{u}}] \mid h]. \end{aligned}$$

To investigate representations in terms of the set (3.1.12) we consider some analogue of Condition 3.4.1.

CONDITION 3.7.1 $\forall x \in \mathbf{X} \setminus \mathbf{Y}\ \exists H_1 \in N_{\boldsymbol{\tau}_{\mathbf{l}}}(x)\ \exists H_2 \in \mathcal{Y}_{\mathbf{l}} : H_1 \cap H_2 = \varnothing$.

PROPOSITION 3.7.3 *Let Condition 3.7.1 be valid. Then* $\forall x \in \mathbf{X} \setminus \mathbf{Y}$ $\exists H_1 \in N_{\boldsymbol{\tau}_{\mathbf{u}}}(x)\ \exists H_2 \in \mathcal{Y}_{\mathbf{u}} : H_1 \cap H_2 = \varnothing$.

The proof follows from (3.7.9) and (3.7.13).

PROPOSITION 3.7.4 *Let Condition 3.7.1 be valid. Then*

$$g^{-1}(\mathbf{Y}) = (\tau^{(3)} - \mathrm{LIM})[s^{-1}[\mathcal{Y}_{\mathbf{l}}] \mid m] = (\tau^{(3)} - \mathrm{LIM})[s^{-1}[\mathcal{Y}_{\mathbf{u}}] \mid m].$$

The proof is an obvious corollary of Proposition 3.4.4. Namely, we use this proposition twice. In the first variant we employ the constructions of Section 3.4 under $\tau^{(1)} = \boldsymbol{\tau}_{\mathbf{l}}$. Such employment is justified by (3.7.10),

(3.7.11) and Condition 3.7.1. From this and Proposition 3.4.4 the coincidence of $g^{-1}(\mathbf{Y})$ and $(\tau^{(3)} - \mathrm{LIM})[s^{-1}[\mathcal{Y}_{\mathrm{l}}] \mid m]$ follows (see (3.4.3) and (3.7.12)). The second variant is direct application of the constructions from Section 3.4 under $\tau^{(1)} = \boldsymbol{\tau}_{\mathbf{u}}$ (see (3.4.3),(3.4.5),(3.7.10),(3.7.12)). In this case, from Propositions 3.4.4 and 3.7.3 we obtain the coincidence of $g^{-1}(\mathbf{Y})$ and $(\tau^{(3)} - \mathrm{LIM})[s^{-1}[\mathcal{Y}_{\mathbf{u}}] \mid m]$. □

THEOREM 3.7.1 *Let Condition 3.7.1 be valid. Moreover, let* $\omega \in C_{\mathbf{ap}}(\mathbf{K}, \tau^{(3)}, \mathbf{H}, \tau^{(2)})$. *Then*

$$\omega^1(g^{-1}(\mathbf{Y})) = (\tau^{(2)} - \mathrm{LIM})[s^{-1}[\mathcal{Y}_{\mathrm{l}}] \mid h] = (\tau^{(2)} - \mathrm{LIM})[s^{-1}[\mathcal{Y}_{\mathbf{u}}] \mid h]. \tag{3.7.16}$$

To prove this we use two variants of Theorem 3.4.1. We consider the case $\tau^{(1)} = \boldsymbol{\tau}_{\mathrm{l}}$ for which it is advisable to employ (3.7.11), (3.7.12) and Condition 3.7.1. Moreover, we consider (in Theorem 3.4.1) the case $\tau^{(1)} = \boldsymbol{\tau}_{\mathbf{u}}$ with using Proposition 3.7.3, (3.4.1) and (3.7.12). As a result we obtain the equality (3.7.16) by combination of the two above variants. The coincidence of the first and second sets in (3.7.16) follows from the first case. The coincidence of the first and third sets of (3.7.16) is obtained from the second particular case.

PROPOSITION 3.7.5 *Let* $\tau^{(2)} \in (\mathrm{top})_0[\mathbf{H}]$ *and Condition 3.7.1 be valid. Then*

$$h^1(s^{-1}(\mathbf{Y})) \subset ((\mathbf{c}\tau^{(3)}, \tau^{(2)}) - \mathrm{LIM})[s^{-1}[\mathcal{Y}_{\mathbf{u}}] \mid m; h]$$
$$\subset ((\mathbf{c}\tau^{(3)}, \tau^{(2)}) - \mathrm{LIM})[s^{-1}[\mathcal{Y}_{\mathrm{l}}] \mid m; h] \subset \omega^1(g^{-1}(\mathbf{Y})).$$

The proof follows from Propositions 3.7.2 and 3.7.4

THEOREM 3.7.2 *Let* $\tau^{(2)} \in (\mathrm{top})_0[\mathbf{H}]$ *and* $\tau^{(3)} \in (\mathbf{c} - \mathrm{top})_{\mathrm{loc}}[\mathbf{K}]$. *Moreover, let Condition 3.7.1 be valid. Then*

$$\begin{aligned} \omega^1(g^{-1}(\mathbf{Y})) &= ((\mathbf{c}\tau^{(3)}, \tau^{(2)}) - \mathrm{LIM})[s^{-1}[\mathcal{Y}_{\mathrm{l}}] \mid m; h] \\ &= ((\mathbf{c}\tau^{(3)}, \tau^{(2)}) - \mathrm{LIM})[s^{-1}[\mathcal{Y}_{\mathbf{u}}] \mid m; h]. \end{aligned} \tag{3.7.17}$$

PROOF. Consider Theorem 3.4.2 in accordance with the two following variants:

1) $\tau^{(1)} = \boldsymbol{\tau}_{\mathrm{l}}$ and consequently $\mathcal{Y} = \mathcal{Y}_{\mathrm{l}}$;

2) $\tau^{(1)} = \boldsymbol{\tau}_{\mathbf{u}}$ and $\mathcal{Y} = \mathcal{Y}_{\mathbf{u}}$.

Here the relations (3.4.3) and (3.7.12) are used. In the case 1) we use also (3.7.11) and Condition 3.7.1. As a consequence we obtain the validity of the corresponding variant of Condition 3.4.1. Hence the two first sets in (3.7.17) coincide. The case 2) is realized as a concrete variant of Theorem 3.4.2. In this case Condition 3.4.1 follows from Proposition 3.7.3. □.

Fix a nonempty set $\mathbf{Q}$ as in Section 3.4. In addition, fix

$$\Lambda_0 \in (\mathrm{UNIF})[\mathbf{Q}; \boldsymbol{\tau}_\mathrm{l} \mid \mathbf{Y}]. \tag{3.7.18}$$

Thus we use a concrete variant of the corresponding definitions from Section 3.4, and set $\tau^{(1)} = \boldsymbol{\tau}_\mathrm{l}$ in these definitions. Let

$$\mathfrak{N}_\mathrm{l} \triangleq \mathcal{U}(\mathbf{Q}, \boldsymbol{\tau}_\mathrm{l}, \mathbf{Y} \mid \Lambda_0). \tag{3.7.19}$$

By (3.7.18) and (3.7.19) we realize the required concrete variant of the constructions from Section 3.4.

CONDITION 3.7.2 $\forall x \in \mathbf{X} \setminus \mathbf{Y}\ \exists \mathbf{q} \in \mathbf{Q} : \Lambda_0(x, \mathbf{q}) \cap \left( \bigcup_{y \in \mathbf{Y}} \Lambda_0(y, \mathbf{q}) \right) = \varnothing$.

Note that Condition 3.7.2 is a version of Condition 3.4.2 corresponding to the case $\tau^{(1)} = \boldsymbol{\tau}_\mathrm{l}$ (compare (3.4.5) and (3.7.11)).

PROPOSITION 3.7.6 *Condition 3.7.2 implies Condition 3.7.1*

The proof follows from Proposition 3.4.6 and the above remark about the connection of Conditions 3.4.2 and 3.7.2.

PROPOSITION 3.7.7 *Let Condition 3.7.2 be valid. Then*

$$\begin{aligned} g^{-1}(\mathbf{Y}) &= (\tau^{(3)} - \mathrm{LIM})[s^{-1}[\mathcal{Y}_\mathrm{l}] \mid m] \\ &= (\tau^{(3)} - \mathrm{LIM})[s^{-1}[\mathcal{Y}_\mathbf{u}] \mid m] \\ &= (\tau^{(3)} - \mathrm{LIM})[s^{-1}[\mathfrak{N}_\mathrm{l}] \mid m]. \end{aligned} \tag{3.7.20}$$

To prove this it is sufficient to compare Propositions 3.4.7, 3.7.4, and 3.7.6. Indeed, Proposition 3.4.7 is used under $\tau^{(1)} = \boldsymbol{\tau}_\mathrm{l}$. In this case a natural combination of (3.7.10) and (3.7.11) is taken instead of (3.4.5). In addition, the family $\mathcal{Y}$ from Section 3.4 is replaced by $\mathcal{Y}_\mathrm{l}$ (see (3.4.3) and (3.7.12)), and the family $\mathfrak{N}$ from Section 3.4 is replaced by $\mathfrak{N}_\mathrm{l}$. As a result, from Proposition 3.4.7 we obtain the coincidence of the first, second, and fourth sets in (3.7.20). Moreover, Proposition 3.7.6 implies the validity of Condition 3.7.1. Therefore from Proposition 3.7.4 we obtain the coincidence of $g^{-1}(\mathbf{Y})$ and the third set of (3.7.20). □

COROLLARY 3.7.1 *Let Condition 3.7.2 be valid. Then* $\forall \mathcal{H} \in \mathcal{B}[\mathbf{F}]$:

$$\left(s^{-1}[\mathfrak{N}_\mathrm{l}] \subset \mathcal{H} \subset s^{-1}[\mathcal{Y}_\mathbf{u}]\right) \Rightarrow \left(g^{-1}(\mathbf{Y}) = (\tau^{(3)} - \mathrm{LIM})[\mathcal{H} \mid m]\right).$$

To prove this, it is sufficient to compare (3.2.8) and Proposition 3.7.7.

PROPOSITION 3.7.8 *Let Condition 3.7.2 be valid. In addition, let* $\mathfrak{N}_{\mathrm{l}} \in \mathcal{B}_{\mathcal{N}}(\mathbf{X})$, $\tau^{(3)} \in (\mathbf{c}_{\mathcal{N}} - \mathrm{top})[\mathbf{K}]$, $\tau^{(2)} \in (\mathcal{D} - \mathrm{top})[\mathbf{H}]$ *and* $\omega \in C_{\mathrm{cl}}(\mathbf{K}, \tau^{(3)}, \mathbf{H}, \tau^{(2)})$. *Then*

$$\omega^1(g^{-1}(\mathbf{Y})) = (\tau^{(2)} - \mathrm{LIM})[s^{-1}[\mathfrak{N}_{\mathrm{l}}] \mid h]. \tag{3.7.21}$$

The proof is a concrete variant of Proposition 3.4.8. Namely, we consider the case $\tau^{(1)} = \boldsymbol{\tau}_{\mathrm{l}}$ and use (3.4.22), (3.7.11), and (3.7.19). Then (3.4.24) gives the required statement (3.7.21).

PROPOSITION 3.7.9 *Let Condition 3.7.2 be valid. In addition, let* $\mathfrak{N}_{\mathrm{l}} \in \mathcal{B}_{\mathcal{N}}(\mathbf{X})$ *and* $\omega \in C_{\mathbf{qp}}(\mathbf{K}, \tau^{(3)}, \mathbf{H}, \tau^{(2)})$. *Then (3.7.21) holds.*

The proof follows from Proposition 3.4.9 under $\tau^{(1)} = \boldsymbol{\tau}_{\mathrm{l}}$.

We restrict ourselves to Propositions 3.7.8 and 3.7.9 when discussing variants of the employment of countably compact model TS for constructing 'universal' versions of AS. Other concrete variants can be obtained on the basis of corresponding statements from Section 3.4 (see, for example, Propositions 3.4.10* and 3.4.11*).

THEOREM 3.7.3 *Let Condition 3.7.2 be valid. In addition, let* $\omega \in C_{\mathbf{ap}}(\mathbf{K}, \tau^{(3)}, \mathbf{H}, \tau^{(2)})$. *Then*

$$\begin{aligned} \omega^1(g^{-1}(\mathbf{Y})) &= (\tau^{(2)} - \mathrm{LIM})[s^{-1}[\mathfrak{Y}_{\mathrm{l}}] \mid h] \\ &= (\tau^{(2)} - \mathrm{LIM})[s^{-1}[\mathfrak{Y}_{\mathbf{u}}] \mid h] \\ &= (\tau^{(2)} - \mathrm{LIM})[s^{-1}[\mathfrak{N}_{\mathrm{l}}] \mid h]. \end{aligned} \tag{3.7.22}$$

PROOF. Condition 3.7.1 is valid by Proposition 3.7.6. Theorem 3.7.1 implies the coincidence of the three first sets in (3.7.22). The coincidence of $g^{-1}(\mathbf{Y})$ and $(\tau^{(3)} - \mathrm{LIM})[s^{-1}[\mathfrak{N}_{\mathrm{l}}] \mid m]$ follows from Proposition 3.7.7. Note that Condition 3.7.2 is a concrete variant of Condition 3.4.2 corresponding to the case $\tau^{(1)} = \boldsymbol{\tau}_{\mathrm{l}}$. Hence by Theorem 3.5.1 we have the coincidence of $\omega^1(g^{-1}(\mathbf{Y}))$ and $(\tau^{(2)} - \mathrm{LIM})[s^{-1}[\mathfrak{N}_{\mathrm{l}}] \mid h]$. Here it is advisable to use (3.7.11). □.

PROPOSITION 3.7.10 *Suppose that Condition 3.7.2 is valid and* $\tau^{(2)} \in (\mathrm{top})_0[\mathbf{H}]$. *Then*

$$\begin{aligned} h^1(s^{-1}(\mathbf{Y})) &\subset ((\mathbf{c}\tau^{(3)}, \tau^{(2)}) - \mathrm{LIM})[s^{-1}[\mathfrak{Y}_{\mathbf{u}}] \mid m; h] \\ &\subset ((\mathbf{c}\tau^{(3)}, \tau^{(2)}) - \mathrm{LIM})[s^{-1}[\mathfrak{Y}_{\mathrm{l}}] \mid m; h] \\ &\subset ((\mathbf{c}\tau^{(3)}, \tau^{(2)}) - \mathrm{LIM})[s^{-1}[\mathfrak{N}_{\mathrm{l}}] \mid m; h] \\ &\subset \omega^1(g^{-1}(\mathbf{Y})). \end{aligned} \tag{3.7.23}$$

**Proof.** Recall that Condition 3.7.1 is valid (see Proposition 3.7.6). Then the statement of Proposition 3.7.5 holds. Moreover, by (3.2.19), (3.7.12), (3.7.18), and (3.7.19) we have

$$\mathfrak{N}_\mathrm{l} \subset \mathcal{Y}_\mathrm{l}. \tag{3.7.24}$$

From (3.2.10) and (3.7.24) we obtain the inclusion

$$((\mathbf{c}\tau^{(3)}, \tau^{(2)}) - \mathrm{LIM})[s^{-1}[\mathcal{Y}_\mathrm{l}] \mid m; h]$$
$$\subset ((\mathbf{c}\tau^{(3)}, \tau^{(2)}) - \mathrm{LIM})[s^{-1}[\mathfrak{N}_\mathrm{l}] \mid m; h].$$

Consider the last inclusion in (3.7.23). To establish it we use Proposition 3.5.1. Recall that Condition 3.7.2 is a version of Condition 3.4.2 corresponding to the case $\tau^{(1)} = \tau_\mathrm{l}$ (see (3.7.10) and (3.7.11)). Hence from Proposition 3.5.1 we obtain the last inclusion in (3.7.23); in addition, (3.4.22) and (3.7.19) should be used. Subsequent arguments are obvious. □

**Theorem 3.7.4** *Let Condition 3.7.2 be valid. Moreover, let* $\tau^{(2)} \in (\mathrm{top})_0[\mathbf{H}]$ *and* $\tau^{(3)} \in (\mathbf{c} - \mathrm{top})_{\mathrm{loc}}[\mathbf{K}]$. *Then*

$$\begin{aligned}\omega^1(g^{-1}(\mathbf{Y})) &= ((\mathbf{c}\tau^{(3)}, \tau^{(2)}) - \mathrm{LIM})[s^{-1}[\mathcal{Y}_\mathrm{u}] \mid m; h] \\ &= ((\mathbf{c}\tau^{(3)}, \tau^{(2)}) - \mathrm{LIM})[s^{-1}[\mathcal{Y}_\mathrm{l}] \mid m; h] \\ &= ((\mathbf{c}\tau^{(3)}, \tau^{(2)}) - \mathrm{LIM})[s^{-1}[\mathfrak{N}_\mathrm{l}] \mid m; h].\end{aligned} \tag{3.7.25}$$

**Proof.** By Proposition 3.7.6, Condition 3.7.1 is valid. Hence from Theorem 3.7.2 we have the coincidence of the first three sets in (3.7.5); in this connection see (3.7.17). Moreover, we again use the natural equivalence of Conditions 3.4.2 and 3.7.2 in the case when, in Section 3.4, $\tau^{(1)} = \tau_\mathrm{l}$. Then by Theorem 3.5.2 we obtain the equality

$$\omega^1(g^{-1}(\mathbf{Y})) = ((\mathbf{c}\tau^{(3)}, \tau^{(2)}) - \mathrm{LIM})[s^{-1}[\mathfrak{N}_\mathrm{l}] \mid m; h]$$

(in the last equality it is advisable to compare (3.4.22) and (3.7.19), and also to use (3.7.11)). □

In the conclusion we note that, in typical cases of the initial setting of the problem of attainability under perturbed constraints, a universal regularization in the form of AS exists when we can construct a respective model TS and auxiliary mappings for which representations similar to (3.4.5) and (3.4.6) take place. In this section we work with some variants of the above relations. We have to state that a rational construction of the space of GE is our main goal. The respective conditions (see (3.4.5), (3.4.6), and (3.7.10)) determine the required structure of this model space. Note that the above concrete variant of rational choice of the space of GE was realized in [29, 30, 32, 35, 40, 41, 45, 47]. On the other hand, for classical problems of control with geometric constraints another concrete scheme of the space of GE was constructed (see, for example, [117]).

## 3.8 CONCLUSION

The schemes constructed for extension of abstract problems of attainability in topological spaces are conceptually rise to [117], where more concrete constructions for control problems and calculus of variations are discussed. At the same time, in [117, 120] we can see some concepts of topology, which, in fact, are realized in the framework of sequential constructions; i.e., sequences of usual solutions are employed for formalization of approximate solutions. The idea of topological extension (compactification) is developed, to a smaller extent, in mathematical programming, where the duality is most used. However, this duality is applied in general in problems of convex programming (see in this connection [65, 79]). On the other hand, the idea of compactification in general requires some correction in extension constructions for extremal problems and problems of attainability. Roughly speaking, the extension with preserving some connections is required here. This idea is subsequently realized in Chapter 3 in general topological terms; to this end (and it is a new point), the authors do not identify the extension of a problem only with the compactification of the space of solutions in the presence of connections, but consider more general procedures going back to [28, 32, 35, 40, 45, 46, 47]. In particular, the idea of local compactifications under perfect and almost perfect mappings is used here. Only consequences of the application of them to the asymptotic realization of effects arising under subsequent relaxation of constraints are investigated. On the whole, the methods of Chapter 3 can be regarded as schemes of 'topological regularization' of origin problems of attainability; the latter can be ill posed and require, in fact, some improvement at the level of mathematical setting.

# Chapter 4

# ELEMENTS OF MEASURE THEORY AND EXTENSION CONSTRUCTIONS

## 4.1 INTRODUCTION

In the previous chapter a topological basis of extension was given. But realization of the general idea above requires constructing the space of GE. It is useful to note that GE are very often defined as measures. In this chapter we also follow this approach. We consider the spaces of measures and employ these spaces as model TS (see Section 3.4). Of course, we use properties of measures selectively. Firstly, we mean questions concerning the identification of measures and linear functionals. In this connection we note the well known Riesz theorem on representation of linear functionals on the space of continuous functions. In some cases it is required to consider such functionals on the space of discontinuous functions. For a respective representation we need to use finitely additive measures (FAM). On this basis the theory of extension for problems with integral constraints was suggested in [32, 35, 45, 46]. Constructions based on the Riesz theorem are used for the extension of control problems with geometric constraints (see, for example, [78, 117, 120]). In this chapter we discuss in detail constructions connected with integration with respect to FAM. In particular, we consider questions connected with the representation of linear continuous functionals on a space of discontinuous functions. This representation is very important for constructing compactifications which are used both in constructions similar to those considered in the previous chapter and in some constructions connected with problems of functional analysis. In particular, we touch upon some questions connected with the problem of universal integrability of bounded functions (see, for example, [3, 22, 92]). We use (0,1)-measures as 'the material' for compactifications.

## 4.2 FINITELY ADDITIVE AND COUNTABLY ADDITIVE MEASURES

In this section we fix a set $E$. Let us consider families of subsets of $E$, i.e., elements of $\mathcal{P}(\mathcal{P}(E))$ (see Section 2.2). If $\mathcal{L} \in 2^{\mathcal{P}(E)}$, then $\mathbf{S}[\mathbb{R}^{\mathcal{L}} \neq \varnothing]$; we call elements of the set of the type $\mathbb{R}^{\mathcal{L}}$ set functions. Among such set functions we select FAM. Moreover, we consider a stronger property of countable additivity of set functions. First we recall various types of measurable spaces (MS) on which FAM and countably additive measures (CAM) are considered. Recall that (see (2.2.10))

$$\begin{aligned}\pi[E] = \{\mathcal{L} \in \mathcal{P}(\mathcal{P}(E)) \mid (\emptyset \in \mathcal{L}) \;\&\; (E \in \mathcal{L}) \\ \&\; (\forall A \in \mathcal{L}\, \forall B \in \mathcal{L} :\; A \cap B \in \mathcal{L})\} \subset 2^{\mathcal{P}(E)}.\end{aligned} \tag{4.2.1}$$

In (4.2.1) we consider (following [21]) the set of all multiplicative families of subsets of $E$ with zero and unit. In the sequel we simply call elements of (4.2.1) multiplicative families.

**Example.** Let $E \triangleq ]0,1[$ and $\mathcal{L}$ be the family of all intervals $]a,b[$, $(a,b) \in [0,1] \times [0,1]$; then $\mathcal{L} \in \pi[E]$. But the family $\mathcal{L}$ has a property which makes employment of it in problems of measure theory difficult. Namely, it is possible to indicate a set $L \in \mathcal{L}$ for which $E \setminus L \notin \mathcal{L}$ and, what is more, a finite partition of $E \setminus L$ by elements of $\mathcal{L}$ is impossible.

In connection with the previous example and also other causes, it is important to introduce the notion of a measurable partition. If $\mathcal{L} \in \pi[E]$, $H \in \mathcal{P}(E)$, and $n \in \mathcal{N}$, then we have (see (2.2.25))

$$\begin{aligned}\Delta_n(H, \mathcal{L}) = \{(L_i)_{i \in \overline{1,n}} \in \mathcal{L}^n \mid (H = \bigcup_{i=1}^{n} L_i) \;\&\; (\forall p \in \overline{1,n} \\ \forall q \in \overline{1,n} \setminus \{p\} :\; L_p \cap L_q = \emptyset)\} \in \mathcal{P}(\mathcal{L}^n).\end{aligned} \tag{4.2.2}$$

**Remark.** Recall that an arbitrary set can be used in place of $E$. In (2.2.25) and (4.2.2) it does not matter for which set $E$ the set $H$ is considered as a subset. Moreover, employment of an element of $\pi[E]$ in place of $\mathcal{L}$ is not necessary. Therefore in the conditions defining (4.2.2) we can impose the following collection of stipulations: $\mathcal{L}$ is a family of sets; $H$ is a set; and $n \in \mathcal{N}$. But we use the form of stipulations relative to conditions of (4.2.2) for the reason of the methodical character. Recall that families of the type $\mathcal{L} \in \pi[E]$ were considered in Section 2.2. But in these constructions we have oriented toward applications to topology. We now make of use of (2.2.10) in measure theory. In this connection the notion of a measurable partition plays a very important role. In fact, we begin to develop another branch of applications of multiplicative families.

Recall the very important notion of a semi-algebra of sets (see, for example, [103, Ch.I]). We have

$$\Pi[E] = \{\mathcal{L} \in \pi[E] | \ \forall L \in \mathcal{L}\ \exists n \in \mathcal{N}: \ \Delta_n(E \setminus L, \mathcal{L}) \neq \emptyset\} \in 2^{\pi[E]} \quad (4.2.3)$$

(we note that the family of intervals from the example above is not element of the set (4.2.3)). Then (see Section 2.2)

$$\begin{aligned}(\mathrm{alg})[E] &= \{\mathcal{L} \in \Pi[E] \mid \forall L \in \mathcal{L}: \ E \setminus L \in \mathcal{L}\} \\ &= \{\mathcal{L} \in \pi[E] \mid \forall L \in \mathcal{L}: \ E \setminus L \in \mathcal{L}\} \in 2^{\Pi[E]}\end{aligned} \quad (4.2.4)$$

is the set of all algebras of subsets of $E$. It is useful to note that $\forall \mathcal{L} \in (\mathrm{alg})[E]\ \forall n \in \mathcal{N}\ \forall (L_i)_{i \in \overline{1,n}} \in \mathcal{L}^n$:

$$\left(\bigcup_{i=1}^{n} L_i \in \mathcal{L}\right) \ \& \ \left(\bigcap_{i=1}^{n} L_i \in \mathcal{L}\right).$$

However, the union of all sets of a sequence in an algebra $\mathcal{L}$ of subsets of $E$ may not be element of $\mathcal{L}$. Therefore we use the set

$$\begin{aligned}(\sigma - \mathrm{alg})[E] &= \left\{\mathcal{L} \in (\mathrm{alg})[E] | \ \forall (L_i)_{i \in \mathcal{N}} \in \mathcal{L}^{\mathcal{N}}: \ \bigcup_{i \in \mathcal{N}} L_i \in \mathcal{L}\right\} \\ &= \left\{\mathcal{L} \in (\mathrm{alg})[E] | \ \forall (L_i)_{i \in \mathcal{N}} \in \mathcal{L}^{\mathcal{N}}: \ \bigcap_{i \in \mathcal{N}} L_i \in \mathcal{L}\right\}.\end{aligned} \quad (4.2.5)$$

In Section 2.2 questions of constructing the algebra and $\sigma$-algebra generated by an arbitrary family of sets were considered. Let $\forall \mathcal{L} \in \pi[E]$ $\forall A \in \mathcal{P}(E)$:

$$\Delta_\infty(A, \mathcal{L}) \triangleq \{(L_i)_{i \in \mathcal{N}} \in \mathcal{L}^{\mathcal{N}} \ |(A = \bigcup_{i \in \mathcal{N}} L_i) \& (\forall p \in \mathcal{N}\ \forall q \in \mathcal{N} \setminus \{p\}: \ L_p \cap L_q = \varnothing)\}. \quad (4.2.6)$$

From (4.2.1), (4.2.3)–(4.2.5) we deduce the following scale of measurable structures:

$$(\sigma - \mathrm{alg})[E] \subset (\mathrm{alg})[E] \subset \Pi[E] \subset \pi[E] \subset 2^{\mathcal{P}(E)}. \quad (4.2.7)$$

We use (4.2.7) when investigating various types of MS. If $\mathcal{L} \in (\sigma - \mathrm{alg})[E]$, then the pair $(E, \mathcal{L})$ is called the standard MS or, more briefly, SMS. Such SMS are usually used in classical measure theory. In Section 2.2 we considered MS with a semi-algebra of sets. The concrete application of families of $\pi[E] \setminus \Pi[E]$ is very bound. But it is advisable to realize some constructions of integration with respect to FAM for MS $(E, \mathcal{L})$, where $\mathcal{L} \in \pi[E]$. Of course, we use the extension procedure for measures (see, for example, [103]).

Suppose that $\forall \mathcal{L} \in \pi[E]$:

$$\Big((\text{add})[\mathcal{L}] \triangleq \{\mu \in \mathbb{R}^{\mathcal{L}} \mid \forall L \in \mathcal{L}\ \forall n \in \mathcal{N}\ \forall (L_i)_{i \in \overline{1,n}} \in \Delta_n(L, \mathcal{L}) : \mu(L) = \sum_{i=1}^{n} \mu(L_i)\}\Big)$$
$$\& \Big((\sigma - \text{add})[\mathcal{L}] \triangleq \{\mu \in \mathbb{R}^{\mathcal{L}} \mid \forall L \in \mathcal{L}\ \forall (L_i)_{i \in \mathcal{N}} \in \Delta_\infty(L, \mathcal{L}) : \Big(\sum_{i=1}^{k} \mu(L_i)\Big)_{k \in \mathcal{N}} \to \mu(L))\}\Big). \tag{4.2.8}$$

In (4.2.8) FAM and CAM on arbitrary multiplicative families are defined. Recall that each topology is a multiplicative family. Hence it is possible to consider FAM and CAM defined on a topology. However, such construction is not sufficiently informative. At the same time, the employment of at least semi-algebra of sets changes the situation cardinally.

We use the notation and stipulations of Section 2.2. From (4.2.8) we have $\forall \mathcal{L} \in \pi[E]$:

$$(\sigma - \text{add})[\mathcal{L}] \subset (\text{add})[\mathcal{L}] \subset \mathbb{R}^{\mathcal{L}}. \tag{4.2.9}$$

In (4.2.8) and (4.2.9) measures with alternating signs are used. Sometimes such measures are called charges (see, for example, [106, 110]). But the case of non-negative measures is most important for our goals. In this connection we suppose that $\forall \mathcal{L} \in \pi[E]$:

$$\begin{aligned}((\text{add})_+[\mathcal{L}] &\triangleq \{\mu \in (\text{add})[\mathcal{L}] \mid \mathbb{O}_{\mathcal{L}} \leqq \mu\}) \\ \&((\sigma - \text{add})_+[\mathcal{L}] &\triangleq \{\mu \in (\sigma - \text{add})[\mathcal{L}] \mid \mathbb{O}_{\mathcal{L}} \leqq \mu\}).\end{aligned} \tag{4.2.10}$$

Elements of the second set in (4.2.10) are measures in the classical sense (if $\mathcal{L} \in (\sigma - \text{alg})[\mathcal{L}]$). Obviously, $\forall \mathcal{L} \in \pi[E]$:

$$(\sigma - \text{add})_+[\mathcal{L}] \subset (\text{add})_+[\mathcal{L}] \subset [0, \infty[^{\mathcal{L}}.$$

The last property follows from (4.2.9). Among more general measures (see (4.2.8)), FAM with the bounded variation are selected. Some notations are required to introduce a corresponding definition. If $\mathcal{L} \in \pi[E]$,

$\mu \in \mathbb{R}^{\mathcal{L}}$, and $L \in \mathcal{L}$, then we denote by $(\mathrm{VAR})_L[\mu]$ the (nonempty) set of all $c \in [0, \infty[$ such that

$$\exists n \in \mathcal{N}\ \exists (L_i)_{i \in \overline{1,n}} \in \Delta_n(L, \mathcal{L}) : c = \sum_{i=1}^{n} |\mu(L_i)|;$$

in particular, $(\mathrm{VAR})_E[\mu]$ is defined. If $(\mathrm{VAR})_E[\mu]$ is a bounded subset of $\mathbb{R}$, then the set function $\mu$ is called the function with bounded variation. It is useful to introduce the space of all such functions. Namely, $\forall \mathcal{L} \in \pi[E]$:

$$(B - \mathrm{var})[\mathcal{L}] \triangleq \{\mu \in \mathbb{R}^{\mathcal{L}} \mid \exists c \in [0, \infty[: (\mathrm{VAR})_E[\mu] \subset [0, c]\}. \quad (4.2.11)$$

The variation as a set function is defined on the space (4.2.11). In particular, the complete variation can be considered. Namely, suppose that $\forall \mathcal{L} \in \pi[E]\ \forall \mu \in (B - \mathrm{var})[\mathcal{L}]$ :

$$V_\mu \triangleq \sup((\mathrm{VAR})_E[\mu]) \in [0, \infty[. \quad (4.2.12)$$

The number (4.2.12) is the complete variation of the set function $\mu$. The following definition is very important for extension theory (see [32, 35, 45, 46]). Assume that $\forall \mathcal{L} \in \pi[E]$:

$$\mathbf{A}(\mathcal{L}) \triangleq (\mathrm{add})[\mathcal{L}] \cap (B - \mathrm{var})[\mathcal{L}]. \quad (4.2.13)$$

In (4.2.13) we have the set of all FAM of bounded variation with the domain $\mathcal{L}$. Note that for $\mathcal{L} \in \pi[E]$, $\mu \in (\mathrm{add})_+[\mathcal{L}]$, and $L \in \mathcal{L}$ the set $(\mathrm{VAR})_L[\mu]$ is $\{\mu(L)\}$. As a consequence we have $\forall \mathcal{L} \in \pi[E]$:

$$(\sigma - \mathrm{add})_+[\mathcal{L}] \subset (\mathrm{add})_+[\mathcal{L}] \subset \mathbf{A}(\mathcal{L}). \quad (4.2.14)$$

If $\mathcal{L} \in \pi[E]$ and $\mu \in (\mathrm{add})_+[\mathcal{L}]$, then $V_\mu = \mu(E)$. We now recall an important representation connected with the well known Hewitt Yosida decomposition; see [66, 106, 110, 118]. We restrict ourselves to the realization of components of this decomposition in the cone of non-negative elements of $\mathbf{A}(\mathcal{L})$, i.e., in the cone $(\mathrm{add})_+[\mathcal{L}]$. Recall that $\mathbf{A}(\mathcal{L})$ is regarded as a subset of $\mathbb{R}^{\mathcal{L}}$ with pointwise linear operations. In addition, for $\mathcal{L} \in \pi[E]$, we have linear subspaces of $\mathbb{R}^{\mathcal{L}}$ in the form $(\mathrm{add})[\mathcal{L}]$, $(\sigma - \mathrm{add})[\mathcal{L}]$ and $(B - \mathrm{var})[\mathcal{L}]$; moreover,

$$\mu \mapsto V_\mu : (B - \mathrm{var})[\mathcal{L}] \to [0, \infty[ \quad (4.2.15)$$

is a semi-norm of $(B - \mathrm{var})[\mathcal{L}]$, and

$$\mu \mapsto V_\mu : \mathbf{A}(\mathcal{L}) \to [0, \infty[ \quad (4.2.16)$$

is a semi-norm of $\mathbf{A}(\mathcal{L})$. If $\mathcal{L} \in \Pi[E]$, then (4.2.15) and (4.2.16) are norms; moreover, (4.2.16) is called the strong norm of $\mathbf{A}(\mathcal{L})$ and, besides, $(\text{add})_+[\mathcal{L}] = \{\mu \in \mathbf{A}(\mathcal{L}) \mid \mathbb{O}_\mathcal{L} \leqq \mu\}$ (this equality is also valid for $\mathcal{L} \in \pi[E]$). Returning to the representation of components in the above cone, we introduce $\forall \mathcal{L} \in \pi[E]$:

$$(\mathbf{p} - \text{add})_+[\mathcal{L}] \triangleq \{\mu \in (\text{add})_+[\mathcal{L}] \mid \forall \nu \in (\sigma - \text{add})_+[\mathcal{L}] : (\nu \leqq \mu) \Rightarrow (\nu = \mathbb{O}_\mathcal{L})\}. \tag{4.2.17}$$

In (4.2.17) non-negative purely FAM on $\mathcal{L}$ are defined. Of course, (4.2.17) is a cone. This cone is usually considered in the case when $\mathcal{L}$ is at least a semi-algebra of sets. In terms of the pair

$$((\sigma - \text{add})_+[\mathcal{L}], (\mathbf{p} - \text{add})_+[\mathcal{L}]) \tag{4.2.18}$$

the Hewitt Yosida decomposition in the cone $(\text{add})_+[\mathcal{L}]$ can be realized. Components of the pair (4.2.18) can be regarded as components of this decomposition in $(\text{add})_+[\mathcal{L}]$. Suppose that $\forall \mathcal{L} \in \pi[E]$:

$$\mathbb{P}(\mathcal{L}) \triangleq \{\mu \in (\text{add})_+[\mathcal{L}] \mid \mu(E) = 1\}. \tag{4.2.19}$$

It is possible to call elements of (4.2.19) finitely additive probabilities (FAP); such interpretation is informative in the case when $\mathcal{L}$ is at least a semi-algebra of sets. In correspondence with (4.2.18) we introduce $\forall \mathcal{L} \in \pi[E]$:

$$(\mathbb{P}_\sigma(\mathcal{L}) \triangleq \mathbb{P}(\mathcal{L}) \cap (\sigma - \text{add})_+[\mathcal{L}])\&(\mathbb{P}_\mathbf{p}(\mathcal{L}) \triangleq \mathbb{P}(\mathcal{L}) \cap (\mathbf{p} - \text{add})_+[\mathcal{L}]). \tag{4.2.20}$$

Elements of the first set in (4.2.20) for $\mathcal{L} \in (\sigma - \text{alg})[E]$ are Kolmogorov probabilities. They play a very important role in probability theory. Let $\forall \mathcal{L} \in \pi[E]$:

$$\mathbb{T}(\mathcal{L}) \triangleq \{\mu \in \mathbb{P}(\mathcal{L}) \mid \forall L \in \mathcal{L} : (\mu(L) = 0) \vee (\mu(L) = 1)\}. \tag{4.2.21}$$

We call elements of the set (4.2.21) two-valued FAP or $(0,1)$-measures. These measures are used in the theory of Boolean algebras [111, 116], non-standard analysis [63] and other fields of mathematics. Of course, in these investigations $\mathcal{L}$ is supposed to be an algebra of sets. We have $\forall \mathcal{L} \in \pi[E]$:

$$\begin{aligned}(\mathbb{T}_\sigma(\mathcal{L}) &\triangleq \mathbb{T}(\mathcal{L}) \cap (\sigma - \text{add})_+[\mathcal{L}] \\ &= \{\mu \in \mathbb{P}_\sigma(\mathcal{L}) \mid \forall L \in \mathcal{L} : (\mu(L) = 0) \vee (\mu(L) = 1)\}) \\ \&(\mathbb{T}_\mathbf{p}(\mathcal{L}) &\triangleq \mathbb{T}(\mathcal{L}) \cap (\mathbf{p} - \text{add})_+[\mathcal{L}] \\ &= \{\mu \in \mathbb{P}_\mathbf{p}(\mathcal{L}) \mid \forall L \in \mathcal{L} : (\mu(L) = 0) \vee (\mu(L) = 1)\}).\end{aligned} \tag{4.2.22}$$

In (4.2.22) we realize components of the Hewitt Yosida decomposition for the set (4.2.21). Note that the simplest example of (0, 1)-measure of the set $\mathbb{T}_\sigma(\mathcal{L})$ (4.2.22) is the well known Dirac measure concentrated at some point of $E$ and defined on $\mathcal{L}$. The Dirac measure itself is usually defined on the $\sigma$-algebra of all subsets of $E$. The above variant connected with (4.2.22) is, strongly speaking, a restriction of the 'true' Dirac measure. It is possible to give other examples of CAM of $\mathbb{T}_\sigma(\mathcal{L})$. In [2] necessary and sufficient conditions for coincidence of $\mathbb{T}_\sigma(\mathcal{L})$ with the set of all Dirac measures were established. Note the very important case $\mathcal{L} = \mathcal{P}(E)$. In particular, the well known measure problem or the problem of the existence of measurable cardinal numbers should be mentioned; see, for example, [71, Sec. 3.11]. The question is of whether $\{\mu \in \mathbb{T}_\sigma(\mathcal{P}(E)) \mid \forall x \in E : \mu(\{x\}) = 0\} \neq \varnothing$ (for some concrete variant of the set $E$)? The set (4.2.21) was used in extension constructions (see, for example, [24, 25, 27, 32, 35, 38, 121]).

Returning to the case of FAM with alternating signs, it is natural to regard the variation of FAM as a set function. If $\mathcal{L} \in \Pi[E]$ and $\mu \in (B-\text{var})[\mathcal{L}]$, then $(\text{VAR})_L[\mu] \in 2^{[0,V_\mu]}$ for every $L \in \mathcal{L}$; as a consequence it is possible to introduce the following real-valued function $v_\mu$:

$$L \mapsto \sup((\text{VAR})_L[\mu]) : \mathcal{L} \to [0, V_\mu]. \tag{4.2.23}$$

So $v_\mu(L)$ coincides with the supremum of $(\text{VAR})_L[\mu]$ for $L \in \mathcal{L}$. From (4.2.12) we have $\forall \mathcal{L} \in \Pi[E]\ \forall \mu \in (B - \text{var})[\mathcal{L}]$:

$$v_\mu(E) = V_\mu. \tag{4.2.24}$$

The relations (4.2.12), (4.2.23) and (4.2.24) characterize the variation as a set function. Obviously, these relations can be used in the case when $\mu \in \mathbf{A}(\mathcal{L})$. So we obtain a corollary of general definitions for FAM with alternating signs.

It is useful to recall some properties connected with the employment of spaces equipped with an algebra of sets, which is traditional for FAM theory. Note that $\forall \mathcal{L} \in (\text{alg})[E]$:

$$\begin{aligned}(\text{add})[\mathcal{L}] = \{\mu \in \mathbb{R}^{\mathcal{L}} \mid &\forall A \in \mathcal{L}\ \forall B \in \mathcal{L} : (A \cap B = \varnothing) \\ &\Rightarrow (\mu(A \cup B) = \mu(A) + \mu(B))\}.\end{aligned} \tag{4.2.25}$$

The known property of strong additivity is connected with (4.2.25): if $\mathcal{L} \in (\text{alg})[E]$, $\mu \in (\text{add})[\mathcal{L}]$, $A \in \mathcal{L}$ and $B \in \mathcal{L}$, then $\mu(A \cup B) + \mu(A \cap B) = \mu(A) + \mu(B)$. The monotonocity of non-negative FAM is connected with this property. Namely, $\forall \mathcal{L} \in (\text{alg})[E]\ \forall A \in \mathcal{L}\ \forall B \in \mathcal{L}$:

$$(A \subset B) \Rightarrow (\forall \mu \in (\text{add})_+[\mathcal{L}] : \mu(A) \leq \mu(B)); \tag{4.2.26}$$

(4.2.26) and many other properties are in natural way extended to the case of FAM defined on a semi-algebra of sets. This is connected with the well known procedure of extension of FAM from a semi-algebra of sets to the algebra generated by this semi-algebra. This algebra is realized very simply (see (2.2.38)). Moreover, it is useful to note that $\forall \mathcal{L} \in \Pi[E]$ $\forall \mu \in (\text{add})[\mathcal{L}]$ $\forall A \in a_E^0(\mathcal{L})$ $\forall m \in \mathcal{N}$ $\forall (L_i)_{i \in \overline{1,m}} \in \Delta_m(A, \mathcal{L})$ $\forall n \in \mathcal{N}$ $\forall (\Lambda_j)_{j \in \overline{1,n}} \in \Delta_n(A, \mathcal{L})$:

$$\sum_{i=1}^{m} \mu(L_i) = \sum_{j=1}^{n} \mu(\Lambda_j). \tag{4.2.27}$$

From (2.2.38) and (4.2.27) we obtain the following definition. Namely, if $\mathcal{L} \in \Pi[E]$ and $\mu \in (\text{add})[\mathcal{L}]$, then $\exists! \nu \in \mathbb{R}^{a_E^0(\mathcal{L})}$ $\forall A \in a_E^0(\mathcal{L})$ $\forall m \in \mathcal{N}$ $\forall (L_i)_{i \in \overline{1,m}} \in \Delta_m(A, \mathcal{L})$:

$$\sum_{i=1}^{m} \mu(L_i) = \nu(A).$$

So, we have established the possibility of some extension of FAM defined on a semi-algebra of sets. If $\mathcal{L} \in \Pi[E]$ and $\mu \in (\text{add})[\mathcal{L}]$, then we set that $\alpha[\mu] \in \mathbb{R}^{a_E^0(\mathcal{L})}$ is by definition the set function such that $\forall A \in a_E^0(\mathcal{L})$ $\forall m \in \mathcal{N}$ $\forall (L_i)_{i \in \overline{1,m}} \in \Delta_m(A, \mathcal{L})$:

$$\alpha[\mu](A) = \sum_{i=1}^{m} \mu(L_i);$$

obviously, $(\alpha[\mu] \mid \mathcal{L}) = \mu$. From (4.2.25) it follows that if $\mathcal{L} \in \Pi[E]$ and $\mu \in (\text{add})[\mathcal{L}]$, then

$$\alpha[\mu] \in (\text{add})[a_E^0(\mathcal{L})]. \tag{4.2.28}$$

If $\mu \in (\text{add})_+[\mathcal{L}]$ in (4.2.28), then $\alpha[\mu] \in (\text{add})_+[a_E^0(\mathcal{L})]$. From (2.2.38) and (4.2.8) we derive the obvious property that if $\mathcal{L} \in \Pi[E]$, $\mu \in (\text{add})[\mathcal{L}]$, and $\nu \in (\text{add})[a_E^0(\mathcal{L})]$, then

$$(\mu = (\nu \mid \mathcal{L})) \Rightarrow (\nu = \alpha[\mu]);$$

so from (4.2.28) it follows that $\alpha[\mu]$ is the unique extension of $\mu$ to $a_E^0(\mathcal{L})$ in the class of FAM. It is useful to note that $\forall \mathcal{L} \in \Pi[E]$ $\forall \mu \in (\sigma\text{--add})[\mathcal{L}]$:

$$\alpha[\mu] \in (\sigma - \text{add})[a_E^0(\mathcal{L})]. \tag{4.2.29}$$

Obviously,

$$\alpha[\mu] \in (\sigma - \text{add})_+[a_E^0(\mathcal{L})]$$

for $\mathcal{L} \in \Pi[E]$ and $\mu \in (\sigma - \text{add})_+[\mathcal{L}]$. This property is an obvious consequence of (4.2.29) and the statement (4.2.28). We use (4.2.28) and (4.2.29) for expansion of some useful properties of FAM to the case $\mathcal{L} \in \Pi[E]$. We consider only a simple example. By (4.2.25) one can verify that

$$v_\mu \in (\text{add})_+[\tilde{\mathcal{L}}]$$

for $\tilde{\mathcal{L}} \in (\text{alg})[E]$ and $\mu \in \mathbf{A}(\tilde{\mathcal{L}})$. Return to the definition of the variation of FAM.

Using (2.2.38) and the definition of FAM (4.2.28), we recall that $\forall \mathcal{L} \in \Pi[E]$ $\forall \mu \in (\text{add})[\mathcal{L}]$ $\forall H \in a_E^0(\mathcal{L})$ $\forall t \in (\text{VAR})_H[\alpha[\mu]]$ $\exists m \in \mathcal{N}$ $\exists (L_i)_{i \in \overline{1,m}} \in \Delta_m(H, \mathcal{L})$:

$$t \leq \sum_{i=1}^{m} |\mu(L_i)|. \tag{4.2.30}$$

As a consequence we obtain (see (4.2.30)) the following useful statement. Namely, $\forall \mathcal{L} \in \Pi[E]$ $\forall \mu \in (\text{add})[\mathcal{L}]$:

$$(\mu \in (B - \text{var})[\mathcal{L}]) \Leftrightarrow (\alpha[\mu] \in (B - \text{var})[a_E^0(\mathcal{L})]). \tag{4.2.31}$$

From (4.2.31) we have $\forall \mathcal{L} \in \Pi[E]$ $\forall \mu \in \mathbf{A}(\mathcal{L})$: $\alpha[\mu] \in \mathbf{A}(a_E^0(\mathcal{L}))$. So we can use the additivity of the variation of FAM defined on an algebra of sets: if $\mathcal{L} \in \Pi[E]$ and $\mu \in \mathbf{A}(\mathcal{L})$, then $v_{\alpha[\mu]} \in (\text{add})_+[a_E^0(\mathcal{L})]$. Consequently, the restriction of $v_{\alpha[\mu]}$ to $\mathcal{L}$ is a non-negative FAM on $\mathcal{L}$. Thus by (4.2.30) we have $\forall \mathcal{L} \in \Pi[E]$ $\forall \mu \in \mathbf{A}(\mathcal{L})$:

$$v_\mu = (v_{\alpha[\mu]} \mid \mathcal{L}) \in (\text{add})_+[\mathcal{L}]. \tag{4.2.32}$$

From (4.2.24) and (4.2.32) we obtain, in particular, that $\forall \mathcal{L} \in \Pi[E]$ $\forall \mu \in \mathbf{A}(\mathcal{L})$: $V_\mu = V_{\alpha[\mu]}$. Since the mapping

$$\mu \mapsto \alpha[\mu] : (\text{add})[\mathcal{L}] \to (\text{add})[a_E^0(\mathcal{L})]$$

is a linear bijection for $\mathcal{L} \in \Pi[E]$ (this follows from the properties that the extension (4.2.28) is unique and $(\text{add})[a_E^0(\mathcal{L})]$ is linear), we conclude that the mapping

$$\mu \mapsto \alpha[\mu] : \mathbf{A}(\mathcal{L}) \to \mathbf{A}(a_E^0(\mathcal{L})) \tag{4.2.33}$$

is the isometric isomorphism of $\mathbf{A}(\mathcal{L})$ onto $\mathbf{A}(a_E^0(\mathcal{L}))$ under the equipment with strong norms of the type (4.2.16). Thus $\mathbf{A}(\mathcal{L})$ and $\mathbf{A}(a_E^0(\mathcal{L}))$ are in essence identified (it is possible to add some reasons concerning the order properties). We note the following well known property. Namely, $\forall \mathcal{L} \in (\text{alg})[E]$ :

$$\mathbf{A}(\mathcal{L}) = (\text{add})[\mathcal{L}] \cap \mathbb{B}(\mathcal{L}).$$

Recall (see Section 2.2) that $\mathbb{B}(\mathcal{L})$ is the set of all bounded functions on $\mathcal{L}$. As for (4.2.32), we note that for $\mathcal{L} \in \Pi[E]$ and $\mu \in \mathbf{A}(\mathcal{L})$ the FAM

$$\left(\mu^+ \triangleq \frac{1}{2}(v_\mu + \mu) \in (\text{add})_+[\mathcal{L}]\right) \& \left(\mu^- \triangleq \frac{1}{2}(v_\mu - \mu) \in (\text{add})_+[\mathcal{L}]\right) \tag{4.2.34}$$

realize the Jordan decomposition of the initial FAM $\mu$:

$$(\mu = \mu^+ - \mu^-) \& (v_\mu = \mu^+ + \mu^-). \tag{4.2.35}$$

As for the decomposition (4.2.34) and (4.2.35), some corollaries should be noted for the traditional case $\mathcal{L} \in (\text{alg})[E]$. Moreover, an important representation in terms of the theory of complete vector lattices is connected with this decomposition (see [19, 106, 109, 110]). We now omit this important consideration.

Returning to the problem of countable additivity of set functions it is advisable to recall the representation in terms of continuity with respect to monotone sequences of sets. If $(A_i)_{i\in\mathcal{N}} \in \mathcal{P}(E)^{\mathcal{N}}$ and $A \in \mathcal{P}(E)$, then, as usual, we set by definition that

$$\begin{gathered}\left(((A_i)_{i\in\mathcal{N}} \downarrow A) \Leftrightarrow ((A = \bigcap_{i\in\mathcal{N}} A_i)\&(\forall j \in \mathcal{N} : A_{j+1} \subset A_j))\right)\\ \& \left(((A_i)_{i\in\mathcal{N}} \uparrow A) \Leftrightarrow ((A = \bigcup_{i\in\mathcal{N}} A_i)\&(\forall j \in \mathcal{N} : A_j \subset A_{j+1}))\right);\end{gathered} \tag{4.2.36}$$

see these definitions, for example, in [103, Ch. I]. The first notion of (4.2.36) is most important in the case $A = \varnothing$. In this connection suppose that $\forall \mathcal{L} \in \pi[E]$:

$$\begin{aligned}(\mathbf{c} - \text{add})[\mathcal{L}] \triangleq \{\mu \in (\text{add})[\mathcal{L}] \,|\forall (L_i)_{i\in\mathcal{N}} \in \mathcal{L}^{\mathcal{N}} : ((L_i)_{i\in\mathcal{N}} \downarrow \varnothing)\\ \Rightarrow ((\mu(L_i))_{i\in\mathcal{N}} \to 0)\}.\end{aligned} \tag{4.2.37}$$

Since $\mu(\varnothing) = 0$ for $\mathcal{L} \in \pi[E]$ and $\mu \in (\text{add})[\mathcal{L}]$, the basic property characterized by (4.2.37) has the sense of sequential continuity at the point $\varnothing$. So (4.2.37) has some analogy with notions of Section 2.6. In addition (see, for example, [103, Ch. I]), $\forall \mathcal{L} \in (\text{alg})[E]$:

$$(\sigma - \text{add})[\mathcal{L}] = (\mathbf{c} - \text{add})[\mathcal{L}]. \tag{4.2.38}$$

Thus CAM are exactly FAM sequentially continuous at the point $\varnothing$. Using (4.2.29) we obtain $\forall \mathcal{L} \in \Pi[E]$:

$$(\sigma - \text{add})[\mathcal{L}] \subset (\mathbf{c} - \text{add})[\mathcal{L}]. \tag{4.2.39}$$

The relation (4.2.39) is an example of extending the properties of FAM and CAM defined on algebras of sets to the case of analogous set functions defined on semi-algebras of sets. It is useful to supplement the relations above by Proposition 2.2.1 which permits us to reduce MS with multiplicative families to MS with semi-algebra of sets.

## 4.3 INTEGRAL REPRESENTATIONS OF LINEAR CONTINUOUS FUNCTIONALS

In this section we consider a very important question. We study the problem of identification of measures and linear continuous functionals. In this connection it is advisable first to note the well known Riesz theorem on representation of the spaces topologically conjugate to the corresponding spaces of continuous functions. But we first investigate some questions connected with the integral representation of spaces topologically conjugate to Banach spaces of discontinuous functions. In this case FAM are very important. We consider spaces of the type $B(S,\Sigma)$ [66, Ch.IV]. If $\mathbf{S}[X \neq \varnothing]$ and $A \in \mathcal{P}(X)$, then we denote by $\chi_A[X]$ the function of $\mathbb{R}^X$ for which

$$(\forall x \in A : \chi_A[X](x) \triangleq 1) \ \& \ (\forall x \in X \setminus A : \chi_A[X](x) \triangleq 0). \tag{4.3.1}$$

In particular, we use (4.3.1) for $X = E$. In other cases it is advisable to use a nonempty family of subsets of $X$ in place of $X$. Note that $\forall_X \ \mathbf{S}[X \neq \varnothing]$:

$$\mathbb{O}_X = \chi_\varnothing[X].$$

We construct natural step functions on the basis of indicators (4.3.1). Suppose that $\mathbf{S}[E \neq \varnothing]$. We use the pointwise linear operations on the space $\mathbb{R}^E$ of all real-valued functions on $E$ (see stipulations of Section 2.2). So, $\forall \mathcal{L} \in \pi[E]$:

$$B_0(E,\mathcal{L}) \triangleq \{f \in \mathbb{R}^E \mid \exists n \in \mathcal{N} \ \exists (\alpha_i)_{i\in\overline{1,n}} \in \mathbb{R}^n \ \exists (L_i)_{i\in\overline{1,n}} \in \Delta_n(E,\mathcal{L}) :$$

$$f = \sum_{i=1}^{n} \alpha_i \chi_{L_i}[E]\} \in 2^{\mathbb{B}(E)}. \tag{4.3.2}$$

It is worth noting that $B_0(E,\mathcal{L})$ (4.3.2) is a linear manifold in $\mathbb{B}(E)$. Moreover, $\forall \mathcal{L} \in \pi[E] \ \forall f \in B_0(E,\mathcal{L}) \ \forall g \in B_0(E,\mathcal{L})$:

$$fg \in B_0(E,\mathcal{L}). \tag{4.3.3}$$

As for (4.3.3), we recall that the product in $\mathbb{R}^E$ is defined pointwise. Omitting a number of other general properties, we note only the fact that the statement $\chi_L[E] \in B_0(E,\mathcal{L})$ can be not true under $\mathcal{L} \in \pi[E]$

and some $L \in \mathcal{L}$. A corresponding example can be easily constructed with using the example of Section 4.2. However, $\forall \mathcal{L} \in \Pi[E]$ $\forall L \in \mathcal{L}$:

$$\chi_L[E] \in B_0(E, \mathcal{L}). \tag{4.3.4}$$

This property is very important. In particular, (4.3.4) is used under constructing extensions in [32, 35, 45, 46]. Note that in the case of MS with a semi-algebra of sets the linear manifold of step functions is the linear span of indicators: from (4.3.2) and (4.3.4) for $\mathcal{L} \in \Pi[E]$ we have the equality

$$B_0(E, \mathcal{L}) \triangleq \{f \in \mathbb{R}^E \mid \exists n \in \mathcal{N}\ \exists (\alpha_i)_{i \in \overline{1,n}} \in \mathbb{R}^n\ \exists (L_i)_{i \in \overline{1,n}} \in \mathcal{L}^n :$$

$$f = \sum_{i=1}^{n} \alpha_i \chi_{L_i}[E]\}. \tag{4.3.5}$$

The relation (4.3.5) can be interpreted as some distinctive 'linear closure' of $\mathcal{L}$ under the imbedding of $\mathcal{L}$ in $\mathbb{R}^E$ by virtue of indicators. We use the property that

$$H \mapsto \chi_H[E] : \mathcal{P}(E) \to \mathbb{R}^E$$

is an injective mapping. In the sequel we equip the linear space $\mathbb{B}(E)$ with the sup-norm $\|\cdot\|$, setting $\forall f \in \mathbb{B}(E)$:

$$\|f\| \triangleq \sup(\{|f(x)| : x \in E\}).$$

We introduce an elementary integral with respect to FAM. We use the property that if $\mathcal{L} \in \pi[E]$, $\mu \in (\mathrm{add})[\mathcal{L}]$, $n \in \mathcal{N}$, $(\alpha_i)_{i \in \overline{1,n}} \in \mathbb{R}^n$, $(L_i)_{i \in \overline{1,n}} \in \Delta_n(E, \mathcal{L})$, $m \in \mathcal{N}$, $(\beta_j)_{j \in \overline{1,m}} \in \mathbb{R}^m$, and $(\Lambda_j)_{j \in \overline{1,m}} \in \Delta_m(E, \mathcal{L})$, then

$$\left(\sum_{i=1}^{n} \alpha_i \chi_{L_i}[E] = \sum_{j=1}^{m} \beta_j \chi_{\Lambda_j}[E]\right)$$

$$\Rightarrow \left(\forall \mu \in (\mathrm{add})[\mathcal{L}] : \sum_{i=1}^{n} \alpha_i \mu(L_i) = \sum_{j=1}^{m} \beta_j \mu(\Lambda_j)\right). \tag{4.3.6}$$

The proof of (4.3.6) uses (as in [21]) only the additivity of $\mu$; see (4.2.8). Owing to (4.3.6) we can properly give the following definition. Namely, if $\mathcal{L} \in \pi[E]$, $\mu \in (\mathrm{add})[\mathcal{L}]$, and $f \in B_0(E, \mathcal{L})$, then

$$\overset{(\mathrm{el})}{\int_E} f \, d\mu \in \mathbb{R} \tag{4.3.7}$$

is the number for which $\forall n \in \mathcal{N}\ \forall (\alpha_i)_{i \in \overline{1,n}} \in \mathbb{R}^n\ \forall (L_i)_{i \in \overline{1,n}} \in \Delta_n(E, \mathcal{L})$ :

$$\left(f = \sum_{i=1}^{n} \alpha_i \chi_{L_i}[E]\right) \Rightarrow \left( {}^{(\mathrm{el})}\!\!\int_E f\, d\mu = \sum_{i=1}^{n} \alpha_i \mu(L_i)\right). \tag{4.3.8}$$

In fact, (4.3.7) and (4.3.8) is another notation of (4.3.6). The notation (4.3.7) is auxiliary. But it is advisable to establish some properties of the elementary integral. We note that for $\mathcal{L} \in \pi[E]$ the dependence

$$(f, \mu) \mapsto {}^{(\mathrm{el})}\!\!\int_E f\, d\mu : B_0(E, \mathcal{L}) \times (\mathrm{add})[\mathcal{L}] \to \mathbb{R} \tag{4.3.9}$$

is a bilinear functional. We use (4.3.9) in the sequel basic definition of an integral. Moreover, $\forall \mathcal{L} \in \pi[E]\ \forall \mu \in \mathbf{A}(\mathcal{L})\ \forall f \in B_0(E, \mathcal{L})$:

$$\left| {}^{(\mathrm{el})}\!\!\int_E f\, d\mu \right| \leq V_\mu \|f\|. \tag{4.3.10}$$

The property (4.3.10) follows from (4.3.8). If $\mu \in \mathbf{A}(\mathcal{L})$ is fixed, then the elementary $\mu$-integral is an uniformly continuous functional on $B_0(E, \mathcal{L})$. In the sequel the symbol $\rightrightarrows$ denotes the uniform convergence in the space of functionals. In particular, we apply this notion in the case $\mathbb{R}^E$. It is obvious that the uniform convergence and the convergence in the sense of sup-norm are equivalent in $\mathbb{B}(E)$. Of course, the convergence in the sense of sup-norm is the usual one in metric space under the supposition that the metric is generated by the sup-norm. Thus we have the usual sequential convergence in a TS. If $\mathcal{L} \in \pi[E]$, then the set $B(E, \mathcal{L})$ is defined as the closure of $B_0(E, \mathcal{L})$ in the Banach space $\mathbb{B}(E)$ with the sup-norm (i.e., in the topology of $\mathbb{B}(E)$ generated by the sup-norm):

$$B(E, \mathcal{L}) = \{f \in \mathbb{B}(E) \mid \exists (f_i)_{i \in \mathcal{N}} \in B_0(E, \mathcal{L})^{\mathcal{N}} : (f_i)_{i \in \mathcal{N}} \rightrightarrows f\}. \tag{4.3.11}$$

Then for $\mathcal{L} \in \pi[E]$ we have (in the form of (4.3.11)) a subspace of the Banach space $\mathbb{B}(E)$: $B(E, \mathcal{L})$ is the linear manifold closed in $\mathbb{B}(E)$ equipped with the sup-norm. Therefore $B(E, \mathcal{L})$ itself equipped with the norm induced from $(\mathbb{B}(E), \|\cdot\|)$ is a Banach space. In addition, $B_0(E, \mathcal{L})$ is a linear manifold in the Banach space $B(E, \mathcal{L})$. From (4.3.10) we extract the following correct definition. Namely, we note that $\forall \mathcal{L} \in \pi[E]$ $\forall \mu \in \mathbf{A}(\mathcal{L})\ \forall f \in B(E, \mathcal{L})\ \exists! c \in \mathbb{R}\ \forall (f_i)_{i \in \mathcal{N}} \in B_0(E, \mathcal{L})^{\mathcal{N}}$:

$$((f_i)_{i \in \mathcal{N}} \rightrightarrows f) \Rightarrow \left( \left( {}^{(\mathrm{el})}\!\!\int_E f_i\, d\mu \right)_{i \in \mathcal{N}} \to c \right). \tag{4.3.12}$$

In (4.3.12) we use the completeness of $(\mathbb{R}, |\cdot|)$. The number $c$ in (4.3.12) is called the definite $\mu$-integral of $f$. This number depends on only $f$ and $\mu$. Therefore we use the traditional designation with the parameters $f$ and $\mu$ in place of $c$. Thus, if $\mathcal{L} \in \pi[E]$, $\mu \in \mathbf{A}(\mathcal{L})$, and $f \in B(E, \mathcal{L})$, then

$$\int_E f \, d\mu \in \mathbb{R} \tag{4.3.13}$$

is the number for which $\forall (f_i)_{i \in \mathcal{N}} \in B_0(E, \mathcal{L})^{\mathcal{N}}$:

$$((f_i)_{i \in \mathcal{N}} \rightrightarrows f) \Rightarrow \left( \left( {}^{(\mathrm{el})}\!\!\int_E f_i \, d\mu \right)_{i \in \mathcal{N}} \to \int_E f \, d\mu \right). \tag{4.3.14}$$

In (4.3.13) and (4.3.14) we realize the property (4.3.12). Note that the definition on the basis of (4.3.13) and (4.3.14)is analogous to that in [110, p. 288]. A more general definition of the integral with respect to FAM is given in [66, Ch.III]. But we restrict ourselves to the more simple version (4.3.13), (4.3.14). The integrals (4.3.13) inherit many properties of elementary integrals. We call elements of (4.3.11) the stratum functions, following conceptually [97]. We also call the integral (4.3.13), (4.3.14) the stratum integral. From (4.3.10) and (4.3.14) we immediately obtain $\forall \mathcal{L} \in \pi[E]\ \forall \mu \in \mathbf{A}(\mathcal{L})\ \forall f \in B(E, \mathcal{L})$:

$$\left| \int_E f \, d\mu \right| \leq V_\mu \|f\|. \tag{4.3.15}$$

Moreover, we note the obvious property of bilinearity. Namely, if $\mathcal{L} \in \pi[E]$, then

$$(f, \mu) \mapsto \int_E f \, d\mu : B(E, \mathcal{L}) \times \mathbf{A}(\mathcal{L}) \to \mathbb{R} \tag{4.3.16}$$

is a bilinear functional. The property (4.3.16) follows from the similar property of the mapping (4.3.9). From (4.3.15) and (4.3.16) we get the important conclusion: if $\mu \in \mathbf{A}(\mathcal{L})$ is fixed, then the mapping

$$f \mapsto \int_E f \, d\mu : B(E, \mathcal{L}) \to \mathbb{R} \tag{4.3.17}$$

is a linear bounded functional on the Banach space $B(E, \mathcal{L})$. In this connection we introduce $\forall \mathcal{L} \in \pi[E]$:

$$\begin{aligned} B^*(E, \mathcal{L}) \triangleq \{x^* \in \mathbb{R}^{B(E,\mathcal{L})} | & (\forall \alpha \in \mathbb{R}\ \forall f \in B(E, \mathcal{L}) : x^*(\alpha f) = \alpha x^*(f)) \\ & \&(\forall f \in B(E, \mathcal{L})\ \forall g \in B(E, \mathcal{L}) : x^*(f+g) = x^*(f) + x^*(g)) \\ & \& \, (\exists c \in [0, \infty[\ \forall f \in B(E, \mathcal{L}) : |x^*(f)| \leq c\|f\|)\}. \end{aligned} \tag{4.3.18}$$

The space topologically conjugate to the Banach space $B(E, \mathcal{L})$ is introduced in (4.3.18). From (4.3.17)and (4.3.18) we have $\forall \mathcal{L} \in \pi[E]$ $\forall \mu \in \mathbf{A}(\mathcal{L})$:

$$\int_E d\mu \triangleq \left(\int_E f\, d\mu\right)_{f \in B(E,\mathcal{L})} \in B^*(E, \mathcal{L}). \tag{4.3.19}$$

We get the corresponding integral functional from $B^*(E, \mathcal{L})$; see (4.3.19). It is useful to recall the natural norm in $B^*(E, \mathcal{L})$, where $\mathcal{L} \in \pi[E]$. If $\mathcal{L} \in \pi[E]$, then we set

$$(\mathbb{U}(\mathcal{L}) \triangleq \{f \in B(E, \mathcal{L}) | \|f\| \leq 1\})\&(\mathbb{U}_\partial(\mathcal{L}) \triangleq \{f \in B(E, \mathcal{L}) | \|f\| = 1\}),$$

obtaining the corresponding unit ball and unit sphere in the Banach space $B(E, \mathcal{L})$. We have $\forall \mathcal{L} \in \pi[E]$ $\forall h \in B^*(E, \mathcal{L})$:

$$\|h\|^* \triangleq \sup(\{|h(f)| : f \in \mathbb{U}(\mathcal{L})\}) \in [0, \infty[.$$

It is known that this construction defines the natural norm in the space topologically conjugate to a Banach space. In particular, it is possible to define the value of this norm at the point (4.3.19). Then $\forall \mathcal{L} \in \pi[E]$ $\forall \mu \in \mathbf{A}(\mathcal{L})$:

$$\left\|\int_E d\mu\right\|^* = \sup\left(\left\{\left|\int_E f\, d\mu\right| : f \in \mathbb{U}(\mathcal{L})\right\}\right).$$

Note that $\mathbb{O}_E = \chi_\varnothing[E] \in \mathbb{U}(\mathcal{L}) \cap B_0(E, \mathcal{L})$ for $\mathcal{L} \in \pi[E]$.

LEMMA 4.3.1 *If* $\mathcal{L} \in \pi[E]$ *and* $\mu \in \mathbf{A}(\mathcal{L})$, *then*

$$V_\mu = \left\|\int_E d\mu\right\|^* = \sup\left(\left\{\left| {}^{(\mathrm{el})}\!\!\int_E f\, d\mu\right| : f \in \mathbb{U}_\partial(\mathcal{L}) \cap B_0(E, \mathcal{L})\right\}\right). \tag{4.3.20}$$

PROOF. Fix $\mathcal{L}$ and $\mu$ in correspondence with the conditions. From (4.3.15) and (4.3.19) we have the inequality

$$\left\|\int_E d\mu\right\|^* \leq V_\mu. \tag{4.3.21}$$

Denote by $a$ the last number in (4.3.20): $a$ is the required supremum over the class of step functions. Then

$$a \leq \left\|\int_E d\mu\right\|^*. \tag{4.3.22}$$

We use the corollary of (4.3.14) that

$$\int_E f d\mu = {}^{(\mathrm{el})}\!\!\int_E f\, d\mu \tag{4.3.23}$$

for $f \in B_0(E, \mathcal{L})$. In the following we use (4.3.23) without additional clarification. So (4.3.22) is true. We recall the definition of the total variation (see (4.2.12), (4.2.15) and (4.2.16)). Let $\varepsilon \in ]0, \infty[$. It is possible to choose $c_\varepsilon \in (\mathrm{VAR})_E[\mu]$ for which $V_\mu - \varepsilon < c_\varepsilon$. Further we choose $n \in \mathcal{N}$ and $(L_i)_{i\in\overline{1,n}} \in \Delta_n(E, \mathcal{L})$ for which

$$c_\varepsilon = \sum_{i=1}^{n} \mid \mu(L_i) \mid . \tag{4.3.24}$$

We use the corresponding definition of Section 4.2. Introduce $(\alpha_i)_{i\in\overline{1,n}} \in \mathbb{R}^n$, setting $\forall k \in \overline{1,n}$:

$$((\mu(L_k) < 0) \Rightarrow (\alpha_k \triangleq -1))\ \&((0 \leq \mu(L_k)) \Rightarrow (\alpha_k \triangleq 1)).$$

Then $\forall j \in \overline{1,n} : \alpha_j \mu(L_j) = \mid \mu(L_j) \mid$. By (4.3.5) we obtain

$$\varphi \triangleq \sum_{i=1}^{n} \alpha_i \chi_{L_i}[E] \in B_0(E, \mathcal{L}).$$

In addition, $\forall j \in \overline{1,n}\ \forall x \in L_j$: $\varphi(x) = \alpha_j$. As a consequence we have $\forall x \in E$: $\mid \varphi(x) \mid = 1$. Hence $\|\varphi\| = 1$, i.e., $\varphi \in \mathbb{U}_\partial(\mathcal{L}) \cap B_0(E, \mathcal{L})$ and

$$\left| {}^{(\mathrm{el})}\!\!\int_E \varphi\, d\mu \right| \leq a. \tag{4.3.25}$$

But by (4.3.8) and (4.3.24) we obtain

$$\left| {}^{(\mathrm{el})}\!\!\int_E \varphi\, d\mu \right| = \left| \sum_{i=1}^{n} \alpha_i \mu(L_i) \right| = \sum_{i=1}^{n} \mid \mu(L_j) \mid = c_\varepsilon.$$

From (4.3.25) we have the inequality

$$V_\mu - \varepsilon < \left| {}^{(\mathrm{el})}\!\!\int_E \varphi\, d\mu \right| \leq a. \tag{4.3.26}$$

Since the choice of $\varepsilon \in ]0, \infty[$ was arbitrary, we conclude that $V_\mu \leq a$. Therefore by (4.3.21) and (4.3.22) we have the required equality (4.3.20).
□

From (4.3.19) and Lemma 4.3.1 we obtain that for $\mathcal{L} \in \pi[E]$ the mapping

$$\mu \mapsto \int_E d\mu: \quad \mathbf{A}(\mathcal{L}) \to B^*(E, \mathcal{L}) \tag{4.3.27}$$

is an isometric inclusion.

THEOREM 4.3.1 *Let $\mathcal{L} \in \Pi[E]$. Then (4.3.27) is the isometric isomorphism of $\mathbf{A}(\mathcal{L})$ with the strong norm (4.2.16) onto $B^*(E, \mathcal{L})$ with the norm*

$$h \mapsto \|h\|^* : B^*(E, \mathcal{L}) \to [0, \infty[. \tag{4.3.28}$$

**Remark.** We call the norm (4.3.28) traditional. Theorem 4.3.1 means the validity of the three following statements:

1) (4.3.27) is a bijection;
2) (4.3.27) is a linear operator;
3) (4.3.27) is an isometry.

The proof is analogous to that of [66, Ch. IV]. The part 1) of the remark above was considered in [21, pp. 65–67]. The part 2) follows from the bilinearity of (4.3.16). The part 3) is realized in correspondence with Lemma 4.3.1. So for $\mathcal{L} \in \Pi[E]$ the spaces $\mathbf{A}(\mathcal{L})$ and $B^*(E, \mathcal{L})$ are identified. This fact is used in extension constructions.

## 4.4 INDEFINITE INTEGRAL

In this brief section we consider a useful example of employing Theorem 4.3.1 similarly to statements of [66, Ch. IV]. We recall (4.3.3) and (4.3.11). Then we have $\forall \mathcal{L} \in \pi[E]$ $\forall u \in B(E, \mathcal{L})$ $\forall v \in B(E, \mathcal{L})$

$$uv = (u(x)v(x))_{x \in E} \in B(E, \mathcal{L}). \tag{4.4.1}$$

As a consequence, for $\mathcal{L} \in \pi[E]$, $\mu \in \mathbf{A}(\mathcal{L})$ and $f \in B(E, \mathcal{L})$ we have the functional

$$u \mapsto \int_E uf \, d\mu : B(E, \mathcal{L}) \to \mathbb{R}. \tag{4.4.2}$$

Obviously, (4.4.2) is a linear functional. In this connection we recall that the linear operations and the product are defined pointwise. Therefore for $\alpha \in \mathbb{R}$, $u \in \mathbb{R}^E$ and $v \in \mathbb{R}^E$ we obtain $(\alpha u) \cdot f = \alpha \cdot (uf)$. In this relation we use the dot as the product symbol for right interpretation of the sequence of realized operations: in the first case we consider the product of $\alpha u \in \mathbb{R}^E$ and $f \in \mathbb{R}^E$; in the second case we deal with the product of the number $\alpha$ and the function $uf$. Moreover, $\forall u \in \mathbb{R}^E$ $\forall v \in \mathbb{R}^E$ $\forall f \in \mathbb{R}^E$: $(u + v) \cdot f = (uf) + (vf)$. From the two above representations we obtain the linearity of (4.4.2); of course, we use the properties of (4.3.16) and (4.3.19). Moreover, we note that by the definition of the sup-norm, $\forall p \in \mathbb{B}(E)\ \forall q \in \mathbb{B}(E)$: $\|pq\| \leq \|p\| \cdot \|q\|$. Therefore by (4.3.15) we obtain $\forall \mathcal{L} \in \pi[E]\ \forall \mu \in \mathbf{A}(\mathcal{L})\ \forall u \in B(E, \mathcal{L})$ $\forall f \in B(E, \mathcal{L})$:

$$\left| \int_E uf\ d\mu \right| \leq V_\mu \cdot \|u\| \cdot \|f\|. \tag{4.4.3}$$

The following corollary can be extracted from (4.4.3). If $\mathcal{L} \in \pi[E]$, $\mu \in \mathbf{A}(\mathcal{L})$ and $f \in B(E, \mathcal{L})$, then $c_f \triangleq V_\mu \|f\| \in [0, \infty[$ has the property

$$\left| \int_E uf\ d\mu \right| \leq c_f \|u\|$$

for $u \in B(E, \mathcal{L})$. Thus (4.4.2) defines a linear bounded functional on the space of stratum functions. Namely, from (4.3.18) we have $\forall \mathcal{L} \in \pi[E]$ $\forall \mu \in \mathbf{A}(\mathcal{L})\ \forall f \in B(E, \mathcal{L})$:

$$\left( \int_E uf\ d\mu \right)_{u \in B(E, \mathcal{L})} \in B^*(E, \mathcal{L}). \tag{4.4.4}$$

In (4.4.2) and (4.4.4) we have action of some FAM from $\mathbf{A}(\mathcal{L})$ defined by $f \in B(E, \mathcal{L})$ and $\mu \in \mathbf{A}(\mathcal{L})$. We construct this measure for the case $\mathcal{L} \in \Pi[E]$, where Theorem 4.3.1 can be used. Namely, if $\mathcal{L} \in \Pi[E]$, then $\exists! \nu \in \mathbf{A}(\mathcal{L})\ \forall u \in B(E, \mathcal{L})$:

$$\int_E uf\ d\mu = \int_E u\ d\nu.$$

This fact is an obvious corollary of Theorem 4.3.1; of course, $\nu = \nu(f, \mu)$. Therefore the following definition is correct. If $\mathcal{L} \in \Pi[E]$, $\mu \in \mathbf{A}(\mathcal{L})$ and $f \in B(E, \mathcal{L})$, then by definition

$$f * \mu \in \mathbf{A}(\mathcal{L}) \tag{4.4.5}$$

is the unique FAM of bounded variation with the property that

$$\forall u \in B(E, \mathcal{L}) : \int_E uf\ d\mu = \int_E u\ d(f * \mu). \tag{4.4.6}$$

We call the FAM (4.4.5), (4.4.6) the indefinite $\mu$-integral of $f$ with respect to $\mu$. As for concrete constructing (4.4.5), we note that $\forall \mathcal{L} \in \Pi[E]$ $\forall f \in B(E, \mathcal{L})\ \forall L \in \mathcal{L}$: $f\chi_L[E] \in B(E, \mathcal{L})$. We use (4.3.4) and (4.4.1). Therefore $\forall \mathcal{L} \in \Pi[E]\ \forall \mu \in \mathbf{A}(\mathcal{L})\ \forall f \in B(E, \mathcal{L})\ \forall L \in \mathcal{L}$:

$$\int_L f\ d\mu \triangleq \int_E f\chi_L[E] d\mu \in \mathbb{R}. \tag{4.4.7}$$

Note that the numbers (4.3.13) and (4.4.7) coincide under $L = E$. Hence we use the common definition (4.4.7) both in the case $L \neq E$ and in the 'previous' case $L = E$. From (4.4.6) we have $\forall \mathcal{L} \in \Pi[E]\ \forall \mu \in \mathbf{A}(\mathcal{L})\ \forall f \in B(E, \mathcal{L})\ \forall L \in \mathcal{L}$:

$$(f * \mu)(L) = \int_L f \, d\mu. \tag{4.4.8}$$

To prove (4.4.8) it is sufficient to suppose $u = \chi_L[E]$ in (4.4.6). From (4.4.5) and (4.4.8) we conclude that $\forall \mathcal{L} \in \Pi[E]\ \forall \mu \in \mathbf{A}(\mathcal{L})\ \forall f \in B(E, \mathcal{L})$:

$$f * \mu = \left( \int_L f \, d\mu \right)_{L \in \mathcal{L}} \in \mathbf{A}(\mathcal{L}).$$

The number (4.4.7) is called the $\mu$-integral of $f$ over the set $L$. Such numbers are values of the FAM (4.4.5). From (4.4.1) and (4.4.6) we have the following useful corollary in the form of the chain of equalities:

$$(p * (f * \mu))(L) = \int_L p \, d(f * \mu) = \int_E p\chi_L[E] d(f * \mu)$$

$$= \int_E (p\chi_L[E]) \cdot f \, d\mu = \int_E (pf) \cdot \chi_L[E] d\mu = \int_L pf \, d\mu = ((pf) * \mu)(L),$$

where $\mathcal{L} \in \Pi[E]$, $\mu \in \mathbf{A}(\mathcal{L})$, $f \in B(E, \mathcal{L})$, $p \in B(E, \mathcal{L})$, and $L \in \mathcal{L}$. Of course, $\forall \mathcal{L} \in \Pi[E]\ \forall \mu \in \mathbf{A}(\mathcal{L})\ \forall f \in B(E, \mathcal{L})\ \forall p \in B(E, \mathcal{L})$:

$$p * (f * \mu) = (pf) * \mu = (fp) * \mu = f * (p * \mu). \tag{4.4.9}$$

In connection with (4.4.9) see [21, pp.110–112]. From (4.4.9) we have the useful representation of the definite integral: if $\mathcal{L} \in \Pi[E]$, $\mu \in \mathbf{A}(\mathcal{L})$, $f \in B(E, \mathcal{L})$, and $p \in B(E, \mathcal{L})$, then

$$\int_E p \, d(f * \mu) = \int_E pf \, d\mu = \int_E f \, d(p * \mu). \tag{4.4.10}$$

In the sequel we consider (4.4.10) in connection with representation of the mathematical expectation when the probability has a density. We now note the particular case when in (4.4.5) and (4.4.6) the function $f$ is an indicator. Then for $\mathcal{L} \in \Pi[E]$, $\mu \in \mathbf{A}(\mathcal{L})$ and $\Lambda \in \mathcal{L}$:

$$\chi_\Lambda[E] * \mu = \left( \int_L \chi_\Lambda[E] \, d\mu \right)_{L \in \mathcal{L}} = (\mu(\Lambda \cap L))_{L \in \mathcal{L}} \in \mathbf{A}(\mathcal{L}). \tag{4.4.11}$$

In addition, we use the known property that

$$\chi_\Lambda[E]\chi_L[E] = \chi_{\Lambda \cap L}[E]$$

for $\Lambda \in \mathcal{P}(E)$ and $L \in \mathcal{P}(E)$. In fact, the relation (4.4.11) defines the restriction of a FAM to a corresponding subspace. From (4.4.11) we obviously have the equality $(\chi_\Lambda[E] * \mu)(\Lambda) = \mu(\Lambda)$ for $\mathcal{L} \in \Pi[E]$, $\mu \in \mathbf{A}(\mathcal{L})$ and $\Lambda \in \mathcal{L}$. If $\mathcal{L} \in (\mathrm{alg})[E]$, then we can consider $(\chi_\Lambda[E]*\mu)(E\setminus\Lambda)$; in this case we have

$$(\chi_\Lambda[E] * \mu)(E \setminus \Lambda) = \mu(\varnothing) = 0.$$

We obtain (in fact) some restriction of the initial FAM to a subspace of MS.

## 4.5 INTEGRATION WITH RESPECT TO NON-NEGATIVE FINITELY ADDITIVE MEASURES

In this section we very briefly consider a concrete variant of statements of Sections 4.3 and 4.4; this variant corresponds to the case when integration is realized with respect to non-negative FAM. In this connection we introduce $\forall \mathcal{L} \in \pi[E]$:

$$\begin{aligned}(B_0^+(E,\mathcal{L}) &\triangleq \{f \in B_0(E,\mathcal{L}) \mid \mathbb{O}_E \leqq f\})\\ \& \ (B^+(E,\mathcal{L}) &\triangleq \{f \in B(E,\mathcal{L}) \mid \mathbb{O}_E \leqq f\}).\end{aligned} \tag{4.5.1}$$

In (4.5.1) we use stipulations of Section 2.2. So we obtain two convex cones of non-negative elements of the linear space above. The relation (4.5.1) implies the following definition. If $\mathcal{L} \in \pi[E]$, then

$$B_+^*(E,\mathcal{L}) \triangleq \{h \in B^*(E,\mathcal{L}) \mid \forall f \in B^+(E,\mathcal{L}) : 0 \leq h(f)\}. \tag{4.5.2}$$

Elements of the set (4.5.2) are traditionally called non-negative functionals from $B^*(E,\mathcal{L})$. In the sequel we suppose that $\mathbb{R}_+^n \triangleq [0,\infty[^n = \{(\alpha_i)_{i\in\overline{1,n}} \in \mathbb{R}^n \mid \forall j \in \overline{1,n} : \alpha_j \in [0,\infty[\}$. From (4.3.2) we have $\forall \mathcal{L} \in \pi[E]$:

$$B_0^+(E,\mathcal{L}) = \{f \in \mathbb{R}^E \mid \exists n \in \mathcal{N}\ \exists (\alpha_i)_{i\in\overline{1,n}} \in \mathbb{R}_+^n\ \exists (L_i)_{i\in\overline{1,n}} \in \Delta_n(E,\mathcal{L}) :$$

$$f = \sum_{i=1}^n \alpha_i \chi_{L_i}[E]\}. \tag{4.5.3}$$

**Remark.** We note that for $f \in B_0^+(E,\mathcal{L})$, $n \in \mathcal{N}$, $(\alpha_i)_{i\in\overline{1,n}} \in \mathbb{R}^n$, and $(L_i)_{i\in\overline{1,n}} \in \Delta_n(E,\mathcal{L})$, under the condition

$$f = \sum_{i=1}^n \alpha_i \chi_{L_i}[E],$$

the inequality $\alpha_k < 0$ can be hold for some $k \in \overline{1,n}$; but for such $k$ the equality $L_k = \varnothing$ must be true. From (4.3.7), (4.3.8) and (4.5.3) we have $\forall_E\, \mathbf{S}[E \neq \varnothing]\ \forall \mathcal{L} \in \pi[E]\ \forall \mu \in (\mathrm{add})_+[\mathcal{L}]\ \forall f \in B_0^+(E, \mathcal{L})$:

$$\overset{(\mathrm{el})}{\int_E} f\, d\mu \in [0, \infty[. \tag{4.5.4}$$

An analogous property holds for integrals of stratum functions. In this connection we note the following simple property: if $\mathcal{L} \in \pi[E]$, $f \in B_0(E, \mathcal{L})$, and $g \in B^+(E, \mathcal{L})$, then the function

$$f_+ \triangleq (\sup(\{f(x); 0\}))_{x \in E}$$

is an element of $B_0^+(E, \mathcal{L})$, and $\|f_+ - g\| \leq \|f - g\|$. As a consequence, for $\mathcal{L} \in \pi[E]$ and $g \in B^+(E, \mathcal{L})$ we can choose some sequence $(g_i)_{i \in \mathcal{N}} : \mathcal{N} \to B_0^+(E, \mathcal{L})$ for which $(g_i)_{i \in \mathcal{N}} \rightrightarrows g$. By (4.2.14), (4.3.13), (4.3.14), and (4.5.4) we have $\forall \mathcal{L} \in \pi[E]\ \forall \mu \in (\mathrm{add})_+[\mathcal{L}]\ \forall f \in B^+(E, \mathcal{L})$:

$$\int_E f\, d\mu \in [0, \infty[. \tag{4.5.5}$$

The relations (4.5.4) and (4.5.5) are represented in terms of functionals in a natural way. Namely, from (4.3.19), (4.5.2), and (4.5.5) we obtain the property that if $\mathcal{L} \in \pi[E]$ and $\mu \in (\mathrm{add})_+[\mathcal{L}]$, then $\int_E d\mu \in B_+^*(E, \mathcal{L})$. Hence we have the imbedding of $(\mathrm{add})_+[\mathcal{L}]$ into $B_+^*(E, \mathcal{L})$.

PROPOSITION 4.5.1 *If $\mathcal{L} \in \Pi[E]$, then the operator*

$$\mu \mapsto \int_E d\mu : (\mathrm{add})_+[\mathcal{L}] \to B_+^*(E, \mathcal{L}) \tag{4.5.6}$$

*is a bijection of* $(\mathrm{add})_+[\mathcal{L}]$ *onto* $B_+^*(E, \mathcal{L})$.

PROOF. Fix $\mathcal{L} \in \Pi[E]$ and denote the operator (4.5.6) by $\varphi$. Let $\tilde{\varphi}$ be the operator (4.3.27); we obviously have

$$\varphi = (\tilde{\varphi} \mid (\mathrm{add})_+[\mathcal{L}]).$$

Recall that $\tilde{\varphi}$ is a bijection. Therefore $\varphi$ is an injective operator: if $\varphi(\mu_1) = \varphi(\mu_2)$ for $\mu_1 \in (\mathrm{add})_+[\mathcal{L}]$ and $\mu_2 \in (\mathrm{add})_+[\mathcal{L}]$, then $\mu_1 = \mu_2$. The mapping $\varphi$ is a surjective operator. Indeed, let $h \in B_+^*(E, \mathcal{L})$. Then $h \in B^*(E, \mathcal{L})$ and $h(f) \in [0, \infty[$ for $f \in B^+(E, \mathcal{L})$. As a consequence

$$\mu \triangleq (h(\chi_L[E]))_{L \in \mathcal{L}} \tag{4.5.7}$$

is a non-negative functional on $\mathcal{L}$. Certainly, $\mu(L) = h(\chi_L[E])$ for $L \in \mathcal{L}$. Using Theorem 4.3.1, choose $\nu \in \mathbf{A}(\mathcal{L})$ for which $h = \int_E d\nu$. Then by (4.3.19), for $L \in \mathcal{L}$ we have

$$\mu(L) = h(\chi_L[E]) = \int_E \chi_L[E] d\nu = \nu(L).$$

So $\mu = \nu$; as a consequence $\mu = \nu \in (\mathrm{add})_+[\mathcal{L}]$ and $h = \int_E d\mu = \varphi(\mu)$. But the choice of $h$ was arbitrary. Therefore $B_+^*(E, \mathcal{L})$ is a subset of the image of $(\mathrm{add})_+[\mathcal{L}]$ under the mapping $\varphi$. Hence $\varphi$ is a surjection. □

Note that for $\mathcal{L} \in \Pi[E]$, $f \in B^+(E, \mathcal{L})$ and $L \in \mathcal{L}$ the property $f\chi_L[E] \in B^+(E, \mathcal{L})$ holds (see (4.5.1)); from (4.4.7) and (4.5.5) it follows that $\int_L f\, d\mu \in [0, \infty[$ for $\mu \in (\mathrm{add})_+[\mathcal{L}]$. As a consequence we can note the corresponding property of non-negativity for the indefinite integral. Namely, $\forall \mathcal{L} \in \Pi[E]\ \forall \mu \in (\mathrm{add})_+[\mathcal{L}]\ \forall f \in B^+(E, \mathcal{L})$:

$$f * \mu = \left( \int_L f\, d\mu \right)_{L \in \mathcal{L}} \in (\mathrm{add})_+[\mathcal{L}]. \tag{4.5.8}$$

As a consequence of (4.5.8) we infer that if $\mathcal{L} \in \Pi[E]$, $f \in B(E, \mathcal{L})$ and $g \in B(E, \mathcal{L})$, then

$$(f \leqq g) \Rightarrow (\forall \mu \in (\mathrm{add})_+[\mathcal{L}] : f * \mu \leqq g * \mu). \tag{4.5.9}$$

Obviously, we can supplement (4.5.8) and (4.5.9). Here we note only the following reasoning of a probability character. If $\mathcal{L} \in \Pi[E]$, $\mu \in (\mathrm{add})_+[\mathcal{L}]$ and $f \in B^+(E, \mathcal{L})$ have the property

$$\int_E f\, d\mu = 1, \tag{4.5.10}$$

then $f * \mu \in \mathbb{P}(\mathcal{L})$, and by (4.4.6) we can consider $f$ as a density of the FAP $f * \mu$. In addition, (4.5.10) is the typical requirement of norming. Note that similar representations can be used in control theory with impulse constraints; in this connection see constructions of [32, 35, 45]. As usual, for $f \in \mathbb{R}^E$ we denote by $\mid f \mid$ the function $x \mapsto \mid f(x) \mid : E \to [0, \infty[$. In conclusion we note that $\forall \mathcal{L} \in \pi[E]\ \forall f \in B(E, \mathcal{L})$:

$$\mid f \mid \in B^+(E, \mathcal{L}).$$

## 4.6 TOPOLOGICAL EQUIPMENT. I

Equipping the space of FAM of bounded variation with the $*$-weak topology is very important in connection with Theorem 4.3.1. We now restrict ourselves to only simplest representations (a detailed consideration is given in [66, Ch.V]). We fix $\mathcal{L} \in \Pi[E]$, where $\mathbf{S}[E \neq \varnothing]$, until the end of this chapter. Let us consider a topological equipment of the set $\mathbf{A}(\mathcal{L})$. We are interested in the weakest topology of $\mathbf{A}(\mathcal{L})$ for which each functional $\left(\int_E f\,d\mu\right)_{\mu \in \mathbf{A}(\mathcal{L})} \in \mathbb{R}^{\mathbf{A}(\mathcal{L})}$, $f \in B(E,\mathcal{L})$, is continuous. We set $\forall \mu \in \mathbf{A}(\mathcal{L})\ \forall K \in \mathrm{Fin}(B(E,\mathcal{L}))\ \forall \varepsilon \in\, ]0,\infty[$:

$$N^*_{\mathcal{L}}(\mu,K,\varepsilon) \triangleq \left\{\nu \in \mathbf{A}(\mathcal{L}) \mid \forall f \in K : \left|\int_E f\,d\mu - \int_E f\,d\nu\right| < \varepsilon\right\}. \quad (4.6.1)$$

Obviously, each set (4.6.1) is not empty. Consider some families of sets of the type (4.6.1). If $\mu \in \mathbf{A}(\mathcal{L})$, then

$$\mathfrak{N}^0_*(\mu \mid \mathcal{L}) \triangleq \{H \in \mathcal{P}(\mathbf{A}(\mathcal{L})) \mid \exists K \in \mathrm{Fin}(B(E,\mathcal{L}))\ \exists \varepsilon \in\, ]0,\infty[: H = N^*_{\mathcal{L}}(\mu,K,\varepsilon)\}; \quad (4.6.2)$$

thus (4.6.2) is the family of all sets (4.6.1) under enumeration of $K \in \mathrm{Fin}(B(E,\mathcal{L}))$ and $\varepsilon \in\, ]0,\infty[$. Moreover, we have

$$\mathfrak{N}_*(\mathcal{L}) \triangleq \bigcup_{\mu \in \mathbf{A}(\mathcal{L})} \mathfrak{N}^0_*(\mu \mid \mathcal{L}) \in (\mathrm{BAS})[\mathbf{A}(\mathcal{L})]. \quad (4.6.3)$$

On the basis of (4.6.3) we can introduce the required topology:

$$\tau_*(\mathcal{L}) \triangleq \{\cup\}(\mathfrak{N}_*(\mathcal{L})); \quad (4.6.4)$$

in this connection see Section 2.6 (for example, see (2.6.2)). Of course, $\mathfrak{N}_*(\mathcal{L}) \subset \tau_*(\mathcal{L})$. Therefore the sets (4.6.1) are open neighborhoods. Namely, $\forall \mu \in \mathbf{A}(\mathcal{L})$:

$$\mathfrak{N}^0_*(\mu \mid \mathcal{L}) \subset N^0_{\tau_*(\mathcal{L})}(\mu). \quad (4.6.5)$$

By using (4.6.4) we can strengthen (4.6.5). In this connection we note that the topology (4.6.4) admits the following representation:

$$\begin{aligned}\tau_*(\mathcal{L}) &= \{G \in \mathcal{P}(\mathbf{A}(\mathcal{L})) \mid \forall \mu \in G\ \exists H \in \mathfrak{N}^0_*(\mu \mid \mathcal{L}) : H \subset G\} \\ &= \{G \in \mathcal{P}(\mathbf{A}(\mathcal{L})) \mid \forall \mu \in G\ \exists K \in \mathrm{Fin}(B(E,\mathcal{L})) \\ &\qquad \exists \varepsilon \in\, ]0,\infty[: N^*_{\mathcal{L}}(\mu,K,\varepsilon) \subset G\}.\end{aligned} \quad (4.6.6)$$

From (4.6.6) and definitions of Section 2.2 we have $\forall \mu \in \mathbf{A}(\mathcal{L})\ \forall G \in N^0_{\tau_*(\mathcal{L})}(\mu)\ \exists H \in \mathfrak{N}^0_*(\mu \mid \mathcal{L})$: $H \subset G$. As a consequence we obtain $\forall \mu \in \mathbf{A}(\mathcal{L})$:

$$\mathfrak{N}^0_*(\mu \mid \mathcal{L}) \in (\mu - \mathrm{bas})[\tau_*(\mathcal{L})]. \quad (4.6.7)$$

In (4.6.7) we use (2.6.8) and the above mentioned corollary of (4.6.6). Thus the local bases of $\tau_*(\mathcal{L})$ are defined by (4.6.1) and (4.6.2). As a consequence $\forall M \in \mathcal{P}(\mathbf{A}(\mathcal{L}))$:

$$\mathrm{cl}(M, \tau_*(\mathcal{L})) = \{\mu \in \mathbf{A}(\mathcal{L}) \mid \forall H \in \mathfrak{N}_*^0(\mu \mid \mathcal{L}) : H \cap M \neq \varnothing\}. \quad (4.6.8)$$

As for Theorem 4.3.1, it is possible to characterize (4.6.4) as the $*$-weak topology of $\mathbf{A}(\mathcal{L})$. In addition,

$$(\mathbf{A}(\mathcal{L}), \tau_*(\mathcal{L})) \quad (4.6.9)$$

is the basic TS (in the sequel). We now note some properties of (4.6.9). If $b \in [0, \infty[$, then

$$U_b(\mathcal{L}) \triangleq \{\mu \in \mathbf{A}(\mathcal{L}) \mid V_\mu \leq b\}.$$

We have introduced the ball in the sense of the strong norm of $\mathbf{A}(\mathcal{L})$. On this basis we introduce the family of all strongly bounded subsets of $\mathbf{A}(\mathcal{L})$. Suppose that

$$\mathbb{B}_*(\mathcal{L}) \triangleq \{H \in \mathcal{P}(\mathbf{A}(\mathcal{L})) \mid \exists c \in [0, \infty[: H \subset U_c(\mathcal{L})\}.$$

From the well known Alaoglu theorem we get the equality

$$(\tau_*(\mathcal{L}) - \mathrm{comp})[\mathbf{A}(\mathcal{L})] = \mathbb{B}_*(\mathcal{L}) \cap \mathcal{F}_{\tau_*(\mathcal{L})}. \quad (4.6.10)$$

In (4.6.10) we have the very useful representation: a subset of $\mathbf{A}(\mathcal{L})$ is $*$-weakly compact iff it is a strongly bounded and $*$-weakly closed subset of $\mathbf{A}(\mathcal{L})$. Obviously, $\forall c \in [0, \infty[$:

$$U_c(\mathcal{L}) \in (\tau_*(\mathcal{L}) - \mathrm{comp})[\mathbf{A}(\mathcal{L})]. \quad (4.6.11)$$

Therefore (4.6.9) is a $\sigma$-compactum: $\mathbf{A}(\mathcal{L})$ is the union of the balls $U_k(\mathcal{L})$, $k \in \mathcal{N}$. From (4.6.11) we have the required statement. Note that by (4.6.1) $\forall K \in \mathrm{Fin}(B(E, \mathcal{L}))\ \forall \varepsilon \in ]0, \infty[$:

$$N_{\mathcal{L}}^*(\mathbb{O}_{\mathcal{L}}, K, \varepsilon) = \{\nu \in \mathbf{A}(\mathcal{L}) \mid \forall f \in K : \left| \int_E f \, d\nu \right| < \varepsilon\}. \quad (4.6.12)$$

By properties of (4.3.16) we get that each set (4.6.12) is convex. As a consequence (4.6.9) is a locally convex TS. Of course, here one must verify that (4.6.9) is a topological vector space (TVS). But the corresponding reasoning is obvious, and we omit it.

Thus (4.6.9) is a locally convex $\sigma$-compactum. We now give some examples of compact sets in the TS (4.6.9). First recall (4.6.11). We note that

$$(\mathrm{add})_+[\mathcal{L}] \in \mathcal{F}_{\tau_*(\mathcal{L})}. \quad (4.6.13)$$

To prove (4.6.13) it is sufficient to compare (4.2.10), (4.6.11), and (4.6.6). From (4.2.19), (4.6.1), (4.6.6), and (4.6.13) we get that

$$\mathbb{P}(\mathcal{L}) \in (\tau_*(\mathcal{L}) - \text{comp})[\mathbf{A}(\mathcal{L})] \tag{4.6.14}$$

(the compactum of all FAP on $\mathcal{L}$). From (4.2.21), (4.6.1), (4.6.6), (4.6.10), and (4.6.14), we have

$$\mathbb{T}(\mathcal{L}) \in (\tau_*(\mathcal{L}) - \text{comp})[\mathbf{A}(\mathcal{L})]. \tag{4.6.15}$$

The property (4.6.15) is very important for constructions analogous to those of nonstandard analysis. Returning to (2.2.47) and (2.2.48), we note that by (2.2.4), (4.6.1) and (4.6.7) $\forall_{\mathbf{D}}\, \mathbf{S}[\mathbf{D} \neq \varnothing]\ \forall \preceq \in (\text{DIR})[\mathbf{D}]$ $\forall h \in \mathbf{A}(\mathcal{L})^{\mathbf{D}}\ \forall \mu \in \mathbf{A}(\mathcal{L})$:

$$\begin{gathered}\left((\mathbf{D}, \preceq, h) \overset{\tau_*(\mathcal{L})}{\to} \mu\right) \\ \Leftrightarrow \left(\forall f \in B(E, \mathcal{L}) : \left(\mathbf{D}, \preceq, \left(\int_E f\, dh(\delta)\right)_{\delta \in \mathbf{D}}\right) \overset{\tau_{\mathbb{R}}}{\to} \int_E f\, d\mu\right).\end{gathered} \tag{4.6.16}$$

Of course, (4.6.16) defines the basic representation of (4.6.9) in terms of convergence. Note that from (4.6.16) we have $\forall_{\mathbf{D}}\, \mathbf{S}[\mathbf{D} \neq \varnothing]\ \forall \preceq \in$ $(\text{DIR})[\mathbf{D}]\ \forall h \in \mathbf{A}(\mathcal{L})^{\mathbf{D}}\ \forall \mu \in \mathbf{A}(\mathcal{L})$:

$$((\mathbf{D}, \preceq, h) \overset{\tau_*(\mathcal{L})}{\to} \mu) \Rightarrow (\forall L \in \mathcal{L} : (\mathbf{D}, \preceq, (h(\delta)(L))_{\delta \in \mathbf{D}}) \overset{\tau_{\mathbb{R}}}{\to} \mu(L)). \tag{4.6.17}$$

Note that the corresponding equivalence is lacking in (4.6.17) (see the appropriate example in [32, pp.93,94]). Thus the statement of the corollary in (4.6.17) defines another type of convergence. In the sequel we will consider this type. Moreover, we will introduce some other topologies of $\mathbf{A}(\mathcal{L})$. These topologies play the essential role in generalized constructions of AS; in this connection see [32, 35, 45, 46].

Let us consider the nonempty set $\mathbb{R}^{\mathcal{L}}$ of all real-valued (r.-v.) functions on $\mathcal{L}$. The set $\mathbf{A}(\mathcal{L})$ is a nonempty subset of $\mathbb{R}^{\mathcal{L}}$. Therefore in the form of $\mathbf{A}(\mathcal{L})$ we obtain a subspace of $\mathbb{R}^{\mathcal{L}}$ under appropriate equipment of the latter with a topology. We now give only two variants of such equipment. Namely, we consider the Tichonoff product of samples of the TS $(\mathbb{R}, \tau_{\mathbb{R}})$ and analogous product of samples of the TS $(\mathbb{R}, \tau_\partial)$. In these cases the index set is defined as $\mathcal{L}$. As a consequence we obtain the spaces

$$(\mathbb{R}^{\mathcal{L}}, \otimes^{\mathcal{L}}(\tau_{\mathbb{R}})), \quad (\mathbb{R}^{\mathcal{L}}, \otimes^{\mathcal{L}}(\tau_\partial)).$$

The corresponding precise definitions were given in Section 2.6. We recall (2.6.42). So we assume that (see (2.6.24))

$$\begin{aligned}&(\tau_{\otimes}(\mathcal{L}) \triangleq \otimes^{\mathcal{L}}(\tau_{\mathbb{R}}) \mid_{\mathbf{A}(\mathcal{L})} \in (\text{top})_0[\mathbf{A}(\mathcal{L})])\\ &\&(\tau_0(\mathcal{L}) \triangleq \otimes^{\mathcal{L}}(\tau_{\partial}) \mid_{\mathbf{A}(\mathcal{L})} \in (\text{top})_0[\mathbf{A}(\mathcal{L})])\end{aligned} \tag{4.6.18}$$

is connected by the following variant of (2.6.42):

$$\tau_{\otimes}(\mathcal{L}) \subset \tau_0(\mathcal{L}). \tag{4.6.19}$$

It is useful to establish the connection of the TS (4.6.9) and

$$(\mathbf{A}(\mathcal{L}), \tau_{\otimes}(\mathcal{L})). \tag{4.6.20}$$

This connection is the supplement of (4.6.19), and it was indicated in [32, p. 80] and [35, p. 45]. Namely,

$$\tau_{\otimes}(\mathcal{L}) \subset \tau_*(\mathcal{L}). \tag{4.6.21}$$

It is useful to recall the natural justification. Returning to (4.6.18) we introduce $\forall \mu \in \mathbf{A}(\mathcal{L})\ \forall \mathcal{K} \in \text{Fin}(\mathcal{L})\ \forall \varepsilon \in ]0, \infty[$:

$$\begin{aligned}\mathbb{N}_{\mathcal{L}}^{\otimes}(\mu, \mathcal{K}, \varepsilon) &= \{\nu \in \mathbf{A}(\mathcal{L}) \mid \forall L \in \mathcal{K} : \mid \mu(L) - \nu(L) \mid < \varepsilon\}\\ &= \mathbb{N}_{\mathcal{L}}(\mu, \mathcal{K}, \varepsilon) \cap \mathbf{A}(\mathcal{L}).\end{aligned} \tag{4.6.22}$$

On the basis of (4.6.22) we construct corresponding families of neighborhoods. If $\mu \in \mathbf{A}(\mathcal{L})$, then

$$\mathfrak{N}_{\mathcal{L}}^{\otimes}(\mu) \triangleq \{\mathbb{N}_{\mathcal{L}}^{\otimes}(\mu, \mathcal{K}, \varepsilon) : (\mathcal{K}, \varepsilon) \in \text{Fin}(\mathcal{L}) \times ]0, \infty[\}. \tag{4.6.23}$$

Compare (4.6.23) and the family $\mathfrak{N}_{\mathcal{L}}^{0}(\mu) \mid_{\mathbf{A}(\mathcal{L})}$ (see Section 2.6). Then for $\mu \in \mathbf{A}(\mathcal{L})$:

$$\mathfrak{N}_{\mathcal{L}}^{0}(\mu) \mid_{\mathbf{A}(\mathcal{L})} = \{\mathbb{N}_{\mathcal{L}}(\mu, \mathcal{K}, \varepsilon) \cap \mathbf{A}(\mathcal{L}) : (\mathcal{K}, \varepsilon) \in \text{Fin}(\mathcal{L}) \times ]0, \infty[\} = \mathfrak{N}_{\mathcal{L}}^{\otimes}(\mu) \tag{4.6.24}$$

(see (2.3.1)). As a consequence we have the inclusion

$$\mathfrak{N}^{\otimes}[\mathcal{L}] \triangleq \bigcup_{\mu \in \mathbf{A}(\mathcal{L})} \mathfrak{N}_{\mathcal{L}}^{\otimes}(\mu) \subset \mathfrak{N}_{\mathcal{L}} \mid_{\mathbf{A}(\mathcal{L})}. \tag{4.6.25}$$

We use (2.3.1), (2.6.28), and (4.6.24) in (4.6.25). Therefore by (2.6.1), (2.6.3), (2.6.6), (2.6.44), (2.6.45), and (4.6.25) we obtain

$$\{\cup\}(\mathfrak{N}^{\otimes}[\mathcal{L}]) \subset \{\cup\}(\mathfrak{N}_{\mathcal{L}}) \mid_{\mathbf{A}(\mathcal{L})}. \tag{4.6.26}$$

From (2.6.29), (4.6.18) and (4.6.26) we have the inclusion

$$\{\cup\}(\mathfrak{N}^{\otimes}[\mathcal{L}]) \subset \tau_{\otimes}(\mathcal{L}). \tag{4.6.27}$$

Returning to (2.6.2), (4.6.22) and (4.6.25), we infer that

$$\mathfrak{N}^{\otimes}[\mathcal{L}] \in (\mathrm{BAS})[\mathbf{A}(\mathcal{L})]. \tag{4.6.28}$$

In (4.6.28) we use the following corollary of (4.6.22): if $\mu_1 \in \mathbf{A}(\mathcal{L})$, $\mu_2 \in \mathbf{A}(\mathcal{L})$, $\mathcal{K}_1 \in \mathrm{Fin}(\mathcal{L})$, $\mathcal{K}_2 \in \mathrm{Fin}(\mathcal{L})$, $\varepsilon_1 \in ]0, \infty[$, $\varepsilon_2 \in ]0, \infty[$, and $\mu \in \mathbb{N}^{\otimes}_{\mathcal{L}}(\mu_1, \mathcal{K}_1, \varepsilon_1) \cap \mathbb{N}^{\otimes}_{\mathcal{L}}(\mu_2, \mathcal{K}_2, \varepsilon_2)$, then $\exists \tilde{\varepsilon} \in ]0, \infty[$:

$$\mathbb{N}^{\otimes}_{\mathcal{L}}(\mu, \mathcal{K}_1 \cup \mathcal{K}_2, \tilde{\varepsilon}) \subset \mathbb{N}^{\otimes}_{\mathcal{L}}(\mu_1, \mathcal{K}_1, \varepsilon_1) \cap \mathbb{N}^{\otimes}_{\mathcal{L}}(\mu_2, \mathcal{K}_2, \varepsilon_2).$$

From (4.6.27) it follows that some topology of $\mathbf{A}(\mathcal{L})$ is realized on the left hand side of (4.6.27). Let $G'_{\otimes} \in \tau_{\otimes}(\mathcal{L})$ and $\mu_{\otimes} \in G'_{\otimes}$. Choose $G' \in \otimes^{\mathcal{L}}(\tau_{\mathbb{R}})$ for which $G'_{\otimes} = G' \cap \mathbf{A}(\mathcal{L})$. Then $\mu_{\otimes} \in G'$. By (2.6.30) we choose $\mathcal{K}' \in \mathrm{Fin}(\mathcal{L})$ and $\varepsilon' \in ]0, \infty[$ for which $\mathbb{N}_{\mathcal{L}}(\mu_{\otimes}, \mathcal{K}', \varepsilon') \subset G'$. Then by (4.6.22) we have

$$\mathbb{N}^{\otimes}_{\mathcal{L}}(\mu_{\otimes}, \mathcal{K}', \varepsilon') \subset G'_{\otimes}.$$

Since the choice of $\mu_{\otimes}$ was arbitrary, we have established that $\forall \mu \in G'_{\otimes}$ $\exists H \in \mathfrak{N}^{\otimes}_{\mathcal{L}}(\mu)$: $H \subset G'_{\otimes}$. From ((2.6.3), (4.6.22), (4.6.24), and (4.6.25) we have $G'_{\otimes} \in \{\cup\}(\mathfrak{N}^{\otimes}[\mathcal{L}])$. Therefore $\tau_{\otimes}(\mathcal{L}) \subset \{\cup\}(\mathfrak{N}^{\otimes}[\mathcal{L}])$; we recall that the choice of $G'_{\otimes}$ was arbitrary. By (4.6.27) we now have $\tau_{\otimes}(\mathcal{L}) = \{\cup\}(\mathfrak{N}^{\otimes}[\mathcal{L}])$, i.e.,

$$\mathfrak{N}^{\otimes}[\mathcal{L}] \in (\tau_{\otimes}(\mathcal{L}) - \mathrm{BAS})_0[\mathbf{A}(\mathcal{L})]. \tag{4.6.29}$$

But from (4.6.1), (4.6.2) and (4.6.22) it follows that

$$\mathbb{N}^{\otimes}_{\mathcal{L}}(\mu, \mathcal{K}, \varepsilon) \in \mathfrak{N}^{0}_{*}(\mu \mid \mathcal{L}) \tag{4.6.30}$$

for $\mu \in \mathbf{A}(\mathcal{L})$, $\mathcal{K} \in \mathrm{Fin}(\mathcal{L})$, and $\varepsilon \in ]0, \infty[$. Indeed, let $\mu \in \mathbf{A}(\mathcal{L})$, $\mathcal{K} \in \mathrm{Fin}(\mathcal{L})$, and $\varepsilon \in ]0, \infty[$. Then

$$\mathbb{K} \triangleq \{\chi_L[E] : L \in \mathcal{K}\} \in \mathrm{Fin}(B(E, \mathcal{L})).$$

Therefore from (4.6.1) we conclude that $N^{*}_{\mathcal{L}}(\mu, \mathbb{K}, \varepsilon)$ is the set of all $\nu \in \mathbf{A}(\mathcal{L})$ such that

$$\forall f \in \mathbb{K} : \left| \int_E f \, d\mu - \int_E f \, d\nu \right| < \varepsilon.$$

Then (see (4.3.4)) the last relation is equivalent to the condition $\mid \mu(L) - \nu(L) \mid < \varepsilon$ for $L \in \mathcal{K}$. This conclusion follows from (4.3.8) and (4.3.14). Then by (4.6.22) we have the equality

$$N^{*}_{\mathcal{L}}(\mu, \mathbb{K}, \varepsilon) = \mathbb{N}^{\otimes}_{\mathcal{L}}(\mu, \mathcal{K}, \varepsilon). \tag{4.6.31}$$

From (4.6.2) and (4.6.31) we obtain (4.6.30). Then by (4.6.23) we have $\forall \mu \in \mathbf{A}(\mathcal{L})$:

$$\mathfrak{N}^{\otimes}_{\mathcal{L}}(\mu) \subset \mathfrak{N}^{0}_{*}(\mu \mid \mathcal{L}). \tag{4.6.32}$$

From (4.6.3), (4.6.25) and (4.6.32) we infer the inclusion $\mathfrak{N}^{\otimes}[\mathcal{L}] \subset \mathfrak{N}_*(\mathcal{L})$. Obviously, from (4.6.4), (4.6.29) and the last inclusion we get (4.6.21). Thus (4.6.21) is established.

PROPOSITION 4.6.1 ( SEE [35, P. 45]) *If* $H \in \mathbb{B}_*(\mathcal{L})$, *then*

$$\tau_*(\mathcal{L}) \mid_H = \tau_{\otimes}(\mathcal{L}) \mid_H = \otimes^{\mathcal{L}}(\tau_{\mathbb{R}}) \mid_H .$$

The proof is obvious. Here (4.3.2), (4.3.11) and (4.3.15) are used. We note only that it is assumed that $H \in \mathbb{B}_*(\mathcal{L}) \setminus \{\varnothing\}$ (the case $H = \varnothing$ is trivial).

Suppose that $\forall H \in \mathcal{P}((\text{add})[\mathcal{L}])$ $\forall \mu \in (\text{add})[\mathcal{L}]$ $\forall K \in \text{Fin}(B_0(E, \mathcal{L}))$ $\forall \varepsilon \in ]0, \infty[$:

$$N^{(\text{el})}_{\mathcal{L}}(\mu, K, \varepsilon \mid H) \triangleq \{\nu \in H \mid \forall f \in K : \mid {}^{(\text{el})}\!\!\int_E f \, d\mu - {}^{(\text{el})}\!\!\int_E f \, d\nu \mid < \varepsilon\}.$$

For $H \in \mathcal{P}((\text{add})[\mathcal{L}])$ and $\mu \in (\text{add})[\mathcal{L}]$ consider the family $\mathfrak{N}^{(\text{el})}(\mu \mid H)$ of all sets $N^{(\text{el})}_{\mathcal{L}}(\mu, K, \varepsilon \mid H)$, $(K, \varepsilon) \in \text{Fin}(B_0(E, \mathcal{L})) \times ]0, \infty[$. Finally, for $H \in 2^{(\text{add})[\mathcal{L}]}$ we have

$$\mathfrak{N}^{(\text{el})}_*(H) \triangleq \bigcup_{\mu \in H} \mathfrak{N}^{(\text{el})}(\mu \mid H) \in (\text{BAS})[H].$$

The corresponding proof is an obvious consequence of definitions from Section 2.5. In addition, for $H \in 2^{(\text{add})[\mathcal{L}]}$ we have

$$\begin{aligned} \tau^{(\text{el})}_*(\mathcal{L} \mid H) &\triangleq \{\cup\}(\mathfrak{N}^{(\text{el})}_*(H)) \\ &= \{X \in \mathcal{P}(H) \mid \forall \mu \in X \ \exists B \in \mathfrak{N}^{(\text{el})}(\mu \mid H) : B \subset X\} \\ &\in (\text{top})[H]. \end{aligned}$$

As a consequence we can use the last notion for $\mathbb{H} \in \mathbb{B}_*(\mathcal{L}) \setminus \{\varnothing\}$:

$$\tau^{(\text{el})}_*(\mathcal{L} \mid \mathbb{H}) = \tau_*(\mathcal{L}) \mid_{\mathbb{H}} . \tag{4.6.33}$$

Proposition 4.6.1 can be easily established on the basis of (4.6.33) (see (4.3.8)).

We recall (4.6.10). From Proposition 4.6.1 we have $\forall K \in (\tau_*(\mathcal{L}) - \text{comp})[\mathbf{A}(\mathcal{L})]$:

$$\tau_*(\mathcal{L}) \mid_K = \tau_{\otimes}(\mathcal{L}) \mid_K \in (\mathbf{c} - \text{top})[K].$$

In particular, from (4.6.11) we have $\forall c \in [0, \infty[$:

$$\tau_*(\mathcal{L}) \mid_{U_c(\mathcal{L})} = \tau_\otimes(\mathcal{L}) \mid_{U_c(\mathcal{L})} \in (\mathbf{c}-\mathrm{top})[U_c(\mathcal{L})]. \tag{4.6.34}$$

The relation (4.6.34) is a version of the well known Alaoglu theorem. As for the TS (4.6.9), it is useful to recall a property of the non-negative cone. Namely, the topologies

$$\begin{aligned}(\tau_*^+(\mathcal{L}) &\triangleq \tau_*(\mathcal{L}) \mid_{(\mathrm{add})_+[\mathcal{L}]}) \\ \& \ (\tau_0^+(\mathcal{L}) &\triangleq \tau_0(\mathcal{L}) \mid_{(\mathrm{add})_+[\mathcal{L}]}) \\ \& \ (\tau_\otimes^+(\mathcal{L}) &\triangleq \tau_\otimes(\mathcal{L}) \mid_{(\mathrm{add})_+[\mathcal{L}]})\end{aligned} \tag{4.6.35}$$

satisfy the following relations:

$$\tau_*^+(\mathcal{L}) = \tau_\otimes^+(\mathcal{L}) \subset \tau_0^+(\mathcal{L}). \tag{4.6.36}$$

Here the last inclusion immediately follows from (4.6.19). Moreover, (4.6.21) implies the inclusion $\tau_\otimes^+(\mathcal{L}) \subset \tau_*^+(\mathcal{L})$. Let $F \in \mathcal{F}_\tau$, where $\tau = \tau_*^+(\mathcal{L})$. Consider an arbitrary net $(\mathbf{D}, \preceq, h)$ in $F$ and $\mu \in (\mathrm{add})_+[\mathcal{L}]$ with the property

$$(\mathbf{D}, \preceq, h) \overset{\tau_\otimes^+(\mathcal{L})}{\longrightarrow} \mu. \tag{4.6.37}$$

By (4.6.37) we have the convergence of $(\mathbf{D}, \preceq, h)$ to $\mu$ in the TS (4.6.20). In particular, from definitions of Section 2.6 (see (2.6.22)) it follows that

$$(\mathbf{D}, \preceq, (h(d)(E))_{d \in \mathbf{D}}) \overset{\tau_\mathbb{R}}{\to} \mu(E). \tag{4.6.38}$$

In this statement we use the obvious convergence of $(\mathbf{D}, \preceq, h)$ to $\mu$ in the TS $(\mathbb{R}^\mathcal{L}, \otimes^\mathcal{L}(\tau_\mathbb{R}))$; see (2.3.9) and (4.6.18). In addition, we use (2.6.27) and (2.6.31). Introduce $b \triangleq \mu(E) + 1 \in ]0, \infty[$. Then by (4.6.38), for some $\delta \in \mathbf{D}$, we have $h(d)(E) \leq b$ under $d \in \mathbf{D}$, $\delta \preceq d$. Suppose that $\mathbf{D}_\delta \triangleq \{d \in \mathbf{D} \mid \delta \preceq d\}$. Then $\delta \in \mathbf{D}_\delta$. So $\mathbf{S}[\mathbf{D}_\delta \neq \varnothing]$. Let $\sqsubseteq \triangleq \preceq \cap (\mathbf{D}_\delta \times \mathbf{D}_\delta)$; we have a new relation. Obviously, $\forall d_1 \in \mathbf{D}_\delta \ \forall d_2 \in \mathbf{D}_\delta$:

$$(d_1 \sqsubseteq d_2) \Leftrightarrow (d_1 \preceq d_2). \tag{4.6.39}$$

We note (2.2.7) in connection with (4.6.39). From (2.2.4) and (2.2.7) we infer that $\mathbf{D}_\delta \in (\preceq -\mathrm{cof})[\mathbf{D}]$. As a consequence of (2.2.8) we obtain $\sqsubseteq \in (\mathrm{DIR})[\mathbf{D}_\delta]$. Assuming $h_\delta \triangleq (h \mid \mathbf{D}_\delta)$, we get the net $(\mathbf{D}_\delta, \sqsubseteq, h_\delta)$ in $U_b(\mathcal{L})$; here we use the definition of a ball and the remark after (4.2.14). From (4.6.37) we have the convergence in the TS (4.6.20); see (2.3.9). Consequently

$$(\mathbf{D}_\delta, \sqsubseteq, h_\delta) \overset{\tau_\otimes(\mathcal{L})}{\longrightarrow} \mu. \tag{4.6.40}$$

Again use (2.3.9) (and (4.6.40)). We get the convergence

$$(\mathbf{D}_\delta, \sqsubseteq, h_\delta) \overset{\tau_\otimes(\mathcal{L})|_{U_b(\mathcal{L})}}{\longrightarrow} \mu. \tag{4.6.41}$$

This convergence is an obvious consequence of (4.6.40). From (4.6.41) with taking (4.6.34) into account we get the convergence of $(\mathbf{D}_\delta, \sqsubseteq, h_\delta)$ in the sense of the topology $\tau_*(\mathcal{L}) \mid_{U_b(\mathcal{L})}$. Returning to (2.3.9), we obtain the convergence

$$(\mathbf{D}_\delta, \sqsubseteq, h_\delta) \overset{\tau_*(\mathcal{L})}{\to} \mu.$$

From (2.3.9) and (4.6.35) it follows that

$$(\mathbf{D}_\delta, \sqsubseteq, h_\delta) \overset{\tau_*^+(\mathcal{L})}{\to} \mu. \tag{4.6.42}$$

But $(\mathbf{D}_\delta, \sqsubseteq, h_\delta)$ is a net in $F$. By properties of $F$ and (4.6.42) we have $\mu \in F$. Thus (see (4.6.37))

$$((\mathbf{D}, \preceq, h) \overset{\tau_\otimes^+(\mathcal{L})}{\longrightarrow} \mu) \Rightarrow (\mu \in F).$$

Since the choice of $(\mathbf{D}, \preceq, h)$ and $\mu$ was arbitrary, we infer that

$$F \in \mathcal{F}_{\tau_\otimes^+(\mathcal{L})}.$$

So we have established that $\mathcal{F}_{\tau_*^+(\mathcal{L})} \subset \mathcal{F}_{\tau_\otimes^+(\mathcal{L})}$. As a consequence $\tau_*^+(\mathcal{L}) \subset \tau_\otimes^+(\mathcal{L})$. Therefore $\tau_*^+(\mathcal{L}) = \tau_\otimes^+(\mathcal{L})$. Using (4.6.19) we obtain (4.6.36). Note that from (4.6.10) and (4.6.13) we have $\forall b \in [0, \infty[$:

$$\begin{aligned} U_b^+(\mathcal{L}) &\triangleq U_b(\mathcal{L}) \cap (\text{add})_+[\mathcal{L}] \\ &= \{\mu \in (\text{add})_+[\mathcal{L}] \mid \mu(E) \le b\} \\ &\in (\tau_*(\mathcal{L}) - \text{comp})[\mathbf{A}(\mathcal{L})]. \end{aligned} \tag{4.6.43}$$

In (4.6.43) we use the property $(\text{add})_+[\mathcal{L}] \in \mathcal{F}_{\tau_*(\mathcal{L})}$; in this connection see (4.6.1) and (4.6.6). So by (2.3.6) for $b \in [0, \infty[$ it follows that

$$\tau_*(\mathcal{L}) \mid_{U_b^+(\mathcal{L})} = \tau_*^+(\mathcal{L}) \mid_{U_b^+(\mathcal{L})} \in (\mathbf{c} - \text{top})[U_b^+(\mathcal{L})]$$

and therefore

$$U_b^+(\mathcal{L}) \in (\tau_*^+(\mathcal{L}) - \text{comp})[(\text{add})_+[\mathcal{L}]]. \tag{4.6.44}$$

On the other hand, for $b \in [0, \infty[$

$$\begin{aligned} &\{\mu \in (\text{add})_+[\mathcal{L}] \mid \mu(E) < b\} \\ =&\{\mu \in \mathbf{A}(\mathcal{L}) \mid \mu(E) < b\} \cap (\text{add})_+[\mathcal{L}] \in \tau_*^+(\mathcal{L}); \end{aligned} \tag{4.6.45}$$

we use (4.6.1) and (4.6.6) in (4.6.45). Certainly, (4.6.45) is a subset of the set (4.6.43). As a consequence we have the highly obvious

PROPOSITION 4.6.2 $\tau_*^+(\mathcal{L}) \in (\mathbf{c} - \mathrm{top})^0_{\mathrm{loc}}[(\mathrm{add})_+[\mathcal{L}]]$.

PROOF. From (2.3.8), (4.6.4) and (4.6.35) we conclude that $\tau_*^+(\mathcal{L}) \in (\mathrm{top})_0[(\mathrm{add})_+[\mathcal{L}]]$. We take into account (2.9.1), (2.9.4) and (2.9.5). Let $\mu_* \in (\mathrm{add})_+[\mathcal{L}]$ and $b_* \triangleq \mu_*(E) + 1$. We have

$$U^+_{b_*}(\mathcal{L}) = \{\mu \in (\mathrm{add})_+[\mathcal{L}] \mid \mu(E) \leq b_*\} \in (\tau_*^+(\mathcal{L}) - \mathrm{comp})[(\mathrm{add})_+[\mathcal{L}]] \tag{4.6.46}$$

(see (4.6.44)). Moreover, by (4.6.45), $\mathbb{S} \triangleq \{\mu \in (\mathrm{add})_+[\mathcal{L}] \mid \mu(E) < b_*\} \in \tau_*^+(\mathcal{L})$. Since $\mu_* \in \mathbb{S}$, we have

$$\mathbb{S} \in N^0_{\tau_*^+(\mathcal{L})}(\mu_*).$$

In addition, $\mathbb{S} \subset U^+_{b_*}(\mathcal{L})$. Therefore from (4.6.46) we obtain that

$$U^+_{b_*}(\mathcal{L}) \in N_{\tau_*^+(\mathcal{L})}(\mu_*) \cap (\tau_*^+(\mathcal{L}) - \mathrm{comp})[(\mathrm{add})_+[\mathcal{L}]].$$

Since the choice of $\mu_*$ was arbitrary, the property

$$\tau_*^+(\mathcal{L}) \in (\mathbf{c} - \mathrm{top})_{\mathrm{loc}}[(\mathrm{add})_+[\mathcal{L}]] \tag{4.6.47}$$

is established (see (2.9.1)). By (2.9.5) and (4.6.47) we obtain the required statement. □

## 4.7 TOPOLOGICAL EQUIPMENT. II

In this brief part (see ([35, Ch. 3]) we consider questions connected with the bounded $*$-weak topology

$$\tau^*_{\mathbb{B}}(\mathcal{L}) \triangleq \{G \in \mathcal{P}(\mathbf{A}(\mathcal{L})) \mid \forall c \in [0, \infty[ : U_c(\mathcal{L}) \cap G \in \tau_*(\mathcal{L}) \mid_{U_c(\mathcal{L})}\} \in (\mathrm{top})[\mathbf{A}(\mathcal{L})]. \tag{4.7.1}$$

From (4.7.1) we have the relation

$$\tau_*(\mathcal{L}) \subset \tau^*_{\mathbb{B}}(\mathcal{L}). \tag{4.7.2}$$

It is useful to note the obvious corollary of (4.6.10) and (4.7.1):

$$\tau^*_{\mathbb{B}}(\mathcal{L}) = \{G \in \mathcal{P}(\mathbf{A}(\mathcal{L})) \mid \forall K \in (\tau_*(\mathcal{L}) - \mathrm{comp})[\mathbf{A}(\mathcal{L})] : K \cap G \in \tau_*(\mathcal{L}) \mid_K\}. \tag{4.7.3}$$

So the topology (4.7.1), (4.7.3) turns $\mathbf{A}(\mathcal{L})$ into a Hausdorff $k$-space [71]. In addition

$$(\tau_*(\mathcal{L}) - \text{comp})[\mathbf{A}(\mathcal{L})] = (\tau^*_{\mathbb{B}}(\mathcal{L}) - \text{comp})[\mathbf{A}(\mathcal{L})] \tag{4.7.4}$$

((4.7.2) and (4.7.3) should be used in (4.7.4)). Moreover, we note that $\forall K \in (\tau_*(\mathcal{L}) - \text{comp})[\mathbf{A}(\mathcal{L})]$:

$$\tau_*(\mathcal{L}) \mid_K = \tau^*_{\mathbb{B}}(\mathcal{L}) \mid_K . \tag{4.7.5}$$

In (4.7.4) and (4.7.5) we have important representations: the topologies connected by virtue of (4.7.2) have the same collection of compact sets. The construction above admits a natural generalization. Namely, let $\forall_X\ \mathbf{S}[X]\ \forall \tau \in (\text{top})[X]$:

$$\mathbf{K}_*[\tau] \triangleq \{G \in \mathcal{P}(X) \mid \forall K \in (\tau - \text{comp})[X] : K \cap G \in \tau \mid_K\} \in (\text{top})[X]; \tag{4.7.6}$$

in (4.7.6) we have the compact-generating topology. It is worth noting that $\forall_X\ \mathbf{S}[X]\ \forall \tau \in (\text{top})_0[X]\ \forall H \in \mathcal{P}(X)$:

$$(\tau \mid_H \in (\mathbf{c} - \text{top})_{\text{loc}}[H]) \Rightarrow (\tau \mid_H = \mathbf{K}_*[\tau] \mid_H). \tag{4.7.7}$$

To prove (4.7.7) it is sufficient to apply constructions of Section 2.9. In connection with (4.7.7) we consider some auxiliary notions. If $\mathbf{S}[X]$, then denote by $(\mathcal{K} - \text{top})[X]$ the family of all $\tau \in (\text{top})[X]$ with the property that

$$\{E \in \mathcal{P}(X) \mid \forall K \in (\tau - \text{comp})[X] : K \cap E \in \mathcal{F}_{\tau|_K}\} = \mathcal{F}_\tau. \tag{4.7.8}$$

In (4.7.8) we have the basic property of the compact-generating topology. In fact, to prove (4.7.7) it suffices to show that

$$(\mathbf{c} - \text{top})^0_{\text{loc}}[X] \subset (\mathcal{K} - \text{top})[X]. \tag{4.7.9}$$

Let $\mathbf{S}[X]$ and $\theta \in (\mathbf{c} - \text{top})^0_{\text{loc}}[X]$. Denote by $\mathcal{M}$ the family on the left hand side of (4.7.8) under the replacement $\tau \to \theta$. Obviously, $\mathcal{F}_\theta \subset \mathcal{M}$. Choose $M \in \mathcal{M}$ and consider $\text{cl}(M, \theta) \in \mathcal{F}_\theta$. Let $\mu \in \text{cl}(M, \theta)$ and $\Omega \in N_\theta(\mu) \cap (\theta - \text{comp})[X]$ (we use (2.9.1)). Then $\Omega \cap M \in \mathcal{F}_{\theta|_\Omega}$. In addition, by (2.3.8)

$$N_{\theta|_\Omega}(\mu) = N_\theta(\mu) \mid_\Omega . \tag{4.7.10}$$

Choose arbitrarily an element $\Lambda$ of the family (4.7.10). Let $\Lambda_0 \in N_\theta(\mu)$ be the set for which $\Lambda = \Omega \cap \Lambda_0$. For some $G_0 \in N^0_\theta(\mu)$ we have $G_0 \subset \Lambda_0$. By the choice of $\Omega$ we can take $G^0 \in N^0_\theta(\mu)$ such that $G^0 \subset \Omega$. Then

$G^0 \cap G_0 \in N_\theta^0(\mu)$ and $\Lambda = \Lambda_0 \cap \Omega \in N_\theta(\mu)$; hence $\Lambda \cap M \neq \varnothing$. Then $\Lambda \cap (\Omega \cap M) \neq \varnothing$. Since the choice of $\Lambda$ was arbitrary, we have $\mu \in \mathrm{cl}(\Omega \cap M, \theta\mid_\Omega)$, i.e., $\mu \in \Omega \cap M$. In particular, $\mu \in M$ and the inclusion $\mathrm{cl}(M, \theta) \subset M$ is established. So $M \in \mathcal{F}_\theta$. We obtain the inclusion $\mathcal{M} \subset \mathcal{F}_\theta$. The inverse inclusion is obvious. Thus $\theta \in (\mathcal{K} - \mathrm{top})[X]$ (see (4.7.8)) □.

The implication (4.7.7) follows from (4.7.9). Since

$$\tau_{\mathbb{B}}^*(\mathcal{L}) = \mathbf{K}_*[\tau_*(\mathcal{L})],$$

we have (see Proposition 4.6.1)

$$\tau_*^+(\mathcal{L}) = \tau_{\mathbb{B}}^*(\mathcal{L})\mid_{(\mathrm{add})_+[\mathcal{L}]}; \tag{4.7.11}$$

in this connection see [35, p. 55].

## 4.8 A CONSTRUCTION OF A DIRECTED SET

In the previous sections several topologies of $\mathbf{A}(\mathcal{L})$ are used. It turns out that it is impossible to obtain an exhausting representation in terms of the convergence of sequences in $\mathbf{A}(\mathcal{L})$ for these topologies. Nets or filters should be used here. Therefore we consider some concrete construction of a directed set with the aim of getting the above mentioned nets. Return to partitions similar to (4.2.2). But we introduce unordered partitions, following [32, p.57]. Suppose that $\forall H \in \mathcal{P}(E)$:

$$\begin{gathered}\mathbf{D}(H, \mathcal{L}) \triangleq \{\mathcal{K} \in \mathrm{Fin}(\mathcal{L}) \mid (H = \bigcup_{L \in \mathcal{K}} L) \& (\forall A \in \mathcal{K}\ \forall B \in \mathcal{K} : \\ (A \cap B) \neq \varnothing) \Rightarrow (A = B))\}\end{gathered} \tag{4.8.1}$$

The basic case is when $H = E$. Hence we consider the nonempty set $\mathbf{D}(E, \mathcal{L})$ (recall that $\{E\} \in \mathbf{D}(E, \mathcal{L})$). We use a particular case of (4.8.1) for constructing an important directed set. Following the stipulation (2.2.1), we introduce the relation $\prec \in \mathcal{P}(\mathbf{D}(E, \mathcal{L}) \times \mathbf{D}(E, \mathcal{L}))$ by the rule: for $\mathcal{U} \in \mathbf{D}(E, \mathcal{L})$ and $\mathcal{V} \in \mathbf{D}(E, \mathcal{L})$,

$$(\mathcal{U} \prec \mathcal{V}) \Leftrightarrow (\forall V \in \mathcal{V}\ \exists U \in \mathcal{U} : V \subset U). \tag{4.8.2}$$

The relation (4.8.2) corresponds to the notion of refinement. We note that $\forall \mathcal{U} \in \mathbf{D}(E, \mathcal{L})\ \forall \mathcal{V} \in \mathbf{D}(E, \mathcal{L})$:

$$\{U \cap V : (U, V) \in \mathcal{U} \times \mathcal{V}\} \in \mathbf{D}(E, \mathcal{L}). \tag{4.8.3}$$

From (2.2.3), (2.2.4), (4.8.2), and (4.8.3) we get the obvious property

$$\prec \in (\mathrm{DIR})[\mathbf{D}(E, \mathcal{L})]. \tag{4.8.4}$$

**Remark.** Certainly, the direction (4.8.4) depends on $\mathcal{L}$, i.e., $\prec = \prec_{\mathcal{L}}$. Therefore in the case when a family $\mathcal{H}$ is used in place of $\mathcal{L}$ (i.e., the MS is replaced), we take the corresponding symbol $\prec_{\mathcal{H}}$ in place of $\prec$. Here we preserve the simplified notation $\prec$ instead of $\prec_{\mathcal{L}}$. Recall that we fixed $\mathcal{L} \in \Pi[E]$ in Section 4.6.

So we have the (nonempty) directed set $(\mathbf{D}(E, \mathcal{L}), \prec)$. In addition, $\forall \mathcal{U} \in \mathbf{D}(E, \mathcal{L})\ \forall \mathcal{V} \in \mathbf{D}(E, \mathcal{L})$:

$$(\mathcal{U} \prec \mathcal{V}) \Leftrightarrow (\forall U \in \mathcal{U} \setminus \{\varnothing\} \exists \mathcal{H} \in \mathbf{D}(U, \mathcal{L}) : \mathcal{H} \subset \mathcal{V}). \tag{4.8.5}$$

See [32, pp. 83,84] in connection with (4.8.2)–(4.8.5). The proof of (4.8.5) is obvious (see (4.8.1)). If

$$h : \mathbf{D}(E, \mathcal{L}) \to \mathbf{A}(\mathcal{L}), \tag{4.8.6}$$

then we have the net $(\mathbf{D}(E, \mathcal{L}), \prec, h)$ in $\mathbf{A}(\mathcal{L})$. The concrete choice of (4.8.6) depends on the problem under consideration. We have

$$\forall \mathcal{H} \in \mathbf{D}(E, \mathcal{L}) : \mathcal{H} \setminus \{\varnothing\} \in \mathbf{D}(E, \mathcal{L}). \tag{4.8.7}$$

As for (4.8.7), we note that $\mathbf{S}[E \neq \varnothing]$ and (4.8.1). Using (4.8.7) we consider a variant of (4.8.6). Namely, relying on the axiom of choice we introduce some function of choice

$$\kappa \in \prod_{L \in \mathcal{L} \setminus \{\varnothing\}} L, \tag{4.8.8}$$

which is defined on the nonempty family $\mathcal{L} \setminus \{\varnothing\}$ (we recall that $E \in \mathcal{L} \setminus \{\varnothing\}$). In addition, for $L \in \mathcal{L} \setminus \{\varnothing\}$, we have $\kappa(L) \in L$. So $\mathcal{L} \setminus \{\varnothing\}$ is domain of $\kappa$ and hence, by (4.8.8), $\mathcal{L} = (\mathcal{L} \setminus \{\varnothing\}) \cup \{\varnothing\}$ is defined in a unique way. If $\mu \in \mathbf{A}(\mathcal{L})$ and $\mathcal{H} \in \mathbf{D}(E, \mathcal{L})$, then it is possible to consider the linear combination of Dirac measures concentrated at the points $\kappa(H) \in H$ with the weights $\mu(H)$, $H \in \mathcal{H} \setminus \{\varnothing\}$. Return to definitions of Section 4.2.

If $x \in E$, then we denote by $\delta_x$ the Dirac measure defined on $\mathcal{P}(E)$ and concentrated at the point $x$: by definition, $\delta_x \in \mathbb{T}_\sigma(\mathcal{P}(E))$ is defined by the rule

$$\forall L \in \mathcal{P}(E) : ((x \in L) \Rightarrow (\delta_x(L) = 1))\&((x \notin L) \Rightarrow (\delta_x(L) = 0)). \tag{4.8.9}$$

Then for $x \in E$ we have the property

$$(\delta_x \mid \mathcal{L}) = (\delta_x(L))_{L \in \mathcal{L}} \in \mathbb{T}_\sigma(\mathcal{L}). \tag{4.8.10}$$

By (4.8.9) and (4.8.10) the natural restriction of the Dirac measure is defined. As a consequence $\forall n \in \mathcal{N}\ \forall (\alpha_i)_{i\in\overline{1,n}} \in \mathbb{R}^n\ \forall (x_i)_{i\in\overline{1,n}} \in E^n$:

$$\sum_{i=1}^{n} \alpha_i(\delta_{x_i} \mid \mathcal{L}) \in (\sigma - \text{add})[\mathcal{L}] \cap (B - \text{var})[\mathcal{L}]. \tag{4.8.11}$$

In (4.8.11) we have a linear combination of the 'restricted' Dirac measures. Using the pointwise definition of linear operations for functionals, we introduce unordered linear combination of Dirac measures. Here we employ unordered sums of numbers and, in fact, obtain sums of the type (4.8.11). Hence for $\mathcal{H} \in \mathbf{D}(E,\mathcal{L})$ and $\mu \in \mathbf{A}(\mathcal{L})$ we have

$$\mathfrak{D}_{\mathcal{L}}(\mu,\kappa,\mathcal{H}) \triangleq \sum_{H\in\mathcal{H}\setminus\{\varnothing\}} \mu(H)(\delta_{\kappa(H)} \mid \mathcal{L}) \in (\sigma - \text{add})[\mathcal{L}] \cap (B - \text{var})[\mathcal{L}]; \tag{4.8.12}$$

of course, for $L \in \mathcal{L}$ we have the relation

$$\mathfrak{D}_{\mathcal{L}}(\mu,\kappa,\mathcal{H})(L) = \sum_{H\in\mathcal{H}\setminus\{\varnothing\}} \mu(H)\delta_{\kappa(H)}(L). \tag{4.8.13}$$

In (4.8.13) we use the ordinary sum of numbers. If $\mu \in \mathbf{A}(\mathcal{L})$, then

$$\mathfrak{D}_{\mathcal{L}}(\mu,\kappa,\cdot) \triangleq (\mathfrak{D}_{\mathcal{L}}(\mu,\kappa,\mathcal{H}))_{\mathcal{H}\in\mathbf{D}(E,\mathcal{L})} \tag{4.8.14}$$

is the mapping from $\mathbf{D}(E,\mathcal{L})$ into $(\sigma - \text{add})[\mathcal{L}] \cap (B - \text{var})[\mathcal{L}]$. In particular, for $\mu \in \mathbf{A}(\mathcal{L})$ we have a net in $\mathbf{A}(\mathcal{L})$ in the form

$$(\mathbf{D}(E,\mathcal{L}), \prec, \mathfrak{D}_{\mathcal{L}}(\mu,\kappa,\cdot)); \tag{4.8.15}$$

see (4.8.4). In the sequel we assume that

$$\mathfrak{M}_*(\mathcal{L}) \triangleq \{\tau_*(\mathcal{L}), \tau_{\mathbb{B}}^*(\mathcal{L})\}.$$

Moreover, let

$$\mathfrak{M}(\mathcal{L}) \triangleq \mathfrak{M}_*(\mathcal{L}) \cup \{\tau_{\otimes}(\mathcal{L}), \tau_0(\mathcal{L})\}. \tag{4.8.16}$$

In (4.8.16) we have a four-element set whose each element is a topology of $\mathbf{A}(\mathcal{L})$.

PROPOSITION 4.8.1 *Let $\mu \in \mathbf{A}(\mathcal{L})$ and $b \triangleq V_\mu$. Then (4.8.15) is a net in $U_b(\mathcal{L})$.*

PROOF. Recall that values of the mapping (4.8.14) are the measures (4.8.12). Let $\mathcal{H} \in \mathbf{D}(E,\mathcal{L})$. From (4.8.1), (4.8.7) and (4.8.12) we have

$$V_{\mathfrak{D}_{\mathcal{L}}(\mu,\kappa,\mathcal{H})} \le \sum_{H\in\mathcal{H}\setminus\{\varnothing\}} \mid \mu(H) \mid V_{(\delta_{\kappa(H)}|\mathcal{L})} = \sum_{H\in\mathcal{H}\setminus\{\varnothing\}} \mid \mu(H) \mid \le V_\mu = b. \tag{4.8.17}$$

In (4.8.17) we use the basic property of the mapping (4.2.15) and other simple properties from Section 4.2. From (4.8.12) for $\mathfrak{D}_{\mathcal{L}}(\mu, \kappa, \mathcal{H}) \in \mathbf{A}(\mathcal{L})$ we obtain that (4.8.17) is fulfilled. From the definition of Section 4.6 we have the required statement. □

THEOREM 4.8.1 *If* $\mu \in \mathbf{A}(\mathcal{L})$, *then* $\forall \tau \in \mathfrak{M}(\mathcal{L})$:

$$(\mathbf{D}(E, \mathcal{L}), \prec, \mathfrak{D}_{\mathcal{L}}(\mu, \kappa, \cdot)) \xrightarrow{\tau} \mu. \tag{4.8.18}$$

PROOF. Fix $\mu \in \mathbf{A}(\mathcal{L})$. At first consider (4.8.18) in the case $\tau = \tau_0(\mathcal{L})$. We use (4.6.18) and (2.6.36). Fix $L \in \mathcal{L}$. If $L = \varnothing$, then $\mu(L) = 0$ and the value of (4.8.12) is also zero. Consider the case $L \neq \varnothing$. Using (4.2.3) we can choose $\mathcal{H} \in \mathbf{D}(E, \mathcal{L})$ for which $L \in \mathcal{H}$. In addition, $L \in \mathcal{H} \setminus \{\varnothing\}$. Let $\tilde{\mathcal{H}} \in \mathbf{D}(E, \mathcal{L})$ have the property $\mathcal{H} \prec \tilde{\mathcal{H}}$. By (4.8.5) we obtain that $\exists \tilde{\mathcal{H}}_1 \in \mathbf{D}(L, \mathcal{L})$: $\tilde{\mathcal{H}}_1 \subset \tilde{\mathcal{H}}$. We fix such partition $\tilde{\mathcal{H}}_1$. So $\tilde{\mathcal{H}}_1 \in \mathbf{D}(L, \mathcal{L})$. In particular, $L$ is the union of all sets $H \in \tilde{\mathcal{H}}_1$. And what is more, $L$ is the union of all sets of the nonempty family $\tilde{\mathcal{H}}_1 \setminus \{\varnothing\}$. In particular, we have $\tilde{\mathcal{H}}_1 \setminus \{\varnothing\} \in \mathrm{Fin}(\tilde{\mathcal{H}} \setminus \{\varnothing\})$. Let $\nu \triangleq \mathfrak{D}_{\mathcal{L}}(\mu, \kappa, \tilde{\mathcal{H}})$. Then

$$\nu(L) = \sum_{H \in \tilde{\mathcal{H}} \setminus \{\varnothing\}} \mu(H) \delta_{\kappa(H)}(L). \tag{4.8.19}$$

Let $\tilde{H} \in \tilde{\mathcal{H}} \setminus \tilde{\mathcal{H}}_1$. Then $\forall H \in \tilde{\mathcal{H}}_1$: $\tilde{H} \neq H$. By (4.8.1) we have $\forall H \in \tilde{\mathcal{H}}_1$: $\tilde{H} \cap H = \varnothing$. But $L$ is the union of all sets $H \in \tilde{\mathcal{H}}_1$. As a consequence we get

$$\tilde{H} \cap L = \tilde{H} \cap (\bigcup_{H \in \tilde{\mathcal{H}}_1} H) = \bigcup_{H \in \tilde{\mathcal{H}}_1} (\tilde{H} \cap H) = \varnothing.$$

Since the choice of $\tilde{H}$ was arbitrary, we have

$$\forall H \in \tilde{\mathcal{H}} \setminus \tilde{\mathcal{H}}_1 : H \cap L = \varnothing. \tag{4.8.20}$$

In addition, $\tilde{\mathcal{H}} \setminus \{\varnothing\} \subset (\tilde{\mathcal{H}}_1 \setminus \{\varnothing\}) \cup (\tilde{\mathcal{H}} \setminus \tilde{\mathcal{H}}_1)$. Consider the following cases:

1) $\tilde{\mathcal{H}} \setminus \{\varnothing\} = \tilde{\mathcal{H}}_1 \setminus \{\varnothing\}$;
2) $\tilde{\mathcal{H}} \setminus \{\varnothing\} \neq \tilde{\mathcal{H}}_1 \setminus \{\varnothing\}$.

In case 1) by (4.8.19) we have the obvious equality

$$\nu(L) = \sum_{H \in \tilde{\mathcal{H}}_1 \setminus \{\varnothing\}} \mu(H) \delta_{\kappa(H)}(L). \tag{4.8.21}$$

In case 2) we have $\tilde{\mathcal{K}} \triangleq (\tilde{\mathcal{H}} \setminus \{\varnothing\}) \setminus (\tilde{\mathcal{H}}_1 \setminus \{\varnothing\}) \in \mathrm{Fin}(\tilde{\mathcal{H}} \setminus \{\varnothing\})$, and by (4.8.19)

$$\nu(L) = \sum_{H \in \tilde{\mathcal{H}}_1 \setminus \{\varnothing\}} \mu(H) \delta_{\kappa(H)}(L) + \sum_{H \in \tilde{\mathcal{K}}} \mu(H) \delta_{\kappa(H)}(L). \tag{4.8.22}$$

In addition, $\forall H \in \tilde{\mathcal{K}} : H \notin \tilde{\mathcal{H}}_1$. Then $\forall H_1 \in \tilde{\mathcal{K}}\ \forall H_2 \in \tilde{\mathcal{H}}_1$: $H_1 \neq H_2$. From (4.8.1) we obtain that $\forall H_1 \in \tilde{\mathcal{K}}\ \forall H_2 \in \tilde{\mathcal{H}}_1$: $H_1 \cap H_2 = \varnothing$. Consequently $\forall H \in \tilde{\mathcal{K}} : H \cap L = \varnothing$. In addition, from (4.8.8) we get the property that $\tilde{\mathcal{K}} \subset \mathcal{L} \setminus \{\varnothing\}$ and hence $\kappa(H) \in E \setminus L$ for $H \in \tilde{\mathcal{K}}$. By (4.8.9) and (4.8.21) we have the following transformation of (4.8.22): in case 2) $\nu(L)$ is also defined by (4.8.21). So (4.8.21) is fulfilled always. For $H \in \tilde{\mathcal{H}}_1 \setminus \{\varnothing\}$ we have $\kappa(H) \in L$; in this connection see (4.8.8). As a consequence of (4.8.9) and (4.8.21) we have the equality

$$\nu(L) = \sum_{H \in \tilde{\mathcal{H}}_1 \setminus \{\varnothing\}} \mu(H) = \sum_{H \in \tilde{\mathcal{H}}_1} \mu(H). \tag{4.8.23}$$

In (4.8.23) we use the obvious corollary of (4.2.8): $\mu(\varnothing) = 0$. By the choice of $\tilde{\mathcal{H}}_1$ and (4.2.8) we have in (4.8.23) the equality $\mu(L) = \nu(L)$. Since the choice of $\tilde{\mathcal{H}}$ was arbitrary, we have established that $\forall \mathcal{H}' \in \mathbf{D}(E, \mathcal{L})$:

$$(\mathcal{H} \prec \mathcal{H}') \Rightarrow (\mathfrak{D}_{\mathcal{L}}(\mu, \kappa, \mathcal{H}')(L) = \mu(L)). \tag{4.8.24}$$

In other words, for $L \neq \varnothing$ the statement (4.8.24) holds for some $\mathcal{H} \in \mathbf{D}(E, \mathcal{L})$. But this statement is obvious for $L = \varnothing$. So $\forall \hat{L} \in \mathcal{L}\ \exists \hat{\mathcal{H}} \in \mathbf{D}(E, \mathcal{L})\ \forall \mathcal{H}' \in \mathbf{D}(E, \mathcal{L})$:

$$(\hat{\mathcal{H}} \prec \mathcal{H}') \Rightarrow (\mathfrak{D}_{\mathcal{L}}(\mu, \kappa, \mathcal{H}')(\hat{L}) = \mu(\hat{L})). \tag{4.8.25}$$

On the basis of (4.8.25) it is possible to establish the convergence

$$(\mathbf{D}(E, \mathcal{L}), \prec, \mathfrak{D}_{\mathcal{L}}(\mu, \kappa, \cdot)) \overset{\otimes^{\mathcal{L}}(\tau_\partial)}{\longrightarrow} \mu. \tag{4.8.26}$$

We use (2.6.38). Choose $\mathbb{H}^* \in N_{\otimes^{\mathcal{L}}(\tau_\partial)}(\mu)$. By (2.6.8) and (2.6.38) for some $\mathcal{K}^* \in \mathrm{Fin}(\mathcal{L})$ we have

$$\mathbb{N}_{\mathcal{L}}^{(\partial)}(\mu, \mathcal{K}^*) \subset \mathbb{H}^*.$$

Further we use (4.8.25) to prove (4.8.26). Indeed, let $n \in \mathcal{N}$ and $(L_i^*)_{i \in \overline{1,n}} \in \mathcal{L}^n$ satisfy the equality $\mathcal{K}^* = \{L_i^* : i \in \overline{1,n}\}$. By (4.8.25) we can choose

$$(\mathcal{H}_i^*)_{i \in \overline{1,n}} : \overline{1,n} \to \mathbf{D}(E, \mathcal{L})$$

for which $\forall k \in \overline{1,n}\ \forall \mathcal{H} \in \mathbf{D}(E, \mathcal{L})$:

$$(\mathcal{H}_k^* \prec \mathcal{H}) \Rightarrow (\mathfrak{D}_{\mathcal{L}}(\mu, \kappa, \mathcal{H})(L_k^*) = \mu(L_k^*)). \tag{4.8.27}$$

Using (2.2.4) and the obvious induction, we infer that $\forall i \in \overline{1,n}$: $\mathcal{H}_i^* \prec \mathcal{H}_*$ for some $\mathcal{H}_* \in \mathbf{D}(E, \mathcal{L})$. Of course, we use (4.8.4) too. Then from (4.8.27) we have $\forall \mathcal{H} \in \mathbf{D}(E, \mathcal{L})$:

$$(\mathcal{H}_* \prec \mathcal{H}) \Rightarrow (\forall L \in \mathcal{K}^* : \mathfrak{D}_{\mathcal{L}}(\mu, \kappa, \mathcal{H})(L) = \mu(L)).$$

This relation means that (see Section 2.6) for $\mathcal{H} \in \mathbf{D}(E, \mathcal{L})$, $\mathcal{H}_* \prec \mathcal{H}$, the inclusion $\mathfrak{D}_{\mathcal{L}}(\mu, \kappa, \mathcal{H}) \in \mathbb{N}_{\mathcal{L}}^{(\partial)}(\mu, \mathcal{K}^*)$ holds. As a consequence $\forall \mathcal{H} \in \mathbf{D}(E, \mathcal{L})$:

$$(\mathcal{H}_* \prec \mathcal{H}) \Rightarrow (\mathfrak{D}_{\mathcal{L}}(\mu, \kappa, \mathcal{H}) \in \mathbb{H}^*).$$

Since the choice of $\mathbb{H}^*$ was arbitrary, we have the convergence (4.8.26). Using (2.3.9) and (4.6.18) we obtain the convergence

$$(\mathbf{D}(E, \mathcal{L}), \prec, \mathfrak{D}_{\mathcal{L}}(\mu, \kappa, \cdot)) \overset{\tau_0(\mathcal{L})}{\longrightarrow} \mu. \tag{4.8.28}$$

From (4.6.19) and (4.8.28) we have the convergence

$$(\mathbf{D}(E, \mathcal{L}), \prec, \mathfrak{D}_{\mathcal{L}}(\mu, \kappa, \cdot)) \overset{\tau_\otimes(\mathcal{L})}{\longrightarrow} \mu. \tag{4.8.29}$$

We now use Proposition 4.8.1. Let $b \triangleq V_\mu$; $b \in [0, \infty[$. Then by (2.3.9) and (4.8.9) we have

$$(\mathbf{D}(E, \mathcal{L}), \prec, \mathfrak{D}_{\mathcal{L}}(\mu, \kappa, \cdot)) \overset{\tau_\otimes(\mathcal{L})|_{U_b(\mathcal{L})}}{\longrightarrow} \mu.$$

Using (4.6.34) we obtain the following convergence

$$(\mathbf{D}(E, \mathcal{L}), \prec, \mathfrak{D}_{\mathcal{L}}(\mu, \kappa, \cdot)) \overset{\tau_*(\mathcal{L})|_{U_b(\mathcal{L})}}{\longrightarrow} \mu. \tag{4.8.30}$$

From (2.3.9) and (4.8.30) it follows that

$$(\mathbf{D}(E, \mathcal{L}), \prec, \mathfrak{D}_{\mathcal{L}}(\mu, \kappa, \cdot)) \overset{\tau_*(\mathcal{L})}{\longrightarrow} \mu. \tag{4.8.31}$$

From (4.6.11) and (4.7.5) we derive the equality

$$\tau_*(\mathcal{L}) \mid_{U_b(\mathcal{L})} = \tau_{\mathbb{B}}^*(\mathcal{L}) \mid_{U_b(\mathcal{L})} .$$

Hence by (4.8.30) we have the convergence of $(\mathbf{D}(E, \mathcal{L}), \prec, \mathfrak{D}_{\mathcal{L}}(\mu, \kappa, \cdot))$ to $\mu$ in the topology $\tau_{\mathbb{B}}^*(\mathcal{L}) \mid_{U_b(\mathcal{L})}$ . From (2.3.9) we get

$$(\mathbf{D}(E, \mathcal{L}), \prec, \mathfrak{D}_{\mathcal{L}}(\mu, \kappa, \cdot)) \overset{\tau_{\mathbb{B}}^*(\mathcal{L})}{\longrightarrow} \mu. \tag{4.8.32}$$

From (4.8.16), (4.8.28), (4.8.29), (4.8.31), and (4.8.32) we obtain the required statement.

Note that by (4.8.12) we have the property that if $\mu \in (\text{add})_+[\mathcal{L}]$, then (4.8.15) is a net in $(\sigma-\text{add})_+[\mathcal{L}]$. Indeed, $\forall \mu \in (\text{add})_+[\mathcal{L}]\ \forall \mathcal{H} \in \mathbf{D}(E, \mathcal{L})$:

$$\mathfrak{D}_{\mathcal{L}}(\mu, \kappa, \mathcal{H}) \in (\sigma - \text{add})_+[\mathcal{L}]. \tag{4.8.33}$$

A variant on the basis of (4.8.33) can be used, in particular, in the case $\mu \in \mathbb{P}(\mathcal{L})$ (for probability settings). The general case of the net (4.8.15) can be realized in settings of the purely impulse control. Let us discuss the last possibility very briefly. Introduce $\forall b \in [0, \infty[$:

$$U_b^0(\mathcal{L}) = \Big\{\mu \in \mathbf{A}(\mathcal{L}) \,|\, \exists n \in \mathcal{N}\ \exists (\alpha_i)_{i \in \overline{1,n}} \in \mathbb{R}^n\ \exists (x_i)_{i \in \overline{1,n}} \in E^n : \\ \Big(\mu = \sum_{i=1}^{n} \alpha_i (\delta_{x_i} \mid \mathcal{L})\Big) \,\&\, \Big(\sum_{i=1}^{n} \mid \alpha_i \mid \leq b\Big)\Big\}. \tag{4.8.34}$$

Obviously, $U_b^0(\mathcal{L}) \subset U_b(\mathcal{L})$ for $b \in [0, \infty[$. In this property we use arguments similar to the proof of Proposition 4.8.1 (in particular, see (4.8.17)). We exploit known properties of semi-norms (see [119, Ch. I]). From (4.8.11) and (4.8.34) we have $\forall b \in [0, \infty[$:

$$U_b^0(\mathcal{L}) \subset (\sigma - \text{add})[\mathcal{L}] \cap (B - \text{var})[\mathcal{L}]. \tag{4.8.35}$$

In addition, from (4.8.12), (4.8.34) and arguments similar to Proposition 4.8.1, for $\mu \in \mathbf{A}(\mathcal{L})$, $b = V_\mu$ and $\mathcal{H} \in \mathbf{D}(E, \mathcal{L})$ we obtain the property

$$\mathfrak{D}_{\mathcal{L}}(\mu, \kappa, \mathcal{H}) \in U_b^0(\mathcal{L}). \tag{4.8.36}$$

In (4.8.35) and (4.8.36) we are able to approximate $\mu \in \mathbf{A}(\mathcal{L})$ in the class of simplest CAM on $\mathcal{L}$ in a natural way. The following statements can be connected with (4.8.33) and (4.8.36). To formulate the first statement we introduce some new notations. The sense of these notations is analogous to (4.8.34).

Recall that, for $n \in \mathcal{N}$, $\mathbb{R}_+^n$ is the set of all $(x_i)_{i \in \overline{1,n}} \in \mathbb{R}^n$ for which $\forall j \in \overline{1,n} : 0 \leq x_j$. Clearly, $\mathbb{R}_+^n$ is the non-negative cone of $\mathbb{R}^n$. Suppose that $\forall n \in \mathcal{N}$:

$$(\text{simpl})[n] \triangleq \{(\alpha_i)_{i \in \overline{1,n}} \in \mathbb{R}_+^n \mid \sum_{i=1}^{n} \alpha_i = 1\}.$$

We note that by (4.2.20) and (4.8.11), $\forall n \in \mathcal{N}$:

$$\mathbb{P}_\sigma^0(\mathcal{L}) \triangleq \{\mu \in \mathbf{A}(\mathcal{L}) \mid \exists n \in \mathcal{N}\ \exists (\alpha_i)_{i \in \overline{1,n}} \in (\text{simpl})[n]\ \exists (x_i)_{i \in \overline{1,n}} \in E^n : \\ \mu = \sum_{i=1}^{n} \alpha_i (\delta_{x_i} \mid \mathcal{L})\} \in 2^{\mathbb{P}_\sigma(\mathcal{L})}. \tag{4.8.37}$$

In (4.8.37) the set of all finitely distributed probabilities is introduced. Elements of (4.8.37) are exactly the simplest Kolmogorov probabilities.

Their role is analogous to that of (4.8.34). Obviously, from (4.8.12) we have $\forall \mu \in \mathbb{P}(\mathcal{L})\ \forall \mathcal{H} \in \mathbf{D}(E, \mathcal{L})$:

$$\mathfrak{D}_{\mathcal{L}}(\mu, \kappa, \mathcal{H}) \in \mathbb{P}^0_\sigma(\mathcal{L}). \tag{4.8.38}$$

In (4.8.38) we use (4.2.8), (4.2.19), (4.8.1) and (4.8.33). In addition, we exploit (for $H = E$) the obvious connection of (4.2.2) and (4.8.1). Namely, if $n \in \mathcal{N}$ and $(L_i)_{i\in\overline{1,n}} \in \Delta_n(E, \mathcal{L})$, then $\{L_i : i \in \overline{1,n}\} \in \mathbf{D}(E, \mathcal{L})$. Analogously, let $\tilde{\mathcal{H}} \in \mathbf{D}(E, \mathcal{L})$. Then, in particular, $\tilde{\mathcal{H}} \in \mathrm{Fin}(\mathcal{L})$ by (4.8.1), and the sets of $\tilde{\mathcal{H}}$ can be bijectively enumerated. Namely, we can choose $k \in \mathcal{N}$ and a bijection [110, p.25] $(H_i)_{i\in\overline{1,k}} \in \tilde{\mathcal{H}}^k$. By (4.8.1) the union of all sets $H_i$, $i \in \overline{1,k}$, coincides with $E$. If $p \in \overline{1,k}$ and $q \in \overline{1,k} \setminus \{p\}$, then $H_p \neq H_q$, since the corresponding enumeration of $\tilde{\mathcal{H}}$ is bijective. Therefore by (4.8.1) we have $H_p \cap H_q = \varnothing$. By (4.2.2), $(H_i)_{i\in\overline{1,k}} \in \Delta_k(E, \mathcal{L})$. Assume that $\mathcal{H} \in \mathbf{D}(E, \mathcal{L})$ corresponds to (4.8.12) and $\tilde{\mathcal{H}} \triangleq \mathcal{H} \setminus \{\varnothing\}$. Then we can use (4.8.7) and the above bijection $(H_i)_{i\in\overline{1,k}} \in \tilde{\mathcal{H}}^k$. By the property of $(H_i)_{i\in\overline{1,k}}$ and (4.2.8) we get

$$\mu(E) = \sum_{i=1}^{k} \mu(H_i) = \sum_{H\in\mathcal{H}\setminus\{\varnothing\}} \mu(H).$$

If $\mu \in \mathbb{P}(\mathcal{L})$, then by (4.2.19) we have $(\mu(H_i))_{i\in\overline{1,k}} \in (\mathrm{simpl})[k]$ and (by (4.8.12) and (4.8.37))

$$\mathfrak{D}_{\mathcal{L}}(\mu, \kappa, \mathcal{H}) = \sum_{i=1}^{k} \mu(H_i)(\delta_{\kappa(H_i)} \mid \mathcal{L}) \in \mathbb{P}^0_\sigma(\mathcal{L}).$$

So (4.8.37) is established. From Theorem 4.8.1 we obtain the following

THEOREM 4.8.2 $\forall \tau \in \mathfrak{M}(\mathcal{L}) : \mathbb{P}(\mathcal{L}) = \mathrm{cl}(\mathbb{P}^0_\sigma(\mathcal{L}), \tau)$.

PROOF. By (4.2.20) and (4.8.37) we have the chain of inclusions $\mathbb{P}^0_\sigma(\mathcal{L}) \subset \mathbb{P}_\sigma(\mathcal{L}) \subset \mathbb{P}(\mathcal{L})$. From (4.6.10) and (4.6.14) we obtain $\mathbb{P}(\mathcal{L}) \in \mathcal{F}_{\tau_*(\mathcal{L})}$. Hence $\mathrm{cl}(\mathbb{P}^0_\sigma(\mathcal{L}), \tau_*(\mathcal{L})) \subset \mathbb{P}(\mathcal{L})$. From (4.7.2) we have $\mathbb{P}(\mathcal{L}) \in \mathcal{F}_\tau$, where $\tau = \tau^*_{\mathbb{B}}(\mathcal{L})$. As a consequence $\mathrm{cl}(\mathbb{P}^0_\sigma(\mathcal{L}), \tau^*_{\mathbb{B}}(\mathcal{L})) \subset \mathbb{P}(\mathcal{L})$. Thus $\forall \tau \in \mathfrak{M}_*(\mathcal{L})$: $\mathrm{cl}(\mathbb{P}^0_\sigma(\mathcal{L}), \tau) \subset \mathbb{P}(\mathcal{L})$. We note that by (2.6.27), (2.6.31) and (4.2.8) we have $\forall_{D_0} \mathbf{S}[D_0 \neq \varnothing]\ \forall \preceq \in (\mathrm{DIR})[D_0]\ \forall h \in (\mathrm{add})[\mathcal{L}]^{D_0}\ \forall \mu \in \mathbb{R}^{\mathcal{L}}$:

$$((D_0, \preceq, h) \overset{\otimes^{\mathcal{L}}(\tau_{\mathbb{R}})}{\longrightarrow} \mu) \Rightarrow (\mu \in (\mathrm{add})[\mathcal{L}]).$$

Therefore $(\mathrm{add})[\mathcal{L}] \in \mathcal{F}_\tau$, where $\tau = \otimes^{\mathcal{L}}(\tau_{\mathbb{R}})$. Hence from (2.6.27), (2.6.31) and (4.2.10) we have $\forall_{D_0} \mathbf{S}[D_0 \neq \varnothing]\ \forall \preceq \in (\mathrm{DIR})[D_0]\ \forall h \in$

$(\mathrm{add})_+[\mathcal{L}]^{D_0}\ \forall \mu \in \mathbb{R}^{\mathcal{L}}$:

$$((D_0, \preceq, h) \overset{\otimes^{\mathcal{L}}(\tau_{\mathbb{R}})}{\longrightarrow} \mu) \Rightarrow (\mu \in (\mathrm{add})_+[\mathcal{L}]).$$

We see that the set $(\mathrm{add})_+[\mathcal{L}]$ is closed in the Tichonoff product of samples $(\mathbb{R}, \tau_{\mathbb{R}})$ with the index set $\mathcal{L}$. Finally, if $(D_0, \preceq, h)$ is a net in $\mathbb{P}(\mathcal{L})$, $\mu \in (\mathrm{add})_+[\mathcal{L}]$, and $(D_0, \preceq, h)$ converges to $\mu$ in the sense of $\otimes^{\mathcal{L}}(\tau_{\mathbb{R}})$, then by (2.6.27) and (2.6.31) we have $\mu(E) = 1$ because $h(d)(E) = 1$ for $d \in D_0$. So $\mathbb{P}(\mathcal{L}) \in \mathcal{F}_\tau$ for $\tau = \otimes^{\mathcal{L}}(\tau_{\mathbb{R}})$, and $\mathbb{P}(\mathcal{L}) \subset \mathbf{A}(\mathcal{L})$. By (2.3.1), (2.3.12) and (4.6.18) we have $\mathbb{P}(\mathcal{L}) \in \mathcal{F}_\tau \big|_{\mathbf{A}(\mathcal{L})}$ and hence $\mathbb{P}(\mathcal{L}) \in \mathcal{F}_{\tau_\otimes(\mathcal{L})}$. From (4.6.19) we obtain the property $\mathbb{P}(\mathcal{L}) \in \mathcal{F}_{\tau_0(\mathcal{L})}$. Hence we have

$$(\mathrm{cl}(\mathbb{P}^0_\sigma(\mathcal{L}), \tau_\otimes(\mathcal{L})) \subset \mathbb{P}(\mathcal{L}))\ \&\ (\mathrm{cl}(\mathbb{P}^0_\sigma(\mathcal{L}), \tau_0(\mathcal{L})) \subset \mathbb{P}(\mathcal{L})).$$

By (4.8.16) we obtain $\forall \tau \in \mathfrak{M}(\mathcal{L})$: $\mathrm{cl}(\mathbb{P}^0_\sigma(\mathcal{L}), \tau) \subset \mathbb{P}(\mathcal{L})$. From (2.3.11), Theorem 4.8.1 and (4.8.38) we have $\forall \tau \in \mathfrak{M}(\mathcal{L})$: $\mathbb{P}(\mathcal{L}) \subset \mathrm{cl}(\mathbb{P}^0_\sigma(\mathcal{L}), \tau)$. □

So we have established the natural statement on approximation. As a corollary we have $\forall \tau \in \mathfrak{M}(\mathcal{L})$:

$$\mathbb{P}(\mathcal{L}) = \mathrm{cl}(\mathbb{P}_\sigma(\mathcal{L}), \tau). \tag{4.8.39}$$

In (4.8.39) we use (4.2.20), (4.8.37) and Theorem 4.8.2.

LEMMA 4.8.1 $\forall b \in [0, \infty[\ \forall \tau \in \mathfrak{M}(\mathcal{L}) : U_b(\mathcal{L}) \in \mathcal{F}_\tau$.

PROOF. Fix $b \in [0, \infty[$. By (4.6.10) and (4.6.11) we have

$$U_b(\mathcal{L}) \in \mathcal{F}_{\tau_*(\mathcal{L})}. \tag{4.8.40}$$

From (4.7.2) and (4.8.40) we get

$$U_b(\mathcal{L}) \in \mathcal{F}_{\tau^*_{\mathbf{B}}(\mathcal{L})}. \tag{4.8.41}$$

From (4.8.40) and (4.8.41) we have $\forall \tau \in \mathfrak{M}_*(\mathcal{L}) : U_b(\mathcal{L}) \in \mathcal{F}_\tau$. Consider the case $\tau = \tau_\otimes(\mathcal{L})$. Let

$$\mu \in \mathbf{A}(\mathcal{L}) \setminus U_b(\mathcal{L}).$$

By the corresponding definition of Section 4.6 we have $b < V_\mu$. So,

$$\varepsilon_b \triangleq \frac{V_\mu - b}{2} \in ]0, \infty[.$$

Using (4.2.12) we fix a number $c_b \in (\mathrm{VAR})_E[\mu]$ such that $V_\mu - \varepsilon_b < c_b$. Now we choose (see definitions of Section 4.2) $n \in \mathcal{N}$ and $(L_i)_{i \in \overline{1,n}} \in$

$\Delta_n(E,\mathcal{L})$ for which $c_b = \sum_{i=1}^{n} \mid \mu(L_i) \mid$. Then for $\varepsilon_b^0 \triangleq \varepsilon_b/n \in ]0,\infty[$ and $\mathcal{K} \triangleq \{L_i : i \in \overline{1,n}\} \in \mathrm{Fin}(\mathcal{L})$ we have the representation

$$\begin{aligned}\mathrm{N}_{\mathcal{L}}^{\otimes}(\mu,\mathcal{K},\varepsilon_b^0) &= \{\nu \in \mathbf{A}(\mathcal{L}) \mid \forall L \in \mathcal{K} : \mid \mu(L) - \nu(L) \mid < \varepsilon_b^0\} \\ &= \{\nu \in \mathbf{A}(\mathcal{L}) \mid \forall i \in \overline{1,n} : \mid \mu(L_i) - \nu(L_i) \mid < \varepsilon_b^0\}.\end{aligned} \tag{4.8.42}$$

Obviously, by (4.6.23) and (4.8.42) we have $\mathrm{N}_{\mathcal{L}}^{\otimes}(\mu,\mathcal{K},\varepsilon_b^0) \in \mathfrak{N}_{\mathcal{L}}^{\otimes}(\mu)$. In particular, $\mathrm{N}_{\mathcal{L}}^{\otimes}(\mu,\mathcal{K},\varepsilon_b^0) \in \mathfrak{N}^{\otimes}[\mathcal{L}]$. Let $\xi \in \mathrm{N}_{\mathcal{L}}^{\otimes}(\mu,\mathcal{K},\varepsilon_b^0)$. Then $\xi \in \mathbf{A}(\mathcal{L})$, and the inequality $\mid \mu(L_i) - \xi(L_i) \mid < \varepsilon_b^0$ holds for $i \in \overline{1,n}$. Hence

$$\sum_{i=1}^{n} \mid \mu(L_i) \mid - n\varepsilon_b^0 < \sum_{i=1}^{n} \mid \xi(L_i) \mid \leq V_\xi.$$

But $n\varepsilon_b^0 = \varepsilon_b$ and hence we obtain $V_\mu - 2\varepsilon_b < c_b - \varepsilon_b = \sum_{i=1}^{n} \mid \mu(L_i) \mid - \varepsilon_b \leq V_\xi$. In other words, we have the inequality $b < V_\xi$, and therefore $\xi \in \mathbf{A}(\mathcal{L}) \setminus U_b(\mathcal{L})$. Since the choice of $\xi$ was arbitrary, we obtain the inclusion

$$\mathrm{N}_{\mathcal{L}}^{\otimes}(\mu,\mathcal{K},\varepsilon_b^0) \subset \mathbf{A}(\mathcal{L}) \setminus U_b(\mathcal{L}).$$

In addition, $\mu \in \mathrm{N}_{\mathcal{L}}^{\otimes}(\mu,\mathcal{K},\varepsilon_b^0)$. This property follows from (4.6.22). We have established that

$$\forall \mu \in \mathbf{A}(\mathcal{L}) \setminus U_b(\mathcal{L})\ \exists B \in \mathfrak{N}^{\otimes}[\mathcal{L}] : (\mu \in B)\&(B \subset \mathbf{A}(\mathcal{L}) \setminus U_b(\mathcal{L})).$$

By (2.6.3) we have the property $\mathbf{A}(\mathcal{L}) \setminus U_b(\mathcal{L}) \in \{\cup\}(\mathfrak{N}^{\otimes}[\mathcal{L}])$. From (4.6.29) we conclude that $\mathbf{A}(\mathcal{L}) \setminus U_b(\mathcal{L}) \in \tau_{\otimes}(\mathcal{L})$. As a consequence we have

$$U_b(\mathcal{L}) \in \mathcal{F}_{\tau_{\otimes}(\mathcal{L})}. \tag{4.8.43}$$

From (4.6.19) and (4.8.43) we get $U_b(\mathcal{L}) \in \mathcal{F}_{\tau_0(\mathcal{L})}$. Thus from (4.8.16), (4.8.40), (4.8.41) and (4.8.43) we have the required statement. □

THEOREM 4.8.3 $\forall b \in [0,\infty[\ \forall \tau \in \mathfrak{M}(\mathcal{L})$: $U_b(\mathcal{L}) = \mathrm{cl}(U_b^0(\mathcal{L}),\tau)$.

PROOF. By (4.8.34) and Lemma 4.8.1, $\forall \tau \in \mathfrak{M}(\mathcal{L})$: $\mathrm{cl}(U_b^0(\mathcal{L}),\tau) \subset U_b(\mathcal{L})$. By (2.3.11), Theorem 4.8.1, (4.8.34) and (4.8.36) we have $\forall \tau \in \mathfrak{M}(\mathcal{L})$: $U_b(\mathcal{L}) \subset \mathrm{cl}(U_b^0(\mathcal{L}),\tau)$. □

It is useful to note that Theorem 4.8.3 can be exploited in extension constructions for the purely impulse control. In these problems the constant $b$ in (4.8.34) defines the corresponding resource constraint on the choice of controls.

Note two circumstances connected with the Alaoglu theorem. From (4.6.11), (4.6.14), Theorems 4.8.2 and 4.8.3 it follows that

$$\begin{aligned}&(\mathbb{P}(\mathcal{L}) = \mathrm{cl}(\mathbb{P}_\sigma^0(\mathcal{L}),\tau_*(\mathcal{L})) \in (\tau_*(\mathcal{L}) - \mathrm{comp})[\mathbf{A}(\mathcal{L})]) \\ &\&(\forall c \in [0,\infty[: U_c(\mathcal{L}) = \mathrm{cl}(U_c^0(\mathcal{L}),\tau_*(\mathcal{L})) \in (\tau_*(\mathcal{L}) - \mathrm{comp})[\mathbf{A}(\mathcal{L})]).\end{aligned} \tag{4.8.44}$$

In (4.8.44) we have two variants of compactifications in the sense of Chapter 3. Obviously, (4.8.44) can be supplemented by other similar statements. In particular, (4.6.44) can by used. But now we do not consider these possibilities. Note only that in Theorem 4.8.1 we have the universal construction of the approximate solution based on the use of the choice function $\kappa$. Corresponding analogues of Theorems 4.8.2 and 4.8.3 can be obtained without direct employment of the 'global' choice function (in this connection see [35, pp. 286,287], where some 'local' choice is used).

## 4.9 WEAKLY ABSOLUTELY CONTINUOUS MEASURES AND QUESTIONS OF APPROXIMATION IN THE CLASS OF INDEFINITE INTEGRALS

Constructions of this section are given in [32, 35, 45, 46] and many other papers. Therefore we consider basic constructions very briefly. It is worth noting some analogues of Theorems 4.8.1–4.8.2. Moreover, the 'collection' (4.8.16) is used here. Finally, it should be mentioned that these constructions have the applied meaning in questions of the impulse control (see extension constructions for control problems in [32, 35, 45] and examples of Chapter 1). We discuss the substantive sense of this question. Let us recall the construction (4.4.5). In this section we fix $\eta \in (\mathrm{add})_+[\mathcal{L}]$. We use $\eta$ in (4.4.5) instead of $\mu$. Denote by $\mathcal{J}_\eta$ the operator

$$f \mapsto f * \eta : B(E, \mathcal{L}) \to \mathbf{A}(\mathcal{L}). \tag{4.9.1}$$

So by (4.9.1) we have $\forall f \in B(E, \mathcal{L})$: $\mathcal{J}_\eta(f) = f * \eta$. If $\mathbb{H} \in \mathcal{P}(B(E, \mathcal{L}))$, then by (2.2.6) $\mathcal{J}_\eta^1(\mathbb{H}) = \{f * \eta : f \in \mathbb{H}\} \in \mathcal{P}(\mathbf{A}(\mathcal{L}))$. In some problems the following question is important: how to represent the set $\mathrm{cl}(\mathcal{J}_\eta^1(\mathbb{H}), \tau)$, where $\tau \in \mathfrak{M}(\mathcal{L})$. Especially it is important to know the set $\mathrm{cl}(\mathcal{J}_\eta^1(\mathbb{H}), \tau_*(\mathcal{L}))$. Elements of this set are analogues of elements of $\mathbb{H}$. In particular, the case $\mathbb{H} \in \mathcal{P}(B^+(E, \mathcal{L}))$ can be considered. This case is advisable in connection with Proposition 4.6.1. We begin the study with this case, following [32, 35, 45, 46]. In addition, as generalized analogues of functions $f \in B^+(E, \mathcal{L})$ we consider so called weakly absolutely continuous (with respect to $\eta$) measures. Here we use the terminology of [106] (we recall that in this section $\mathbf{S}[E \neq \varnothing]$ and $\mathcal{L} \in \Pi[E]$). In addition,

$$(E, \mathcal{L}, \eta) \tag{4.9.2}$$

is (generally speaking) a nonstandard measure space. Let

$$(\mathrm{add})^+[\mathcal{L};\eta] \triangleq \{\mu \in (\mathrm{add})_+[\mathcal{L}] \mid \forall L \in \mathcal{L} : (\eta(L) = 0) \Rightarrow (\mu(L) = 0)\}. \tag{4.9.3}$$

Note that (4.9.3) is the cone in $\mathbf{A}(\mathcal{L})$ and

$$\begin{aligned}\mathbf{A}_\eta[\mathcal{L}] &\triangleq \{\mu - \nu : \mu \in (\mathrm{add})^+[\mathcal{L};\eta], \nu \in (\mathrm{add})^+[\mathcal{L};\eta]\} \\ &= \{\mu \in \mathbf{A}(\mathcal{L}) \mid v_\mu \in (\mathrm{add})^+[\mathcal{L};\eta]\} \\ &= \{\mu \in \mathbf{A}(\mathcal{L}) \mid (\mu^+ \in (\mathrm{add})^+[\mathcal{L};\eta]) \;\&\; (\mu^- \in (\mathrm{add})^+[\mathcal{L};\eta])\} \\ &= \{\mu \in \mathbf{A}(\mathcal{L}) \mid \forall L \in \mathcal{L} : (\eta(L) = 0) \Rightarrow (\mu(L) = 0)\}.\end{aligned} \tag{4.9.4}$$

In connection with (4.9.3) and (4.9.4) see [32, Ch. 4], [35, Ch. 3], [45, 46], and many other papers. Of course, $\mathbf{A}_\eta[\mathcal{L}]$ is a linear subspace of $\mathbf{A}(\mathcal{L})$; this subspace is generated by the cone $(\mathrm{add})^+[\mathcal{L};\eta]$ (see (4.9.4)). From (4.9.3) and (4.9.4) we have

$$(\mathrm{add})^+[\mathcal{L};\eta] = \mathbf{A}_\eta[\mathcal{L}] \cap (\mathrm{add})_+[\mathcal{L}]. \tag{4.9.5}$$

We operate with $\mu \in \mathbf{A}_\eta[\mathcal{L}]$ as with GE in problems for which the space (4.9.2) is fixed.

**Remark.** Note that $\mathbf{A}_\eta[\mathcal{L}]$ is the band or component of the complete vector lattice $\mathbf{A}(\mathcal{L})$ with the point-wise order induced from $\mathbb{R}^{\mathcal{L}}$ and defined in terms of $\leqq$. In addition, $(\mathrm{add})^+[\mathcal{L};\eta]$ (4.9.5) is the corresponding generating cone. It is worth noting that the disjunct band

$$\mathbf{A}_\eta^{(\mathbf{d})}[\mathcal{L}] \triangleq \{\mu \in \mathbf{A}(\mathcal{L}) \mid \forall \nu \in \mathbf{A}_\eta[\mathcal{L}] : v_\mu - (v_\mu - v_\nu)^+ = \mathbb{O}_{\mathcal{L}}\}$$

is generated by the cone

$$\begin{aligned}(\mathrm{add})^+_{\mathbf{D}}[\mathcal{L};\eta] \triangleq \{\mu \in (\mathrm{add})_+[\mathcal{L}] \mid &\forall \nu \in (\mathrm{add})^+[\mathcal{L};\eta] : \\ &(\nu \leqq \mu) \Rightarrow (\nu = \mathbb{O}_{\mathcal{L}})\}.\end{aligned}$$

Indeed, the linear space $\mathbf{A}_\eta^{(\mathbf{d})}[\mathcal{L}]$ is such that $\mathbf{A}_\eta^{(\mathbf{d})}[\mathcal{L}] \cap (\mathrm{add})_+[\mathcal{L}] = (\mathrm{add})^+_{\mathbf{D}}[\mathcal{L};\eta]$ and hence

$$\mathbf{A}_\eta^{(\mathbf{d})}[\mathcal{L}] = \{\mu - \nu : \mu \in (\mathrm{add})^+_{\mathbf{D}}[\mathcal{L};\eta], \nu \in (\mathrm{add})^+_{\mathbf{D}}[\mathcal{L};\eta]\}.$$

The pair $(\mathbf{A}_\eta[\mathcal{L}], \mathbf{A}_\eta^{(\mathbf{d})}[\mathcal{L}])$ is generated by the natural decomposition of $\mathbf{A}(\mathcal{L})$ into the ordered direct sum. This decomposition is similar to that of Hewitt Yosida. In this connection see [80, 118]. The Hewitt Yosida decomposition [118] is the most known example. The decomposition on the basis of $(\mathbf{A}_\eta[\mathcal{L}], \mathbf{A}_\eta^{(\mathrm{d})}[\mathcal{L}])$ is another one ( see the example in [32, p.89]).

Consider a highly general approximate scheme [32, 35, 45, 46], setting

$$\mathcal{L}_0[\eta] \triangleq \{L \in \mathcal{L} \mid \eta(L) \neq 0\}. \tag{4.9.6}$$

We interpret (4.9.6) as the $\eta$-nonzero part of $\mathcal{L}$. If $\mu \in \mathbf{A}(\mathcal{L})$, then $\theta_\eta[\mu]$ is by definition the mapping for which

$$\left(\forall L \in \mathcal{L}_0[\eta] : \theta_\eta[\mu](L) = \frac{\mu(L)}{\eta(L)}\right) \& \left(\forall L \in \mathcal{L} \setminus \mathcal{L}_0[\eta] : \theta_\eta[\mu](L) = 0\right). \tag{4.9.7}$$

The definition (4.9.7) corresponds to [35, p.244]. Recall that by (4.8.1) $\forall \mathcal{H} \in \mathbf{D}(E, \mathcal{L})$ $\forall x \in E$ $\exists! H \in \mathcal{H}$: $x \in H$. Hence the following definition is correct: if $\mu \in \mathbf{A}(\mathcal{L})$ and $\mathcal{H} \in \mathbf{D}(E, \mathcal{L})$, then $\Theta_\eta[\mu; \mathcal{H}] \in B_0(E, \mathcal{L})$ is by definition the functional (on E) such that

$$\forall L \in \mathcal{H}\ \forall x \in L : \Theta_\eta[\mu; \mathcal{H}](x) = \theta_\eta[\mu](L). \tag{4.9.8}$$

In other words, we use the representation

$$\Theta_\eta[\mu; \mathcal{H}] = \sum_{L \in \mathcal{H}} \theta_\eta[\mu](L) \chi_L[E]. \tag{4.9.9}$$

The relations (4.9.8) and (4.9.9) are equivalent. We can consider (4.9.9) as the basic definition. Of course, $\forall \mu \in \mathbf{A}_\eta[\mathcal{L}]$ $\forall L \in \mathcal{L}$:

$$\mu(L) = \theta_\eta[\mu](L)\eta(L). \tag{4.9.10}$$

In (4.9.10) we use (4.9.7) and the corollary of (4.9.4) that if $\mu \in \mathbf{A}_\eta[\mathcal{L}]$ and $L \in \mathcal{L} \setminus \mathcal{L}_0[\eta]$, then $\eta(L) = 0$ and $\mu(L) = 0$. From (4.3.8), (4.3.14), (4.8.1), (4.9.9), and (4.9.10) we have

$$\begin{aligned} {}^{(\mathrm{el})}\!\!\int_E \Theta_\eta[\mu; \mathcal{H}] d\eta &= \int_E \Theta_\eta[\mu; \mathcal{H}] d\eta = (\Theta_\eta[\mu; \mathcal{H}] * \eta)(E) \\ &= \sum_{L \in \mathcal{H}} \theta_\eta[\mu](L)\eta(L) = \sum_{L \in \mathcal{H}} \mu(L) = \mu(E). \end{aligned} \tag{4.9.11}$$

Obviously, (4.9.11) is satisfied (in particular) for $\mu \in (\mathrm{add})^+[\mathcal{L}; \eta]$. In addition, $\forall \mu \in (\mathrm{add})^+[\mathcal{L}; \eta]$ $\forall \mathcal{H} \in \mathbf{D}(E, \mathcal{L})$: $\Theta_\eta[\mu; \mathcal{H}] \in B_0^+(E, \mathcal{L})$. Here we use (4.9.7).

PROPOSITION 4.9.1 *If* $\mu \in \mathbf{A}_\eta[\mathcal{L}]$ *and* $\mathcal{H} \in \mathbf{D}(E, \mathcal{L})$, *then*

$$\Theta_\eta[\mu; \mathcal{H}] = \Theta_\eta[\mu^+; \mathcal{H}] - \Theta_\eta[\mu^-; \mathcal{H}]. \tag{4.9.12}$$

PROOF. We use (4.2.35). Fix $\mu \in \mathbf{A}_\eta[\mathcal{L}]$ and $\mathcal{H} \in \mathbf{D}(E, \mathcal{L})$. Then from (4.9.4) it follows that $\mu \in \mathbf{A}(\mathcal{L})$ and

$$(\mu^+ \in (\text{add})^+[\mathcal{L};\eta]) \;\&\; (\mu^- \in (\text{add})^+[\mathcal{L};\eta]).$$

By (4.9.6) and (4.9.7) we have $\forall L \in \mathcal{L}_0[\eta]$:

$$\theta_\eta[\mu](L) = \frac{\mu^+(L) - \mu^-(L)}{\eta(L)} = \theta_\eta[\mu^+](L) - \theta_\eta[\mu^-](L). \tag{4.9.13}$$

From (4.9.7) it follows that $\forall L \in \mathcal{L} \setminus \mathcal{L}_0[\eta]$:

$$\theta_\eta[\mu](L) = 0 = \theta_\eta[\mu^+](L) - \theta_\eta[\mu^-](L).$$

Then by (4.9.13) we obtain that $\theta_\eta[\mu] = \theta_\eta[\mu^+] - \theta_\eta[\mu^-]$. As a consequence, from (4.9.9) we have

$$\begin{aligned}\Theta_\eta[\mu;\mathcal{H}] &= \sum_{L\in\mathcal{H}} \theta_\eta[\mu^+](L)\chi_L[E] - \sum_{L\in\mathcal{H}} \theta_\eta[\mu^-](L)\chi_L[E] \\ &= \Theta_\eta[\mu^+;\mathcal{H}] - \Theta_\eta[\mu^-;\mathcal{H}].\end{aligned}$$

The relation (4.9.12) is established. □

So by Proposition 4.9.1 for $\mu \in \mathbf{A}_\eta[\mathcal{L}]$ and $\mathcal{H} \in \mathbf{D}(E,\mathcal{L})$ we have $\Theta_\eta[\mu^+;\mathcal{H}] \in B_0^+(E,\mathcal{L})$, $\Theta_\eta[\mu^-;\mathcal{H}] \in B_0^+(E,\mathcal{L})$, and the relation (4.9.12) holds. Recall the simple property (see [35, p.56]) that if $f \in B(E,\mathcal{L})$ and $L \in \mathcal{L}$, then

$$v_{f*\eta}(L) = \int_L |\, f \,|\, d\eta = (|\, f \,| *\eta)(L).$$

In other words, $\forall f \in B(E,\mathcal{L})$:

$$v_{f*\eta} = |\, f \,| *\eta. \tag{4.9.14}$$

In particular, from (4.2.4) and (4.9.14) for $f \in B(E,\mathcal{L})$ we have

$$V_{f*\eta} = \int_E |\, f \,|\, d\eta. \tag{4.9.15}$$

PROPOSITION 4.9.2 *Let* $\mu \in \mathbf{A}_\eta[\mathcal{L}]$, $\mathcal{H} \in \mathbf{D}(E,\mathcal{L})$, *and* $b \triangleq V_\mu$. *Then* $\Theta_\eta[\mu;\mathcal{H}] * \eta \in U_b(\mathcal{L})$.

PROOF. We use (4.9.15) and Proposition 4.9.1. In addition,

$$\begin{aligned}\int_E |\, \Theta_\eta[\mu;\mathcal{H}] \,|\, d\eta &\le \int_E |\, \Theta_\eta[\mu^+;\mathcal{H}] \,|\, d\eta + \int_E |\, \Theta_\eta[\mu^-;\mathcal{H}] \,|\, d\eta \\ &= \mu^+(E) + \mu^-(E) = v_\mu(E) = V_\mu = b.\end{aligned}$$

Of course, we also use (4.2.35) and (4.9.11). By (4.9.15) and definitions of Section 4.6 we obtain that $\Theta_\eta[\mu; \mathcal{H}] * \eta \in U_b(\mathcal{L})$. $\square$

In the sequel we use the stipulation that for $\mu \in \mathbf{A}(\mathcal{L})$:

$$\Theta_\eta[\mu; \cdot] * \eta \triangleq (\Theta_\eta[\mu; \mathcal{H}] * \eta)_{\mathcal{H} \in \mathbf{D}(E, \mathcal{L})}. \tag{4.9.16}$$

It is worth noting that (4.9.16) is an element of $\mathbf{A}(\mathcal{L})^{\mathbf{D}(E,\mathcal{L})}$. The following assertion is obvious.

PROPOSITION 4.9.3 *If $f \in B(E, \mathcal{L})$, then $f * \eta \in \mathbf{A}_\eta[\mathcal{L}]$.*

The proof is very simple. For $L \in \mathcal{L} \setminus \mathcal{L}_0[\eta]$ we have $\eta(L) = 0$ and

$$\begin{aligned} | (f * \eta)(L) | &= \left| \int_L f \, d\eta \right| = \left| \int_E f \chi_L[E] \, d\eta \right| \leq \int_E | f | \chi_L[E] \, d\eta \\ &\leq \|f\| \int_E \chi_L[E] \, d\eta = \|f\| \eta(L) = 0. \end{aligned}$$

We use (4.5.9) and the property that $\eta$ is non-negative. So $\forall L \in \mathcal{L} \setminus \mathcal{L}_0[\eta]$: $(f * \eta)(L) = 0$. From (4.4.5), (4.9.4) and (4.9.6) we have $f * \eta \in \mathbf{A}_\eta[\mathcal{L}]$. $\square$

From (4.9.16) and Proposition 4.9.3 it follows that if $\mu \in \mathbf{A}_\eta[\mathcal{L}]$, then

$$(\mathbf{D}(E, \mathcal{L}), \prec, \Theta_\eta[\mu; \cdot] * \eta) \tag{4.9.17}$$

is a net in $\mathbf{A}_\eta[\mathcal{L}]$; simultaneously (4.9.17) is a net in $U_b(\mathcal{L})$ for $b \triangleq V_\mu$. Recall that by (4.9.16) for $\mu \in \mathbf{A}_\eta[\mathcal{L}]$ the mapping $\Theta_\eta[\mu; \cdot] * \eta$ is the function

$$\mathcal{H} \mapsto \Theta_\eta[\mu; \mathcal{H}] * \eta : \mathbf{D}(E, \mathcal{L}) \to \mathbf{A}_\eta[\mathcal{L}].$$

THEOREM 4.9.1 ( SEE [35, CH. 3]) *Let $\mu \in \mathbf{A}_\eta[\mathcal{L}]$. Then $\forall \tau \in \mathfrak{M}(\mathcal{L})$:*

$$(\mathbf{D}(E, \mathcal{L}), \prec, \Theta_\eta[\mu; \cdot] * \eta) \xrightarrow{\tau} \mu. \tag{4.9.18}$$

PROOF. Assume that $b \triangleq V_\mu$. Then (4.9.17) is a net in $\mathbf{A}_\eta[\mathcal{L}] \cap U_b(\mathcal{L})$. At first we establish (4.9.18) in the case $\tau = \tau_0(\mathcal{L})$. Fix $\mathbb{L} \in \mathcal{L}$. Using (4.8.1) and the property that $\mathcal{L} \in \Pi[E]$, we choose $\mathcal{K} \in \mathbf{D}(E, \mathcal{L})$ for which $\mathbb{L} \in \mathcal{K}$. Let $\mathcal{M} \in \mathbf{D}(E, \mathcal{L})$ be such that $\mathcal{K} \prec \mathcal{M}$. If $\mathbb{L} = \varnothing$, then $\mu(\mathbb{L}) = \mu(\varnothing) = 0 = (\Theta_\eta[\mu; \mathcal{M}] * \eta)(\varnothing) = (\Theta_\eta[\mu; \mathcal{M}] * \eta)(\mathbb{L})$. Consider the case $\mathbb{L} \neq \varnothing$. So, let $\mathbb{L} \in \mathcal{K} \setminus \{\varnothing\}$. Then by (4.8.5) we have $\tilde{\mathcal{M}} \subset \mathcal{M}$ for some $\tilde{\mathcal{M}} \in \mathbf{D}(\mathbb{L}, \mathcal{L})$. From (4.2.8) it follows that $\forall \nu \in (\text{add})[\mathcal{L}]$:

$$\nu(\mathbb{L}) = \sum_{L \in \tilde{\mathcal{M}}} \nu(L).$$

In particular, $\mu(\mathbb{L})$ is the sum of all numbers $\mu(L)$, $L \in \tilde{\mathcal{M}}$. Analogously

$$(\Theta_\eta[\mu;\mathcal{M}] * \eta)(\mathbb{L}) = \sum_{L\in\tilde{\mathcal{M}}} (\Theta_\eta[\mu;\mathcal{M}] * \eta)(L). \tag{4.9.19}$$

We use (4.9.9). Then for $\tilde{L} \in \tilde{\mathcal{M}}$ we have (see (4.4.7) and (4.4.8))

$$\begin{aligned}(\Theta_\eta[\mu;\mathcal{M}] * \eta)(\tilde{L}) &= \int_{\tilde{L}} \Theta_\eta[\mu;\mathcal{M}]d\eta = \int_E \Theta_\eta[\mu;\mathcal{M}]\chi_{\tilde{L}}d\eta \\ &= \sum_{L\in\mathcal{M}} \theta_\eta[\mu](L) \int_E \chi_L[E]\chi_{\tilde{L}}[E]d\eta \\ &= \sum_{L\in\mathcal{M}} \theta_\eta[\mu](L) \int_E \chi_{L\cap\tilde{L}}[E]d\eta \\ &= \sum_{L\in\mathcal{M}} \theta_\eta[\mu](L)\eta(L\cap\tilde{L}).\end{aligned} \tag{4.9.20}$$

If $L \in \mathcal{M}\setminus\tilde{\mathcal{M}}$, then $\forall L' \in \tilde{\mathcal{M}} : L \cap L' = \varnothing$. In particular, $\forall L \in \mathcal{M}\setminus\tilde{\mathcal{M}}$: $L \cap \tilde{L} = \varnothing$. In this connection see (4.8.1). From (4.2.8) and (4.9.20) we have

$$(\Theta_\eta[\mu;\mathcal{M}] * \eta)(\tilde{L}) = \sum_{L\in\tilde{\mathcal{M}}} \theta_\eta[\mu](L)\eta(L\cap\tilde{L}) = \theta_\eta[\mu](\tilde{L})\eta(\tilde{L}).$$

Indeed, by (4.8.1) $\forall L \in \tilde{\mathcal{M}}\setminus\{\tilde{L}\}$: $L\cap\tilde{L} = \varnothing$. But the choice of $\tilde{L}$ was arbitrary. We obtain that (see (4.9.19))

$$(\Theta_\eta[\mu;\mathcal{M}] * \eta)(\mathbb{L}) = \sum_{L\in\tilde{\mathcal{M}}} \theta_\eta[\mu](L)\eta(L).$$

This means (see (4.9.10)) the validity of the following representation:

$$(\Theta_\eta[\mu;\mathcal{M}] * \eta)(\mathbb{L}) = \sum_{L\in\tilde{\mathcal{M}}} \mu(L) = \mu(\mathbb{L}).$$

Thus we always have the equality

$$(\Theta_\eta[\mu;\mathcal{M}] * \eta)(\mathbb{L}) = \mu(\mathbb{L}).$$

Recall that the choice of $\mathcal{M}$ with the property $\mathcal{K} \prec \mathcal{M}$ was arbitrary. Therefore we have

$$\begin{aligned}&\forall L \in \mathcal{L}\ \exists \mathcal{K}_L \in \mathbf{D}(E,\mathcal{L})\ \forall \mathcal{H} \in \mathbf{D}(E,\mathcal{L}) : (\mathcal{K}_L \prec \mathcal{H}) \\ &\qquad \Rightarrow (\Theta_\eta[\mu;\mathcal{H}] * \eta)(L) = \mu(L)).\end{aligned} \tag{4.9.21}$$

By (4.9.21) and the basic property of directed sets we have (see (2.2.4) and (4.8.4)) $\forall \mathbb{K} \in \mathrm{Fin}(\mathcal{L})\ \exists \mathfrak{K} \in \mathbf{D}(E, \mathcal{L})\ \forall \mathcal{H} \in \mathbf{D}(E, \mathcal{L})$:

$$(\mathfrak{K} \prec \mathcal{H}) \Rightarrow (\forall L \in \mathbb{K} : (\Theta_\eta[\mu; \mathcal{H}] * \eta)(L) = \mu(L)).$$

In other words, (see Section 2.6) $\forall \mathbb{K} \in \mathrm{Fin}(\mathcal{L})\ \exists \mathfrak{K} \in \mathbf{D}(E, \mathcal{L})\ \forall \mathcal{H} \in \mathbf{D}(E, \mathcal{L})$:

$$(\mathfrak{K} \prec \mathcal{H}) \Rightarrow (\Theta_\eta[\mu; \mathcal{H}] * \eta \in \mathbb{N}_{\mathcal{L}}^{(\partial)}(\mu, \mathbb{K})).$$

By definitions of Section 2.6 we have

$$\forall T \in \mathfrak{N}_{\mathcal{L}}^{(\partial)}(\mu)\ \exists \mathfrak{K} \in \mathbf{D}(E, \mathcal{L})\ \forall \mathcal{H} \in \mathbf{D}(E, \mathcal{L}) :$$

$$(\mathfrak{K} \prec \mathcal{H}) \Rightarrow (\Theta_\eta[\mu; \mathcal{H}] * \eta \in T).$$

From (2.6.8) and (2.6.38) we obtain the convergence

$$(\mathbf{D}(E, \mathcal{L}), \prec, \Theta_\eta[\mu; \cdot] * \eta) \overset{\otimes^{\mathcal{L}}(\tau_\partial)}{\longrightarrow} \mu.$$

By (2.3.9) and (4.6.18) we have

$$(\mathbf{D}(E, \mathcal{L}), \prec, \Theta_\eta[\mu; \cdot] * \eta) \overset{\tau_0(\mathcal{L})}{\longrightarrow} \mu. \tag{4.9.22}$$

From (4.6.19) and (4.9.22) we get the convergence

$$(\mathbf{D}(E, \mathcal{L}), \prec, \Theta_\eta[\mu; \cdot] * \eta) \overset{\tau_\otimes(\mathcal{L})}{\longrightarrow} \mu. \tag{4.9.23}$$

Introduce $\tau_\otimes^{(b)}[\mathcal{L}] \triangleq \tau_\otimes(\mathcal{L}) \mid_{U_b(\mathcal{L})}$. From (4.6.34) it follows that

$$\tau_\otimes^{(b)}[\mathcal{L}] = \tau_*(\mathcal{L}) \mid_{U_b(\mathcal{L})} . \tag{4.9.24}$$

From (2.3.9) and (4.9.23) we obtain

$$(\mathbf{D}(E, \mathcal{L}), \prec, \Theta_\eta[\mu; \cdot] * \eta) \overset{\tau_\otimes^{(b)}[\mathcal{L}]}{\longrightarrow} \mu. \tag{4.9.25}$$

From (2.3.9), (4.9.24) and (4.9.25) we have the convergence

$$(\mathbf{D}(E, \mathcal{L}), \prec, \Theta_\eta[\mu; \cdot] * \eta) \overset{\tau_*(\mathcal{L})}{\longrightarrow} \mu. \tag{4.9.26}$$

Recall (4.7.5). From (4.6.11), (4.7.5) and (4.9.24) it follows that

$$\tau_*(\mathcal{L}) \mid_{U_b(\mathcal{L})} = \tau_{\mathbb{B}}^*(\mathcal{L}) \mid_{U_b(\mathcal{L})} = \tau_\otimes^{(b)}[\mathcal{L}]$$

and by properties of (4.9.17) we have the convergence

$$(\mathbf{D}(E, \mathcal{L}), \prec, \Theta_\eta[\mu; \cdot] * \eta) \overset{\tau_{\mathbb{B}}^*(\mathcal{L})\mid_{U_b(\mathcal{L})}}{\longrightarrow} \mu.$$

From (2.3.9) we get the required convergence

$$(\mathbf{D}(E,\mathcal{L}), \prec, \Theta_\eta[\mu;\cdot] * \eta) \stackrel{\tau_{\mathbb{B}}^*(\mathcal{L})}{\longrightarrow} \mu. \tag{4.9.27}$$

From (4.8.16), (4.9.22), (4.9.23), (4.9.26) and (4.9.27) we have the statement of the theorem. □

It is useful to compare Theorems 4.8.1 and 4.9.1. We have a multi-topology property. Using Theorem 4.9.1 it is possible to establish many statements about the density of 'usual' controls in the corresponding subspace of $\mathbf{A}_\eta[\mathcal{L}]$. In this connection we note the following statement.

PROPOSITION 4.9.4 $\forall \tau \in \mathfrak{M}(\mathcal{L}) : \mathbf{A}_\eta[\mathcal{L}] \in \mathcal{F}_\tau$.

PROOF. Let $\mu \in \mathrm{cl}(\mathbf{A}_\eta[\mathcal{L}], \tau_\otimes(\mathcal{L}))$. Then $\mu \in \mathbf{A}(\mathcal{L})$ and by (2.3.11) we can choose $\mathbf{S}[\mathbf{D} \neq \varnothing]$, $\preceq \in (\mathrm{DIR})[\mathbf{D}]$ and $f \in \mathbf{A}_\eta[\mathcal{L}]^{\mathbf{D}}$ for which

$$(\mathbf{D}, \preceq, f) \stackrel{\tau_\otimes(\mathcal{L})}{\longrightarrow} \mu.$$

From (2.3.9) and (4.6.18) we have the convergence

$$(\mathbf{D}, \preceq, f) \stackrel{\otimes^{\mathcal{L}}(\tau_{\mathbb{R}})}{\longrightarrow} \mu. \tag{4.9.28}$$

From (2.6.27), (2.6.31) and (4.9.28) we infer that $\forall L \in \mathcal{L}$:

$$(\mathbf{D}, \preceq, (f(d)(L))_{d \in \mathbf{D}}) \stackrel{\tau_{\mathbb{R}}}{\longrightarrow} \mu(L). \tag{4.9.29}$$

Note that by (4.9.4) and (4.9.6) we have $\forall d \in \mathbf{D}\ \forall L \in \mathcal{L} \setminus \mathcal{L}_0[\eta]$: $f(d)(L) = 0$. From (4.9.29) we obtain $\forall L \in \mathcal{L} \setminus \mathcal{L}_0[\eta]$: $\mu(L) = 0$. We have $\mu \in \mathbf{A}_\eta[\mathcal{L}]$ (see (4.9.4) and (4.9.6)). Thus $\mathrm{cl}(\mathbf{A}_\eta[\mathcal{L}], \tau_\otimes(\mathcal{L})) \subset \mathbf{A}_\eta[\mathcal{L}]$. Therefore

$$\mathrm{cl}(\mathbf{A}_\eta[\mathcal{L}], \tau_\otimes(\mathcal{L})) = \mathbf{A}_\eta[\mathcal{L}]. \tag{4.9.30}$$

From (4.6.19) we have the inclusion $\mathrm{cl}(\mathbf{A}_\eta[\mathcal{L}], \tau_0(\mathcal{L})) \subset \mathbf{A}_\eta[\mathcal{L}]$. Hence $\mathbf{A}_\eta[\mathcal{L}] = \mathrm{cl}(\mathbf{A}_\eta[\mathcal{L}], \tau_0(\mathcal{L}))$. From (4.6.21) and (4.9.30) we have the inclusion $\mathrm{cl}(\mathbf{A}_\eta[\mathcal{L}], \tau_*(\mathcal{L})) \subset \mathbf{A}_\eta[\mathcal{L}]$. As a consequence we obtain the equality

$$\mathbf{A}_\eta[\mathcal{L}] = \mathrm{cl}(\mathbf{A}_\eta[\mathcal{L}], \tau_*(\mathcal{L})). \tag{4.9.31}$$

From (4.7.2) and (4.9.31) it follows that $\mathrm{cl}(\mathbf{A}_\eta[\mathcal{L}], \tau_{\mathbb{B}}^*(\mathcal{L})) \subset \mathbf{A}_\eta[\mathcal{L}]$. Hence $\mathbf{A}_\eta[\mathcal{L}] = \mathrm{cl}(\mathbf{A}_\eta[\mathcal{L}], \tau_{\mathbb{B}}^*(\mathcal{L}))$. By (4.8.16) we get the required statement. □

THEOREM 4.9.2 ([35, CH.3]) $\forall \tau \in \mathfrak{M}(\mathcal{L}) : \mathbf{A}_\eta[\mathcal{L}] = \mathrm{cl}(\{f * \eta : f \in B_0(E,\mathcal{L})\}, \tau) = \mathrm{cl}(\{f * \eta : f \in B(E,\mathcal{L})\}, \tau)$.

The proof is realized by combination of (2.3.11), Proposition 4.9.3, Theorem 4.9.1 and Proposition 4.9.4.

From Lemma 4.8.1 and Proposition 4.9.4 we have $\forall \tau \in \mathfrak{M}(\mathcal{L})$:

$$U_b(\mathcal{L}) \cap \mathbf{A}_\eta[\mathcal{L}] \in \mathcal{F}_\tau. \tag{4.9.32}$$

THEOREM 4.9.3 ([35, CH.3]) $\forall b \in [0, \infty[\ \forall \tau \in \mathfrak{M}(\mathcal{L})$:

$$\begin{aligned} U_b(\mathcal{L}) \cap \mathbf{A}_\eta[\mathcal{L}] &= \mathrm{cl}(\{f * \eta : f \in B_0(E, \mathcal{L}), \int_E |f| \, d\eta \le b\}, \tau) \\ &= \mathrm{cl}(\{f * \eta : f \in B(E, \mathcal{L}), \int_E |f| \, d\eta \le b\}, \tau). \end{aligned}$$

PROOF. Fix $b \in [0, \infty[$. From (4.9.15) and Proposition 4.9.3 we have $\forall f \in B(E, \mathcal{L})$:

$$\left(\int_E |f| \, d\eta \le b\right) \Rightarrow (f * \eta \in U_b(\mathcal{L}) \cap \mathbf{A}_\eta[\mathcal{L}]). \tag{4.9.33}$$

In this connection see definitions of Section 4.6. From (4.9.32) and (4.9.33) we obtain $\forall \tau \in \mathfrak{M}(\mathcal{L})$:

$$\begin{aligned} &\mathrm{cl}(\{f * \eta : f \in B_0(E, \mathcal{L}), \int_E |f| \, d\eta \le b\}, \tau) \\ &\subset \mathrm{cl}(\{f * \eta : f \in B(E, \mathcal{L}), \int_E |f| \, d\eta \le b\}, \tau) \\ &\subset U_b(\mathcal{L}) \cap \mathbf{A}_\eta[\mathcal{L}]. \end{aligned} \tag{4.9.34}$$

From (4.9.15), Propositions 4.9.2 and 4.9.3 it follows that

$$\Theta_\eta[\mu; \mathcal{H}] \in B_0(E, \mathcal{L}) : \int_E |\Theta_\eta[\mu; \mathcal{H}]| \, d\eta \le b$$

for $\mu \in U_b(\mathcal{L}) \cap \mathbf{A}_\eta[\mathcal{L}]$ and $\mathcal{H} \in \mathbf{D}(E, \mathcal{L})$. Therefore from Theorem 4.9.1 we have $\forall \tau \in \mathfrak{M}(\mathcal{L})$:

$$U_b(\mathcal{L}) \cap \mathbf{A}_\eta[\mathcal{L}] \subset \mathrm{cl}(\{f * \eta : f \in B_0(E, \mathcal{L}), \int_E |f| \, d\eta \le b\}, \tau). \tag{4.9.35}$$

From (4.9.34) and (4.9.35) we get the required statement. □

It should be noted that a number of other density properties similar to Theorems 4.9.2 and 4.9.3 are known; see, for example, [35, pp.56–60]. We restrict ourselves to consideration of constructions in the non-negative cone of the space of FAM. We note that

$$\forall \tau \in \mathfrak{M}(\mathcal{L}) : (\mathrm{add})_+[\mathcal{L}] \in \mathcal{F}_\tau. \tag{4.9.36}$$

Indeed, from (4.6.13) and (4.7.2) we get that $\forall \tau \in \mathfrak{M}_*(\mathcal{L}) : (\mathrm{add})_+[\mathcal{L}] \in \mathcal{F}_\tau$. From (2.6.31) and (4.2.10) we have $\forall_T \mathbf{S}[T \neq \varnothing]\ \forall \preceq \in (\mathrm{DIR})[T]$ $\forall f \in (\mathrm{add})_+[\mathcal{L}]^T\ \forall \mu \in \mathbf{A}(\mathcal{L})$:

$$((T, \preceq, f) \overset{\otimes^{\mathcal{L}}(\tau_{\mathbb{R}})}{\longrightarrow} \mu) \Rightarrow (\mu \in (\mathrm{add})_+[\mathcal{L}]).$$

By (2.3.9) and (4.6.18) we obtain $\forall_T \mathbf{S}[T \neq \varnothing] \forall \preceq \in (\mathrm{DIR})[T]\ \forall f \in (\mathrm{add})_+[\mathcal{L}]^T\ \forall \mu \in \mathbf{A}(\mathcal{L})$:

$$((T, \preceq, f) \overset{\tau_\otimes(\mathcal{L})}{\longrightarrow} \mu) \Rightarrow (\mu \in (\mathrm{add})_+[\mathcal{L}]).$$

This property means that $(\mathrm{add})_+[\mathcal{L}] \in \mathcal{F}_{\tau_\otimes(\mathcal{L})}$. From (4.6.19) we have $(\mathrm{add})_+[\mathcal{L}] \in \mathcal{F}_{\tau_0(\mathcal{L})}$. Thus we get (4.9.36). From (4.9.5), (4.9.36) and Proposition 4.9.4 we have $\forall \tau \in \mathfrak{M}(\mathcal{L})$:

$$(\mathrm{add})^+[\mathcal{L}; \eta] \in \mathcal{F}_\tau. \tag{4.9.37}$$

THEOREM 4.9.4 ([35, CH.3]) *$\forall \tau \in \mathfrak{M}(\mathcal{L})$:*

$$\begin{aligned}(\mathrm{add})^+[\mathcal{L}; \eta] &= \mathrm{cl}(\{f * \eta : f \in B^+(E, \mathcal{L})\}, \tau)\\ &= \mathrm{cl}(\{f * \eta : f \in B_0^+(E, \mathcal{L})\}, \tau).\end{aligned}$$

PROOF. From Proposition 4.9.3, (4.5.8) and (4.9.5) it follows that $\forall f \in B^+(E, \mathcal{L}) : f * \eta \in (\mathrm{add})^+[\mathcal{L}; \eta]$. Therefore

$$\{f * \eta : f \in B^+(E, \mathcal{L})\} \subset (\mathrm{add})^+[\mathcal{L}; \eta].$$

Hence from (4.9.37) we obtain $\forall \tau \in \mathfrak{M}(\mathcal{L})$:

$$\mathrm{cl}(\{f * \eta : f \in B^+(E, \mathcal{L})\}, \tau) \subset (\mathrm{add})^+[\mathcal{L}; \eta]. \tag{4.9.38}$$

Let $\mu \in (\mathrm{add})^+[\mathcal{L}; \eta]$. Using (4.9.5), consider the net (4.9.17). In addition, $\Theta_\eta[\mu, \cdot] \in B_0^+(E, \mathcal{L})^{\mathbf{D}(E,\mathcal{L})}$. Consequently (4.9.17) is a net in the set $\{f * \eta : f \in B_0^+(E, \mathcal{L})\}$. By Theorem 4.9.1 we have (see (2.3.11)) $\forall \tau \in \mathfrak{M}(\mathcal{L})$: $\mu \in \mathrm{cl}(\{f * \eta : f \in B_0^+(E, \mathcal{L})\}, \tau)$. Since the choice of $\mu$ was arbitrary, we conclude that $\forall \tau \in \mathfrak{M}(\mathcal{L})$:

$$(\mathrm{add})^+[\mathcal{L}; \eta] \subset \mathrm{cl}(\{f * \eta : f \in B_0^+(E, \mathcal{L})\}, \tau). \tag{4.9.39}$$

From (4.9.38) and (4.9.39) we have the required statement. □

We note a simple corollary of the property $\{f * \eta : f \in B^+(E, \mathcal{L})\} \in \mathcal{P}((\mathrm{add})_+[\mathcal{L}])$. Namely, from Theorem 4.9.4 and Proposition 4.6.1 we get that $\tau_*^+(\mathcal{L}) \in (\mathbf{c} - \mathrm{top})^0_{\mathrm{loc}}[(\mathrm{add})_+[\mathcal{L}]]$ has the property

$$\begin{aligned}(\mathrm{add})^+[\mathcal{L}; \eta] &= \mathrm{cl}(\{f * \eta : f \in B^+(E, \mathcal{L})\}, \tau_*^+(\mathcal{L}))\\ &= \mathrm{cl}(\{f * \eta : f \in B_0^+(E, \mathcal{L})\}, \tau_*^+(\mathcal{L})).\end{aligned} \tag{4.9.40}$$

To prove (4.9.40) it is sufficient to compare (2.3.13) and (4.6.35). We get the very interesting possibility of the extension of informative problems in the class of non-negative functions with employing appropriate locally compact TS. In connection with this possibility we recall constructions of Section 3.2. The last scheme was considered in [29, 30, 32, 35, 45] and other publications.

Return to (4.6.13) and consider the sets $U_b^+(\mathcal{L}) \cap \mathbf{A}_\eta[\mathcal{L}] = U_b^+(\mathcal{L}) \cap (\text{add})^+[\mathcal{L};\eta]$, $b \in [0,\infty[$. By (4.6.43), Proposition 4.8.1 and (4.9.36) we have $\forall b \in [0,\infty[\ \forall \tau \in \mathfrak{M}(\mathcal{L})$:

$$U_b^+(\mathcal{L}) \in \mathcal{F}_\tau. \tag{4.9.41}$$

THEOREM 4.9.5 ([35, CH.3]) *$\forall b \in [0,\infty[\ \forall \tau \in \mathfrak{M}(\mathcal{L})$:*

$$\begin{aligned} U_b(\mathcal{L}) \cap (\text{add})^+[\mathcal{L};\eta] &= U_b^+(\mathcal{L}) \cap \mathbf{A}_\eta[\mathcal{L}] = U_b^+(\mathcal{L}) \cap (\text{add})^+[\mathcal{L};\eta] \\ &= \text{cl}(\{f * \eta : f \in B^+(E,\mathcal{L}), \int_E f\, d\eta \le b\}, \tau) \\ &= \text{cl}(\{f * \eta : f \in B_0^+(E,\mathcal{L}), \int_E f\, d\eta \le b\}, \tau). \end{aligned}$$

PROOF. Fix $b \in [0,\infty[$. By (4.9.37) and (4.9.41) we have $\forall \tau \in \mathfrak{M}(\mathcal{L})$:

$$U_b^+(\mathcal{L}) \cap (\text{add})^+[\mathcal{L};\eta] \in \mathcal{F}_\tau. \tag{4.9.42}$$

On the other hand, for $f \in B^+(E,\mathcal{L})$ with the property $\int_E f\, d\eta \le b$ we get (see (4.5.8) and (4.6.43)) that $f * \eta \in U_b^+(\mathcal{L})$. Using Theorem 4.9.4 we obtain the property $f * \eta \in U_b^+(\mathcal{L}) \cap (\text{add})^+[\mathcal{L};\eta]$ for such function $f$. We have established that $\forall \tau \in \mathfrak{M}(\mathcal{L})$:

$$\text{cl}(\{f * \eta : f \in B^+(E,\mathcal{L}), \int_E f\, d\eta \le b\}, \tau) \subset U_b^+(\mathcal{L}) \cap (\text{add})^+[\mathcal{L};\eta] \tag{4.9.43}$$

(see (4.9.42)). Recall that for $\mu \in U_b^+(\mathcal{L}) \cap (\text{add})^+[\mathcal{L};\eta]$ and $\mathcal{H} \in \mathbf{D}(E,\mathcal{L})$ we have $\Theta_\eta[\mu;\mathcal{H}] \in B_0^+(E,\mathcal{L})$ and

$$\int_E \Theta_\eta[\mu;\mathcal{H}]d\eta = (\Theta_\eta[\mu;\mathcal{H}] * \eta)(E) = \mu(E) \le b$$

(see (4.6.43) and (4.9.11)). Using (4.5.8) and (4.6.43) we obtain (from the last relation) that

$$\Theta_\eta[\mu;\mathcal{H}] * \eta \in \{f * \eta : f \in B_0^+(E,\mathcal{L}), \int_E f\, d\eta \le b\}.$$

From Theorem 4.9.1 it follows that $\forall \tau \in \mathfrak{M}(\mathcal{L})$:

$$U_b^+(\mathcal{L}) \cap (\text{add})^+[\mathcal{L};\eta] \subset \text{cl}(\{f * \eta : f \in B_0^+(E,\mathcal{L}), \int_E f\, d\eta \le b\}, \tau). \tag{4.9.44}$$

From (4.9.43) and (4.9.44) we get the required statement. □

THEOREM 4.9.6 ([35, CH.3]) $\forall b \in [0, \infty[\ \forall \tau \in \mathfrak{M}(\mathcal{L})$:

$$\begin{aligned}&\{\mu \in (\mathrm{add})^+[\mathcal{L};\eta] \mid \mu(E) = b\}\\ &= \mathrm{cl}(\{f * \eta : f \in B^+(E,\mathcal{L}), \int_E f\, d\eta = b\}, \tau)\\ &= \mathrm{cl}(\{f * \eta : f \in B_0^+(E,\mathcal{L}), \int_E f\, d\eta = b\}, \tau).\end{aligned}$$

PROOF. Fix $b \in [0, \infty[$. First we note that

$$\begin{aligned}Q &\triangleq \{\mu \in (\mathrm{add})^+[\mathcal{L};\eta] \mid \mu(E) = b\}\\ &= (\mathrm{add})^+[\mathcal{L};\eta] \cap \{\mu \in \mathbf{A}(\mathcal{L}) \mid \mu(E) = b\} \in \mathcal{F}_{\tau_\otimes(\mathcal{L})}.\end{aligned}$$

Indeed, by (2.6.30) we have $\{\mu \in \mathbb{R}^{\mathcal{L}} \mid \mu(E) \neq b\} = \mathbb{R}^{\mathcal{L}} \setminus \{\mu \in \mathbb{R}^{\mathcal{L}} \mid \mu(E) = b\} \in \otimes^{\mathcal{L}}(\tau_{\mathbb{R}})$. Therefore the set $P \triangleq \{\mu \in \mathbb{R}^{\mathcal{L}} \mid \mu(E) = b\}$ is closed in $(\mathbb{R}^{\mathcal{L}}, \otimes^{\mathcal{L}}(\tau_{\mathbb{R}}))$. As a consequence, from (2.3.12) and (4.6.18) it follows that

$$\{\mu \in \mathbf{A}(\mathcal{L}) \mid \mu(E) = b\} = P \cap \mathbf{A}(\mathcal{L}) \in \mathcal{F}_{\tau_\otimes(\mathcal{L})}. \tag{4.9.45}$$

From (4.9.37) and (4.9.45) we obtain that $Q \in \mathcal{F}_{\tau_\otimes(\mathcal{L})}$. Then by (4.6.19) $Q \in \mathcal{F}_{\tau_0(\mathcal{L})}$. By (4.6.21) we have $Q \in \mathcal{F}_{\tau_*(\mathcal{L})}$. From (4.7.2) we obtain $Q \in \mathcal{F}_{\tau^*_{\mathbf{B}}(\mathcal{L})}$. So

$$\forall \tau \in \mathfrak{M}(\mathcal{L}) : Q \in \mathcal{F}_\tau. \tag{4.9.46}$$

By (4.4.8) and Theorem 4.9.4 we have

$$\begin{aligned}&\{f * \eta : f \in B_0^+(E,\mathcal{L}), \int_E f\, d\eta = b\}\\ &\subset \{f * \eta : f \in B^+(E,\mathcal{L}), \int_E f\, d\eta = b\} \subset Q.\end{aligned} \tag{4.9.47}$$

From (4.9.46) and (4.9.47) we obtain $\forall \tau \in \mathfrak{M}(\mathcal{L})$:

$$\begin{aligned}&\mathrm{cl}(\{f * \eta : f \in B_0^+(E,\mathcal{L}), \int_E f\, d\eta = b\}, \tau)\\ &\subset \mathrm{cl}(\{f * \eta : f \in B^+(E,\mathcal{L}), \int_E f\, d\eta = b\}, \tau) \subset Q.\end{aligned} \tag{4.9.48}$$

Let $\mu \in Q$. Then $\mu \in (\mathrm{add})^+[\mathcal{L};\eta]$. In addition, $\mu(E) = b$. Consider the net (4.9.17). Then for $\mathcal{H} \in \mathbf{D}(E,\mathcal{L})$ we have (see (4.9.11))

$$\Theta_\eta[\mu;\mathcal{H}] \in B_0^+(E,\mathcal{L}) : \int_E \Theta_\eta[\mu;\mathcal{H}] d\eta = b.$$

So (4.9.17) is a net in the set $Q_1 \triangleq \{f * \eta : f \in B_0^+(E, \mathcal{L}), \int_E f \, d\eta = b\}$. By Theorem 4.9.1 we have $\mu \in \mathrm{cl}(Q_1, \tau)$ for $\tau \in \mathfrak{M}(\mathcal{L})$. Since the choice of $\mu$ was arbitrary, we have $\forall \tau \in \mathfrak{M}(\mathcal{L})$: $Q \subset \mathrm{cl}(Q_1, \tau)$. From (4.9.48) we get for any $\tau \in \mathfrak{M}(\mathcal{L})$ the required chain of equalities

$$\begin{aligned} Q &= \mathrm{cl}(\{f * \eta : f \in B_0^+(E, \mathcal{L}), \int_E f \, d\eta = b\}, \tau) \\ &= \mathrm{cl}(\{f * \eta : f \in B^+(E, \mathcal{L}), \int_E f \, d\eta = b\}, \tau). \quad \square \end{aligned}$$

Note the important particular case of Theorem 4.9.6 when $b = 1$. By (4.2.19)

$$\begin{aligned} \mathbb{P}_\eta(\mathcal{L}) &\triangleq \mathbb{P}(\mathcal{L}) \cap (\mathrm{add})^+[\mathcal{L}; \eta] = \mathbb{P}(\mathcal{L}) \cap \mathbf{A}_\eta[\mathcal{L}] \\ &= \{\mu \in (\mathrm{add})^+[\mathcal{L}; \eta] \mid \mu(E) = 1\} \end{aligned} \tag{4.9.49}$$

has the property that $\forall \tau \in \mathfrak{M}(\mathcal{L})$:

$$\begin{aligned} \mathbb{P}_\eta(\mathcal{L}) &\triangleq \mathrm{cl}(\{f * \eta : f \in B^+(E, \mathcal{L}), \int_E f \, d\eta = 1\}, \tau) \\ &= \mathrm{cl}(\{f * \eta : f \in B_0^+(E, \mathcal{L}), \int_E f \, d\eta = 1\}, \tau). \end{aligned} \tag{4.9.50}$$

Let us discuss (4.9.49) and (4.9.50). In (4.9.49) we have the set of all FAP weakly continuous with respect to the space (4.9.2). Then by (4.9.50) we get the important approximate property in the class of densities (of FAP) with respect to (4.9.2).

## 4.10 TWO-VALUED MEASURES AND ULTRAFILTERS OF MEASURABLE SPACES

In this section we return to some constructions of Section 2.4. We deal with the pair $(\mathbb{T}(\mathcal{L}), \mathbb{F}_0^*(\mathcal{L}))$. Such pair is investigated in [35, Ch. 7] in detail. Moreover, the theory of Boolean algebras considers this pair in the case of the MS with an algebra of sets; see, for example, [111, 116]. We treat $\mathbb{T}(\mathcal{L})$ as a natural set of generalized elements. In addition, the set $E$ plays the role of the space of 'usual' solutions. Recall that $\mathcal{L} \in \Pi[E]$ is fixed. Embedding the set $E$ into $\mathbb{T}(\mathcal{L})$, we strive to realize the density property with respect to each topology of $\mathfrak{M}(\mathcal{L})$. We recall the very useful Proposition 2.6.6, in which ultrafilters of MS with algebra of sets are considered. This and several other circumstances make the given case basic. Statements of Section 2.4 concerning the extension of ultrufilters of MS permit us to realize the desired expansion of useful properties to other MS. We use (4.3.1). Then by Proposition 2.2.6, (4.2.25) and (4.3.1) we have the property that if $\mathcal{A} \in (\mathrm{alg})[E]$, then

$$\forall \mathcal{H} \in \mathbb{F}_0^*(\mathcal{A}) : \chi_{\mathcal{H}}[\mathcal{A}] \in \mathbb{T}(\mathcal{A}). \tag{4.10.1}$$

The auxiliary case (4.10.1) is very useful. In reality, the following property holds: if $\mathcal{A} \in (\mathrm{alg})[E]$, then

$$\mathbb{T}(\mathcal{A}) = \{\chi_{\mathcal{H}}[\mathcal{A}] : \mathcal{H} \in \mathbb{F}_0^*(\mathcal{A})\}. \tag{4.10.2}$$

The property (4.10.2) is well known. But we very briefly discuss the scheme of the proof. Denote by $\mathbb{T}_0^*$ the relation on the right hand side of (4.10.2). Let $\mu \in \mathbb{T}(\mathcal{A})$ and $\mathcal{H} \triangleq \{A \in \mathcal{A} \mid \mu(A) = 1\}$. Then by (4.2.8) and (4.2.25) we have $\varnothing \notin \mathcal{H}$ and $E \in \mathcal{H}$ (see (4.2.19) and (4.2.21)). So $\mathcal{H} \in 2^{\mathcal{A}}$. Let $A_1 \in \mathcal{H}$ and $A_2 \in \mathcal{H}$. Consider $A_1 \cap A_2 \in \mathcal{A}$. We have $\mu(E \setminus A_1) = \mu(E \setminus A_2) = 0$ and $E \setminus (A_1 \cap A_2) = (E \setminus A_1) \cup (E \setminus A_2)$. Using the natural monotonocity of $\mu$, we conclude that $\mu(A_1 \cap A_2) = 1$, i.e., $A_1 \cap A_2 \in \mathcal{H}$. Of course, $\forall H \in \mathcal{H}\ \forall A \in \mathcal{A}$: $(H \subset A) \Rightarrow (A \in \mathcal{H})$. Thus $\mathcal{H} \in \mathbb{F}^*(\mathcal{A})$ by (2.2.45). Let $\mathfrak{G} \in \mathbb{F}^*(\mathcal{A})$ and $\mathcal{H} \subset \mathfrak{G}$. Let $\mathfrak{G} \setminus \mathcal{H} \neq \varnothing$. Choose $G \in \mathfrak{G} \setminus \mathcal{H}$. Then $G \in \mathcal{A} \setminus \{\varnothing\}$ and $\mu(G) = 0$. As a consequence $E \setminus G \in \mathcal{H}$ by (4.2.25). Therefore $E \setminus G \in \mathfrak{G}$ and $G \cap (E \setminus G) = \varnothing$. But $\mathfrak{G}$ is an $\mathcal{A}$-filter and $G \cap (E \setminus G) \in \mathfrak{G}$ (see (2.2.45)). This is impossible. The contradiction means that $\mathfrak{G} \setminus \mathcal{H} = \varnothing$. We have $\mathcal{H} \in \mathbb{F}_0^*(\mathcal{A})$. As a result $\mu \in \mathbb{T}_0^*$. From (4.10.1) it follows that $\mathbb{T}_0^* \subset \mathbb{T}(\mathcal{A})$. So (4.10.2) is established.

We recall (2.4.4). Then $\forall \mathcal{M} \in \pi[E]\ \forall \mathcal{E} \in \pi_0[E; \mathcal{M}]\ \forall \mathcal{X} \in \mathbb{F}^*(\mathcal{E})$:

$$\psi[\mathcal{M}; \mathcal{X}] \cap \mathcal{E} = \{C \in \mathcal{E} \mid \exists G \in \mathcal{X} : G \subset C\} = \mathcal{X}. \tag{4.10.3}$$

In connection with constructions of Section 2.4 we note the following property: if $\tilde{\mathcal{L}} \in \pi[E]$, $\mathcal{M} \in \pi_0[E; \tilde{\mathcal{L}}]$ and $\mathcal{Y} \in \mathbb{F}^*(\tilde{\mathcal{L}})$, then $\mathcal{Y} \cap \mathcal{M} \in \mathbb{F}^*(\mathcal{M})$. This property is connected with (2.4.5).

We note that (4.10.2) holds for $\mathcal{A} = a_E^0(\mathcal{L})$:

$$\mathbb{T}(a_E^0(\mathcal{L})) = \{\chi_{\mathcal{H}}[a_E^0(\mathcal{L})] : \mathcal{H} \in \mathbb{F}_0^*(a_E^0(\mathcal{L}))\}. \tag{4.10.4}$$

By (2.4.2) and Proposition 2.4.3 we have $\forall \mathcal{U} \in \mathbb{F}_0^*(\mathcal{L})$:

$$\psi[a_E^0(\mathcal{L}); \mathcal{U}] = \{A \in a_E^0(\mathcal{L}) \mid \exists U \in \mathcal{U} : U \subset A\} \in \mathbb{F}_0^*(a_E^0(\mathcal{L})). \tag{4.10.5}$$

By the corollary of Proposition 2.4.3 we have

$$\forall \mathcal{U} \in \mathbb{F}_0^*(\mathcal{L})\ \exists \mathcal{V} \in \mathbb{F}_0^*(a_E^0(\mathcal{L})) : \mathcal{U} = \mathcal{V} \cap \mathcal{L}. \tag{4.10.6}$$

In this connection see (4.10.3) and (4.10.5).

PROPOSITION 4.10.1 *Let* $\mathcal{U} \in \mathbb{F}_0^*(a_E^0(\mathcal{L}))$. *Then* $\mathcal{U} \cap \mathcal{L} \in \mathbb{F}_0^*(\mathcal{L})$.

PROOF. By the previous properties we have $\mathcal{V} \triangleq \mathcal{U} \cap \mathcal{L} \in \mathbb{F}^*(\mathcal{L})$. Note that by Proposition 2.2.3 the inclusion $\mathcal{V} \subset \mathcal{W}$ holds for some $\mathcal{W} \in \mathbb{F}_0^*(\mathcal{L})$. By (4.10.6) choose $\mathcal{F} \in \mathbb{F}_0^*(a_E^0(\mathcal{L}))$ for which $\mathcal{W} = \mathcal{F} \cap \mathcal{L}$. Using (4.10.4) we get

$$\chi_{\mathcal{U}}[a_E^0(\mathcal{L})] \in \mathbb{T}(a_E^0(\mathcal{L})).$$

On the other hand, $\chi_{\mathcal{V}}[\mathcal{L}] \leqq \chi_{\mathcal{W}}[\mathcal{L}]$. But

$$\chi_{\mathcal{V}}[\mathcal{L}] = (\chi_{\mathcal{U}}[a_E^0(\mathcal{L})] \mid \mathcal{L}) \in \mathbb{T}(\mathcal{L}). \tag{4.10.7}$$

In addition, $\chi_{\mathcal{F}}[a_E^0(\mathcal{L}]) \in \mathbb{T}(a_E^0(\mathcal{L}))$ and as a consequence we have the property

$$\chi_{\mathcal{W}}[\mathcal{L}] = (\chi_{\mathcal{F}}[a_E^0(\mathcal{L}] \mid \mathcal{L}) \in \mathbb{T}(\mathcal{L}). \tag{4.10.8}$$

We recall that $\chi_{\mathcal{V}}[\mathcal{L}] \leqq \chi_{\mathcal{W}}[\mathcal{L}]$. Then by definitions of Section 4.2 we have

$$\alpha[\chi_{\mathcal{V}}[\mathcal{L}]] \leqq \alpha[\chi_{\mathcal{W}}[\mathcal{L}]].$$

From (4.10.7) and (4.10.8) we get

$$\chi_{\mathcal{U}}[a_E^0(\mathcal{L})] \leqq \chi_{\mathcal{F}}[a_E^0(\mathcal{L})].$$

Then $\mathcal{U} \subset \mathcal{F}$. As a result (see (2.2.51)) $\mathcal{U} = \mathcal{F}$ and $\mathcal{U} \cap \mathcal{L} = \mathcal{V} = \mathcal{F} \cap \mathcal{L} = \mathcal{W} \in \mathbb{F}_0^*(\mathcal{L})$. □

PROPOSITION 4.10.2 $\mathbb{T}(\mathcal{L}) = \{\chi_{\mathcal{H}}[\mathcal{L}] : \mathcal{H} \in \mathbb{F}_0^*(\mathcal{L})\}$.

PROOF. We use the obvious property that $\alpha[\mu] \in \mathbb{T}(a_E^0(\mathcal{L}))$ for $\mu \in \mathbb{T}(\mathcal{L})$. Therefore by (4.10.2) we have for $\mu \in \mathbb{T}(\mathcal{L})$ the equality

$$\mu = (\chi_{\mathcal{H}}[a_E^0(\mathcal{L}] \mid \mathcal{L}) = \chi_{\mathcal{H} \cap \mathcal{L}}[\mathcal{L}],$$

where $\mathcal{H} \in \mathbb{F}_0^*(a_E^0(\mathcal{L}))$. In addition, by Proposition 4.10.1 we obtain that the set

$$\Omega \triangleq \{\chi_{\mathcal{H}}[\mathcal{L}] : \mathcal{H} \in \mathbb{F}_0^*(\mathcal{L})\}$$

has the property $\mathbb{T}(\mathcal{L}) \subset \Omega$. Choose arbitrarily $\omega \in \Omega$. Let $\mathcal{F} \in \mathbb{F}_0^*(\mathcal{L})$ have the property $\omega = \chi_{\mathcal{F}}[\mathcal{L}]$. By (4.10.6) we choose $\Lambda \in \mathbb{F}_0^*(a_E^0(\mathcal{L}))$ with the property $\mathcal{F} = \Lambda \cap \mathcal{L}$. Then

$$\omega = \chi_{\Lambda \cap \mathcal{L}}[\mathcal{L}] = (\chi_{\Lambda}[a_E^0(\mathcal{L}] \mid \mathcal{L}) \in \mathbb{T}(\mathcal{L}).$$

In this case we use (4.10.1), (4.10.4) and the obvious property of restrictions of two-valued measures. So $\Omega \subset \mathbb{T}(\mathcal{L})$ and as a consequence $\mathbb{T}(\mathcal{L}) = \Omega$. □

Recall that $\mathbb{F}_0^*(\mathcal{L}) \neq \varnothing$ by Proposition 2.2.2. Moreover, we note an obvious property following from Proposition 2.2.4. Namely, if $\mathcal{A} \in (\text{alg})[E]$, then $\forall \mathcal{H} \in \mathbb{F}_0^*(\mathcal{A})\ \forall m \in \mathcal{N}\ \forall (L_i)_{i\in\overline{1,m}} \in \mathcal{A}^m$:

$$\left(\bigcup_{i=1}^{m} L_i \in \mathcal{H}\right) \Rightarrow (\exists j \in \overline{1,m} : L_j \in \mathcal{H}).$$

This useful property is extended to the case of MS with semi-algebra of sets. We note that the mapping

$$\mathcal{H} \mapsto \chi_{\mathcal{H}}[\mathcal{L}] : \mathbb{F}_0^*(\mathcal{L}) \to \mathbb{T}(\mathcal{L}) \tag{4.10.9}$$

is a bijection. This property of (4.10.9) supplements Proposition 4.10.2. Moreover,

$$\mathcal{U} \mapsto \psi[a_E^0(\mathcal{L}); \mathcal{U}] : \mathbb{F}_0^*(\mathcal{L}) \to \mathbb{F}_0^*(a_E^0(\mathcal{L})) \tag{4.10.10}$$

is a bijection. So we may not distinguish $\mathbb{F}_0^*(\mathcal{L})$, $\mathbb{T}(\mathcal{L})$, $\mathbb{F}_0^*(a_E^0(\mathcal{L}))$ and $\mathbb{T}(a_E^0(\mathcal{L}))$. In this connection we note that

$$\mathcal{H} \mapsto \mathcal{H} \cap \mathcal{L} : \mathbb{F}_0^*(a_E^0(\mathcal{L})) \to \mathbb{F}_0^*(\mathcal{L})$$

is the bijection for which the corresponding inverse bijection coincides with (4.10.10). The above reasoning shows that the known properties of (0, 1)-measures on an algebra of sets are easily extended to the case of such measures on a semi-algebra of sets. The last case is suitable for applications. Note that natural semi-algebras of sets arise directly in many cases, while the construction of the generated algebra of sets requires some efforts. For example, recall the well known property of the product of MS when we get the semi-algebra (of sets) under employing measurable rectangles. In the sequel we discuss the case of $(E, \mathcal{L})$ with $\mathcal{L} \in \Pi[E]$.

It is obvious that

$$\mathbb{T}(\mathcal{L}) \in \mathcal{F}_{\tau_*(\mathcal{L})}. \tag{4.10.11}$$

This property follows from definitions (see (4.2.19), (4.2.21), (4.6.21) and Theorem 4.8.2). As a consequence, by (4.6.10) and (4.10.11) we have the property

$$\mathbb{T}(\mathcal{L}) \in (\tau_*(\mathcal{L}) - \text{comp})[\mathbf{A}(\mathcal{L})] \tag{4.10.12}$$

(here it is useful to use (4.6.14)). Obviously, from (4.8.10) and (4.10.11) we get

$$\text{cl}(\{(\delta_x \mid \mathcal{L}) : x \in E\}, \tau_*(\mathcal{L})) \subset \mathbb{T}(\mathcal{L}). \tag{4.10.13}$$

In reality we have the equality in (4.10.13) (see [35, p.306]). We will give a scheme of proof, paying special attention to constructing a specific net.

For this we consider an analogue of the filter (2.2.47). We realize the construction (2.2.47), using only measurable sets. We have $\forall_{\mathbf{D}}\, \mathbf{S}[\mathbf{D} \neq \varnothing]$ $\forall \preceq \in (\mathrm{DIR})[\mathbf{D}]\ \forall f \in E^{\mathbf{D}}$:

$$(\mathcal{L} - \mathrm{ASS})[\mathbf{D}; \preceq; h] \triangleq \{L \in \mathcal{L} \,|\, \exists d \in \mathbf{D}\ \forall \delta \in \mathbf{D} : (d \preceq \delta) \Rightarrow (f(\delta) \in L)\} \in \mathbb{F}^*(\mathcal{L}). \tag{4.10.14}$$

We use (4.10.14) similarly to (2.2.47). Recall that for each ordered pair $z$ we denote by $\mathrm{pr}_1(z)$ and $\mathrm{pr}_2(z)$ the corresponding components of $z$: if $z = (u, v)$ for some objects $u$ and $v$, then $\mathrm{pr}_1(z) = u$ and $\mathrm{pr}_2(z) = v$. We apply this stipulation to the case of the Cartesian product of $E$ and some family of subsets of $E$. Suppose that $\forall \mathcal{H} \in 2^{\mathcal{P}(E)}$:

$$\tilde{D}(E, \mathcal{H}) \triangleq \{z \in E \times \mathcal{H} \mid \mathrm{pr}_1(z) \in \mathrm{pr}_2(z)\}. \tag{4.10.15}$$

If $\mathcal{H} \in 2^{(2^E)}$, then $\mathbf{S}[\tilde{D}(E, \mathcal{H}) \neq \varnothing]$. Note that $\mathbb{F}^*(\mathcal{L}) \subset 2^{(2^E)}$. In particular, (4.10.15) can be realized in the case $\mathcal{H} \in \mathbb{F}^*(\mathcal{L})$. The scheme inverse (in some sense) to (4.10.14) can be constructed on this basis. Let us consider it. Note that if $\mathcal{H} \in \mathbb{F}^*(\mathcal{L})$, then we have $\mathbf{S}[E \times \mathcal{H} \neq \varnothing]$; in addition, $\forall x \in E : (x, E) \in \tilde{D}(E, \mathcal{H})$. Thus $\forall \mathcal{H} \in \mathbb{F}^*(\mathcal{L})$:

$$\mathbf{S}[\tilde{D}(E, \mathcal{H}) \neq \varnothing].$$

We now consider a natural construction of the direction defined on (4.10.15) in the case when $\mathcal{H}$ is a filter of $\mathcal{L}$. For this we investigate the general case of (4.10.15). For $\mathcal{H} \in 2^{\mathcal{P}(E)}$ we introduce the set

$$(\mathcal{H} - \mathrm{dir})[E] \triangleq \{z \in \tilde{D}(E, \mathcal{H}) \times \tilde{D}(E, \mathcal{H}) \mid \mathrm{pr}_2(\mathrm{pr}_2(z)) \subset \mathrm{pr}_2(\mathrm{pr}_1(z))\};$$

so if $x_1 \in E$, $H_1 \in \mathcal{H}$, $x_2 \in E$, $H_2 \in \mathcal{H}$ have the properties $(x_1, H_1) \in \tilde{D}(E, \mathcal{H})$ and $(x_2, H_2) \in \tilde{D}(E, \mathcal{H})$, then (see (2.2.1))

$$((x_1, H_1)\ (\mathcal{H} - \mathrm{dir})[E]\ (x_2, H_2)) \Leftrightarrow (H_2 \subset H_1).$$

Note the following obvious property: if $\mathcal{H} \in \mathbb{F}^*(\mathcal{L})$, then

$$(\mathcal{H} - \mathrm{dir})[E] \in (\mathrm{DIR})[\tilde{D}(E, \mathcal{H})] \tag{4.10.16}$$

(in reality (4.10.16) is true already for $\mathcal{L} \in \pi[E]$ and $\mathcal{H} \in \mathbb{F}^*(\mathcal{L})$). So by (4.10.16) for $\mathcal{H} \in \mathbb{F}^*(\mathcal{L})$ we have the nonempty directed set

$$(\tilde{D}(E, \mathcal{H}), (\mathcal{H} - \mathrm{dir})[E]). \tag{4.10.17}$$

On this base we realize (see [25]) the required type of a net. Indeed, $\forall \mathcal{H} \in 2^{(2^E)}$:

$$(\mathcal{H} - \mathrm{pr})[E] \triangleq (\mathrm{pr}_1(z))_{z \in \tilde{D}(E, \mathcal{H})} \in E^{\tilde{D}(E, \mathcal{H})}. \tag{4.10.18}$$

For $\mathcal{H} \in \mathbb{F}^*(\mathcal{L})$, by combination of (4.10.17) and (4.10.18) we obtain the required net in $E$:

$$(\tilde{D}(E,\mathcal{H}), (\mathcal{H}-\text{dir})[E], (\mathcal{H}-\text{pr})[E]). \tag{4.10.19}$$

PROPOSITION 4.10.3 *If $\mathcal{H} \in \mathbb{F}^*(\mathcal{L})$ and the net $(\mathbf{D}, \preceq, f)$ in (4.10.14) coincides with (4.10.19), then (4.10.14) coincides with $\mathcal{H}$:*

$$\mathcal{H} = (\mathcal{L}-\text{ASS})[\tilde{D}(E,\mathcal{H}); (\mathcal{H}-\text{dir})[E]; (\mathcal{H}-\text{pr})[E]].$$

The proof is similar to arguments used in general topology (see, for example, [71, Sect. 1.6]). We omit this proof. Of course, Proposition 4.10.3 defines some inversion of (4.10.14).

**Remark.** If $\mathcal{L} \subset \mathcal{P}(E)$, then (4.10.14) and Proposition 4.10.3 correspond to constructions of Section 2.2 (see (2.2.47) and (2.2.48)). Then for $X = E$ we have the coincidence of (2.2.47) and (4.10.14). Therefore (2.2.48) has the sense of convergence of the corresponding filter. Namely, if $(X, \tau)$ is a TS, $x \in X$ and $\mathcal{H} \in \mathbb{F}^*(\mathcal{P}(X))$, then we define (see [71, Sect. 1.6]) the $\tau$-convergence of $\mathcal{H}$ at the point $x$ as the property $N_\tau(x) \subset \mathcal{H}$. Then (2.2.48) corresponds to the convergence of the filter associated with the net used in (2.2.48).

Note that a net in $\mathbb{T}_\sigma(\mathcal{L})$ can be defined on the basis of the net (4.10.19). For this it is required to use the immersion of $E$ into $\mathbb{T}_\sigma(\mathcal{L})$ in terms of Dirac measures (see (4.8.10)). We use this approach somewhat later. Now we compare the relative topologies of $\mathbb{T}(\mathcal{L})$ generated by the topologies of the set (4.8.16). In addition, we take into account (4.10.2). But first we note some circumstances connected with topological properties. From (4.10.12) and the statements of Section 4.6 we obtain

$$\tau^*_{(0,1)}[\mathcal{L}] \triangleq \tau_*(\mathcal{L}) \mid_{\mathbb{T}(\mathcal{L})} = \tau_\otimes(\mathcal{L}) \mid_{\mathbb{T}(\mathcal{L})} = \otimes^{\mathcal{L}}(\tau_{\mathbb{R}}) \mid_{\mathbb{T}(\mathcal{L})} \tag{4.10.20}$$

(see (4.6.18)). Moreover, from (4.7.5) and (4.10.12) we have

$$\tau^*_{(0,1)}[\mathcal{L}] = \tau^*_{\mathbb{B}}(\mathcal{L}) \mid_{\mathbb{T}(\mathcal{L})} . \tag{4.10.21}$$

From (4.6.16) and (4.10.20) we get the inclusion $\tau^*_{(0,1)}[\mathcal{L}] \subset \tau_0(\mathcal{L}) \mid_{\mathbb{T}(\mathcal{L})}$. In reality we have the equality here (see [35, p.305] and [34]). By [35, p.305] and (4.10.20) we obtain that

$$\tau^*_{(0,1)}[\mathcal{L}] = \tau_0(\mathcal{L}) \mid_{\mathbb{T}(\mathcal{L})} . \tag{4.10.22}$$

From (4.8.16) and (4.10.20)–(4.10.22) we have the very suitable property that $\forall \tau \in \mathfrak{M}(\mathcal{L})$:

$$\tau^*_{(0,1)}[\mathcal{L}] = \tau \mid_{\mathbb{T}(\mathcal{L})} . \tag{4.10.23}$$

As for (4.10.23), we note the discussion in [35, p. 306]. Thus from (4.10.12) and (4.10.20)–(4.10.23) we infer that

$$\tau^*_{(0,1)}[\mathcal{L}] \in (\mathbf{c}-\text{top})[\mathbb{T}(\mathcal{L})] \tag{4.10.24}$$

defines the highly universal subspace in the form of the compactum

$$(\mathbb{T}(\mathcal{L}), \tau^*_{(0,1)}[\mathcal{L}]). \tag{4.10.25}$$

With this universal representation in terms of (4.10.24) and (4.10.25) it is advisable to connect the approximate construction based on (4.10.19) and Proposition 4.10.3. Denote by $\tilde{\delta}^0[\mathcal{L}]$ the mapping from $E$ into $\mathbb{T}_\sigma(\mathcal{L})$ for which $\forall x \in E$:

$$\tilde{\delta}^0[\mathcal{L}](x) \triangleq (\delta_x \mid \mathcal{L}). \tag{4.10.26}$$

In particular, by (4.10.26) we introduce the immersion $\tilde{\delta}^0[\mathcal{L}] \in \mathbb{T}(\mathcal{L})^E$.

PROPOSITION 4.10.4 *Let $\mathcal{H} \in \mathbb{F}^*_0(\mathcal{L})$. Then*

$$(\tilde{D}(E,\mathcal{H}), (\mathcal{H}-\text{dir})[E], \tilde{\delta}^0[\mathcal{L}] \circ (\mathcal{H}-\text{pr})[E]) \overset{\tau^*_{(0,1)}[\mathcal{L}]}{\rightarrow} \chi_{\mathcal{H}}[\mathcal{L}].$$

The proof uses (4.10.9), (4.10.10) and their corollaries. Moreover, it is required to take into account the basic property of ultrafilters of an algebra of sets (see Proposition 2.2.4). We note that this property is extended to the basic case of $\mathcal{L}$. Namely,

$$\forall \mathcal{H} \in \mathbb{F}^*_0(\mathcal{L})\ \forall n \in \mathcal{N}\ \forall (L_i)_{i\in\overline{1,n}} \in \Delta_n(E,\mathcal{L})\ \exists! j \in \overline{1,n} : L_j \in \mathcal{H}. \tag{4.10.27}$$

To prove (4.10.27) it is sufficient to use transformation of the type (4.10.10). From Proposition 4.10.4 we have

PROPOSITION 4.10.5 *If $\mathcal{H} \in \mathbb{F}^*_0(\mathcal{L})$, then $\forall \tau \in \mathfrak{M}(\mathcal{L})$:*

$$(\tilde{D}(E,\mathcal{H}), (\mathcal{H}-\text{dir})[E], \tilde{\delta}^0[\mathcal{L}] \circ (\mathcal{H}-\text{pr})[E]) \overset{\tau}{\rightarrow} \chi_{\mathcal{H}}[\mathcal{L}].$$

The proof is realized by combination of (2.3.9) and Proposition 4.10.4; moreover, we use the property that for $\mathcal{H} \in \mathbb{F}^*_0(\mathcal{L})$ we have the net in $\mathbb{T}(\mathcal{L})$ in the form

$$(\tilde{D}(E,\mathcal{H}), (\mathcal{H}-\text{dir})[E], \tilde{\delta}^0[\mathcal{L}] \circ (\mathcal{H}-\text{pr})[E]). \tag{4.10.28}$$

COROLLARY 4.10.1 ([35, P.306]) $\forall \tau \in \mathfrak{M}(\mathcal{L}) : \mathbb{T}(\mathcal{L}) = \text{cl}(\{(\delta_x \mid \mathcal{L}) : x \in E\}, \tau)$.

PROOF. Let $\mu \in \mathbb{T}(\mathcal{L})$. Using Proposition 4.10.2 we choose $\mathcal{H} \in \mathbb{F}^*_0(\mathcal{L})$ with the property $\mu = \chi_{\mathcal{H}}[\mathcal{L}]$. Then by Proposition 4.10.5 we have the 'universal' convergence $\forall \tau \in \mathfrak{M}(\mathcal{L})$:

$$(\tilde{D}(E,\mathcal{H}), (\mathcal{H}-\text{dir})[E], \tilde{\delta}^0[\mathcal{L}] \circ (\mathcal{H}-\text{pr})[E]) \overset{\tau}{\rightarrow} \mu,$$

where (4.10.28) is a net in the set $\{(\delta_x \mid \mathcal{L}) : x \in E\}$ (see (4.10.26)). By (2.3.11) we have $\forall \tau \in \mathfrak{M}(\mathcal{L})$: $\mu \in \mathrm{cl}(\{(\delta_x \mid \mathcal{L}) : x \in E\}, \tau)$. So the inclusion

$$\mathbb{T}(\mathcal{L}) \subset \mathrm{cl}(\{(\delta_x \mid \mathcal{L}) : x \in E\}, \tau) \tag{4.10.29}$$

holds for $\tau \in \mathfrak{M}(\mathcal{L})$. Using (4.10.13) and (4.10.29) we get

$$\mathbb{T}(\mathcal{L}) = \mathrm{cl}(\{(\delta_x \mid \mathcal{L}) : x \in E\}, \tau_*(\mathcal{L})). \tag{4.10.30}$$

By (4.7.2) and (4.10.13) we have $\mathrm{cl}(\{(\delta_x \mid \mathcal{L}) : x \in E\}, \tau_{\mathbb{B}}^*(\mathcal{L})) \subset \mathbb{T}(\mathcal{L})$. As a consequence, from (4.10.29) we have the equality

$$\mathbb{T}(\mathcal{L}) = \mathrm{cl}(\{(\delta_x \mid \mathcal{L}) : x \in E\}, \tau_{\mathbb{B}}^*(\mathcal{L})). \tag{4.10.31}$$

Recall a property of the space (4.6.20). By (2.6.29) we have $\mathfrak{N}_{\mathcal{L}} \subset \otimes^{\mathcal{L}}(\tau_{\mathbb{R}})$. Then $\forall \mu \in \mathbb{R}^{\mathcal{L}} : \mathfrak{N}_{\mathcal{L}}^0(\mu) \subset \otimes^{\mathcal{L}}(\tau_{\mathbb{R}})$; we use (2.6.28). As a consequence $\forall \mu \in \mathbf{A}(\mathcal{L})\ \forall \mathcal{K} \in \mathrm{Fin}(\mathcal{L})\ \forall \varepsilon \in ]0, \infty[$:

$$\mathbb{N}_{\mathcal{L}}(\mu, \mathcal{K}, \varepsilon) \in \otimes^{\mathcal{L}}(\tau_{\mathbb{R}}).$$

Using (4.6.22) we get that for $\mu \in \mathbf{A}(\mathcal{L})$, $\mathcal{K} \in \mathrm{Fin}(\mathcal{L})$ and $\varepsilon \in ]0, \infty[$:

$$\mathbb{N}_{\mathcal{L}}^{\otimes}(\mu, \mathcal{K}, \varepsilon) \in \tau_{\otimes}(\mathcal{L}). \tag{4.10.32}$$

Hence if $\mu \in \mathbf{A}(\mathcal{L}) \setminus (\mathrm{add})_+[\mathcal{L}]$, then for some $\mathcal{K} \in \mathrm{Fin}(\mathcal{L})$ and $\varepsilon \in ]0, \infty[$ we have

$$\mathbb{N}_{\mathcal{L}}^{\otimes}(\mu, \mathcal{K}, \varepsilon) \subset \mathbf{A}(\mathcal{L}) \setminus (\mathrm{add})_+[\mathcal{L}]. \tag{4.10.33}$$

We use here the property that

$$(\mathrm{add})_+[\mathcal{L}] = \{\mu \in \mathbf{A}(\mathcal{L}) \mid \mathbb{O}_{\mathcal{L}} \leqq \mu\}$$

(see (4.2.10) and (4.2.14)). From (4.10.32) and (4.10.33) we have $\forall \mu \in \mathbf{A}(\mathcal{L}) \setminus (\mathrm{add})_+[\mathcal{L}]\ \exists G \in \tau_{\otimes}(\mathcal{L})$: $(\mu \in G)\&(G \subset \mathbf{A}(\mathcal{L}) \setminus (\mathrm{add})_+[\mathcal{L}])$. Then $\mathbf{A}(\mathcal{L}) \setminus (\mathrm{add})_+[\mathcal{L}] \in \tau_{\otimes}(\mathcal{L})$, since this set is the union of some family of open sets (see (2.2.21)). We obtain that

$$(\mathrm{add})_+[\mathcal{L}] \in \mathcal{F}_{\tau_{\otimes}(\mathcal{L})}. \tag{4.10.34}$$

Hence by (2.3.4) and (4.6.35) we have

$$\begin{aligned} \mathcal{F}_{\tau_{\otimes}^+(\mathcal{L})} = \mathcal{F}_{\tau_*^+(\mathcal{L})} &= \mathcal{F}_{\tau_{\otimes}(\mathcal{L})} \cap \mathcal{P}((\mathrm{add})_+[\mathcal{L}]) \\ &= \{H \in \mathcal{F}_{\tau_{\otimes}(\mathcal{L})} \mid H \subset (\mathrm{add})_+[\mathcal{L}]\}. \end{aligned} \tag{4.10.35}$$

From (2.3.12) and (4.10.11) it follows that $\mathbb{T}(\mathcal{L}) = \mathbb{T}(\mathcal{L}) \cap (\mathrm{add})_+[\mathcal{L}] \in \mathcal{F}_{\tau_*^+(\mathcal{L})}$. Using (4.10.35) we get $\mathbb{T}(\mathcal{L}) \in \mathcal{F}_{\tau_{\otimes}(\mathcal{L})}$. Taking into account (4.2.22) and (4.8.10) we have

$$\mathrm{cl}(\{(\delta_x \mid \mathcal{L}) : x \in E\}, \tau_{\otimes}(\mathcal{L})) \subset \mathbb{T}(\mathcal{L}).$$

Therefore by (4.8.16) and (4.10.29) we obtain

$$\mathbb{T}(\mathcal{L}) = \mathrm{cl}(\{(\delta_x \mid \mathcal{L}) : x \in E\}, \tau_{\otimes}(\mathcal{L})). \tag{4.10.36}$$

From (4.6.19) and (4.10.36) it follows that

$$\mathrm{cl}(\{(\delta_x \mid \mathcal{L}) : x \in E\}, \tau_0(\mathcal{L})) \subset \mathbb{T}(\mathcal{L}).$$

By (4.8.16) and (4.10.29) we obtain the equality

$$\mathbb{T}(\mathcal{L}) = \mathrm{cl}(\{(\delta_x \mid \mathcal{L}) : x \in E\}, \tau_0(\mathcal{L})). \tag{4.10.37}$$

From (4.10.30), (4.10.31), (4.10.36) and (4.10.37) we have the required statement. □

## 4.11 SOME GENERALIZATIONS. I

The common construction using a compact space of GE was realized in the three last sections. Namely, we have some set **F**, $\mathbf{F} \neq \varnothing$, of 'usual' solutions. This set admits some immersion $m$ in $\mathbf{A}(\mathcal{L})$. In addition, the image $m^1(\mathbf{F})$ is a subset of some compactum in the TS (4.6.9). And what is more, this image has the density property in the compactum above (we can not require this property to be fulfillment; see the corresponding arguments in the previous chapter). Finally, this density is universal with respect to the topologies of $\mathfrak{M}(\mathcal{L})$ (4.8.16). In this connection, Theorems 4.8.2, 4.8.3, 4.9.3, 4.9.5, 4.9.6 and Corollary 4.10.1 should be compare. Moreover, (4.8.44) and (4.9.50) should be noted. The natural construction of an approximate solution net was realized in the all three variants; see Theorems 4.8.1, 4.9.1 and Proposition 4.10.5. Note that in the last statements the concrete approximate solutions-nets were constructed. It is natural to connect this fact with constructions of the previous chapter. Namely, we equip the space of 'usual' solutions with some constraints. Then we introduce some weakenings of these constraints. As a result a problem of asymptotic attainability arises. A corresponding modification of constructions of Section 3.1 (see (3.1.6)–(3.1.9)) was given in [32, 35, 45, 46] and other investigations; in addition, the space of GE is defined as a subset of $\mathbf{A}(\mathcal{L})$. Note that in [32, 35, 45, 46] mainly the scheme of immersion (in $\mathbf{A}(\mathcal{L})$) similar to the construction of Section 4.9 was investigated. In addition, several essential statements were obtained. It is possible to select two directions of these statements. Firstly, in [32, 35, 45, 46] the construction of generalized problems was suggested. This construction was realized in the class of weakly absolutely continuous measures; see the set $\mathbf{A}_\eta[\mathcal{L}]$ of Section 4.9. Other direction of the investigation concerns the property of asymptotic non-sensitivity under perturbation of a part of constraints.

These two directions are connected. Namely, the corresponding generalized problem has the important property of universality with respect to different variants of the weakening of integral constraints. Of course, this universality takes place in some range of the types of weakening of constraints. But this range is sufficiently broad. In addition, a representation of the above mentioned property of asymptotic non-sensitivity can be obtained from results of Chapter 3. We do not once more consider constructions similar to those of [32, 35, 45, 46]. We investigate a type of generalized problems realized in $\mathbf{A}(\mathcal{L})$. In our constructions we are oriented towards statements of the three last sections. In addition, we use the general scheme of Section 3.1.

We suppose that $\mathbf{K} \in (\tau_*(\mathcal{L}) - \text{comp})[\mathbf{A}(\mathcal{L})] \setminus \{\varnothing\}$ and, in correspondence with (3.1.6), introduce an imbedding of $\mathbf{F}$ into $\mathbf{K}$. But for the reason of methodical character we denote this mapping by $\mathbf{m}$. So $\mathbf{m} \in \mathbf{K}^{\mathbf{F}}$. We have

$$\tau^*_{\mathbf{K}}(\mathcal{L}) \triangleq \tau_*(\mathcal{L}) \,\big|_{\mathbf{K}} \in (\mathbf{c} - \text{top})[\mathbf{K}]. \tag{4.11.1}$$

Moreover, in this section we suppose that $(\mathbf{D}, \sqsubseteq)$ is a fixed directed set. Thus $\mathbf{S}[\mathbf{D} \neq \varnothing]$ and $\sqsubseteq \in (\text{DIR})[\mathbf{D}]$. In addition, we suppose that

$$(\xi_\mu)_{\mu \in \mathbf{K}} : \mathbf{K} \to \mathbf{F}^{\mathbf{D}}. \tag{4.11.2}$$

So for $\mu \in \mathbf{K}$ we have a net in $\mathbf{F}$ in the form $(\mathbf{D}, \sqsubseteq, \xi_\mu)$. We get that for $\mu \in \mathbf{K}$ the triplet $(\mathbf{D}, \sqsubseteq, \mathbf{m} \circ \xi_\mu)$ is a net in $\mathbf{K}$ and, in particular, in $\mathbf{A}(\mathcal{L})$. Let $\forall \mu \in \mathbf{K}\ \forall \tau \in \mathfrak{M}(\mathcal{L})$:

$$(\mathbf{D}, \sqsubseteq, \mathbf{m} \circ \xi_\mu) \xrightarrow{\tau} \mu. \tag{4.11.3}$$

By (4.11.1)-(4.11.3) we define a variant of compactification of the set $\mathbf{F}$. As for (4.11.3), we note Theorems 4.8.1, 4.9.1 and Proposition 4.10.5. So we have several concrete variants of the given general scheme.

Return to (4.11.1)-(4.11.3). This construction is considered in correspondence with (3.1.6). We fix $r \in \mathcal{N}$ and suppose $\mathcal{R} \triangleq \mathbb{R}^r$. Moreover, let $\mathbf{S}[\Gamma \neq \varnothing]$. We use elements of $\Gamma$ as distinctive indices. In this section we suppose that the set $\mathbf{X}$ of Section 3.1 coincides with $\mathcal{R}^\Gamma$. So

$$\mathbf{X} = \mathcal{R}^\Gamma. \tag{4.11.4}$$

Of course, we have the operators $s$ and $h$ in correspondence with (3.1.1). We postulate the properties of these operators in terms of the above mentioned immersion. Let

$$(S_{i,\gamma})_{(i,\gamma) \in \overline{1,r} \times \Gamma} : \overline{1,r} \times \Gamma \to B(E, \mathcal{L}) \tag{4.11.5}$$

be a generalized matriciant. We suppose that the mapping $g$ in (3.1.6) has the form

$$\mu \mapsto \left(\left(\int_E S_{i,\gamma} d\mu\right)_{i\in\overline{1,r}}\right)_{\gamma\in\Gamma} : \mathbf{K} \to \mathcal{R}^\Gamma. \tag{4.11.6}$$

Then in terms of (4.11.4) and (4.11.6) we have $g \in \mathbf{X}^\mathbf{K}$ and $\forall \mu \in \mathbf{K}$ $\forall \gamma \in \Gamma$:

$$g(\mu)(\gamma) = \left(\int_E S_{i,\gamma} d\mu\right)_{i\in\overline{1,r}}. \tag{4.11.7}$$

Recall that by (3.1.9) $s = g \circ \mathbf{m}$. Using (4.11.4), we construct a topological equipment of $\mathbf{X}$. Employing constructions of Chapter 3, we introduce two comparable topologies of $\mathbf{X}$ (4.11.4). First we equip $\mathbf{X}$ with the topology of pointwise convergence. Namely, we introduce the topology $\tau_\mathbb{R}^{(r)} \in (\mathrm{top})[\mathcal{R}]$ of coordinate-wise convergence in $\mathcal{R}$. It is possible, for example, to take the metric

$$\rho_r : \mathcal{R} \times \mathcal{R} \to [0, \infty[$$

for which $\forall (u_i)_{i\in\overline{1,r}} \in \mathcal{R}$ $\forall (v_i)_{i\in\overline{1,r}} \in \mathcal{R}$:

$$\rho_r((u_i)_{i\in\overline{1,r}}, (v_i)_{i\in\overline{1,r}}) \triangleq \sup(\{\,| u_i - v_i | : i \in \overline{1,r}\}).$$

Of course, $\rho_r \in (\mathrm{Dist})[\mathcal{R}]$. We recall that by (2.7.4) $\forall (u_i)_{i\in\overline{1,r}} \in \mathcal{R}$ $\forall \varepsilon \in ]0, \infty[$:

$$\mathbf{B}^0_{\rho_r}((u_i)_{i\in\overline{1,r}}, \varepsilon) = \{(v_i)_{i\in\overline{1,r}} \in \mathcal{R} \mid \rho_r((u_i)_{i\in\overline{1,r}}, (v_i)_{i\in\overline{1,r}}) < \varepsilon\}.$$

We define $\tau_\mathbb{R}^{(r)}$ by the requirement

$$\tau_\mathbb{R}^{(r)} \triangleq \tau^\natural_{\rho_r}(\mathcal{R}) \in (\mathrm{top})_0[\mathcal{R}]. \tag{4.11.8}$$

In (4.11.8) we use (2.7.5) and (2.7.7). Hence we consider the topology

$$\otimes^\Gamma(\tau_\mathbb{R}^{(r)}) \in (\mathrm{top})_0[\mathcal{R}^\Gamma] \tag{4.11.9}$$

(see (2.6.22) and (2.6.24)). It is useful to choose some natural basis of the topology (4.11.9). Of course, the construction (2.6.23) can be used. But in this case we can apply a more simple variant. For (4.11.4) we use here a scheme similar to (2.6.30). If $u \in \mathbf{X}$, $K \in \mathrm{Fin}(\Gamma)$ and $\varepsilon \in ]0, \infty[$, then we suppose that

$$\mathrm{N}_\Gamma^{(r)}(u, K, \varepsilon) \triangleq \{v \in \mathcal{R}^\Gamma \mid \forall \gamma \in K : v(\gamma) \in \mathbf{B}^0_{\rho_r}(u(\gamma), \varepsilon)\}. \tag{4.11.10}$$

We use (4.11.4) and (4.11.10) by analogy with (2.6.30).

PROPOSITION 4.11.1 *If* $u_* \in \mathbf{X}$, $K_* \in \mathrm{Fin}(\Gamma)$ *and* $\varepsilon_* \in ]0,\infty[$, *then* $\mathbb{N}_\Gamma^{(r)}(u_*, K_*, \varepsilon_*) \in N^0_{\otimes^\Gamma(\tau_{\mathbb{R}}^{(r)})}(u_*)$. *Moreover, the following equality holds:*

$$\otimes^\Gamma(\tau_{\mathbb{R}}^{(r)}) = \{G \in \mathcal{P}(\mathbf{X}) \mid \forall u \in G\ \exists K \in \mathrm{Fin}(\Gamma)\ \exists \varepsilon \in ]0,\infty[ : \mathbb{N}_\Gamma^{(r)}(u, K, \varepsilon) \subset G\}.$$

PROOF. Consider (2.6.23) under $X = \Gamma$, $Y = \mathcal{R}$ and $\tau = \tau_{\mathbb{R}}^{(r)}$. By (2.6.21) and (4.11.4) we have $\forall \gamma \in \Gamma$:

$$(\gamma - \mathrm{proj})[\Gamma; \mathcal{R}] = (f(\gamma))_{f \in \mathbf{X}} \in \mathcal{R}^{\mathbf{X}}.$$

Recall that

$$(\mathrm{Proj})[\Gamma; \mathcal{R}] = \{(\gamma - \mathrm{proj})[\Gamma; \mathcal{R}] : \gamma \in \Gamma\} = \{(f(\gamma))_{f \in \mathbf{X}} : \gamma \in \Gamma\}. \tag{4.11.11}$$

From (2.6.23) we obtain the following useful representation:

$$\beta^0_\otimes(\Gamma, \mathcal{R}, \tau_{\mathbb{R}}^{(r)}) = \beta^0(\mathcal{R}^\Gamma \mid \tau_{\mathbb{R}}^{(r)}, (\mathrm{Proj})[\Gamma; \mathcal{R}]) \in (\otimes^\Gamma(\tau_{\mathbb{R}}^{(r)}) - \mathrm{BAS})_0[\mathbf{X}] \tag{4.11.12}$$

is the family of all sets $T \in \mathcal{P}(\mathbf{X})$ for which

$$\exists n \in \mathcal{N}\ \exists (h_i)_{i \in \overline{1,n}} \in (\mathrm{Proj})[\Gamma; \mathcal{R}]^n\ \exists (G_i)_{i \in \overline{1,n}} \in (\tau_{\mathbb{R}}^{(r)})^n :$$

$$T = \bigcap_{i=1}^n h_i^{-1}(G_i).$$

Using (4.11.11) we obtain that $\beta^0_\otimes(\Gamma, \mathcal{R}, \tau_{\mathbb{R}}^{(r)})$ is the family of all sets $T \in \mathcal{P}(\mathbf{X})$ for which

$$\exists n \in \mathcal{N}\ \exists (\gamma_i)_{i \in \overline{1,n}} \in \Gamma^n\ \exists (G_i)_{i \in \overline{1,n}} \in (\tau_{\mathbb{R}}^{(r)})^n :$$

$$T = \bigcap_{i=1}^n \{f \in \mathbf{X} \mid f(\gamma_i) \in G_i\} = \{f \in \mathbf{X} \mid \forall i \in \overline{1,n} : f(\gamma_i) \in G_i\}.$$

So we have established that

$$\beta^0_\otimes(\Gamma, \mathcal{R}, \tau_{\mathbb{R}}^{(r)}) = \Big\{T \in \mathcal{P}(\mathbf{X}) \mid \exists n \in \mathcal{N}\ \exists (\gamma_i)_{i \in \overline{1,n}} \in \Gamma^n\ \exists (G_i)_{i \in \overline{1,n}} \in (\tau_{\mathbb{R}}^{(r)})^n : T = \{f \in \mathbf{X} \mid \forall i \in \overline{1,n} : f(\gamma_i) \in G_i\}\Big\}. \tag{4.11.13}$$

Recall (4.11.12). By (2.6.3) and (2.6.6) we have the property

$$\begin{aligned}\otimes^{\Gamma}(\tau_{\mathbb{R}}^{(r)}) &= \{\cup\}(\beta_{\otimes}^{0}(\Gamma,\mathcal{R},\tau_{\mathbb{R}}^{(r)})) \\ &= \{G \in \mathcal{P}(\mathbf{X}) \mid \forall f \in G\ \exists B \in \beta_{\otimes}^{0}(\Gamma,\mathcal{R},\tau_{\mathbb{R}}^{(r)}) : \\ &\qquad (f \in B)\ \&\ (B \subset G)\}.\end{aligned} \tag{4.11.14}$$

Introduce the family

$$\mathfrak{G} \triangleq \{G \in \mathcal{P}(\mathbf{X}) \mid \forall u \in G\ \exists K \in \mathrm{Fin}(\Gamma)\ \exists \varepsilon \in ]0,\infty[: \mathbb{N}_{\Gamma}^{(r)}(u,K,\varepsilon) \subset G\}. \tag{4.11.15}$$

Compare the families (4.11.14) and (4.11.15). Recall that by (2.7.6) and (4.11.8) $\forall (u_i)_{i\in\overline{1,r}} \in \mathcal{R}\ \forall \varepsilon \in ]0,\infty[$:

$$\mathbf{B}_{\rho_r}^{0}((u_i)_{i\in\overline{1,r}},\varepsilon) \in \tau_{\mathbb{R}}^{(r)}. \tag{4.11.16}$$

By (2.7.5) and (4.11.8) we have $\forall H \in \tau_{\mathbb{R}}^{(r)}\ \forall x \in H\ \exists \varepsilon \in ]0,\infty[$:

$$\mathbf{B}_{\rho_r}^{0}(x,\varepsilon) \subset H.$$

Then from (4.11.10) and (4.11.13) it follows that

$$\forall T \in \beta_{\otimes}^{0}(\Gamma,\mathcal{R},\tau_{\mathbb{R}}^{(r)})\ \forall f \in T\ \exists K \in \mathrm{Fin}(\Gamma)\ \exists \varepsilon \in ]0,\infty[: \mathbb{N}_{\Gamma}^{(r)}(f,K,\varepsilon) \subset T. \tag{4.11.17}$$

Moreover, from (4.11.10) and (4.11.13) we obtain the property that

$$\mathbb{N}_{\Gamma}^{(r)}(f,K,\varepsilon) \in \beta_{\otimes}^{0}(\Gamma,\mathcal{R},\tau_{\mathbb{R}}^{(r)}) \tag{4.11.18}$$

for each $f \in \mathbf{X}$, $K \in \mathrm{Fin}(\Gamma)$ and $\varepsilon \in ]0,\infty[$. Recall that by (2.6.6) and (4.11.12) we have $\beta_{\otimes}^{0}(\Gamma,\mathcal{R},\tau_{\mathbb{R}}^{(r)}) \in (\mathrm{BAS})[\mathbf{X}]$. Then $\beta_{\otimes}^{0}(\Gamma,\mathcal{R},\tau_{\mathbb{R}}^{(r)}) \subset \{\cup\}(\beta_{\otimes}^{0}(\Gamma,\mathcal{R},\tau_{\mathbb{R}}^{(r)}))$ and by (4.11.14) $\beta_{\otimes}^{0}(\Gamma,\mathcal{R},\tau_{\mathbb{R}}^{(r)}) \subset \otimes^{\Gamma}(\tau_{\mathbb{R}}^{(r)})$. By (2.2.14), (4.11.10) and (4.11.18), for $f \in \mathbf{X}$, $K \in \mathrm{Fin}(\Gamma)$ and $\varepsilon \in ]0,\infty[$ we obtain

$$\mathbb{N}_{\Gamma}^{(r)}(f,K,\varepsilon) \in N^{0}_{\otimes^{\Gamma}(\tau_{\mathbb{R}}^{(r)})}(f). \tag{4.11.19}$$

In (4.11.19) we have the first part of the required statement. From (4.11.10), (4.11.14), (4.11.15) and (4.11.18) we obtain the inclusion

$$\mathfrak{G} \subset \otimes^{\Gamma}(\tau_{\mathbb{R}}^{(r)}). \tag{4.11.20}$$

Let $\Omega \in \otimes^{\Gamma}(\tau_{\mathbb{R}}^{(r)})$. Then by (4.11.14) $\Omega \in \mathcal{P}(\mathbf{X})$ and

$$\forall f \in \Omega\ \exists B \in \beta_{\otimes}^{0}(\Gamma,\mathcal{R},\tau_{\mathbb{R}}^{(r)}) : (f \in B)\&(B \subset \Omega). \tag{4.11.21}$$

Fix $\omega \in \Omega$. Using (4.11.21) we choose $B_\omega \in \beta^0_\otimes(\Gamma, \mathcal{R}, \tau^{(r)}_\mathbb{R})$ for which

$$(\omega \in B_\omega)\&(B_\omega \subset \Omega). \tag{4.11.22}$$

Using (4.11.13) we choose $n \in \mathcal{N}$, $(\gamma_i)_{i\in\overline{1,n}} \in \Gamma^n$ and $(G_i)_{i\in\overline{1,n}} \in (\tau^{(r)}_\mathbb{R})^n$ with the property

$$B_\omega = \{f \in \mathbf{X} \mid \forall i \in \overline{1,n} : f(\gamma_i) \in G_i\}. \tag{4.11.23}$$

Since $G_1 \in \tau^{(r)}_\mathbb{R}, \ldots, G_n \in \tau^{(r)}_\mathbb{R}$ and by (4.11.22) $\forall i \in \overline{1,n} : \omega(\gamma_i) \in G_i$ we have for some $\varepsilon' \in ]0, \infty[$, the inclusion $\mathbf{B}^0_{\rho_r}(\omega(\gamma_i), \varepsilon') \subset G_i$ for each $i \in \overline{1,n}$. Let $K' = \{\gamma_i : i \in \overline{1,n}\}$. Using (4.11.10), introduce

$$\mathbb{N}^{(r)}_\Gamma(\omega, K', \varepsilon') = \{v \in \mathbf{X} \mid \forall \gamma \in K' : v(\gamma) \in \mathbf{B}^0_{\rho_r}(\omega(\gamma), \varepsilon')\}.$$

For $v \in \mathbb{N}^{(r)}_\Gamma(\omega, K', \varepsilon')$ we have $\forall i \in \overline{1,n} : v(\gamma_i) \in \mathbf{B}^0_{\rho_r}(\omega(\gamma_i), \varepsilon')$. As a consequence $v(\gamma_i) \in G_i$ for $i \in \overline{1,n}$. Then by (4.11.22) and (4.11.23) we have

$$\mathbb{N}^{(r)}_\Gamma(\omega, K', \varepsilon') \subset B_\omega \subset \Omega.$$

Since the choice of $\omega$ was arbitrary, by (4.11.15) we have the property $\Omega \in \mathfrak{G}$. So

$$\otimes^\Gamma(\tau^{(r)}_\mathbb{R}) \subset \mathfrak{G}. \tag{4.11.24}$$

From (4.11.20) and (4.11.24) we have the required equality $\otimes^\Gamma(\tau^{(r)}_\mathbb{R}) = \mathfrak{G}$.
□

It is useful to introduce a variant of neighborhood-valued mapping of Section 3.2. Namely, in (3.2.15) we consider the case $X = \mathbf{X}$, $Q = \mathrm{Fin}(\Gamma)\times]0, \infty[$ and $M = \mathbf{X}$. For this case we define a variant of the mapping $\Lambda$ (3.2.18) by (4.11.10). Consider the set $\mathfrak{Q}_\mathbf{X}(\mathrm{Fin}(\Gamma)\times]0, \infty[, \otimes^\Gamma(\tau^{(r)}_\mathbb{R}))$. Then by Proposition 4.11.1

$$\begin{aligned}\Lambda^{(r)}_\Gamma &\triangleq (\mathbb{N}^{(r)}_\Gamma(\mathrm{pr}_1(z), \mathrm{pr}_1(\mathrm{pr}_2(z)), \mathrm{pr}_2(\mathrm{pr}_2(z))))_{z\in\mathbf{X}\times(\mathrm{Fin}(\Gamma)\times]0,\infty[)}\\ &\in \mathfrak{Q}_\mathbf{X}(\mathrm{Fin}(\Gamma)\times]0, \infty[, \otimes^\Gamma(\tau^{(r)}_\mathbb{R})).\end{aligned} \tag{4.11.25}$$

From (4.11.25) we have the mapping

$$\Lambda^{(r)}_\Gamma : \mathbf{X} \times (\mathrm{Fin}(\Gamma)\times]0, \infty[) \to \mathcal{P}(\mathbf{X})$$

such that

$$\Lambda^{(r)}_\Gamma(f, (K, \varepsilon)) = \mathbb{N}^{(r)}_\Gamma(f, K, \varepsilon)$$

for $f \in \mathbf{X}$, $K \in \mathrm{Fin}(\Gamma)$ and $\varepsilon \in ]0,\infty[$. The mapping (4.11.25) has the following useful property:

$$\Lambda_\Gamma^{(r)} \in (\mathrm{UNIF})[\mathrm{Fin}(\Gamma)\times]0,\infty[, \otimes^\Gamma(\tau_{\mathbb{R}}^{(r)}) \mid \mathbf{X}]. \tag{4.11.26}$$

To verify this we use (3.2.18), fixing $q_1 \in \mathrm{Fin}(\Gamma)\times]0,\infty[$ and $q_2 \in \mathrm{Fin}(\Gamma)\times]0,\infty[$. Then $K_1 \triangleq \mathrm{pr}_1(q_1) \in \mathrm{Fin}(\Gamma)$, $\varepsilon_1 \triangleq \mathrm{pr}_2(q_1) \in ]0,\infty[$, $K_2 \triangleq \mathrm{pr}_1(q_2) \in \mathrm{Fin}(\Gamma)$ and $\varepsilon_2 \triangleq \mathrm{pr}_2(q_2) \in ]0,\infty[$. Suppose that $q_3 \triangleq (K_1 \cup K_2, \inf(\{\varepsilon_1,\varepsilon_2\}))$. Of course, $q_3 \in \mathrm{Fin}(\Gamma)\times]0,\infty[$. Then by (4.11.10) for $f \in \mathbf{X}$ we have

$$\begin{aligned}\Lambda_\Gamma^{(r)}(f,q_3) &= \mathbb{N}_\Gamma^{(r)}(f, K_1 \cup K_2, \inf(\{\varepsilon_1,\varepsilon_2\}))\\ &\subset \mathbb{N}_\Gamma^{(r)}(f,K_1,\varepsilon_1) \cap \mathbb{N}_\Gamma^{(r)}(f,K_2,\varepsilon_2)\\ &= \Lambda_\Gamma^{(r)}(f,q_1) \cap \Lambda_\Gamma^{(r)}(f,q_2).\end{aligned}$$

So (4.11.26) is established. This permits us to use Proposition 3.2.1.

We obtain a variant of the topological equipment of $\mathbf{X}$. For this equipment we use the weakening of constraints on the basis of (3.2.21). Recall that a simple sufficient condition of asymptotic non-sensitivity under the weakening of integral constraints was established in [32, 35, 45, 46]. This condition means the graduatedness of a part of components of the mapping (4.11.5). We will not repeat constructions of [32, 35, 45, 46] connected with the extension of integral constraints. Instead of this we consider a highly natural concrete variant of general constructions of Chapter 3. In this connection we introduce one more topology of $\mathbf{X}$ (4.11.4). We set

$$\Gamma_0 \triangleq \{\gamma \in \Gamma \mid \forall i \in \overline{1,r} : S_{i,\gamma} \in B_0(E,\mathcal{L})\}. \tag{4.11.27}$$

We call $\Gamma_0$ (4.11.27) the set of graduatedness. In terms of $\Gamma_0$ we introduce a new topological structure. Namely, if $u \in \mathbf{X}$, $K \in \mathrm{Fin}(\Gamma)$ and $\varepsilon \in ]0,\infty[$, then

$$\begin{aligned}\overset{\circ}{\mathbb{N}}{}_\Gamma^{(r)}(u,K,\varepsilon) \triangleq \{v \in \mathcal{R}^\Gamma \mid &(\forall \gamma \in K \cap \Gamma_0 : u(\gamma) = v(\gamma))\\ &\&(\forall \gamma \in K \setminus \Gamma_0 : v(\gamma) \in \mathbf{B}^0_{\rho_r}(u(\gamma),\varepsilon))\}.\end{aligned} \tag{4.11.28}$$

Of course, (4.11.28) is a subset of the set (4.11.10). Introduce the family

$$\begin{aligned}\mathcal{T}_r^0[\Gamma] \triangleq \{G \in \mathcal{P}(\mathbf{X}) \mid &\forall u \in G\ \exists K \in \mathrm{Fin}(\Gamma)\ \exists \varepsilon \in ]0,\infty[:\\ &\overset{\circ}{\mathbb{N}}{}_\Gamma^{(r)}(u,K,\varepsilon) \subset G\}.\end{aligned} \tag{4.11.29}$$

Comparing (4.11.10) and (4.11.28) we get (see Proposition 4.11.1 and (4.11.29)) the property

$$\otimes^{\Gamma}(\tau_{\mathbb{R}}^{(r)}) \subset \mathcal{T}_r^0[\Gamma]. \qquad (4.11.30)$$

From (4.11.28) we infer that if $u \in \mathbf{X}$, $K_1 \in \mathrm{Fin}(\Gamma)$, $\varepsilon_1 \in ]0, \infty[$, $K_2 \in \mathrm{Fin}(\Gamma)$ and $\varepsilon_2 \in ]0, \infty[$, then for $K_1 \cup K_2 \in \mathrm{Fin}(\Gamma)$ and $\inf(\{\varepsilon_1, \varepsilon_2\}) \in ]0, \infty[$ we get

$$\overset{\circ\,(r)}{\mathbb{N}_\Gamma}(u, K_1 \cup K_2, \inf(\{\varepsilon_1, \varepsilon_2\})) \subset \overset{\circ\,(r)}{\mathbb{N}_\Gamma}(u, K_1, \varepsilon_1) \cap \overset{\circ\,(r)}{\mathbb{N}_\Gamma}(u, K_2, \varepsilon_2).$$

Then from (4.11.4) and (4.11.29) we have the following important statement:

$$\mathcal{T}_r^0[\Gamma] \in (\mathrm{top})_0[\mathcal{R}^\Gamma]; \qquad (4.11.31)$$

in (4.11.31) we use (4.11.9) and (4.11.30). Thus by (4.11.31) we construct the new TS

$$(\mathbf{X}, \mathcal{T}_r^0[\Gamma]) = (\mathcal{R}^\Gamma, \mathcal{T}_r^0[\Gamma]). \qquad (4.11.32)$$

Of course, (4.11.32) is a Hausdorff space. Return to (4.11.28). From (2.7.2), (2.7.4) and (4.11.28) the useful property follows: if $u \in \mathbf{X}$, $K \in \mathrm{Fin}(\Gamma)$, $\varepsilon \in ]0, \infty[$ and $v \in \overset{\circ\,(r)}{\mathbb{N}_\Gamma}(u, K, \varepsilon)$, then it is possible to choose $\varepsilon_1 \in ]0, \varepsilon]$ for which

$$\overset{\circ\,(r)}{\mathbb{N}_\Gamma}(v, K, \varepsilon_1) \subset \overset{\circ\,(r)}{\mathbb{N}_\Gamma}(u, K, \varepsilon).$$

As a consequence we have $\forall u \in \mathbf{X}\ \forall K \in \mathrm{Fin}(\Gamma)\ \forall \varepsilon \in ]0, \infty[$:

$$\overset{\circ\,(r)}{\mathbb{N}_\Gamma}(u, K, \varepsilon) \in N^0_{\mathcal{T}_r^0[\Gamma]}(u). \qquad (4.11.33)$$

We use (2.2.14) in (4.11.33). And what is more, from (2.2.14), (2.6.8), (4.11.29) and (4.11.33) we get $\forall u \in \mathbf{X}$:

$$\begin{aligned} \overset{\circ\,(r)}{\mathfrak{N}}[u] &\triangleq \{\overset{\circ\,(r)}{\mathbb{N}_\Gamma}(u, \mathrm{pr}_1(z), \mathrm{pr}_2(z)) : z \in \mathrm{Fin}(\Gamma) \times ]0, \infty[\} \\ &\in (u - \mathrm{bas})[\mathcal{T}_r^0[\Gamma]]. \end{aligned} \qquad (4.11.34)$$

In (4.11.34) we have a highly useful property from the point of view of representation of the Moore Smith convergence in the TS (4.11.32).

**Remark.** The space (4.11.32) is analogous to TS considered in [39, 45]; in particular, see [39, pp. 350,351]. In addition, the corresponding constructions of [39, 45] are connected with investigation of the case $r = 1$. In [39, p.351] the natural homeomorph of the space similar to

(4.11.32) was indicated. We do not now consider the structure of such homeomorph, referring to [39,45].

Recall (4.11.6) and (4.11.7). We consider the important property of continuity. Of course,

$$\tau^0_{\mathbf{K}}(\mathcal{L}) \triangleq \tau_0(\mathcal{L}) \mid_{\mathbf{K}} \in (\mathrm{top})_0[\mathbf{K}]. \tag{4.11.35}$$

In addition, from (4.6.19), (4.11.35) and properties of the type (4.6.34) we have

$$\tau^*_{\mathbf{K}}(\mathcal{L}) = \tau_\otimes(\mathcal{L}) \mid_{\mathbf{K}} \subset \tau^0_{\mathbf{K}}(\mathcal{L}). \tag{4.11.36}$$

Consider the space of mappings from the Hausdorff TS

$$(\mathbf{K}, \tau^0_{\mathbf{K}}(\mathcal{L})) \tag{4.11.37}$$

into (4.11.32). Of course, (4.11.37) is an auxiliary TS, whose role is defined by (4.11.36).

PROPOSITION 4.11.2 *The following property holds:*

$$g \in C(\mathbf{K}, \tau^0_{\mathbf{K}}(\mathcal{L}), \mathbf{X}, \mathcal{T}^0_r[\Gamma]).$$

PROOF. We use the representation of continuity based on (2.5.4). Let $(D_0, \preceq, \varphi)$ is a net in $\mathbf{K}$; then $\mathbf{S}[D_0 \neq \varnothing]$, $\preceq \in (\mathrm{DIR})[D_0]$ and $\varphi \in \mathbf{K}^{D_0}$. Moreover, let $\mu_0 \in \mathbf{K}$ and

$$(D_0, \preceq, \varphi) \overset{\tau^0_{\mathbf{K}}(\mathcal{L})}{\to} \mu_0. \tag{4.11.38}$$

Note that, in particular, $(D_0, \preceq, \varphi)$ is a net in $\mathbf{A}(\mathcal{L})$. By (2.3.9) and (4.11.35) we obtain that

$$(D_0, \preceq, \varphi) \overset{\tau_0(\mathcal{L})}{\to} \mu_0. \tag{4.11.39}$$

Note several properties connected with the TS (4.6.9). Namely, from (4.11.36) and (4.11.38) we have the convergence

$$(D_0, \preceq, \varphi) \overset{\tau^*_{\mathbf{K}}(\mathcal{L})}{\to} \mu_0. \tag{4.11.40}$$

In (4.11.40) we take into account (2.2.14) and (2.2.48). Using (2.3.9), (4.11.1) and (4.11.40) we get the convergence

$$(D_0, \preceq, \varphi) \overset{\tau_*(\mathcal{L})}{\to} \mu_0. \tag{4.11.41}$$

Finally, by (2.3.9) and (4.6.18) we have

$$(D_0, \preceq, \varphi) \overset{\otimes^{\mathcal{L}}(\tau_\partial)}{\to} \mu_0. \tag{4.11.42}$$

Of course, in (4.11.42) we consider $(D_0, \preceq, \varphi)$ as a net in $\mathbb{R}^{\mathcal{L}}$ and interpret $\mu$ as an element of $\mathbb{R}^{\mathcal{L}}$. From (4.6.1), (4.6.2), (4.6.5), (4.6.7) and (4.6.42) we have $\forall h \in B(E, \mathcal{L})$:

$$\left(D_0, \preceq, \left(\int_E h \, d\varphi(\delta)\right)_{\delta \in D_0}\right) \stackrel{\mathbb{R}}{\to} \int_E h \, d\mu_0. \tag{4.11.43}$$

We supplement (4.11.43) by the following simple corollary of (4.11.42). Namely, from (2.6.29) and (4.11.42) we have $\forall L \in \mathcal{L} \; \exists d_1 \in D_0 \; \forall d_2 \in D_0$:

$$(d_1 \preceq d_2) \Rightarrow (\varphi(d_2)(L) = \mu_0(L)). \tag{4.11.44}$$

From (4.3.2), (4.3.8) and (4.11.44) we obtain that $\forall h \in B_0(E, \mathcal{L}) \; \exists \delta_1 \in D_0 \; \forall \delta_2 \in D_0$:

$$(\delta_1 \preceq \delta_2) \Rightarrow \left( {}^{(\mathrm{el})}\!\!\int_E h \, d\varphi(\delta_2) = {}^{(\mathrm{el})}\!\!\int_E h \, d\mu_0 \right).$$

Using (4.3.11) and (4.3.14) we get

$$\begin{aligned} &\forall h \in B_0(E, \mathcal{L}) \; \exists \delta_1 \in D_0 \; \forall \delta_2 \in D_0 : (\delta_1 \preceq \delta_2) \\ &\Rightarrow \left( \int_E h \, d\varphi(\delta_2) = \int_E h \, d\mu_0 \right). \end{aligned} \tag{4.11.45}$$

We use a natural combination of (4.11.43) and (4.11.45). Consider $g(\mu_0) \in \mathbf{X}$. Then $g(\mu_0) : \Gamma \to \mathcal{R}$. Choose an arbitrary neighborhood $\mathbb{H}_0 \in N_{\mathcal{T}_r^0[\Gamma]}(g(\mu_0))$. Using (2.6.8) and (4.11.34), choose $K^\natural \in \mathrm{Fin}(\Gamma)$ and $\varepsilon^\natural \in ]0, \infty[$ for which

$$\overset{\circ}{\mathbb{N}}{}_\Gamma^{(r)}(g(\mu_0), K^\natural, \varepsilon^\natural) \subset \mathbb{H}_0. \tag{4.11.46}$$

Recall that by (4.11.28) we have the equality

$$\begin{aligned} \overset{\circ}{\mathbb{N}}{}_\Gamma^{(r)}(g(\mu_0), K^\natural, \varepsilon^\natural) = \{ v \in \mathbf{X} \mid (\forall \gamma \in K^\natural \cap \Gamma_0 : g(\mu_0)(\gamma) = v(\gamma)) \\ \&(\forall \gamma \in K^\natural \setminus \Gamma_0 : v(\gamma) \in \mathbf{B}^0_{\rho_r}(g(\mu_0)(\gamma), \varepsilon^\natural))\}. \end{aligned}$$

Using (2.7.4) and the definition of the metric $\rho_r$, we get

$$\begin{aligned} &\overset{\circ}{\mathbb{N}}{}_\Gamma^{(r)}(g(\mu_0), K^\natural, \varepsilon^\natural) \\ &= \{ v \in \mathbf{X} \mid (\forall \gamma \in K^\natural \cap \Gamma_0 : g(\mu_0)(\gamma) = v(\gamma)) \\ &\quad \&(\forall \gamma \in K^\natural \setminus \Gamma_0 : \sup(\{ \mid g(\mu_0)(\gamma)(i) - v(\gamma)(i) \mid : i \in \overline{1, r}\}) < \varepsilon^\natural)\}. \end{aligned} \tag{4.11.47}$$

We recall that by (4.11.7) the following representation holds. Namely, $\forall \gamma \in \Gamma$:

$$\left(\forall \delta \in D_0 : (g \circ \varphi)(\delta)(\gamma) = g(\varphi(\delta))(\gamma) = \left(\int_E S_{i,\gamma} d\varphi(\delta)\right)_{i \in \overline{1,r}}\right)$$

$$\& \left(g(\mu_0)(\gamma) = \left(\int_E S_{i,\gamma} d\mu_0\right)_{i \in \overline{1,r}}\right). \tag{4.11.48}$$

From (4.11.27) we have $\forall \gamma \in \Gamma_0\ \forall i \in \overline{1,r}$: $S_{i,\gamma} \in B_0(E, \mathcal{L})$. Therefore from (4.11.45) and (4.11.48) it follows that $\forall \gamma \in \Gamma_0\ \exists \delta_1 \in D_0\quad \forall \delta_2 \in D_0$:

$$(\delta_1 \preceq \delta_2) \Rightarrow ((g \circ \varphi)(\delta_2)(\gamma) = g(\mu_0)(\gamma)). \tag{4.11.49}$$

Of course, we use (2.2.4) in (4.11.49). Moreover, from (4.11.5) and (4.11.43) we have $\forall \gamma \in \Gamma_0\ \exists \delta' \in D_0\quad \forall \delta' \in D_0$:

$$(\delta' \preceq \delta') \Rightarrow \left(\forall i \in \overline{1,r} : \left|\int_E S_{i,\gamma} d\varphi(\delta') - \int_E S_{i,\gamma} d\mu_0\right| < \varepsilon^\natural\right). \tag{4.11.50}$$

But by (4.11.48) we obtain that $\forall \gamma \in \Gamma\ \forall i \in \overline{1,r}$:

$$\left(\forall \delta \in D_0 : (g \circ \varphi)(\delta)(\gamma)(i) = g(\varphi(\delta))(\gamma)(i) = \int_E S_{i,\gamma} d\varphi(\delta)\right)$$

$$\& \left(g(\mu_0)(\gamma)(i) = \int_E S_{i,\gamma} d\mu_0\right).$$

Therefore by (4.11.48) and (4.11.50) we have $\forall \gamma \in \Gamma\ \exists \delta' \in D_0\quad \forall \delta' \in D_0$:

$$(\delta' \preceq \delta') \Rightarrow \left(\forall i \in \overline{1,r} : |g(\mu_0)(\gamma)(i) - (g \circ \varphi)(\delta')(\gamma)(i)| < \varepsilon^\natural\right). \tag{4.11.51}$$

Consider the combination of (4.11.47), (4.11.49) and (4.11.51). Since $K^\natural \cap \Gamma_0$ and $K^\natural \setminus \Gamma_0$ are finite sets, we infer that $\exists \overline{\delta} \in D_0\ \forall \delta \in D_0$:

$$(\overline{\delta} \preceq \delta) \Rightarrow ((g \circ \varphi)(\delta) \in \overset{\circ}{\mathbb{N}}{}_\Gamma^{(r)} (g(\mu_0), K^\natural, \varepsilon^\natural)).$$

By (4.11.46) we obtain that $\exists \overline{\delta} \in D_0\ \forall \delta \in D_0$:

$$(\overline{\delta} \preceq \delta) \Rightarrow ((g \circ \varphi)(\delta) \in \mathbb{H}_0). \tag{4.11.52}$$

Since the choice of $\mathbb{H}_0$ was arbitrary, we get the convergence

$$(D_0, \preceq, g \circ \varphi) \overset{\mathcal{T}_r^0[\Gamma]}{\to} g(\mu_0). \tag{4.11.53}$$

So (4.11.38) implies (4.11.53). We have established the (local) continuity of $g$ at the point $\mu_0$ in the sense of topologies $\tau^0_{\mathbf{K}}(\mathcal{L})$ and $\mathcal{T}^0_r[\Gamma]$. But the choice of $\mu_0$ was arbitrary. Hence the function $g$ is continuous at each point of the domain. Thus $g$ is a continuous function in the sense of the TS (4.11.37) and (4.11.32). □

PROPOSITION 4.11.3 *The property* $g \in C(\mathbf{K}, \tau^*_{\mathbf{K}}(\mathcal{L}), \mathbf{X}, \otimes^{\Gamma}(\tau^{(r)}_{\mathbb{R}}))$ *is valid.*

PROOF. Let $(D_0, \preceq, \varphi)$ is a net in $\mathbf{K}$. So, $\mathbf{S}[D_0 \neq \varnothing]$, $\preceq \in (\mathrm{DIR})[D_0]$ and $\varphi \in \mathbf{K}^{D_0}$. Let $\mu_0 \in \mathbf{K}$. Finally, suppose that

$$(D_0, \preceq, \varphi) \overset{\tau^*_{\mathbf{K}}(\mathcal{L})}{\to} \mu_0. \tag{4.11.54}$$

Using the definition of $\mathbf{K}$, we infer that $(D_0, \preceq, \varphi)$ is a net in $\mathbf{A}(\mathcal{L})$. By (2.3.9), (4.11.1) and (4.11.54) we have the convergence

$$(D_0, \preceq, \varphi) \overset{\tau_*(\mathcal{L})}{\to} \mu_0. \tag{4.11.55}$$

From (4.6.2), (4.6.7) and (4.11.55) we get that $\forall h \in B(E, \mathcal{L})$:

$$\left(D_0, \preceq, \left(\int_E h\, d\varphi(\delta)\right)_{\delta \in D_0}\right) \overset{\tau_{\mathbb{R}}}{\to} \int_E h\, d\mu_0. \tag{4.11.56}$$

From (4.11.7) and (4.11.56) we have $\forall \gamma \in \Gamma$:

$$\left(D_0, \preceq, \left(\rho_r\left(\left(\int_E S_{i,\gamma} d\varphi(\delta)\right)_{i \in \overline{1,r}}, \left(\int_E S_{i,\gamma} d\mu_0\right)_{i \in \overline{1,r}}\right)\right)_{\delta \in D_0}\right) \overset{\tau_{\mathbb{R}}}{\to} 0. \tag{4.11.57}$$

In (4.11.57) we use the definition of the metric $\rho_r$. From (4.11.7) and (4.11.57) we obtain that $\forall \gamma \in \Gamma$:

$$(D_0, \preceq, (\rho_r((g \circ \varphi)(\delta)(\gamma), g(\mu_0)(\gamma)))_{\delta \in D_0}) \overset{\tau_{\mathbb{R}}}{\to} 0.$$

Therefore by (4.11.10) we have $\forall K \in \mathrm{Fin}(\Gamma)$ $\forall \varepsilon \in ]0, \infty[$ $\exists \delta' \in D_0$ $\forall \delta' \in D_0$:

$$(\delta' \preceq \delta') \Rightarrow ((g \circ \varphi)(\delta') \in \mathbb{N}^{(r)}_{\Gamma}(g(\mu_0), K, \varepsilon)). \tag{4.11.58}$$

But by (2.6.8) and Proposition 4.11.1 we have

$$\{\mathbb{N}^{(r)}_{\Gamma}(g(\mu_0), K, \varepsilon)) : (K, \varepsilon) \in \mathrm{Fin}(\Gamma) \times ]0, \infty[\} \in (g(\mu_0) - \mathrm{bas})[\otimes^{\Gamma}(\tau^{(r)}_{\mathbb{R}})].$$

Therefore from (4.11.58) it follows that (see (2.6.8))

$$(D_0, \preceq, g \circ \varphi) \overset{\otimes^{\Gamma}(\tau^{(r)}_{\mathbb{R}})}{\to} g(\mu_0). \tag{4.11.59}$$

So (4.11.54) implies (4.11.59). Since the choice of $(D_0, \preceq, \varphi)$ and $\mu_0$ was arbitrary, by (2.5.4) we have the required property $g \in C(\mathbf{K}, \tau^*_{\mathbf{K}}(\mathcal{L}), \mathbf{X}, \otimes^{\Gamma}(\tau^{(r)}_{\mathbb{R}}))$. □

In conclusion of the section we note simple corollaries of (4.11.3).

PROPOSITION 4.11.4 $\mathbf{K} = \mathrm{cl}(\mathbf{m}^1(\mathbf{F}), \tau^*_{\mathbf{K}}(\mathcal{L})) = \mathrm{cl}(\mathbf{m}^1(\mathbf{F}), \tau^0_{\mathbf{K}}(\mathcal{L}))$.

PROOF. From (2.3.9), (4.11.1), (4.11.3) and (4.11.35) we have $\forall \mu \in \mathbf{K}$ $\forall \tau \in \{\tau^*_{\mathbf{K}}(\mathcal{L}); \tau^0_{\mathbf{K}}(\mathcal{L})\}$:

$$(\mathbf{D}, \sqsubseteq, \mathbf{m} \circ \xi_\mu) \xrightarrow{\tau} \mu.$$

Hence by (2.3.11) and (4.11.2) we have the required density properties of $\mathbf{m}^1(\mathbf{F})$ in $\mathbf{K}$.

## 4.12 SOME GENERALIZATIONS. II

We continue the constructing of the previous section, following notations of Section 4.11.

So we return to the constructions of Chapter 3, restricting ourselves to consideration of the set $g^{-1}(\mathbf{Y})$ of Section 3.1. Later on we return to the problem connected with determining the set (3.1.12). In addition, we use the property (4.11.30).

We fix the set $\mathbf{Y} \in \mathcal{F}_{\otimes^{\Gamma}(\tau^{(r)}_{\mathbb{R}})}$ and consider the constraints

$$s(f) \in \mathbf{Y} \tag{4.12.1}$$

on the choice of $f \in \mathbf{F}$. In connection with (4.12.1) we note the following obvious

PROPOSITION 4.12.1 $\forall u \in \mathbf{X} \setminus \mathbf{Y}\ \exists z \in \mathrm{Fin}(\Gamma) \times ]0, \infty[$:

$$\Lambda^{(r)}_{\Gamma}(u, z) \cap \left( \bigcup_{y \in \mathbf{Y}} \Lambda^{(r)}_{\Gamma}(y, z) \right) = \varnothing.$$

PROOF. Fix $u \in \mathbf{X} \setminus \mathbf{Y}$. Since $\mathbf{X} \setminus \mathbf{Y} \in \otimes^{\Gamma}(\tau^{(r)}_{\mathbb{R}})$ (see (4.11.9)), by Proposition 4.11.1 we have the property that for some $K' \in \mathrm{Fin}(\Gamma)$ and $\varepsilon' \in ]0, \infty[$ the inclusion $\mathbb{N}^{(r)}_{\Gamma}(u, K', \varepsilon') \subset \mathbf{X} \setminus \mathbf{Y}$ holds. Suppose that

$$(K \triangleq K') \ \&(\varepsilon = \varepsilon'/2).$$

Let $z \triangleq (K, \varepsilon)$. Consider the sets $\Lambda^{(r)}_{\Gamma}(u, z) = \mathbb{N}^{(r)}_{\Gamma}(u, K, \varepsilon)$ and

$$V \triangleq \bigcup_{y \in \mathbf{Y}} \Lambda^{(r)}_{\Gamma}(y, z). \tag{4.12.2}$$

Then $\Lambda_\Gamma^{(r)}(u,z)\cap V = \varnothing$. Indeed, suppose the contrary: $\Lambda_\Gamma^{(r)}(u,z)\cap V \neq \varnothing$. Choose $v \in \Lambda_\Gamma^{(r)}(u,z)\cap V$. By (4.11.10) $v \in \mathbf{X}$ has the property $v(\gamma) \in \mathbb{B}^0_{\rho_r}(u(\gamma),\varepsilon)$ for $\gamma \in K$. This means the following inequality: $\rho_r(u(\gamma),v(\gamma)) < \varepsilon$ for $\gamma \in K$ (see (2.7.4)). Consider the inclusion $v \in V$. Using (4.12.2) we choose $y_1 \in \mathbf{Y}$ for which $v \in \Lambda_\Gamma^{(r)}(y_1,z)$. By (4.11.25) we have

$$v \in \mathbb{N}_\Gamma^{(r)}(y_1,K,\varepsilon). \tag{4.12.3}$$

From(4.11.10) and (4.12.3) it follows that $v(\gamma) \in \mathbb{B}^0_{\rho_r}(y_1(\gamma),\varepsilon)$ for $\gamma \in K$. By (2.7.4) we obtain that $\forall\gamma \in K$: $\rho_r(y_1(\gamma),v(\gamma)) < \varepsilon$. Then by the triangle inequality we have $\forall\gamma \in K$:

$$\begin{aligned}\rho_r(y_1(\gamma),u(\gamma)) &\leq \rho_r(y_1(\gamma),v(\gamma)) + \rho_r(v(\gamma),u(\gamma))\\ &= \rho_r(y_1(\gamma),v(\gamma)) + \rho_r(u(\gamma),v(\gamma))\\ &< 2\varepsilon = \varepsilon'.\end{aligned} \tag{4.12.4}$$

By (2.7.2), (2.7.4) and (4.12.4) we have $\forall\gamma \in K'$: $y_1(\gamma) \in \mathbf{B}^0_{\rho_r}(u(\gamma),\varepsilon')$. Using (4.11.10) we get that $y_1 \in \mathbb{N}_\Gamma^{(r)}(u,K',\varepsilon')$. Then $y_1 \in \mathbf{X}\setminus\mathbf{Y}$, which is impossible. The obtained contradiction shows that $\Lambda_\Gamma^{(r)}(u,z)\cap V = \varnothing$. From (4.12.2) we have the required statement. □

We recall that by (3.1.9) we have the equality $s = g \circ \mathbf{m}$.

COROLLARY 4.12.1 *The following equality holds:*

$$\begin{aligned}&(\tau^*_{\mathbf{K}}(\mathcal{L}) - \mathrm{LIM})[s^{-1}[\mathbb{N}_{\otimes^\Gamma(\tau_{\mathbb{R}}^{(r)})}[\mathbf{Y}]] \mid \mathbf{m}]\\ =&(\tau^*_{\mathbf{K}}(\mathcal{L}) - \mathrm{LIM})[s^{-1}[\mathcal{U}(\mathrm{Fin}(\Gamma)\times]0,\infty[,\otimes^\Gamma(\tau_{\mathbb{R}}^{(r)}),\mathbf{Y} \mid \Lambda_\Gamma^{(r)})] \mid \mathbf{m}].\end{aligned} \tag{4.12.5}$$

PROOF. We use Propositions 3.2.1, 4.11.3, 4.12.1 and (4.11.26). In addition, we take into account the following concrete variant of Proposition 3.2.1: $X = \mathbf{F}$, $Q = \mathrm{Fin}(\Gamma)\times]0,\infty[$, $U = \mathbf{X}$, $V = \mathbf{K}$, $\tau_1 = \otimes^\Gamma(\tau_{\mathbb{R}}^{(r)})$, $\tau_2 = \tau^*_{\mathbf{K}}(\mathcal{L})$, $\mathbf{M} = \mathbf{Y}$, $\Lambda = \Lambda_\Gamma^{(r)}$, $\alpha = s$, and $\beta = \mathbf{m}$. Then (4.12.5) follows from Proposition 4.12.1. □

We use Proposition 3.2.4. But first we recall Proposition 4.11.4.

PROPOSITION 4.12.2 *The following equality holds:*

$$(\tau^*_{\mathbf{K}}(\mathcal{L}) - \mathrm{LIM})[s^{-1}[\mathbb{N}_{\otimes^\Gamma(\tau_{\mathbb{R}}^{(r)})}[\mathbf{Y}]] \mid \mathbf{m}] = g^{-1}(\mathbf{Y}).$$

PROOF. We use the following concrete variant of Proposition 3.2.4: $X = \mathbf{F}$, $U = \mathbf{X}$, $V = \mathbf{K}$, $\tau_1 = \otimes^\Gamma(\tau_{\mathbb{R}}^{(r)})$, $\tau_2 = \tau^*_{\mathbf{K}}(\mathcal{L})$, $\mathbf{M} = \mathbf{Y}$, $p = \mathbf{m}$ and

$q = g$. By Proposition 4.11.4 we have the corresponding analogue of (3.2.37), because $\mathbf{K} = \mathrm{cl}(\mathbf{m}^1(\mathbf{F}), \tau^*_{\mathbf{K}}(\mathcal{L})) = \mathrm{cl}(p^1(X), \tau_2)$.

Note that the corresponding variant of (3.2.44) is correct too. Indeed, consider Proposition 4.12.1. By (3.2.15) and (4.11.25) we have $\forall y \in \mathbf{X}$ $\forall K \in \mathrm{Fin}(\Gamma)$ $\forall \varepsilon \in ]0, \infty[$:

$$\Lambda^{(r)}_\Gamma(y, (K, \varepsilon)) = \mathbb{N}^{(r)}_\Gamma(y, K, \varepsilon) \in N_{\otimes^\Gamma(\tau^{(r)}_{\mathbb{R}})}(y). \tag{4.12.6}$$

Moreover, by (3.2.16), (3.2.18) and (4.11.25) we have $\forall K \in \mathrm{Fin}(\Gamma)$ $\forall \varepsilon \in ]0, \infty[$:

$$\bigcup_{y \in \mathbf{Y}} \Lambda^{(r)}_\Gamma(y, (K, \varepsilon)) \in \mathbb{N}_{\otimes^\Gamma(\tau^{(r)}_{\mathbb{R}})}[\mathbf{Y}]. \tag{4.12.7}$$

Then by Proposition 4.12.1, (4.12.6) and (4.12.7) we get that

$$\forall u \in \mathbf{X} \setminus \mathbf{Y}\ \exists H_1 \in N_{\otimes^\Gamma(\tau^{(r)}_{\mathbb{R}})}(u)\ \exists H_2 \in \mathbb{N}_{\otimes^\Gamma(\tau^{(r)}_{\mathbb{R}})}[\mathbf{Y}] : H_1 \cap H_2 = \varnothing. \tag{4.12.8}$$

We obtain the required concrete variant of (3.2.44). By (3.2.45) we have the basic statement of this proposition.

From Corollary 4.12.1 and Proposition 4.12.2 we have

$$g^{-1}(\mathbf{Y}) = (\tau^*_{\mathbf{K}}(\mathcal{L}) - \mathrm{LIM})[s^{-1}[\mathcal{U}(\mathrm{Fin}(\Gamma) \times ]0, \infty[, \otimes^\Gamma(\tau^{(r)}_{\mathbb{R}}), \mathbf{Y} | \Lambda^{(r)}_\Gamma)] \mid \mathbf{m}]. \tag{4.12.9}$$

Along with (4.12.9) we consider a similar representation connected with another topological equipment of $\mathbf{X}$ and $\mathbf{K}$. We use the topologies $\mathcal{T}^0_r[\Gamma]$ and $\tau^0_{\mathbf{K}}(\mathcal{L})$, respectively. Moreover, we employ Proposition 4.11.2.

PROPOSITION 4.12.3 *The following equality holds:*

$$(\tau^0_{\mathbf{K}}(\mathcal{L}) - \mathrm{LIM})[s^{-1}[\mathbb{N}_{\mathcal{T}^0_r[\Gamma]}[\mathbf{Y}]] \mid \mathbf{m}] = g^{-1}(\mathbf{Y}). \tag{4.12.10}$$

PROOF. Return to Proposition 3.2.4, setting in its conditions: $X = \mathbf{F}$, $U = \mathbf{X}$, $V = \mathbf{K}$, $\tau_1 = \mathcal{T}^0_r[\Gamma]$, $\tau_2 = \tau^0_{\mathbf{K}}(\mathcal{L})$, $\mathbf{M} = \mathbf{Y}$, $p = \mathbf{m}$ and $q = g$. By Proposition 4.11.2 we have the property $q \in C(V, \tau_2, U, \tau_1)$. By Proposition 4.11.4 we have $V = \mathrm{cl}(p^1(X), \tau_2)$. Finally, we use arguments similar to the relations (4.12.6) and (4.12.7) in the proof of Proposition 4.12.2. By (4.11.26) we obtain that (4.12.8) is correct. But by (4.11.30) and (4.12.8) we have

$$\forall u \in \mathbf{X} \setminus \mathbf{Y}\ \exists H_1 \in N_{\mathcal{T}^0_r[\Gamma]}(u)\ \exists H_2 \in \mathbb{N}_{\mathcal{T}^0_r[\Gamma]}[\mathbf{Y}] : H_1 \cap H_2 = \varnothing.$$

As a consequence, by Proposition 3.2.4 we obtain (4.12.10). □

Joining (4.12.9), Propositions 4.12.2 and 4.12.3, we get

$$\begin{aligned} g^{-1}(\mathbf{Y}) &= (\tau^*_{\mathbf{K}}(\mathcal{L}) - \mathrm{LIM})[s^{-1}[\mathcal{U}(\mathrm{Fin}(\Gamma)\times]0,\infty[, \otimes^{\Gamma}(\tau^{(r)}_{\mathbb{R}}), \mathbf{Y} \mid \Lambda^{(r)}_{\Gamma})] \mid \mathbf{m}] \\ &= (\tau^*_{\mathbf{K}}(\mathcal{L}) - \mathrm{LIM})[s^{-1}[\mathbb{N}_{\otimes^{\Gamma}(\tau^{(r)}_{\mathbf{R}})}[\mathbf{Y}]] \mid \mathbf{m}] \\ &= (\tau^0_{\mathbf{K}}(\mathcal{L}) - \mathrm{LIM})[s^{-1}[\mathbb{N}_{\mathcal{T}^0_r[\Gamma]}[\mathbf{Y}]] \mid \mathbf{m}]. \end{aligned} \tag{4.12.11}$$

In (4.12.11) we have the first concrete variant of statements of Chapter 3 about the asymptotic non-sensitivity under perturbation of a part of constraints. Let us discuss some questions connected with (4.12.11). It is useful to note that $g^{-1}(\mathbf{Y})$ determines two 'polar' versions of asymptotic solutions. The first one is characterized by the second expression in (4.12.11). Here some not large collection of neighborhoods is used for weakening of the **Y**-constraints. And what is more, these neighborhoods themselves correspond to a weaker topology of the set **X** (see(4.11.30)). The following principle is realized here. Not many 'large' neighborhoods of **Y** are employed for weakening of the **Y**-constraints. In addition, the topology $\tau^*_{\mathbf{K}}(\mathcal{L})$ satisfying the relation (4.11.36) is used in the second expression of (4.12.11). Different topology $\tau^0_{\mathbf{K}}(\mathcal{L})$ is used in the last relation of (4.12.11). In addition, in this expression all neighborhoods in the TS (4.11.32) are taken for weakening of the **Y**-constraints; in this connection we recall (4.11.30). The different principle is realized here: a lot of 'small' neighborhoods of **Y** are used for weakening of the **Y**-constraints.

We supplement (4.12.11) by some simple statements. From (4.11.30) it follows that

$$\mathbb{N}^0_{\otimes^{\Gamma}(\tau^{(r)}_{\mathbf{R}})}[\mathbf{Y}] \subset \mathbb{N}^0_{\mathcal{T}^0_r[\Gamma]}[\mathbf{Y}]. \tag{4.12.12}$$

This property directly follows from (2.2.12). Using (2.2.13) and (4.12.12), we get that

$$\mathbb{N}_{\otimes^{\Gamma}(\tau^{(r)}_{\mathbf{R}})}[\mathbf{Y}] \subset \mathbb{N}_{\mathcal{T}^0_r[\Gamma]}[\mathbf{Y}]. \tag{4.12.13}$$

From (3.2.8) and (4.12.13) we have the inclusion

$$\begin{aligned} &(\tau^*_{\mathbf{K}}(\mathcal{L}) - \mathrm{LIM})[s^{-1}[\mathbb{N}_{\mathcal{T}^0_r[\Gamma]}[\mathbf{Y}]] \mid \mathbf{m}] \\ &\subset (\tau^*_{\mathbf{K}}(\mathcal{L}) - \mathrm{LIM})[s^{-1}[\mathbb{N}_{\otimes^{\Gamma}(\tau^{(r)}_{\mathbf{R}})}[\mathbf{Y}]] \mid \mathbf{m}]. \end{aligned} \tag{4.12.14}$$

By (4.11.36) we get $\forall H \in \mathcal{P}(\mathbf{K})$: $\mathrm{cl}(H, \tau^0_{\mathbf{K}}(\mathcal{L})) \subset \mathrm{cl}(H, \tau^*_{\mathbf{K}}(\mathcal{L}))$. As a consequence, from (3.2.8) we obtain that $\forall \mathcal{H} \in \mathcal{B}[\mathbf{F}]$:

$$(\tau^0_{\mathbf{K}}(\mathcal{L}) - \mathrm{LIM})[\mathcal{H} \mid \mathbf{m}] \subset (\tau^*_{\mathbf{K}}(\mathcal{L}) - \mathrm{LIM})[\mathcal{H} \mid \mathbf{m}]. \tag{4.12.15}$$

Hence from (4.12.11), (4.12.14) and (4.12.15) we have the inclusion

$$\begin{aligned} g^{-1}(\mathbf{Y}) &\subset (\tau^*_{\mathbf{K}}(\mathcal{L}) - \mathrm{LIM})[s^{-1}[\mathbb{N}_{\mathcal{T}^0_r[\Gamma]}[\mathbf{Y}]] \mid \mathbf{m}] \\ &\subset (\tau^*_{\mathbf{K}}(\mathcal{L}) - \mathrm{LIM})[s^{-1}[\mathbb{N}_{\otimes^\Gamma(\tau^{(r)}_{\mathbf{R}})}[\mathbf{Y}]] \mid \mathbf{m}] = g^{-1}(\mathbf{Y}). \end{aligned}$$

We get the equality

$$g^{-1}(\mathbf{Y}) = (\tau^*_{\mathbf{K}}(\mathcal{L}) - \mathrm{LIM})[s^{-1}[\mathbb{N}_{\mathcal{T}^0_r[\Gamma]}[\mathbf{Y}]] \mid \mathbf{m}]. \tag{4.12.16}$$

Thus we have established that $\forall \tau \in \{\tau^*_{\mathbf{K}}(\mathcal{L}); \tau^0_{\mathbf{K}}(\mathcal{L})\}$:

$$g^{-1}(\mathbf{Y}) = (\tau - \mathrm{LIM})[s^{-1}[\mathbb{N}_{\mathcal{T}^0_r[\Gamma]}[\mathbf{Y}]] \mid \mathbf{m}]. \tag{4.12.17}$$

From (3.2.8), (4.12.11) and (4.12.13) it follows that

$$\begin{aligned} g^{-1}(\mathbf{Y}) &= (\tau^0_{\mathbf{K}}(\mathcal{L}) - \mathrm{LIM})[s^{-1}[\mathbb{N}_{\mathcal{T}^0_r[\Gamma]}[\mathbf{Y}]] \mid \mathbf{m}] \\ &\subset (\tau^0_{\mathbf{K}}(\mathcal{L}) - \mathrm{LIM})[s^{-1}[\mathbb{N}_{\otimes^\Gamma(\tau^{(r)}_{\mathbf{R}})}[\mathbf{Y}]] \mid \mathbf{m}]. \end{aligned}$$

Using (4.12.11) and (4.12.15) we obtain that

$$\begin{aligned} g^{-1}(\mathbf{Y}) &\subset (\tau^0_{\mathbf{K}}(\mathcal{L}) - \mathrm{LIM})[s^{-1}[\mathbb{N}_{\otimes^\Gamma(\tau^{(r)}_{\mathbf{R}})}[\mathbf{Y}]] \mid \mathbf{m}] \\ &\subset (\tau^*_{\mathbf{K}}(\mathcal{L}) - \mathrm{LIM})[s^{-1}[\mathbb{N}_{\otimes^\Gamma(\tau^{(r)}_{\mathbf{R}})}[\mathbf{Y}]] \mid \mathbf{m}] = g^{-1}(\mathbf{Y}). \end{aligned}$$

Thus we have established the equality

$$g^{-1}(\mathbf{Y}) = (\tau^0_{\mathbf{K}}(\mathcal{L}) - \mathrm{LIM})[s^{-1}[\mathbb{N}_{\otimes^\Gamma(\tau^{(r)}_{\mathbf{R}})}[\mathbf{Y}]] \mid \mathbf{m}].$$

As a consequence we have $\forall \tau \in \{\tau^*_{\mathbf{K}}(\mathcal{L}); \tau^0_{\mathbf{K}}(\mathcal{L})\}$:

$$g^{-1}(\mathbf{Y}) = (\tau - \mathrm{LIM})[s^{-1}[\mathbb{N}_{\otimes^\Gamma(\tau^{(r)}_{\mathbf{R}})}[\mathbf{Y}]] \mid \mathbf{m}]. \tag{4.12.18}$$

Consider an analogue of the second expression of (4.12.11). By (4.11.36), (4.12.11) and (4.12.15) we have

$$(\tau^0_{\mathbf{K}}(\mathcal{L}) - \mathrm{LIM})[s^{-1}[\mathcal{U}(\mathrm{Fin}(\Gamma)\times]0,\infty[, \otimes^\Gamma(\tau^{(r)}_{\mathbb{R}}), \mathbf{Y}|\Lambda^{(r)}_\Gamma)] \mid \mathbf{m}] \subset g^{-1}(\mathbf{Y}). \tag{4.12.19}$$

We recall that by (3.2.21), (4.11.26) and (4.12.7) we have the inclusion

$$\mathcal{U}(\mathrm{Fin}(\Gamma)\times]0,\infty[, \otimes^\Gamma(\tau^{(r)}_{\mathbb{R}}), \mathbf{Y} \mid \Lambda^{(r)}_\Gamma) \subset \mathbb{N}_{\otimes^\Gamma(\tau^{(r)}_{\mathbf{R}})}[\mathbf{Y}].$$

As a consequence, by (3.2.8) we obtain that

$$\begin{aligned} &(\tau^0_{\mathbf{K}}(\mathcal{L}) - \mathrm{LIM})[s^{-1}[\mathbb{N}_{\otimes^\Gamma(\tau^{(r)}_{\mathbf{R}})}[\mathbf{Y}]] \mid \mathbf{m}] \\ &\subset (\tau^0_{\mathbf{K}}(\mathcal{L}) - \mathrm{LIM})[s^{-1}[\mathcal{U}(\mathrm{Fin}(\Gamma)\times]0,\infty[, \otimes^\Gamma(\tau^{(r)}_{\mathbb{R}}), \mathbf{Y}|\Lambda^{(r)}_\Gamma)] \mid \mathbf{m}]. \end{aligned} \tag{4.12.20}$$

From (4.12.18) and (4.12.20) we get the inclusion

$$g^{-1}(\mathbf{Y}) \subset (\tau^0_{\mathbf{K}}(\mathcal{L}) - \mathrm{LIM})[s^{-1}[\mathcal{U}(\mathrm{Fin}(\Gamma)\times]0,\infty[,\otimes^{\Gamma}(\tau^{(r)}_{\mathbb{R}}),\mathbf{Y}|\Lambda^{(r)}_{\Gamma})] \mid \mathbf{m}]. \tag{4.12.21}$$

Thus by (4.12.19) and (4.12.21) we obtain the equality

$$(\tau^0_{\mathbf{K}}(\mathcal{L}) - \mathrm{LIM})[s^{-1}[\mathcal{U}(\mathrm{Fin}(\Gamma)\times]0,\infty[,\otimes^{\Gamma}(\tau^{(r)}_{\mathbb{R}}),\mathbf{Y}|\Lambda^{(r)}_{\Gamma})] \mid \mathbf{m}] = g^{-1}(\mathbf{Y}).$$

Using (4.12.11) we have $\forall \tau \in \{\tau^*_{\mathbf{K}}(\mathcal{L}); \tau^0_{\mathbf{K}}(\mathcal{L})\}$:

$$g^{-1}(\mathbf{Y}) = (\tau - \mathrm{LIM})[s^{-1}[\mathcal{U}(\mathrm{Fin}(\Gamma)\times]0,\infty[,\otimes^{\Gamma}(\tau^{(r)}_{\mathbb{R}}),\mathbf{Y}|\Lambda^{(r)}_{\Gamma})] \mid \mathbf{m}]. \tag{4.12.22}$$

From (4.12.17), (4.12.18) and (4.12.22) we have the statement about the 'broad' universality of the set $g^{-1}(\mathbf{Y})$. Of course, the statement is connected with the topological regularization of the admissible set. But this statement can be extended to the case of the attainable sets.

We introduce a nonempty set $\mathbf{H}$ and a topology $\boldsymbol{\tau} \in (\mathrm{top})_0[\mathbf{H}]$. We use the TS $(\mathbf{H}, \boldsymbol{\tau})$ in place of $(\mathbf{H}, \tau^{(2)})$ of Chapter 3. Moreover, we have (see (3.1.1)) a mapping $h \in \mathbf{H}^{\mathbf{F}}$. Finally, in correspondence with the general statement of Chapter 3 we have

$$\omega \in C(\mathbf{K}, \tau^*_{\mathbf{K}}(\mathcal{L}), \mathbf{H}, \boldsymbol{\tau}), \tag{4.12.23}$$

for which $h = \omega \circ \mathbf{m}$. Here we choose $\tau^*_{\mathbf{K}}(\mathcal{L})$ from the set $\{\tau^*_{\mathbf{K}}(\mathcal{L}); \tau^0_{\mathbf{K}}(\mathcal{L})\}$; in addition, we use arguments which are employed in practical problems and, sometimes, convenient in theoretical respect. First we give the following natural

**Example.** Consider a variant of (4.12.23), setting that $\mathbf{H} = \mathbb{R}^n$ for some $n \in \mathcal{N}$. Certainly, we suppose here that $\boldsymbol{\tau}$ is the natural topology of coordinate-wise convergence in $\mathbb{R}^n$. Fix $m \in \mathcal{N}$ and

$$(\tilde{\xi}_i)_{i\in\overline{1,m}} : \overline{1,m} \to B(E, \mathcal{L}).$$

Finally, we fix the mapping

$$\zeta : \mathbb{R}^m \to \mathbb{R}^n.$$

We suppose that $\zeta$ is the operator continuous relative to the topologies of coordinate-wise convergence of the spaces $\mathbb{R}^m$ and $\mathbb{R}^n$, respectively. Let $\omega$ be defined by the condition

$$\omega(\mu) \triangleq \zeta\left(\left(\int_E \tilde{\xi}_i \, d\mu\right)_{i\in\overline{1,m}}\right) \tag{4.12.24}$$

for $\mu \in \mathbf{K}$. So $\omega$ can be regarded as the mapping

$$\mu \mapsto \zeta\Big(\int_E \tilde{\xi}_1 \, d\mu, \ldots, \int_E \tilde{\xi}_m \, d\mu\Big) : \mathbf{K} \to \mathbb{R}^n. \tag{4.12.25}$$

The variant (4.12.24), (4.12.25) is similar to the construction of [35, p.74] and is typical for many applied settings of control problems. In particular, the scheme for extension of the problem of constructing the attainability domain in a linear system (see Chapter 1) can be considered in terms of (4.12.24) and (4.12.25). Of course, (4.12.23) is correct in this case, since in (4.12.24), (4.12.25) we have the 'universal' superposition of two continuous mappings.

The choice of $\tau^*_{\mathbf{K}}(\mathcal{L})$ in (4.12.23) convenient in theoretical respect is connected with the following circumstance: (4.12.23) is a continuous mapping from a compact space into a Hausdorff space. Recall that in this case

$$\omega \in C_{\mathbf{ap}}(\mathbf{K}, \tau^*_{\mathbf{K}}(\mathcal{L}), \mathbf{H}, \boldsymbol{\tau}); \tag{4.12.26}$$

see (2.8.7). In addition, (4.12.26) implies several useful properties of the considered problem. We recall that by the scheme of Section 3.1 the replacement

$$h^1(s^{-1}(\mathbf{Y})) \to \omega^1(g^{-1}(\mathbf{Y}))$$

is a concrete effect of the extension. From (4.12.26) we have the following statement.

THEOREM 4.12.1 *The following equalities hold:*

$$\begin{aligned}
\omega^1(g^{-1}(\mathbf{Y})) &= (\boldsymbol{\tau} - \mathrm{LIM})[s^{-1}[\mathbb{N}_{\mathcal{T}^0_r[\Gamma]}[\mathbf{Y}]] \mid h] \\
&= (\boldsymbol{\tau} - \mathrm{LIM})[s^{-1}[\mathbb{N}_{\otimes^\Gamma(\tau^{(r)}_{\mathbb{R}})}[\mathbf{Y}]] \mid h] \\
&= (\boldsymbol{\tau} - \mathrm{LIM})[s^{-1}[\mathcal{U}(\mathrm{Fin}(\Gamma) \times ]0, \infty[, \otimes^\Gamma(\tau^{(r)}_{\mathbb{R}}), \mathbf{Y} \mid \Lambda^{(r)}_\Gamma)] \mid h].
\end{aligned}$$

The proof is an obvious corollary of the general statements of Chapter 3. In this connection see (4.12.26) and Theorem 3.5.1.

We note that by Proposition 4.11.3 and the definition of $\mathbf{Y}$ we have $g^{-1}(\mathbf{Y}) \in \mathcal{F}_\tau$ for $\tau = \tau^*_{\mathbf{K}}(\mathcal{L})$; see (2.5.2). From (4.11.1) it follows that $g^{-1}(\mathbf{Y}) \in (\tau^*_{\mathbf{K}}(\mathcal{L}) - \mathrm{comp})[\mathbf{K}]$. Thus

$$\tau^*_{\mathbf{K}}(\mathcal{L}) \mid_{g^{-1}(\mathbf{Y})} \in (\mathbf{c} - \mathrm{top})[g^{-1}(\mathbf{Y})].$$

Using (4.11.1) we get that $\tau_*(\mathcal{L}) \mid_{g^{-1}(\mathbf{Y})} \in (\mathbf{c} - \mathrm{top})[g^{-1}(\mathbf{Y})]$. As a consequence

$$g^{-1}(\mathbf{Y}) \in (\tau_*(\mathcal{L}) - \mathrm{comp})[\mathbf{A}(\mathcal{L})].$$

Returning to the compactness of $g^{-1}(\mathbf{Y})$ in the sense of (4.11.1), we note that (4.12.23) implies the property

$$\omega^1(g^{-1}(\mathbf{Y})) \in (\boldsymbol{\tau} - \mathrm{comp})[\mathbf{H}]. \tag{4.12.27}$$

By (4.12.27) and Theorem 4.12.1 we have the universal compact AS. Some constructions of Sections 3.5 and 3.6 can be used for this AS. We restrict ourselves to consideration of Proposition 3.6.1 and its corollaries. To apply this proposition we note some analogues of (4.12.27).

PROPOSITION 4.12.4 *If* $U \in \mathcal{P}(\mathbf{F})$, *then* $\mathrm{cl}(h^1(U), \boldsymbol{\tau}) \in (\boldsymbol{\tau} - \mathrm{comp})[\mathbf{H}]$.

PROOF. Fix $U \in \mathcal{P}(\mathbf{F})$. Then $h^1(U) = \omega^1(\mathbf{m}^1(U))$ and hence

$$\mathrm{cl}(h^1(U), \boldsymbol{\tau}) = \mathrm{cl}(\omega^1(\mathbf{m}^1(U)), \boldsymbol{\tau}) = \omega^1(\mathrm{cl}(\mathbf{m}^1(U), \tau^*_{\mathbf{K}}(\mathcal{L}))). \tag{4.12.28}$$

In (4.12.28) we use the fact that $\omega \in C_{\mathrm{cl}}(\mathbf{K}, \tau^*_{\mathbf{K}}(\mathcal{L}), \mathbf{H}, \boldsymbol{\tau})$, following from (2.8.2) and (4.12.26); as a consequence (2.8.4) implies that

$$\omega^1(\mathrm{cl}(V, \tau^*_{\mathbf{K}}(\mathcal{L}))) = \mathrm{cl}(\omega^1(V), \boldsymbol{\tau}) \tag{4.12.29}$$

for $V \in \mathcal{P}(\mathbf{K})$. In (4.12.29) we suppose that $V = \mathbf{m}^1(U)$. Then we obtain (4.12.28). We have

$$\mathrm{cl}(\mathbf{m}^1(U), \tau^*_{\mathbf{K}}(\mathcal{L})) \in \mathcal{F}_{\tau^*_{\mathbf{K}}(\mathcal{L})}. \tag{4.12.30}$$

From (4.11.1) and (4.12.30) it follows that

$$\mathrm{cl}(\mathbf{m}^1(U), \tau^*_{\mathbf{K}}(\mathcal{L})) \in (\tau^*_{\mathbf{K}}(\mathcal{L}) - \mathrm{comp})[\mathbf{K}]. \tag{4.12.31}$$

From (4.11.23) and (4.12.31) we obtain that $\omega^1(\mathrm{cl}(\mathbf{m}^1(U), \tau^*_{\mathbf{K}}(\mathcal{L}))) \in (\boldsymbol{\tau} - \mathrm{comp})[\mathbf{H}]$. From (4.12.28) the required property $\mathrm{cl}(h^1(U), \boldsymbol{\tau}) \in (\boldsymbol{\tau} - \mathrm{comp})[\mathbf{H}]$ follows. □

From Proposition 4.12.4 we have $\forall \mathcal{U} \in \mathcal{B}[\mathbf{F}]\ \forall U \in \mathcal{U}$:

$$\mathrm{cl}(h^1(U), \boldsymbol{\tau}) \in (\boldsymbol{\tau} - \mathrm{comp})[\mathbf{H}]. \tag{4.12.32}$$

Note that the families

$$s^{-1}[\mathbb{N}_{\mathcal{T}^0_r[\Gamma]}[\mathbf{Y}]], \quad s^{-1}[\mathbb{N}_{\otimes^\Gamma(\tau^{(r)}_{\mathbb{R}})}[\mathbf{Y}]],$$

$$s^{-1}[\mathcal{U}(\mathrm{Fin}(\Gamma) \times ]0, \infty[, \otimes^\Gamma(\tau^{(r)}_{\mathbb{R}}), \mathbf{Y} \mid \Lambda^{(r)}_\Gamma)]$$

can be used in place of $\mathcal{U}$. From (4.12.32) and Theorem 4.12.1 we get several natural concrete variants of (3.6.4):

$$\left(\mathbb{N}_{\boldsymbol{\tau}}[\omega^1(g^{-1}(\mathbf{Y}))] \subset \bigcup_{U \in \mathbb{N}_{\mathcal{T}^0_r[\Gamma]}[\mathbf{Y}]} \mathbb{N}_{\boldsymbol{\tau}}[\mathrm{cl}(h^1(s^{-1}(U)), \boldsymbol{\tau})]\right)$$

$$\& \left( \mathbb{N}_{\boldsymbol{\tau}}[\omega^1(g^{-1}(\mathbf{Y}))] \subset \bigcup_{U \in \mathbb{N}_{\otimes^{\Gamma}(\tau_{\mathbf{R}}^{(r)})}[\mathbf{Y}]} \mathbb{N}_{\boldsymbol{\tau}}[\mathrm{cl}(h^1(s^{-1}(U)), \boldsymbol{\tau})] \right)$$

$$\& \left( \mathbb{N}_{\boldsymbol{\tau}}[\omega^1(g^{-1}(\mathbf{Y}))] \subset \bigcup_{U \in \mathcal{U}(\mathrm{Fin}(\Gamma) \times ]0,\infty[, \otimes^{\Gamma}(\tau_{\mathbf{R}}^{(r)}), \mathbf{Y} | \Lambda_{\Gamma}^{(r)})} \mathbb{N}_{\boldsymbol{\tau}}[\mathrm{cl}(h^1(s^{-1}(U)), \boldsymbol{\tau})] \right). \tag{4.12.33}$$

In (4.12.33) we have the natural possibility of the neighborhood realization of the AS $\omega^1(g^{-1}(\mathbf{Y}))$.

In conclusion of this chapter we consider the question of the structure of approximate solution nets realizing AS (4.12.27). Recall that $(\mathbf{D}, \sqsubseteq)$ and the mapping (4.11.2) define some class of nets in $\mathbf{F}$. For brevity we denote by $\Xi$ the mapping (4.11.2).So

$$\Xi : \mathbf{K} \to \mathbf{F}^{\mathbf{D}}$$

is defined by the equality $\Xi(\mu) \triangleq \xi_\mu$ for $\mu \in \mathbf{K}$. Consider the set

$$\Xi^1(g^{-1}(\mathbf{Y})) = \{\xi_\mu : \mu \in g^{-1}(\mathbf{Y})\} \in \mathcal{P}(\mathbf{F}^{\mathbf{D}}). \tag{4.12.34}$$

Then for $\tilde{\xi} \in \Xi^1(g^{-1}(\mathbf{Y}))$ we have the net $(\mathbf{D}, \sqsubseteq, \tilde{\xi})$ in $\mathbf{F}$.

PROPOSITION 4.12.5 *If $\mu \in g^{-1}(\mathbf{Y})$, then the net $(\mathbf{D}, \sqsubseteq, \xi_\mu)$ in $\mathbf{F}$ has the following properties:*

$$\Big( s^{-1}[\mathbb{N}_{\mathcal{T}_r^0[\Gamma]}[\mathbf{Y}]] \cup s^{-1}[\mathbb{N}_{\otimes^{\Gamma}(\tau_{\mathbf{R}}^{(r)})}[\mathbf{Y}]]$$
$$\subset (\mathbf{F} - \mathrm{ass})[\mathbf{D}; \sqsubseteq; \xi_\mu]\Big) \,\&\, \Big( (\mathbf{D}, \sqsubseteq, h \circ \xi_\mu) \xrightarrow{\boldsymbol{\tau}} \omega(\mu) \Big).$$

PROOF. Fix $\mu \in g^{-1}(\mathbf{Y})$. Then $\xi_\mu = \Xi(\mu) \in \mathbf{F}^{\mathbf{D}}$ and hence $(\mathbf{D}, \sqsubseteq, \xi_\mu)$ is a net in $\mathbf{F}$. And what is more, by (4.11.3) we have the 'universal' convergence

$$\forall \tau \in \mathfrak{M}(\mathcal{L}) : (\mathbf{D}, \sqsubseteq, \mathbf{m} \circ \xi_\mu) \xrightarrow{\tau} \mu. \tag{4.12.35}$$

In addition, $\mathbf{m} \circ \xi_\mu \in \mathbf{K}^{\mathbf{D}}$. Then by (4.8.16) and (4.12.35) we have

$$(\mathbf{D}, \sqsubseteq, \mathbf{m} \circ \xi_\mu) \xrightarrow{\tau_*(\mathcal{L})} \mu.$$

Using (2.3.9) and (4.11.1) we obtain the convergence

$$(\mathbf{D}, \sqsubseteq, \mathbf{m} \circ \xi_\mu) \xrightarrow{\tau_{\mathbf{K}}^*(\mathcal{L})} \mu.$$

By (2.5.4) and (4.12.22) we have the property

$$(\mathbf{D}, \sqsubseteq, \omega \circ (\mathbf{m} \circ \xi_\mu)) \xrightarrow{\boldsymbol{\tau}} \omega(\mu).$$

But by the above representation of $h$ we get $\omega \circ (\mathbf{m} \circ \xi_\mu) = \omega \circ \mathbf{m} \circ \xi_\mu = (\omega \circ \mathbf{m}) \circ \xi_\mu = h \circ \xi_\mu$, and

$$(\mathbf{D}, \sqsubseteq, h \circ \xi_\mu) \overset{\mathcal{T}}{\rightarrow} \omega(\mu). \qquad (4.12.36)$$

In (4.12.36) we have the second part of our proposition. Consider the proof of the first part. In this connection we recall that by (4.8.16) and (4.12.35) the convergence

$$(\mathbf{D}, \sqsubseteq, \mathbf{m} \circ \xi_\mu) \overset{\tau_0(\mathcal{L})}{\rightarrow} \mu \qquad (4.12.37)$$

holds. By (2.3.9) and (4.11.36) we have

$$(\mathbf{D}, \sqsubseteq, \mathbf{m} \circ \xi_\mu) \overset{\tau^0_{\mathbf{K}}(\mathcal{L})}{\rightarrow} \mu.$$

From Proposition 4.11.2 and (2.5.4) it follows that

$$(\mathbf{D}, \sqsubseteq, g \circ (\mathbf{m} \circ \xi_\mu)) \overset{\mathcal{T}^0_r[\Gamma]}{\rightarrow} g(\mu).$$

But $g \circ (\mathbf{m} \circ \xi_\mu) = g \circ \mathbf{m} \circ \xi_\mu = (g \circ \mathbf{m}) \circ \xi_\mu = s \circ \xi_\mu$. As a consequence

$$(\mathbf{D}, \sqsubseteq, s \circ \xi_\mu) \overset{\mathcal{T}^0_r[\Gamma]}{\rightarrow} g(\mu). \qquad (4.12.38)$$

In addition, $g(\mu) \in \mathbf{Y}$. Therefore

$$\mathbb{N}_{\mathcal{T}^0_r[\Gamma]}[\mathbf{Y}] \subset N_{\mathcal{T}^0_r[\Gamma]}(g(\mu)).$$

From (2.2.48) and (4.12.38) we have

$$\mathbb{N}_{\mathcal{T}^0_r[\Gamma]}[\mathbf{Y}] \subset (\mathbf{X} - \text{ass})[\mathbf{D}; \sqsubseteq; s \circ \xi_\mu]. \qquad (4.12.39)$$

From (2.2.47), (2.5.8) and (4.12.39) we get

$$s^{-1}[\mathbb{N}_{\mathcal{T}^0_r[\Gamma]}[\mathbf{Y}]] \subset (\mathbf{F} - \text{ass})[\mathbf{D}; \sqsubseteq; \xi_\mu]. \qquad (4.12.40)$$

We supplement (4.12.40) by some analogues connected with Proposition 4.11.3. We recall the convergence of $(\mathbf{D}, \sqsubseteq, \mathbf{m} \circ \xi_\mu)$ in the sense of $\tau^*_{\mathbf{K}}(\mathcal{L})$. From Proposition 4.11.3 it follows that

$$(\mathbf{D}, \sqsubseteq, g \circ (\mathbf{m} \circ \xi_\mu)) \overset{\otimes^\Gamma(\tau^{(r)}_{\mathbb{R}})}{\rightarrow} g(\mu).$$

As a consequence we get the convergence

$$(\mathbf{D}, \sqsubseteq, s \circ \xi_\mu)) \overset{\otimes^\Gamma(\tau^{(r)}_{\mathbb{R}})}{\rightarrow} g(\mu). \qquad (4.12.41)$$

By the choice of $\mu$ we have the inclusion

$$\mathbb{N}_{\otimes^{\Gamma}(\tau_{\mathbb{R}}^{(r)})}[\mathbf{Y}] \subset N_{\otimes^{\Gamma}(\tau_{\mathbb{R}}^{(r)})}(g(\mu)).$$

Therefore from (2.2.48) and (4.12.41) we get

$$\mathbb{N}_{\otimes^{\Gamma}(\tau_{\mathbb{R}}^{(r)})}[\mathbf{Y}] \subset (\mathbf{X} - \mathrm{ass})[\mathbf{D}; \sqsubseteq; s \circ \xi_{\mu}]. \tag{4.12.42}$$

From (2.2.47), (2.5.8) and (4.12.42) it follows that

$$s^{-1}[\mathbb{N}_{\otimes^{\Gamma}(\tau_{\mathbb{R}}^{(r)})}[\mathbf{Y}]] \subset (\mathbf{F} - \mathrm{ass})[\mathbf{D}; \sqsubseteq; \xi_{\mu}].$$

Using (4.12.40) we obtain the inclusion

$$s^{-1}[\mathbb{N}_{\mathcal{T}_r^0[\Gamma]}[\mathbf{Y}]] \cup s^{-1}[\mathbb{N}_{\otimes^{\Gamma}(\tau_{\mathbb{R}}^{(r)})}[\mathbf{Y}]] \subset (\mathbf{F} - \mathrm{ass})[\mathbf{D}; \sqsubseteq; \xi_{\mu}]. \tag{4.12.43}$$

From (4.12.36) and (4.12.43) we get the required statement. □

COROLLARY 4.12.2 *For $\mu \in g^{-1}(\mathbf{Y})$ we have*

$$s^{-1}[\mathcal{U}(\mathrm{Fin}(\Gamma) \times ]0, \infty[, \otimes^{\Gamma}(\tau_{\mathbb{R}}^{(r)}), \mathbf{Y} \mid \Lambda_{\Gamma}^{(r)})] \subset (\mathbf{F} - \mathrm{ass})[\mathbf{D}; \sqsubseteq; \xi_{\mu}].$$

PROOF. We recall (4.11.27). Combining (3.2.19) and (4.11.26), we obtain that

$$\mathcal{U}(\mathrm{Fin}(\Gamma) \times ]0, \infty[, \otimes^{\Gamma}(\tau_{\mathbb{R}}^{(r)}), \mathbf{Y} \mid \Lambda_{\Gamma}^{(r)}) \subset \mathbb{N}_{\otimes^{\Gamma}(\tau_{\mathbb{R}}^{(r)})}[\mathbf{Y}].$$

As a consequence, by (2.5.8) and Proposition 4.12.5 we have

$$\begin{aligned} s^{-1}[\mathcal{U}(\mathrm{Fin}(\Gamma) \times ]0, \infty[, \otimes^{\Gamma}(\tau_{\mathbb{R}}^{(r)}), \mathbf{Y} \mid \Lambda_{\Gamma}^{(r)})] &\subset s^{-1}[\mathbb{N}_{\otimes^{\Gamma}(\tau_{\mathbb{R}}^{(r)})}[\mathbf{Y}]] \\ &\subset (\mathbf{F} - \mathrm{ass})[\mathbf{D}; \sqsubseteq; \xi_{\mu}]. \end{aligned}$$

Note that owing to Proposition 4.12.4 and Corollary 4.12.2 we can regard (4.12.34) as some sufficient set of approximate solutions. Such solutions are defined as the triplets $(\mathbf{D}, \sqsubseteq, \hat{\xi})$, $\hat{\xi} \in \Xi^1(g^{-1}(\mathbf{Y}))$. Varying $\hat{\xi}$, we realize all the AS $\omega^1(g^{-1}(\mathbf{Y}))$ under the validity of the corresponding 'asymptotic constraints'. In fact, (4.12.34) defines the concrete type of approximate solution nets realizing the 'universal' AS $\omega^1(g^{-1}(\mathbf{Y}))$. Recall that three concrete variants of (4.12.34) can be extracted from the definitions of Sections 4.8–4.10 (see, in particular, Theorems 4.8.1, 4.9.1 and Proposition 4.10.28). Certainly, using (4.12.34) we do not exclude employment of other types of approximate solutions. For example, in many cases it is sufficient to use only sequences of usual solutions (see [35, p. 38]. Note only that Proposition 4.12.4 and Corollary 4.12.2 define a highly general variant of asymptotic realization of the AS $\omega^1(g^{-1}(\mathbf{Y}))$. This variant is coordinated with the conditions determining the considered class of problems. In this connection it is useful to return to (4.11.3).

Note the following fact. If $(\mathbf{D}, \sqsubseteq)$, where $\mathbf{S}[\mathbf{D} \neq \varnothing]$ and $\sqsubseteq \in (\mathrm{DIR})[\mathbf{D}]$, and the mapping (4.11.2) are fixed, then (4.11.3) will be valid for any $\mu \in \mathbf{K}$ and $\tau \in \mathfrak{M}(\mathcal{L})$ if

$$\forall \mu \in \mathbf{K} : (\mathbf{D}, \sqsubseteq, \mathbf{m} \circ \xi_\mu)) \overset{\tau_0(\mathcal{L})}{\to} \mu. \qquad (4.12.44)$$

Indeed, let us suppose that only (4.12.44) is valid. We use the stipulations (4.11.36) and (4.11.37). More exactly, we follow (4.11.36) and, by properties of the type (4.6.34), we have by the definition of $\tau^*_{\mathbf{K}}(\mathcal{L})$ that (see (4.11.1))

$$\tau^*_{\mathbf{K}}(\mathcal{L}) = \tau_*(\mathcal{L}) \mid_{\mathbf{K}} .$$

Fix $\mu \in \mathbf{K}$. Then from (2.3.9), (4.11.36) and (4.12.44) we get the convergence

$$(\mathbf{D}, \sqsubseteq, \mathbf{m} \circ \xi_\mu) \overset{\tau^0_{\mathbf{K}}(\mathcal{L})}{\to} \mu.$$

Using (4.6.19) and (4.11.37) we obtain that

$$(\mathbf{D}, \sqsubseteq, \mathbf{m} \circ \xi_\mu) \overset{\tau^*_{\mathbf{K}}(\mathcal{L})}{\to} \mu.$$

From (2.3.9) and (4.11.37) it follows that

$$((\mathbf{D}, \sqsubseteq, \mathbf{m} \circ \xi_\mu) \overset{\tau_*(\mathcal{L})}{\to} \mu) \;\&\; ((\mathbf{D}, \sqsubseteq, \mathbf{m} \circ \xi_\mu) \overset{\tau_\otimes(\mathcal{L})}{\to} \mu).$$

Finally, by (4.7.5) we have the equality

$$\tau^*_{\mathbf{K}}(\mathcal{L}) = \tau_*(\mathcal{L}) \mid_{\mathbf{K}} = \tau^*_{\mathbb{B}}(\mathcal{L}) \mid_{\mathbf{K}} .$$

Therefore we obtain the obvious convergence

$$(\mathbf{D}, \sqsubseteq, \mathbf{m} \circ \xi_\mu) \overset{\tau^*_{\mathbb{B}}(\mathcal{L})|_{\mathbf{K}}}{\longrightarrow} \mu.$$

Again use (2.3.9). As a consequence we obtain the convergence

$$(\mathbf{D}, \sqsubseteq, \mathbf{m} \circ \xi_\mu) \overset{\tau^*_{\mathbb{B}}(\mathcal{L})}{\longrightarrow} \mu.$$

Taking into account (4.8.16), (4.12.44) and the consequent properties, we get (4.11.3) for any topology of $\mathfrak{M}(\mathcal{L})$. Since the choice of $\mu$ was arbitrary, we obtain the important property that (4.12.44) is sufficient for realization of the basic approximate condition.

## 4.13 A PARTICULAR CASE OF CONTROL PROBLEM

We return to the problem of control by a mass point under integral constraints, which was considered in Chapter 1. So we examine the concrete differential equation (1.2.1). Constraints (1.2.4) are regarded as basic ones. For clarity we consider the first part of (1.2.5); i.e., we investigate the attainability domain relative to geometric coordinates under constraints (1.2.4). For simplicity we preserve the stipulation of Section 1.2 with respect to $m_1, \ldots, m_r$ and $f$. In fact, for this case the corresponding asymptotic setting is given in Chapter 1. The most important moment is connected with the asymptotic equivalence of constraints (1.2.14) and (1.2.20). Here we use (1.2.14) as a variant of the scheme used in Section 4.11 and connected with the mapping (4.11.25). Constraint (1.2.14) is of interest from the practical point of view. Of course, we use some singularities of the **Y**-constraint connected with the phase constraint of Section 1.2.

Under the conditions of Section 1.2 we introduce the following particular case of the MS $(E, \mathcal{L})$ of Section 4.2. In this section we suppose that

$$E \triangleq I = [0, \vartheta_0[$$

and $\mathcal{L} \triangleq \{[\mathrm{pr}_1(z), \mathrm{pr}_2(z)[ : z \in I_0 \times I_0\}$, where $I_0 = [0, \vartheta_0]$. Thus $(E, \mathcal{L}) = (I, \mathcal{L})$ is the space pointer. In correspondence with Section 4.9 we fix $\eta \in (\mathrm{add})_+[\mathcal{L}]$. We choose the simplest variant, setting $\eta([a, b[) = b - a$ for $a \in I_0$ and $b \in I_0$ with $a \leq b$. Thus $\eta$ is the length function. We do not use the corresponding $\sigma$-additivity property for $\eta$ (of course, we can regard $\eta$ as the trace of the Lebesgue measure). So we have the MS $(E, \mathcal{L}, \eta)$ of Section 4.9. In particular, we will consider the variant of constructions of Sections 4.11 and 4.12.

It is worth noting that in our case the set $F$ of Section 1.2 coincides with $B_0(E, \mathcal{L}) = B_0(I, \mathcal{L})$, where $I = [0, \vartheta_0[$. We fix $c \in [0, \infty[$ as some recourse constant. So the number $c$ defines the available fuel. Recall that the phase constraints is defined by the mapping

$$t \mapsto N_t : I_0 \to \mathcal{K}_r \tag{4.13.1}$$

(see Section 1.2). In addition, the mapping(4.13.1) is continuous in the sense of the Hausdorff metric of $\mathcal{K}_r$. So (4.13.1) is a 'good' compact-valued mapping. Observe that

$$\mathcal{K}_r = (\tau_{\mathbb{R}}^{(r)} - \mathrm{comp})[\mathbb{R}^r] \setminus \{\varnothing\}.$$

Using (4.13.1) we introduce the set-valued mapping

$$t \mapsto \tilde{N}_t : I_0 \to \mathcal{K}_r \tag{4.13.2}$$

by the relation (1.2.18). Then we arrive at the constraints of the type (1.2.20). We use (4.13.2) for representation of the following conditions: if $K \in \mathrm{Fin}(I_0)$ and $\alpha \in ]0, \infty[$, then we consider the requirement

$$\left(\int_0^t (t-\tau)m(\tau)f(\tau)d\tau\right)_{t\in K} \in \prod_{t\in K} \tilde{N}_t^{[\alpha]}. \tag{4.13.3}$$

In (1.2.20) we have the equivalent variant of (4.13.3). On the other hand, in (1.2.20) we have some analogue of the constraint defined in Section 4.12 in terms of (4.12.7). To consider this analogue we give a concrete variant of the constructions of Sections 4.11 and 4.12 for the above MS $(E, \mathcal{L}, \eta)$.

Suppose that $\mathcal{R} = \mathbb{R}^r$ as in Section 4.11. Moreover, in this section assume that $\Gamma = I_0 = [0, \vartheta_0]$. As in the general part, $\mathbf{X}$ is the set of all mappings from $I_0 = \Gamma$ into $\mathcal{R}$. The concrete variant of the generalized matriciant (4.11.5) is defined as follows:

$$S_{i,\gamma} = \left((\gamma - \tau)m_i(\tau)\chi_{[0,\gamma[}(\tau)\right)_{\tau\in I}, \tag{4.13.4}$$

where $i \in \overline{1,r}$ and $\gamma \in I_0$. In connection with (4.13.4) recall that all integrals of vector functions in the integral representations of Section 1.2 are defined coordinate-wise. Note that by (4.11.7) and (4.13.4) we can give the corresponding variant of the generalized operator $g$. But we first note that the set $\mathbf{K}$ is defined here as the ball $U_c(\mathcal{L})$: $\mathbf{K} = U_c(\mathcal{L})$. In addition, we observe that $\mathbf{A}_\eta[\mathcal{L}] = \mathbf{A}(\mathcal{L})$ for our space $(E, \mathcal{L}, \eta)$ (see (4.9.4)). Therefore in our case we 'automatically' have

$$\mathbf{K} = U_c(\mathcal{L}) \cap \mathbf{A}_\eta[\mathcal{L}].$$

We now can introduce the concrete variant of (4.11.6):

$$g : U_c(\mathcal{L}) \to \mathcal{R}^\Gamma.$$

By (4.13.4) we have $\forall \mu \in U_c(\mathcal{L})\ \forall \gamma \in I_0$:

$$g(\mu)(\gamma) = \left(\int_{[0,\gamma[} (\gamma - \tau)m_i(\tau)\mu(d\tau)\right)_{i\in\overline{1,r}}. \tag{4.13.5}$$

Interpreting the vector integral in (4.13.5) as the integral of a vector function of Section 1.2, we obtain for $\mu \in U_c(\mathcal{L})$ and $t \in I_0$ the equality

$$g(\mu)(t) = \int_{[0,t[} (t-\tau)m(\tau)\mu(d\tau). \tag{4.13.6}$$

In (4.13.6) we have a natural analogue of the relations used in (1.2.20) and (1.2.21). The integral of the vector function in (4.13.6) is defined coordinate-wise. We can use the Hausdorff topology (4.11.9) and the neighborhood-valued mapping (4.11.25). Also it is possible to consider the topology (4.11.32). But we are not interested in this construction, although Proposition 4.11.2 remains valid. In our constructions Proposition 4.11.3 is more important. We define the set $\mathbf{Y}$ of Section 4.12 as follows: $\mathbf{Y} = \tilde{\mathbb{N}}$ (see Section 1.2). In this part we note that $\forall t \in I_0 : \tilde{N}_t \in \mathcal{K}_r$. In particular, for $t \in I_0$ the set $\tilde{N}_t$ is closed in the space $(\mathcal{R}, \tau_{\mathbb{R}}^{(r)})$. As a consequence

$$\tilde{\mathbb{N}} = \prod_{t \in I_0} \tilde{N}_t \in \mathcal{F}_{\otimes^{\Gamma}(\tau_{\mathbb{R}}^{(r)})}. \tag{4.13.7}$$

From (4.13.7) we get the basic requirement connected with (4.12.1). Chiefly we consider the asymptotic version of our problem. But in connection with (4.13.7) the consideration of the non-perturbed problem is of interest. By (4.13.6) we use the generalized constraints in the form

$$\left(\int_{[0,t[} (t-\tau)m(\tau)\mu(d\tau)\right)_{t \in I_0} \in \tilde{\mathbb{N}}. \tag{4.13.8}$$

Note a variant of the mapping $\mathbf{m}$ as the natural restriction of the mapping (4.9.1). Of course, we keep in mind Theorem 4.9.3. Thus we here suppose that $\mathbf{m}$ is the mapping from $\mathbf{F}$, defined as the set of all $f \in F$ such that

$$\int_0^{\vartheta_0} | f(t) | \, dt = \int_I | f | \, d\eta \leq c,$$

into $U_c(\mathcal{L}) = \mathbf{K}$ with the values $f * \eta$; so $\mathbf{m}$ is

$$f \mapsto f * \eta : \mathbf{F} \to U_c(\mathcal{L}).$$

Then $s$ is the mapping from $\mathbf{F}$ into $\mathbf{X} = \mathbb{R}^{\Gamma}$ for which

$$\begin{aligned} s(f) = g(\mathbf{m}(f)) &= \left(\int_{[0,t[} (t-\tau)m(\tau)f(\tau)\eta(d\tau)\right)_{t \in I_0} \\ &= \left(\int_0^t (t-\tau)m(\tau)f(\tau)d\tau\right)_{t \in I_0}. \end{aligned} \tag{4.13.9}$$

Here we use (4.4.6) and the identification of $F$ and $B_0(E, \mathcal{L})$ in our case. Then (4.12.1) is the condition

$$s(f) = \left(\int_0^t (t-\tau)m(\tau)f(\tau)d\tau\right)_{t \in I_0} \in \tilde{\mathbb{N}}. \tag{4.13.10}$$

Certainly, (4.12.1) is realized as the non-perturbed constraint of Section 12.1. Condition (4.13.8) has the sense of some regularization. Note that (4.13.10) is equivalent to the following requirement (see Section 1.2):

$$\forall t \in I_0 : y_f(t) \in N_t. \tag{4.13.11}$$

In (4.13.11) we use (1.2.18) and (4.13.7). We see that the requirement (4.13.8) (recall that (4.13.10) is a concrete variant of (4.12.1)) is equivalent to the traditional (in control theory) phase constraint. It is logical to consider (4.13.8), where $\mu \in U_c(\mathcal{L})$, as the correct extension of the problem of constructing the attainability domain under the resource $c$-constraint and the phase constraint (4.13.11). But here it is useful to give the natural analogue of (4.13.8) in terms of some generalized trajectories of (1.2.1). In addition, we use $\mu \in U_c(\mathcal{L})$ and the natural stipulation: the $\mu$-integral of a vector function with components from $B(I, \mathcal{L})$ is defined component-wise. Namely, we suppose that for $t \in I_0$

$$\tilde{y}_\mu(t) \triangleq y_0 + \dot{y}_0 t + \int_{[0,t[} (t-\tau) m(\tau) \mu(d\tau), \tag{4.13.12}$$

and, in a similar way, we assume

$$\tilde{y}_\mu^{(v)}(t) \triangleq \dot{y}_0 + \int_{[0,t[} m(\tau) \mu(d\tau). \tag{4.13.13}$$

We interpret (4.13.12) as the generalized $\mu$-trajectory and (4.13.13) as the generalized $\mu$-velocity. We use only (4.13.12). By (1.2.18) and (4.13.12) we get that (4.13.8) is equivalent to the following condition on the choice of $\mu \in U_c(\mathcal{L})$:

$$\forall t \in I_0 : \tilde{y}_\mu(t) \in N_t. \tag{4.13.14}$$

It is useful to supplement (4.3.13) by the following property: if $f \in \mathbf{F}$ and $\mu = f * \eta$, then $\mu \in U_c(\mathcal{L})$ and

$$\tilde{y}_\mu = y_f. \tag{4.13.15}$$

This statement follows from Theorem 4.9.3 and (4.4.6); moreover, see relations of Section 1.2. Finally, along with (4.3.14) we note the natural density property. Namely, consider $\nu \in U_c(\mathcal{L})$. Then for our case of $(E, \mathcal{L}, \eta)$ we have $\nu \in U_c(\mathcal{L}) \cap \mathbf{A}_\eta[\mathcal{L}]$. Therefore we can consider the net

$$(\mathbf{D}(I, \mathcal{L}), \prec, \Theta_\eta[\nu; \cdot])$$

in $B_0(I, \mathcal{L})$. From (4.4.8), (4.9.15) and Proposition 4.9.2 we have $\forall \mathcal{H} \in \mathbf{D}(I, \mathcal{L}) : \Theta_\eta[\nu; \mathcal{H}] \in \mathbf{F}$. Of course, $\Theta_\eta[\nu; \mathcal{H}] * \eta \in U_c(\mathcal{L})$ (see Proposition 4.9.2) and by (4.13.15)

$$\tilde{y}_{\Theta_\eta[\nu;\mathcal{H}]*\eta} = y_{\Theta_\eta[\nu;\mathcal{H}]}. \tag{4.13.16}$$

By Theorem 4.9.1 we have the convergence

$$(\mathbf{D}(I,\mathcal{L}),\prec,\Theta_\eta[\nu;\cdot]*\eta) \overset{\tau_*(\mathcal{L})}{\longrightarrow} \nu. \tag{4.13.17}$$

Taking into account (4.13.17) and the definition of $\tau_*(\mathcal{L})$, we get the convergence

$$\left(\mathbf{D}(I,\mathcal{L}),\prec,\left(\int_{[0,t[}(t-\tau)m(\tau)\Theta_\eta[\nu;\mathcal{H}](\tau)\eta(d\tau)\right)_{\mathcal{H}\in\mathbf{D}(I,\mathcal{L})}\right) \overset{\tau_{\mathbb{R}}^{(r)}}{\longrightarrow}$$

$$\int_{[0,t[}(t-\tau)m(\tau)\nu(d\tau),$$

where $t \in I_0$. Of course, we take into account (4.4.6). Using (4.13.12) and representations of Section 1.2, we get the following convergence for $t \in I_0$:

$$(\mathbf{D}(I,\mathcal{L}),\prec,(y_{\Theta_\eta[\nu;\mathcal{H}]}(t))_{\mathcal{H}\in\mathbf{D}(I,\mathcal{L})}) \overset{\tau_{\mathbb{R}}^{(r)}}{\longrightarrow} \tilde{y}_\nu(t).$$

So we have the convergence

$$(\mathbf{D}(I,\mathcal{L}),\prec,(y_{\Theta_\eta[\nu;\mathcal{H}]})_{\mathcal{H}\in\mathbf{D}(I,\mathcal{L})}) \overset{\otimes^{I_0}(\tau_{\mathbb{R}}^{(r)})}{\longrightarrow} \tilde{y}_\nu. \tag{4.13.18}$$

We obtain the natural way of realization of $\tilde{y}_\nu$. From (4.13.18) we have

$$\{\tilde{y}_\mu : \mu \in U_c(\mathcal{L})\} \subset \mathrm{cl}\left(\{y_f : f \in \mathbf{F}\}, \otimes^{I_0}(\tau_{\mathbb{R}}^{(r)})\right). \tag{4.13.19}$$

Let us supplement (4.3.19). Note that by (4.13.12), the mapping

$$\mu \mapsto \tilde{y}_\mu : U_c(\mathcal{L}) \to \mathbf{X}$$

is continuous in the sense of the topologies $\tau_*(\mathcal{L})\,\big|_{U_c(\mathcal{L})}$ and $\otimes^{I_0}(\tau_{\mathbb{R}}^{(r)})$. The first topology is compact and the second one is a Hausdorff topology. Hence the set on the left hand side of (4.13.19) is closed. Using (4.13.15), we obtain the inclusion inverse to (4.13.19). In reality,

$$\{\tilde{y}_\mu : \mu \in U_c(\mathcal{L})\} = \mathrm{cl}\left(\{y_f : f \in \mathbf{F}\}, \otimes^{I_0}(\tau_{\mathbb{R}}^{(r)})\right). \tag{4.13.20}$$

The relation (4.13.20) permits us to consider (4.13.14) as an extension of (4.13.11). This conclusion is proved by analogy with [117, Chapters III,IV]. Namely, we introduce (as in [117]) the corresponding generalized system based on (4.13.12) and (4.13.13). Moreover, using the closedness of $\mathbf{Y} = \widetilde{\mathbb{N}}$, we replace (4.13.11) by (4.13.14). Such a way is typical for control theory. But it is useful to consider this replacement in terms of constructions of Sections 4.11 and 4.12. Here we return to (4.13.8). Recall that (4.13.8) is equivalent to (4.13.14). Hence we can consider some weakening of the constraint (4.13.8). In this connection we return to (4.12.7). Owing to (4.11.10) and (4.12.6) we get for $K \in \mathrm{Fin}(\Gamma)$ and $\varepsilon \in ]0, \infty[$

$$\begin{aligned}\bigcup_{y \in \mathbf{Y}} \Lambda_\Gamma^{(r)}(y, (K, \varepsilon)) &= \{v \in \mathbf{X} \mid \exists y \in \mathbf{Y}\ \forall \gamma \in K : v(\gamma) \in \mathbf{B}^0_{\rho_r}(y(\gamma), \varepsilon)\} \\ &= \{v \in \mathbf{X} \mid \exists y \in \mathbf{Y}\ \forall \gamma \in K\ \forall i \in \overline{1, r} : \\ &\qquad \mid y(\gamma)(i) - v(\gamma)(i) \mid < \varepsilon\}\end{aligned} \tag{4.13.21}$$

(recall (4.11.27); then $\mathcal{U}(\mathrm{Fin}(\Gamma) \times ]0, \infty[, \otimes^\Gamma(\tau_{\mathbb{R}}^{(r)}), \mathbf{Y} \mid \Lambda_\Gamma^{(r)})$ is the family of all sets (4.13.21) under enumeration of $(K, \varepsilon) \in \mathrm{Fin}(\Gamma) \times ]0, \infty[$ ). In connection with Theorem 4.12.1 it is useful to obtain a concrete variant of (4.13.21) corresponding to realization of $\mathbf{Y}$ in the form of (4.13.7). And what is more, it is important for us to consider sets of the type

$$s^{-1}\left(\bigcup_{y \in \mathbf{Y}} \Lambda_\Gamma^{(r)}(y, (K, \varepsilon))\right), \tag{4.13.22}$$

where $K \in \mathrm{Fin}(\Gamma)$ and $\varepsilon \in ]0, \infty[$. In this connection we recall sets of the type $F(K, \varepsilon)$ from Section 1.2. In Section 1.2 and here we use different norms of $\mathcal{R} = \mathbb{R}^r$. But this circumstance is unessential (we use constructions of Section 1.2 only as an reference point). So we consider the sets (4.13.21) and (4.13.22). Recall that by (2.7.13) and (2.7.16) we have $\forall t \in I_0\ \forall \varepsilon \in [0, \infty[$:

$$\begin{aligned}\mathbf{B}^0_{\rho_r}[\tilde{N}_t; \varepsilon] &= \{v \in \mathcal{R} \mid (\rho_r - \inf)[v; \tilde{N}_t] < \varepsilon\} \\ &= \{v \in \mathcal{R} \mid \exists w \in \tilde{N}_t : \rho_r(v, w) < \varepsilon\} \\ &= \{v \in \mathcal{R} \mid \exists w \in \tilde{N}_t\ \forall i \in \overline{1, r} : \mid v(i) - w(i) \mid < \varepsilon\}.\end{aligned} \tag{4.13.23}$$

From (4.13.21) and (4.13.23) we immediately get

$$\bigcup_{y \in \mathbf{Y}} \Lambda_\Gamma^{(r)}(y, (K, \varepsilon)) \subset \{x \in \mathbf{X} \mid (x \mid K) \in \prod_{t \in K} \mathbf{B}^0_{\rho_r}[\tilde{N}_t; \varepsilon]\}. \tag{4.13.24}$$

On the other hand, if $x \in \mathbf{X}$ has the property

$$(x \mid K) \in \prod_{t \in K} \mathbf{B}^0_{\rho_r}[\tilde{N}_t; \varepsilon], \tag{4.13.25}$$

then we can choose $u \in \prod_{t \in K} \tilde{N}_t$ for which $\forall \gamma \in K\ \forall i \in \overline{1,r}$:

$$| u(\gamma)(i) - x(\gamma)(i) | < \varepsilon.$$

Since $\tilde{\mathbb{N}} \neq \varnothing$, we choose $w \in \tilde{\mathbb{N}}$ and replace the values $w(\gamma)$ by $u(\gamma)$ for $\gamma \in K$. As a result we obtain the new function $\tilde{w} \in \tilde{\mathbb{N}}$ for which $u = (\tilde{w} \mid K)$; then $\forall \gamma \in K\ \forall i \in \overline{1,r}$:

$$| \tilde{w}(\gamma)(i) - x(\gamma)(i) | < \varepsilon.$$

Using (4.13.21) we get the inclusion

$$x \in \bigcup_{y \in \mathbf{Y}} \Lambda_\Gamma^{(r)}(y, (K, \varepsilon)).$$

Since the choice of $x$ with the property (4.13.25) was arbitrary, we have (see (4.13.24))

$$\bigcup_{y \in \mathbf{Y}} \Lambda_\Gamma^{(r)}(y, (K, \varepsilon)) = \{x \in \mathbf{X} \mid (x \mid K) \in \prod_{t \in K} \mathbf{B}^0_{\rho_r}[\tilde{N}_t; \varepsilon]\}. \tag{4.13.26}$$

Recall that in (4.13.26) $K \in \operatorname{Fin}(I_0)$ and $\varepsilon \in ]0, \infty[$ are arbitrary. From the representation of $s$ and (4.13.26) we have

$$s^{-1}\left(\bigcup_{y \in \mathbf{Y}} \Lambda_\Gamma^{(r)}(y, (K, \varepsilon))\right)$$

$$= \left\{ f \in \mathbf{F} \mid \forall t \in K : \int_0^t (t - \tau) m(\tau) f(\tau) d\tau \in \mathbf{B}^0_{\rho_r}[\tilde{N}_t; \varepsilon] \right\}, \tag{4.13.27}$$

where $K \in \operatorname{Fin}(I_0)$ and $\varepsilon \in ]0, \infty[$. So in (4.13.27) we have the required concrete variant of the employment of the family

$$\begin{aligned} &\mathcal{U}(\operatorname{Fin}(\Gamma) \times ]0, \infty[, \otimes^\Gamma(\tau_{\mathbb{R}}^{(r)}), \mathbf{Y} \mid \Lambda_\Gamma^{(r)}) \\ =&\mathcal{U}(\operatorname{Fin}(I_0) \times ]0, \infty[, \otimes^{I_0}(\tau_{\mathbb{R}}^{(r)}), \tilde{\mathbb{N}} \mid \Lambda_{I_0}^{(r)}) \end{aligned} \tag{4.13.28}$$

used in Theorem 4.12.1. Namely, (4.13.28) realizes the follwing property: $s^{-1}[\mathcal{U}(\operatorname{Fin}(I_0) \times ]0, \infty[, \otimes^{I_0}(\tau_{\mathbb{R}}^{(r)}), \tilde{\mathbb{N}} \mid \Lambda_{I_0}^{(r)})]$ is the family of all sets (4.13.27) under the enumeration of $(K, \varepsilon) \in \operatorname{Fin}(I_0) \times ]0, \infty[$. Here it is required to give the concrete definition of $\omega$. Recall that we consider the problem of constructing the attainability domain. Therefore it is advisable to define $h \in \mathbf{H}^{\mathbf{F}}$, where $(\mathbf{H}, \boldsymbol{\tau}) = (\mathcal{R}, \tau_{\mathbb{R}}^{(r)})$, by the following rule:

$$h(f) \triangleq y_f(\vartheta_0) = y_0 + \dot{y}_0 \vartheta_0 + \int_0^{\vartheta_0} (\vartheta_0 - \tau) m(\tau) f(\tau) d\tau. \tag{4.13.29}$$

Owing to (4.13.29) we can easily construct $\omega \in \mathcal{R}^{U_c(\mathcal{L})}$:

$$\omega(\mu) \triangleq \tilde{y}_\mu(\vartheta_0) = y_0 + \dot{y}_0\vartheta_0 + \int_I (\vartheta_0 - \tau)m(\tau)\mu(d\tau). \qquad (4.13.30)$$

From (4.13.15), (4.13.29) and (4.13.30) we have the equality $h = \omega \circ \mathbf{m}$. By the definition of $\tau_*(\mathcal{L})$ we conclude that (4.12.22) is valid. In addition, $(\mathbf{H}, \boldsymbol{\tau})$ is a Hausdorff space. Then (4.12.25) holds and Theorem 4.12.1 is true. Note only the coincidence of $\omega^1(g^{-1}(\mathbf{Y}))$ with the AS corresponding to the family (4.13.28). In addition, by (4.13.14) and (4.13.30) this AS is the generalized attainability domain: $\omega^1(g^{-1}(\mathbf{Y})) = \omega^1(g^{-1}(\tilde{\mathbb{N}}))$ is the set of all points $\tilde{y}_\mu(\vartheta_0)$ such that $\mu \in U_c(\mathcal{L})$ and (4.13.14) holds. Thus, using FAM for constructing GE, we realize AS for a very interesting (from the practical point of view) variant of the weakening of phase constraints (see (4.13.27) and its analogue realized in terms of $y_f$ and $(N_t)_{t \in I_0}$). But it is very interesting to obtain another result examined in Section 1.2 and connected with conditions of the type (1.2.4). These conditions are not considered in Sections 4.11 and 4.12 but, at the informative level, are discussed in Chapter 1.

Now it is advisable to repeat some elements of this setting, using unessential modifications connected with the basic constructions of this chapter. If $\varepsilon \in \,]0, \infty[$, then we suppose (in this section) that

$$\mathbf{F}_\varepsilon \triangleq \{f \in \mathbf{F} \mid \forall t \in I_0 : y_f(t) \in \mathbf{B}^0_{\rho_r}[N_t; \varepsilon]\}. \qquad (4.13.31)$$

We consider the family $\mathfrak{F}$ of all sets $\mathbf{F}_\varepsilon$, $\varepsilon \in \,]0, \infty[$. Obviously (see (4.13.31)),

$$\mathfrak{F} \in \mathcal{B}[\mathbf{F}]. \qquad (4.13.32)$$

It is worth noting that $\mathfrak{F} \in \mathcal{B}_{\mathcal{N}}(\mathbf{F})$. In addition, (4.13.31) and (4.13.32) are similar to asymptotic constructions connected with the constraints of the type (1.2.6). These weakened versions of phase constraints are very typical for many practical control problems. With (4.13.31) we connect the idea of approximate observation of phase constraints. We consider the connection of the constraints defined in terms of (4.13.27), (4.13.28) and those defined in terms of (4.13.31), (4.13.32).

Note that from (1.2.18) and other definitions of Section 1.2 it follows that $\forall \varepsilon \in \,]0, \infty[$:

$$\mathbf{F}_\varepsilon = \{f \in \mathbf{F} \mid \forall t \in I_0 : \int_0^t (t - \tau)m(\tau)f(\tau)d\tau \in \mathbf{B}^0_{\rho_r}[\tilde{N}_t; \varepsilon]\}. \quad (4.13.33)$$

To prove (4.13.33) it is useful to take into account several simplest properties connected with the considered concrete variant of (2.7.16),

(2.7.17). Consider some properties of such a type, fixing $t \in I_0$. By (1.2.18) we have the coincidence of $\tilde{N}_t$ and the set $\{z-(y_0+t\dot{y}_0) : z \in N_t\}$. In addition, $N_t \in \mathcal{K}_r$. Namely, $N_t$ is a nonempty compactum in the sense of $\tau_{\mathbb{R}}^{(r)}$. The metric $\rho_r$ was defined in Section 4.11. Note the general property (2.7.11). From the triangle inequality we get several known simple estimates. If $u \in \mathcal{R}$, $v \in \mathcal{R}$ and $w \in \mathcal{R}$, then

$$\big|\, \rho_r(u,v) - \rho_r(u,w) \,\big| \leq \rho_r(v,w).$$

Moreover, if such $u$, $v$ and $w$ are given, then $\rho_r(u+w, v+w) = \rho_r(u,v)$ (in fact, we deal with a normed space). Let $u \in \mathbf{B}^0_{\rho_r}[\tilde{N}_t;\varepsilon]$. Then we can choose $z \in N_t$ for which

$$\rho_r(u, z - (y_0 + t\dot{y}_0)) < \varepsilon.$$

We use (2.7.13) and (2.7.16). But in this case we obtain

$$\rho_r((y_0 + t\dot{y}_0) + u, z) < \varepsilon.$$

As a consequence we have (see (2.7.13)) the inequality

$$(\rho_r - \inf)[(y_0 + t\dot{y}_0) + u; N_t] < \varepsilon.$$

This means (see (2.7.16)) the property $(y_0 + t\dot{y}_0) + u \in \mathbf{B}^0_{\rho_r}[N_t;\varepsilon]$. So we have established that if $u' \in \mathbf{B}^0_{\rho_r}[\tilde{N}_t;\varepsilon]$, then $(y_0 + t\dot{y}_0) + u' \in \mathbf{B}^0_{\rho_r}[N_t;\varepsilon]$. Conversely, if $v \in \mathbf{B}^0_{\rho_r}[N_t;\varepsilon]$, then $\rho_r(v,\tilde{z}) < \varepsilon$ for some $\tilde{z} \in N_t$ and as a consequence

$$\rho_r(v - (y_0 + t\dot{y}_0), \tilde{z} - (y_0 + t\dot{y}_0)) < \varepsilon.$$

From (2.7.13), (2.7.16) and the last inequality we get

$$(\rho_r - \inf)[v - (y_0 + t\dot{y}_0); \tilde{N}_t] < \varepsilon$$

and hence

$$v - (y_0 + t\dot{y}_0) \in \mathbf{B}^0_{\rho_r}[\tilde{N}_t;\varepsilon].$$

Since the choice of $v$ was arbitrary, we have the property that if $v' \in \mathbf{B}^0_{\rho_r}[N_t;\varepsilon]$, then

$$v' - (y_0 + t\dot{y}_0) \in \mathbf{B}^0_{\rho_r}[\tilde{N}_t;\varepsilon].$$

The two last properties are used in (4.13.31) to verify (4.13.33). We now compare (4.13.31) and the set on the right hand side of (4.13.33). In addition, we use the representation of Section 1.2 following from the Cauchy formula. Namely, from (4.13.31) for $\varepsilon \in \,]0,\infty[$ we have $\forall f \in \mathbf{F}_\varepsilon\ \forall t \in I_0$:

$$\int_0^t (t-\tau)m(\tau)f(\tau)d\tau \in \mathbf{B}^0_{\rho_r}[\tilde{N}_t;\varepsilon]. \qquad (4.13.34)$$

Then $\mathbf{F}_\varepsilon$ (4.13.31) is a subset of the set on the right hand side of (4.13.33). Let $f'$ be a point of the last set: $f' \in \mathbf{F}$ and (4.13.34) is satisfied for $t \in I_0$ and $f = f'$. Then for

$$u' = \int_0^t (t-\tau)m(\tau)f'(\tau)d\tau,$$

we have $u' \in \mathbf{B}^0_{\rho_r}[\tilde{N}_t;\varepsilon]$ and as a consequence (see Section 1.2) $y_{f'}(t) \in \mathbf{B}^0_{\rho_r}[N_t;\varepsilon]$. So $f' \in \mathbf{F}_\varepsilon$ (see (4.13.31)). Since the choice of $f'$ was arbitrary, we get (4.13.33). Thus (4.13.33) is established.

Hence for $\varepsilon \in\, ]0,\infty[$ and $K \in \mathrm{Fin}(I_0)$, from (4.13.27) and (4.13.33), we have the obvious inclusion

$$\mathbf{F}_\varepsilon \subset s^{-1}\left(\bigcup_{y\in\mathbf{Y}} \Lambda_\Gamma^{(r)}(y(K,\varepsilon))\right). \tag{4.13.35}$$

On the other hand, if $\varepsilon \in\, ]0,\infty[$, then by continuity of the compact-valued mapping

$$t \mapsto N_t : I_0 \to \mathcal{K}_r, \tag{4.13.36}$$

for some $\varepsilon' \in ]0,\varepsilon[$ and $K' \in \mathrm{Fin}(I_0)$ we have the inclusion

$$s^{-1}\left(\bigcup_{y\in\mathbf{Y}} \Lambda_\Gamma^{(r)}(y,(K',\varepsilon'))\right) \subset \mathbf{F}_\varepsilon. \tag{4.13.37}$$

The proof is analogous to the reasoning of Chapter 1 (see the relation (1.3.2)). Indeed, fix $\varepsilon \in ]0,\infty[$. Since the mapping (4.13.36) from the compact space $I_0$ with the metric-modulus into a metric space is continuous, it is uniformly continuous [66, Ch. 1]. Hence for this $\varepsilon$ there exists $\delta \in\, ]0,\infty[$ such that $\forall t_1 \in I_0\ \forall t_2 \in I_0$:

$$(|t_1 - t_2| < \delta) \Rightarrow (N_{t_1} \in \mathbf{B}^0_{\rho_r}[N_{t_2};\varepsilon/3])\&(N_{t_2} \in \mathbf{B}^0_{\rho_r}[N_{t_1};\varepsilon/3]).$$

Choose $n' \in \mathcal{N}$ such that

$$n' > \max\left\{\frac{3c_3\vartheta_0}{\varepsilon}, \frac{\vartheta_0}{\delta}\right\}.$$

It is easily checked that (4.13.37) is valid for $\varepsilon' = \varepsilon/3$ and $K' = \{i\vartheta_0/n' : i \in \overline{0,n'}\}$.

Returning to Section 1.3 and taking into account the equivalence of the Euclidean metric and the metric $\rho_r$, from Theorem 4.12.1 and (4.13.29) we conclude that (see (1.3.8))

$$\begin{aligned}\mathrm{Att}_1 =&\,\omega^1(g^{-1}(\widetilde{\mathbb{N}}))\\ =&(\tau_{\mathbb{R}}^{(r)} - \mathrm{LIM})[s^{-1}[\mathcal{U}(\mathrm{Fin}(I_0)\times]0,\infty[,\otimes^{I_0}(\tau_{\mathbb{R}}^{(r)}),\widetilde{\mathbb{N}} \mid \Lambda_{I_0}^{(r)})] \mid h].\end{aligned} \tag{4.13.38}$$

Note that points of the AS (4.13.38) are realized as limits of nets. However, taking into account Proposition 1.3.1 and the equality (1.3.11), we can get a more convenient representation of this AS, admitting sequential realization of the points. Namely, let $\forall n \in \mathcal{N}$ :

$$\mathbf{F}_n = \{f \in \mathbf{F} \mid \forall t \in K_n : \int_0^t (t-\tau)m(\tau)f(\tau)d\tau \in \mathbf{B}^0_{\rho_r}[\tilde{N}_t; \frac{1}{n}]\}, \tag{4.13.39}$$

where $K_n = \left\{\frac{i\vartheta_0}{n} : i \in \overline{0,n}\right\}$ (see (1.3.9)). In other words,

$$\mathbf{F}_n = s^{-1}\left(\bigcup_{y\in\mathbf{Y}} \Lambda_{I_0}^{(r)}(y,(K_n,\frac{1}{n}))\right).$$

Note that $\mathbf{F}_n$ is defined as $F_n$ from Section 1.3, but different metrics in the space of attainable elements are used in their definitions. Of course, this difference is unessential. Analogously to Proposition 1.3.1 we can state that the sequence $(\mathbf{F}_n)_{n\in\mathcal{N}}$ is fundamental for the family $\mathcal{U}(\mathrm{Fin}(I_0)\times]0,\infty[,\otimes^{I_0}(\tau_{\mathbb{R}}^{(r)}),\widetilde{\mathbb{N}} \mid \Lambda_{I_0}^{(r)})$. Consequently we have

$$\begin{aligned}&(\tau_{\mathbb{R}}^{(r)} - \mathrm{LIM})[s^{-1}[\mathcal{U}(\mathrm{Fin}(I_0)\times]0,\infty[,\otimes^{I_0}(\tau_{\mathbb{R}}^{(r)}),\widetilde{\mathbb{N}} \mid \Lambda_{I_0}^{(r)})] \mid h]\\ &= (\tau_{\mathbb{R}}^{(r)} - \mathrm{LIM})[\{\mathbf{F}_n : n\in\mathcal{N}\} \mid h] = \bigcap_{n\in\mathcal{N}} \mathrm{cl}(h^1(\mathbf{F}_n),\tau_{\mathbb{R}}^{(r)}).\end{aligned} \tag{4.13.40}$$

From (4.13.38) and (4.13.40) it follows that in the case under consideration it is sufficient to use sequences for realization of points of the attraction set.

**Remark.** In this problem the sequential realization of the attraction set is possible only for the attainability domain because the space of attainable elements (here $\mathbb{R}^r$) satisfies the first axiom of countability (see [35, p. 38], [99, 100]) . As for the space of admissible elements (here $\mathbf{A}(\mathcal{L})$), it does not satisfy this axiom. In this case the presence of a fundamental countable system of sets in the family determining the relaxed constraints of the problem is not sufficient for the possibility of sequential realization of points. Consider the following example. (This example is taken from [107, Ch. 4,Problem 12], but as distinct from this work, the interpretation of it is given in terms of FAM).

Let $\mathcal{L} = \mathcal{P}(\mathcal{N})$. Then $\mathbb{B}(\mathcal{N}) = B(\mathcal{N}, \mathcal{P}(\mathcal{N}))$ is the set of all bounded sequences. Consider the sequence of Dirac measures

$$(\delta_n)_{n\in\mathcal{N}} \in \mathbb{T}(\mathcal{P}(\mathcal{N}))^{\mathcal{N}}; \tag{4.13.41}$$

recall that $\delta_n$ is the Dirac measure defined on the family of all subsets of the positive integers and concentrated at the point $n$, $n \in \mathcal{N}$. Note that the set $\mathbb{T}(\mathcal{P}(\mathcal{N}))$ is compact in $(\mathbf{A}(\mathcal{P}(\mathcal{N})), \tau_*(\mathcal{P}(\mathcal{N})))$ (see (4.10.12)). However, this sequence has no convergent subsequences. Let us show this. Suppose the contrary; i.e., there exist $(n_k)_{k\in\mathcal{N}} \in \mathbf{N}$; i.e., $(n_k)_{k\in\mathcal{N}} \in \mathcal{N}^{\mathcal{N}}$ with the property $n_s < n_{s+1}$, $s \in \mathcal{N}$, and a measure $\mu^* \in \mathbb{T}(\mathcal{P}(\mathcal{N}))$ such that

$$(\delta_{n_k})_{k\in\mathcal{N}} \overset{\tau_*(\mathcal{P}(\mathcal{N}))}{\longrightarrow} \mu^*. \tag{4.13.42}$$

Note that for any $\mathbf{a} = (a_i)_{i\in\mathcal{N}} \in B(\mathcal{N}, \mathcal{P}(\mathcal{N}))$ we have

$$\int_{\mathcal{N}} \mathbf{a}\, d\delta_n = a_n. \tag{4.13.43}$$

From (4.13.42) and (4.13.43) it follows that for any $\mathbf{a} \in B(\mathcal{N}, \mathcal{P}(\mathcal{N}))$ the sequence $(a_{n_k})_{k\in\mathcal{N}}$ must converge. But we can take the sequence $\tilde{\mathbf{a}} \in B(\mathcal{N}, \mathcal{P}(\mathcal{N}))$ defined as follows:

$$(\forall i \in \mathcal{N} : \tilde{\mathbf{a}}(n_{2i}) \triangleq 1) \ \&\ (\forall s \in \mathcal{N} \setminus \{n_{2i} : i \in \mathcal{N}\} : \tilde{\mathbf{a}}(s) \triangleq 0).$$

Then the subsequence $(\tilde{\mathbf{a}}_{n_k})_{k\in\mathcal{N}} \in B(\mathcal{N}, \mathcal{P}(\mathcal{N})$, whose elements are equal to 1 for even $k$ and to 0 for odd $k$, is not convergent. The contradiction obtained proves our statement. Thus there is no convergent subsequence for the sequence (4.13.41). However, from the compactness of the set $\mathbb{T}(\mathcal{P}(\mathcal{N}))$ it follows that there exists a convergent subnet for this sequence; i.e., there exist a directed set $(\mathbf{D}, \preceq)$, $\mathbf{D} \neq \varnothing$, an isotone operator $\rho \in (\mathrm{Isot})[\mathbf{D}; \preceq]$ (see (2.2.50)), and a measure $\mu^0 \in \mathbb{T}(\mathcal{P}(\mathcal{N}))$ such that

$$(\mathbf{D}, \preceq, (\delta_{\rho(d)})_{d\in\mathbf{D}}) \overset{\tau_*(\mathcal{P}(\mathcal{N}))}{\longrightarrow} \mu^0.$$

By the way, note that from (4.10.31) we get

$$\mathbb{T}(\mathcal{P}(\mathcal{N})) = \mathrm{cl}(\{\delta_n : n \in \mathcal{N}\}, \tau_*(\mathcal{P}(\mathcal{N}))),$$

which means that the set $\mathbb{T}(\mathcal{P}(\mathcal{N}))$ is separable. □

As for the generalized admissible set $g^{-1}(\widetilde{\mathbb{N}})$, we can get another representation along with the equality (4.12.22). Namely, consider the family whose elements are sets of the form

$$\mathbf{F}_\xi(\varepsilon) \triangleq \left\{ f \in \mathbf{F} \mid \int_0^\xi (\xi - \tau) m(\tau) f(\tau) d\tau \in \mathbf{B}^0_{\rho_r}[\tilde{N}_\xi; \varepsilon] \right\}$$
$$= s^{-1}\left( \bigcup_{y \in \mathbf{Y}} \Lambda^{(r)}_{I_0}(y, (\{\xi\}, \varepsilon)) \right). \tag{4.13.44}$$

That is, we consider only one-element sets of $I_0$ instead of the family of all finite subsets of $I_0$. Obviously, from (4.13.28) and (4.13.44) we have $\forall K \in \mathrm{Fin}(I_0)\ \forall \varepsilon \in ]0, \infty[$:

$$s^{-1}\left( \bigcup_{y \in \mathbf{Y}} \Lambda^{(r)}_{I_0}(y, (K, \varepsilon)) \right) = \bigcap_{\xi \in K} \mathbf{F}_\xi(\varepsilon).$$

In addition, from (4.13.34) and (4.13.44) it follows that $\forall \varepsilon \in ]0, \infty[$:

$$\mathbf{F}_\varepsilon = \bigcap_{\xi \in I_0} \mathbf{F}_\xi(\varepsilon).$$

THEOREM 4.13.1 *The set of admissible generalized controls has the following representation:*

$$g^{-1}(\widetilde{\mathbb{N}}) = \bigcap_{\xi \in I_0} \bigcap_{\varepsilon \in\, ]0,\infty[} \mathrm{cl}(\mathbf{m}^1(\mathbf{F}_\xi(\varepsilon)), \tau^*_{\mathbf{K}}(\mathcal{L})). \tag{4.13.45}$$

PROOF. For brevity denote by $\mathfrak{F}$ the family

$$s^{-1}[\mathcal{U}(\mathrm{Fin}(I_0) \times ]0, \infty[, \otimes^{I_0}(\tau^{(r)}_{\mathbb{R}}) \widetilde{\mathbb{N}} | \Lambda^{(r)}_{I_0})].$$

In correspondence with (4.12.22) we have

$$g^{-1}(\widetilde{\mathbb{N}}) = (\tau^*_{\mathbf{K}}(\mathcal{L}) - \mathrm{LIM})[\mathfrak{F} \mid \mathbf{m}]. \tag{4.13.46}$$

However, by (4.13.44)

$$\{F_\xi(\varepsilon) : (\xi, \varepsilon) \in I_0 \times ]0, \infty[\} \subset \mathfrak{F},$$

and hence

$$(\tau^*_{\mathbf{K}}(\mathcal{L}) - \mathrm{LIM})[\mathfrak{F} \mid \mathbf{m}] \subset \bigcap_{\xi \in I_0} \bigcap_{\varepsilon \in\, ]0,\infty[} \mathrm{cl}(\mathbf{m}^1(\mathbf{F}_\xi(\varepsilon)), \tau^*_{\mathbf{K}}(\mathcal{L})),$$

which implies, by (4.13.46), the inclusion

$$g^{-1}(\widetilde{\mathbb{N}}) \subset \bigcap_{\xi \in I_0} \bigcap_{\varepsilon \in\, ]0,\infty[} \mathrm{cl}(\mathbf{m}^1(\mathbf{F}_\xi(\varepsilon)), \tau^*_{\mathbf{K}}(\mathcal{L})).$$

Let now $\mu \in \bigcap_{\xi \in I_0} \bigcap_{\varepsilon \in ]0,\infty[} \mathrm{cl}(\mathbf{m}^1(\mathbf{F}_\xi(\varepsilon)), \tau^*_{\mathbf{K}}(\mathcal{L}))$. This means that $\forall \xi \in I_0\ \forall \varepsilon \in ]0, \infty[$:

$$\mu \in \mathrm{cl}(\mathbf{m}^1(\mathbf{F}_\xi(\varepsilon)), \tau^*_{\mathbf{K}}(\mathcal{L})). \qquad (4.13.47)$$

Fix any $\xi \in I_0$ and $\varepsilon \in ]0, \infty[$. Then (4.13.47) holds and in accordance with (2.3.11) we have

$$\exists_{\mathbf{D}}\, \mathbf{S}[\mathbf{D} \neq \varnothing]\ \exists \preceq \in (\mathrm{DIR})[\mathbf{D}]\ \exists z \in F_\xi(\varepsilon)^{\mathbf{D}} : (\mathbf{D}, \preceq, \mathbf{m} \circ z) \overset{\tau^*_{\mathbf{K}}(\mathcal{L})}{\longrightarrow} \mu.$$

By the continuity of $g$ (see Proposition 4.11.3) we get the convergence

$$(\mathbf{D}, \preceq, g \circ \mathbf{m} \circ z) \overset{\otimes^{I_0}(\tau^{(r)}_{\mathbb{R}})}{\longrightarrow} g(\mu). \qquad (4.13.48)$$

Taking into account (4.13.9) and (4.13.44), we conclude that $\forall d \in \mathbf{D}$:

$$g(\mathbf{m}(z(d)))(\xi) = \int_0^\xi (\xi - \tau) m(\tau) z(d)(\tau) d\tau \in \mathbf{B}^0_{\rho_r}[\tilde{N}_\xi; \varepsilon]. \qquad (4.13.49)$$

On the other hand, from (4.13.48) we get

$$(\mathbf{D}, \preceq, (g(\mathbf{m}(z(d)))(\xi))_{d \in \mathbf{D}}) \overset{\tau^{(r)}_{\mathbb{R}}}{\longrightarrow} g(\mu)(\xi). \qquad (4.13.50)$$

From (4.13.49) and (4.13.50) we obtain

$$g(\mu)(\xi) \in \mathrm{cl}(\mathbf{B}^0_{\rho_r}[\tilde{N}_\xi; \varepsilon], \tau^{(r)}_{\mathbb{R}}),$$

which means, by (2.7.16), the inclusion

$$g(\mu)(\xi) \in \mathbf{B}_{\rho_r}[\tilde{N}_\xi; \varepsilon]. \qquad (4.13.51)$$

Since (4.13.51) is valid for any $\varepsilon \in ]0, \infty[$, we get

$$g(\mu)(\xi) \in \tilde{N}_\xi. \qquad (4.13.52)$$

But $\xi \in I_0$ was also chosen arbitrarily. Hence (4.13.52) is valid for all $\xi \in I_0$, which implies, by (4.13.7), the inclusion

$$g(\mu) \in \widetilde{\mathbb{N}}.$$

This means that $\mu \in g^{-1}(\widetilde{\mathbb{N}})$, and consequently we have the required inverse inclusion

$$\bigcap_{\xi \in I_0} \bigcap_{\varepsilon \in ]0,\infty[} \mathrm{cl}(\mathbf{m}^1(\mathbf{F}_\xi(\varepsilon)), \tau^*_{\mathbf{K}}(\mathcal{L})) \subset g^{-1}(\widetilde{\mathbb{N}}),$$

which completes the proof of the theorem. □

Note that the representation (4.13.45) essentially differs from the general constructions of Chapter 3 because the family $\{F_\xi(\varepsilon) : (\xi, \varepsilon) \in I_0 \times ]0, \infty[\}$ is not semi-multiplicative; i.e., $\{F_\xi(\varepsilon) : (\xi, \varepsilon) \in I_0 \times ]0, \infty[\} \notin \mathcal{B}[\mathbf{F}]$, as is required in the definition of an attraction set (see (3.2.8)). But in the case of the problem under consideration the set $\mathbf{Y}$, which determines the constraints of the problem (here the set $\widetilde{\mathbb{N}}$), is represented as the product of sets (see (4.13.7)). Namely this peculiarity of the problem gives the representation (4.13.45).

# 4.14 AN ASYMPTOTIC CONSTRUCTION FOR SOME UNBOUNDED PROBLEMS OF ATTAINABILITY

In this section we consider a possibility of extension for a non-compactificable problem of attainability. We use the scheme of Section 4.9. Moreover, we consider a new type of AS (see [28, 98]). In this connection we recall that 'unbounded' problems of attainability were considered in [29, 30, 35, 39]. In particular, attraction sets connected with a bounded convergence are investigated in these works. It is useful also to recall the general constructions of Chapter 3, which are concerned with compactificable approximate solutions. Roughly speaking, special types of AS are studied in [29, 30, 35, 39]. For points of these AS the possibility of approximate realization in the class of 'strongly bounded' solution nets is postulated. But simultaneously, variants of unbounded asymptotic realization are possible. Such a possibility is assumed in [29, 30, 35, 39]. The case when only 'bounded approximate realization' is possible is considered in [28, 98]. In this section we describe very briefly this approach. Moreover, we note a typical singularity of control problems. Very often in these problems programs of control are vector functions. Therefore corresponding GE are defined as vector measures. Of course, we consider only simplest variants of such vector measures. In our constructions we use finite collections of scalar FAM; here we follow [35, Sect. 3.8]. In addition, we use vector measures only in this section. We strive to realize an analogous possibility for corresponding modifications of other extension procedures considered in this book. But we first give a very simple example for scalar controls. Following Chapter 1 and the previous section, consider the problem of control by a unit mass point

$$\dot{x}_1(t) = x_2(t), \quad \dot{x}_2(t) = f(t) \tag{4.14.1}$$

on the time interval $[0, 1]$. The corresponding initial conditions are supposed to be equal to zero: $x_1(0) = x_2(0) = 0$. Assume that only non-negative piecewise constant and continuous from the right r.-v. functions on $[0,1[$ can be used in (4.14.1). Denote by $F$ the set of all such functions treated as controls. Moreover, we postulate the following boundary conditions:

$$x_1(1) = z_* = \frac{1}{2}. \tag{4.14.2}$$

Of course, here $x(\cdot) = x_f(\cdot)$ is defined by the familiar Cauchy formula:

$$x_{f,1}(1) = \int_0^1 (1-t)f(t)\,dt, \quad x_{f,2}(1) = \int_0^1 f(t)\,dt, \quad f \in F. \tag{4.14.3}$$

We can transform (4.14.2) in terms of (4.14.3), obtaining the moment constraint. Consider possible variants of realization of the point $z_*$. For the control $\hat{f} \in F$ satisfying the condition $\hat{f}(t) \equiv 1$ we have the equality

$$z_* = x_{\hat{f},1}(1) = \int_0^1 (1-t)f(t)\,dt$$

(see 4.14.3)). Therefore we get an obvious possibility for the 'bounded' asymptotic realization of $z_*$; for this, the stationary sequence $(\hat{f}_i)_{i\in\mathcal{N}}$ in $F$ with $\hat{f}_i \equiv \hat{f}$ should be taken. In addition, the resource of each control of $(\hat{f}_i)_{i\in\mathcal{N}}$ is unit. Note that $(\hat{f}_i)_{i\in\mathcal{N}}$ can be regarded as a concrete variant of asymptotic realization of $z_*$. Indeed, the variant of precise realization of $z_*$ is simultaneously a variant of the 'bounded' asymptotic realization.

Consider another variant. For $\delta \in ]0,1[$, define $f^{(\delta)} \in F$ by the following rule:

$$f^{(\delta)}(t) \triangleq \begin{cases} 0, & t \in [0, 1-\delta[, \\ 1/\delta^2, & t \in [1-\delta, 1[. \end{cases}$$

For $q \in \mathcal{N}$ define $f_q \in F$ by the rule

$$f_q \triangleq f^{(\frac{1}{2q})}. \tag{4.14.4}$$

So we have the sequence $(f_k)_{k\in\mathcal{N}}$ in $F$. For $k \in \mathcal{N}$ consider

$$\begin{aligned} x_{f_k,1}(1) &= \int_{1-1/2k}^1 (1-t)4k^2\,dt = 4k^2 \int_{1-1/2k}^1 (1-t)\,dt \\ &= 4k^2 \int_0^{1/2k} \xi\,d\xi = 4k^2 \frac{\xi^2}{2}\Big|_0^{\frac{1}{2k}} = \frac{1}{2} = z_*. \end{aligned}$$

Note that

$$\int_0^1 |\, f_k(t) \,|\, dt = \int_0^1 f_k(t)\,dt = 4k^2\,\frac{1}{2k} = 2k.$$

So we have got a concrete variant of unbounded asymptotic realization of $z_*$. We use the constraints (4.14.2) and obtain many variants of admissible controls. And what is more, here we have very different variants of the precise realization of $z_*$. Of course, we can consider $z_*$ both as a point of the 'non-perturbed' attainability domain and as a point of the corresponding AS. Thus we get a pathology of the precise realization of attainable elements.

Consider another example, fixing the scalar control system

$$\dot{x}(t) = f(t), \quad x(0) = 0, \tag{4.14.5}$$

where $t \in [0,1]$. We preserve the set $F$ of the previous example as the set of all possible controls. Let the choice of $f \in F$ be subject to the constraint

$$\int_0^1 t f(t)\,dt \leq 0. \tag{4.14.6}$$

Introduce the function $g : \mathbb{R} \to \mathbb{R}$ such that

$$g(x) \triangleq \begin{cases} 2\mid x - 1/2 \mid, & x \in ]-\infty, 1], \\ 1/x, & x \in [1, \infty[. \end{cases}$$

We obtain the following functional $\mathbf{g}$:

$$f \mapsto g\left(\int_0^1 f(t)\,dt\right) : F \to \mathbb{R}. \tag{4.14.7}$$

In other words, by (4.14.7) we have $\mathbf{g}(f) = g(x_f(1))$. So we consider the simplest linear system with a nonlinear instantaneous transformer. Take the point $z^* = 0$. Introduce two (sequential) approximate solutions:

$$\left((f_k)_{k\in\mathcal{N}} \in F^{\mathcal{N}}\right) \;\&\; \left((f^{(k)})_{k\in\mathcal{N}} \in F^{\mathcal{N}}\right).$$

Consider the first sequence. Suppose that for $m \in \mathcal{N}$ the control $f_m \in F$ is defined as follows:

$$f_m(t) \triangleq \begin{cases} m/2, & t \in [0, 1/m[, \\ 0, & t \in [1/m, 1[. \end{cases} \tag{4.14.8}$$

We obviously have

$$\int_0^1 t f_m(t)\,dt = \int_0^{1/m} t f_m(t)\,dt \leq \frac{1}{m}\int_0^{1/m} f_m(t)\,dt = \frac{1}{2m}.$$

So $(f_k)_{k\in\mathcal{N}}$ is a concrete approximate solution (see Chapter 3). In addition, $x_{f_m}(1) = \int_0^{1/m} f_m(t)\,dt = \frac{1}{2}$. Therefore $\mathbf{g}(f_k) \equiv 0$. We obtain a concrete bounded variant of the asymptotic realization of $z^*$. Consider another approximate solution. Namely, for $k \in \mathcal{N}$ suppose that $f^{(k)} \in F$ is defined as follows:

$$f^{(k)}(t) \triangleq \begin{cases} k^3, & t \in [0, 1/k^2[, \\ 0, & t \in [1/k^2, 1[. \end{cases}$$

Then for such control $f^{(k)}$ we have the property

$$\int_0^1 tf^{(k)}(t)\, dt = \int_0^{1/k^2} tf^{(k)}(t)\, dt \leq \frac{1}{k^2} \int_0^{1/k^2} f^{(k)}(t)\, dt = \frac{1}{k}. \quad (4.14.9)$$

From (4.14.9) we get the natural property of an approximate solution. In addition, for $k \in \mathcal{N}$

$$x_{f^{(k)}}(1) = \int_0^{1/k^2} f^{(k)}(t)\, dt = k.$$

As a consequence we have the convergence

$$\left(\mathbf{g}(f^{(k)})\right)_{k \in \mathcal{N}} \to 0. \quad (4.14.10)$$

From (4.14.9) and (4.14.10) we get a concrete variant of the unbounded asymptotic realization of $z^*$. So we have two different possibilities again.

It is of interest to consider some general conditions under which all points of AS admit only the bounded asymptotic realization (we strive to exclude the possibility of an unbounded asymptotic realization as undesirable). Of course, this section contains several singularities which show that general settings of Chapter 3 can be developed in very different directions. Let $\forall_X\, \mathbf{S}[X]\ \forall_M\, \mathbf{S}[M \neq \varnothing]\ \forall \tau \in (\mathrm{top})[X]\ \forall \preceq \in (\mathrm{DIR})[M]$ $\forall h \in X^M$:

$$(\tau - \mathrm{cl})[M; \preceq; h] \triangleq \{x \in X \mid \forall Q \in N_\tau(x) : h^{-1}(Q) \in (\preceq - \mathrm{cof})[M]\}. \quad (4.14.11)$$

In (4.14.11) we have the set of all limit points of an arbitrary net in TS. We introduce a more simple designation for finite products of TS. We use (2.6.22). If $\mathbf{S}[X]$, $\tau \in (\mathrm{top})[X]$ and $m \in \mathcal{N}$, then we use the notation $\otimes^m[\tau]$ instead of $\otimes^{\overline{1,m}}(\tau)$; so $(X^m, \otimes^m[\tau])$ is the TS called $m$-multiple product of samples of $(X, \tau)$.

We return to the general notation of Section 4.6, fixing a nonempty set $E$ and $\mathcal{L} \in \Pi[E]$. Moreover, as in Section 4.9 we fix $\eta \in (\mathrm{add})_+[\mathcal{L}]$ and consider the cone (4.9.3). Let (in this section)

$$\tau_\eta^*(\mathcal{L}) \triangleq \tau_*(\mathcal{L}) \mid_{(\mathrm{add})^+[\mathcal{L};\eta]} = \tau_*^+(\mathcal{L}) \mid_{(\mathrm{add})^+[\mathcal{L};\eta]} = \tau_\otimes^+(\mathcal{L}) \mid_{(\mathrm{add})^+[\mathcal{L};\eta]}; \quad (4.14.12)$$

so by (4.14.12) we obtain the nonempty TS

$$((\mathrm{add})^+[\mathcal{L};\eta], \tau^*_\eta(\mathcal{L})). \tag{4.14.13}$$

Of course, (4.14.13) is a subspace of the TS (4.6.9). Suppose that $r \in \mathcal{N}$ and

$$(\mathrm{add})^+_r[\mathcal{L};\eta] \triangleq (\mathrm{add})^+[\mathcal{L};\eta]^r. \tag{4.14.14}$$

We equip the nonempty set (4.14.14) with the natural topology of locally compact space. Namely, using (2.9.5), (4.14.12) and Proposition 4.6.1, we easily obtain the property

$$\otimes^r[\tau^*_\eta(\mathcal{L})] \in (\mathbf{c}-\mathrm{top})^0_{\mathrm{loc}}[(\mathrm{add})^+_r[\mathcal{L};\eta]].$$

Consider the corresponding locally compact TS

$$\left((\mathrm{add})^+_r[\mathcal{L};\eta], \otimes^r[\tau^*_\eta(\mathcal{L})]\right). \tag{4.14.15}$$

In (4.14.14) and (4.14.15) we have the usual product of samples of TS. Introduce

$$B^+_{0,r}[E;\mathcal{L}] \triangleq B^+_0(E,\mathcal{L})^r. \tag{4.14.16}$$

In (4.14.16) we act like in (4.14.14). We consider $f \in B^+_{0,r}[E;\mathcal{L}]$ as a function from $\overline{1,r}$ into $B^+_0(E,\mathcal{L})$. So the corresponding collections are regarded as functions on $\overline{1,r}$. Then by analogy with [35, p. 62] we have the following property of density:

$$(\mathrm{add})^+_r[\mathcal{L};\eta] = \mathrm{cl}(\{(f_i * \eta)_{i\in\overline{1,r}} : (f_i)_{i\in\overline{1,r}} \in B^+_{0,r}[E;\mathcal{L}]\}, \otimes^r[\tau^*_\eta(\mathcal{L})]). \tag{4.14.17}$$

Of course, (4.14.17) follows and from (4.9.14). We do not consider other properties of density, referring the reader to [35, Sect. 3.8].

Suppose $\forall_Q\, \mathbf{S}[Q \neq \varnothing]\ \forall \preceq \in (\mathrm{DIR})[Q]$:

$$\mathbb{B}_r(Q,\preceq,E,\mathcal{L},\eta) \triangleq \{g \in B^+_{0,r}[E;\mathcal{L}]^Q \mid \exists p \in Q\ \exists c \in ]0,\infty[\ \forall q \in Q : (p \preceq q) \Rightarrow (\sum_{i=1}^{r} \int_E g(q)(i) d\eta \le c)\}. \tag{4.14.18}$$

We consider each element of (4.14.18) as some essentially bounded approximate solution net. We now investigate a very general setting of the problem of asymptotic attainability.

Fix $\mathfrak{F} \in \beta\left[B^+_{0,r}[E;\mathcal{L}]\right]$ (see Chapter 3). We regard $\mathfrak{F}$ as distinctive asymptotic constraints. Moreover, we use the TS $(\mathbb{H},\theta)$, $\mathbb{H} \neq \varnothing$, and the mapping

$$w \in C((\mathrm{add})^+_r[\mathcal{L};\eta], \otimes^r[\tau^*_\eta(\mathcal{L})], \mathbb{H}, \theta). \tag{4.14.19}$$

We use (4.14.19) as the generalized goal mapping. Introduce the 'real' goal mapping

$$W \triangleq \left(w\left(\left(f_i * \eta\right)_{i\in\overline{1,r}}\right)\right)_{(f_i)_{i\in\overline{1,r}}\in B^+_{0,r}[E;\mathcal{L}]} \in \mathbb{H}^{B^+_{0,r}[E;\mathcal{L}]}. \tag{4.14.20}$$

Very often in concrete problems the goal mapping (4.14.20) is given previously. But in our theoretical constructions the previous definition of (4.14.19) is useful, since such approach permits us to investigate very different applied problems. Therefore we use the passage $w \to W$ realized in (4.14.20); of course,

$$W\left((f_i)_{i\in\overline{1,r}}\right) = w\left((f_i * \eta)_{i\in\overline{1,r}}\right)$$

for $(f_i)_{i\in\overline{1,r}} \in B^+_{0,r}[E;\mathcal{L}]$. There exist many concrete versions of (4.14.20). In particular, we can consider the following case. Fix $m \in \mathcal{N}$, $k \in \mathcal{N}$, a matriciant

$$(H_{i,j})_{(i,j)\in\overline{1,k}\times\overline{1,r}} : \overline{1,k}\times\overline{1,r} \to B(E,\mathcal{L}),$$

and a continuous mapping $g$ from $\mathbb{R}^k$ into $\mathbb{R}^m$. Let $W$ be as follows:

$$W\left((f_i)_{i\in\overline{1,r}}\right) = g\left(\left(\sum_{j=1}^{r}\int_E H_{i,j} f_j d\eta\right)_{i\in\overline{1,k}}\right), \tag{4.14.21}$$

where $(f_i)_{i\in\overline{1,r}} \in B^+_{0,r}[E;\mathcal{L}]$. Indeed, introduce $w$ by the rule:

$$w\left((\mu_i)_{i\in\overline{1,r}}\right) = g\left(\left(\sum_{j=1}^{r}\int_E H_{i,j}\, d\mu_j\right)_{i\in\overline{1,k}}\right) \tag{4.14.22}$$

for $(\mu_i)_{i\in\overline{1,r}} \in (\mathrm{add})^+_r[\mathcal{L};\eta]$. To represent (4.14.21) in terms of (4.14.20) and (4.14.22) we use the basic property of indefinite $\eta$-integrals. Indeed, by (4.4.5) and (4.14.22) we have $\forall (f_i)_{i\in\overline{1,r}} \in B^+_{0,r}[E;\mathcal{L}]$:

$$w\left((f_i * \eta)_{i\in\overline{1,r}}\right) = g\left(\left(\sum_{j=1}^{r}\int_E H_{i,j} f_j\, d\eta\right)_{i\in\overline{1,k}}\right) = W\left((f_i)_{i\in\overline{1,r}}\right).$$

This equality realizes (4.14.20). So by (4.14.21) we have a very typical concrete variant of (4.14.20).

Returning to the general case, we introduce some designations. Suppose that $\forall_T\, \mathbf{S}[T \neq \varnothing]\ \forall \preceq \in (\mathrm{DIR})[T]\ \forall \omega \in \mathbb{H}$:

$$(\mathrm{AS})[T;\preceq;\omega] \triangleq \{\alpha \in B^+_{0,r}[E;\mathcal{L}]^T \mid (\forall U \in \mathfrak{F}\ \exists m \in T\ \forall t \in T : (m \preceq t) \Rightarrow (\alpha(t) \in U))\&\ ((T,\preceq, W\circ\alpha) \xrightarrow{\theta} \omega)\}. \tag{4.14.23}$$

On the basis of (4.14.23) we introduce the usual AS (see Chapter 3)

$$\begin{aligned}\mathbf{AC} &\triangleq \{\omega \in \mathbb{H} \mid \exists_T\, \mathbf{S}[T \neq \varnothing]\ \exists \preceq \in (\mathrm{DIR})[T] : (\mathrm{AS})[T;\preceq;\omega] \neq \varnothing\}\\ &= (\theta - \mathrm{LIM})[\mathfrak{F} \mid W]\end{aligned} \tag{4.14.24}$$

and the new AS (based on (4.14.18))

$$\begin{aligned}\mathbf{BAC} \triangleq \{\omega \in \mathbf{AC} \,|\forall_T\, &\mathbf{S}[T \neq \varnothing]\ \forall \preceq \in (\mathrm{DIR})[T] :\\ &(\mathrm{AS})[T;\preceq;\omega] \subset \mathbb{B}_r(T,\preceq,E,\mathcal{L},\eta)\}.\end{aligned} \tag{4.14.25}$$

The last AS corresponds to the 'asymptotic' setting for which only integrally bounded approximate solutions are assumed.

Denote by $\mathbf{I}$ the mapping

$$(f_i)_{i\in\overline{1,r}} \mapsto (f_i * \eta)_{i\in\overline{1,r}} : B^+_{0,r}[E;\mathcal{L}] \to (\mathrm{add})^+_r[\mathcal{L};\eta].$$

THEOREM 4.14.1 *If* $\omega \in \mathbf{AC}$, *then the following equivalence holds:*

$$\begin{aligned}(\omega \in \mathbf{BAC}) \Leftrightarrow (\forall_T\, \mathbf{S}[T \neq \varnothing]\ \forall \preceq \in (\mathrm{DIR})[T]\ \forall h \in (\mathrm{AS})[T;\preceq;\omega] :\\ (\otimes^r[\tau^*_\eta(\mathcal{L})] - \mathrm{cl})[T;\preceq;\mathbf{I}\circ h] \neq \varnothing).\end{aligned} \tag{4.14.26}$$

The proof is very obvious. We give only a very brief scheme. If $\omega \in \mathbf{BAC}$, then the second statement of (4.14.26) holds by the compactness (in the TS (4.14.15)) of the sets

$$\Sigma^+_r[b;\eta] \triangleq \{(\mu_i)_{i\in\overline{1,r}} \in (\mathrm{add})^+_r[\mathcal{L};\eta] \mid \sum_{i=1}^{r} \mu_i(E) \leq b\} \quad (b \in [0,\infty[).$$

It is useful to compare (4.14.18) and $\Sigma^+_r[b;\eta]$, $b \geq 0$.

Let the last statement of (4.14.26) be valid. Then

$$\exists_T\, \mathbf{S}[T \neq \varnothing]\ \exists \preceq \in (\mathrm{DIR})[T] : (\mathrm{AS})[T;\preceq;\omega] \neq \varnothing. \tag{4.14.27}$$

We use (4.14.27). Let $\mathbf{S}[T \neq \varnothing]$ and $\preceq \in (\mathrm{DIR})[T]$ be such that $(\mathrm{AS})[T;\preceq;\omega] \neq \varnothing$. Fix $\varphi \in (\mathrm{AS})[T;\preceq;\omega]$. Then

$$\varphi \in \mathbb{B}_r(T,\preceq,E,\mathcal{L},\eta). \tag{4.14.28}$$

Indeed, suppose the contrary: (4.14.28) is not valid. Then by (4.14.18) $\forall p \in T\ \forall b \in\ ]0,\infty[\ \exists q \in T$:

$$(p \preceq q)\ \&\ (b < \sum_{i=1}^{r} \int_E \varphi(q)(i)d\eta). \tag{4.14.29}$$

Return to (4.14.29). We have $\forall z \in T\times]0,\infty[$:

$$\Phi_z \triangleq \{q \in T \mid (\mathrm{pr}_1(z) \preceq q) \,\&(\mathrm{pr}_2(z) < \sum_{i=1}^{r} \int_E \varphi(q)(i)d\eta)\} \in 2^T.$$

So we get the mapping

$$z \mapsto \Phi_z : T\times]0,\infty[ \to 2^T.$$

Using the axiom of choice we choose

$$\psi \in \prod_{z\in T\times]0,\infty[} \Phi_z. \tag{4.14.30}$$

Then for $\psi \in T^{T\times]0,\infty[}$ we get the property that if $z \in T\times]0,\infty[$, then

$$(\mathrm{pr}_1(z) \preceq \psi(z)) \,\&\, (\mathrm{pr}_2(z) < \sum_{i=1}^{r} \int_E (\varphi \circ \psi)(z)(i)d\eta). \tag{4.14.31}$$

On the basis of $\psi$ (4.14.30) we will construct a new net realizing the point $\omega$. For this, we equip $T\times]0,\infty[$ with the traditional construction of the directed product. Namely, we introduce the binary relation $\sqsubseteq$ in $T\times]0,\infty[$, setting by definition for $z' \in T\times]0,\infty[$ and $z' \in T\times]0,\infty[$:

$$(z' \sqsubseteq z') \Leftrightarrow ((\mathrm{pr}_1(z') \preceq \mathrm{pr}_1(z')) \,\&\, (\mathrm{pr}_2(z') \leq (\mathrm{pr}_2(z')). \tag{4.14.32}$$

Of course, $\sqsubseteq \in (\mathrm{DIR})[T\times]0,\infty[\,]$ and $(T\times]0,\infty[, \sqsubseteq, \psi)$ is a net in $T$. As a consequence, in the form of

$$(T\times]0,\infty[, \sqsubseteq, \varphi\circ\psi) \tag{4.14.33}$$

we have a net in $B^+_{0,r}[E;\mathcal{L}]$. By the choice of $\varphi$ we get

$$\varphi\circ\psi \in (\mathrm{AS})[T\times]0,\infty[, \sqsubseteq, \omega]. \tag{4.14.34}$$

Indeed, $\varphi \circ \psi : T\times]0,\infty[\to B^+_{0,r}[E;\mathcal{L}]$. Fix $U \in \mathfrak{F}$ and choose (using the basic property of $\varphi$) $m \in T$ for which $\forall t \in T$:

$$(m \preceq t) \Rightarrow (\varphi(t) \in U). \tag{4.14.35}$$

If $b \in ]0,\infty[$, then $\beta \triangleq (m, b) \in T\times]0,\infty[$ and by (4.14.31) and (4.14.32) for $z \in T\times]0,\infty[$ with the property $\beta \sqsubseteq z$ we have

$$(m \preceq \mathrm{pr}_1(z)) \,\&\, (\mathrm{pr}_1(z) \preceq \psi(z)),$$

which owing to (4.14.35) implies

$$(\varphi\circ\psi)(z) = \varphi(\psi(z)) \in U.$$

Of course, we use (4.14.31) too. We get the property

$$\forall U' \in \mathfrak{F}\ \exists z' \in T\times]0,\infty[\ \forall z \in T\times]0,\infty[: (z' \sqsubseteq z) \Rightarrow ((\varphi \circ \psi)(z) \in U'). \tag{4.14.36}$$

Let $\Lambda \in N_\theta(\omega)$. By the choice of $\varphi$ we have

$$\Lambda \in (\mathbb{H} - \text{ass})[T; \preceq; W \circ \varphi].$$

Choose $t_* \in T$ such that for $t \in T$ with the property $t_* \preceq t$ the inclusion $(W \circ \varphi)(t) = W(\varphi(t)) \in \Lambda$ holds. Fix now $b_* \in ]0,\infty[$ and consider $z_* \triangleq (t_*, b_*) \in T\times]0,\infty[$. Choose $\tilde{z} \in T\times]0,\infty[$ with the property $z_* \sqsubseteq \tilde{z}$. Then from (4.14.32) it follows that $t_* \preceq \text{pr}_1(\tilde{z})$ and from (4.14.31) we have $\text{pr}_1(z) \preceq \psi(\tilde{z})$. As a consequence $(\varphi \circ \psi)(\tilde{z}) \in B^+_{0,r}[E;\mathcal{L}]$ has the properties $(\varphi \circ \psi)(\tilde{z}) = \varphi(\psi(\tilde{z}))$ and $t_* \preceq \psi(\tilde{z})$. Hence

$$(W \circ \varphi \circ \psi)(\tilde{z}) = (W \circ \varphi)(\psi(\tilde{z})) \in \Lambda.$$

Since the choice of $z$ was arbitrary, we have $\forall z \in T\times]0,\infty[$:

$$(z_* \sqsubseteq z) \Rightarrow ((W \circ \varphi \circ \psi)(z) \in \Lambda).$$

Since the choice of $\Lambda$ was arbitrary, we get

$$N_\theta(\omega) \subset (\mathbb{H} - \text{ass})[T\times]0,\infty[; \sqsubseteq; W \circ \varphi \circ \psi].$$

In other words, we have established the convergence

$$(T\times]0,\infty[, \sqsubseteq, W \circ \varphi \circ \psi) \xrightarrow{\theta} \omega.$$

From (4.14.23) and (4.14.36) we get (4.14.34). Thus (4.14.33) is a net realizing the point $\omega$. Then

$$(\otimes^r[\tau^*_\eta(\mathcal{L})] - \text{cl})[T\times]0,\infty[; \sqsubseteq; \mathbf{I} \circ \varphi \circ \psi] \neq \varnothing. \tag{4.14.37}$$

Let $\nu$ be an element of the set on the left hand side of (4.14.37). Hence $\nu$ is the limit point of the net $(T\times]0,\infty[, \sqsubseteq, \mathbf{I}\circ\varphi\circ\psi)$, and $\nu \in (\text{add})^+_r[\mathcal{L};\eta]$. In addition, $\forall Q \in N_{\otimes^r[\tau^*_\eta(\mathcal{L})]}(\nu)$:

$$(\mathbf{I} \circ \varphi \circ \psi)^{-1}(Q) \in (\sqsubseteq - \text{cof})[T\times]0,\infty[\,]. \tag{4.14.38}$$

We use (4.14.38) for some special choice of $Q$. Note that

$$a_0 \triangleq \sum_{i=1}^{r} \nu(i)(E) \in [0,\infty[.$$

Recall that for $i \in \overline{1,r}$

$$\mathbb{Q}_i \triangleq \{\xi \in (\text{add})^+[\mathcal{L};\eta] \mid |\nu(i)(E) - \xi(E)| < 1/r\} \in \tau^*_\eta(\mathcal{L}).$$

As a consequence we have

$$\prod_{i=1}^{r} \mathbb{Q}_i \in N^0_{\otimes^r[\tau^*_\eta(\mathcal{L})]}(\nu).$$

Here we use (4.14.12). It is worth noting that

$$\prod_{i=1}^{r} \mathbb{Q}_i \subset \{\mu \in (\mathrm{add})^+_r[\mathcal{L};\eta] \mid \sum_{i=1}^{r} \mu(i)(E) < a_0 + 1\}. \qquad (4.14.39)$$

Denote by $\tilde{\mathbb{Q}}$ the set on the right hand side of (4.14.39). By (4.14.38) we have

$$(\mathbf{I} \circ \varphi \circ \psi)^{-1}(\tilde{\mathbb{Q}}) \in (\sqsubseteq - \mathrm{cof})[T\times]0, \infty[\,]. \qquad (4.14.40)$$

Consider a natural combination of (4.14.39) and (4.14.40). From (2.2.7) and (4.14.40) we get that

$$(\mathbf{I} \circ \varphi \circ \psi)^{-1}(\tilde{\mathbb{Q}}) \in \mathcal{P}([T\times]0, \infty[\,)$$

has the following property:

$$\forall z' \in T\times]0, \infty[\; \exists z' \in (\mathbf{I} \circ \varphi \circ \psi)^{-1}(\tilde{\mathbb{Q}}) : z' \sqsubseteq z'. \qquad (4.14.41)$$

Fix $\mathbf{t} \in T$ and consider the point

$$\mathbf{z} \triangleq (\mathbf{t}, a_0 + 1) \in T\times]0, \infty[.$$

From (4.14.41) it follows that $\mathbf{z} \sqsubseteq \lambda$ for some $\lambda \in (\mathbf{I} \circ \varphi \circ \psi)^{-1}(\tilde{\mathbb{Q}})$. Introduce $\lambda_1 \triangleq \mathrm{pr}_1(\lambda) \in T$ and $\lambda_2 \triangleq \mathrm{pr}_2(\lambda) \in ]0, \infty[$. By (4.14.32) we have

$$(\mathbf{t} \preceq \lambda_1) \;\&\; (a_0 + 1 \leq \lambda_2). \qquad (4.14.42)$$

By the choice of $\lambda$ we have $(\mathbf{I} \circ \varphi \circ \psi)(\lambda) \in \tilde{\mathbb{Q}}$. In addition, $\lambda \in T\times]0, \infty[$. Then by the definition of $\mathbf{I}$ we get

$$((\varphi \circ \psi)(\lambda)(i) * \eta)_{i \in \overline{1,r}} \in \tilde{\mathbb{Q}}.$$

From (4.14.39) we obtain

$$\sum_{i=1}^{r} ((\varphi \circ \psi)(\lambda)(i) * \eta)(E) = \sum_{i=1}^{r} \int_E (\varphi \circ \psi)(\lambda)(i)\, d\eta < a_0 + 1. \qquad (4.14.43)$$

By (4.14.31) and (4.14.42) we derive the inequality

$$a_0 + 1 \leq \lambda_2 = \mathrm{pr}_2(\lambda) < \sum_{i=1}^{r} \int_E (\varphi \circ \psi)(\lambda)(i)\, d\eta,$$

which contradicts to (4.14.43). This contradiction proves (4.14.28). Thus

$$(\mathrm{AS})[T;\preceq;\omega] \subset \mathbb{B}_r(T,\preceq,E,\mathcal{L},\eta).$$

Since the choice of $T$ and $\preceq$ was arbitrary, we have

$$\forall_T\, \mathbf{S}[T\neq\varnothing]\ \forall \preceq\in (\mathrm{DIR})[T] : (\mathrm{AS})[T;\preceq;\omega] \subset \mathbb{B}_r(T,\preceq,E,\mathcal{L},\eta).$$

Since $\omega\in\mathbf{AC}$, from (4.14.25) we get the inclusion $\omega\in\mathbf{BAC}$. So the validity of the second statement of (4.14.26) implies the validity of its first statement. □

**A special case.** In the sequel (until the end of this section) we consider a special case when $(\mathbb{H},\theta)$ is a metrizable space. Let

$$\rho:\mathbb{H}\times\mathbb{H}\to[0,\infty[$$

be the metric of $\mathbb{H}$ generating the topology $\theta$. So $(\mathbb{H},\rho)$ is a metric space.

CONDITION 4.14.1 $\exists\omega^0\in\mathbb{H}\ \forall a\in\,]0,\infty[\ \exists b\in[0,\infty[\ \forall\mu\in(\mathrm{add})_r^+[\mathcal{L};\eta]$:

$$\left(b<\sum_{i=1}^{r}\mu(i)(E)\right)\Rightarrow(a\le\rho(\omega^0,w(\mu))).$$

This condition was used (in fact) in [35, Chapter 4] for other goal.

THEOREM 4.14.2 *If Condition 4.14.1 is valid, then* $\mathbf{AC}=\mathbf{BAC}$.

PROOF. Let Condition 4.14.1 be valid. From (4.14.25) we get the inclusion $\mathbf{BAC}\subset\mathbf{AC}$. Let $\omega\in\mathbf{AC}$. Then $\omega\in\mathbb{H}$. Choose a directed set $(T,\preceq)$, $T\neq\varnothing$, with the property $(\mathrm{AS})[T;\preceq;\omega]\neq\varnothing$. So $\mathbf{S}[T\neq\varnothing]$ and $\preceq\in(\mathrm{DIR})[T]$. Let $h\in(\mathrm{AS})[T;\preceq;\omega]$. Then $(T,\preceq,h)$ is a net in $B^+_{0,r}[E;\mathcal{L}]$ and

$$h:T\to B^+_{0,r}[E;\mathcal{L}].$$

By (4.14.23) we have the following two statements:

$$(\forall U\in\mathfrak{F}\ \exists m\in T\ \forall t\in T:(m\preceq t)\Rightarrow(h(t)\in U))\ \&\ ((T,\preceq,W\circ h)\xrightarrow{\theta}\omega). \tag{4.14.44}$$

But $\theta=\tau_\rho^\natural(\mathbb{H})$ (see Section 2.7). From (4.14.44) we get the convergence

$$(T,\preceq,(\rho(W(h(t)),\omega))_{t\in T})\xrightarrow{\tau_{\mathbb{R}}}0.$$

Therefore for some $t^0\in T$ we have $\forall t\in T$:

$$(t^0\preceq t)\Rightarrow(\rho(W(h(t)),\omega)<1). \tag{4.14.45}$$

Using Condition 4.14.1 we choose $\omega^0 \in \mathbb{H}$ for which $\forall a \in ]0, \infty[\ \exists b \in [0, \infty[\ \forall \mu \in (\mathrm{add})_r^+[\mathcal{L}; \eta]$:

$$\left(b < \sum_{i=1}^{r} \mu(i)(E)\right) \Rightarrow (a \leq \rho(\omega^0, w(\mu))).$$

We can use the number $\rho(\omega^0, \omega) + 1$ instead of $a$. Let $b^0 \in [0, \infty[$ be a number for which $\forall \mu \in (\mathrm{add})_r^+[\mathcal{L}; \eta]$:

$$(b^0 < \sum_{i=1}^{r} \mu(i)(E) \Rightarrow ((\rho(\omega^0, \omega) + 1 \leq \rho(\omega^0, w(\mu))). \tag{4.14.46}$$

In (4.14.46) we use the case

$$\mu = (h(t)(i) * \eta)_{i \in \overline{1,r}},$$

where $t \in T$; then

$$\sum_{i=1}^{r} \mu(i)(E) = \sum_{i=1}^{r} (h(t)(i) * \eta)(E) = \sum_{i=1}^{r} \int_E h(t)(i) d\eta,$$

and by (4.14.20) we get the following representation:

$$W(h(t)) = (W \circ h)(t) = w((h(t)(i) * \eta)_{i \in \overline{1,r}}) \in \mathbb{H}. \tag{4.14.47}$$

From (4.14.45) it follows that $\forall t \in T$:

$$((t^0 \preceq t) \Rightarrow (\rho(W(h(t)), \omega^0) < \rho(\omega^0, \omega) + 1). \tag{4.14.48}$$

From (4.14.46) and (4.14.47) we have the implication

$$\left(b^0 < \sum_{i=1}^{r} (h(t)(i) * \eta)(E)\right) \Rightarrow (\rho(\omega^0, \omega) + 1 \leq \rho(W(h(t)), \omega^0)). \tag{4.14.49}$$

Owing to (4.14.48) and (4.14.49) we get for $t \in T$ with the property $t^0 \preceq t$ the inequality

$$\sum_{i=1}^{r} \int_E h(t)(i) d\eta \leq b^0.$$

As a consequence we have the property

$$h \in \mathbb{B}_r(T, \preceq, E, \mathcal{L}, \eta).$$

Since the choice of $h$ was arbitrary, we have the inclusion

$$(\mathrm{AS})[T; \preceq; \omega] \subset \mathbb{B}_r(T, \preceq, E, \mathcal{L}, \eta).$$

But the choice of $T$ and $\preceq$ was arbitrary too. Hence by (4.14.25) we get $\omega \in \mathbf{BAC}$. Thus $\mathbf{AC} \subset \mathbf{BAC}$. □

**Some examples.** Consider a variant of linear control systems with discontinuous coefficients. Let $p \in \mathcal{N}$, $\mathbb{H} = \mathbb{R}^p$ and $\theta = \tau_{\mathbb{R}}^{(p)}$. In addition, $(\mathbb{H}, \theta) = (\mathbb{R}^p, \tau_{\mathbb{R}}^{(p)})$ is a concrete metrizable TS. Fix $\omega^0 \in \mathbb{R}^p$ and the matriciant

$$(i,j) \mapsto M_{i,j} : \overline{1,p} \times \overline{1,r} \to B^+(E, \mathcal{L})$$

for which

$$\exists \alpha \in ]0, \infty[\ \forall j \in \overline{1,r}\ \exists i \in \overline{1,p}\ \forall x \in E : \alpha \leq M_{i,j}(x). \tag{4.14.50}$$

Let the operator $W$ be defined as

$$f \mapsto \omega^0 + \left( \sum_{j=1}^{r} \int_E M_{i,j} f(j) d\eta \right)_{i \in \overline{1,p}} : B_{0,r}^{+}[E; \mathcal{L}] \to \mathbb{R}^p. \tag{4.14.51}$$

We introduce the mapping $w$ corresponding to the operator (4.14.51) as follows:

$$\mu \mapsto \omega^0 + \left( \sum_{j=1}^{r} \int_E M_{i,j} d\mu(j) \right)_{i \in \overline{1,p}} : (\mathrm{add})_r^+[\mathcal{L}; \eta] \to \mathbb{R}^p. \tag{4.14.52}$$

From (4.14.51) and (4.14.52) we have the representation (4.14.20). We now define $\rho$ as the metric of $\mathbb{H}$ generated by the sup-norm: for $(x_i')_{i \in \overline{1,p}} \in \mathbb{R}^p$ and $(x_i')_{i \in \overline{1,p}} \in \mathbb{R}^p$,

$$\rho((x_i')_{i \in \overline{1,p}}, (x_i')_{i \in \overline{1,p}}) \triangleq \sup(\{ \mid x_i' - x_i' \mid : i \in \overline{1,p}\}).$$

Return to (4.14.52). Note that for $\mu \in (\mathrm{add})_r^+[\mathcal{L}; \eta]$ we have

$$\rho(\omega^0, w(\mu)) = \sup\left( \left\{ \left| \sum_{j=1}^{r} \int_E M_{i,j} d\mu(j) \right| : i \in \overline{1,p} \right\} \right). \tag{4.14.53}$$

Using (4.14.50), fix $\alpha \in ]0, \infty[$ with the property

$$\forall j \in \overline{1,r}\ \exists i \in \overline{1,p}\ \forall x \in E : \alpha \leq M_{i,j}(x). \tag{4.14.54}$$

Consider (4.14.53). If $a \in ]0, \infty[$, then we suppose $b_a = pa/\alpha$. Choose arbitrarily $\mu \in (\mathrm{add})_r^+[\mathcal{L}; \eta]$ with the property $b_a < \sum_{j=1}^{r} \mu(j)(E)$. Then $a/\alpha < \mu(k)(E)$ for some $k \in \overline{1,r}$. From (4.14.54) for some $q \in \overline{1,p}$ we get $\forall x \in E : \alpha \leq M_{q,k}(x)$. As a consequence we obviously have

$$a < \alpha\mu(k)(E) \leq \int_E M_{q,k} d\mu(k) \leq \sum_{j=1}^{r} \int_E M_{q,j} d\mu(j) \leq \rho(\omega^0, w(\mu)).$$

Here Condition 4.14.1 is satisfied. Note that in (4.14.51) we have a model of the Cauchy formula in the theory of linear controlled differential equation. In (4.14.52) we have a model of the extended Cauchy formula. In this part we obtain an analogy with the constructions of the previous section. Of course, it is possible to give many examples of concrete linear controlled systems for which the basic condition (4.14.50) holds. In particular, this condition can be realized in the problem of investigating the attainability domain for a unit mass point in the whole phase space. But now we give only the following example.

Suppose that $p = r$ and $\vartheta_0 \in ]0, \infty[$. Consider the system

$$\dot{x}(t) = Ax(t) + f(t) \tag{4.14.55}$$

that functions in the phase space $\mathbb{R}^p$ on the time interval $[0, \vartheta_0]$; an initial condition is given: $x(0) = x_0 \in \mathbb{R}^p$. Suppose that $f = f(\cdot)$ is a non-negative (component-wise) vector function; in addition, $f$ is assumed to be an element of $B^+_{0,r}[E; \mathcal{L}]$ under $E = [0, \vartheta_0[$ and the simplest variant of $\mathcal{L}$: $\mathcal{L}$ is $\{[u, v[ : 0 \leq u \leq v \leq \vartheta_0\}$. Let $A$ be the diagonal $(r \times r)$-matrix with diagonal elements $\lambda_1 \in \mathbb{R}, \ldots, \lambda_r \in \mathbb{R}$. In this case the matriciant is defined as the fundamental matrix of solutions for the homogeneous system $\dot{x} = Ax$:

$$\dot{x}_1 = \lambda_1 x_1, \ldots, \dot{x}_p = \lambda_p x_p.$$

We have a very simple homogeneous system corresponding to (4.14.55). It follows that $\forall i \in \overline{1,p}$:

$$\alpha_i \triangleq \exp\left(- \mid \lambda_i \mid \vartheta_0\right) \in ]0, \infty[;$$

let $\alpha \triangleq \inf(\{\alpha_i : i \in \overline{1,p}\})$. Then $\alpha \in ]0, \infty[$ corresponds to (4.14.50). Indeed, $\forall i \in \overline{1,p}\ \forall t \in [0, \vartheta_0] : \alpha \leq \alpha_i \leq \exp \lambda_i(\vartheta_0 - t)$.

## 4.15 CONCLUSION

The main contents of Chapter 4 are connected with applying FAM as generalized elements in the constructions of Chapter 3. For the matter of that, this chapter continues the investigations [28, 32, 35, 40, 45, 47].

However, there is a new important element, which to a considerable extent is reflected in Sections 4.11 and 4.12. The question is of a peculiar logical interconnection of rather different schemes of extension with using FAM, which earlier were considered singly with applying special methods. In connection with general constructions of the FAM theory we note the works [92]–[96], [106, 110, 118, 121]; especially, we should recall the monograph [66, Chapters 3, 4], where a harmonic theory on integration with respect to FAM is suggested and important properties of spaces of FAM are established (as for the earlier investigations on the theory of integration with respect to FAM, we should note [72, 76]). At the present time, many remarkable results are obtained in the FAM theory, but we restrict ourselves to only the most close ones. In this connection, de Finetti's work [73] on the theory of finitely additive probabilities should be also noted; in addition, see [64, 74]. It seems that the statements of the FAM theory naturally supplement the classic (Lebesgue) measure theory. In particular, such a combination is useful for the probability theory, where impressive achievements of the Kolmogorov theory is complimented with constructions using finitely additive probability (for example, in the theory of random processes).

# Chapter 5

# COMPACTIFICATIONS AND PROBLEMS OF INTEGRATION

## 5.1 INTRODUCTION

In this chapter we continue to consider the effects arising when employing extensions in the class of FAM. But we now discuss some applications of extensions to problems of 'pure' mathematics. We investigate the problems of universal integrability of bounded functions and universal measurability of sets. In these questions the effects of compactifications in the class of FAM are also presented. Of course, here we strive to attain other goals. We pay the basic attention to corresponding constructions of functional analysis. However, in the most essential questions the constructions considered below have much in common with constructions of the applied character given in the previous chapters.

## 5.2 INTEGRATION IN THE SENSE OF DARBOUX

We consider a natural scheme of integration. This scheme differs from the corresponding definition of Section 4.3 (see (4.3.13), (4.3.14)). We call this scheme the Darboux scheme although it can be regarded as the Riemann scheme. In connection with the considered problem we note the investigations [3, 20, 22, 92].

Let us return to the construction of the elementary integral in Section 4.3. This construction (see (4.3.7), (4.3.8)) is very natural. However, this scheme can be extended to a more general class of bounded functions in not only the way (4.3.13), (4.3.14).

In this chapter we fix a nonempty set $E$ and $\mathcal{L} \in \Pi[E]$ (see (4.2.3)) unless otherwise stipulated. Thus we consider the MS $(E, \mathcal{L})$ with the fixed semi-algebra of sets. If $n \in \mathcal{N}$ and $(L_i)_{i \in \overline{1,n}} \in \Delta_n(E, \mathcal{L})$, then we set

$$\mathfrak{K}[(L_i)_{i\in\overline{1,n}}] \triangleq \{k \in \overline{1,n} \mid L_k \neq \varnothing\}; \tag{5.2.1}$$

if $x \in E$, then $\exists! j \in \mathfrak{K}[(L_i)_{i\in\overline{1,n}}] : x \in L_j$. If $f \in \mathbb{B}(E)$, $n \in \mathcal{N}$ and $(L_i)_{i\in\overline{1,n}} \in \Delta_n(E,\mathcal{L})$, then $\underline{\mathbb{G}}[f;(L_i)_{i\in\overline{1,n}}] \in \mathbb{R}^E$ is defined by the following rule:

$$\underline{\mathbb{G}}[f;(L_i)_{i\in\overline{1,n}}](x) \triangleq \inf(\{f(y) : y \in L_k\}) \tag{5.2.2}$$

for $k \in \mathfrak{K}[(L_i)_{i\in\overline{1,n}}]$ and $x \in L_k$. Of course, (5.2.2) is correct owing to (5.2.1). Analogously, for $f \in \mathbb{B}(E)$, $n \in \mathcal{N}$ and $(L_i)_{i\in\overline{1,n}} \in \Delta_n(E,\mathcal{L})$ we define $\overline{\mathbb{G}}[f;(L_i)_{i\in\overline{1,n}}] \in \mathbb{R}^E$ by the rule:

$$\overline{\mathbb{G}}[f;(L_i)_{i\in\overline{1,n}}](x) \triangleq \sup(\{f(y) : y \in L_k\})$$

for $k \in \mathfrak{K}[(L_i)_{i\in\overline{1,n}}]$ and $x \in L_k$. Of course, $\forall f \in \mathbb{B}(E)$ $\forall n \in \mathcal{N}$ $\forall (L_i)_{i\in\overline{1,n}} \in \Delta_n(E,\mathcal{L})$:

$$(\underline{\mathbb{G}}[f;(L_i)_{i\in\overline{1,n}}] \leqq f) \,\&(f \leqq \overline{\mathbb{G}}[f;(L_i)_{i\in\overline{1,n}}]). \tag{5.2.3}$$

We note that the step minorant and step majorant are used in (5.2.3) because $\forall f \in \mathbb{B}(E)$ $\forall n \in \mathcal{N}$ $\forall (L_i)_{i\in\overline{1,n}} \in \Delta_n(E,\mathcal{L})$:

$$(\underline{\mathbb{G}}[f;(L_i)_{i\in\overline{1,n}}] \in B_0(E,\mathcal{L})) \,\&(\overline{\mathbb{G}}[f;(L_i)_{i\in\overline{1,n}}] \in B_0(E,\mathcal{L})). \tag{5.2.4}$$

By (4.3.11) we obtain $\forall s \in B(E,\mathcal{L})$ $\forall \varepsilon \in ]0,\infty[$ $\exists n \in \mathcal{N}$ $\exists (L_i)_{i\in\overline{1,n}} \in \Delta_n(E,\mathcal{L})$:

$$(\|\underline{\mathbb{G}}[s;(L_i)_{i\in\overline{1,n}}] - s\| < \varepsilon) \,\&(\|\overline{\mathbb{G}}[s;(L_i)_{i\in\overline{1,n}}] - s\| < \varepsilon). \tag{5.2.5}$$

As for (5.2.4) and (5.2.5), it is advisable to introduce the monotone uniform convergence of sequences in $\mathbb{B}(E)$. We set $\forall (f_i)_{i\in\mathcal{N}} \in \mathbb{B}(E)^{\mathcal{N}}$ $\forall f \in \mathbb{B}(E)$:

$$\begin{aligned}&\Big(((f_i)_{i\in\mathcal{N}} \upuparrows f) \Leftrightarrow ((f_i)_{i\in\mathcal{N}} \rightrightarrows f) \,\&\, (\forall j \in \mathcal{N} : f_j \leqq f_{j+1})\Big)\\ \&&\Big(((f_i)_{i\in\mathcal{N}} \downdownarrows f) \Leftrightarrow ((f_i)_{i\in\mathcal{N}} \rightrightarrows f) \,\&\, (\forall j \in \mathcal{N} : f_{j+1} \leqq f_j)\Big).\end{aligned} \tag{5.2.6}$$

Any sequence in $B_0(E,\mathcal{L})$ can be used instead of $(f_i)_{i\in\mathcal{N}}$ in (5.2.6). By (5.2.3), (5.2.5) and (5.2.6) we have $\forall f \in B(E,\mathcal{L})$:

$$\begin{aligned}&\big(\exists (f_i)_{i\in\mathcal{N}} \in B_0(E,\mathcal{L})^{\mathcal{N}} : (f_i)_{i\in\mathcal{N}} \upuparrows f\big)\\ \&&\Big(\exists (\tilde{f}_i)_{i\in\mathcal{N}} \in B_0(E,\mathcal{L})^{\mathcal{N}} : (f_i)_{i\in\mathcal{N}} \downdownarrows f\Big).\end{aligned} \tag{5.2.7}$$

A weaker property holds for arbitrary bounded functions on $E$. Namely, $\forall f \in \mathbb{B}(E)$:

$$\Big(B^0_{\leqq}(E,\mathcal{L},f) \triangleq \{\tilde{f} \in B_0(E,\mathcal{L}) \mid \tilde{f} \leqq f\} \in 2^{B_0(E,\mathcal{L})}\Big)$$
$$\& \Big(B^0_{\geqq}(E,\mathcal{L},f) \triangleq \{\hat{f} \in B_0(E,\mathcal{L}) \mid f \leqq \hat{f}\} \in 2^{B_0(E,\mathcal{L})}\Big). \qquad (5.2.8)$$

From (5.2.8) we obtain $\forall \mu \in (\mathrm{add})_+[\mathcal{L}]\ \forall f \in \mathbb{B}(E)$:

$$\Big(\{\tilde{f} * \mu : \tilde{f} \in B^0_{\leqq}(E,\mathcal{L},f)\} \in 2^{\mathbf{A}(\mathcal{L})}\Big)$$
$$\& \Big(\{\hat{f} * \mu : \hat{f} \in B^0_{\geqq}(E,\mathcal{L},f)\} \in 2^{\mathbf{A}(\mathcal{L})}\Big). \qquad (5.2.9)$$

The lower and upper Darboux integrals can be introduced on the basis of (5.2.9). We consider the construction of [20, 22] using ordinal properties of $\mathbf{A}(\mathcal{L})$. However, beforehand we note some simplest ordinal properties of such a type.

## 5.3 SOME ORDINAL PROPERTIES OF FINITELY ADDITIVE MEASURES OF BOUNDED VARIATION

We consider $\mathbf{A}(\mathcal{L})$ with the order defined in Section 2.2. Namely, $\mathbf{A}(\mathcal{L})$ is regarded as a subspace of $\mathbb{R}^{\mathcal{L}}$ with the pointwise order. Then $\forall \mu \in \mathbf{A}(\mathcal{L})\ \forall \nu \in \mathbf{A}(\mathcal{L})$:

$$(\mu \leqq \nu) \Leftrightarrow (\forall L \in \mathcal{L} : \mu(L) \leq \nu(L)). \qquad (5.3.1)$$

Certainly, (5.3.1) defines the natural order of $\mathbf{A}(\mathcal{L})$. We recall properties of this ordered space very briefly. By (2.2.3),

$$\leqq_{\mathcal{L}} \triangleq \{z \in \mathbf{A}(\mathcal{L}) \times \mathbf{A}(\mathcal{L}) \mid \mathrm{pr}_1(z) \leqq \mathrm{pr}_2(z)\} \in (\mathrm{Ord})_0[\mathcal{L}];$$

we deal with $(\mathbf{A}(\mathcal{L}), \leqq_{\mathcal{L}})$. Of course, $\forall \mu \in \mathbf{A}(\mathcal{L})\ \forall \nu \in \mathbf{A}(\mathcal{L})$:

$$(\mu \leqq_{\mathcal{L}} \nu) \Leftrightarrow (\mu \leqq \nu).$$

We use the traditional definitions of minorants and majorants. In this connection we set $\forall U \in \mathcal{P}(\mathbf{A}(\mathcal{L}))\ \forall V \in \mathcal{P}(\mathbf{A}(\mathcal{L}))$:

$$\Big((\leqq_{\mathcal{L}} - \mathrm{Ma})_U[V] \triangleq \{\mu \in U \mid \forall \nu \in V : \nu \leqq \mu\}\Big)$$
$$\& \Big((\leqq_{\mathcal{L}} - \mathrm{Mi})_U[V] \triangleq \{\mu \in U \mid \forall \nu \in V : \mu \leqq \nu\}\Big). \qquad (5.3.2)$$

Moreover, if $H \in \mathcal{P}(\mathbf{A}(\mathcal{L}))$, then

$$\begin{aligned}&\Big((\leqq_{\mathcal{L}} - \mathrm{Ma})[H] \triangleq (\leqq_{\mathcal{L}} - \mathrm{Ma})_{\mathbf{A}(\mathcal{L})}[H]\Big)\\ \& &\Big((\leqq_{\mathcal{L}} - \mathrm{Mi})[H] \triangleq (\leqq_{\mathcal{L}} - \mathrm{Mi})_{\mathbf{A}(\mathcal{L})}[H]\Big)\\ \& &\Big((\leqq_{\mathcal{L}} - \mathrm{Ma})^0[H] \triangleq (\leqq_{\mathcal{L}} - \mathrm{Ma})_H[H]\Big)\\ \& &\Big((\leqq_{\mathcal{L}} - \mathrm{Mi})^0[H] \triangleq (\leqq_{\mathcal{L}} - \mathrm{Mi})_H[H]\Big).\end{aligned} \tag{5.3.3}$$

In (5.3.2) and (5.3.3) we define the corresponding sets of minorants and majorants; moreover, for each set $H$, $H \subset \mathbf{A}(\mathcal{L})$, we introduce the sets of all greatest and smallest elements of $H$. The greatest and smallest elements are defined in a unique way when $H$ is a nonempty set; certainly, the existence of such elements is supposed. Thus $\forall H \in \mathcal{P}(\mathbf{A}(\mathcal{L}))$:

$$\big((\leqq_{\mathcal{L}} - \mathrm{Ma})^0[H] \neq \varnothing\big) \Rightarrow \big(\exists! \mu \in H : (\leqq_{\mathcal{L}} - \mathrm{Ma})^0[H] = \{\mu\}\big). \tag{5.3.4}$$

Analogously, if $H \in \mathcal{P}(\mathbf{A}(\mathcal{L}))$, then

$$\big((\leqq_{\mathcal{L}} - \mathrm{Mi})^0[H] \neq \varnothing\big) \Rightarrow \big(\exists! \mu \in H : (\leqq_{\mathcal{L}} - \mathrm{Mi})^0[H] = \{\mu\}\big). \tag{5.3.5}$$

We use (5.3.4) and (5.3.5) for determining the supremum and infimum. Let

$$(\mathrm{SUP})[\mathcal{L}] \triangleq \{H \in \mathcal{P}(\mathbf{A}(\mathcal{L})) \mid (\leqq_{\mathcal{L}} - \mathrm{Mi})^0[(\leqq_{\mathcal{L}} - \mathrm{Ma})[H]\,] \neq \varnothing\}. \tag{5.3.6}$$

From (5.3.5) and (5.3.6) we have $\forall H \in (\mathrm{SUP})[\mathcal{L}]\ \exists! \mu \in (\leqq_{\mathcal{L}} - \mathrm{Ma})[H]$:

$$(\leqq_{\mathcal{L}} - \mathrm{Mi})^0[(\leqq_{\mathcal{L}} - \mathrm{Ma})[H]\,] = \{\mu\}. \tag{5.3.7}$$

Using (5.3.7) we define for $H \in (\mathrm{SUP})[\mathcal{L}]$ the unique element

$$(\mathcal{L} - \sup)[H] \in (\leqq_{\mathcal{L}} - \mathrm{Ma})[H] \tag{5.3.8}$$

for which

$$(\leqq_{\mathcal{L}} - \mathrm{Mi})^0[(\leqq_{\mathcal{L}} - \mathrm{Ma})[H]\,] = \{(\mathcal{L} - \sup)[H]\}. \tag{5.3.9}$$

In (5.3.8) and (5.3.9) we have the usual definition of the supremum of a set. From (5.3.6) it follows that $\forall \mu \in \mathbf{A}(\mathcal{L})$:

$$\{\mu\} \in (\mathrm{SUP})[\mathcal{L}].$$

Hence we have a nonempty family in (5.3.6). If $H \in (\mathrm{SUP})[\mathcal{L}]$, then $(\mathcal{L} - \sup)[H] \in \mathbf{A}(\mathcal{L})$ possesses the following property:

$$(\forall \mu \in H : \mu \leqq (\mathcal{L} - \sup)[H])\ \&\ (\forall \nu \in (\leqq_{\mathcal{L}} - \mathrm{Ma})[H] : (\mathcal{L} - \sup)[H] \leqq \nu).$$

Moreover, we set

$$(\mathrm{INF})[\mathcal{L}] \triangleq \{H \in \mathcal{P}(\mathbf{A}(\mathcal{L})) \mid (\leqq_{\mathcal{L}} - \mathrm{Ma})^0[(\leqq_{\mathcal{L}} - \mathrm{Mi})[H]\,] \neq \varnothing\}. \tag{5.3.10}$$

Of course, $\forall \mu \in \mathbf{A}(\mathcal{L})$: $\{\mu\} \in (\mathrm{INF})[\mathcal{L}]$. Therefore $\mathbf{S}[(\mathrm{INF})[\mathcal{L}] \neq \varnothing]$. From (5.3.4) we obtain that for $H \in (\mathrm{INF})[\mathcal{L}]$ $\exists! \mu \in (\leqq_{\mathcal{L}} - \mathrm{Mi})[H]$:

$$(\leqq_{\mathcal{L}} - \mathrm{Ma})^0[(\leqq_{\mathcal{L}} - \mathrm{Mi})[H]\,] = \{\mu\}. \tag{5.3.11}$$

We use (5.3.4) and (5.3.10) in (5.3.11). By (5.3.11) we introduce the traditional definition. Namely, if $H \in (\mathrm{INF})[\mathcal{L}]$, then we define the unique element

$$(\mathcal{L} - \inf)[H] \in (\leqq_{\mathcal{L}} - \mathrm{Mi})[H] \tag{5.3.12}$$

for which

$$(\leqq_{\mathcal{L}} - \mathrm{Ma})^0[(\leqq_{\mathcal{L}} - \mathrm{Mi})[H]\,] = \{(\mathcal{L} - \inf)[H]\}. \tag{5.3.13}$$

From (5.3.13) it follows that for $H \in (\mathrm{INF})[\mathcal{L}]$ the FAM $(\mathcal{L} - \inf)[H] \in \mathbf{A}(\mathcal{L})$ possesses the following property:

$$(\forall \mu \in H : (\mathcal{L} - \inf)[H] \leqq \mu) \ \&\ (\forall \nu \in (\leqq_{\mathcal{L}} - \mathrm{Mi})[H] : \nu \leqq (\mathcal{L} - \inf)[H]). \tag{5.3.14}$$

Thus the supremum and infimum are defined for some subsets of $\mathbf{A}(\mathcal{L})$. With due of (4.2.34) and (4.2.35) we can verify that $\forall \mu \in \mathbf{A}(\mathcal{L})$ $\forall \nu \in \mathbf{A}(\mathcal{L})$:

$$\mu + (\nu - \mu)^+ \in (\leqq_{\mathcal{L}} - \mathrm{Mi})^0[(\leqq_{\mathcal{L}} - \mathrm{Ma})[\{\mu; \nu\}]\,]. \tag{5.3.15}$$

The proof of (5.3.15) uses the extension (4.2.33) and obvious properties of this extension. From (5.3.6) and (5.3.15) we have $\forall \mu \in \mathbf{A}(\mathcal{L})$ $\forall \nu \in \mathbf{A}(\mathcal{L})$:

$$\{\mu; \nu\} \in (\mathrm{SUP})[\mathcal{L}]. \tag{5.3.16}$$

From (5.3.9), (5.3.15) and (5.3.16) we obtain that $\forall \mu \in \mathbf{A}(\mathcal{L})$ $\forall \nu \in \mathbf{A}(\mathcal{L})$:

$$(\mathcal{L} - \sup)[\{\mu; \nu\}] = \mu + (\nu - \mu)^+. \tag{5.3.17}$$

We do not consider properties of the above mentioned ordered structure of $\mathbf{A}(\mathcal{L})$. We restrict ourselves to only a brief summary; in this connection see [19, 109].

Let

$$(\uparrow - \mathrm{str})[\mathcal{L}] \triangleq \{H \in \mathcal{P}(\mathbf{A}(\mathcal{L})) \mid \forall \mu \in H \forall \nu \in H : (\mathcal{L} - \sup)[\{\mu; \nu\}] \in H\}. \tag{5.3.18}$$

In (5.3.18) we consider the family of all upper semi-lattice subsets of $\mathbf{A}(\mathcal{L})$. It is useful to note the case of subsets of $(\mathrm{add})_+[\mathcal{L}]$. Let

$$\begin{aligned}(\uparrow -\mathrm{str})'_+[\mathcal{L}] &\triangleq (\uparrow -\mathrm{str})[\mathcal{L}] \cap 2^{(\mathrm{add})_+[\mathcal{L}]} \\ &= \{H \in 2^{(\mathrm{add})_+[\mathcal{L}]} \mid \forall \mu \in H\ \forall \nu \in H : \mu + (\nu - \mu)^+ \in H\}.\end{aligned} \quad (5.3.19)$$

Elements of (5.3.19) are exactly nonempty upper semi-lattice subsets of $(\mathrm{add})_+[\mathcal{L}]$. Let

$$(\uparrow -\mathrm{str})^+_{\mathbb{B}}[\mathcal{L}] \triangleq \{H \in (\uparrow -\mathrm{str})'_+[\mathcal{L}] \mid (\leqq_{\mathcal{L}} -\mathrm{Ma})_{(\mathrm{add})_+[\mathcal{L}]}[H] \neq \varnothing\}. \quad (5.3.20)$$

In (5.3.20) we add the requirement of ordered boundedness to the condition in (5.3.19). From (5.3.20) we have $\forall H \in (\uparrow -\mathrm{str})^+_{\mathbb{B}}[\mathcal{L}]\ \forall L \in \mathcal{L}$ $\exists c \in [0, \infty[$:

$$\{\mu(L) : \mu \in H\} \in 2^{[0,c]}.$$

This property follows from (5.3.2). In addition, $\forall H \in (\uparrow -\mathrm{str})^+_{\mathbb{B}}[\mathcal{L}]$ $\forall L \in \mathcal{L}$:

$$\sup(\{\mu(L) : \mu \in H\}) \in [0, \infty[.$$

PROPOSITION 5.3.1 *If* $H \in (\uparrow -\mathrm{str})^+_{\mathbb{B}}[\mathcal{L}]$, *then*

$$(\sup(\{\mu(L) : \mu \in H\}))_{L \in \mathcal{L}} \in (\mathrm{add})_+[\mathcal{L}]. \quad (5.3.21)$$

The corresponding proof is obvious. In addition, it is useful beforehand to consider the case of an algebra of sets. Later on we use properties of the extension (4.2.28). We give an example of the element of (5.3.20). Namely, $\forall H \in 2^{(\mathrm{add})_+[\mathcal{L}]}$:

$$(\leqq_{\mathcal{L}} -\mathrm{Mi})_{(\mathrm{add})_+[\mathcal{L}]}[H] \in (\uparrow -\mathrm{str})^+_{\mathbb{B}}[\mathcal{L}]. \quad (5.3.22)$$

Thus the construction of Proposition 5.3.1 (see (5.3.21)) can be applied to the set on the left hand side of (5.3.22). We note the following simple property: if $H \in 2^{(\mathrm{add})_+[\mathcal{L}]}$, then

$$(\leqq_{\mathcal{L}} -\mathrm{Ma})_{(\mathrm{add})_+[\mathcal{L}]}[H] = (\leqq_{\mathcal{L}} -\mathrm{Ma})[H].$$

From (5.3.2), (5.3.3) and Proposition 5.3.1 it follows that for $H \in (\uparrow -\mathrm{str})^+_{\mathbb{B}}[\mathcal{L}]$, the FAM (5.3.21) is an element of $(\leqq_{\mathcal{L}} -\mathrm{Ma})[H]$. In reality, each element of the last set is the majorant with respect to the FAM (5.3.21). Hence the FAM (5.3.21) is an element of the set $(\leqq_{\mathcal{L}} -\mathrm{Mi})^0[(\leqq_{\mathcal{L}} -\mathrm{Ma})[H]\,]$. Then by (5.3.6) $(\uparrow -\mathrm{str})^+_{\mathbb{B}}[\mathcal{L}] \subset (\mathrm{SUP})[\mathcal{L}]$, and by (5.3.8) $\forall H \in (\uparrow -\mathrm{str})^+_{\mathbb{B}}[\mathcal{L}]$: $(\mathcal{L}-\sup)[H] \in (\leqq_{\mathcal{L}} -\mathrm{Ma})[H]$. Hence from (5.3.8) we obtain that

$$(\mathcal{L} - \sup)[H] = (\sup(\{\mu(L) : \mu \in H\}))_{L \in \mathcal{L}} \quad (5.3.23)$$

for $H \in (\uparrow - \mathrm{str})^+_{\mathbb{B}}[\mathcal{L}]$. In (5.3.23) we have a simple example of determining the supremum in a constructive way. It is possible to apply the scheme (5.3.23) to the case (5.3.22). Note that by (5.3.16) we have

$$\mathrm{Fin}(\mathbf{A}(\mathcal{L})) \subset (\mathrm{SUP})[\mathcal{L}]. \tag{5.3.24}$$

To prove (5.3.24) it is advisable to use the obvious induction on the cardinality of $K \in \mathrm{Fin}(\mathbf{A}(\mathcal{L}))$; certainly, we use (5.3.7). In the sequel we consider corresponding analogues of (5.3.24) for (5.3.12). Simplest facts of the vector lattice theory are used here. Note that by (5.3.1) the order $\leqq_{\mathcal{L}}$ and the (pointwise) linear operations in $\mathbf{A}(\mathcal{L})$ are coordinated. Namely, $\forall \mu \in \mathbf{A}(\mathcal{L})\ \forall \nu \in \mathbf{A}(\mathcal{L})$:

$$(\mu \leqq \nu) \Rightarrow ((\forall \xi \in \mathbf{A}(\mathcal{L}) : \mu + \xi \leqq \nu + \xi)\ \&\ (\forall \alpha \in ]0, \infty[: \alpha\mu \leqq \alpha\nu)). \tag{5.3.25}$$

Moreover, by (5.3.16) we obtain $\forall \mu \in \mathbf{A}(\mathcal{L})\ \forall \nu \in \mathbf{A}(\mathcal{L})$:

$$\{\mu; \nu\} \subset (\mathrm{SUP})[\mathcal{L}] \cap (\mathrm{INF})[\mathcal{L}]. \tag{5.3.26}$$

From (5.3.25) it follows that $\forall \mu \in \mathbf{A}(\mathcal{L})\ \forall \nu \in \mathbf{A}(\mathcal{L})\ \forall \xi \in \mathbf{A}(\mathcal{L})$:

$$(\mathcal{L} - \sup)[\{\mu + \xi; \nu + \xi\}] = (\mathcal{L} - \sup)[\{\mu; \nu\}] + \xi.$$

Using (5.3.25) and (5.3.26) we obtain $\forall \mu \in \mathbf{A}(\mathcal{L})\ \forall \nu \in \mathbf{A}(\mathcal{L})$:

$$(\mathcal{L} - \inf)[\{\mu; \nu\}] = -(\mathcal{L} - \sup)[\{-\mu; -\nu\}]. \tag{5.3.27}$$

From (5.3.27) and some simplest facts of the vector lattice theory it follows that $\forall \mu \in \mathbf{A}(\mathcal{L})\ \forall \nu \in \mathbf{A}(\mathcal{L})$:

$$\mu + \nu = (\mathcal{L} - \sup)[\{\mu; \nu\}] + (\mathcal{L} - \inf)[\{\mu; \nu\}].$$

For $\mu \in \mathbf{A}(\mathcal{L})$ we have $v_\mu = (\mathcal{L} - \sup)[\{\mu; -\mu\}]$. A number of other similar properties of $\mathbf{A}(\mathcal{L})$ are considered in [19, 20, 32, 35]. Note that $\forall \alpha \in ]0, \infty[\ \forall \mu \in \mathbf{A}(\mathcal{L})\ \forall \nu \in \mathbf{A}(\mathcal{L})$:

$$\begin{aligned} &((\mathcal{L} - \sup)[\{\alpha\mu; \alpha\nu\}] = \alpha \cdot (\mathcal{L} - \sup)[\{\mu; \nu\}]) \\ &\&\ ((\mathcal{L} - \inf)[\{\alpha\mu; \alpha\nu\}] = \alpha \cdot (\mathcal{L} - \inf)[\{\mu; \nu\}]). \end{aligned}$$

For $\alpha \in ]0, \infty[$ and $\mu \in \mathbf{A}(\mathcal{L})$ we have $v_{\alpha\mu} = |\alpha| v_\mu$. Certainly, for $\mu \in \mathbf{A}(\mathcal{L})$ we have $(\mathcal{L} - \inf)[\{\mu^+; \mu^-\}] = \mathbb{O}_{\mathcal{L}}$ and $\forall \nu_1 \in (\mathrm{add})_+[\mathcal{L}]\ \forall \nu_2 \in (\mathrm{add})_+[\mathcal{L}]$:

$$((\mu = \nu_1 - \nu_2)\&((\mathcal{L} - \inf)[\{\nu_1; \nu_2\}] = \mathbb{O}_{\mathcal{L}})) \Rightarrow ((\nu_1 = \mu^+)\&(\nu_2 = \mu^-)).$$

The last relation characterizes the useful property of the Jordan decomposition. If $\mu \in \mathbf{A}(\mathcal{L})$ and $\nu \in \mathbf{A}(\mathcal{L})$, we set

$$[\mu; \nu]^{(0)} \triangleq \{\xi \in \mathbf{A}(\mathcal{L}) \mid (\mu \leqq \xi)\&(\xi \leqq \nu)\}; \tag{5.3.28}$$

if $\mu \in (\mathrm{add})_+[\mathcal{L}]$ and $\nu \in (\mathrm{add})_+[\mathcal{L}]$, then

$$[\mathbb{O}_{\mathcal{L}};\mu+\nu]^{(0)} = \{\mathrm{pr}_1(z)+\mathrm{pr}_2(z) : z \in [\mathbb{O}_{\mathcal{L}};\mu]^{(0)} \times [\mathbb{O}_{\mathcal{L}};\nu]^{(0)}\}.$$

In (5.3.28) we have the order interval corresponding to $\mu$ and $\nu$. Note that $\forall \mu \in \mathbf{A}(\mathcal{L})\ \forall \nu \in \mathbf{A}(\mathcal{L})$:

$$(\mu \leqq \nu) \Leftrightarrow ((\mu^+ \leqq \nu^+)\&(\nu^- \leqq \mu^-)).$$

The following notions are important. Suppose that

$$\left(\mathbb{B}^{\uparrow}_{\leqq}(\mathcal{L}) \triangleq \{H \in 2^{\mathbf{A}(\mathcal{L})} \mid (\leqq_{\mathcal{L}} - \mathrm{Ma})[H] \neq \varnothing\}\right)$$
$$\& \left(\mathbb{B}^{\downarrow}_{\leqq}(\mathcal{L}) \triangleq \{H \in 2^{\mathbf{A}(\mathcal{L})} \mid (\leqq_{\mathcal{L}} - \mathrm{Mi})[H] \neq \varnothing\}\right).$$

Sets from $\mathbb{B}^{\uparrow}_{\leqq}(\mathcal{L})$ (from $\mathbb{B}^{\downarrow}_{\leqq}(\mathcal{L})$) are called upper (lower) orderly bounded. The very important property of the lattice $(\mathbf{A}(\mathcal{L}), \leqq_{\mathcal{L}})$ consists in the following:

$$(\mathbb{B}^{\uparrow}_{\leqq}(\mathcal{L}) \subset (\mathrm{SUP})[\mathcal{L}])\&(\mathbb{B}^{\downarrow}_{\leqq}(\mathcal{L}) \subset (\mathrm{INF})[\mathcal{L}]). \tag{5.3.29}$$

So, $(\mathbf{A}(\mathcal{L}), \leqq_{\mathcal{L}})$ is a complete vector lattice. The proof of (5.3.29) uses the construction (5.3.23). We have

$$(\forall U \in \mathbb{B}^{\uparrow}_{\leqq}(\mathcal{L}) : (\mathcal{L} - \sup)[U] \in \mathbf{A}(\mathcal{L}))$$
$$\& (\forall V \in \mathbb{B}^{\downarrow}_{\leqq}(\mathcal{L}) : (\mathcal{L} - \inf)[V] \in \mathbf{A}(\mathcal{L})).$$

Note the simple topological property: if $\mu \in \mathbf{A}(\mathcal{L})$ and $\nu \in \mathbf{A}(\mathcal{L})$, then

$$[\mu;\nu]^{(0)} \in (\tau_*(\mathcal{L}) - \mathrm{comp})[\mathbf{A}(\mathcal{L})]. \tag{5.3.30}$$

As a supplement of (5.3.24), we note that $\mathrm{Fin}(\mathbf{A}(\mathcal{L})) \subset (\mathrm{SUP})[\mathcal{L}] \cap (\mathrm{INF})[\mathcal{L}]$. Therefore $\forall K \in \mathrm{Fin}(\mathbf{A}(\mathcal{L}))$:

$$((\mathcal{L} - \sup)[K] \in \mathbf{A}(\mathcal{L}))\ \&\ ((\mathcal{L} - \inf)[K] \in \mathbf{A}(\mathcal{L})).$$

## 5.4 THE INDEFINITE DARBOUX INTEGRAL

We return to the constructions of Section 4.2. We use (5.3.29) for the required representations of the sets (5.2.9). Namely, $\forall \mu \in (\mathrm{add})_+[\mathcal{L}]$ $\forall f \in \mathbb{B}(E)$:

$$\begin{aligned}&\left(\{\tilde{f} * \mu : \tilde{f} \in B^0_{\leq}(E,\mathcal{L},f)\} \in \mathbb{B}^{\uparrow}_{\leqq}(\mathcal{L})\right)\\ \&&\left(\{\hat{f} * \mu : \hat{f} \in B^0_{\geq}(E,\mathcal{L},f)\} \in \mathbb{B}^{\downarrow}_{\leqq}(\mathcal{L})\right).\end{aligned} \tag{5.4.1}$$

The proof of (5.4.1) is very simple (see (5.2.8)). From (5.4.1) we have $\forall \mu \in (\mathrm{add})_+[\mathcal{L}]\ \forall f \in \mathbb{B}(E)$:

$$\begin{aligned}&\left((f * \mu)_{\mathbf{d}} \triangleq (\mathcal{L} - \sup)[\{\tilde{f} * \mu : \tilde{f} \in B^0_{\leq}(E, \mathcal{L}, f)\}] \in \mathbf{A}(\mathcal{L})\right)\\ \& &\left((f * \mu)^{\mathbf{d}} \triangleq (\mathcal{L} - \inf)[\{\hat{f} * \mu : \hat{f} \in B^0_{\geq}(E, \mathcal{L}, f)\}] \in \mathbf{A}(\mathcal{L})\right).\end{aligned} \tag{5.4.2}$$

We call the two FAM in (5.4.2) the lower and upper indefinite Darboux $\mu$-integral of $f$. It is obvious that $\forall \mu \in (\mathrm{add})_+[\mathcal{L}]\ \forall f \in \mathbb{B}(E)$:

$$(f * \mu)_{\mathbf{d}} \leqq (f * \mu)^{\mathbf{d}}.$$

Using (5.3.30) we introduce the set-valued Darboux integral

$$[(f * \mu)_{\mathbf{d}}; (f * \mu)^{\mathbf{d}}]^{(0)} \in (\tau_*(\mathcal{L}) - \mathrm{comp})[\mathbf{A}(\mathcal{L})] \setminus \{\varnothing\} \tag{5.4.3}$$

for $\mu \in (\mathrm{add})_+[\mathcal{L}]$ and $f \in \mathbb{B}(E)$. By (5.2.7) we obtain that $\forall \mu \in (\mathrm{add})_+[\mathcal{L}]\ \forall f \in B(E, \mathcal{L})$:

$$(f * \mu)_{\mathbf{d}} = (f * \mu)^{\mathbf{d}} = f * \mu. \tag{5.4.4}$$

It is useful to consider functions $f \in \mathbb{B}(E)$ possessing the property $(f * \mu)_{\mathbf{d}} = (f * \mu)^{\mathbf{d}}$ for $\mu \in (\mathrm{add})_+[\mathcal{L}]$. This coincidence has the sense of universal integrability. We will concern this question. At first, we note that such universal integrability and the integrability of a bounded function in the class of all non-negative CAM are (generally speaking) different notions.

**Example.** Let $E = \mathcal{N}$. Assume that $\mathcal{L}_1 \triangleq \overline{\{\mathrm{pr}_1(z), \mathrm{pr}_2(z) : z \in \mathcal{N} \times \mathcal{N}\}}$ and $\mathcal{L}_2 \triangleq \{\overrightarrow{n, \infty} : n \in \mathcal{N}\}$. Then $\forall \mu \in (\sigma - \mathrm{add})_+[\mathcal{L}]\ \forall f \in \mathbb{B}(E)$: $(f * \mu)_{\mathbf{d}} = (f * \mu)^{\mathbf{d}}$. To prove this, we use (4.2.39). Introduce $\nu \in (\mathbf{p} - \mathrm{add})_+[\mathcal{L}]$ (see [19, 20, 22]) by the rule: $(\forall L \in \mathcal{L}_1 : \nu(L) \triangleq 0)\ \&(\forall L \in \mathcal{L}_2 : \nu(L) \triangleq 1)$. Then $\forall f \in \mathbb{B}(\mathcal{N})$:

$$((f * \mu)_{\mathbf{d}} = \underline{\lim}_{k \to \infty} f(k)\ \nu)\ \&((f * \mu)^{\mathbf{d}} = \overline{\lim}_{k \to \infty} f(k)\ \nu).$$

Hence it is possible to choose $f \in \mathbb{B}(E)$ for which

$$(f * \nu)_{\mathbf{d}} \neq (f * \nu)^{\mathbf{d}}.$$

And what is more, in this example $B(E, \mathcal{L})$ is the set of all universally integrable functions $f \in \mathbb{B}(E)$. For a more detailed consideration of this example, see [19, 20, 22].

Note the following obvious property: if $\mu \in (\mathrm{add})[\mathcal{L}]$, $\nu \in (\mathrm{add})[\mathcal{L}]$, then

$$((\mu \leqq \nu)\&(\mu(E) = \nu(E))) \Rightarrow (\mu = \nu).$$

As a corollary, from (5.4.3) it follows that

$$\begin{aligned}&\{f \in \mathbb{B}(E) \mid \forall \mu \in H : (f * \mu)_{\mathbf{d}} = (f * \mu)^{\mathbf{d}}\}\\ =&\{f \in \mathbb{B}(E) \mid \forall \mu \in H : (f * \mu)_{\mathbf{d}}(E) = (f * \mu)^{\mathbf{d}}(E)\}\end{aligned} \tag{5.4.5}$$

for $H \in \mathcal{P}((\text{add})_+[\mathcal{L}])$.

To investigate the conditions of universal integrability of bounded functions we use (5.4.5). Consider some obvious representations of the numbers

$$(f * \mu)_{\mathbf{d}}(E) \in \mathbb{R}, (f * \mu)^{\mathbf{d}}(E) \in \mathbb{R}.$$

But beforehand it is advisable to note some general properties. In particular, $\forall \mu \in \mathbf{A}(\mathcal{L})$:

$$(\forall \alpha \in \mathbb{R}\, \forall f \in B(E, \mathcal{L}) : (\alpha f) * \mu = \alpha \cdot (f * \mu))$$

$$\& \ (\forall f \in B(E, \mathcal{L})\, \forall g \in B(E, \mathcal{L}) : (f + g) * \mu = (f * \mu) + (g * \mu)). \tag{5.4.6}$$

To prove (5.4.6) we use simplest properties of the integrals of Chapter 4. Moreover, if $\mu \in (\text{add})_+[\mathcal{L}]$ and $f \in B_0(E, \mathcal{L})$, then for the function $f_+ \in B_0^+(E, \mathcal{L})$ defined by the rule $f_+(x) \triangleq \sup(\{f(x); 0\})$ (under $x \in E$) we have $f_+ * \mu = (f * \mu)^+$. Suppose that $\forall f \in \mathbb{R}^E\ \forall g \in \mathbb{R}^E$:

$$\text{Sup}(\{f; g\}) \triangleq (\sup(\{f(x); g(x)\}))_{x \in E}.$$

We obtain a function of $\mathbb{R}^E$. Of course, $\forall f \in B_0(E, \mathcal{L})\ \forall g \in B_0(E, \mathcal{L})$: $\text{Sup}(\{f; g\}) \in B_0(E, \mathcal{L})$. We note that this property implies the equality

$$(\mathcal{L} - \sup)[\{f * \mu; \mathbb{O}_{\mathcal{L}}\}] = \text{Sup}(\{f; \mathbb{O}_E\}) * \mu$$

for $\mu \in (\text{add})_+[\mathcal{L}]$ and $f \in B_0(E, \mathcal{L})$. As a corollary we obtain that $\forall \mu \in (\text{add})_+[\mathcal{L}]\ \forall f \in B_0(E, \mathcal{L})\ \forall g \in B_0(E, \mathcal{L})$:

$$(\mathcal{L} - \sup)[\{f * \mu; g * \mu\}] = \text{Sup}(\{f; g\}) * \mu. \tag{5.4.7}$$

This property can be extended to the case $f \in B(E, \mathcal{L})$ and $g \in B(E, \mathcal{L})$, but we do not consider this construction. We note that $\forall f \in B(E, \mathcal{L})$:

$$(\forall \alpha \in \mathbb{R}\, \forall \mu \in \mathbf{A}(\mathcal{L}) : f * (\alpha \mu) = \alpha \cdot (f * \mu))$$

$$\& \ (\forall \mu \in \mathbf{A}(\mathcal{L})\, \forall \nu \in \mathbf{A}(\mathcal{L}) : f * (\mu + \nu) = (f * \mu) + (f * \nu)). \tag{5.4.8}$$

The property (5.4.8) follows from the obvious statements of Section 4.3. In (5.4.8) we have a useful supplement of (5.4.6). From (5.4.7) we obtain $\forall \mu \in (\text{add})_+[\mathcal{L}]\ \forall f \in \mathbb{B}(E)$:

$$\{\tilde{f} * \mu : \tilde{f} \in B^0_{\leq}(E, \mathcal{L}, f)\} \in (\uparrow - \text{str})[\mathcal{L}]. \tag{5.4.9}$$

Let us consider some analogues of (5.4.9) for definite integrals. First we note that $\forall \mu \in (\mathrm{add})_+[\mathcal{L}]\ \forall f \in \mathbb{B}(E)\ \forall \tilde{f} \in B^0_{\leq}(E,\mathcal{L},f)\ \forall \hat{f} \in B^0_{\geq}(E,\mathcal{L},f)$:

$$\overset{(\mathrm{el})}{\int_E} \tilde{f}\, d\mu \leq \overset{(\mathrm{el})}{\int_E} \hat{f}\, d\mu. \tag{5.4.10}$$

Simplest properties of the integrals from Section 4.3 are used for checking (5.4.10). From (5.2.8) and (5.4.10) we have $\forall \mu \in (\mathrm{add})_+[\mathcal{L}]\ \forall f \in \mathbb{B}(E)$:

$$\left(\exists c \in \mathbb{R} : \left\{ \overset{(\mathrm{el})}{\int_E} \tilde{f}\, d\mu : \tilde{f} \in B^0_{\leq}(E,\mathcal{L},f) \right\} \in 2^{]-\infty,c]}\right)$$

$$\& \left(\exists \tilde{c} \in \mathbb{R} : \left\{ \overset{(\mathrm{el})}{\int_E} \hat{f}\, d\mu : \hat{f} \in B^0_{\geq}(E,\mathcal{L},f) \right\} \in 2^{[\tilde{c},\infty[}\right).$$

As a consequence we obtain that $\forall \mu \in (\mathrm{add})_+[\mathcal{L}]\ \forall f \in \mathbb{B}(E)$:

$$\left(\sup\left(\left\{ \overset{(\mathrm{el})}{\int_E} \tilde{f}\, d\mu : \tilde{f} \in B^0_{\leq}(E,\mathcal{L},f) \right\}\right) \in \mathbb{R}\right)$$

$$\& \left(\inf\left(\left\{ \overset{(\mathrm{el})}{\int_E} \hat{f}\, d\mu : \hat{f} \in B^0_{\geq}(E,\mathcal{L},f) \right\}\right) \in \mathbb{R}\right).$$

PROPOSITION 5.4.1 *$\forall \mu \in (\mathrm{add})_+[\mathcal{L}]\ \forall f \in \mathbb{B}(E)$:*

$$(f * \mu)_{\mathbf{d}}(E) = \sup\left(\left\{ \overset{(\mathrm{el})}{\int_E} \tilde{f}\, d\mu : \tilde{f} \in B^0_{\leq}(E,\mathcal{L},f) \right\}\right).$$

We omit a highly obvious proof. Note only that we use (5.3.23) and the natural compatibility of the order $\leqq_{\mathcal{L}}$ and the linear operations in $\mathbf{A}(\mathcal{L})$ (see (5.3.1)).

It is possible to verify that $\forall f \in \mathbb{B}(E)$:

$$B^0_{\leq}(E,\mathcal{L},-f) = \{-\hat{f} : \hat{f} \in B^0_{\geq}(E,\mathcal{L},f)\}. \tag{5.4.11}$$

With due of (5.4.11) one can obtain the corresponding analogue of Proposition 5.4.1. Namely, $\forall \mu \in (\mathrm{add})_+[\mathcal{L}]\ \forall f \in \mathbb{B}(E)$:

$$((-f) * \mu)_{\mathbf{d}} = -(f * \mu)^{\mathbf{d}}.$$

As a corollary, Proposition 5.4.1 and (5.4.11) imply the following property: if $\mu \in (\mathrm{add})_+[\mathcal{L}]$ and $f \in \mathbb{B}(E)$, then

$$(f * \mu)^{\mathbf{d}}(E) = \inf\left(\left\{ \overset{(\mathrm{el})}{\int_E} \hat{f}\, d\mu : \hat{f} \in B^0_{\geq}(E,\mathcal{L},f) \right\}\right). \tag{5.4.12}$$

From Proposition 5.4.1 and (5.4.12) it follows that for $H \in 2^{(\mathrm{add})_+[\mathcal{L}]}$

$$\left\{f \in \mathbb{B}(E) \mid \forall \mu \in H : (f * \mu)_{\mathbf{d}} = (f * \mu)^{\mathbf{d}}\right\}$$
$$= \left\{f \in \mathbb{B}(E) \mid \forall \mu \in H : \sup\left(\left\{ {}^{(\mathrm{el})}\!\!\int_E \tilde{f}\, d\mu : \tilde{f} \in B^0_{\leq}(E, \mathcal{L}, f)\right\}\right)\right.$$
$$\left. = \inf\left(\left\{ {}^{(\mathrm{el})}\!\!\int_E \hat{f}\, d\mu : \hat{f} \in B^0_{\geq}(E, \mathcal{L}, f)\right\}\right)\right\}. \tag{5.4.13}$$

We use (4.10.12) in the subsequent constructions. Namely, (4.10.12) can be exploited for $H$ in (5.4.13).

PROPOSITION 5.4.2 *If $f \in B_0^+(E, \mathcal{L})$ and $\varepsilon \in ]0, \infty[$, then*

$$\left\{\mu \in \mathbb{T}(\mathcal{L}) \mid \varepsilon \leq {}^{(\mathrm{el})}\!\!\int_E f\, d\mu\right\} \in \mathcal{F}_{\tau_*(\mathcal{L})}. \tag{5.4.14}$$

PROOF. Denote the set on the left hand side of (5.4.14) by $\Omega$. In addition, introduce the set

$$W \triangleq \left\{\mu \in \mathbf{A}(\mathcal{L}) \mid \left| {}^{(\mathrm{el})}\!\!\int_E f\, d\mu\right| < \varepsilon\right\}.$$

Then $\Omega = \mathbb{T}(\mathcal{L}) \cap (\mathbf{A}(\mathcal{L}) \setminus W)$. But $W \in \mathfrak{N}^0_*(\mathbb{O}_{\mathcal{L}} \mid \mathcal{L})$; see (4.6.1) and (4.6.2). As a consequence we obtain $W \in \tau_*(\mathcal{L})$ and $\mathbf{A}(\mathcal{L}) \setminus W \in \mathcal{F}_{\tau_*(\mathcal{L})}$. From (4.6.10) and (4.10.12) it follows that $\mathbb{T}(\mathcal{L}) \in \mathcal{F}_{\tau_*(\mathcal{L})}$ and hence $\Omega \in \mathcal{F}_{\tau_*(\mathcal{L})}$. □

It should be noted that $\forall H \in 2^{B_0(E,\mathcal{L})}$:

$$(\forall f \in H\ \forall g \in H\ \exists s \in H : (s \leqq f) \& (s \leqq g))$$
$$\Rightarrow (\forall m \in \mathcal{N}\ \forall (h_i)_{i \in \overline{1,m}} \in H^m\ \exists h \in H\ \forall j \in \overline{1,m} : h \leqq h_j).$$

PROPOSITION 5.4.3 *$\forall f \in \mathbb{B}(E) \setminus B(E, \mathcal{L})\ \exists \mu \in \mathbb{T}(\mathcal{L})$:*

$$\sup\left(\left\{ {}^{(\mathrm{el})}\!\!\int_E \tilde{f}\, d\mu : \tilde{f} \in B^0_{\leq}(E, \mathcal{L}, f)\right\}\right)$$
$$< \inf\left(\left\{ {}^{(\mathrm{el})}\!\!\int_E \hat{f}\, d\mu : \hat{f} \in B^0_{\geq}(E, \mathcal{L}, f)\right\}\right). \tag{5.4.15}$$

PROOF. Fix $f \in \mathbb{B}(E) \setminus B(E, \mathcal{L})$. Note that $\forall \tilde{f} \in B^0_{\leq}(E, \mathcal{L}, f)\ \forall \hat{f} \in B^0_{\geq}(E, \mathcal{L}, f)$: $\hat{f} - \tilde{f} \in B_0^+(E, \mathcal{L})$. It is obvious that

$$\{h \in B_0(E, \mathcal{L}) \mid \|h - f\| \leq \kappa\} = \varnothing \tag{5.4.16}$$

for some $\kappa \in ]0, \infty[$. From (5.4.16) we have the following simple corollary:

$$\forall f \in B_0(E, \mathcal{L}) \; \exists x \in E : \kappa < |h(x) - f(x)|. \tag{5.4.17}$$

Therefore (5.4.17) is true for $h \in B^0_{\leqq}(E, \mathcal{L}, f) \cup B^0_{\geqq}(E, \mathcal{L}, f)$. Consequently we obtain

$$\begin{aligned} &\left(\forall \tilde{f} \in B^0_{\leqq}(E, \mathcal{L}, f) \; \exists x \in E : \kappa < f(x) - \tilde{f}(x)\right) \\ \& &\left(\forall \hat{f} \in B^0_{\geqq}(E, \mathcal{L}, f) \; \exists x' \in E : \kappa < \hat{f}(x') - f(x')\right). \end{aligned} \tag{5.4.18}$$

From (5.2.8) and (5.4.18) it follows that

$$\forall \tilde{f} \in B^0_{\leqq}(E, \mathcal{L}, f) \; \forall \hat{f} \in B^0_{\geqq}(E, \mathcal{L}, f) \; \exists x \in E : \kappa < \hat{f}(x) - \tilde{f}(x).$$

In connection with the last property we introduce the set $B^0(E, \mathcal{L}, f) \triangleq \{\mathrm{pr}_2(z) - \mathrm{pr}_1(z) : z \in B^0_{\leqq}(E, \mathcal{L}, f) \times B^0_{\geqq}(E, \mathcal{L}, f)\}$. Then $\forall g \in B^0(E, \mathcal{L}, f)$ $\exists x \in E : \kappa < g(x)$. For $\mu \in \mathbb{T}(\mathcal{L})$ we denote the numbers on the left hand and right hand sides of (5.4.15) by $a_0(\mu)$ and $a^0(\mu)$, respectively. We have $a_0(\mu) \leq a^0(\mu)$. Let us consider the topology $\tau^*_{(0,1)}[\mathcal{L}]$ (4.10.20). Moreover, we use (4.10.24). Let

$$\mathcal{F}^*_{(0,1)} \triangleq \mathcal{F}_\tau \big|_{\tau = \tau^*_{(0,1)}[\mathcal{L}]} .$$

We obtain a family of closed sets in the compact TS (4.10.25). In addition, $\forall g \in B^0(E, \mathcal{L}, f)$:

$$\Pi_g \triangleq \left\{ \mu \in \mathbb{T}(\mathcal{L}) \mid \kappa \leq {}^{(\mathrm{el})}\!\!\int_E g \, d\mu \right\} \in \mathcal{F}^*_{(0,1)}.$$

From (5.2.8) it follows that $\forall \tilde{f} \in B^0(E, \mathcal{L}, f)$ $\forall \tilde{g} \in B^0(E, \mathcal{L}, f)$ $\exists \tilde{s} \in B^0(E, \mathcal{L}, f)$:

$$(\tilde{s} \leqq \tilde{f}) \; \& \; (\tilde{s} \leqq \tilde{g}).$$

We use the obvious property considered in front of this proposition. As a consequence we have $\forall m \in \mathcal{N}$ $\forall (h_i)_{i \in \overline{1,m}} \in B^0(E, \mathcal{L}, f)^m$ $\exists h \in B^0(E, \mathcal{L}, f)$ $\forall j \in \overline{1,m} : h \leqq h_j$. Then $\forall m \in \mathcal{N}$ $\forall (g_i)_{i \in \overline{1,m}} \in B^0(E, \mathcal{L}, f)^m$ $\exists g \in B^0(E, \mathcal{L}, f)$:

$$\Pi_g \subset \bigcap_{i=1}^{m} \Pi_{g_i}. \tag{5.4.19}$$

Moreover, $\Pi_g \neq \varnothing$ for $g \in B^0(E, \mathcal{L}, f)$. In addition, each set $\Pi_g$, $g \in B^0(E, \mathcal{L}, f)$, contains some Dirac measure. By (5.4.19) we obtain that

$\forall m \in \mathcal{N}\ \forall (g_i)_{i \in \overline{1,m}} \in B^0(E, \mathcal{L}, f)^m$:

$$\bigcap_{i=1}^{m} \Pi_{g_i} \neq \varnothing. \tag{5.4.20}$$

Consider the (nonempty) family

$$\mathbb{F} \triangleq \{\Pi_g : g \in B^0(E, \mathcal{L}, f)\} \in 2^{\mathcal{F}^*_{(0,1)}}.$$

From (5.4.20) we obviously have $\mathbb{F} \in \mathbf{Z}[\mathcal{F}^*_{(0,1)}]$. Using (2.2.21), (4.10.20) and (4.10.24) we obtain the property

$$\bigcap_{g \in B^0(E,\mathcal{L},f)} \Pi_g = \bigcap_{F \in \mathbb{F}} F \neq \varnothing.$$

Choose an element $\overline{\eta}$ of the intersection of all sets $\Pi_g$, $g \in B^0(E, \mathcal{L}, f)$. Then $\forall g \in B^0(E, \mathcal{L}, f)$:

$$\kappa \leq {}^{(\mathrm{el})}\int_E g \, d\overline{\eta}.$$

As a consequence we obtain the inequality $a_0(\overline{\eta}) + \kappa \leq a^0(\overline{\eta})$. In particular, $a_0(\overline{\eta}) < a^0(\overline{\eta})$. □

COROLLARY 5.4.1 $\forall f \in \mathbb{B}(E) \setminus B(E, \mathcal{L})\ \exists \mu \in (\mathrm{add})_+[\mathcal{L}]$:

$$\sup\left(\left\{ {}^{(\mathrm{el})}\int_E \tilde{f}\, d\mu : \tilde{f} \in B^0_{\leq}(E, \mathcal{L}, f) \right\}\right)$$
$$< \inf\left(\left\{ {}^{(\mathrm{el})}\int_E \hat{f}\, d\mu : \hat{f} \in B^0_{\geq}(E, \mathcal{L}, f) \right\}\right).$$

The proof is obvious. Using Propositions 5.4.1, 5.4.3 and (5.4.12) we obtain that $\forall f \in \mathbb{B}(E) \setminus B(E, \mathcal{L})\ \exists \mu \in \mathbb{T}(\mathcal{L})$: $(f * \mu)_{\mathbf{d}}(E) < (f * \mu)^{\mathbf{d}}(E)$. By (5.4.4) and (5.4.5) we have

$$\begin{aligned} B(E, \mathcal{L}) &= \{f \in \mathbb{B}(E) \mid \forall \mu \in \mathbb{T}(\mathcal{L}) : (f * \mu)_{\mathbf{d}} = (f * \mu)^{\mathbf{d}}\} \\ &= \{f \in \mathbb{B}(E) \mid \forall \mu \in \mathbb{T}(\mathcal{L}) : (f * \mu)_{\mathbf{d}}(E) = (f * \mu)^{\mathbf{d}}(E)\} \\ &= \{f \in \mathbb{B}(E) \mid \forall \mu \in (\mathrm{add})_+[\mathcal{L}] : (f * \mu)_{\mathbf{d}} = (f * \mu)^{\mathbf{d}}\} \\ &= \{f \in \mathbb{B}(E) \mid \forall \mu \in (\mathrm{add})_+[\mathcal{L}] : (f * \mu)_{\mathbf{d}}(E) = (f * \mu)^{\mathbf{d}}(E)\}. \end{aligned} \tag{5.4.21}$$

It should be noted that (5.4.21) is compatible with the statements of [92, 95]. Thus $B(E, \mathcal{L})$ exhausts our possibilities in the question of universal integrability. In addition, (5.4.4) defines a concrete form of the corresponding 'universal' integral. Recall that the basic fact used in the proof of (5.4.21) (see Proposition 5.4.3) is connected with the natural compactification of the set of Dirac measures. In essence, we apply Corollary 4.10.1. In fact, we use a distinctive realization of extension in constructions of the pure mathematics.

Let us consider some corollaries. It is useful to note that $\forall \mu \in (\text{add})_+[\mathcal{L}]\ \forall f \in \mathbb{B}(E)\ \forall L \in \mathcal{L}$:

$$((f * \mu)_{\mathbf{d}}(L) = ((f\chi_L[E]) * \mu)_{\mathbf{d}}(E))\&((f * \mu)^{\mathbf{d}}(L) = ((f\chi_L[E]) * \mu)^{\mathbf{d}}(E)).$$

This property is similar to (4.4.7) and (4.4.8). We get the 'usual' property of the indefinite integral. Returning to (5.4.21), we obtain that Proposition 5.4.1 and (5.4.12) imply the following statements.

THEOREM 5.4.1 *The set $B(E, \mathcal{L})$ consists of all $f \in \mathbb{B}(E)$ for which $\forall \mu \in (\text{add})_+[\mathcal{L}]$:*

$$\sup\left(\left\{ {}^{(\text{el})}\!\!\int_E \tilde{f}\, d\mu : \tilde{f} \in B^0_{\leq}(E, \mathcal{L}, f)\right\}\right)$$
$$= \inf\left(\left\{ {}^{(\text{el})}\!\!\int_E \hat{f}\, d\mu : \hat{f} \in B^0_{\geq}(E, \mathcal{L}, f)\right\}\right).$$

THEOREM 5.4.2 *The set $B(E, \mathcal{L})$ consists of all $f \in \mathbb{B}(E)$ for which $\forall \mu \in \mathbb{T}(\mathcal{L})$:*

$$\sup\left(\left\{ {}^{(\text{el})}\!\!\int_E \tilde{f}\, d\mu : \tilde{f} \in B^0_{\leq}(E, \mathcal{L}, f)\right\}\right)$$
$$= \inf\left(\left\{ {}^{(\text{el})}\!\!\int_E \hat{f}\, d\mu : \hat{f} \in B^0_{\geq}(E, \mathcal{L}, f)\right\}\right).$$

These theorems are compatible with the statements of [92]. It should be noted that two-valued measures are 'indicators' of the universal integrability. Constructions of this section and, in particular, Theorem 5.4.2 have many useful developments. We note the representations of [3, 23] connected with the universal integrability with respect to measures-products. In these representations $\mathbb{T}(\mathcal{L})$ plays a very important role. On the basis of (5.4.3) the construction of set-valued integration in the sense of Darboux was developed. Integration with respect to CAM and purely FAM was considered, in particular, in [26]. We do not touch upon these interesting questions in this monograph.

## 5.5 UNIVERSAL MEASURABILITY

In this section we suppose that $\mathcal{L} \in (\text{alg})[E]$. Thus the MS $(E, \mathcal{L})$ with an algebra of sets is considered. Note that in this case

$$\forall A_1 \in \mathcal{L}\, \forall A_2 \in \mathcal{L} : A_1 \setminus A_2 = A_1 \cap (E \setminus A_2) \in \mathcal{L}.$$

We introduce the family of the Jordan measurable subsets of $E$, setting $\forall \mu \in (\text{add})_+[\mathcal{L}]$:

$$\begin{aligned}\mathcal{J}[\mu] \triangleq \{H \in \mathcal{P}(E) \,|\forall \varepsilon \in ]0, \infty[\ \exists L_1 \in \mathcal{L}\, \exists L_2 \in \mathcal{L} : \\ (L_1 \subset H \subset L_2)\ \&\ (\mu(L_2 \setminus L_1) < \varepsilon)\}.\end{aligned} \tag{5.5.1}$$

We consider (5.5.1) under enumeration of $\mu$. We call a set $H \in \mathcal{P}(E)$ universally Jordan measurable iff $H$ is an element of the intersection of all families (5.5.1). In this connection see [22].

PROPOSITION 5.5.1 *The family of all universally Jordan measurable sets coincides with $\mathcal{L}$:*

$$\mathcal{L} = \bigcap_{\mu \in (\text{add})_+[\mathcal{L}]} \mathcal{J}[\mu]. \tag{5.5.2}$$

PROOF. We have $\mathcal{L} \subset \mathcal{J}[\mu]$, $\mu \in (\text{add})_+[\mathcal{L}]$. Therefore the family $\mathbb{J}$ on the right hand side of (5.5.2) has the property $\mathcal{L} \subset \mathbb{J}$. Let $\Gamma \in \mathbb{J}$. Then by (5.5.1) $\Gamma \in \mathcal{P}(E)$ has the property that $\forall \mu \in (\text{add})_+[\mathcal{L}]\ \forall \varepsilon \in ]0, \infty[$ $\exists L_1 \in \mathcal{L}\ \exists L_2 \in \mathcal{L}$:

$$(L_1 \subset \Gamma \subset L_2)\ \&\ (\mu(L_2 \setminus L_1) < \varepsilon). \tag{5.5.3}$$

Consider the function $\chi_\Gamma[E] \in \mathbb{B}(E)$ of the type (4.3.1). Then $\forall \mu \in (\text{add})_+[\mathcal{L}]$:

$$(\chi_\Gamma[L] * \mu)_{\mathbf{d}}(E) = (\chi_\Gamma[L] * \mu)^{\mathbf{d}}(E).$$

Indeed, fix $\eta \in (\text{add})_+[\mathcal{L}]$ and $\varepsilon \in ]0, \infty[$. Using (5.5.3), choose $L_1 \in \mathcal{L}$ and $L_2 \in \mathcal{L}$ with the property (5.5.3) under $\mu = \eta$. From (5.2.8) it follows that

$$\Big(\chi_{L_1}[E] \in B^0_{\leq}(E, \mathcal{L}, \chi_\Gamma[E])\Big)\ \&\Big(\chi_{L_2}[E] \in B^0_{\geq}(E, \mathcal{L}, \chi_\Gamma[E])\Big). \tag{5.5.4}$$

In addition, we use Proposition 4.5.1. Then

$$\begin{aligned}\eta(L_1) = {}^{(\text{el})}\!\!\int_E \chi_{L_1}[E]\, d\eta \leq {}^{(\text{el})}\!\!\int_E \chi_{L_2}[E]\, d\eta = \eta(L_2) \\ = \eta(L_1) + \eta(L_2 \setminus L_1) < \eta(L_1) + \varepsilon.\end{aligned}$$

As a corollary, for $f \triangleq \chi_\Gamma[E]$ the following inequalities hold:

$$(f * \eta)^{\mathbf{d}}(E) \leq {}^{(\mathrm{el})}\!\!\int_E \chi_{L_2}[E]\, d\eta < {}^{(\mathrm{el})}\!\!\int_E \chi_{L_1}[E]\, d\eta + \varepsilon \leq (f * \eta)_{\mathbf{d}}(E) + \varepsilon. \tag{5.5.5}$$

Proposition 5.4.1, (5.4.12) and (5.5.4) are used in (5.5.5). Thus $(f * \eta)^{\mathbf{d}}(E) - (f * \eta)_{\mathbf{d}}(E) < \varepsilon$. Since the choice of $\varepsilon$ was arbitrary, we obtain the inequality

$$(f * \eta)^{\mathbf{d}}(E) - (f * \eta)_{\mathbf{d}}(E) \leq 0.$$

From (5.4.3) it follows that $(f * \eta)^{\mathbf{d}}(E) = (f * \eta)_{\mathbf{d}}(E)$. Since the choice of $\eta$ was arbitrary, we have the required property of the universal integrability of $\chi_\Gamma[E]$. From (5.4.21) we obtain that

$$\chi_\Gamma[E] \in B(E, \mathcal{L}). \tag{5.5.6}$$

The property $\Gamma \in \mathcal{L}$ follows from (4.3.5), (4.3.11), and (5.5.6). Indeed, if $\Gamma = \varnothing$, we immediately have $\Gamma \in \mathcal{L}$. Let $\Gamma \neq \varnothing$. Using (4.3.11) we choose $f_0 \in B_0(E, \mathcal{L})$ for which

$$\|f_0 - \chi_\Gamma[E]\| \leq \frac{1}{4}. \tag{5.5.7}$$

Choose $n \in \mathcal{N}$, $(\alpha_i)_{i \in \overline{1,n}} \in \mathbb{R}^n$ and $(L_i)_{i \in \overline{1,n}} \in \Delta_n(E, \mathcal{L})$ for which

$$f_0 = \sum_{i=1}^{n} \alpha_i \chi_{L_i}.$$

By (5.5.7) $\mathbb{K} \triangleq \{i \in \overline{1,n} \mid |\alpha_i - 1| \leq 1/4\} \in 2^{\overline{1,n}}$. Using the definitions from Section 4.2, we obtain the property

$$\mathbb{L} \triangleq \bigcup_{i \in \mathbb{K}} L_i \in \mathcal{L}. \tag{5.5.8}$$

If $u \in \Gamma$ then $\chi_\Gamma[E](u) = 1$, and for some $p \in \overline{1,n}$ we have $u \in L_p$. Then $f_0(u) = \alpha_p$ and by (5.5.7)

$$|\alpha_p - 1| = |f_0(u) - 1| = |f_0(u) - \chi_\Gamma[E](u)| \leq \|f_0 - \chi_\Gamma[E]\| \leq 1/4.$$

Thus $p \in \mathbb{K}$ and hence $u \in \mathbb{L}$. The inclusion $\Gamma \subset \mathbb{L}$ is established. Consider $v \in \mathbb{L}$. Choose (see (5.5.8)) $q \in \overline{1,n}$ for which $v \in L_q$; then $|\alpha_q - 1| \leq 1/4$. In particular, $v \in E$. By (5.5.7) we have

$$|\alpha_q - \chi_\Gamma[E](v)| = |f_0(v) - \chi_\Gamma[E](v)| \leq 1/4. \tag{5.5.9}$$

In addition, $(\chi_\Gamma[E](v) = 0) \vee (\chi_\Gamma[E](v) = 1)$. If $\chi_\Gamma[E](v) = 0$, then $|\alpha_q| \leq 1/4$ by (5.5.9) and as a consequence

$$|\alpha_q - 1| = |1 - \alpha_q| = 1 - \alpha_q \geq 3/4 > 1/4.$$

Thus $\chi_\Gamma[E](v) = 1$ and $v \in \Gamma$. The inclusion $\mathbb{L} \subset \Gamma$ is established. As a result $\Gamma = \mathbb{L}$. By (5.5.8) we obtain the inclusion $\Gamma \in \mathcal{L}$. So $(\Gamma \neq \varnothing) \Rightarrow (\Gamma \in \mathcal{L})$. We have $\Gamma \in \mathcal{L}$ always. Consequently we have the inclusion $\mathbb{J} \subset \mathcal{L}$. The relation (5.5.2) is established. □

Consider some corollaries of Proposition 5.5.1. Note that $\forall A \in \mathcal{P}(E)$ $\forall B \in \mathcal{P}(E)$:

$$(A\Delta B = (A \setminus B) \cup (B \setminus A)) \ \& \ ((A \setminus B) \cap (B \setminus A) = \varnothing).$$

Obviously, $A\Delta B \in \mathcal{L}$ for $A \in \mathcal{L}$ and $B \in \mathcal{L}$. On the other hand, it is of interest to consider the case when $A \in \mathcal{L}$ but, generally speaking, it may be that $B \notin \mathcal{L}$.

For $\mu \in (\mathrm{add})_+[\mathcal{L}]$ we denote by $\mu^{(*)}$ the function from $\mathcal{P}(E)$ into $[0, \infty[$ for which $\forall H \in \mathcal{P}(E)$:

$$\mu^{(*)}(H) \triangleq \inf(\{\mu(L) : L \in \mathcal{L},\ H \subset L\}). \tag{5.5.10}$$

See the definition of the type (5.5.10), for example, in [66, Ch.III]. Consider the following simple statement.

PROPOSITION 5.5.2 *If* $\mu \in (\mathrm{add})_+[\mathcal{L}]$, *then*

$$\mathcal{J}[\mu] = \{H \in \mathcal{P}(E) \mid \forall \varepsilon \in ]0, \infty[\ \exists L \in \mathcal{L} : \mu^{(*)}(L\Delta H) < \varepsilon\}. \tag{5.5.11}$$

The proof is obvious. Fix $\mu \in (\mathrm{add})_+[\mathcal{L}]$ and denote by $\mathbb{H}_\mu$ the family on the right hand side of (5.5.11). Recall that $\mu$ is an isotone mapping from $\mathcal{L}$ into $[0, \infty[$. We obtain (see (5.5.1) and (5.5.10)) that $\mathcal{J}[\mu] \subset \mathbb{H}_\mu$. Let $H^* \in \mathbb{H}_\mu$. Fix $\varepsilon \in ]0, \infty[$ and choose $L^* \in \mathcal{L}$ for which

$$\mu^{(*)}(L^*\Delta H^*) < \varepsilon/2.$$

Using (5.5.10) we choose $L^{**} \in \mathcal{L}$ with the properties

$$((L^*\Delta H^* \subset L^{**}) \ \& \ (\mu(L^{**}) < \varepsilon).$$

In addition, $L^* \setminus H^* \subset L^{**}$ and $H^* \setminus L^* \subset L^{**}$. Let

$$(L_1 \triangleq L^* \setminus L^{**}) \ \& \ (L_2 \triangleq L^* \cup L^{**}).$$

Let $x \in L_1$. Then $x \in L^*$ and $x \notin L^* \setminus H^*$. So $x \in H^*$. We have $L_1 \subset H^*$. Let $y \in H^*$. Then $y \in L^*$ or $y \in H^* \setminus L^*$, where $H^* \setminus L^* \subset L^{**}$. We have $y \in L_2$. Thus $H^* \subset L_2$. Certainly, $L_1 \in \mathcal{L}$ and $L_2 \in \mathcal{L}$ possess the properties

$$(L_1 \subset H^*) \,\&\, (H^* \subset L_2). \tag{5.5.12}$$

Let $z \in L_2 \setminus L_1$. Then, by the definition of $L_2$, $z \in L^*$ or $z \in L^{**}$. On the other hand, by the definition of $L_1$, $z \notin L^*$ or $z \in L^{**}$. If $z \notin L^{**}$, then we have a contradiction. Therefore $z \in L^{**}$. Thus we have $L_2 \setminus L_1 \subset L^{**}$ and

$$\mu(L_2 \setminus L_1) \leq \mu(L^{**}) < \varepsilon. \tag{5.5.13}$$

Since the choice of $\varepsilon$ was arbitrary, from (5.5.12) and (5.5.13) we have $H^* \in \mathcal{J}[\mu]$ (see (5.5.1)). So $\mathbb{H}_\mu \subset \mathcal{J}[\mu]$ and consequently $\mathcal{J}[\mu] = \mathbb{H}_\mu$. □

We note that from (5.5.10) it follows that $\mu^{(*)} : \mathcal{P}(E) \to [0, \mu(E)]$ for $\mu \in (\text{add})_+[\mathcal{L}]$. Consider the case of FAP. Then for $G \in \mathcal{P}(E)$ the number set

$$\{\inf_{L \in \mathcal{L}} \nu^{(*)}(L \Delta G) : \nu \in \mathbb{P}(\mathcal{L})\}$$

is a nonempty subset of $[0, 1]$; therefore

$$\sup_{\nu \in \mathbb{P}(\mathcal{L})} \inf_{L \in \mathcal{L}} \nu^{(*)}(L \Delta G) \in [0, 1]$$

is well defined.

PROPOSITION 5.5.3 *The following equality holds:*

$$\mathcal{L} = \{G \in \mathcal{P}(E) \mid \sup_{\nu \in \mathbb{P}(\mathcal{L})} \inf_{L \in \mathcal{L}} \nu^{(*)}(L \Delta G) = 0\}. \tag{5.5.14}$$

PROOF. Denote by $\mathbb{L}$ the set on the right hand side of (5.5.14). By (5.5.10) and Proposition 5.5.2 we have $\mathcal{L} \subset \mathcal{J}[\mu]$ for $\mu \in (\text{add})_+[\mathcal{L}]$. Moreover, $\forall \mu \in (\text{add})_+[\mathcal{L}]$: $(\mu^{(*)} \mid \mathcal{L}) = \mu$. Therefore $\forall \tilde{L} \in \mathcal{L}$: $\inf(\{\mu^{(*)}(L \Delta \tilde{L}) : L \in \mathcal{L}\}) = 0$. As a result $\mathcal{L} \subset \mathbb{L}$. Let $A \in \mathbb{L}$. Then $A \in \mathcal{P}(E)$. In addition,

$$\inf_{L \in \mathcal{L}} \nu^{(*)}(L \Delta A) = 0$$

for each $\nu \in \mathbb{P}(\mathcal{L})$. Consequently $\forall \nu \in \mathbb{P}(\mathcal{L})$ $\forall \varepsilon \in ]0, \infty[$ $\exists L \in \mathcal{L}$: $\nu^{(*)}(L \Delta A) < \varepsilon$. We obtain that $\forall \mu \in \mathbb{P}(\mathcal{L})$: $A \in \mathcal{J}[\mu]$; see Proposition 5.5.2. From (5.5.1) we have $A \in \mathcal{J}[\mathbb{O}_\mathcal{L}]$. Then $\forall \mu \in (\text{add})_+[\mathcal{L}]$ : $A \in \mathcal{J}[\mu]$. We use (5.5.1). From Proposition 5.5.1 we have $A \in \mathcal{L}$. Thus $\mathbb{L} \subset \mathcal{L}$ and hence $\mathcal{L} = \mathbb{L}$. □

Let us consider one more question concerned with the problem of universal measurability. In this connection we recall the known definition of the absolutely measurable set (see [103, Ch. I]). If for each $\mu \in \mathbb{P}_\sigma(\mathcal{L})$,

$\mathcal{L}_\mu^*$ is the completion of $\mathcal{L}$ with respect to $\mu$, then the intersection of all families $\mathcal{L}_\mu^*$, $\mu \in (\sigma - \text{add})_+[\mathcal{L}]$, is the set of all absolutely measurable sets. We consider the corresponding precise definitions and introduce a scheme meaning a distinctive extension of the problem on measurability.

For $\mu \in (\text{add})_+[\mathcal{L}]$ denote by $\mathfrak{N}_\mu$ the family of all sets $L \in \mathcal{L}$ for which $\mu(L) = 0$; then $\varnothing \in \mathfrak{N}_\mu$, and hence $\mathfrak{N}_\mu \in 2^{\mathcal{L}}$. Introduce null-sets, setting that, for $\mu \in (\text{add})_+[\mathcal{L}]$, $\mathfrak{N}_\mu^*$ is the family of all $H \in \mathcal{P}(E)$ with the property that $\exists L \in \mathfrak{N}_\mu : H \subset L$. Then it is possible to introduce the notion of an $\mu$-extension of $\mathcal{L}$. This definition is standard, but we use it in the case when $\mu$ is a FAM. Thus $\forall \mu \in (\text{add})_+[\mathcal{L}]$:

$$\mathcal{L}_\mu^* \triangleq \{\text{pr}_1(z) \cup \text{pr}_2(z) : z \in \mathcal{L} \times \mathfrak{N}_\mu^*\}. \qquad (5.5.15)$$

In other words, for $\mu \in (\text{add})_+[\mathcal{L}]$ we define $\mathcal{L}_\mu^*$ as the family of all sets $L \cup H$, where $L \in \mathcal{L}$ and $H \in \mathfrak{N}_\mu^*$.

We note the following obvious property:

$$\forall \mu \in (\text{add})_+[\mathcal{L}] : \mathcal{L}_\mu^* \subset \mathcal{J}[\mu]. \qquad (5.5.16)$$

Indeed, fix $\mu \in (\text{add})_+[\mathcal{L}]$ and choose $L^* \in \mathcal{L}_\mu^*$. Using (5.5.15), take $L \in \mathcal{L}$ and $H \in \mathfrak{N}_\mu^*$ for which $L^* = L \cup H$. Then for some $H_0 \in \mathfrak{N}_\mu$ we have the inclusion $H \subset H_0$. Hence $L \subset L^* \subset L \cup H_0$; in addition, $L \in \mathcal{L}$ and $L \cup H_0 \in \mathcal{L}$. Moreover, $\mu(H_0) = 0$. Note that (see Section 4.2)

$$\mu(L \cup H_0) \leq \mu(L \cup H_0) + \mu(L \cap H_0) = \mu(L) + \mu(H_0) = \mu(L) \leq \mu(L \cup H_0).$$

Here we use (4.2.26). Thus $\mu(L \cup H_0) = \mu(L)$. In addition, $(L \cup H_0) \setminus L \subset H_0$ and by (4.2.26) we have the inequality

$$\mu((L \cup H_0) \setminus L) \leq \mu(H_0) = 0.$$

By (5.5.1) we obtain the property $L^* \in \mathcal{J}[\mu]$. We have (5.5.16). Moreover, from (5.5.15) it follows that $\forall \mu \in (\text{add})_+[\mathcal{L}] : \mathcal{L} \subset \mathcal{L}_\mu^*$. We obtain that

$$\mathcal{L} \subset \bigcap_{\mu \in (\text{add})_+[\mathcal{L}]} \mathcal{L}_\mu^* \subset \bigcap_{\mu \in (\text{add})_+[\mathcal{L}]} \mathcal{J}[\mu].$$

From Proposition 5.5.1 it follows that

$$\mathcal{L} = \bigcap_{\mu \in (\text{add})_+[\mathcal{L}]} \mathcal{L}_\mu^*. \qquad (5.5.17)$$

In (5.5.2) and (5.5.17) we have useful statements about universal measurability. It is essential that the intersection is taken over all non-negative FAM. Generally speaking, another family is obtained when the intersection in (5.5.2) is taken over all non-negative CAM. See the corresponding example in [22, p. 83]. The effect attained in (5.5.2) and (5.5.17) is connected with the 'good' conditions of compactness (see (4.6.10)) in the sense of the $*$-weak topology of $\mathbf{A}(\mathcal{L})$.

## 5.6 ASYMPTOTIC VERSIONS OF THE PROBABILITY UNIFORM DISTRIBUTION

A distinctive employment of the constructions of Chapter 3 will be considered here. We suppose that $\mathcal{L} \in \Pi[E]$. We use (4.8.9) and (4.8.10). In addition, (4.8.11) is exploited under the condition $\alpha_1 = \ldots = \alpha_n$. As a matter of fact, we follow (4.8.13), using only a particular case. For $K \in \mathrm{Fin}(E)$ we denote by $|K|$ the cardinality of the set $K$; of course, $|K| \in \mathcal{N}$. In addition, $\forall K \in \mathrm{Fin}(E)$:

$$\frac{1}{|K|} \sum_{x \in K} (\delta_x \mid \mathcal{L}) \in \mathbb{P}_\sigma(\mathcal{L}). \tag{5.6.1}$$

By virtue of (5.6.1) we construct a mapping from the nonempty family $\mathrm{Fin}(E)$ into $\mathbb{P}_\sigma(\mathcal{L})$, $\mathbb{P}_\sigma(\mathcal{L}) \subset \mathbb{P}(\mathcal{L})$. We recall (see Chapter 4) that $\mathbb{P}(\mathcal{L})$ can be equipped with some compact topology. Suppose

$$\mathbf{h} : \mathrm{Fin}(E) \to \mathbb{P}_\sigma(\mathcal{L}) \tag{5.6.2}$$

is defined by the condition that $\forall K \in \mathrm{Fin}(E)$:

$$\mathbf{h}(K) \triangleq \frac{1}{|K|} \sum_{x \in K} (\delta_x \mid \mathcal{L}). \tag{5.6.3}$$

The mapping $\mathbf{h}$ (5.6.2), (5.6.3) can be regarded as the imbedding into $\mathbb{P}(\mathcal{L})$ with the topology

$$\tau^*_{\mathbb{P}}(\mathcal{L}) \triangleq \tau_*(\mathcal{L}) \mid_{\mathbb{P}(\mathcal{L})} \in (\mathbf{c}-\mathrm{top})[\mathbb{P}(\mathcal{L})].$$

Note that from(5.6.1)–(5.6.3) and (4.8.37) it follows that

$$\mathbf{h} : \mathrm{Fin}(E) \to \mathbb{P}^0_\sigma(\mathcal{L}).$$

Using definitions from Section 2.2 we introduce

$$\mathfrak{K} \triangleq \{(\mathrm{Fin})[E \mid K] : K \in \mathrm{Fin}(E)\} \in \mathcal{B}_0[\mathrm{Fin}(E)]. \tag{5.6.4}$$

We use (3.2.1) and (3.2.2). From (3.2.8) it follows that

$$\begin{aligned}(\tau^*_{\mathbb{P}}(\mathcal{L}) - \mathrm{LIM})[\mathfrak{K} \mid \mathbf{h}] &= \bigcap_{K \in \mathrm{Fin}(E)} \mathrm{cl}(\mathbf{h}^1((\mathrm{Fin})[E \mid K]), \tau^*_{\mathbb{P}}(\mathcal{L})) \\ &= \bigcap_{U \in \mathfrak{K}} \mathrm{cl}(\mathbf{h}^1(U), \tau^*_{\mathbb{P}}(\mathcal{L})).\end{aligned} \tag{5.6.5}$$

In (5.6.4) and (5.6.5) we have some concrete version of the construction from Section 3.2. Obviously, the family $\{\mathrm{cl}(\mathbf{h}^1(U), \tau^*_{\mathbb{P}}(\mathcal{L})) : U \in \mathfrak{K}\}$ has the property of centrality. In connection with (5.6.5) we note that $\forall U \in \mathfrak{K}$:

$$\mathrm{cl}(\mathbf{h}^1(U), \tau^*_{\mathbb{P}}(\mathcal{L})) = \mathrm{cl}(\mathbf{h}^1(U), \tau_*(\mathcal{L})) \cap \mathbb{P}(\mathcal{L}) = \mathrm{cl}(\mathbf{h}^1(U), \tau_*(\mathcal{L})). \quad (5.6.6)$$

We use (4.6.10) and (4.6.14) in (5.6.6): $\mathbb{P}(\mathcal{L}) \in \mathcal{F}_{\tau_*(\mathcal{L})}$ and $\mathbf{h}^1(U) \subset \mathbb{P}(\mathcal{L})$ for $U \in \mathfrak{K}$. Then from (5.6.5) and (5.6.6) it follows that

$$(\tau^*_{\mathbb{P}}(\mathcal{L}) - \mathrm{LIM})[\mathfrak{K} \mid \mathbf{h}] = \bigcap_{U \in \mathfrak{K}} \mathrm{cl}(\mathbf{h}^1(U), \tau_*(\mathcal{L})).$$

Return to (5.6.5). Note that

$$\mathbb{H} \triangleq \{\mathrm{cl}(\mathbf{h}^1(U), \tau^*_{\mathbb{P}}(\mathcal{L})) : U \in \mathfrak{K}\} \in \mathbf{Z}[\mathcal{F}_{\tau^*_{\mathbb{P}}(\mathcal{L})}]. \quad (5.6.7)$$

Indeed, let $n \in \mathcal{N}$ and $(H_i)_{i \in \overline{1,n}} \in \mathbb{H}^n$. Choose $(U_i)_{i \in \overline{1,n}} \in \mathfrak{K}^n$ for which $\forall j \in \overline{1,n} : H_j = \mathrm{cl}(\mathbf{h}^1(U_j), \tau^*_{\mathbb{P}}(\mathcal{L}))$. Using (5.6.4) we take

$$(K_i)_{i \in \overline{1,n}} : \overline{1,n} \to \mathrm{Fin}(E)$$

such that $\forall j \in \overline{1,n} : U_j = (\mathrm{Fin})[E \mid K_j]$. By definitions of Section 2.2 we have

$$(U_i)_{i \in \overline{1,n}} : \overline{1,n} \to 2^{\mathrm{Fin}(E)}.$$

In addition,

$$U_j = \{K \in \mathrm{Fin}(E) \mid K_j \subset K\}$$

for $j \in \overline{1,n}$. Consider the sets $\mathbb{K} \triangleq \bigcup_{i=1}^n K_i \in \mathrm{Fin}(E)$ and

$$\mathbb{U} \triangleq (\mathrm{Fin})[E \mid \mathbb{K}] = \{K \in \mathrm{Fin}(E) \mid \mathbb{K} \subset K\} \in \mathfrak{K}.$$

By the definition of $\mathbb{K}$ we obtain $\mathbb{U} \subset \bigcap_{i=1}^n U_i$. Then

$$\mathrm{cl}(\mathbf{h}^1(\mathbb{U}), \tau^*_{\mathbb{P}}(\mathcal{L})) \subset \bigcap_{i=1}^n H_i. \quad (5.6.8)$$

But $\mathrm{cl}(\mathbf{h}^1(\mathbb{U}), \tau^*_{\mathbb{P}}(\mathcal{L})) \in \mathbb{H}$. Since $\mathbb{K} \in \mathbb{U}$, we have $\mathbf{h}^1(\mathbb{U}) \neq \varnothing$. From (5.6.8) it follows that $\bigcap_{i=1}^n H_i \neq \varnothing$. Since the choice of $n, H_1, \ldots, H_n$ was arbitrary, from (2.2.20) we have (5.6.7). By (2.2.21), (2.3.6), (4.6.14), and (5.6.7) we obtain that

$$(\tau^*_{\mathbb{P}}(\mathcal{L}) - \mathrm{LIM})[\mathfrak{K} \mid \mathbf{h}] = \bigcap_{H \in \mathbb{H}} H \neq \varnothing. \quad (5.6.9)$$

In (5.6.9) we have the statement about the existence of 'asymptotically uniform' distributions. It is advisable to use (3.2.8) in (5.6.5) and (5.6.9). Note that (5.6.5) is a concrete variant of AS. But we employ this AS in constructions of 'pure' mathematics. Arbitrary finite subsets of $E$ were used in (5.6.4) and (5.6.5). However, some analogues with employing only subfamilies of $\mathrm{Fin}(K)$ can be considered. Note that the corresponding bases of $\mathrm{Fin}(K)$ can be used in (5.6.5). Let $\mathbf{F}_0 \in 2^{\mathrm{Fin}(E)}$ have the cofinality property:

$$\forall P \in \mathrm{Fin}(E) : (\mathrm{Fin})[E \mid P] \cap \mathbf{F}_0 \neq \varnothing. \tag{5.6.10}$$

In place of (5.6.4) we take the family

$$\mathfrak{K}_0 \triangleq \{(\mathrm{Fin})[E \mid K] \mid K \in \mathbf{F}_0\}. \tag{5.6.11}$$

From (5.6.4) and (5.6.11) we have the inclusion $\mathfrak{K}_0 \subset \mathfrak{K}$. Then $\varnothing \notin \mathfrak{K}_0$. If $K_1 \in \mathbf{F}_0$ and $K_2 \in \mathbf{F}_0$, then by (5.6.10) we have

$$(\mathrm{Fin})[E \mid K_1 \cup K_2] \cap \mathbf{F}_0 \neq \varnothing. \tag{5.6.12}$$

If $K^0$ is an arbitrary element of the set on the left hand side of (5.6.12), then $K^0 \in \mathbf{F}_0$ and $K_1 \cup K_2 \subset K^0$. Therefore for $K \in (\mathrm{Fin})[E \mid K^0]$ we have the inclusions $K \in (\mathrm{Fin})[E \mid K_1]$ and $K \in (\mathrm{Fin})[E \mid K_2]$. Thus we have established that $\forall K_1 \in \mathbf{F}_0 \ \forall K_2 \in \mathbf{F}_0 \ \exists K \in \mathbf{F}_0$:

$$(\mathrm{Fin})[E \mid K] \subset (\mathrm{Fin})[E \mid K_1] \cap (\mathrm{Fin})[E \mid K_2].$$

By (3.2.1), (3.2.2) and (5.6.11) we obtain

$$\mathfrak{K}_0 \in \mathcal{B}_0[\mathrm{Fin}(E)].$$

Consider the set $(\tau^*_{\mathbb{P}}(\mathcal{L}) - \mathrm{LIM})[\mathfrak{K}_0 \mid \mathbf{h}]$. By (3.2.8) we have the equality

$$(\tau^*_{\mathbb{P}}(\mathcal{L}) - \mathrm{LIM})[\mathfrak{K}_0 \mid \mathbf{h}] = (\tau^*_{\mathbb{P}}(\mathcal{L}) - \mathrm{LIM})[\mathfrak{K} \mid \mathbf{h}]. \tag{5.6.13}$$

Indeed, from (5.6.4) and (5.6.11) we obtain the inclusion

$$(\tau^*_{\mathbb{P}}(\mathcal{L}) - \mathrm{LIM})[\mathfrak{K} \mid \mathbf{h}] \subset (\tau^*_{\mathbb{P}}(\mathcal{L}) - \mathrm{LIM})[\mathfrak{K}_0 \mid \mathbf{h}].$$

Choose arbitrarily $\mu \in (\tau^*_{\mathbb{P}}(\mathcal{L}) - \mathrm{LIM})[\mathfrak{K}_0 \mid \mathbf{h}]$. Then $\mu \in \mathbb{P}(\mathcal{L})$. In addition, by (3.2.8) and (5.6.11) we have $\forall K \in \mathbf{F}_0$:

$$\mu \in \mathrm{cl}(\mathbf{h}^1((\mathrm{Fin})[E \mid K]), \tau^*_{\mathbb{P}}(\mathcal{L})). \tag{5.6.14}$$

Let $P \in \mathrm{Fin}(E)$. Using (5.6.10) we choose $Q \in (\mathrm{Fin})[E \mid P] \cap \mathbf{F}_0$. Then (see Section 2.2) $Q \in \mathrm{Fin}(E)$ and $P \subset Q$. As a consequence

$$(\mathrm{Fin})[E \mid Q] \subset (\mathrm{Fin})[E \mid P].$$

Hence we obtain the inclusion

$$\mathrm{cl}(\mathbf{h}^1((\mathrm{Fin})[E \mid Q]), \tau_{\mathbb{P}}^*(\mathcal{L})) \subset \mathrm{cl}(\mathbf{h}^1((\mathrm{Fin})[E \mid P]), \tau_{\mathbb{P}}^*(\mathcal{L})). \quad (5.6.15)$$

But by (5.6.14) $\mu \in \mathrm{cl}(\mathbf{h}^1((\mathrm{Fin})[E \mid Q]), \tau_{\mathbb{P}}^*(\mathcal{L}))$. Then (see (5.6.15))

$$(\tau_{\mathbb{P}}^*(\mathcal{L}) - \mathrm{LIM})[\mathfrak{K}_0 \mid \mathbf{h}] \subset \mathrm{cl}(\mathbf{h}^1((\mathrm{Fin})[E \mid P]), \tau_{\mathbb{P}}^*(\mathcal{L})).$$

From (5.6.5) we have the inclusion

$$(\tau_{\mathbb{P}}^*(\mathcal{L}) - \mathrm{LIM})[\mathfrak{K}_0 \mid \mathbf{h}] \subset (\tau_{\mathbb{P}}^*(\mathcal{L}) - \mathrm{LIM})[\mathfrak{K} \mid \mathbf{h}].$$

As a consequence we obtain (5.6.13).

**Example.** Let $E \triangleq \mathcal{N}$, $\mathcal{L}_1 \triangleq \{\overline{p,q} : p \in \mathcal{N}, q \in \mathcal{N}\}$, $\mathcal{L}_2 \triangleq \{\overline{n,\infty} : n \in \mathcal{N}\}$, $\mathcal{L} \triangleq \mathcal{L}_1 \cup \mathcal{L}_2$, $\mathbf{F}_0 \triangleq \{\overline{1,n} : n \in \mathcal{N}\}$ (here (5.6.10) is valid). The family $\mathfrak{K}_0$ is realized by (5.6.11). In addition, the equality (5.6.13) holds. We have

$$(\tau_{\mathbb{P}}^*(\mathcal{L}) - \mathrm{LIM})[\mathfrak{K} \mid \mathbf{h}] = \bigcap_{n \in \mathcal{N}} \mathrm{cl}(\mathbf{h}^1((\mathrm{Fin})[E \mid \overline{1,n}\,]), \tau_{\mathbb{P}}^*(\mathcal{L})). \quad (5.6.16)$$

We use the property $\mathfrak{K}_0 = \{(\mathrm{Fin})[E \mid \overline{1,n}\,] : n \in \mathcal{N}\}$. Of course, it is possible to use (3.2.8), considering (5.6.16) as AS. From (3.2.8) and (5.6.16) it follows that

$$\begin{aligned}&(\tau_{\mathbb{P}}^*(\mathcal{L}) - \mathrm{LIM})[\mathfrak{K} \mid \mathbf{h}]\\ =&\{\mu \in \mathbb{P}(\mathcal{L}) \mid \exists_{\mathbf{D}}\, \mathbf{S}[\mathbf{D} \neq \varnothing]\ \exists \preceq \in (\mathrm{DIR})[\mathbf{D}]\ \exists g \in \mathrm{Fin}(E)^{\mathbf{D}} :\\ &\qquad (\forall n \in \mathcal{N} : (\mathrm{Fin})[E \mid \overline{1,n}\,] \in (\mathrm{Fin}(E) - \mathrm{ass})[\mathbf{D}; \preceq; g])\\ &\qquad \& \ ((\mathbf{D}, \preceq, \mathbf{h} \circ g) \overset{\tau_{\mathbb{P}}^*(\mathcal{L})}{\longrightarrow} \mu)\}.\end{aligned} \quad (5.6.17)$$

Consider the most obvious case of a sequence in $\mathrm{Fin}(E) = \mathrm{Fin}(\mathcal{N})$. Denote by $\xi$ the sequence

$$k \mapsto \overline{1,k} : \mathcal{N} \to \mathrm{Fin}(E);$$

thus $\forall s \in \mathcal{N} : \xi(s) \triangleq \overline{1,s}$. We equip $\mathcal{N}$ with the usual order, i.e., with the direction in $\mathcal{N}$. So $(\mathbf{D}, \preceq)$ is $\mathcal{N}$ with the usual order. Then $\forall n \in \mathcal{N}$ $\forall s \in \overline{n,\infty}$: $\xi(s) \in (\mathrm{Fin})[E \mid \overline{1,n}\,]$. Hence $\forall n \in \mathcal{N}$:

$$(\mathrm{Fin})[E \mid \overline{1,n}\,] \in (\mathrm{Fin}(E) - \mathrm{ass})[\mathbf{D}; \preceq; \xi]. \quad (5.6.18)$$

Moreover, the sequence $\mathbf{h} \circ \xi$,

$$k \mapsto \mathbf{h}(\overline{1,k}) : \mathcal{N} \to \mathbb{P}_{\sigma}^0(\mathcal{L}),$$

converges to the following FAM. Namely, in accordance with [32, p. 91] we introduce the function $\delta_\infty : \mathcal{L} \to \{0; 1\}$ as follows:

$$(\forall L \in \mathcal{L}_1 : \delta_\infty(L) \triangleq 0) \; \& \; (\forall L \in \mathcal{L}_2 : \delta_\infty(L) \triangleq 1).$$

Then $\delta_\infty \in \mathbb{T}_{\mathbf{p}}(\mathcal{L})$; see (4.2.22). We have $\forall k \in \mathcal{N}$:

$$\mathbf{h}(\overline{1,k}) = \frac{1}{k}\sum_{i=1}^{k}(\delta_i \mid \mathcal{L}).$$

This relation means that for $k \in \mathcal{N}$ and $L \in \mathcal{L}$ we have

$$\mathbf{h}(\overline{1,k})(L) = \frac{1}{k}\sum_{i=1}^{k}\delta_i(L).$$

For $L \in \mathcal{L}_1$ we have the convergence

$$(\mathbf{h}(\overline{1,k})(L))_{k\in\mathcal{N}} \to 0. \tag{5.6.19}$$

Let $n \in \mathcal{N}$. Consider the case $L = \overrightarrow{n,\infty}$. For $k \in \overrightarrow{n,\infty}$ we have

$$(\forall i \in \overline{1,n} \setminus \{n\} : \delta_i(\overrightarrow{n,\infty}) = 0) \& (\forall i \in \overrightarrow{n,\infty} : \delta_i(\overrightarrow{n,\infty}) = 1).$$

Hence

$$\mathbf{h}(\overline{1,k})(\overrightarrow{n,\infty}) = \frac{k-n+1}{k}.$$

As a result we obtain the convergence

$$(\mathbf{h}(\overline{1,k})(\overrightarrow{n,\infty}))_{k\in\mathcal{N}} \to 1.$$

Since the choice of $n$ was arbitrary, we have $\forall L \in \mathcal{L}_2$:

$$(\mathbf{h}(\overline{1,k})(L))_{k\in\mathcal{N}} \to 1. \tag{5.6.20}$$

From (5.6.19) and (5.6.20) it follows that the sequence $(\mathbf{h}(\overline{1,k})(L))_{k\in\mathcal{N}} \in \mathbb{R}^{\mathcal{N}}$ converges to $\delta_\infty(L)$. In other words, we obtain the convergence

$$(\mathbf{h}(\overline{1,k}))_{k\in\mathcal{N}} \overset{\otimes^{\mathcal{L}}(\tau_{\mathbb{R}})}{\longrightarrow} \delta_\infty. \tag{5.6.21}$$

We use (2.6.31) in (5.6.21). Note that $\mathbf{h} \circ \xi \in \mathbb{P}(\mathcal{L})^{\mathcal{N}}$. Therefore from (5.6.21) we have the convergence

$$(\mathbf{h}(\overline{1,k}))_{k\in\mathcal{N}} \overset{\tau_\otimes(\mathcal{L})|_{\mathbb{P}(\mathcal{L})}}{\longrightarrow} \delta_\infty;$$

we use (2.3.9) and (4.6.18). From Proposition 4.6.1 and the definition of $\tau_{\mathbb{P}}^*(\mathcal{L})$ we obtain the convergence

$$(\mathbf{h}(\overline{1,k}))_{k\in\mathcal{N}} \overset{\tau_{\mathbb{P}}^*(\mathcal{L})}{\longrightarrow} \delta_\infty. \tag{5.6.22}$$

From (5.6.17), (5.6.18), and (5.6.22) we have

$$\delta_\infty \in (\tau_{\mathbb{P}}^*(\mathcal{L}) - \mathrm{LIM})[\mathfrak{K} \mid \mathbf{h}]. \tag{5.6.23}$$

Choose an arbitrary element

$$\mu \in (\tau_{\mathbb{P}}^*(\mathcal{L}) - \mathrm{LIM})[\mathfrak{K} \mid \mathbf{h}]. \tag{5.6.24}$$

By (5.6.17) and (5.6.24) we have $\mu \in \mathbb{P}(\mathcal{L})$. Choose a net $(\mathbf{D}, \preceq, g)$ in $\mathrm{Fin}(E)$ with the properties

$$(\forall n \in \mathcal{N} : (\mathrm{Fin})[E \mid \overline{1,n}] \in (\mathrm{Fin}(E) - \mathrm{ass})[\mathbf{D}; \preceq; g])$$

$$\& \; ((\mathbf{D}, \preceq, \mathbf{h} \circ g) \overset{\tau_{\mathbb{P}}^*(\mathcal{L})}{\longrightarrow} \mu). \tag{5.6.25}$$

The first condition in (5.6.25) means that $\mu = \delta_\infty$. Indeed, fix $\tilde{L} \in \mathcal{L}_1$. By the first statement of (5.6.25) we have

$$\forall \varepsilon \in ]0, \infty[\; \exists d_0 \in \mathbf{D} \; \forall d \in \mathbf{D} : (d_0 \preceq d) \Rightarrow ((\mathbf{h} \circ g)(d)(\tilde{L}) < \varepsilon).$$

Indeed, choose $n_0 \in \mathcal{N}$ with the property $\tilde{L} = \overline{1, n_0}$. Then for some $d_0 \in \mathbf{D}$ we have the property $\overline{1, n_0} \subset g(d)$, and

$$(\mathbf{h} \circ g)(d)(\tilde{L}) = \frac{n_0}{|g(d)|} \tag{5.6.26}$$

for each $d \in \mathbf{D}$, $d_0 \preceq d$. We use (5.6.3). But from the first statement of (5.6.25) it follows that $\forall N \in \mathcal{N} \; \exists d' \in \mathbf{D} \; \forall d'' \in \mathbf{D}$ :

$$(d' \preceq d'') \Rightarrow (N \leq |g(d'')|). \tag{5.6.27}$$

Using the definition of a directed set from Section 2.2 and taking (5.6.26), (5.6.27) into account, we obtain $\forall N \in \mathcal{N} \; \exists d_* \in \mathbf{D} \; \forall d \in \mathbf{D}$ :

$$(d_* \preceq d) \Rightarrow \left((\mathbf{h} \circ g)(d)(\tilde{L}) \leq \frac{n_0}{N}\right).$$

We have the required property of convergence. This property can be represented as follows:

$$\left(\mathbf{D}, \preceq, ((\mathbf{h} \circ g)(d)(\tilde{L}))_{d\in\mathbf{D}}\right) \overset{\mathbb{R}}{\longrightarrow} 0.$$

Since the choice of $\tilde{L}$ was arbitrary, we have $\forall L \in \mathcal{L}_1$:

$$(\mathbf{D}, \preceq, ((\mathbf{h} \circ g)(d)(L))_{d \in \mathbf{D}}) \xrightarrow{\tau_{\mathbb{R}}} 0. \qquad (5.6.28)$$

Now consider $L^* \in \mathcal{L}_2$. Let $n^* \in \mathcal{N}$ be the number for which $L^* = \overrightarrow{n^*, \infty}$. By (5.6.25) there exists $d^* \in \mathbf{D}$ for which $\forall d \in \mathbf{D}$:

$$(d^* \preceq d) \Rightarrow (\overline{1, n^*} \subset g(d)).$$

From (5.6.3) it follows that

$$(\mathbf{h} \circ g)(d)(L^*) = \frac{1}{|g(d)|} |g(d) \cap L^*| \qquad (5.6.29)$$

for $d \in \mathbf{D}$, $d^* \preceq d$. Let us estimate the value on the right hand side of (5.6.29). For each $d \in \mathbf{D}$, $d^* \preceq d$, we have

$$|g(d) \cap L^*| \geq |g(d)| - n^*.$$

From (5.6.29), for $d \in \mathbf{D}$, $d^* \preceq d$, we obtain the inequality

$$1 - \frac{n^*}{|g(d)|} \leq (\mathbf{h} \circ g)(d)(L^*). \qquad (5.6.30)$$

Using the first statement of (5.6.25) and taking (5.6.3), (5.6.30) into account, we obtain the convergence

$$(\mathbf{D}, \preceq, ((\mathbf{h} \circ g)(d)(L^*))_{d \in \mathbf{D}}) \xrightarrow{\tau_{\mathbb{R}}} 1.$$

Since the choice of $L^*$ was arbitrary, we have $\forall L \in \mathcal{L}_2$:

$$(\mathbf{D}, \preceq, ((\mathbf{h} \circ g)(d)(L))_{d \in \mathbf{D}}) \xrightarrow{\tau_{\mathbb{R}}} 1. \qquad (5.6.31)$$

From (5.6.28) and (5.6.31) it follows that $\forall L \in \mathcal{L}$:

$$(\mathbf{D}, \preceq, ((\mathbf{h} \circ g)(d)(L))_{d \in \mathbf{D}}) \xrightarrow{\tau_{\mathbb{R}}} \delta_\infty(L). \qquad (5.6.32)$$

The convergence (5.6.32) means the following property (see [35, p. 35]):

$$(\mathbf{D}, \preceq, \mathbf{h} \circ g) \xrightarrow{\otimes^{\mathcal{L}}(\tau_{\mathbb{R}})} \delta_\infty. \qquad (5.6.33)$$

Note that by (5.6.2) we have

$$\mathbf{h} \circ g : \mathbf{D} \to \mathbb{P}(\mathcal{L}).$$

Hence from (2.3.9) and (5.6.33) we obtain the convergence

$$(\mathbf{D}, \preceq, \mathbf{h} \circ g) \xrightarrow{\otimes^{\mathcal{L}}(\tau_{\mathbb{R}})|_{\mathbb{P}(\mathcal{L})}} \delta_\infty.$$

As a consequence, from (4.6.18) it follows that

$$(\mathbf{D}, \preceq, \mathbf{h} \circ g) \stackrel{\tau_\otimes(\mathcal{L})|_{\mathbb{P}(\mathcal{L})}}{\longrightarrow} \delta_\infty. \qquad (5.6.34)$$

In (5.6.34) we use the natural thransitivity of the operation of the passage to a subspace of a TS. Taking into account (4.6.14) and the relation preceding to (4.6.34), we obtain

$$\tau_{\mathbb{P}}^*(\mathcal{L}) = \tau_\otimes(\mathcal{L}) \,|_{\mathbb{P}(\mathcal{L})} \,. \qquad (5.6.35)$$

From (5.6.33) and (5.6.35) we have the convergence

$$(\mathbf{D}, \preceq, \mathbf{h} \circ g) \stackrel{\tau_{\mathbb{P}}^*(\mathcal{L})}{\longrightarrow} \delta_\infty. \qquad (5.6.36)$$

Then from (5.6.25) and (5.6.36) we obtain the required equality $\mu = \delta_\infty$. We use the property that each net in a Hausdorff TS has not more than one limit. Since the choice (5.6.24) was arbitrary, we have

$$(\tau_{\mathbb{P}}^*(\mathcal{L}) - \mathrm{LIM})[\mathfrak{K} \mid \mathbf{h}] \subset \{\delta_\infty\}.$$

Using (5.6.23) we obtain the equality

$$(\tau_{\mathbb{P}}^*(\mathcal{L}) - \mathrm{LIM})[\mathfrak{K} \mid \mathbf{h}] = \{\delta_\infty\}. \qquad (5.6.37)$$

From (5.6.3) and (5.6.37) we have the highly universal representation of $\delta_\infty$ as some uniform distribution on the semi-algebra of intervals of $\mathcal{N}$.
□

## 5.7 THE PURELY FINITELY ADDITIVE VERSION OF UNIFORM DISTRIBUTION ON THE FAMILY OF ALL SUBSETS OF THE INTERVAL [0,1]

In this section we investigate a question of principle of measure theory. We deal with the traditional space $[0, 1]$ considered as a subspace of $\mathbb{R}$ with the usual $|\cdot|$-topology. Our aim lies in constructing analogues of the uniform distribution (the latter is defined as the trace of the Lebesgue measure). For simplicity, we consider only the closed interval $\mathbf{I} \triangleq [0, 1]$. Thus we suppose here that $E \triangleq \mathbf{I}$. We equip $\mathbf{I}$ with different measurable structures. In particular, we consider $\mathcal{J} \triangleq \mathcal{P}(\mathbf{I}) \in (\sigma - \mathrm{alg})[\mathbf{I}]$ and the corresponding space $(\mathbf{I}, \mathcal{J})$ as some all embracing space. In addition, we consider

$$\begin{aligned} &(\mathcal{I}_1 \triangleq \{[\mathrm{pr}_1(z), \mathrm{pr}_2(z)[: z \in \mathbf{I} \times \mathbf{I}\}) \\ \& &(\mathcal{I}_2 \triangleq \{[\mathrm{pr}_1(z), \mathrm{pr}_2(z)] : z \in \mathbf{I} \times \mathbf{I}\}) \\ \& &(\mathcal{I}_3 \triangleq \{]\,\mathrm{pr}_1(z), \mathrm{pr}_2(z)] : z \in \mathbf{I} \times \mathbf{I}\}) \\ \& &(\mathcal{I}_4 \triangleq \{]\,\mathrm{pr}_1(z), \mathrm{pr}_2(z)[: z \in \mathbf{I} \times \mathbf{I}\}). \end{aligned} \qquad (5.7.1)$$

Then $\mathcal{I} \triangleq \bigcup_{j=1}^{4} \mathcal{I}_j \in \Pi[\mathbf{I}]$, and the pair $(\mathbf{I}, \mathcal{I})$ is the simplest measurable space. For

$$\mathfrak{B}_0 \triangleq \sigma_{\mathbf{I}}^0(\mathcal{I}) \in (\sigma - \text{alg})[\mathbf{I}]$$

we have the following chain of inclusions: $\mathcal{I} \subset \mathfrak{B}_0 \subset \mathcal{J}$. Let us introduce $\mathbf{l} : \mathcal{I} \to [0,1]$, setting

$$(\mathbf{l}(\varnothing) = 0) \ \& \ (\forall I \in \mathcal{I} \setminus \{\varnothing\} : \mathbf{l}(I) \triangleq \sup(I) - \inf(I)).$$

Hence $\mathbf{l} \in \mathbb{P}(\mathcal{I})$. We consider a FAP $\mu \in \mathbb{P}(\mathcal{J})$ satisfying the condition $(\mu \mid \mathcal{I}) = \mathbf{l}$ and some additional properties. In particular, for the set $Q$ of all rational numbers from $\mathbf{I}$ we require the equality $\mu(Q) = 1$ to be held. To investigate such extensions of $\mathbf{l}$, we employ methods similar to those used when compactifying the spaces of 'usual' solutions. Thus we use constructions of the finitely additive measure theory.

Certainly, $\mathbf{l}$ has the property of countable additivity. Then it is possible to use the well known scheme of the Lebesgue extension. However, preserving the countable additivity property, we can not realize the corresponding extension of $\mathbf{l}$ to $\mathcal{J}$. Of course, the known Lebesgue measure has zero value on the set $Q$. Therefore we consider constructions of a 'non-Lebesgue' extension of the given CAM. Note that a lot of possibilities exist for such an extension. The Tarskiı theorem [110, p.292] realizes these possibilities. But in this section we consider some special extensions of $\mathbf{l}$. Namely, we investigate extensions $\mu \in \mathbb{P}(\mathcal{J})$ of $\mathbf{l}$ with the property $\mu(Q) = 1$. In the sequel we use different concrete variants of the semi-algebra $\mathcal{L}$. In particular, we examine the following two cases: $\mathcal{L} = \mathcal{I}$ and $\mathcal{L} = \mathcal{J}$.

Let us consider some approximate construction. If $n \in \mathcal{N}$, $(a_i)_{i \in \overline{1,n}} : \overline{1,n} \to [0, \infty[$, and $(t_i)_{i \in \overline{1,n}} \in \mathbf{I}^n$, then by (4.8.10) we have

$$\sum_{i=1}^{n} a_i \delta_{t_i} \in (\sigma - \text{add})_+[\mathcal{J}].$$

Suppose that

$$\forall k \in \mathcal{N} \ \forall i \in \overline{0,k} : t_i^{(k)} \triangleq i/k. \tag{5.7.2}$$

We construct some grids on $\mathbf{I}$ in terms of (5.7.2). In particular, we set $\forall k \in \mathcal{N}$

$$I_k \triangleq \{t_i^{(k)} : i \in \overline{0,k}\}.$$

In addition, we obtain the obvious property

$$Q = \bigcup_{k \in \mathcal{N}} I_k. \tag{5.7.3}$$

We use (5.7.3) for constructing some simple probabilities on $\mathcal{J}$. Namely, we have $\forall k \in \mathcal{N}$:

$$\mu^{(k)} \triangleq \frac{1}{k} \sum_{i=1}^{k} \delta_{t_i^{(k)}} \in \mathbb{P}_\sigma(\mathcal{J}). \tag{5.7.4}$$

Note an important property of the measures (5.7.4). From (5.7.3) and (5.7.4) we obtain

$$\forall k \in \mathcal{N} : \mu^{(k)}(\mathbf{I}) = \mu^{(k)}(Q) = \mu^{(k)}(I_k) = 1. \tag{5.7.5}$$

Thus by (5.7.5) all elements of the sequence $(\mu^{(k)})_{k \in \mathcal{N}}$ are localized on $Q$. From (5.7.4) we have

$$\forall k \in \mathcal{N} : \nu^{(k)} \triangleq (\mu^{(k)} \mid \mathcal{I}) \in \mathbb{P}_\sigma(\mathcal{I}). \tag{5.7.6}$$

By (5.7.6) we have introduced a sequence in $\mathbb{P}_\sigma(\mathcal{I})$.

PROPOSITION 5.7.1 $(\nu^{(k)})_{k \in \mathcal{N}} \stackrel{\tau_\otimes(\mathcal{I})}{\longrightarrow} \mathbb{1}$.

PROOF. Choose arbitrarily $S \in \mathcal{I}$. We have the sequence

$$(\mu^{(k)}(S))_{k \in \mathcal{N}} = (\nu^{(k)}(S))_{k \in \mathcal{N}}$$

in $\mathbb{R}$. In addition,

$$(\nu^{(k)}(S))_{k \in \mathcal{N}} \to \mathbb{1}(S). \tag{5.7.7}$$

To verify (5.7.7) we consider several possible cases. If $S = \varnothing$, then (5.7.7) is obviously true.

Let $S \neq \varnothing$. First we consider the simplest case when $S = \{s\}$ for some $s \in \mathbf{I}$. Then $\mathbb{1}(S) = 0$. Note that $\forall k \in \mathcal{N}\ \forall i \in \overline{0,k}\ \forall j \in \overline{0,k} \setminus \{i\}$: $t_i^{(k)} \neq t_j^{(k)}$. From (5.7.4) we have

$$\forall k \in \mathcal{N} : \mu^{(k)}(S) = \mu^{(k)}(\{s\}) \leq \frac{1}{k}. \tag{5.7.8}$$

Indeed, the case $\{i \in \overline{0,k} \mid t_i^{(k)} = s\} \neq \varnothing$ is possible for $k \in \mathcal{N}$. Then by (5.7.4) we have

$$\mu^{(k)}(S) = \frac{1}{k}.$$

On the other hand, the case $\{i \in \overline{0,k} \mid t_i^{(k)} = s\} = \varnothing$ is also possible for $k \in \mathcal{N}$. Then we have (see (5.7.4)) $\mu^{(k)}(S) = 0$. Thus (5.7.8) is true. From (5.7.8) we have the convergence (5.7.7). Hence we obtain the convergence (5.7.7) in the case $\{s \in \mathbf{I} : S = \{s\}\} \neq \varnothing$.

Consider the case when $\forall s \in \mathbf{I} : S \neq \{s\}$. Of course, we preserve the supposition $S \neq \varnothing$. Since $S \in \mathcal{I} \setminus \{\varnothing\}$, we have the following chain of inclusions for some $a \in \mathbf{I}$ and $b \in \mathbf{I}$:

$$]a, b[ \subset S \subset [a, b]. \tag{5.7.9}$$

Moreover, $a < b$ because $S$ is not a singleton. By (5.7.9), always

$$\mathfrak{l}(S) = b - a \in ]0, \infty[.$$

Choose $N \in \mathcal{N}$ with the property $1/N < b - a = \mathfrak{l}(S)$. Fix $k \in \overrightarrow{N, \infty}$. Suppose that

$$]a, b[ \cap I_k = \varnothing. \tag{5.7.10}$$

Since $\theta \triangleq (a + b)/2 \in ]a, b[$, we have $\Theta \triangleq \{i \in \overline{1, k} \mid \theta \leq t_i^{(k)}\} \neq \varnothing$. We take into account the inclusion $]a, b[ \subset ]0, 1[$. Then $m \triangleq \inf(\Theta) \in \Theta$ and $t_{m-1}^{(k)} \in I_k$. In addition, $t_{m-1}^{(k)} < \theta$. Here it is advisable to examine the cases when $m - 1 = 0$ and $m - 1 \in \overline{1, k}$. We have $\theta \leq t_m^{(k)}$. But $t_m^{(k)} \in I_k$. Hence $\theta \neq t_m^{(k)}$ by (5.7.10). Thus we have

$$t_{m-1}^{(k)} < \theta < t_m^{(k)}. \tag{5.7.11}$$

Recall that $\theta \in ]a, b[$. If $a < t_{m-1}^{(k)}$, then $t_{m-1}^{(k)} \in ]a, b[$ by (5.7.11), and hence

$$t_{m-1}^{(k)} \in ]a, b[ \cap I_k.$$

The last property is impossible. Therefore $t_{m-1}^{(k)} \leq a$. If $t_m^{(k)} < b$, then $t_m^{(k)} \in ]a, b[$ by (5.7.11), and hence

$$t_m^{(k)} \in ]a, b[ \cap I_k.$$

We again obtain a contradiction. Therefore $b \leq t_m^{(k)}$. As a consequence

$$t_{m-1}^{(k)} \leq a < b \leq t_m^{(k)}.$$

Then $\mathfrak{l}(S) = b - a \leq t_m^{(k)} - t_{m-1}^{(k)} = 1/k \leq 1/N$, which contradicts the choice of $N$. Thus (5.7.10) is impossible. Consequently

$$\forall k \in \overrightarrow{N, \infty} : ]a, b[ \cap I_k \neq \varnothing. \tag{5.7.12}$$

Recall that $b - a \leq 1 - a \leq 1$. Hence we have the equality $1/N < 1$. Consequently $1 < N$ or, in other words, $2 \leq N$. Then $N - 1 \in \mathcal{N}$.

We note that $\forall k \in \mathcal{N}$:

$$(P_k \triangleq \{i \in \overline{1, k} \mid a < t_i^{(k)}\} \neq \varnothing) \& (Q_k \triangleq \{i \in \overline{1, k} \mid b \leq t_i^{(k)}\} \neq \varnothing).$$

It follows that $\forall k \in \mathcal{N}$:

$$(\alpha_k \triangleq \inf(P_k) \in P_k) \;\&\; (\tilde{\beta}_k \triangleq \inf(Q_k) \in Q_k).$$

We note that $0 \leq a < t_{\alpha_k}^{(k)}$ for $k \in \mathcal{N}$. If $\alpha_k = 1$, then

$$a \in [t_{\alpha_k - 1}^{(k)}, t_{\alpha_k}^{(k)}[. \tag{5.7.13}$$

In the case $\alpha_k \geq 2$ we have $\alpha_k - 1 < \alpha_k$ and $\alpha_k - 1 \notin P_k$; therefore $t_{\alpha_k - 1}^{(k)} \leq a$. Thus we again have (5.7.13) for $\alpha_k \geq 2$. By (5.7.2) we obtain

$$\forall k \in \mathcal{N} : a \in \left[ t_{\alpha_k}^{(k)} - \frac{1}{k}, t_{\alpha_k}^{(k)} \right[. \tag{5.7.14}$$

Assume that $\forall k \in \mathcal{N} : \beta_k \triangleq \tilde{\beta}_k - 1$. Fix $k \in \mathcal{N}$. Recall that $b \leq t_i^{(k)} \mid_{i=\tilde{\beta}_k}$. In addition, we have the equality $t_{\beta_k}^{(k)} < b$. We use the property $0 \leq a < b$. Hence we automatically have $t_{\beta_k}^{(k)} < b$ for $\tilde{\beta}_k = 1$ (see (5.7.2)). If $\tilde{\beta}_k \geq 2$, then $\beta_k \in \overline{1,k} \setminus Q_k$, and consequently the required inequality is realized by the definition of $Q_k$. So we have

$$b \in \left] t_{\beta_k}^{(k)}, t_{\beta_k}^{(k)} + \frac{1}{k} \right]. \tag{5.7.15}$$

From (5.7.14) and (5.7.15) we have the two chains of inequalities

$$\left( t_{\alpha_k}^{(k)} - \frac{1}{k} \leq a < t_{\alpha_k}^{(k)} \right) \;\&\; \left( t_{\beta_k}^{(k)} < b \leq t_{\beta_k}^{(k)} + \frac{1}{k} \right). \tag{5.7.16}$$

Recall that $a < b$. Hence we have the inequality

$$a < t_i^{(k)} \mid_{i=\tilde{\beta}_k}.$$

Since $\tilde{\beta}_k \in \overline{1,k}$, we obtain $\tilde{\beta}_k \in P_k$. By the definition of $\alpha_k$ we obtain the inequality $\alpha_k \leq \tilde{\beta}_k$. In reality, $\alpha_k \leq \beta_k$ for $k \in \overrightarrow{N, \infty}$. Indeed, suppose that $\beta_k < \alpha_k$ for $k \in \overrightarrow{N, \infty}$. Then $\beta_k \leq \alpha_k - 1$ and hence (see (5.7.2))

$$t_{\beta_k}^{(k)} = \frac{\beta_k}{k} \leq \frac{\alpha_k - 1}{k} = \frac{\alpha_k}{k} - \frac{1}{k} = t_{\alpha_k}^{(k)} - \frac{1}{k}.$$

Using (5.7.16) we obtain the inequalities

$$t_{\beta_k}^{(k)} \leq a < b \leq t_{\beta_k}^{(k)} + \frac{1}{k} = t_i^{(k)} \mid_{i=\tilde{\beta}_k}.$$

Then we arrive at the following inequalities:

$$b - a \leq (t^{(k)}_{\beta_k} + \frac{1}{k}) - a \leq \frac{1}{k}.$$

But $1/k \leq 1/N < b-a = \mathbf{l}(S)$. We have an obvious contradiction. Thus $\alpha_k \leq \beta_k$. Moreover, $\beta_k = \tilde{\beta}_k - 1 \leq k-1$. We obtain that $\alpha_k \in \overline{1,k}$ and $\beta_k \in \overline{\alpha_k, k-1}$ for $k \in \overrightarrow{N,\infty}$. From (5.7.2) and (5.7.16) we have $\forall k \in \overrightarrow{N,\infty}\ \forall i \in \overline{\alpha_k, \beta_k}$:

$$\left(a < t_i^{(k)}\right) \ \&\ \left(t_i^{(k)} < b\right).$$

Therefore from (5.7.9) it follows that $\forall k \in \overrightarrow{N,\infty}\ \forall i \in \overline{\alpha_k, \beta_k}$:

$$t_i^{(k)} \in S. \tag{5.7.17}$$

If $k \in \mathcal{N}$ and $i \in \overline{0, \alpha_k - 1}$, then

$$t_i^{(k)} \leq (t^{(k)}_{\alpha_k} - \frac{1}{k}) \leq a. \tag{5.7.18}$$

If $i \neq \alpha_k - 1$ in (5.7.18), then $i < \alpha_k - 1$ and by (5.7.2) we have $t_i^{(k)} < a$; hence by (5.7.9) $t_i^{(k)} \notin S$. Thus

$$\forall k \in \mathcal{N}\ \forall i \in \overline{0, \alpha_k - 1} \setminus \{\alpha_k - 1\} : t_i^{(k)} \notin S. \tag{5.7.19}$$

Let $k \in \mathcal{N}$ and $i \in \overline{\tilde{\beta}_k, k}$. From (5.7.16) we have

$$b \leq t^{(k)}_{\beta_k} + \frac{1}{k} = t^{(k)}_{\tilde{\beta}_k} \leq t_i^{(k)}. \tag{5.7.20}$$

Suppose that $i \neq \{\tilde{\beta}_k\}$. Then $\tilde{\beta}_k < i$ and, by (5.7.2) and (5.7.20), $b < t_i^{(k)}$. From (5.7.9) it follows that $t_i^{(k)} \notin S$. Thus we have

$$\forall k \in \mathcal{N}\ \forall i \in \overline{\beta_k + 1, k} \setminus \{\beta_k + 1\} : t_i^{(k)} \notin S. \tag{5.7.21}$$

We use (5.7.17), (5.7.19), and (5.7.21). Then $\forall k \in \overrightarrow{N,\infty}$:

$$\begin{aligned}\overline{\alpha_k, \beta_k} \subset \{i \in \overline{1,k} \mid t_i^{(k)} \in S\} \subset \{i \in \overline{0,k} \mid t_i^{(k)} \in S\} \\ \subset \overline{\alpha_k - 1, \beta_k + 1} = \overline{\alpha_k, \beta_k} \cup \{\alpha_k - 1\} \cup \{\beta_k + 1\}.\end{aligned}$$

As a result $\forall k \in \overrightarrow{N,\infty}$:

$$\begin{aligned}\{i \in \overline{1,k} \mid \delta_{t_i^{(k)}}(S) \neq 0\} \setminus \overline{\alpha_k, \beta_k} = \{i \in \overline{1,k} \mid \delta_{t_i^{(k)}}(S) = 1\} \setminus \overline{\alpha_k, \beta_k} \\ \subset \{\alpha_k - 1; \beta_k + 1\}.\end{aligned} \tag{5.7.22}$$

Consider a natural combination of (5.7.4) and (5.7.22). Recall that the case $\alpha_k - 1 \notin \overline{1,k}$ is possible for $k \in \overrightarrow{N,\infty}$. For simplicity, we fix $k \in \overrightarrow{N,\infty}$ in the sequel designations. Compare the two following numbers:

$$(\mu^{(k)}(S) \in \mathbb{R}) \ \& \ \left(\frac{1}{k} \sum_{i=\alpha_k}^{\beta_k} \delta_{t_i^{(k)}}(S) \in \mathbb{R}\right).$$

Note that

$$\mu^{(k)}(S) = \frac{1}{k} \sum_{i=1}^{k} \delta_{t_i^{(k)}}(S).$$

Recall (5.7.22), setting

$$M \triangleq \{i \in \overline{1,k} \mid \delta_{t_i^{(k)}}(S) \neq 0\} = \{i \in \overline{1,k} \mid \delta_{t_i^{(k)}}(S) = 1\}.$$

From (5.7.17) we have the inclusion $\overline{\alpha_k, \beta_k} \subset M$. Moreover, from (5.7.4) we have

$$\mu^{(k)}(S) = \frac{1}{k} \sum_{i \in M} \delta_{t_i^{(k)}}(S) = \frac{|M|}{k}. \tag{5.7.23}$$

In addition,

$$\frac{\beta_k - \alpha_k + 1}{k} = \frac{1}{k} \sum_{i=\alpha_k}^{\beta_k} \delta_{t_i^{(k)}}(S) \leq \mu^{(k)}(S). \tag{5.7.24}$$

We supplement (5.7.23) and (5.7.24) by (5.7.22). Namely, from (5.7.22) we obtain

$$M \setminus \overline{\alpha_k, \beta_k} \subset \{\alpha_k - 1; \beta_k + 1\}. \tag{5.7.25}$$

If $\alpha_k - 1 = 0$, then by the definition of $M$ we have

$$M \setminus \overline{\alpha_k, \beta_k} \subset \{\beta_k + 1\}.$$

Hence by (5.7.23) we obtain the inequalities

$$\mu^{(k)}(S) \leq \frac{1}{k} \sum_{i=\alpha_k}^{\beta_k} \delta_{t_i^{(k)}}(S) + \frac{1}{k}\delta_{t_{\beta_k+1}^{(k)}}(S) \leq \frac{1}{k} \sum_{i=\alpha_k}^{\beta_k} \delta_{t_i^{(k)}}(S) + \frac{1}{k}.$$

Thus for $\alpha_k - 1 = 0$ we have

$$\frac{\beta_k - \alpha_k + 1}{k} \leq \mu^{(k)}(S) \leq \frac{\beta_k - \alpha_k + 1}{k} + \frac{1}{k}. \tag{5.7.26}$$

So we have investigated the case $\alpha_k - 1 = 0$ with taking (5.7.25) into account. Consider now the case $\alpha_k - 1 \in \overline{1, k-1}$. From (5.7.3) and (5.7.25) it follows that

$$\begin{aligned} \mu^{(k)}(S) &= \frac{1}{k}\sum_{i=\alpha_k}^{\beta_k} \delta_{t_i^{(k)}}(S) + \frac{1}{k}\delta_{t^{(k)}_{\alpha_k - 1}}(S) + \frac{1}{k}\delta_{t^{(k)}_{\beta_k+1}}(S) \\ &\leq \frac{1}{k}\sum_{i=\alpha_k}^{\beta_k} \delta_{t_i^{(k)}}(S) + \frac{2}{k}. \end{aligned} \tag{5.7.27}$$

So we have (5.7.27) for $\alpha_k - 1 \in \overline{1, k-1}$. But $(\alpha_k - 1 = 0) \vee (\alpha_k - 1 \in \overline{1, k-1})$. Hence from (5.7.24), (5.7.26) and (5.7.27) we always have

$$\frac{\beta_k - \alpha_k + 1}{k} \leq \mu^{(k)}(S) \leq \frac{\beta_k - \alpha_k + 1}{k} + \frac{2}{k}.$$

As a consequence we obtain

$$\left|\mu^{(k)}(S) - \frac{\beta_k - \alpha_k + 1}{k}\right| \leq \frac{2}{k}. \tag{5.7.28}$$

Recall that (5.7.2) implies the equalities

$$\left(t^{(k)}_{\alpha_k} = \frac{\alpha_k}{k}\right) \ \& \ \left(t^{(k)}_{\beta_k} = \frac{\beta_k}{k}\right).$$

From (5.7.28) we have the following estimate

$$\mid \mu^{(k)}(S) - (t^{(k)}_{\beta_k} - t^{(k)}_{\alpha_k}) \mid \leq \frac{3}{k}. \tag{5.7.29}$$

By (5.7.16) we obtain

$$t^{(k)}_{\beta_k} - t^{(k)}_{\alpha_k} < b - a \leq t^{(k)}_{\beta_k} - t^{(k)}_{\alpha_k} + \frac{2}{k}.$$

As a consequence we have the inequality

$$\mid (t^{(k)}_{\beta_k} - t^{(k)}_{\alpha_k}) - (b-a) \mid = (b-a) - (t^{(k)}_{\beta_k} - t^{(k)}_{\alpha_k}) \leq \frac{2}{k}.$$

Taking (5.7.29) into account we obtain

$$\mid \mu^{(k)}(S) - (b-a) \mid \leq \frac{5}{k}.$$

Since the choice of $k$ was arbitrary, we have the convergence (5.7.7). Thus the convergence (5.7.7) holds for the case $(S \neq \varnothing)\&(\forall s \in \mathbf{I} : S \neq \{s\})$. Hence (5.7.7) holds always. As a consequence

$$\forall S \in \mathcal{I} : (\nu^{(k)}(S))_{k \in \mathcal{N}} \to \mathbf{l}(S).$$

From this and (2.6.27), (2.6.31) we have the convergence

$$(\nu^{(k)})_{k\in\mathcal{N}} \overset{\otimes^{\mathcal{I}}(\tau_{\mathbb{R}})}{\longrightarrow} \mathbb{1}. \tag{5.7.30}$$

Therefore by (2.3.9), (4.6.18) and (5.7.30) we obtain the required statement. □

By (5.7.4) we have constructed the sequence

$$(\mu^{(k)})_{k\in\mathcal{N}} : \mathcal{N} \to \mathbb{P}(\mathcal{J}). \tag{5.7.31}$$

Now we consider (5.7.31) as a sequence in the compact space

$$(\mathbb{P}(\mathcal{J}), \tau_{\mathbb{P}}^{*}(\mathcal{J})); \tag{5.7.32}$$

here we use a concrete variant of the corresponding definition from the previous section (recall that $\tau_{\mathbb{P}}^{*}(\mathcal{J}) \in (\mathbf{c} - \text{top})_0[\mathbb{P}(\mathcal{J})]$; see (2.2.22)). Introduce the set

$$\Omega \triangleq \{\omega \in \mathbb{P}_{\mathbf{p}}(\mathcal{J}) \mid ((\omega \mid \mathcal{I}) = \mathbb{1}) \ \&\ (\omega(Q) = 1)\}. \tag{5.7.33}$$

In (5.7.33) we consider the purely FAP on $\mathcal{J}$ with two special properties.

THEOREM 5.7.1 *The set* $\Omega$ *is not empty and* $\Omega = \{\omega \in \mathbb{P}(\mathcal{J}) \mid ((\omega \mid \mathcal{I}) = \mathbb{1}) \ \&\ (\omega(Q) = 1)\}$.

PROOF. Let $\tilde{\Omega} \triangleq \{\omega \in \mathbb{P}(\mathcal{J}) \mid ((\omega \mid \mathcal{I}) = \mathbb{1}) \ \&\ (\omega(Q) = 1)\}$. Then $\Omega \subset \tilde{\Omega}$. Consider the sequence (5.7.31) in the compact TS (5.7.32). By (2.2.24) we have

$$\tau_{\mathbb{P}}^{*}(\mathcal{J}) \in (\mathbf{c}_{\mathcal{N}} - \text{top})[\mathbb{P}(\mathcal{J})].$$

From (2.3.14) and the definition of $\tau_{\mathbb{P}}^{*}(\mathcal{J})$ it follows that

$$\mathbb{P}(\mathcal{J}) \in (\mathcal{N} - \tau_{*}(\mathcal{J}) - \text{comp})[\mathbf{A}(\mathcal{J})].$$

By (2.3.22) there exist $\eta \in \mathbb{P}(\mathcal{J})$, a directed set $(\mathbf{D}, \preceq)$, $\mathbf{D} \neq \varnothing$, and an operator $\varphi \in (\text{isot})[\mathbf{D}; \preceq]$ such that

$$\left(\mathbf{D}, \preceq, (\mu^{(\varphi(d))})_{d\in\mathbf{D}}\right) \overset{\tau_{*}(\mathcal{J})}{\longrightarrow} \eta. \tag{5.7.34}$$

Of course, $\mathbf{S}[\mathbf{D} \neq \varnothing]$ and $\preceq \in (\text{DIR})[\mathbf{D}]$. In addition, by (5.7.31)

$$d \mapsto \mu^{(\varphi(d))} : \mathbf{D} \to \mathbb{P}(\mathcal{J}).$$

From (4.6.21) and (5.7.34) it follows that $\forall L \in \mathcal{J}$:

$$\left(\mathbf{D}, \preceq, (\mu^{(\varphi(d))}(L))_{d\in\mathbf{D}}\right) \overset{\tau_{\mathbb{R}}}{\rightarrow} \eta(L). \tag{5.7.35}$$

In (5.7.35) we can consider the case $L \in \mathcal{I}$. Namely, $\forall S \in \mathcal{I}$:

$$\left(\mathbf{D}, \preceq, (\mu^{(\varphi(d))}(S))_{d\in\mathbf{D}}\right) \overset{\mathbb{R}}{\to} \eta(S). \tag{5.7.36}$$

By (5.7.6) we have $\forall S \in \mathcal{I}$: $(\mu^{(k)}(S))_{k\in\mathcal{N}} = (\nu^{(k)}(S))_{k\in\mathcal{N}}$. Hence by properties of $\varphi$ we obtain

$$\left(\mathbf{D}, \preceq, (\mu^{(\varphi(d))}(S))_{d\in D}\right) \overset{\mathbb{R}}{\to} \mathbb{1}(S). \tag{5.7.37}$$

We use Proposition 5.7.1. From (5.7.36) and (5.7.37) we have

$$(\eta \mid \mathcal{I}) = \mathbb{1}.$$

Now consider (5.7.35) for $L = Q$. Using (5.7.5) and (5.7.35) we obtain the equality $\eta(Q) = 1$. We have

$$\eta \in \mathbb{P}(\mathcal{J}) : ((\eta \mid \mathcal{I}) = \mathbb{1})\&(\eta(Q) = 1).$$

Note that $\eta \in \tilde{\Omega}$. Of course, $\mathbf{S}[\tilde{\Omega} \neq \varnothing]$. Choose arbitrarily $\tilde{\eta} \in \tilde{\Omega}$. Then $\tilde{\eta} \in \mathbb{P}(\mathcal{J})$ has the properties $((\tilde{\eta} \mid \mathcal{I}) = \mathbb{1})\&(\tilde{\eta}(Q) = 1)$.

Let $\nu \in (\sigma - \mathrm{add})_+[\mathcal{J}]$ be such that $\nu \leqq \tilde{\eta}$, i.e., $\forall H \in \mathcal{J}$: $0 \le \nu(H) \le \tilde{\eta}(H)$. In addition, $\tilde{\eta}(\mathbf{I} \setminus Q) = \tilde{\eta}(\mathbf{I}) - \tilde{\eta}(Q) = 0$. Therefore $\nu(\mathbf{I} \setminus Q) = 0$ and as a consequence $\forall S \in \mathcal{J}$:

$$\nu(S) = \nu(S\cap Q) + \nu(S\setminus Q) = \nu(S\cap Q) + \nu(S\cap(\mathbf{I}\setminus Q)) = \nu(S\cap Q) \le \nu(Q). \tag{5.7.38}$$

Let

$$(r_i)_{i\in\mathcal{N}} : \mathcal{N} \to Q$$

be a bijection from $\mathcal{N}$ onto $Q$. Thus $(\{r_i\})_{i\in\mathcal{N}} \in \Delta_\infty(Q, \mathcal{J})$. For $j \in \mathcal{N}$ we have $\{r_j\} \in \mathcal{I}$ and $\tilde{\eta}(\{r_j\}) = \mathbb{1}(\{r_j\}) = \nu(\{r_j\}) = 0$. But

$$\left(\sum_{i=1}^{k} \nu(\{r_i\})\right)_{k\in\mathcal{N}} \to \nu(Q).$$

As a result $\nu(Q) = 0$. Then from (5.7.38) it follows that $\forall S \in \mathcal{J}$ : $\nu(S) = 0$. Consequently we obtain $\nu = \mathbb{O}_{\mathcal{J}}$ under the condition $\nu \leqq \tilde{\eta}$. In other words, we have the implication

$$(\nu \leqq \tilde{\eta}) \Rightarrow (\nu = \mathbb{O}_{\mathcal{J}}).$$

Since the choice of $\nu$ was arbitrary, by (4.2.17) we obtain

$$\tilde{\eta} \in (\mathbf{p} - \mathrm{add})_+[\mathcal{J}].$$

But $\tilde{\eta}$ is a FAP. Therefore by (4.2.20) we have $\tilde{\eta} \in \mathbb{P}_{\mathbf{p}}(\mathcal{J})$. Using (5.7.33) and the suppositions with respect to $\tilde{\eta}$, we obtain $\tilde{\eta} \in \Omega$. Since the choice of $\tilde{\eta}$ was arbitrary, we have the inclusion $\tilde{\Omega} \subset \Omega$. Hence the equality $\tilde{\Omega} = \Omega$ is established.

PROPOSITION 5.7.2 $\Omega \in (\tau_*(\mathcal{J}) - \text{comp})[\mathbf{A}(\mathcal{J})]$.

PROOF. The topology $\tau^*_{\mathbb{P}}(\mathcal{J})$ coincides with

$$\tau_{\otimes}(\mathcal{J}) \mid_{\mathbb{P}(\mathcal{J})} = \otimes^{\mathcal{J}}(\tau_{\mathbb{R}}) \mid_{\mathbb{P}(\mathcal{J})} .$$

This property follows from Proposition 4.6.1. Let $(\mathbf{D}, \preceq, h)$ be a net in $\Omega$ and $\mu$ be a measure in $\mathbf{A}(\mathcal{J})$ for which

$$\left(\mathbf{D}, \preceq, h\right) \overset{\tau_*(\mathcal{J})}{\longrightarrow} \mu. \tag{5.7.39}$$

Here $(\mathbf{D}, \preceq)$ is a nonempty directed set and $h \in \Omega^{\mathbf{D}}$. In particular,

$$h : \mathbf{D} \to \mathbb{P}(\mathcal{J}). \tag{5.7.40}$$

By (4.6.10) and (4.6.14) we have $\mathbb{P}(\mathcal{J}) \in \mathcal{F}_{\tau_*(\mathcal{J})}$. Then from (5.7.39) and (5.7.40) we obtain the property $\mu \in \mathbb{P}(\mathcal{J})$. From (2.3.9) and (5.7.39) we have the convergence

$$\left(\mathbf{D}, \preceq, h\right) \overset{\tau^*_{\mathbb{P}}(\mathcal{J})}{\longrightarrow} \mu.$$

By (2.3.9) we obtain

$$\left(\mathbf{D}, \preceq, h\right) \overset{\otimes^{\mathcal{J}}(\tau_{\mathbb{R}})}{\longrightarrow} \mu.$$

Then for $S \in \mathcal{J}$ we have the convergence

$$\left(\mathbf{D}, \preceq, (h(d)(S))_{d \in D}\right) \overset{\tau_{\mathbb{R}}}{\to} \mu(S). \tag{5.7.41}$$

If $S \in \mathcal{I}$, then by (5.7.33) we have $\forall d \in \mathbf{D} : h(d)(S) = \mathbf{1}(S)$; therefore by (5.7.41) $\mu(S) = \mathbf{1}(S)$. So $(\mu \mid \mathcal{I}) = \mathbf{1}$. Moreover, from (5.7.41) we have the convergence

$$\left(\mathbf{D}, \preceq, (h(d)(Q))_{d \in D}\right) \overset{\tau_{\mathbb{R}}}{\to} \mu(Q). \tag{5.7.42}$$

But by (5.7.33) we have $\forall d \in \mathbf{D} : h(d)(Q) = 1$. From (5.7.42) we obtain that $\mu(Q) = 1$. Thus $\mu \in \mathbb{P}(\mathcal{J})$ has the properties

$$((\mu \mid \mathcal{I}) = \mathbf{1}) \ \& \ (\mu(Q) = 1).$$

From (5.7.33) we have $\mu \in \Omega$. Since the choice of $(\mathbf{D}, \preceq, h)$ and $\mu$ was arbitrary, we obtain $\Omega \in \mathcal{F}_{\tau_*(\mathcal{J})}$. But $\Omega \subset \mathbb{P}(\mathcal{J})$ and as a consequence $\Omega \in \mathbb{B}_*(\mathcal{J})$. From (4.6.10) we have $\Omega \in (\tau_*(\mathcal{J}) - \text{comp})[\mathbf{A}(\mathcal{J})]$. □

PROPOSITION 5.7.3 *The set $\Omega$ is convex.*

The proof is obvious (see Theorem 5.7.1). So we have established that $\Omega$ is a nonempty convex compactum in the TS $(\mathbf{A}(\mathcal{J}), \tau_*(\mathcal{J}))$. We note some simple properties of $\Omega$.

1) The image of $\mathcal{J}$ under each function $\omega \in \Omega$ is the interval $\mathbf{I}$. Indeed, if $\omega \in \Omega$ and $\xi \in \mathbf{I}$, then for $[0,\xi] \in \mathcal{I}$ we have

$$\xi = \mathbf{l}([0,\xi]) = \omega([0,\xi]) \in \omega^1(\mathcal{I});$$

hence $\xi \in \omega^1(\mathcal{J})$. Thus $\mathbf{I} \subset \omega^1(\mathcal{J})$ for $\omega \in \Omega$; conversely, if $\zeta \in \omega^1(\mathcal{J})$ and $L \in \mathcal{J}$ has the property $\zeta = \omega(L)$, then $0 \leq \zeta = \omega(L) \leq \omega(\mathbf{I}) = 1$. So we have $\forall \omega \in \Omega:$

$$\omega^1(\mathcal{J}) = \omega^1(\mathcal{I}) = [0,1].$$

2) Each FAP $\omega \in \Omega$ has no property of the bounded invariance: it is possible to indicate two sets $S_1 \in \mathcal{J}$, $S_2 \in \mathcal{J}$ and a number $\alpha \in \mathbf{I}$ for which $S_2 = \{s + \alpha : s \in S_1\}$ and $\omega(S_1) \neq \omega(S_2)$. Indeed, let $\omega \in \Omega$, $S_1 \triangleq Q \cap [0,1/2]$ and $\alpha \in (\mathbf{I} \setminus Q) \cap\, ]0,1/2[$. Then $\omega(S_1) = \omega(S_1) + \omega([0,1/2] \setminus Q) = \omega(S_1) + \omega([0,1/2] \setminus S_1) = \omega([0,1/2]) = \mathbf{l}([0,1/2]) = 1/2$ because $0 \leq \omega([0,1/2] \setminus Q) \leq \omega(\mathbf{I} \setminus Q) = 0$. Consider the set $S_2 \triangleq \{s + \alpha : s \in S_1\}$. Then $\forall s \in S_1 : 0 < s + \alpha < 1$. As a consequence $S_2 \subset ]0,1[$ and then $S_2 \in \mathcal{J}$. Therefore $\omega(S_2) \in [0,1]$. Let $\mathcal{N}_0 \triangleq \{0\} \cup \mathcal{N}$; moreover, let $s \in S_1$, $m \in \mathcal{N}_0$, and $n \in \mathcal{N}$ be such that $s = m/n$. Suppose that $s + \alpha \in Q$. In addition, $s + \alpha \neq 0$. Choose $p \in \mathcal{N}$ and $q \in \mathcal{N}$ for which $s + \alpha = p/q$. So

$$\alpha = \frac{p}{q} - \frac{m}{n} = \frac{pn - mq}{nq}. \tag{5.7.43}$$

But this is impossible. Since $\alpha > 0$ and hence $pn - mq \in \mathcal{N}$ and $nq \in \mathcal{N}$. Moreover, $\alpha < 1$, and hence $pn - mq < nq$. So by (5.7.43) we have $\alpha \in Q$, which contradicts the choice of $\alpha$. Thus $s + \alpha \notin Q$. We obtain that $S_2 \subset ]0,1[ \setminus Q \subset \mathbf{I} \setminus Q$. Therefore $0 \leq \omega(S_2) \leq \omega(\mathbf{I} \setminus Q) = 0$, since by (5.7.33), $1 = \omega(\mathbf{I}) = \omega(Q) + \omega(\mathbf{I} \setminus Q) = 1 + \omega(\mathbf{I} \setminus Q)$. Thus $\omega(S_2) = 0$. Finally, we have

$$(\omega(S_1) = 1/2) \ \& \ (\omega(S_2) = 0),$$

which proves property 2).

3) It is possible to verify that

$$(\nu^{(k)})_{k \in \mathcal{N}} \rightrightarrows \mathbf{l}.$$

This property reinforces Proposition 5.7.1. □

We introduce the set $\mathbf{L}$ of all limit points of the sequence (5.7.4) in the TS

$$(\mathbf{A}(\mathcal{J}), \tau_*(\mathcal{J})). \tag{5.7.44}$$

Namely, we assume that

$$\mathbf{L} \triangleq \Big\{ \mu_* \in \mathbf{A}(\mathcal{J}) \mid \exists_{\mathbf{D}}\, \mathbf{S}[\mathbf{D} \neq \varnothing]\ \exists \preceq \in (\mathrm{DIR})[\mathbf{D}]\ \exists \varphi \in (\mathrm{isot})[D; \preceq] :$$

$$\Big(D, \preceq, (\mu^{(\varphi(d))})_{d \in D}\Big) \overset{\tau_*(\mathcal{J})}{\longrightarrow} \mu_* \Big\}. \tag{5.7.45}$$

We use (2.2.50) in (5.7.45). Recall that $\tau_{\mathbb{P}}^*(\mathcal{J})$ is a topology of a compact subspace of the TS (5.7.44).

PROPOSITION 5.7.4 $\mathbf{L} \subset \Omega$.

PROOF. Choose arbitrarily $\mu^* \in \mathbf{L}$. Then $\mu^* \in \mathbf{A}(\mathcal{J})$. By (5.7.45) for some $\mathbf{S}[\mathbf{D} \neq \varnothing]$, $\preceq \in (\mathrm{DIR})[\mathbf{D}]$, and $\varphi \in (\mathrm{isot})[D; \preceq]$ we have the convergence

$$\Big(\mathbf{D}, \preceq, (\mu^{(\varphi(d))})_{d \in \mathbf{D}}\Big) \overset{\tau_*(\mathcal{J})}{\longrightarrow} \mu^*. \tag{5.7.46}$$

We use (5.7.46) by analogy with (5.7.34). Note that (5.7.34) implies $\mu^{(\varphi(d))} \in \mathbb{P}(\mathcal{J})$ for $d \in \mathbf{D}$. Using the compactness of the TS (5.7.32), we obtain

$$\mathbb{P}(\mathcal{J}) \in \mathcal{F}_{\tau_*(\mathcal{J})} \tag{5.7.47}$$

(see (4.6.10) and (4.6.14)). From (5.7.46) and (5.7.47) it follows that $\mu^* \in \mathbb{P}(\mathcal{J})$. Moreover, by Proposition 5.7.1 and the properties of $\varphi$ we obtain that

$$\Big(\mathbf{D}, \preceq, (\nu^{(\varphi(d))})_{d \in \mathbf{D}}\Big) \overset{\tau_\otimes(\mathcal{I})}{\longrightarrow} \mathbb{1}.$$

As a consequence we have the convergence

$$\Big(\mathbf{D}, \preceq, (\mu^{(\varphi(d))}(L))_{d \in \mathbf{D}}\Big) \overset{\tau_{\mathbb{R}}}{\longrightarrow} \mathbb{1}(L) \tag{5.7.48}$$

for each $L \in \mathcal{I}$. In (5.7.48) we use (2.3.9), (2.6.31), and (4.6.18). But from (4.6.21) and (5.7.46) we have the convergence

$$\Big(\mathbf{D}, \preceq, (\mu^{(\varphi(d))})_{d \in \mathbf{D}}\Big) \overset{\tau_\otimes(\mathcal{J})}{\longrightarrow} \mu^*. \tag{5.7.49}$$

From (2.3.9), (4.6.18), and (5.7.49) it follows that

$$\Big(\mathbf{D}, \preceq, (\mu^{(\varphi(d))})_{d \in \mathbf{D}}\Big) \overset{\otimes^{\mathcal{J}}(\tau_{\mathbb{R}})}{\longrightarrow} \mu^*.$$

As a result, for $L \in \mathcal{J}$ we have

$$\left(\mathbf{D}, \preceq, (\mu^{(\varphi(d))}(L))_{d\in\mathbf{D}}\right) \xrightarrow{\mathbb{T}\mathbb{R}} \mu^*(L). \tag{5.7.50}$$

We now use (2.6.31). We can take $L \in \mathcal{I}$ in (5.7.50). Then by (5.7.48) and (5.7.50) we have the equality

$$(\mu^* \mid \mathcal{I}) = \mathbf{1}. \tag{5.7.51}$$

Moreover, we can take $L = Q$ in (5.7.50). Then

$$\left(\mathbf{D}, \preceq, (\mu^{(\varphi(d))}(Q))_{d\in\mathbf{D}}\right) \xrightarrow{\mathbb{T}\mathbb{R}} \mu^*(Q).$$

By (5.7.5) we obtain the equality $\mu^*(Q) = 1$. So $\mu^* \in \mathbb{P}(\mathcal{J})$ has the properties (see (5.7.51)) $((\mu^* \mid \mathcal{I}) = \mathbf{1})\ \&\ (\mu^*(Q) = 1)$. By Theorem 5.7.1 $\mu^* \in \Omega$. Since the choice of $\mu^*$ was arbitrary, the required inclusion is established. □

PROPOSITION 5.7.5 *The set* $Q' \triangleq \bigcup_{k\in\mathcal{N}} I_{2k-1} \in \mathcal{P}(Q)$ *possesses the following property:*

$$(\exists \mu_1 \in \mathbf{L} : \mu_1(Q') = 1)\ \&(\exists \mu_2 \in \mathbf{L} : \mu_2(Q') = 1/2).$$

PROOF. Introduce two mappings in $\mathcal{N}$:

$$\left(\varphi_1 = (2k-1)_{k\in\mathcal{N}} \in \mathcal{N}^{\mathcal{N}}\right) \& \left(\varphi_2 = (2(2k-1))_{k\in\mathcal{N}} \in \mathcal{N}^{\mathcal{N}}\right). \tag{5.7.52}$$

Since $\varphi_1$ is a real-valued function, we have $\varphi_2 = 2\varphi_1$. Suppose that $\forall k \in \mathcal{N}$:

$$(\varsigma^{(k)} \triangleq \mu^{(\varphi_1(k))})\ \&\ (\eta^{(k)} \triangleq \mu^{(\varphi_2(k))}).$$

From (5.7.52) we see that $(\varsigma^{(k)})_{k\in\mathcal{N}}$ and $(\eta^{(k)})_{k\in\mathcal{N}}$ are subsequences of $(\mu^{(k)})_{k\in\mathcal{N}}$. Fix $n \in \mathcal{N}$. Consider

$$I_{\varphi_1(n)} = I_{2n-1} = \{t_i^{(\varphi_1(n))} : i \in \overline{0, \varphi_1(n)}\} \in \mathcal{P}(Q').$$

Using (5.7.5) we obtain that

$$\varsigma^{(n)}(I_{\varphi_1(n)}) = \mu^{(\varphi_1(n))}(I_{\varphi_1(n)}) = 1,$$

and as a consequence we have

$$1 = \varsigma^{(n)}(I_{\varphi_1(n)}) \leq \varsigma^{(n)}(Q') \leq \varsigma^{(n)}(Q) = \mu^{(\varphi_1(n))}(Q) = 1.$$

So $\zeta^{(n)}(Q') = 1$. Consider the number

$$\eta^{(n)}(Q') = \mu^{(\varphi_2(n))}(Q') = \frac{1}{\varphi_2(n)} \sum_{i=1}^{\varphi_2(n)} \delta_{t_i^{(\varphi_2(n))}}(Q') \in [0,1].$$

Return to (5.7.2). Introduce the sets

$$\left(\Xi^{(n)} \triangleq \left\{i \in \overline{1,\varphi_2(n)} \mid \frac{i}{\varphi_2(n)} \in Q'\right\} = \left\{i \in \overline{1,\varphi_2(n)} \mid t_i^{(\varphi_2(n))} \in Q'\right\}\right)$$
$$\& \left( \Xi_1^{(n)} \triangleq \left\{2j : j \in \overline{1,\varphi_1(n)}\right\}\right). \tag{5.7.53}$$

For $k \in \overline{1,\varphi_1(n)}$ we have $2k \in \overline{2,\varphi_2(n)}$ and

$$t_{2k}^{(\varphi_2(n))} = \frac{2k}{2(2n-1)} = \frac{k}{2n-1} = t_k^{(\varphi_1(n))} = t_k^{(2n-1)} \in I_{2n-1},$$

since $\varphi_1(n) = 2n-1$. Hence for $l \in \Xi_1^{(n)}$ we have $l \in \overline{1,\varphi_2(n)}$ and $t_l^{(\varphi_2(n))} \in Q'$. By (5.7.53)

$$\Xi_1^{(n)} \subset \Xi^{(n)}. \tag{5.7.54}$$

In reality, we have the equality in (5.7.54). Indeed, suppose the contrary: $\Xi^{(n)} \setminus \Xi_1^{(n)} \neq \varnothing$. Choose $s \in \Xi^{(n)} \setminus \Xi_1^{(n)}$. From (5.7.53) it follows that $s \in \overline{1,\varphi_2(n)}$ and

$$t_s^{(\varphi_2(n))} = \frac{s}{\varphi_2(n)} \in Q'. \tag{5.7.55}$$

But $s \neq 2j$ for $j \in \overline{1,\varphi_1(n)}$. Consequently $s \in \overline{1,\varphi_2(n)} \setminus \{2j : j \in \overline{1,\varphi_1(n)}\}$, where $\varphi_2(n) = 2\varphi_1(n)$. So we have

$$\Xi_2^{(n)} \triangleq \{j \in \overline{1,\varphi_1(n)} \mid s \leq 2j\} \neq \varnothing.$$

In addition, $p \triangleq \inf(\Xi_2^{(n)}) \in \Xi_2^{(n)}$. Then $p \in \overline{1,\varphi_1(n)}$ and $s \leq 2p$. But $p-1 \in \overline{0,\varphi_1(n)-1}$, and what is more, $p-1 \notin \Xi_2^{(n)}$. Then

$$(p-1=0) \vee (2(p-1) < s).$$

The case $p-1=0$ is very simple. Indeed, then $s \in \mathcal{N}$, $s \leq 2$ and (by the choice of $s$) $s \neq 2$. Therefore $s = 1 = 2p-1$. Consider the case $2(p-1) < s$. Then $2p-1 \leq s \leq 2p$. But by the choice of $s$ we have $s \neq 2p$. Therefore $s \leq 2p-1$. As a consequence, $s = 2p-1$ always. Thus

$$p \in \overline{1,\varphi_1(n)} : s = 2p-1.$$

Using (5.7.55) we choose $m \in \mathcal{N}$ for which $t_s^{(\varphi_2(n))} \in I_{2m-1}$. Therefore for some $r \in \overline{0, 2m-1}$ we have $t_s^{(\varphi_2(n))} = t_r^{(2m-1)}$. By (5.7.55) we have $t_s^{(\varphi_2(n))} \neq 0$. Therefore by (5.7.2) $r \neq 0$. So $r \in \overline{1, 2m-1}$. In addition, from (5.7.2) we have

$$\frac{2p-1}{\varphi_2(n)} = \frac{r}{\varphi_1(m)}.$$

In other words, $r\varphi_2(n) = (2p-1)(2m-1)$ or $2r\varphi_1(n) = (2p-1)(2m-1) = 4pm - 2p - 2m + 1$. Then we have

$$1/2 = r\varphi_1(n) + p + m - 2pm. \tag{5.7.56}$$

We have an integer number on the right hand side of (5.7.56), which is impossible. The contradiction means that $\Xi^{(n)} \subset \Xi_1^{(n)}$ and consequently (see (5.7.54))

$$\Xi_1^{(n)} = \Xi^{(n)}. \tag{5.7.57}$$

From (5.7.53) and (5.7.57) we have $\forall i \in \overline{1, \varphi_2(n)}$:

$$\left(\frac{i}{\varphi_2(n)} \in Q'\right) \Leftrightarrow \left(i \in \Xi_1^{(n)}\right).$$

Therefore for $i \in \overline{1, \varphi_2(n)}$, we obtain the equivalence

$$\left(i \in \Xi_1^{(n)}\right) \Leftrightarrow \left(\delta_{\frac{i}{\varphi_2(n)}}(Q') = 1\right).$$

Recall that $\Xi_1^{(n)} \subset \overline{1, \varphi_2(n)}$. Hence

$$\left\{i \in \overline{1, \varphi_2(n)} \mid \delta_{\frac{i}{\varphi_2(n)}}(Q') = 1\right\}$$
$$= \left\{i \in \overline{1, \varphi_2(n)} \mid \delta_{\frac{i}{\varphi_2(n)}}(Q') \neq 0\right\} = \Xi_1^{(n)}.$$

In this case we have the equality

$$\eta^{(n)}(Q') = \frac{\mid \Xi_1^{(n)} \mid}{\varphi_2(n)}, \tag{5.7.58}$$

where $\mid \Xi_1^{(n)} \mid$ is the cardinality of $\Xi_1^{(n)}$. Of course, we use (5.7.2) in (5.7.58). From (5.7.53) it follows that $\mid \Xi_1^{(n)} \mid = \varphi_1(n) = 2n - 1$. Therefore by (5.7.58) we have the equality

$$\eta^{(n)}(Q') = \frac{\varphi_1(n)}{\varphi_2(n)} = \frac{1}{2}.$$

Since the choice of $n \in \mathcal{N}$ was arbitrary, we have established that

$$\forall k \in \mathcal{N} : (\zeta^{(k)}(Q') = 1) \;\&\; (\eta^{(k)}(Q') = 1/2). \tag{5.7.59}$$

Recall that from (5.7.4) we have

$$\left( (\zeta^{(k)})_{k \in \mathcal{N}} \in \mathbb{P}(\mathcal{J})^{\mathcal{N}} \right) \;\&\; \left( (\eta^{(k)})_{k \in \mathcal{N}} \in \mathbb{P}(\mathcal{J})^{\mathcal{N}} \right). \tag{5.7.60}$$

We use (2.3.21), (2.3.22), and (4.6.14). By (5.7.60) we can choose two (nonempty) directed sets $(\mathbb{T}, \preceq)$ and $(\mathbb{S}, \ll)$, operators $\psi_1 \in (\text{isot})[\mathbb{T}; \preceq]$ and $\psi_2 \in (\text{isot})[\mathbb{S}; \ll]$, measures $\zeta \in \mathbb{P}(\mathcal{J})$ and $\eta \in \mathbb{P}(\mathcal{J})$ for which

$$\left( (\mathbb{T}, \preceq, (\zeta^{(\psi_1(t))})_{t \in \mathbb{T}}) \xrightarrow{\tau_*(\mathcal{J})} \zeta \right) \;\&\; \left( (\mathbb{S}, \ll, (\eta^{(\psi_2(s))})_{s \in \mathbb{S}}) \xrightarrow{\tau_*(\mathcal{J})} \eta \right). \tag{5.7.61}$$

Compare (5.7.59) and (5.7.61). From (4.6.21) and (5.7.61) it follows that

$$\left( (\mathbb{T}, \preceq, (\zeta^{(\psi_1(t))})_{t \in \mathbb{T}}) \xrightarrow{\tau_\otimes(\mathcal{J})} \zeta \right) \;\&\; \left( (\mathbb{S}, \ll, (\eta^{(\psi_2(s))})_{s \in \mathbb{S}}) \xrightarrow{\tau_\otimes(\mathcal{J})} \eta \right).$$

As a consequence we have

$$\left( (\mathbb{T}, \preceq, (\zeta^{(\psi_1(t))}(Q'))_{t \in \mathbb{T}}) \xrightarrow{\tau_\mathbb{R}} \zeta(Q') \right)$$

$$\&\; \left( (\mathbb{S}, \ll, (\eta^{(\psi_2(s))}(Q'))_{s \in \mathbb{S}}) \xrightarrow{\tau_\mathbb{R}} \eta(Q') \right).$$

From (5.7.59) we obtain the equalities

$$(\zeta(Q') = 1) \;\&\; (\eta(Q') = 1/2).$$

Note that $\varphi_1 \circ \psi_1 \in \mathcal{N}^{\mathbb{T}}$ and $\varphi_2 \circ \psi_2 \in \mathcal{N}^{\mathbb{S}}$. Moreover,

$$(\forall t \in \mathbb{T} : (\varphi_1 \circ \psi_1)(t) = \varphi_1(\psi_1(t)) = 2\psi_1(t) - 1)$$

$$\&\; (\forall s \in \mathbb{S} : (\varphi_2 \circ \psi_2)(s) = \varphi_2(\psi_2(s)) = 2\varphi_1(\psi_2(s)) = 2(2\psi_2(s) - 1)).$$

From (2.2.50) it follows that $\{t \in \mathbb{T} \mid k \leq \psi_1(t)\} \neq \varnothing$ for $k \in \mathcal{N}$. Let $t' \in \mathbb{T}$ be such that $k \leq \psi_1(t')$ and consequently

$$k \leq 2k - 1 = (k - 1) + k$$
$$\leq (\psi_1(t') - 1) + \psi_1(t') = 2\psi_1(t') - 1$$
$$= \varphi_1(\psi_1(t')) = (\varphi_1 \circ \psi_1)(t').$$

On the other hand, for $t_1 \in \mathbb{T}$ and $t_2 \in \mathbb{T}$ we have

$$(t_1 \preceq t_2) \Rightarrow (\psi_1(t_1) \leq \psi_1(t_2)).$$

From the definition of $\varphi_1$ it follows that if $t_1 \in \mathbb{T}$, $t_2 \in \mathbb{T}$, and $t_1 \preceq t_2$, then

$$(\varphi_1 \circ \psi_1)(t_1) = \varphi_1(\psi_1(t_1)) \leq \varphi_1(\psi_1(t_2)) = (\varphi_1 \circ \psi_1)(t_2).$$

By (2.2.50) we have

$$\varphi_1 \circ \psi_1 \in (\text{isot})[\mathbb{T}; \preceq]. \tag{5.7.62}$$

Moreover, by the definition of $\varphi_1$ we obtain that $\forall t \in \mathbb{T}$:

$$\zeta^{(\psi_1(t))} = \mu^{(\varphi_1(\psi_1(t)))} = \mu^{((\varphi_1 \circ \psi_1)(t))}.$$

Consequently from (5.7.61) we have the convergence

$$(\mathbb{T}, \preceq, (\mu^{((\varphi_1 \circ \psi_1)(t))})_{t \in \mathbb{T}}) \stackrel{\tau_*(\mathcal{J})}{\longrightarrow} \zeta.$$

The inclusion $\zeta \in \mathbf{L}$ follows from (5.7.45) and (5.7.62). The property $\eta \in \mathbf{L}$ is established in an analogous way. Thus $Q'$ possesses the required property with $\mu_1 = \zeta$ and $\mu_2 = \eta$.

COROLLARY 5.7.1 *The set $\Omega$ contains a nonempty set with the continuum cardinality.*

PROOF. Choose $\mu_1 \in \mathbf{L}$ and $\mu_2 \in \mathbf{L}$ for which

$$(\mu_1(Q') = 1) \ \& \ (\mu_2(Q') = 1/2).$$

In particular, $\mu_1 \in \Omega$ and $\mu_2 \in \Omega$ (see Proposition 5.7.4). From Proposition 5.7.3 we obtain that $\forall \alpha \in [0, 1]$:

$$\eta_\alpha \triangleq \mu_2 + \alpha(\mu_1 - \mu_2) = \alpha\mu_1 + (1 - \alpha)\mu_2 \in \Omega. \tag{5.7.63}$$

Consider the mapping $\rho \triangleq (\eta_\alpha)_{\alpha \in [0,1]} \in \Omega^{[0,1]}$. Let $\Omega_1 \triangleq \rho^1([0,1])$. In other words, $\rho$ is a mapping from $[0, 1]$ onto $\Omega_1 \in \mathcal{P}(\Omega)$. Let $p \in [0, 1]$ and $q \in [0, 1] \setminus \{p\}$. Then $\eta_p(Q') - \eta_q(Q') = (p - q)(\mu_1(Q') - \mu_2(Q')) = (p - q)/2 \neq 0$. Consequently $\rho(p) = \eta_p \neq \eta_q = \rho(q)$. So $\rho$ is a bijection from $[0, 1]$ onto $\Omega_1$, and $\Omega_1$ is the required subset of $\Omega$: the cardinality of $\Omega_1$ is continuum. □

Let $\mathcal{A} \triangleq a^0_{\mathbf{I}}(\mathcal{I})$. By the definition from Section 2.2 we have $\mathcal{A} \in (\text{alg})[\mathbf{I}]$. And what is more, by (2.2.38) we obtain that

$$\mathcal{A} = \{H \in \mathcal{P}(\mathbf{I}) \mid \exists n \in \mathcal{N} : \Delta_n(H, \mathcal{I}) \neq \varnothing\}.$$

Assume that $\mathbf{l}_0 \triangleq \alpha[\mathbf{l}]$. Of course, $\mathbf{l}_0 \in \mathbb{P}(\mathcal{A})$. On the basis of $(\mathbf{I}, \mathcal{A}, \mathbf{l}_0)$ we realize the construction traditional for measure theory. Namely, we introduce the function

$$\mathbf{l}_0^{(*)} : \mathcal{J} \to [0, 1] \tag{5.7.64}$$

by the rule

$$\mathbf{l}_0^{(*)}(H) \triangleq \inf(\{\mathbf{l}_0(A) : A \in \mathcal{A}, H \subset A\}) \tag{5.7.65}$$

for $H \in \mathcal{J}$. From (5.7.64) and (5.7.65) we have $(\mathbf{l}_0^{(*)} \mid \mathcal{A}) = \mathbf{l}_0$. In addition,

$$\begin{aligned}\mathcal{G}_0^{(*)} &\triangleq \{H \in \mathcal{J} \mid \forall \varepsilon \in ]0, \infty[\ \exists A_1 \in \mathcal{A}\ \exists A_2 \in \mathcal{A} : \\ &\qquad (A_1 \subset H \subset A_2)\ \&\ (\mathbf{l}_0(A_2 \setminus A_1) < \varepsilon)\} \\ &= \{H \in \mathcal{J} \mid \forall A \in \mathcal{J} : \mathbf{l}_0^{(*)}(A) = \mathbf{l}_0^{(*)}(A \cap H) + \mathbf{l}_0^{(*)}(A \setminus H)\} \\ &= \{H \in \mathcal{J} \mid \mathbf{l}_0^{(*)}(H) + \mathbf{l}_0^{(*)}(\mathbf{I} \setminus H) = 1\}.\end{aligned} \tag{5.7.66}$$

In connection with (5.7.66) see constructions of [66, Section 3.1]. From (5.7.66) we have

$$\mathcal{G}_0^{(*)} \in (\text{alg})[\mathbf{I} \mid \mathcal{A}]. \tag{5.7.67}$$

Recall the known statement (see [82, p. 281]): if $\nu \in (\text{add})_+[\mathcal{G}_0^{(*)}]$, then

$$(\mathbf{l}_0 = (\nu \mid \mathcal{A})) \Rightarrow ((\mathbf{l}_0^{(*)} \mid \mathcal{G}_0^{(*)}) = \nu). \tag{5.7.68}$$

By Theorem 5.7.1 and (5.7.67) we have $\forall \omega \in \Omega$:

$$(\omega \mid \mathcal{G}_0^{(*)}) \in \mathbb{P}(\mathcal{G}_0^{(*)}).$$

Moreover, for $\omega \in \Omega$ we obtain that $\omega_{\mathcal{A}} \triangleq (\omega \mid \mathcal{A}) \in (\text{add})_+[\mathcal{A}]$ has the property (see Theorem 5.7.1)

$$(\omega_{\mathcal{A}} \mid \mathcal{I}) = \mathbf{l} = (\mathbf{l}_0 \mid \mathcal{I});$$

as a consequence we have $\omega_{\mathcal{A}} = \mathbf{l}_0$. We have established that

$$\forall \omega \in \Omega : (\omega \mid \mathcal{A}) = \mathbf{l}_0 = (\mathbf{l}_0^{(*)} \mid \mathcal{A}). \tag{5.7.69}$$

By (5.7.68) and (5.7.69) we obtain

$$\forall \omega \in \Omega : (\mathbf{l}_0^{(*)} \mid \mathcal{G}_0^{(*)}) = (\omega \mid \mathcal{G}_0^{(*)}). \tag{5.7.70}$$

In addition, we use the following transitivity property: for $\omega \in \Omega$ and $\omega^{(*)} \triangleq (\omega \mid \mathcal{G}_0^{(*)}) \in (\text{add})_+[\mathcal{G}_0^{(*)}]$ the equality $(\omega \mid \mathcal{A}) = (\omega^{(*)} \mid \mathcal{A})$ holds. Since $\Omega \neq \varnothing$, by (5.7.70) we have $(\mathbf{l}_0^{(*)} \mid \mathcal{G}_0^{(*)}) \in \mathbb{P}(\mathcal{G}_0^{(*)})$.

Denote by $\tau_0^1$ the natural $|\cdot|$-topology of $\mathbf{I}$ and assume that (see (2.2.32))

$$\mathfrak{B} \triangleq \mathcal{B}_\sigma(\tau_0^1);$$

then $\mathfrak{B} \in (\sigma - \text{alg})[\mathbf{I}]$ is the algebra of Borel subsets of $\mathbf{I}$. It is possible to check the equality

$$\mathfrak{B} = \mathcal{B}_\sigma(\tau_{\mathbb{R}}\,|_{\mathbf{I}}) = \mathcal{B}_\sigma(\tau_{\mathbb{R}})\,|_{\mathbf{I}} = \sigma_{\mathbf{I}}^0(\mathcal{I}) = \mathfrak{B}_0.$$

Here we use constructions of the type (2.5.17), (2.5.18). Of course, $\mathfrak{B} \in (\text{alg})[\mathbf{I} \mid \mathcal{I}]$. Therefore, $\mathcal{A} \in 2^{\mathfrak{B}}$. Denote by $\lambda$ the Borel Lebesgue measure : $\lambda \in (\sigma - \text{add})_+[\mathfrak{B}]$ and $(\lambda \mid \mathcal{I}) = 1$. In particular, $\lambda \in \mathbb{P}(\mathfrak{B})$. Introduce $\lambda^{(*)} : \mathcal{J} \to [0,1]$ by the natural rule, setting for $H \in \mathcal{J}$

$$\lambda^{(*)}(H) \triangleq \inf(\{\lambda(B) : B \in \mathfrak{B},\ H \subset B\}). \tag{5.7.71}$$

In addition, $\mathcal{A} \subset \mathfrak{B}$ and by (5.7.65) and (5.7.71) we have

$$\lambda^{(*)} \leqq \mathbb{1}_0^{(*)}. \tag{5.7.72}$$

Introduce $(\text{NUL})[\lambda] \triangleq \{B \in \mathfrak{B} \mid \lambda(B) = 0\}$ and (in these terms)

$$(\text{NUL})^*[\lambda] \triangleq \{H \in \mathcal{J} \mid \exists M \in (\text{NUL})[\lambda] : H \subset M\}.$$

Since $(\mathbf{I}, \mathfrak{B}, \lambda)$ is the standard measure space, we have

$$\begin{aligned}\mathfrak{B}^{(*)} &\triangleq \{H \in \mathcal{J} \mid \forall A \in \mathcal{J} : \lambda^{(*)}(A) = \lambda^{(*)}(A \cap H) + \lambda^{(*)}(A \setminus H)\}\\ &= \{H \in \mathcal{J} \mid \lambda^{(*)}(H) + \lambda^{(*)}(\mathbf{I} \setminus H) = 1\}\\ &= \{B \cup N^* : (B, N^*) \in \mathfrak{B} \times (\text{NUL})^*[\lambda]\} \in (\sigma - \text{alg})[\mathbf{I}].\end{aligned}$$

In addition, the Lebesgue measure $\lambda_* \in (\sigma - \text{add})_+[\mathfrak{B}^{(*)}]$ is defined by the rule: for $M \in \mathfrak{B}^{(*)}$ and $B \in \mathfrak{B}$ the implication

$$(\exists S \in (\text{NUL})^*[\lambda] : M = B \cup S) \Rightarrow (\lambda_*(M) = \lambda(B))$$

holds. Moreover, $\lambda_* = (\lambda^{(*)} \mid \mathfrak{B}^{(*)})$. By (5.7.66) and (5.7.72) we have $\mathcal{G}_0^{(*)} \subset \mathfrak{B}^{(*)}$. Moreover,

$$(\lambda_* \mid \mathcal{G}_0^{(*)}) = (\lambda^{(*)} \mid \mathcal{G}_0^{(*)}) = (\mathbb{1}_0^{(*)} \mid \mathcal{G}_0^{(*)}). \tag{5.7.73}$$

Here we use (5.7.66) and (5.7.72). From (5.7.70) and (5.7.72) we have $\forall \omega \in \Omega$:

$$(\omega \mid \mathcal{G}_0^{(*)}) = (\lambda_* \mid \mathcal{G}_0^{(*)}). \tag{5.7.74}$$

The relation (5.7.74) is a very natural property of the algebra of Jordan measurable sets.

Consider some obvious corollaries of (5.7.74). In particular, for the Cantor perfect set $P_0$ [102, p. 50] we have $\forall \omega \in \Omega : \omega(P_0) = \lambda_*(P_0) = 0$.

Here we establish an obvious property $P_0 \in \mathcal{G}_0^{(*)}$ by using the well known representation of $\mathbf{I} \setminus P_0$ via some countable disjunct union of intervals with the unit sum of their lengths. Note some other simple properties. By (5.7.33),we have $\forall \omega \in \Omega : \omega(Q) = 1$. So we have a countable subset of $\mathbf{I}$, for which the corresponding values of FAM from $\Omega$ coincide with unit. But the following statement is true.

PROPOSITION 5.7.6 *Suppose that $(x_i)_{i\in\mathcal{N}}$ is a convergent sequence in $\mathbf{I}$: $(x_i)_{i\in\mathcal{N}} \in \mathbf{I}^{\mathcal{N}}$ and*

$$\exists x^* \in \mathbf{I} : (x_i)_{i\in\mathcal{N}} \to x^*. \tag{5.7.75}$$

*Then $\forall \mu \in \Omega$: $\mu(\{x_i : i \in \mathcal{N}\}) = 0$.*

PROOF. Let $X \triangleq \{x_i : i \in \mathcal{N}\}$. Then $X \in \mathcal{J}$, and what is more, $X \in \mathcal{G}_0^{(*)}$. Indeed, $\forall m \in \mathcal{N}$: $X_m \triangleq \{x_i : i \in \overline{1,m}\} \in \mathcal{A}$. Fix the number $x^* \in \mathbf{I}$ for which the convergence (5.7.75) holds. Fix $\varepsilon \in ]0,\infty[$ and choose $n \in \mathcal{N}$ for which

$$\forall i \in \overline{n,\infty} : |x_i - x^*| < \varepsilon/2.$$

It is obvious that

$$Y \triangleq ]x^* - \varepsilon/2, x^* + \varepsilon/2[ \cap \mathbf{I} \in \mathcal{I},$$

hence by the definition of $\mathcal{A}$ we have $Y \in \mathcal{A}$. Of course, $\mathbf{l}(Y) = \mathbf{l}_0(Y) < \varepsilon$, and $X_n \cup Y \in \mathcal{A}$ by the properties of algebras of sets. Moreover, $X_n \subset X \subset X_n \cup Y$ and

$$(X_n \cup Y) \setminus X_n \subset Y.$$

As a consequence we get the inequality

$$\mathbf{l}_0((X_n \cup Y) \setminus X_n) \leq \mathbf{l}_0(Y) < \varepsilon.$$

Since the choice of $\varepsilon$ was arbitrary, by (5.7.66) we obtain that $X \in \mathcal{G}_0^{(*)}$. From (5.7.66) it easily follows that $\mathbf{l}_0^{(*)}(X) = 0$ (in the previous reasoning it is sufficient to note that $\mathbf{l}_0(X_n \cup Y) \leq \mathbf{l}_0(X_n) + \mathbf{l}_0(Y) = \mathbf{l}_0(Y) < \varepsilon$, since $X_n = \bigcup_{i=1}^{n}\{x_i\}$). □

**Remark.** In connection with (5.7.74) we note the natural cause of the phenomenon that $\omega(Q) = 1$ for $\omega \in \Omega$. Namely: $Q \notin \mathcal{G}_0^{(*)}$. Indeed, let $A_1 \in \mathcal{A}$ have the property $A_1 \subset Q$. Then $A_1$ is a set not more than countable. On the other hand, by the definition of $\mathcal{A}$ we have $\Delta_m(A_1, \mathcal{I}) \neq \varnothing$ for some $m \in \mathcal{N}$. Choose $(I_j)_{j\in\overline{1,m}} \in \Delta_m(A_1, \mathcal{I})$. For $k \in \overline{1,m}$ we have $I_k \in \mathcal{I}$ and $I_k \subset A_1$. As a consequence $I_k$ is a set not more than countable. Suppose that $\mathbf{l}(I_k) > 0$. Then $I_k$ is a continuum (see (5.7.1)): one can choose $a \in \mathbf{I}$ and $b \in \mathbf{I}$ with $a < b$ for which $[a, b] \subset I_k$; in addition,

$$t \mapsto a + t(b - a) : \mathbf{I} \to [a, b]$$

is a bijection of $\mathbf{I}$ onto $[a, b]$. Hence $I_k$ is uncountable. We get a contradiction. Therefore $\mathbf{l}(I_k) = 0$. Then $\forall i \in \overline{1, m} : \mathbf{l}(I_i) = 0$. By the definition of $\mathbf{l}_0$ we have $\mathbf{l}_0(A_1) = \sum_{i=1}^{m} \mathbf{l}(I_i) = 0$. So we see that

$$\forall A \in \mathcal{A} : (A \subset Q) \Rightarrow (\mathbf{l}_0(A) = 0). \tag{5.7.76}$$

Let $A_2 \in \mathcal{A} : Q \subset A_2$. Then $\mathbf{I} \setminus A_2 \subset \mathbf{I} \setminus Q$. In addition, $\mathbf{I} \setminus A_2 \in \mathcal{A}$. Therefore for some $s \in \mathcal{N}$ we have

$$\Delta_s(\mathbf{I} \setminus A_2, \mathcal{I}) \neq \varnothing.$$

Let $(\Lambda_j)_{j \in \overline{1,s}} \in \Delta_s(\mathbf{I} \setminus A_2, \mathcal{I})$. Choose arbitrarily $r \in \overline{1, s}$. Then $\Lambda_r \in \mathcal{I}$ and

$$\Lambda_r \subset \mathbf{I} \setminus A_2 \subset \mathbf{I} \setminus Q.$$

Suppose that $\mathbf{l}(\Lambda_r) \in ]0, \infty[$. Using (5.7.1) we can choose two numbers $a_0 \in \mathbf{I}$ and $b_0 \in \mathbf{I}$ with the properties

$$(a_0 < b_0) \ \& \ (]a_0, b_0[ \subset \Lambda_r).$$

Introduce $\kappa_0 \triangleq b_0 - a_0 \in ]0, \infty[$ and $c_0 \triangleq (a_0 + b_0)/2 \in ]a_0, b_0[$, obtaining the representation

$$]a_0, b_0[ = ]c_0 - \kappa/2, c_0 + \kappa/2[.$$

Taking into account the density of $Q$ in $\mathbf{I}$ in the sense of $|\cdot|$, we can choose $t^\natural \in Q$ for which $| t^\natural - c_0 | < \kappa/2$. Then $t^\natural \in ]a_0, b_0[$ and as a consequence $t^\natural \in \Lambda_r \cap Q$. So $\Lambda_r \cap Q \neq \varnothing$, which is impossible. The contradiction obtained implies that $\mathbf{l}(\Lambda_r) = 0$. Since the choice of $r$ was arbitrary, we get $\forall i \in \overline{1, s} : \mathbf{l}(\Lambda_i) = 0$. So $\mathbf{l}_0(\mathbf{I} \setminus A_2) = \sum_{i=1}^{s} \mathbf{l}(\Lambda_i) = 0$. Therefore $\mathbf{l}_0(A_2) = 1 - \mathbf{l}_0(\mathbf{I} \setminus A_2) = 1$. We have

$$\forall A \in \mathcal{A} : (Q \subset A) \Rightarrow (\mathbf{l}_0(A) = 1). \tag{5.7.77}$$

By (5.7.76) and (5.7.77), for $A^{(1)} \in \mathcal{A}$ and $A^{(2)} \in \mathcal{A}$ with the property $A^{(1)} \subset Q \subset A^{(2)}$ we get

$$\mathbf{l}_0(A^{(2)} \setminus A^{(1)}) = \mathbf{l}_0(A^{(2)} \setminus A^{(1)}) + \mathbf{l}_0(A^{(1)}) = \mathbf{l}_0(A^{(2)}) = 1.$$

Hence $Q \notin \mathcal{G}_0^{(*)}$ (see (5.7.66)). This well known property is given only for completeness of the account.

In conclusion we note some very simple properties of integrals with respect to FAM $\omega \in \Omega$. We follow the definitions of Chapter 4 for $E = \mathbf{I}$. From (5.7.74) it follows that $\forall \omega \in \Omega \ \forall f \in B(\mathbf{I}, \mathcal{G}_0^{(*)})$:

$$\int_{\mathbf{I}} f \, d\omega = \int_{\mathbf{I}} f \, d(\omega \mid \mathcal{G}_0^{(*)}) = \int_{\mathbf{I}} f \, d(\lambda_* \mid \mathcal{G}_0^{(*)}) = \int_{\mathbf{I}} f \, d\lambda_*. \qquad (5.7.78)$$

In addition, for the set $C(\mathbf{I}) = C([0,1])$ of all continuous real-valued functions on $\mathbf{I}$, we have the inclusions $C(\mathbf{I}) \subset B(\mathbf{I}, \mathcal{I}) \subset B(\mathbf{I}, \mathcal{G}_0^{(*)})$ and hence by (5.7.78) we obtain for $\omega \in \Omega$ and $f \in C(\mathbf{I})$ the equality

$$\int_{\mathbf{I}} f \, d\omega = (R) \int_0^1 f(t) \, dt,$$

where the Riemann integral of $f$ is used on the right hand side. So for 'good' functions on $\mathbf{I}$ the $\omega$-integrals coincide with the natural Riemann and Lebesgue integrals.

## 5.8 CONCLUSION

The update measure theory actively uses tools of topology. In particular, it applies compactness similarly to the theory of extension. In connection with the universal integrability we note first of all the work [92]. The integration with respect to FAM is considered in the works of Maharam and her school (see, in particular, [95]). Also observe the investigations [3, 22] on the universal integrability of bounded functions. It is worth noting that the finitely additive version of uniform distribution on a unit interval of the real line, which was considered in [52, 53], is connected in a natural way with the constructions of extension of FAM (see [93, 106, 110] and other). The peculiarity of the proposed construction is the presence of connections (requirements). The extension of the countably additive (in this case) measure on the family of sets of a simple structure must satisfy these requirements. Here we can see an analogy with extension constructions in Chapter 3, where connections are present also. The purely finitely additive versions of uniform distribution that we obtain are defined on the family of all subsets of the unit interval. In a sense, they are asymptotic of finite weighed sums of Dirac measures.

Chapter 6

# NON-ANTICIPATING PROCEDURES OF CONTROL AND ITERATION METHODS FOR CONSTRUCTING THEM

## 6.1 INTRODUCTION

In the previous chapters different extension constructions for natural problems of attainability are considered. These problems are connected with extremal problems; this connection is discussed in detail in [32, 35] and many other investigations. In fact, the constructions of Chapters 3 and 4 are concerned with the employment of FAM for extensions and can be regarded as constructions of the asymptotic optimization. But very often the natural optimization is complicated by noise. These difficulties arise in many dynamic problems. So it is possible to consider the problems of control under the action of a noise. The given direction in control theory, being the important part of this theory, is called differential games theory. The structure of differential games is defined by the well known theorem about an alternative of Krasovskii and Subbotin. The above mentioned alternative was established in the class of very non-regular positional strategies. In addition, two players participate in the game. One player commands a useful control. The other player forms a hindrance. For constructing their own controls the players use feedback. But in the problems (of guidance and evasion) non-linear and discontinuous laws of feedback are required. The realization of such a feedback in the form of positional strategies requires a distinctive extension relative to the formations of trajectories. Namely, the so called constructive motions of Krasovskii are required here. These motions are uniform limits of the natural stepwise motions. This construction is considered (for example) in [88]. The above mentioned positional formalization of Krasovskii is equivalent to the formalization of the game control in the class of set-valued quasi-strategies. The notion of 'quasi-strategies' was used in [70, 108, 115] and in many other investigations. In addition, the 'usual' quasi-strategies are defined as some responses to realization of the hindrance control with the special property of the non-anticipating reaction. Arbitrary responses were called pseudo-strategies (see [70]). In [5, 6, 8, 11, 85, 86] and other investigations the set-valued quasi-strategies on the spaces of generalized controls (defined as measures) are considered. In fact, in these investigations the game methods of extensions are used. Employment of constructions of set-valued mappings and extensions permits us to bring closer the concepts of pseudo-strategies and quasi-strategies. The last circumstance is connected with the familiar method of programmed iterations (MPI) used for solving differential games. Namely, the variant of MPI considered in [6] realizes the extremal (with respect to an order) fixed point of the operator of programmed absorption. In terms of this fixed point the corresponding set-valued extended quasi-strategy solving the game problem of guidance or evasion is defined as some set-valued pseudo-strategy. Such approach was considered in [6, 14] and other investigations. In [112, Chapters IV,V] the detailed account of MPI and employment of MPI for constructing set-valued quasi-strategies are given. We also note the investigations [9, 10, 12, 49, 62, 67, 68, 69] connected with MPI. The construction of MPI was used for solving differential games with the restricted number of switchings and corrections (see [15, 16, 101]).

Later MPI was used for investigation of dynamic problems of another nature. In [113, 114] the scheme of MPI was applied for constructing the generalized solutions of the Hamilton Jacobi equations. Finally, in [36, 37, 42, 50] MPI was used for constructing the extremal hereditary multi-selector of set-valued mappings in functional spaces. We note that the last problem admits several concrete variants connected with the employment of FAM (see, for example, [37, 50]). In addition, in this new version of MPI the constructions similar to extensions are used. In this chapter we consider some such constructions, accentuating topological schemes using compactness, countable compactness, and sequential compactness. Note that sometimes in concrete settings the corresponding properties of the type of compactness (for the bundles of trajectories used) are absent. In these cases it is possible to realize some extension construction for the corresponding set-valued mapping (response). Such a scheme was realized in [31]. Other variants of 'extended' set-valued non-anticipating mappings are studied in [5, 6, 8, 10, 11, 14, 85, 86, 112].

## 6.2 A VERY SIMPLE EXAMPLE

We consider a simple example, using the informative variant of the presentation. Suppose that the simplest scalar control system

$$\dot{x} = u + v, \quad |u| \leqslant 1, |v| \leqslant 1, \tag{6.2.1}$$

functions on the time interval $[0,1]$. Let $x(0) = 0$, $u \in [-1,1]$ be the control parameter and $v \in [-1,1]$ be the hindrance. Of course, $u = u(t)$ and $v = v(t)$. For simplicity, we use $u(\cdot)$ and $v(\cdot)$ only from the set $\mathbb{C}$ of all piecewise constant, continuous from the right functions from $[0,1[$ into $[-1,1]$. Consider the given process of control as a game. Namely, we interpret the control $u(\cdot) \in \mathbb{C}$ as realization of a control function of the first player, and the hindrance $v(\cdot) \in \mathbb{C}$ as the analogous realization of the second player. Suppose that the goal of the first player is to realize the equality $x(1) = 0$. The goal of the second player is opposite. It is know that the given problem is solved in the framework of the formalization of differential game proposed by Krasovskii (see [87, 88]). In this case the set $[0,1] \times \{0\}$ plays the role of the stable bridge of Krasovskii, and the corresponding extremal (with respect to the given bridge) positional strategy guarantees the attainment of the goal of the first player. So it is required to determine the feedback control $U : \mathbb{R} \longrightarrow [-1,1]$ in the form:

$$U(x) \triangleq \begin{cases} 1, & x < 0, \\ -1, & x > 0, \\ 0, & x = 0. \end{cases}$$

However, it is possible to define $U(0) \in [-1,1]$ arbitrarily. Under the employment of $U$ in the discrete scheme with a small step of the partition of $[0,1]$ the first player then guarantees the finish $x(1) \approx 0$. When using the constructive motions of Krasovskii, the strategy $U$ guarantees $x(1) = 0$ under realization of arbitrary hindrances $v(\cdot) \in \mathbb{C}$. Recall [10] that quasi-strategies and positional strategies are equivalent (of course, the corresponding conditions of some informational compatibility are required to be satisfied).

Consider the solution in the class $Q$ of responses $L$ from $\mathbb{C}$ into $2^{\mathbb{C}}$. Among them we select non-anticipating responses. It is natural to introduce the 'broad' programmed reaction to action of hindrances. Let

$$m : \mathbb{C} \longrightarrow 2^{\mathbb{C}}$$

be the mapping for which, under $v(\cdot) \in \mathbb{C}$, $m(v(\cdot))$ is the set of all $u(\cdot) \in \mathbb{C}$ such that

$$\int_0^1 u(t)dt + \int_0^1 v(t)dt = 0.$$

In other words, each control $u(\cdot) \in \mathbb{C}$ for which the trajectory $x(\cdot)$ generated by $(u(\cdot), v(\cdot))$ has the property $x(1) = 0$ is an element of $m(v(\cdot))$. It is very logical to realize $u(\cdot) \in m(v(\cdot))$ by the following rule: $u(t) \triangleq -v(t)$ for $t \in [0,1[$. This variant of $u(\cdot) \in m(v(\cdot))$ guarantees the property $x(t) \equiv 0$. But other controls are contained in the set $m(v(\cdot))$ for $v(\cdot) \in \mathbb{C}$ . Let $\mathbf{O} \triangleq \mathbb{O}_{[0,1[}$ (see Section 2.2); in addition, $\mathbf{O} \in \mathbb{C}$ and $\mathbf{O}(t) \equiv 0$. Introduce $u_*(\cdot) \in \mathbb{C}$ by the following rule:

$$u_*(t) \triangleq \begin{cases} 1, & t \in [0, 1/2[, \\ -1, & t \in [1/2, 1[. \end{cases}$$

Then $u_*(\cdot) \in m(\mathbf{O})$. But consider the possibility when, under the initial choice $v(\cdot) = \mathbf{O}$ at the time moment 1/2, the player forming $v(\cdot)$ replaces own zero control, setting $v(t) = 1$ for $t \in [1/2, 1[$. The control $v^0(\cdot) \in \mathbf{C}$ is realized as follows:

$$v^0(t) \triangleq \begin{cases} 0, & t \in [0, 1/2[, \\ 1, & t \in [1/2, 1[. \end{cases}$$

But for $u(\cdot) \in \mathbb{C}$ with the property $(u_*(\cdot) \mid [0, 1/2[) = (u(\cdot) \mid [0, 1/2[)$ we have

$$\int_0^1 u(t)dt + \int_0^1 v^0(t)dt = \int_0^{1/2} u_*(t)dt + \int_{1/2}^1 u(t)dt + \int_{1/2}^1 v^0(t)dt$$

$$\geq \int_0^{1/2} u_*(t)dt = \frac{1}{2}.$$

So the set $m(v^0(\cdot))$ does not contain the controls $u(\cdot) \in \mathbb{C}$ for which $(u_*(\cdot) \mid [0, 1/2[) = (u(\cdot) \mid [0, 1/2[)$. We see that $m(\mathbf{O})$ contains 'superfluous' controls. Defining the mapping $\mathbf{n} : \mathbb{C} \longrightarrow \mathbb{C}$ as $\mathbf{n}(v(\cdot))(t) \triangleq -v(t)$, we obtain the non-anticipating (one-valued) mapping with the property $\mathbf{n}(v(\cdot)) \in m(v(\cdot))$ for $v(\cdot) \in \mathbb{C}$. So $\mathbf{n}$ is a selector of $m$. In addition, $\mathbf{n}$ has the following useful property. Namely, if $v(\cdot) \in \mathbb{C}$, then $\{\mathbf{n}(v(\cdot))\}$ is the set of all controls $u(\cdot) \in m(v(\cdot))$ for which $\forall v'(\cdot) \in \mathbb{C}\ \forall t \in [0, 1[$:

$$((v(\cdot) \mid [0, t[) = (v'(\cdot) \mid [0, t[))$$

$$\Rightarrow (\exists u'(\cdot) \in m(v'(\cdot)) : (u(\cdot) \mid [0, t[) = (u'(\cdot) \mid [0, t[)). \tag{6.2.2}$$

When constructing the obvious set-valued analogue of $\mathbf{n}$ in the form $\tilde{\mathbf{n}} : \mathbb{C} \to \mathcal{P}(\mathbb{C})$ for which $\tilde{\mathbf{n}}(v(\cdot)) = \{\mathbf{n}(v(\cdot))\}$, one can consider the transformation

$$m \longrightarrow \tilde{\mathbf{n}} \tag{6.2.3}$$

(see (6.2.2)) as a result of some 'programmed' mapping acting in the spaces of set-valued mappings. Below we consider such mapping in very general cases. In addition, (6.2.3) is used as the basic operator defining the corresponding iterated procedure.

## 6.3 SOME SPACES OF SET-VALUED MAPPINGS

In this section we consider a very general setting of the problem of constructing a developing system. In addition we investigate transformations of functions of an abstract argument. The example of the previous section defines, in fact, the scheme of application of the general theory to problems of control. Other applications also can be considered. For example, it is possible to investigate some problems of the control expansion of a FAM under the unexpected expansion of the other FAM.

We introduce designations used in theoretical constructions of this chapter. If $H$ is a nonempty set, then $\mathbb{I}_H \in H^H$ is def the identical mapping of $H$: for $h \in H$ we set $\mathbb{I}_H(h) \triangleq h$; moreover, for $L \in H^H$ we (as usual) introduce the sequence

$$(L^k)_{k \in \mathcal{N}_0} : \mathcal{N}_0 \longrightarrow H^H \tag{6.3.1}$$

for which: $(L^0 \triangleq \mathbb{I}_H)\&(\forall k \in N : L^k = L^0 L^{k-1})$. In this chapter designations of the type $L^k$, where $L$ is a mapping and $k \in \mathcal{N}_0$, are regarded only in the above mentioned sense. If $A$ and $B$ are sets, then $\mathbb{M}(A, B) \triangleq \mathcal{P}(B)^A$ can be used in place of $H$ in (6.3.1); for $\alpha \in \mathbb{M}(A, B)^{\mathbb{M}(A,B)}$ we derive $\alpha^\infty \in \mathbb{M}(A, B)^{\mathbb{M}(A,B)}$, setting for $\zeta \in \mathbb{M}(A, B)$ and $a \in A$

$$\alpha^\infty(\zeta)(a) \triangleq \bigcap_{k \in \mathcal{N}_0} \alpha^k(\zeta)(a).$$

Introduce some natural order and convergence. If A and B are sets, $\alpha \in \mathbb{M}(A, B)$ and $\beta \in \mathbb{M}(A, B)$, then (in the given chapter)

$$(\alpha \sqsubseteq \beta) \Longleftrightarrow (\forall x \in A : \alpha(x) \subset \beta(x));$$

if $(T_i)_{i \in \mathcal{N}} : \mathcal{N} \to \mathbb{M}(A, B)$ and $T \in \mathbb{M}(A, B)$, then (see (4.2.36)) def:

$$((T_i)_{i \in \mathcal{N}} \Downarrow T) \Longleftrightarrow (\forall a \in A : (T_i(a))_{i \in \mathcal{N}} \downarrow T(a)).$$

Finally, for arbitrary sets $A$ and $B$ we suppose

$$\mathcal{M}[A; B] \triangleq \{\alpha \in \mathbb{M}(A, B)^{\mathbb{M}(A,B)} \mid \forall H \in \mathbb{M}(A, B) : \alpha(H) \sqsubseteq H\};$$

if $\beta \in \mathcal{M}[A; B]$ and $\zeta \in \mathbb{M}(A, B)$, then

$$(\beta^k(\zeta))_{k \in \mathcal{N}} \Downarrow \beta^\infty(\zeta). \tag{6.3.2}$$

The last relation defines the basic (for this chapter) variants of the limit passage in spaces of set-valued mappings (for (6.3.2) we have

$$(\beta^0(\varsigma) = \varsigma)\&(\forall k \in \mathcal{N} : \beta^k(\varsigma) = \beta(\beta^{k-1}(\varsigma)));$$

so we obtain an iterated procedure in the space of set-valued mappings).

Fix three nonempty sets $X$, $Y$, and $\Upsilon$. We consider some subsets of $Y^X$ and $\Upsilon^X$. In the example of the previous section the interval $[0, 1[$ is used for $X$.

Moreover, fix two (nonempty) sets $\Omega \in 2^{(\Upsilon^X)}$ and $Z \in 2^{(Y^X)}$. We suppose that $w \in \Omega$ determines the realization of uncertain factors. Consider the construction of $z \in Z$ as a reaction to $w \in \Omega$; namely, we investigate the formation procedure for reaction $z = z(w)$. We consider two types of such a reaction. Namely, we admit arbitrary reactions and the reactions having some special property. This property has the sense of non-anticipation.

Fix some nonempty family $\mathcal{X}$ of nonempty subsets of $X$. In terms of $\mathcal{X}$ we introduce a very general analogue of non-anticipating mappings. In addition, we consider the case of set-valued mappings. Namely, the mapping

$$\beta : \Omega \longrightarrow \mathcal{P}(Z)$$

for which $\forall w_1 \in \Omega\ \forall w_2 \in \Omega\ \forall A \in \mathcal{X}$ :

$$((w_1 \mid A) = (w_2 \mid A)) \Rightarrow (\{(z \mid A) : z \in \beta(w_1)\} = \{(z \mid A) : z \in \beta(w_2)\}) \tag{6.3.3}$$

we call non-anticipating in the generalized sense. The sense of (6.3.3) was illustrated in the previous section. In the following we consider an equivalent representation of the property (6.3.3). Namely, we interpret (6.3.3) as the property of fixed point.

Introduce some auxiliary designations. Suppose that $\Sigma \triangleq \mathcal{P}(\Omega)$, $\Sigma_0 \triangleq 2^\Omega = \Sigma \setminus \{\varnothing\}$, $\mathbb{X} \triangleq \mathcal{P}(X)$, $\mathfrak{X} \triangleq 2^X = \mathbb{X} \setminus \{\varnothing\}$ and $\mathbb{Z} \triangleq \mathcal{P}(Z)$. For $T \in \Sigma_0$ we consider mappings of $\mathbb{M}(T, Z) = \mathbb{Z}^T$; we call these mappings set-valued ones. In the following, for simplicity, we use the obvious representation for $\mathcal{M}[T; Z]$ :

$$\mathcal{M}[T; Z] = \{\alpha \in \mathcal{Z}_T \mid \forall H \in \mathbb{M}(T, Z) : \alpha(H) \sqsubseteq H\}, \tag{6.3.4}$$

where

$$\mathcal{Z}_T \triangleq \mathbb{M}(T, Z)^{\mathbb{M}(T,Z)}.$$

For $\mathcal{Z} \triangleq \mathcal{Z}_\Omega$ we have the important particular case of (6.3.4):

$$\mathcal{M}[\Omega; Z] = \{\alpha \in \mathcal{Z} \mid \forall H \in \mathbb{M}(\Omega, Z) : \alpha(H) \sqsubseteq H\}. \tag{6.3.5}$$

We use (6.3.4) in some local version of the basic variant of MPI. Conversely, (6.3.5) is used in the 'global' version of the given direct variant of MPI. The corresponding local iterated procedures should be coalesced to obtain a global iterated procedure.

For interpretation of (6.3.3) we introduce a special 'programmed' operator. But first we give an important definition.

If $T \in \Sigma_0$, then suppose

$$\mathcal{Z}_T^0 \triangleq \{\mathbb{H} \in \mathcal{Z}_T \mid \mathbb{H} \circ \mathbb{H} = \mathbb{H}\};$$

let $\mathcal{Z}^0 \triangleq \mathcal{Z}_\Omega^0$. Elements of the sets $\mathcal{Z}_T^0$, $T \in \Sigma_0$, and $\mathcal{Z}^0$ are idempotent operators. These operators are not suitable for constructing iterated processes. But by means of such operators we can construct corresponding basic 'programmed' operators. If $T \in \Sigma_0$, $A \in \mathfrak{X}$ and $w \in \Omega$, then we suppose

$$(\mathrm{Ge})[T; w \mid A] \triangleq \{t \in T \mid (w \mid A) = (t \mid A)\}; \qquad (6.3.6)$$

of course, this definition is informative for $w \in T$. In the last case we have in (6.3.6) the $T$-germ of the function $(w \mid A) : A \longrightarrow \Upsilon$. Of course, $\forall A \in \mathfrak{X}\ \forall w \in \Omega$

$$\Omega_0(w \mid A) \triangleq (\mathrm{Ge})[\Omega; w \mid A] = \{\nu \in \Omega \mid (w \mid A) = (\nu \mid A)\} \in \Sigma_0. \quad (6.3.7)$$

In (6.3.7) we obtain the complete germ of $(w \mid A)$. Moreover, we have $\forall A \in \mathfrak{X}\ \forall z \in Z$ :

$$Z_0(z \mid A) \triangleq \{\tilde{z} \in Z \mid (z \mid A) = (\tilde{z} \mid A)\} \in 2^Z. \qquad (6.3.8)$$

In (6.3.8) we have the 'complete' germ of another type. In terms of (6.3.6) and (6.3.8) we introduce a local auxiliary operator (see [42]). Namely, if $T \in \Sigma_0$ and $A \in \mathfrak{X}$, then

$$\tilde{\gamma}_T[A] \in \mathcal{M}[T; Z] \qquad (6.3.9)$$

is defined by the following rule: if $\zeta \in \mathbb{M}(T, Z)$ and $t \in T$, then

$$\tilde{\gamma}_T[A](\zeta)(t) \triangleq \{z \in \zeta(t) \mid \forall s \in (\mathrm{Ge})[T; t \mid A] : Z_0(z \mid A) \cap \zeta(s) \neq \varnothing\}. \qquad (6.3.10)$$

Of course, (6.3.9) defines an element of $\mathcal{Z}_T$. But we use (6.3.4). In terms of (6.3.9), (6.3.10) we construct the local 'programmed' game operator. If $T \in \Sigma_0$, then

$$\gamma_T \in \mathcal{M}[T; Z] \qquad (6.3.11)$$

is defined by the condition that $\forall \zeta \in \mathbb{M}(T, Z)\ \forall t \in T$ :

$$\gamma_T(\zeta)(t) \triangleq \bigcap_{A \in \mathcal{X}} \tilde{\gamma}_T[A](\zeta)(t)$$
$$= \{z \in \zeta(t) \mid \forall A \in \mathcal{X}\ \forall s \in (\mathrm{Ge})[T; t \mid A] : Z_0(z \mid A) \cap \zeta(s) \neq \varnothing\}. \qquad (6.3.12)$$

The mapping (6.3.11) (6.3.12) has the following property. Fix $t \in T$ as the basic program of realization of uncertain factors. But the 'nature' has the following possibilities. Namely, beginning the process of realization of a function of $T$, the 'nature' chooses a set $A \in \mathcal{X}$ for which $t(x), x \in A$, is formed. For point $x \in X \setminus A$ the 'nature' has the possibility to realize $\tilde{t}(x) \in \Upsilon, \tilde{t}(x) \neq t(x)$. Of course, the constraint $\tilde{t} \in T$ is satisfied. The investigator should form $z \in \zeta(t)$, taking into account each possible replacement $t \longrightarrow \tilde{t}$; in addition, under such replacement the investigator must realize the response replacement $z \longrightarrow \tilde{z}$, where $\tilde{z} \in \zeta(\tilde{t})$ and $(z \mid A) = (\tilde{z} \mid A)$.

We select separately the case $T = \Omega$ (see (6.3.11) and (6.3.12)). For this we introduce special designations. Namely, $\forall A \in \mathfrak{X}$ :

$$\Gamma_A \triangleq \gamma_\Omega[A] \in \mathcal{M}[\Omega; Z]. \tag{6.3.13}$$

In addition, from (6.3.10) we have $\forall A \in \mathfrak{X}\ \forall \zeta \in \mathbb{M}(\Omega, Z)\ \forall w \in \Omega$ :

$$\Gamma_A(\zeta)(w) = \{z \in \zeta(w) \mid \forall s \in \Omega_0(w \mid A) : Z_0(z \mid A) \cap \zeta(s) \neq \varnothing\}. \tag{6.3.14}$$

We use (6.3.13) and (6.3.14) for constructing the basic 'global' operator with respect to (6.3.12). Namely, the operator

$$\Gamma \triangleq \gamma_\Omega \in \mathcal{M}[\Omega; Z] \tag{6.3.15}$$

is defined by the following rule: $\forall \zeta \in \mathbb{M}(\Omega, Z)\ \forall w \in \Omega$

$$\begin{aligned}\Gamma(\zeta)(w) &\triangleq \bigcap_{A \in \mathcal{X}} \Gamma_A(\zeta)(w) \\ &= \{z \in \zeta(w) \,|\, \forall A \in \mathcal{X}\ \forall \nu \in \Omega_0(w \,|\, A) : Z_0(z \,|\, A) \cap \zeta(\nu) \neq \varnothing\}.\end{aligned} \tag{6.3.16}$$

PROPOSITION 6.3.1 *Let $\beta \in \mathbb{M}(\Omega, Z)$. Then $\beta$ is a non-anticipating (in the sense of (6.3.3)) mapping iff $\beta = \Gamma(\beta)$.*

PROOF. Let $\beta$ is a non-anticipating mapping: for $w_1 \in \Omega, w_2 \in \Omega$ and $A \in \mathcal{X}$ the implication (6.3.3) is true. Compare $\beta$ and $\Gamma(\beta)$. Fix $w_1 \in \Omega$ and $z_1 \in \beta(w_1)$. Let $A_1 \in \mathcal{X}$ and $\nu \in \Omega_0(w_1 \mid A_1)$. By (6.3.7) we have $\nu_1 \in \Omega$ and $(w_1 \mid A_1) = (\nu_1 \mid A_1)$. Using (6.3.3) we obtain

$$\{(z \mid A_1) : z \in \beta(w_1)\} = \{(z \mid A_1) : z \in \beta(\nu_1)\}. \tag{6.3.17}$$

Since $(z_1 \mid A_1)$ is an element of the first set in (6.3.17), it is possible to choose $z^{(1)} \in \beta(\nu_1)$ for which

$$(z_1 \mid A_1) = (z^{(1)} \mid A_1). \tag{6.3.18}$$

In addition, $z^{(1)} \in Z$. By (6.3.8) and (6.3.18) we have $z^{(1)} \in Z_0(z_1 \mid A_1) \bigcap \beta(\nu_1)$. Since the choice of $A_1$ and $\nu_1$ was arbitrary, the following statement is established. Namely,

$$\forall A \in \mathcal{X}\ \forall \nu \in \Omega_0(w_1 \mid A) : Z_0(z_1 \mid A) \cap \beta(\nu) \neq \varnothing.$$

By (6.3.16) we have $z_1 \in \Gamma(\beta)(w_1)$. But the choice of $z_1$ was arbitrary. Therefore $\beta(w_1) \subset \Gamma(\beta)(w_1)$. Using (6.3.16) we obtain that $\beta(w_1) = \Gamma(\beta)(w_1)$. Since the choice of $w_1$ was arbitrary, $\beta = \Gamma(\beta)$. So if $\beta$ is a non-anticipating (in the generalized sense) mapping, then $\beta$ is a fixed point of $\Gamma$. Let $\beta = \Gamma(\beta)$. Choose $w_1 \in \Omega$, $w_2 \in \Omega$, and $A \in \mathcal{X}$ for which the premise of the implication (6.3.3) is true. Consider $Z_1 \triangleq \{(z \mid A) : z \in \beta(w_1)\}$ and $Z_2 \triangleq \{(z \mid A) : z \in \beta(w_2)\}$. Let $z' \in Z_1$ and $z_1' \in \beta(w_1)$ have the property $z' = (z_1' \mid A)$. Then $z_1' \in \Gamma(\beta)(w_1)$. By the choice of $w_1$ and $w_2$ we have (see (6.3.3)) $w_2 \in \Omega_0(w_1 \mid A)$; here we use (6.3.7). So $A \in \mathcal{X}$ and $w_2 \in \Omega_0(w_1 \mid A)$. By (6.3.16) we have $z_1' \in \beta(w_1)$ and $Z_0(z_1' \mid A) \bigcap \beta(w_2) \neq \varnothing$. Let $z_2' \in Z_0(z_1' \mid A) \bigcap \beta(w_2)$. Then by (6.3.8) we have

$$z' = (z_1' \mid A) = (z_2' \mid A) \in Z_2.$$

Since the choice of $z'$ was arbitrary, the inclusion $Z_1 \subset Z_2$ is established. The inverse inclusion is established in a similar way. So $Z_1 = Z_2$. In other words, (6.3.3) is valid. Since the choice of $w_1, w_2$ and $A$ was arbitrary, $\beta$ is a non-anticipating mapping. We have the following statement: if $\beta$ is a fixed point of $\Gamma$, then $\beta$ is a non-anticipating mapping. □

We have established the following important property:

$$\begin{aligned}\mathbb{N} &\triangleq \{\zeta \in \mathbb{M}(\Omega, Z) \mid \zeta = \Gamma(\zeta)\} \\ &= \{\zeta \in \mathbb{M}(\Omega, Z) \mid \forall w_1 \in \Omega\ \forall w_2 \in \Omega\ \forall A \in \mathcal{X} : ((w_1 \mid A) = (w_2 \mid A)) \\ &\qquad \Rightarrow (\{(z \mid A) : z \in \beta(w_1)\} = \{(z \mid A) : z \in \beta(w_2)\})\}.\end{aligned} \tag{6.3.19}$$

In the form of (6.3.19) we have the representation of the non-anticipation property in terms of the fixed point. We supplement (6.3.19) by the following local definition: if $T \in \Sigma_0$, then

$$\mathfrak{N}[T] \triangleq \{\zeta \in \mathbb{M}(T, Z) \mid \zeta = \gamma_T(\zeta)\}. \tag{6.3.20}$$

Of course, from (6.3.19) and (6.3.20) we have $\mathbb{N} = \mathfrak{N}[\Omega]$. The fixed points of operators (6.3.9), (6.3.13) are connected with (6.3.19), (6.3.20). But it is useful first to note the following property (for a more detailed discussion see in [43]).

PROPOSITION 6.3.2 *If $T \in \Sigma_0$ and $A \in \mathfrak{X}$, then $\tilde{\gamma}_T[A] \in \mathcal{Z}_T^0$.*

PROOF. Fix $T \in \Sigma_0, A \in \mathfrak{X}$ and $\alpha \in \mathbb{M}(T, Z)$. Then by (6.3.9) $\beta \triangleq \tilde{\gamma_T}[A](\alpha) \in \mathbb{M}(T, Z)$. In addition, by (6.3.4) and (6.3.9) we have $\beta \sqsubseteq \alpha$. Consider the mapping

$$\tilde{\gamma_T}[A](\beta) = (\tilde{\gamma_T}[A] \circ \tilde{\gamma_T}[A])(\alpha) \in \mathbb{M}(T, Z).$$

Using (6.3.4) and (6.3.9) we obtain that $\tilde{\gamma_T}[A](\beta) \sqsubseteq \beta$. Fix $w \in T$. Then

$$\tilde{\gamma_T}[A](\beta)(w) \subset \beta(w). \tag{6.3.21}$$

Let $\varphi \in \beta(w)$. Then $\varphi \in \alpha(w)$ and

$$\forall s \in (\mathrm{Ge})[T; w \mid A] : Z_0(\varphi \mid A) \cap \alpha(s) \neq \varnothing. \tag{6.3.22}$$

Moreover, from(6.3.10) we have the equality

$$\tilde{\gamma_T}[A](\beta)(w) = \{z \in \beta(w) \mid \forall s \in (\mathrm{Ge})[T; w \mid A] : Z_0(z \mid A) \cap \beta(s) \neq \varnothing\}. \tag{6.3.23}$$

Choose arbitrarily $\tilde{w} \in (\mathrm{Ge})[T; w \mid A]$. By (6.3.22) we have

$$Z_0(\varphi \mid A) \cap \alpha(\tilde{w}) \neq \varnothing.$$

Let $\tilde{\varphi} \in Z_0(\varphi \mid A) \cap \alpha(\tilde{w})$. Then $\tilde{\varphi} \in Z$ has the property

$$(\varphi \mid A) = (\tilde{\varphi} \mid A). \tag{6.3.24}$$

But in reality $\tilde{\varphi} \in \beta(\tilde{w})$. To prove this statement recall that by (6.3.10)

$$\beta(\tilde{w}) = \{z \in \alpha(\tilde{w}) \mid \forall s \in (\mathrm{Ge})[T; \tilde{w} \mid A] : Z_0(z \mid A) \cap \alpha(s) \neq \varnothing\}. \tag{6.3.25}$$

Let $\hat{w} \in (\mathrm{Ge})[T; \tilde{w} \mid A]$. Then by (6.3.6) $\hat{w} \in T$ and $(\tilde{w} \mid A) = (\hat{w} \mid A)$. But by the choice of $\tilde{w}$ we have (see (6.3.6)) $(w \mid A) = (\tilde{w} \mid A)$. Then $(w \mid A) = (\hat{w} \mid A)$ and as a consequence (see (6.3.6))

$$\hat{w} \in (\mathrm{Ge})[T; w \mid A].$$

By (6.3.22) we obtain that $Z_0(\varphi \mid A) \cap \alpha(\hat{w}) \neq \varnothing$. Choose $\hat{\varphi} \in Z_0(\varphi \mid A) \bigcap \alpha(\hat{w})$. Then $\hat{\varphi} \in Z$ and

$$(\varphi \mid A) = (\hat{\varphi} \mid A).$$

By (6.3.24) we have the equality $(\tilde{\varphi} \mid A) = (\hat{\varphi} \mid A)$. From (6.3.8) we have $\hat{\varphi} \in Z_0(\tilde{\varphi} \mid A)$. Therefore $\hat{\varphi} \in Z_0(\tilde{\varphi} \mid A) \cap \alpha(\hat{w})$. Since the choice of $\hat{w}$ was arbitrary, we obtain that $\forall s \in (\mathrm{Ge})[T; \tilde{w} \mid A] : Z_0(\tilde{\varphi} \mid A) \bigcap \alpha(s) \neq \varnothing$. From (6.3.25) we have the required property $\tilde{\varphi} \in \beta(\tilde{w})$. By the choice of $\tilde{\varphi}$ we obtain that $Z_0(\varphi \mid A) \cap \beta(\tilde{w}) \neq \varnothing$. But the choice of $\tilde{w}$ was arbitrary. Therefore we have established the property

$$\forall s \in (\mathrm{Ge})[T; w \mid A] : Z_0(\varphi \mid A) \cap \beta(s) \neq \varnothing.$$

By (6.3.23) we obtain that $\varphi \in \tilde{\gamma_T}[A](\beta)(w)$. Since the choice of $\varphi$ was arbitrary, we have the inclusion inverse with respect to (6.3.21). As a corollary, from (6.3.21) it follows that $\beta(w) = \tilde{\gamma_T}[A](\beta)(w)$. But the choice of $w$ was arbitrary. So $\beta = \tilde{\gamma_T}[A](\beta)$ or

$$\tilde{\gamma_T}[A](\alpha) = (\tilde{\gamma_T}[A] \circ \tilde{\gamma_T}[A])(\alpha).$$

Since the choice of $\alpha$ was arbitrary, we have $\tilde{\gamma_T}[A] = \tilde{\gamma_T}[A] \circ \tilde{\gamma_T}[A]$. $\square$

COROLLARY 6.3.1 $\forall A \in \mathfrak{X} : \Gamma_A \in \mathcal{Z}^0$.

Introduce in consideration fixed points of the operators (6.3.9). Let $\forall T \in \Sigma_0 \, \forall A \in \mathfrak{X}$ :

$$(\mathrm{f.\,p.})[A \mid T] \triangleq \{\zeta \in \mathbb{M}(T, Z) \mid \zeta = \tilde{\gamma_T}[A](\zeta)\}. \qquad (6.3.26)$$

From Proposition 6.3.2 and (6.3.26) we have the following obvious statement.

PROPOSITION 6.3.3 *If $T \in \Sigma_0$ and $A \in \mathfrak{X}$, then*

$$(\mathrm{f.\,p.})[A \mid T] = \{\tilde{\gamma_T}[A](H) : H \in \mathbb{M}(T, Z)\}.$$

In particular, by (6.3.13) we obtain $\forall A \in \mathfrak{X}$ :

$$\begin{aligned}(\mathrm{f.\,p.})[A] \triangleq (\mathrm{f.\,p.})[A \mid \Omega] &= \{\zeta \in \mathbb{M}(\Omega, Z) \mid \zeta = \Gamma_A(\zeta)\} \\ &= \{\Gamma_A(H) : H \in \mathbb{M}(\Omega, Z)\}.\end{aligned}$$

In addition, for $T \in \Sigma_0$ the set $\mathfrak{N}[T]$ (6.3.20) is the intersection of all sets $(\mathrm{f.\,p.})[A \mid T], A \in \mathcal{X}$. As a corollary, $\mathbb{N}$ is the intersection of all sets $(\mathrm{f.\,p.})[A]$, $A \in \mathcal{X}$.

PROPOSITION 6.3.4 *If $T \in \Sigma_0$, $\zeta \in \mathbb{M}(T, Z)$ and $w \in T$, then $\gamma_T(\zeta)(w)$ is the set of all $z \in \zeta(w)$ for which*

$$\forall A \in \mathcal{X} : (z \mid A) \in \bigcap_{s \in (\mathrm{Ge})[T; w \mid A]} \{(f \mid A) : f \in \zeta(s)\}. \qquad (6.3.27)$$

PROOF. Fix $T \in \Sigma_0, \zeta \in \mathbb{M}(T, Z)$ and $w \in T$. Let $\Lambda$ be the set of all $z \in \zeta(w)$ with the property (6.3.27). Compare $\gamma_T(\zeta)(w)$ and $\Lambda$. Let $z_1 \in \gamma_T(\zeta)(w)$. Then by (6.3.12) $z_1 \in \zeta(w)$, and for all $A \in \mathcal{X}$ and

$s \in (\mathrm{Ge})[T; w \mid A]$ the statement $Z_0(z_1 \mid A) \bigcap \zeta(s) \neq \varnothing$ takes place. In other words (see (6.3.8)),

$$\forall A \in \mathcal{X}\ \forall s \in (\mathrm{Ge})[T; w \mid A]\ \exists z_2 \in \zeta(s) : (z_1 \mid A) = (z_2 \mid A).$$

From the last property we have (6.3.27) for $z = z_1$. Therefore $z_1 \in \Lambda$. So $\gamma_T(\zeta)(w) \subset \Lambda$. Let $\varphi \in \Lambda$. For $\varphi \in \zeta(w)$ we have

$$\forall A \in \mathcal{X}\ \forall s \in (\mathrm{Ge})[T; w \mid A]\ \exists f \in \zeta(s) : (\varphi \mid A) = (f \mid A). \qquad (6.3.28)$$

The given property follows from (6.3.27). Using (6.3.8) and (6.3.28) we obtain that

$$\forall A \in \mathcal{X}\ \forall s \in (\mathrm{Ge})[T; w \mid A] : Z_0(\varphi \mid A) \cap \zeta(s) \neq \varnothing.$$

From (6.3.12) we have $\varphi \in \gamma_T(\zeta)(w)$. The inclusion $\Lambda \subset \gamma_T(\zeta)(w)$ is established. So $\gamma_T(\zeta)(w) = \Lambda$.

**Remark.** In Proposition 6.3.4 we have some analogy with the basic operator of [7].

As a corollary we note that by (6.3.7), (6.3.15), and Proposition 6.3.4 $\forall \zeta \in \mathbb{M}(\Omega, Z) \forall w \in \Omega$ :

$$\Gamma(\zeta)(w) = \{z \in \zeta(w) \mid \forall A \in \mathcal{X} : (z \mid A) \in \bigcap_{s \in \Omega_0(w|A)} \{(f \mid A) : f \in \zeta(s)\}\}. \qquad (6.3.29)$$

**Particular case.** In control theory the following concrete variants of $\mathcal{X}$ are used. Namely, the interval $[t_0, \vartheta_0[, t_0 < \vartheta_0$, is employed in place of $X$, and it is supposed that $\mathcal{X} = \{[t_0, t[ : t \in ]t_0, \vartheta_0[\}$. Such construction was considered in Section 6.2. Similar variants are used in the theory of differential games (see, for example, [70, 108, 115]). Note only the following variant. Consider the two control systems

$$\dot{p} = f_1(t, p) + B_1(t)u, \quad u \in P, \qquad (6.3.30)$$

and

$$\dot{q} = f_2(t, q) + B_2(t)v, \quad v \in Q. \qquad (6.3.31)$$

The interval $[t_0, \vartheta_0] (t_0 < \vartheta_0)$ and the initial states $p(t_0) = p_0$ and $q(t_0) = q_0$ are given. Suppose that $p \in \mathbb{R}^n$ and $q \in \mathbb{R}^n$, where $n \in \mathcal{N}$ is the dimension of the phase space for (6.3.30) and (6.3.31). Suppose that $P$ and $Q$ are compacta in finite-dimensional arithmetical spaces. Let $f_1$ and $f_2$ be continuous functions from $[t_0, \vartheta_0] \times \mathbb{R}^n$ into $\mathbb{R}^n$. Finally, $B_1(\cdot)$ and $B_2(\cdot)$ are $n \times n$ continuous matriciants (i.e., matrix-valued functions) defined on $[t_0, \vartheta_0]$. With respect to (6.3.30) and (6.3.31) we postulate several known properties (see [88]). Namely, we require the local Lipschits conditions for $f_1$ and $f_2$ with respect to the phase variable. Moreover, we require the traditional condition of the sublinear growth with respect to the phase variable; see, for example, [87, p. 38]. Denote by $\mathcal{U}$ and $\mathcal{V}$ the sets of Borel functions from $[t_0, \vartheta_0]$ into P and Q respectively; the corresponding definition of the Borel measurability see in Chapter 2, (2.5.25). We consider the Caratheodory solutions of systems (6.3.30) and (6.3.31). So we obtain two bundles $\mathbf{P}$ and $\mathbf{Q}$ of trajectories of systems (6.3.30) and (6.3.31), respectively: $\mathbf{P}$ is the set of all solutions of (6.3.30) under the given initial conditions $p(t_0) = p_0$ when enumerating all controls of $\mathcal{U}$. The definition of $\mathbf{Q}$ is analogous and is connected with (6.3.31). We have the two compacta $\mathbf{P}$ and $\mathbf{Q}$ in the space $C_n([t_0, \vartheta_0])$ of all continuous $n$-vector functions on $[t_0, \vartheta_0]$ with the topology of uniform convergence. Suppose that the goal of the first player controlling system (6.3.30) is realization of the condition

$$\exists t \in [t_0, \vartheta_0] : (p(t), q(t)) \in M, \tag{6.3.32}$$

where $M$ is a given closed subset of $\mathbb{R}^n \times \mathbb{R}^n$. In fact, the set $M$ is a subset of $\mathbb{R}^{2n}$. We suppose that the second player controls system (6.3.31) and has the goal inverse with respect to (6.3.32). Of course, we restrict ourselves only to the problem of control by system (6.3.30) in the class of set-valued quasi-strategies.

Consider the following variant of such quasi-strategies. Namely, we suppose that it is possible to construct $p = p(\cdot)$ as a response to $q = q(\cdot)$. We admit employment of set-valued mappings. In addition, for such responses we require the non-anticipation property. Namely, let $\alpha \in \mathcal{P}(\mathbf{P})^{\mathbf{Q}}$; then we call this response non-anticipating if $\forall q_1 \in \mathbf{Q}\ \forall q_2 \in \mathbf{Q}\ \forall t \in [t_0, \vartheta_0]$ :

$$((q_1 \mid [t_0, t]) = (q_2 \mid [t_0, t]))$$
$$\Rightarrow (\{(p \mid [t_0, t]) : p \in \alpha(q_1)\} = \{(p \mid [t_0, t]) : p \in \alpha(q_2)\}). \tag{6.3.33}$$

In (6.3.33) it is useful to note (6.3.19). The sense of the property (6.3.33) is clarified in Section 6.2. We note that mappings with the property (6.3.33) must be coordinated with the goal (6.3.32). In this connection we introduce a special goal mapping. Namely, let

$$\zeta : \mathbf{Q} \longrightarrow \mathcal{P}(\mathbf{P}) \tag{6.3.34}$$

be defined by the following rule: for $q \in \mathbf{Q}$ the set $\zeta(q)$ is by definition the set of all $p \in \mathbf{P}$ for which (6.3.32) is true. As in the example of Section 6.2, the set-valued mapping (6.3.34) is, generally speaking, not non-anticipating. But we construct some iterated procedure such that the passage from $\zeta$ to some set-valued mapping $\alpha$ satisfies (6.3.33) and the property

$$\forall q \in \mathbf{Q} : \alpha(q) \subset \zeta(q).$$

The important property $\alpha(q) \neq \varnothing$ $(q \in \mathbf{Q})$ can be lacking. This is a separate question.

We note that the given informative setting can be included in the general construction of Section 6.3. In addition, it should be supposed that $X = [t_0, \vartheta_0], Y = \Upsilon = \mathbb{R}^n, \Omega = \mathbf{Q}, Z = \mathbf{P}, \mathcal{X} = \{[t_0, t] : t \in [t_0, \vartheta_0]\}$.

## 6.4 THE HEREDITARY MULTI-SELECTORS: GENERAL PROPERTIES

First we consider some definitions of the basic structures. If $T \in \Sigma_0, \alpha \in \mathbb{M}(T, Z)$ and $\beta \in \mathbb{M}(T, Z)$, then (see Section 6.3)

$$(\alpha \sqsubseteq \beta) \iff (\forall t \in T : \alpha(t) \subset \beta(t)). \tag{6.4.1}$$

Later the symbol $\sqsubseteq$ (see (6.4.1)) is used only in this sense. Of course, the case $T = \Omega$ can be considered in (6.4.1). Let $\forall T \in \Sigma_0$ $\forall \alpha \in \mathbb{M}(T, Z)$ :

$$S_T(\alpha) \triangleq \{\beta \in \mathbb{M}(T, Z) \mid \beta \sqsubseteq \alpha\}. \tag{6.4.2}$$

Elements of $S_T(\alpha)$ are multi-selectors of $\alpha$. Suppose, for simplicity,

$$\forall \alpha \in \mathbb{M}(\Omega, Z) : \mathbf{S}[\alpha] \triangleq S_\Omega(\alpha).$$

If $T \in \Sigma_0$ and $\alpha \in \mathbb{M}(T, Z)$, then

$$\mathfrak{N}_0[T; \alpha] \triangleq \mathfrak{N}[T] \cap S_T(\alpha) \neq \varnothing. \tag{6.4.3}$$

In fact, the mapping $\beta \in \mathbb{M}(T, Z)$ for which $\beta(t) \equiv \varnothing$ is an element of the set on the left hand side of (6.4.3). We have $\forall \alpha \in \mathbb{M}(\Omega, Z)$ :

$$\mathbb{N}_0[\alpha] \triangleq \mathfrak{N}_0[\Omega; \alpha] = \mathbb{N} \cap \mathbf{S}[\alpha]. \tag{6.4.4}$$

We use the last set when determining the greatest hereditary multi-selectors of $\alpha$. In addition, it is advisable to consider this question more widely. If $T \in \Sigma_0$ and $\mathbf{A} \in 2^{\mathbb{M}(T,Z)}$, then

$$\bigvee_{\alpha \in \mathbf{A}}^{[T]} \alpha \in \mathbb{M}(T, Z) \tag{6.4.5}$$

is defined by the following condition: $\forall w \in T$

$$\left(\bigvee_{\alpha \in \mathbf{A}}^{[T]} \alpha\right)(w) \triangleq \bigcup_{\alpha \in \mathbf{A}} \alpha(w). \tag{6.4.6}$$

Of course, the order of $\mathbb{M}(T,Z)$ defined in terms of $\sqsubseteq$ has the following property: the mapping (6.4.5), (6.4.6) is the $\sqsubseteq$-supremum of **A**. If in (6.4.5) $\mathbf{A} \in 2^{\mathfrak{N}[T]}$, then (6.4.5) is an element of $\mathfrak{N}[T]$, where $T \in \Sigma_0$. Therefore $\forall T \in \Sigma_0\ \forall \alpha \in \mathbb{M}(T,Z)$ :

$$(T - \mathrm{Na})[\alpha] \triangleq \bigvee_{\varsigma \in \mathfrak{N}_0[T;\alpha]}^{[T]} \varsigma \in \mathfrak{N}_0[T;\alpha]. \tag{6.4.7}$$

The mapping (6.4.7) has a very important property. Namely, if $T \in \Sigma_0$ and $\alpha \in \mathbb{M}(T,Z)$, then $(T - \mathrm{Na})[\alpha]$ is the greatest (in the sense of $\sqsubseteq$) element of $\mathfrak{N}_0[T;\alpha]$. Note the important particular case: if $\alpha \in \mathbb{M}(\Omega, Z)$, then

$$(\mathrm{na})[\alpha] \triangleq (\Omega - \mathrm{Na})[\alpha] \in \mathrm{N}_0[\alpha] \tag{6.4.8}$$

has the property $\forall \beta \in \mathrm{N}_0[\alpha]$:

$$\beta \sqsubseteq (\mathrm{na})[\alpha]. \tag{6.4.9}$$

So the hereditary multi-selectors of each set-valued mapping exist always. And what is more (see (6.4.7)–(6.4.9)), always the greatest hereditary multi-selectors exist. It is useful to introduce the infinum of nonempty subsets of $\mathbb{M}(T,Z)$. If $T \in \Sigma_0$ and $\mathbf{A} \in 2^{\mathbb{M}(T,Z)}$, then

$$\bigwedge_{\alpha \in \mathbf{A}}^{[T]} \alpha \in \mathbb{M}(T,Z) \tag{6.4.10}$$

is by definition the mapping such that $\forall w \in T$ :

$$\left(\bigwedge_{\alpha \in \mathbf{A}}^{[T]} \alpha\right)(w) \triangleq \bigcap_{\alpha \in \mathbf{A}} \alpha(w). \tag{6.4.11}$$

Of course, in (6.4.10), (6.4.11) the infinum of **A** in the sense of $\sqsubseteq$ is defined. If $\mathbf{A} \in 2^{\mathbb{M}(\Omega,Z)}$, then, for simplicity, suppose

$$\left(\bigvee_{\alpha \in \mathbf{A}} \alpha \triangleq \bigvee_{\alpha \in \mathbf{A}}^{[\Omega]} \alpha\right) \& \left(\bigwedge_{\alpha \in \mathbf{A}} \alpha \triangleq \bigwedge_{\alpha \in \mathbf{A}}^{[\Omega]} \alpha\right). \tag{6.4.12}$$

From (6.4.11), (6.4.12) we have $\forall \mathbf{A} \in 2^{\mathbb{M}(\Omega,Z)}\ \forall w \in \Omega$ :

$$\left(\bigwedge_{\alpha \in \mathbf{A}} \alpha\right)(w) = \bigcap_{\alpha \in \mathbf{A}} \alpha(w).$$

In the following section we consider some other operations on the spaces of set-valued mappings.

## 6.5 SEWING OF SET-VALUED MAPPINGS

In this section we consider some questions of comparison of local and 'global' set-valued mappings. Our constructions are based on the sewing operation. But employment of this operation requires good blocks for local procedures. We select an important property having the sense of the distinctive invariance of the corresponding block. Let

$$\mathbb{H} \triangleq \{H \in \Sigma \mid \forall w \in H : \bigcup_{A \in \mathcal{X}} \Omega_0(w \mid A) \subset H\}. \tag{6.5.1}$$

In (6.5.1) the above mentioned principle of an invariance is realized. The family (6.5.1) is not empty. Namely,

$$(\varnothing \in \mathbb{H})\&(\Omega \in \mathbb{H}). \tag{6.5.2}$$

Introduce $\forall T \in \Sigma_0 \ \forall \zeta \in \mathbb{M}(T, Z) : (\mathrm{DOM})[\zeta] \triangleq \{w \in T \mid \zeta(w) \neq \varnothing\}$. Then we have $\forall \zeta \in \mathbb{N}$ :

$$(\mathrm{DOM})[\zeta] \in \mathbb{H}. \tag{6.5.3}$$

The proof of (6.5.3) is an obvious corollary of (6.3.19). We note some other properties of $\mathbb{H}$ (6.5.1). At first, $\mathbb{H}$ is an algebra of subsets of $\Omega$ ($\mathbb{H} \in (\mathrm{alg})[\Omega]$). Moreover, we have $\mathbb{H} \in (\mathrm{top})[\Omega]$. We note the following property: if $w \in \Omega, A \in \mathfrak{X}$ and $\tilde{w} \in \Omega_0(w \mid A)$, then

$$\Omega_0(w \mid A) = \Omega_0(\tilde{w} \mid A).$$

Introduce the family $\mathbb{H}_0 \triangleq \mathbb{H} \setminus \{\varnothing\}, \mathbb{H}_0 \subset \Sigma_0$. Then by (6.5.2) we have $\Omega \in \mathbb{H}_0$. So $\mathbf{S}[\mathbb{H}_0 \neq \varnothing]$. Let $\mathbf{H}_0$ be by definition the set of all $\mathcal{U} \in 2^{\mathbb{H}_0}$ such that

$$\left(\Omega = \bigcup_{U \in \mathcal{U}} U\right) \& \left(\forall U_1 \in \mathcal{U} \ \forall U_2 \in \mathcal{U} \setminus \{U_1\} : U_1 \cap U_2 = \varnothing\right).$$

So $\mathbf{H}_0$ is the set of all non-trivial partitions of $\Omega$ by sets of $\mathbb{H}$. Of course, $\{\Omega\} \in \mathbf{H}_0$. Therefore, $\mathbf{S}[\mathbf{H}_0 \neq \varnothing]$. For partitions from $\mathbf{H}_0$ we make the sewing of partial set-valued mappings. Namely, if $\mathcal{U} \in \mathbf{H}_0$ and $(\alpha_U)_{U \in \mathcal{U}} \in \prod_{U \in \mathcal{U}} \mathbb{M}(U, Z)$, then

$$\square_{U \in \mathcal{U}} \alpha_U \in \mathbb{M}(\Omega, Z) \tag{6.5.4}$$

is by definition the mapping from $\Omega$ into $\mathcal{P}(Z)$ such that

$$\forall P \in \mathcal{U} \ \forall w \in P : (\square_{U \in \mathcal{U}} \alpha_U)(w) \triangleq \alpha_P(w). \tag{6.5.5}$$

In (6.5.4) and (6.5.5) we have the required sewing operation. A more simple variant of (6.5.4), (6.5.5) is realized when sewing two set-valued mappings. If $U \in H_0 \setminus \{\Omega\}$, then $\{U; \Omega \setminus U\} \in \mathbf{H}_0$; for $\mu \in \mathbb{M}(U, Z)$ and $\nu \in \mathbb{M}(\Omega \setminus U, Z)$

$$\mu \Box \nu \in \mathbb{M}(\Omega, Z)$$

is by definition the mapping such that

$$(\forall w \in U : (\mu \Box \nu)(w) \triangleq \mu(w))\&(\forall w \in \Omega \setminus U : (\mu \Box \nu)(w) \triangleq \nu(w)).$$

Of course, we have a particular case of (6.5.4) and (6.5.5). Note some useful circumstance: for $\mathcal{U} \in \mathbf{H}_0$ and $(\alpha_U)_{U \in \mathcal{U}} \in \prod_{U \in \mathcal{U}} \mathfrak{N}[U]$, (6.5.4) is an element of $\mathbb{N}$. If $\mathcal{U} \in \mathbf{H}_0, \zeta \in \mathbb{M}(\Omega, Z)$ and $(\alpha_U)_{U \in \mathcal{U}} \in \prod_{U \in \mathcal{U}} \mathfrak{N}_0[U; (\zeta \mid U)]$, then

$$\Box_{U \in \mathcal{U}} \alpha_U \in \mathbb{N}_0[\zeta]. \tag{6.5.6}$$

Note a particular case of (6.5.6): if $\zeta \in \mathbb{M}(\Omega, Z), U \in \mathbb{H}_0 \setminus \{\Omega\}, \alpha \in \mathfrak{N}_0[U; (\zeta \mid U)]$ and $\beta \in \mathfrak{N}_0[\Omega \setminus U; (\zeta \mid \Omega \setminus U)]$, then $\alpha \Box \beta \in \mathbb{N}_0[\zeta]$. From (6.5.6) it follows that for $\alpha \in \mathbb{M}(\Omega, Z)$ and $T \in \mathbb{H}_0$

$$(T - \mathrm{Na})[(\alpha \mid T)] = ((\mathrm{na})[\alpha] \mid T). \tag{6.5.7}$$

In (6.5.7) the requirement $T \in \mathbb{H}_0$ is essential. From (6.5.7) we have $\forall \zeta \in \mathbb{M}(\Omega, Z)\ \forall \mathcal{U} \in \mathbf{H}_0$ :

$$(\mathrm{na})[\zeta] = \Box_{U \in \mathcal{U}} (U - \mathrm{Na})[(\zeta \mid U)]. \tag{6.5.8}$$

Of course, if $\zeta \in \mathbb{M}(\Omega, Z), U \in \mathbb{H}_0 \setminus \{\Omega\}$ and $V \triangleq \Omega \setminus U$, then

$$(\mathrm{na})[\zeta] = (U - \mathrm{Na})[(\zeta \mid U)] \Box (V - \mathrm{Na})[(\zeta \mid V)]. \tag{6.5.9}$$

The sewing operation (6.5.8), (6.5.9) defines the very general approach to the problem of employment of local iterated procedures. We use (6.5.8) and (6.5.9) when constructing the parallel version of the global iterated procedure. In this connection see [42, 43].

## 6.6 SOME PROPERTIES OF UNIVERSAL FIXED POINTS

In this very brief section we consider a development of the representation of the set (6.3.26), which is based on Proposition 6.3.2. Namely, we supplement Proposition 6.3.3 by analogues connected with multiselectors. Suppose that $\forall T \in \Sigma_0\ \forall \alpha \in \mathbb{M}(T, Z)\ \forall A \in \mathfrak{X}$ :

$$(\mathrm{f.p.} - \alpha)[A \mid T] \triangleq (\mathrm{f.p.})[A \mid T] \cap S_T(\alpha).$$

PROPOSITION 6.6.1 *For $T \in \Sigma_0$ and $\alpha \in \mathbb{M}(T, Z)$, the following equality takes place:* $\mathfrak{N}_0[T; \alpha] = \bigcap_{A \in \mathcal{X}} (\mathrm{f.p.} - \alpha)[A \mid T]$.

The corresponding proof is obvious (see (6.3.16), (6.3.20), (6.4.3)). Using Proposition 6.3.3 we obtain that $\forall T \in \Sigma_0 \; \forall \alpha \in \mathbb{M}(T, Z) \; \forall A \in \mathfrak{X}$ :

$$(\text{f. p.} -\alpha)[A \mid T] = \{\tilde{\gamma}_T[A](\mathcal{H}) : \mathcal{H} \in S_T(\alpha)\}. \tag{6.6.1}$$

From (6.6.1) we obtain a natural property of the set-valued mapping (6.4.7). Namely, $\forall T \in \Sigma_0 \; \forall \alpha \in \mathbb{M}(T, Z)$ :

$$(T - \text{Na})[\alpha] \in \bigcap_{A \in \mathcal{X}} \{\tilde{\gamma}_T[A](\mathcal{H}) : \mathcal{H} \in S_T(\alpha)\}. \tag{6.6.2}$$

Of course, the mapping on the left hand side of (6.6.2) is extremal with respect to the set on the right hand side. It is useful to take into account the different additional information. In addition, in (6.6.2) it is natural to consider the case $\alpha = (\zeta \mid T)$, where $\zeta \in \mathbb{M}(\Omega, Z)$. In particular, we discuss the case $T = \Omega$. For $\zeta \in \mathbb{M}(\Omega, Z)$ and $A \in \mathfrak{X}$ we suppose

$$(\text{f. p.} -\zeta)[A] \triangleq (\text{f. p.} -\zeta)[A \mid \Omega]; \tag{6.6.3}$$

by (6.6.3) and definitions of Section 6.4 we have

$$(\text{f. p.} -\zeta)[A] = (\text{f. p.})[A] \cap \mathbf{S}[\zeta].$$

Note an obvious corollary of Proposition 6.6.1. Namely, $\forall \zeta \in \mathbb{M}(\Omega, Z)$ :

$$\mathbb{N}_0[\zeta] = \bigcap_{A \in \mathcal{X}} (\text{f. p.} -\zeta)[A]. \tag{6.6.4}$$

On the other hand, for $\zeta \in \mathbb{M}(\Omega, Z)$ and $A \in \mathfrak{X}$ we have

$$(\text{f. p.} -\zeta)[A] = \{\Gamma_A(\mathcal{H}) : \mathcal{H} \in \mathbf{S}[\zeta]\}; \tag{6.6.5}$$

of course, (6.6.5) is the obvious corollary of (6.6.1). From (6.6.4) and (6.6.5) we have $\forall \zeta \in \mathbb{M}(\Omega, Z)$ :

$$(\text{na})[\zeta] \in \bigcap_{A \in \mathcal{X}} \{\Gamma_A(\mathcal{H}) : \mathcal{H} \in \mathbf{S}[\zeta]\}. \tag{6.6.6}$$

Of course, it is possible to add a more general statement (see (6.6.4) and (6.6.5)) to (6.6.6): if $\zeta \in \mathbb{M}(\Omega, Z)$, then

$$\mathbb{N}_0[\zeta] = \bigcap_{A \in \mathcal{X}} \{\Gamma_A(\mathcal{H}) : \mathcal{H} \in \mathbf{S}[\zeta]\}. \tag{6.6.7}$$

In connection with (6.6.6) and (6.6.7) we discuss a variant of constructing $(\text{na})[\zeta]$. Namely, for $\zeta \in \mathbb{M}[\Omega, Z]$ we consider the mapping $\mathcal{D} \in \mathbb{M}(\Omega, Z)$ for which

$$\forall A \in \mathcal{X} \; \exists \mathcal{H}_A \in \mathbf{S}[\zeta] : \mathcal{D} = \Gamma_A(\mathcal{H}_A).$$

In other words, we consider the intersection on the right hand side of (6.6.7). The given constructing is realized without iterated process. For such set-valued mapping $\mathcal{D}$ we have $\mathcal{D} \in \mathbb{N}_0[\zeta]$. It is advisable to find the $\sqsubseteq$- greatest mapping $\mathcal{D}$ of the above mentioned type. Of course, in this scheme it is useful to reduce enumeration of set-valued mappings by replacement of $\mathbf{S}[\zeta]$ by some its subset. For this it is possible to use some a priori information about $(\mathrm{na})[\zeta]$. We consider this question in a more general form. Namely, for $T \in \Sigma_0, \alpha \in \mathbb{M}(T, Z)$ and $A \in \mathfrak{X}$ we denote by $\mathfrak{M}(T, \alpha, A)$ the family of all sets $\mathbb{P} \in \mathcal{P}(\mathbb{M}(T, Z))$ such that $\forall P \in \mathbb{P} \bigcap S_T(\alpha) : \tilde{\gamma}_T[A](P) \in \mathbb{P}$. So we use some invariance of sets in the space of set-valued mappings. Then $\forall T \in \Sigma_0\ \forall \alpha \in \mathbb{M}(T, Z)\ \forall A \in \mathfrak{X}\ \forall \mathbb{P} \in \mathfrak{M}(T, \alpha, A):$

$$\mathbb{P} \cap (\text{f.p.} -\alpha)[A \mid T] = \{\tilde{\gamma}_T[A](\mathcal{H}) : \mathcal{H} \in \mathbb{P} \cap S_T(\alpha)\}. \tag{6.6.8}$$

We note the important particular case of (6.6.8) when $T = \Omega$. Let $\forall \alpha \in \mathbb{M}(\Omega, Z)\ \forall A \in \mathfrak{X}:$

$$\mathfrak{M}[\alpha; A] \triangleq \mathfrak{M}(\Omega, \alpha, A). \tag{6.6.9}$$

From (6.6.8) and (6.6.9) we obtain that $\forall \zeta \in \mathbb{M}(\Omega, Z)\ \forall A \in \mathfrak{X}\ \forall \mathbb{P} \in \mathfrak{M}[\zeta; A]:$

$$\mathbb{P} \cap (\text{f.p.} -\zeta)[A] = \{\Gamma_A(\mathcal{H}) : \mathcal{H} \in \mathbb{P} \cap \mathbf{S}[\zeta]\}. \tag{6.6.10}$$

In connection with possible application of (6.6.9) and (6.6.10) we note some statements of [42], [43] concerning transformations of set-valued mappings in topological spaces. We will return to properties similar to (6.6.8), (6.6.10).

## 6.7 ITERATIVE PROCEDURES

We return to operators of Section 6.3. We keep in mind $\gamma_T, T \in \Sigma_0$, and $\Gamma$; see (6.3.11) and (6.3.15). Of course, for $T \in \Sigma_0$ we have the sequence

$$(\gamma_T^k)_{k \in \mathcal{N}_0} : \mathcal{N}_0 \longrightarrow \mathcal{M}[T; Z] \tag{6.7.1}$$

defined by (6.3.1). Analogously,

$$\forall T \in \Sigma_0 : \gamma_T^\infty \in \mathcal{M}[T; Z] \tag{6.7.2}$$

(see Section 6.3). Of course, from (6.7.1) and (6.7.2) we have $\forall T \in \Sigma_0\ \forall \zeta \in \mathbb{M}(T, Z):$

$$(\gamma_T^k(\zeta))_{k \in \mathcal{N}} \Downarrow \gamma_T^\infty(\zeta). \tag{6.7.3}$$

Select the important particular case $T = \Omega$. Namely, by (6.3.1) and (6.3.15) we obtain the sequence

$$(\Gamma^k)_{k \in \mathcal{N}_0} : \mathcal{N}_0 \longrightarrow \mathcal{M}[\Omega; Z] \tag{6.7.4}$$

and the operator $\Gamma^\infty \in \mathcal{M}[\Omega; Z]$; in addition, $\forall \zeta \in \mathbb{M}[\Omega, Z]$ :

$$(\Gamma^k(\zeta))_{k\in\mathcal{N}} \Downarrow \Gamma^\infty(\zeta). \tag{6.7.5}$$

In terms of (6.7.1) and (6.7.4) it is possible to introduce natural iterated procedures for which (6.7.3) and (6.7.4) characterize the corresponding effect of convergence. In this connection we recall (6.3.2) and the following interpretation of Section 6.3. We note only the variant of employment of (6.7.4). Namely, if $\zeta \in \mathbb{M}(\Omega, Z)$ is some initial set-valued mapping, then by (6.7.4) we introduce the iterated sequence

$$(\Gamma^k(\zeta))_{k\in\mathcal{N}_0} : \mathcal{N}_0 \longrightarrow \mathbb{M}(\Omega, Z) \tag{6.7.6}$$

for which the following representation is correct:

$$(\Gamma^0(\zeta) = \zeta)\&(\forall k \in \mathcal{N} : \Gamma^k(\zeta) = \Gamma(\Gamma^{k-1}(\zeta))). \tag{6.7.7}$$

In (6.7.6), (6.7.7) we have the global iterated process with the initial element $\zeta$. The basic question is connected with comparison of $\Gamma^\infty(\zeta)$ (realized by (6.7.5)–(6.7.7)) and $(\mathrm{na})[\zeta]$. Of course, for local iterated procedures defined in terms of (6.7.1) and (6.7.2) similar questions arise. We consider some general conditions sufficient for the equality $\Gamma^\infty(\zeta) = (\mathrm{na})[\zeta]$ and for the analogous equalities for local iterated procedures and their limits. We construct some subspaces of spaces of all set-valued mappings of the corresponding type.

Namely, for $T \in \Sigma_0$ we introduce three families $\mathfrak{Z}_T, \mathbf{Z}_T$ and $\mathfrak{C}_T$ of subspaces of $\mathbb{M}(T, Z)$, setting

$$\begin{aligned}
&(\mathfrak{Z}_T \triangleq \{\mathbf{Q} \in \mathcal{P}(\mathbb{M}(T, Z)) \mid \forall Q \in \mathbf{Q} : \gamma_T(Q) \in \mathbf{Q}\}) \\
&\& (\mathbf{Z}_T \triangleq \{\mathfrak{h} \in \mathcal{P}(\mathbb{M}(T, Z)) \mid \forall (\mathcal{H}_i)_{i\in\mathcal{N}} \in \mathfrak{h}^{\mathcal{N}}\ \forall \mathcal{H} \in \mathbb{M}(T, Z) : \\
&\qquad ((\mathcal{H}_i)_{i\in\mathcal{N}} \Downarrow \mathcal{H}) \Rightarrow (\mathcal{H} \in \mathfrak{h})\}) \\
&\& (\mathfrak{C}_T \triangleq \{\mathbf{U} \in \mathcal{P}(\mathbb{M}(T, Z)) \mid \forall (U_j)_{j\in\mathcal{N}} \in \mathbf{U}^{\mathcal{N}}\ \forall U \in \mathbb{M}(T, Z) : \\
&\qquad ((U_j)_{j\in\mathcal{N}} \Downarrow U) \Rightarrow ((\gamma_T(U_j))_{j\in\mathcal{N}} \Downarrow \gamma_T(U))\}).
\end{aligned} \tag{6.7.8}$$

Of course, it is possible to call $\mathbf{Q} \in \mathfrak{Z}_T$ (see (6.7.8)) an invariant subspace of $\gamma_T$. The set $\mathfrak{H} \in \mathbf{Z}_T$ is a sequentially closed subset of $\mathbb{M}(T, Z)$. Finally, $\mathbf{U} \in \mathfrak{Z}_T$ has the sense of the set of sequential continuity of $\gamma_T$. Suppose that

$$(\mathfrak{Z} \triangleq \mathfrak{Z}_\Omega) \& (\mathbf{Z} \triangleq \mathbf{Z}_\Omega) \& (\mathfrak{C} \triangleq \mathfrak{C}_\Omega). \tag{6.7.9}$$

We note some obvious properties. If $T \in \Sigma_0, \mathbf{Q} \in \mathfrak{Z}_T$ and $\alpha \in \mathbf{Q}$, then by (6.7.8) we have

$$(\gamma_T^k(\alpha))_{k\in\mathcal{N}_0} : \mathcal{N}_0 \longrightarrow \mathbf{Q};$$

if, moreover, $\mathbf{Q} \in \mathbf{Z}_T$, then $\gamma_T^\infty \in \mathbf{Q}$ (see (6.7.3)). As a corollary, for $\mathbf{Q} \in \mathfrak{Z}$ and $\alpha \in \mathbf{Q}$ we have

$$(\Gamma^k(\alpha))_{k \in \mathcal{N}_0} : \mathcal{N}_0 \longrightarrow \mathbf{Q},$$

and, under the additional condition $\mathbf{Q} \in \mathbf{Z}$, by (6.7.5) we have $\Gamma^\infty(\alpha) \in \mathbf{Q}$. Using the obvious property of the $\sqsubseteq$ – monotonicity of $\gamma_T, T \in \Sigma_0$, and $\Gamma$, we have the following obvious

PROPOSITION 6.7.1 *If* $T \in \Sigma_0$ *and* $\alpha \in \mathbb{M}(T, Z)$, *then* $(T - \mathrm{Na})[\alpha] = (T - \mathrm{Na})[\gamma_T^\infty(\alpha)]$.

**Scheme of the proof.** By $\sqsubseteq$ – monotonicity of $\gamma_T$ we have (see (6.3.20), (6.4.3)) the property $(T - \mathrm{Na})[\alpha] \in \mathfrak{N}_0[T; \gamma_T^\infty(\alpha)]$. As a consequence, from (6.4.7) we have

$$(T - \mathrm{Na})[\alpha] \sqsubseteq (T - \mathrm{Na})[\gamma_T^\infty(\alpha)]. \tag{6.7.10}$$

On the other hand, we get by (6.4.3) and (6.4.7) the chain of 'inequalities' $(T - \mathrm{Na})[\gamma_T^\infty(\alpha)] \sqsubseteq \gamma_T^\infty(\alpha) \sqsubseteq \alpha$. Therefore $(T - \mathrm{Na})[\gamma_T^\infty(\alpha)] \sqsubseteq \alpha$. Moreover, $(T - \mathrm{Na})[\gamma_T^\infty(\alpha)] \in \mathfrak{N}[T]$. Hence we have

$$(T - \mathrm{Na})[\gamma_T^\infty(\alpha)] \in \mathfrak{N}_0[T; \alpha];$$

see (6.4.3). As a corollary, by (6.4.7) we obtain the property

$$(T - \mathrm{Na})[\gamma_T^\infty(\alpha)] \sqsubseteq (T - \mathrm{Na})[\alpha].$$

Using (6.7.10) we have the required statement. □

COROLLARY 6.7.1 *If* $\zeta \in \mathbb{M}[\Omega, Z]$, *then* $(\mathrm{na})[\zeta] = (\mathrm{na})[\Gamma^\infty(\zeta)]$.

For the proof it is sufficient to use (6.3.15). We note that in Proposition 6.7.1 and in Corollary 6.7.1 we have some estimates of the required solutions; these estimates are realized by iterated procedures. Often these estimates are converted into equalities.

PROPOSITION 6.7.2 *If* $T \in \Sigma_0, \mathbf{Q} \in \mathfrak{Z}_T \bigcap \mathfrak{C}_T$ *and* $\alpha \in \mathbf{Q}$, *then* $\gamma_T^\infty(\alpha) = (T - \mathrm{Na})[\alpha]$.

COROLLARY 6.7.2 *If* $\mathbf{Q} \in \mathfrak{Z} \bigcap \mathfrak{C}$ *and* $\alpha \in \mathbf{Q}$, *then* $\Gamma^\infty(\alpha) = (\mathrm{na})[\alpha]$.

The corresponding proof easily follows from (6.7.8) and (6.7.9). We omit this obvious reasoning.

PROPOSITION 6.7.3 *Let* $T \in \Sigma_0$ *and* $\mathbb{P} \in \mathfrak{Z}_T \bigcap \mathbf{Z}_T \bigcap \mathfrak{C}_T$. *Then the dependence*

$$\alpha \mapsto (T - \mathrm{Na})[\alpha] : \mathbb{M}(T, Z) \longrightarrow \mathfrak{N}[T] \tag{6.7.11}$$

*is the mapping sequentially continuous on* $\mathbb{P}$ : *if* $(\zeta_i)_{i\in\mathcal{N}} \in \mathbb{P}^{\mathcal{N}}$ *and* $\zeta \in \mathbb{M}(T,Z)$, *then* $((\zeta_i)_{i\in\mathcal{N}} \Downarrow \zeta) \Rightarrow (((T-\mathrm{Na})[\zeta_i])_{i\in\mathcal{N}} \Downarrow (T-\mathrm{Na})[\zeta])$.

PROOF. Fix $(\zeta_i)_{i\in\mathcal{N}} \in \mathbb{P}^{\mathcal{N}}$ and $\zeta \in \mathbb{M}(T,Z)$. Suppose that

$$(\zeta_i)_{i\in\mathcal{N}} \Downarrow \zeta. \tag{6.7.12}$$

Then by (6.7.8) and (6.7.12) we have $\zeta \in \mathbb{P}$. In (6.7.12) we consider the convergence in $\mathbb{P}$. Moreover, we have

$$((\gamma_T^{\infty}(\zeta_j))_{j\in\mathcal{N}} \in \mathbb{P}^{\mathcal{N}})\&(\gamma_T^{\infty}(\zeta) \in \mathbb{P}). \tag{6.7.13}$$

By Proposition 6.7.2 we have for values in (6.7.13) the properties

$$(\forall j \in \mathcal{N} : \gamma_T^{\infty}(\zeta_j) = (T-\mathrm{Na})[\zeta_j])\&(\gamma_T^{\infty}(\zeta) = (T-\mathrm{Na})[\zeta]). \tag{6.7.14}$$

From (6.7.12) it follows that

$$\forall w \in T : (\zeta_i(w))_{i\in\mathcal{N}} \downarrow \zeta(w);$$

see Section 6.3. Then, in particular, we have $\forall j \in \mathcal{N} : \zeta_{j+1} \sqsubseteq \zeta_j$. As a result, by (6.7.14) we have

$$\forall j \in \mathcal{N} : (T-\mathrm{Na})[\zeta_{j+1}] \sqsubseteq (T-\mathrm{Na})[\zeta_j]; \tag{6.7.15}$$

see (6.4.2), (6.4.3) and (6.4.7). Then by (6.7.15) $\forall w \in T\ \forall j \in \mathcal{N}$ :

$$(T-\mathrm{Na})[\zeta_{j+1}](w) \subset (T-\mathrm{Na})[\zeta_j](w). \tag{6.7.16}$$

Introduce the mapping $\beta \in \mathbb{M}(T,Z)$ by the rule: $\forall w \in T$

$$\beta(w) \triangleq \bigcap_{i\in\mathcal{N}} (T-\mathrm{Na})[\zeta_i](w). \tag{6.7.17}$$

From (6.7.16) and (6.7.17) we obtain that $\forall w \in T$ :

$$((T-\mathrm{Na})[\zeta_i])_{i\in\mathcal{N}} \downarrow \beta(w).$$

Therefore by definitions of Section 6.3 we have the convergence

$$((T-\mathrm{Na})[\zeta_i])_{i\in\mathcal{N}} \Downarrow \beta. \tag{6.7.18}$$

From (6.7.8), (6.7.13), (6.7.14), and (6.7.18) we have $\beta \in \mathbb{P}$. And what is more, we get

$$\beta \in \mathfrak{N}_0[T;\zeta]. \tag{6.7.19}$$

Of course, in (6.7.19) we use the property that $\gamma_T((T-\mathrm{Na})[\zeta_i]) = (T-\mathrm{Na})[\zeta_i]$ for $i \in \mathcal{N}$; see (6.3.20), (6.4.3) and (6.4.7). Moreover, we use

(6.7.13), (6.7.14) and the properties of $\mathbb{P}$ (see (6.7.8)). From (6.7.19) we have

$$\beta \sqsubseteq (T - \mathrm{Na})[\zeta].$$

But by the choice of $(\zeta_i)_{i\in\mathcal{N}}$ we get $\forall i \in \mathcal{N}$ :

$$(T - \mathrm{Na})[\zeta] \sqsubseteq (T - \mathrm{Na})[\zeta_i]. \quad (6.7.20)$$

From (6.7.18) and (6.7.20) we obtain that $(T-\mathrm{Na})[\zeta] \sqsubseteq \beta$. As a corollary we have $(T - \mathrm{Na})[\zeta] = \beta$. From (6.7.18) we have

$$((T - \mathrm{Na})[\zeta_i])_{i\in\mathcal{N}} \Downarrow (T - \mathrm{Na})[\zeta]. \quad (6.7.21)$$

So (6.7.12) $\Rightarrow$ (6.7.21). The required statement is established. □

COROLLARY 6.7.3 *If* $\mathbb{P} \in \mathfrak{Z}\bigcap\mathbf{Z}\bigcap\mathfrak{C}$, *then the operator*

$$\alpha \mapsto (\mathrm{na})[\alpha] : \mathbb{M}(\Omega, Z) \to \mathrm{N}$$

*has the property of sequential continuity on* $\mathbb{P}$ : *if* $(\zeta_i)_{i\in\mathcal{N}} \in \mathbb{P}^{\mathcal{N}}$ *and* $\zeta \in \mathbb{M}(\Omega.Z)$, *then* $((\zeta_i)_{i\in\mathcal{N}} \Downarrow \zeta) \Rightarrow (((\mathrm{na})[\zeta_i])_{i\in\mathcal{N}} \Downarrow (\mathrm{na})[\zeta])$.

The interesting question about some non-sequential analogues of the two last statements is considered later.

## 6.8 PARALLEL ITERATIVE PROCEDURES

In Proposition 6.7.2 we have some sufficient conditions for convergence of local iterated processes. Simultaneously in Corollary 6.7.2 we have the basic (global) iterated process. It is natural to consider the question of the realization of the basic iterated process by the sewing of local procedures. This question is connected with properties of restrictions of set-valued mappings defined on $\Omega$ and generated by $\Gamma$-iterations. In the general case we have

$$\forall \alpha \in \mathbb{M}(\Omega, Z)\ \forall T \in \Sigma_0\ \forall k \in \mathcal{N}_0 : (\Gamma^k(\alpha) \mid T) \sqsubseteq \gamma_T^k((\alpha \mid T)). \quad (6.8.1)$$

In addition, the equality in (6.8.1) may not hold. Moreover, from (6.8.1) we have $\forall \alpha \in \mathbb{M}(\Omega, Z)\ \forall T \in \Sigma_0 : (\Gamma^\infty(\alpha) \mid T) \sqsubseteq \gamma_T^\infty((\alpha \mid T))$.

PROPOSITION 6.8.1 *Let* $\zeta \in \mathbb{M}(\Omega, Z), U \in \mathbb{H}_0$ *and* $\zeta_U \triangleq (\zeta \mid U)$. *Then* $(\Gamma(\zeta) \mid U) = \gamma_U(\zeta_U)$.

For the proof it is sufficient to use (6.3.12), (6.3.16), and (6.5.1).

COROLLARY 6.8.1 *If* $\alpha \in \mathbb{M}(\Omega, Z), U \in \mathbb{H}_0$ *and* $\alpha_U \triangleq (\alpha \mid U)$, *then* $\forall k \in \mathcal{N}_0 : (\Gamma^k(\alpha) \mid U) = \gamma_U^k(\alpha_U)$.

COROLLARY 6.8.2 *If* $\alpha \in \mathbb{M}(\Omega, Z), U \in \mathbb{H}_0$ *and* $\alpha_U \triangleq (\alpha \mid U)$, *then*

$$(\Gamma^\infty(\alpha) \mid U) = \gamma_U^\infty(\alpha_U).$$

The proof of the two above corollaries is obvious. Now we obtain the following important statement.

THEOREM 6.8.1 *Let* $\alpha \in \mathbb{M}(\Omega, Z)$, $\mathcal{U} \in \mathbf{H}_0$, *and* $\forall U \in \mathcal{U} : \alpha_U \triangleq (\alpha \mid U)$. *Then*

$$(\forall k \in \mathcal{N}_0 : \Gamma^k(\alpha) = \Box_{U \in \mathcal{U}} \gamma_U^k(\alpha_U))\&(\Gamma^\infty(\alpha) = \Box_{U \in \mathcal{U}} \gamma_U^\infty(\alpha_U)).$$

The proof of this theorem is obvious (see Corollaries 6.8.1 and 6.8.2). We only discuss Theorem 6.8.1. Fix $\alpha \in \mathbb{M}(\Omega, Z)$ and $\mathcal{U} \in \mathbf{H}_0$. We have the system $(\alpha_U)_{U \in \mathcal{U}}$ of restrictions of $\alpha$. For each $U \in \mathcal{U}$ we have the iterated process

$$(\alpha_U^{(0)} = \alpha_U)\&(\forall s \in \mathcal{N} : \alpha_U^{(s)} = \gamma_U(\alpha_U^{(s-1)})); \tag{6.8.2}$$

then $\alpha_U^{(k)} = \gamma_U^k(\alpha_U), k \in \mathcal{N}$, and the limit of the sequence defined by (6.8.2) is $\alpha_U^{(\infty)} = \gamma_U^\infty(\alpha_U) \in \mathbb{M}(U, Z)$. On the other hand, we have $(\alpha_k)_{k \in \mathcal{N}_0}$ defined as

$$(\alpha_0 = \alpha)\&(\forall s \in \mathcal{N} : \alpha_s = \Gamma(\alpha_{s-1})); \tag{6.8.3}$$

the sequence defined by (6.8.3) is $(\Gamma^s(\alpha))_{s \in \mathcal{N}_0}$. Of course, the last sequence has the limit $\alpha_\infty \triangleq \Gamma^\infty(\alpha) \in \mathbb{M}(\Omega, Z)$. By Theorem 6.8.1 we have the following properties:

$$(\forall k \in \mathcal{N}_0 : \alpha_k = \Box_{U \in \mathcal{U}} \alpha_U^{(k)})\&(\alpha_\infty = \Box_{U \in \mathcal{U}} \alpha_U^{(\infty)}). \tag{6.8.4}$$

Note the natural particular case.

THEOREM 6.8.2 *Let* $\alpha^* \in \mathbb{M}(\Omega, Z), U_1 \in \mathbb{H}_0 \setminus \{\Omega\}, U_2 \triangleq \Omega \setminus U_1, \alpha_1^* \triangleq (\alpha^* \mid U_1)$, *and* $\alpha_2^* \triangleq (\alpha^* \mid U_2)$. *Then*

$$(\forall k \in \mathcal{N}_0 : \Gamma^k(\alpha^*) = \gamma_{U_1}^k(\alpha_1^*)\Box\gamma_{U_2}^k(\alpha_2^*))\&(\Gamma^\infty(\alpha^*) = \gamma_{U_1}^\infty(\alpha_1^*)\Box\gamma_{U_2}^\infty(\alpha_2^*)).$$

The most interesting cases are connected with the situation when the limit of $\Gamma$-iterations is (na)$[\alpha]$ for some $\alpha \in \mathbb{M}(\Omega, Z)$. Note the following

PROPOSITION 6.8.2 *Let* $\alpha \in \mathbb{M}(\Omega, Z)$ *and* $\mathcal{U} \in \mathbf{H}_0$. *Then*

$$(\forall U \in \mathcal{U} : \gamma_U^\infty((\alpha \mid U)) = (U - \mathrm{Na})[(\alpha \mid U)]) \Rightarrow (\Gamma^\infty(\alpha) = (\mathrm{na})[\alpha]).$$

COROLLARY 6.8.3 *Let* $\alpha^* \in \mathbb{M}(\Omega, Z)$, $U_1 \in \mathbb{H}_0 \backslash \{\Omega\}$, $U_2 \triangleq \Omega \backslash U_1, \alpha_1^* \triangleq (\alpha^* \mid U_1))$, *and* $\alpha_2^* \triangleq (\alpha^* \mid U_2)$. *Then*

$$(((U_1 - \mathrm{Na})[\alpha_1^*] = \gamma_{U_1}^{\infty}(\alpha_1^*)) \;\&\; ((U_2 - \mathrm{Na})[\alpha_2^*] = \gamma_{U_2}^{\infty}(\alpha_2^*)))$$
$$\Rightarrow (\Gamma^{\infty}(\alpha^*) = (\mathrm{na})[\alpha^*]).$$

The proof is realized by obvious combination of (6.5.8) and Theorem 6.8.1 and, moreover, (6.5.9) and Theorem 6.8.2, respectively. We note two obvious corollaries.

THEOREM 6.8.3 *Let* $\alpha \in \mathbb{M}(\Omega, Z)$, $\mathcal{U} \in \mathbf{H}_0$, *and*

$$(Q_U)_{U \in \mathcal{U}} \in \prod_{U \in \mathcal{U}} (\mathfrak{Z}_U \bigcap \mathfrak{C}_U).$$

*Moreover, let* $\forall U \in \mathcal{U} : (\alpha \mid U) \in Q_U$. *Then* $(\mathrm{na})[\alpha] = \Gamma^{\infty}(\alpha)$.

COROLLARY 6.8.4 *If* $\alpha^* \in \mathbb{M}(\Omega, Z), U_1 \in \mathbb{H}_0 \setminus \{\Omega\}, U_2 \triangleq \Omega \setminus U_1, Q_1 \in \mathfrak{Z}_{U_1} \bigcap \mathfrak{C}_{U_1}, Q_2 \in \mathfrak{Z}_{U_2} \bigcap \mathfrak{C}_{U_2}, \alpha_1^* \triangleq (\alpha^* \mid U_1) \in Q_1$, *and* $\alpha_2^* \triangleq (\alpha^* \mid U_2) \in Q_2$, *then* $\Gamma^{\infty}(\alpha^*) = (\mathrm{na})[\alpha^*]$.

For the proof it is sufficient to use the obvious combination of (6.5.8), Proposition 6.7.2 and Theorem 6.8.1.

Note that from Proposition 6.7.1 it follows that for $T \in \Sigma_0$ and $\alpha \in \mathbb{M}(T, Z)$,

$$(\mathrm{DOM})[(T - \mathrm{Na})[\alpha]] \subset (\mathrm{DOM})[\gamma_T^{\infty}(\alpha)]. \qquad (6.8.5)$$

In particular, by (6.8.5) we obtain the statement: if $\zeta \in \mathbb{M}(\Omega, Z)$, then

$$(\mathrm{DOM})[(\mathrm{na})[\zeta]] \subset (\mathrm{DOM})[\Gamma^{\infty}(\zeta)]. \qquad (6.8.6)$$

From (6.8.6) we have the following useful statement. Namely, for the existence of $\beta \in \mathbb{N}_0[\zeta]$ with the property $(\mathrm{DOM})[\beta] = \Omega$ it is necessary that

$$(\mathrm{DOM})[\Gamma^{\infty}(\zeta)] = \Omega. \qquad (6.8.7)$$

From (6.8.7) we obtain that the investigation of iterated procedures of the type (6.8.3) has meaning in the general case of the initial set-valued mapping, not depending on the validity of the conditions defined in (6.7.8). Therefore constructing the procedures of the type (6.8.3) and the corresponding limits is an important problem having an independent meaning.

**Remark.** We discuss a simple procedure of determining the hereditary multi-selector of the goal mapping; this procedure is realized without iterations. First we note that $\forall T \in \Sigma_0 \; \forall \mathcal{H} \in \mathbb{M}(T, Z)$ :

$$(\exists H_0 \in \mathcal{P}(Z) \; \forall t \in T : \mathcal{H}(t) = H_0) \Rightarrow (\mathcal{H} \in \mathfrak{N}[T]). \qquad (6.8.8)$$

We use (6.8.8) for constructing piece-wise constant non-anticipating set-valued mappings. Really, let $\alpha \in \mathbb{M}(\Omega, Z)$ and $\mathcal{U} \in \mathbf{H}_0$. If $U \in \mathcal{U}$, then we define $\mu_U \in \mathbb{M}(U, Z)$ as the following constant set-valued mapping: for $u \in U$ suppose that

$$\mu_U(u) \triangleq \bigcap_{v \in U} \alpha(v); \tag{6.8.9}$$

from (6.8.8) and (6.8.9) we have the property $\mu_U \in \mathfrak{N}_0[U; (\alpha \mid U)]$. In other words,

$$(\mu_U)_{U \in \mathcal{U}} \in \prod_{U \in \mathcal{U}} \mathfrak{N}_0[U; (\alpha \mid U)]. \tag{6.8.10}$$

Using (6.5.6) and (6.8.10) we obtain that the statement

$$\mu^* \triangleq \square_{U \in \mathcal{U}} \mu_U \in \mathbb{N}_0[\alpha]. \tag{6.8.11}$$

holds. In (6.8.11) we have only a variant of the non-anticipating multi-selector of $\alpha$. Of course, it is possible that $\mu^* \neq (\mathrm{na})[\alpha]$. And what is more, it is possible that $(\mathrm{DOM})[\mu^*] \neq \Omega$. In connection with (6.8.11) it is worth noting that the most refined partition of $\Omega$ should be used in place of $\mathcal{U}$. In addition, in the last section we consider some natural condition sufficient for the existence of such a partition.

## 6.9 FACTOR SPACE OF UNCERTAIN ACTIONS

We now consider a particular case of the general setting. This case always occurs in problems of the theory of differential games. So in this section we suppose that the following condition is valid.

CONDITION 6.9.1 *The family $\mathcal{X}$ is a filter base of $X$, i.e., $\forall A \in \mathcal{X}\ \forall B \in \mathcal{X}\ \exists C \in \mathcal{X} : C \subset A \cap B$.*

Thus $\mathcal{X} \in \mathcal{B}_0[X]$ in this section. Now we obviously have

$$\mathfrak{G} \triangleq \{\bigcup_{E \in \mathcal{X}} \Omega_0(w \mid E) : w \in \Omega\} \in \mathbf{H}_0. \tag{6.9.1}$$

**Remark.** Note only an argument connected with the proof of (6.9.1). We define $\mathfrak{G}$ by the relation in (6.9.1). Consider an arbitrary $\mathbb{G} \in \mathfrak{G}$. Choose $w \in \Omega$ for which $\mathbb{G}$ is the union of all sets $\Omega_0(w \mid E), E \in \mathcal{X}$. Fix $\mu \in \mathbb{G}$. Then for some $E' \in \mathcal{X}$ we get $\mu \in \Omega_0(w \mid E')$, i.e., $\mu \in \Omega$ with the property $(\mu \mid E') = (w \mid E')$. Choose arbitrarily

$$\nu \in \bigcup_{A \in \mathcal{X}} \Omega_0(\mu \mid A). \tag{6.9.2}$$

Let $E' \in \mathcal{X}$ be the set for which $(\mu \mid E') = (\nu \mid E')$. Using Condition 6.9.1 we choose $E_0 \in \mathcal{X}$ for which $E_0 \subset E' \cap E'$. Then

$$(\mu \mid E_0) = (\nu \mid E_0) = (w \mid E_0)$$

and as a corollary we obtain that $\nu \in \Omega_0(w \mid E_0)$. Of course, $\nu \in \mathbb{G}$. Since the choice of (6.9.2) was arbitrary, we have

$$\bigcup_{A \in \mathcal{X}} \Omega_0(\mu \mid A) \subset \mathbb{G}.$$

But the choice of $\mu$ was arbitrary too. Therefore by (6.5.1) we have $\mathbb{G} \in \mathbb{H}$. In addition, $w \in \mathbb{G}$. As a corollary $\mathbb{G} \in \mathbb{H}_0$. Since the choice of $\mathbb{G}$ was arbitrary, we get

$$\mathfrak{G} \in 2^{\mathbb{H}_0}.$$

From (6.3.7) and (6.9.1) we infer that $\Omega$ is the union of all sets $G \in \mathfrak{G}$. Let $G_1 \in \mathfrak{G}$ and $G_2 \in \mathfrak{G} \setminus \{G_1\}$. Then

$$G_1 \cap G_2 = \varnothing. \tag{6.9.3}$$

Indeed, by contradiction, let $G_1 \bigcap G_2 \neq \varnothing$. Choose $w_1 \in \Omega$ and $w_2 \in \Omega$ for which

$$\left(G_1 = \bigcup_{E \in \mathcal{X}} \Omega_0(w_1 \mid E)\right) \& \left(G_2 = \bigcup_{E \in \mathcal{X}} \Omega_0(w_2 \mid E)\right). \tag{6.9.4}$$

Choose $\eta \in G_1 \cap G_2$. Owing to (6.9.4) we choose $E_1 \in \mathcal{X}$ and $E_2 \in \mathcal{X}$ for which $\eta \in \Omega_0(w_1 \mid E_1)$ and $\eta \in \Omega_0(w_2 \mid E_2)$. Using Condition 6.9.1 we choose $E_3 \in \mathcal{X}$ for which $E_3 \subset E_1 \cap E_2$. Then $(w_1 \mid E_3) = (\eta \mid E_3) = (w_2 \mid E_3)$. Hence $w_1 \in \Omega_0(w_2 \mid E_3)$ and $w_2 \in \Omega_0(w_1 \mid E_3)$. But $G_1 \in \mathbb{H}$ and $G_2 \in \mathbb{H}$; these properties were established previously. Now by (6.9.4) we have $w_2 \in G_1$ and $w_1 \in G_2$. Therefore by (6.5.1)

$$\left(\bigcup_{A \in \mathcal{X}} \Omega_0(w_2 \mid A) \subset G_1\right) \& \left(\bigcup_{A \in \mathcal{X}} \Omega_0(w_1 \mid A) \subset G_2\right). \tag{6.9.5}$$

From (6.9.4) and (6.9.5) we obtain $(G_2 \subset G_1)\&(G_1 \subset G_2)$. So, $G_1 = G_2$. But by the choice of $G_1$ and $G_2$ we have $G_1 \neq G_2$. We get the obvious contradiction, which proves (6.9.3). We have established that $\forall U_1 \in \mathfrak{G}\ \forall U_2 \in \mathfrak{G} \setminus \{U_1\} : U_1 \cap U_2 = \varnothing$. So $\mathfrak{G} \in \mathbf{H}_0$; (6.9.1) is established. □

We note that from (6.9.1) we have the property $\mathfrak{G} \in 2^{\mathbb{H}_0}$. We recall that $(\Omega, \mathbb{H})$ is a $TS$. From (6.5.1) and (6.9.1) we infer that $\mathfrak{G}$ is a base of the given $TS$.

We equip $\mathbf{H}_0$ with the order $\prec$ defined by the refinement: for $\mathcal{U} \in \mathbf{H}_0$ and $\mathcal{V} \in \mathbf{H}_0$, by definition,

$$(\mathcal{U} \prec \mathcal{V}) \Longleftrightarrow (\forall V \in \mathcal{V}\ \exists U \in \mathcal{U} : V \subset U); \tag{6.9.6}$$

$\prec \in (\mathrm{Ord})_0[\mathbf{H}_0]$ (we omit a very obvious proof of this property; in addition, we use the following properties: 1) for each $G \in \mathfrak{G}$ we have $G \neq \varnothing$; 2) for $G_1 \in \mathfrak{G}$ and $G_2 \in \mathfrak{G}$ with $G_1 \bigcap G_2 \neq \varnothing$ the equality $G_1 = G_2$ holds). In this chapter we use the symbol $\prec$ only in the sense of (6.9.6).

PROPOSITION 6.9.1 *The partition (6.9.1) is the greatest element of* $(\mathbf{H}_0, \prec)$; *namely,* $\prec \in (\mathrm{Ord})_0[\mathbf{H}_0]$ *has the property*

$$\forall \mathcal{U} \in \mathbf{H}_0 : \mathcal{U} \prec \mathfrak{G}. \tag{6.9.7}$$

PROOF. Fix $\mathcal{U} \in \mathbf{H}_0$. Then $\mathcal{U} \in 2^{\mathbb{H}_0}$; therefore $\mathbf{S}[\mathcal{U} \neq \varnothing]$ and $\mathcal{U} \subset \mathbb{H}_0$. In particular, $\mathcal{U} \subset \mathbb{H}$. By (6.5.1) we have

$$\forall U \in \mathcal{U}\ \forall w \in U : \bigcup_{A \in \mathcal{X}} \Omega_0(w \mid A) \subset U. \tag{6.9.8}$$

Let $\mathbb{G} \in \mathfrak{G}$. By (6.9.1), for some $w_* \in \Omega$ we have

$$\mathbb{G} = \bigcup_{E \in \mathcal{X}} \Omega_0(w_* \mid E). \tag{6.9.9}$$

By definition of $\mathbf{H}_0$ we obtain that $w_* \in U_*$ for some $U_* \in \mathcal{U}$. From (6.9.8) and (6.9.9) we have the inclusion $\mathbb{G} \subset U_*$. So $\forall A \in \mathfrak{G}\ \exists B \in \mathcal{U} : A \subset B$. From (6.9.6) we have $\mathcal{U} \prec \mathfrak{G}$. Since the choice of $\mathcal{U}$ was arbitrary, the relation (6.9.7) is established. □

Introduce the following definition: if $w_1 \in \Omega$ and $w_2 \in \Omega$, then suppose that

$$(w_1 \sim w_2) \Longleftrightarrow (\exists A \in \mathcal{X} : (w_1 \mid A) = (w_2 \mid A)). \tag{6.9.10}$$

From (6.9.1) and (6.9.10) we have the obvious property: $\forall w_1 \in \Omega\ \forall w_2 \in \Omega$

$$(w_1 \sim w_2) \Longleftrightarrow (\exists G \in \mathfrak{G} : (w_1 \in G)\&(w_2 \in G)). \tag{6.9.11}$$

From (6.9.1) and (6.9.11) we obtain that (6.9.10) defines the equivalence on $\Omega$. And what is more, (6.9.1) is the factor space for this equivalence. Elements of $\mathfrak{G}$ are called blocks; each block of $\mathfrak{G}$ is an element of $\mathbb{H}_0$.

PROPOSITION 6.9.2 *If* $\mathbb{G} \in \mathfrak{G}$ *and* $\zeta \in \mathbb{M}(\mathbb{G}, Z)$, *then* $((\mathrm{DOM})[\gamma_{\mathbb{G}}(\zeta)] \neq \varnothing) \Rightarrow ((\mathrm{DOM})[\zeta] = \mathbb{G})$.

PROOF. Fix $\mathbb{G} \in \mathfrak{G}$ and $\zeta \in \mathbb{M}(\mathbb{G}, Z)$. Of course, $\mathbb{G} \in \mathbb{H}_0$; therefore $\mathbb{G} \in \mathbb{H}$ and $\mathbb{G} \neq \varnothing$. So $\mathbf{S}[\mathbb{G} \neq \varnothing]$. We use (6.5.1). Then $\mathbb{G} \in \Sigma_0$ and

$$\forall w \in \mathbb{G} : \bigcup_{A \in \mathcal{X}} \Omega_0(w \mid A) \subset \mathbb{G}. \tag{6.9.12}$$

Let $(\mathrm{DOM})[\gamma_{\mathbb{G}}(\zeta)] \neq \varnothing$. Choose

$$\mu \in (\mathrm{DOM})[\gamma_{\mathbb{G}}(\zeta)]. \tag{6.9.13}$$

Since $\gamma_{\mathbb{G}}(\zeta) \in \mathbb{M}(\mathbb{G}, Z)$, we have for $\mu \in \mathbb{G}$ the property

$$\mathbf{S}[\gamma_{\mathbb{G}}(\zeta)(\mu) \neq \varnothing]. \tag{6.9.14}$$

Choose (see (6.9.14)) $\phi \in \gamma_{\mathbb{G}}(\zeta)(\mu)$. Then $\phi \in \zeta(\mu)$. In addition,

$$\forall A \in \mathcal{X}\ \forall s \in (\mathrm{Ge})[\mathbb{G}; \mu \mid A] : Z_0(\phi \mid A) \cap \zeta(s) \neq \varnothing. \tag{6.9.15}$$

We recall that the inclusion $(\mathrm{DOM})[\zeta] \subset \mathbb{G}$ takes place. Let $\nu \in \mathbb{G}$. Then $(\mu \in \mathbb{G})\&(\nu \in \mathbb{G})$. From (6.9.10) and (6.9.11) we derive the equality

$$(\mu \mid A_*) = (\nu \mid A_*) \tag{6.9.16}$$

for some $A_* \in \mathcal{X}$. But from (6.3.6) and (6.9.16) we have $\nu \in (\mathrm{Ge})[\mathbb{G}; \mu \mid A_*]$. By (6.9.15) we obtain that $Z_0(\phi \mid A_*) \cap \zeta(\nu) \neq \varnothing$. Then $\zeta(\nu) \neq \varnothing$ and hence $\nu \in (\mathrm{DOM})[\zeta]$. Since the choice of $\nu$ was arbitrary, we have the inclusion $\mathbb{G} \subset (\mathrm{DOM})[\zeta]$, which implies the equality $(\mathrm{DOM})[\zeta] = \mathbb{G}$. So the implication

$$((\mathrm{DOM})[\gamma_{\mathbb{G}}(\zeta)] \neq \varnothing) \Rightarrow ((\mathrm{DOM})[\zeta] = \mathbb{G})$$

is established. The proof is completed. □

COROLLARY 6.9.1 $\forall \mathbf{G} \in \mathfrak{G}\ \forall \zeta \in \mathfrak{N}[\mathbf{G}]$:

$$((\mathrm{DOM})[\zeta] \neq \varnothing) \Rightarrow ((\mathrm{DOM})[\zeta] = \mathbf{G}).$$

The proof uses (6.3.20) and Proposition 6.9.2.

PROPOSITION 6.9.3 *If* $\alpha \in \mathbb{M}(\Omega, Z)$ *and* $\mathbb{G} \in \mathfrak{G}$, *then*

$$((\mathrm{DOM})[\Gamma(\alpha)] \cap \mathbb{G} \neq \varnothing) \Rightarrow (\mathbb{G} \subset (\mathrm{DOM})[\alpha]). \tag{6.9.17}$$

PROOF. Fix $\alpha \in \mathbb{M}(\Omega, Z)$ and $\mathbb{G} \in \mathfrak{G}$ for which the premise of (6.9.17) takes place:

$$(\mathrm{DOM})[\Gamma(\alpha)] \cap \mathbb{G} \neq \varnothing.$$

For $\beta \triangleq (\alpha \mid \mathbb{G}) \in \mathbb{M}(\mathbb{G}, Z)$, by Proposition 6.8.1 we have the equality

$$(\Gamma(\alpha) \mid \mathbb{G}) = \gamma_{\mathbb{G}}(\beta). \tag{6.9.18}$$

Of course, in (6.9.18) we use the obvious property $\mathbb{G} \in \mathbb{H}_0$. Then by (6.9.18)

$$\begin{aligned}(\mathrm{DOM})[\Gamma(\alpha)] \cap \mathbb{G} &= \{w \in \Omega \mid \Gamma(\alpha)(w) \neq \varnothing\} \cap \mathbb{G} \\ &= \{w \in \mathbb{G} \mid \Gamma(\alpha)(w) \neq \varnothing\} \\ &= \{w \in \mathbb{G} \mid (\Gamma(\alpha) \mid \mathbb{G})(w) \neq \varnothing\} = (\mathrm{DOM})[\gamma_{\mathbb{G}}(\beta)].\end{aligned}$$

By the basic supposition we have $(\mathrm{DOM})[\gamma_{\mathbb{G}}(\beta)] \neq \varnothing$. Now by Proposition 6.9.2 we obtain that $(\mathrm{DOM})[\beta] = \mathbb{G}$. As a result $(\mathrm{DOM})[\alpha] \bigcap \mathbb{G} = \mathbb{G}$ and $\mathbb{G} \subset (\mathrm{DOM})[\alpha]$. □

COROLLARY 6.9.2 *If $\alpha \in \mathbb{M}(\Omega, Z), \mathbb{G} \in \mathfrak{G}$ and $k \in \mathcal{N}_0$, then*

$$(\mathbb{G} \cap (\mathrm{DOM})[\Gamma^{k+1}(\alpha)] \neq \varnothing) \Rightarrow (\mathbb{G} \subset (\mathrm{DOM})[\Gamma^k(\alpha)]).$$

The proof is obvious: under the conditions of this corollary $\Gamma^{k+1}(\alpha) = \Gamma(\Gamma^k(\alpha))$; therefore it is possible to use Proposition 6.9.3 with $\alpha$ replaced by $\Gamma^k(\alpha)$.

PROPOSITION 6.9.4 *If $\alpha \in \mathbb{M}(\Omega, Z)$ and $\mathbb{G} \in \mathfrak{G}$, then*

$$(\mathbb{G} \cap (\mathrm{DOM})[\Gamma^\infty(\alpha)] \neq \varnothing) \Rightarrow \left(\mathbb{G} \subset \bigcap_{k \in \mathcal{N}_0} (\mathrm{DOM})[\Gamma^k(\alpha)]\right).$$

PROOF. Fix $\alpha \in \mathbb{M}(\Omega, Z)$ and $\mathbb{G} \in \mathfrak{G}$. Then by (6.7.5) we have $\forall k \in \mathcal{N}_0 : \Gamma^\infty(\alpha) \sqsubseteq \Gamma^k(\alpha)$. Therefore $\forall k \in \mathcal{N}_0$ :

$$(\mathrm{DOM})[\Gamma^\infty(\alpha)] \subset (\mathrm{DOM})[\Gamma^k(\alpha)]. \tag{6.9.19}$$

Let $\mathbb{G} \bigcap (\mathrm{DOM})[\Gamma^\infty(\alpha)] \neq \varnothing$. Then by (6.9.19) we obtain that

$$\forall k \in \mathcal{N}_0 : \mathbb{G} \cap (\mathrm{DOM})[\Gamma^{k+1}(\alpha)] \neq \varnothing. \tag{6.9.20}$$

From (6.9.20) and Corollary 6.9.2 we get $\forall k \in \mathcal{N}_0 : \mathbb{G} \subset (\mathrm{DOM})[\Gamma^k(\alpha)]$. □

Of course, from (6.4.4), (6.4.8), Corollary 6.7.1, and Proposition 6.9.4 we get that $\forall \alpha \in \mathbb{M}(\Omega, Z)\ \forall \mathbb{G} \in \mathfrak{G}$ :

$$(\mathbb{G} \cap (\mathrm{DOM})[(\mathrm{na})[\alpha]] \neq \varnothing) \Rightarrow \left(\mathbb{G} \subset \bigcap_{k \in \mathcal{N}_0} (\mathrm{DOM})[\Gamma^k(\alpha)]\right).$$

But it is possible to make the last property more precise. Namely, the following statement is valid.

PROPOSITION 6.9.5 *If* $\alpha \in \mathbb{M}(\Omega, Z)$ *and* $\mathbb{G} \in \mathfrak{G}$, *then*

$$((\mathrm{DOM})[(\mathrm{na})[\alpha]] \cap \mathbb{G} \neq \varnothing) \Longleftrightarrow (\mathbb{G} \subset (\mathrm{DOM})[(\mathrm{na})[\alpha]]).$$

PROOF. Fix $\alpha \in \mathbb{M}(\Omega, Z)$ and $\mathbb{G} \in \mathfrak{G}$. Then by (6.4.4) and (6.4.8) we obtain that $(\mathrm{na})[\alpha] \in \mathbb{N}$. Since $\mathbb{G} \in \mathbb{H}_0$, by (6.5.7) we have the equality

$$(\mathbb{G} - \mathrm{Na})[(\alpha \mid \mathbb{G})] = ((\mathrm{na})[\alpha] \mid \mathbb{G}).$$

From (6.4.3) and (6.4.7) we obtain that $((\mathrm{na})[\alpha] \mid \mathbb{G}) \in \mathfrak{N}[\mathbb{G}]$. Then by Corollary 6.9.1 we have

$$((\mathrm{DOM})[((\mathrm{na})[\alpha] \mid \mathbb{G})] \neq \varnothing) \Longleftrightarrow ((\mathrm{DOM})[((\mathrm{na})[\alpha] \mid \mathbb{G})] = \mathbb{G}). \quad (6.9.21)$$

But on the other hand, the following equality holds:

$$(\mathrm{DOM})[((\mathrm{na})[\alpha] \mid \mathbb{G})] = (\mathrm{DOM})[(\mathrm{na})[\alpha]] \cap \mathbb{G}.$$

Therefore by (6.9.21) we have the equivalence

$$((\mathrm{DOM})[(\mathrm{na})[\alpha]] \cap \mathbb{G} \neq \varnothing) \Longleftrightarrow (\mathbb{G} = (\mathrm{DOM})[(\mathrm{na})[\alpha]] \cap \mathbb{G}).$$

The proof is completed. □

By Proposition 6.9.4 we get that factorization on the basis of (6.9.1), (6.9.10), and (6.9.11) is compatible with the heredity formulated in terms of $\mathcal{X}$.

## 6.10 THE CASE OF SET-VALUED MAPPINGS IN A TOPOLOGICAL SPACE

Here we consider the general case of the family $\mathcal{X}$. Namely, now we refuse Condition 6.9.1. But we suppose that $\tau \in (\mathrm{top})[Y]$ is fixed, and we consider the TS $(Y, \tau)$. Then

$$\otimes^X(\tau) \in (\mathrm{top})[Y^X] \quad (6.10.1)$$

and in the form $(Y^X, \otimes^X(\tau))$ we have the Tichonoff product of samples of $(Y, \tau)$ with the index set $X$. Then it is possible to treat $Z$ as the corresponding subspace of the given Tichonoff product. Hence

$$\vartheta \triangleq \otimes^X(\tau) \mid_Z \in (\mathrm{top})[Z] \quad (6.10.2)$$

is the topology induced by the topology (6.10.1) in $Z$. Of course,

$$(Z, \vartheta) \quad (6.10.3)$$

defined by (6.10.2) is a subspace of the above mentioned Tichonoff product. In this section (6.10.2) and (6.10.3) define the basic equipment of $Z$. In this connection we introduce some special designations. Namely, in this section we use the following stipulation.

Suppose that $\mathbb{F} \triangleq \mathcal{F}_\vartheta$ (see Chapter 2). Moreover, we use the notion of sequentially closed set in the TS (6.10.3). In this section we assume that

$$\mathfrak{F} \triangleq \mathcal{F}_{\mathrm{seq}}[\theta]$$

(see the definition (2.3.19)). Of course, elements of $\mathfrak{F}$ are exactly fixed points of the natural operator of sequential closure in the TS (6.10.3). Certainly, $\mathbb{F} \subset \mathfrak{F}$. We consider compact, countably compact, and sequentially compact sets in the TS (6.10.3). Let

$$(\mathbb{K} \triangleq (\vartheta - \mathrm{comp})[Z]) \,\&\, (\mathcal{K} \triangleq (\vartheta - \mathrm{comp}_{\mathrm{seq}})[Z])$$
$$\&\, (\mathbb{C} \triangleq (\mathcal{N} - \vartheta - \mathrm{comp})[Z]); \tag{6.10.4}$$

in connection with (6.10.4) see Chapter 2. We obtain that

$$\mathbb{T} \triangleq \mathbb{C} \cap \mathbb{F} \tag{6.10.5}$$

is the family of all closed countably compact (in (6.10.3)) subsets of $Z$. For $T \in \Sigma_0$ we have

$$(\mathbb{K}^T \subset \mathbb{M}(T,Z))\&(\mathcal{K}^T \subset \mathbb{M}(T,Z))\&(\mathbb{C}^T \subset \mathbb{M}(T,Z))\&(\mathbb{T}^T \subset \mathbb{M}(T,Z)).$$

Of course, $\mathbb{K} \cup \mathcal{K} \subset \mathbb{C}$. Moreover,

$$(\tau \in (\mathrm{top})_0[Y]) \Rightarrow ((\mathbb{K} \subset \mathbb{F})\&(\mathcal{K} \subset \mathfrak{F})\&(\mathbb{C} \subset \mathfrak{F})). \tag{6.10.6}$$

In connection with (6.10.6) see [71, Sect. 3.10]. We note that

$$\forall U \in \Sigma_0 : \mathbb{T}^U \in \mathbf{Z}_U. \tag{6.10.7}$$

We supplement (6.10.7) by the following implication:

$$(\tau \in (\mathrm{top})_0[Y]) \Rightarrow ((\mathbb{K}^U \in \mathbf{Z}_U)\&(\mathcal{K}^U \in \mathbf{Z}_U)). \tag{6.10.8}$$

In (6.10.8) we use (6.7.8) and (6.10.6).

PROPOSITION 6.10.1 *Let* $(Y,\tau)$ *be a Hausdorff space:* $\tau \in (\mathrm{top})_0[Y]$. *Then* $\forall U \in \Sigma_0 : (\mathbb{K}^U \in \mathfrak{G}_U)\&(\mathcal{K}^U \in \mathfrak{G}_U)$.

PROOF. Fix $U \in \Sigma_0$. Consider the set $\mathbb{K}^U \in \mathcal{P}(\mathbb{M}(U,Z))$. Fix $\zeta \in \mathbb{K}^U$. Then $\gamma_U(\zeta) \in \mathbb{M}(U,Z)$. We use (6.3.12). Fix $w \in U$. Then

$$\gamma_U(\zeta)(w) = \{z \in \zeta(w) \mid \forall A \in \mathcal{X}\ \forall s \in (\mathrm{Ge})[U; w \mid A] : Z_0(z \mid A) \cap \zeta(s) \neq \varnothing\}. \tag{6.10.9}$$

Choose arbitrarily $f \in \text{cl}(\gamma_U(\zeta)(w), \theta)$. Using the Birkhoff theorem we choose a net $(\mathcal{D}, \preceq, \phi)$ in $\gamma_U(\zeta)(w)$ for which

$$(\mathcal{D}, \preceq, \phi) \xrightarrow{\theta} f. \tag{6.10.10}$$

Here $\mathbf{S}[\mathcal{D} \neq \varnothing]$, $\preceq \in (\text{DIR})[\mathcal{D}]$ and $\phi \in \gamma_U(\zeta)(w)^{\mathcal{D}}$. We recall that (see (6.10.6)) $\zeta(w) \in \mathbb{F}$. Therefore by (6.10.10) we have $f \in \zeta(w)$. Fix $A \in \mathcal{X}$ and $\tilde{w} \in (\text{Ge})[U; w \mid A]$. From (6.3.6) we have $\tilde{w} \in \Omega$ and $(w \mid A) = (\tilde{w} \mid A)$. Since $\phi(d) \in \gamma_U(\zeta)(w), d \in \mathcal{D}$, from (6.10.9) we obtain that

$$\forall d \in \mathcal{D} : Z_0(\phi(d) \mid A) \cap \zeta(\tilde{w}) \neq \varnothing.$$

We use the axiom of choice. Let

$$\tilde{\phi} \in \prod_{d \in \mathcal{D}} (Z_0(\phi(d) \mid A) \cap \zeta(\tilde{w})). \tag{6.10.11}$$

From (6.10.11) we obtain that

$$\tilde{\phi} : \mathcal{D} \longrightarrow \zeta(\tilde{w})$$

has the following property. Namely, from (6.3.8) and (6.10.11) it follows that

$$\forall d \in \mathcal{D} : (\phi(d) \mid A) = (\tilde{\phi}(d) \mid A). \tag{6.10.12}$$

Of course, $\tilde{\phi} \in Z^{\mathcal{D}}$. And what is more, $\zeta(\tilde{w}) \in \mathbb{K}$. Then (see Chapter 2) it is possible to choose $f_* \in \zeta(\tilde{w})$, a nonempty set $\Delta$, a direction $\ll \in (\text{DIR})[\Delta]$, and $l \in (\text{Isot})[\Delta; \ll; \mathcal{D}; \preceq]$ for which the convergence

$$(\Delta, \ll, \tilde{\phi} \circ l) \xrightarrow{\theta} f_*$$

holds. From (6.10.2) and (6.10.13) we get the convergence

$$(\Delta, \ll, \tilde{\phi} \circ l) \xrightarrow{\otimes^X(\tau)} f_*. \tag{6.10.13}$$

As a result we have $\forall x \in X$:

$$(\Delta, \ll, ((\tilde{\phi} \circ l)(\delta)(x))_{\delta \in \Delta}) \xrightarrow{\tau} f_*(x). \tag{6.10.14}$$

From (6.10.10) we have the convergence $(\Delta, \ll, \phi \circ l) \xrightarrow{\theta} f$. Therefore from (6.10.2) we obtain that

$$(\Delta, \ll, \phi \circ l) \xrightarrow{\otimes^X(\tau)} f. \tag{6.10.15}$$

In (6.10.15) we use the construction similar to (6.10.13). Finally, from (6.10.15) we have $\forall x \in X$ :

$$(\Delta, \ll, ((\phi \circ l)(\delta)(x))_{\delta \in \Delta}) \xrightarrow{\tau} f(x). \tag{6.10.16}$$

But from (6.10.12) we infer that $\forall x \in A$ :

$$((\phi \circ l)(\delta)(x))_{\delta \in \Delta} = ((\tilde{\phi} \circ l)(\delta)(x))_{\delta \in \Delta}.$$

Therefore from (6.10.14) and (6.10.16) we have $\forall x \in A : f(x) = f_*(x)$. Of course, we use the basic property of $(Y, \tau)$. As a result $(f \mid A) = (f_* \mid A)$. By (6.3.8) we have

$$f_* \in Z_0(f \mid A) \cap \zeta(\tilde{w}).$$

Therefore $\mathbf{S}[Z_0(f \mid A) \cap \zeta(\tilde{w}) \neq \varnothing]$. Since the choice of $A$ and $\tilde{w}$ was arbitrary, we have

$$\forall \tilde{A} \in \mathcal{X}\ \forall s \in (\mathrm{Ge})[U; w \mid \tilde{A}] : Z_0(f \mid \tilde{A}) \cap \zeta(s) \neq \varnothing. \qquad (6.10.17)$$

Since $f \in \zeta(w)$, by (6.3.12) and (6.10.17) we obtain that

$$f \in \gamma_U(\zeta)(w).$$

So the inclusion $\mathrm{cl}(\gamma_U(\zeta)(w), \theta) \subset \gamma_U(\zeta)(w)$ is established. Therefore $\gamma_U(\zeta)(w) = \mathrm{cl}(\gamma_U(\zeta)(w), \theta) \in \mathbb{F}$. Since $\gamma_U(\zeta)(w) \subset \zeta(w)$ and $\zeta(w) \in \mathbb{K}$, we have $\gamma_U(\zeta)(w) \in \mathbb{K}$. We have established the property $\gamma_U(\zeta) \in \mathbb{K}^U$. Since the choice of $\zeta$ was arbitrary, by (6.7.8) we have $\mathbb{K}^U \in \mathfrak{Z}_U$. The property $\mathcal{K}^U \in \mathfrak{Z}_U$ is established in a similar way and even easier. □

PROPOSITION 6.10.2 *Let $(Y, \tau)$ be a $T_1$-space: $\tau \in (\mathcal{D} - \mathrm{top})[Y]$. Then*

$$\forall U \in \Sigma_0 : \mathbb{T}^U \in \mathbf{Z}_U \cap \mathfrak{C}_U.$$

PROOF. We use (6.10.7). Fix $U \in \Sigma_0$. Consider $\mathbb{T}^U$. Let $(\zeta_i)_{i \in \mathcal{N}} : \mathcal{N} \to \mathbb{T}^U$ and $\zeta \in \mathbb{M}(U, Z)$. Suppose that

$$(\zeta_i)_{i \in \mathcal{N}} \Downarrow \zeta. \qquad (6.10.18)$$

Consider the sequence $(\gamma_U(\zeta_i))_{i \in \mathcal{N}} : \mathcal{N} \longrightarrow \mathbb{M}(U, Z)$. We have $\gamma_U(\zeta) \in \mathbb{M}(U, Z)$. From (6.10.18) and definitions of Section 3 we get that $\forall w \in U$ :

$$(\zeta_i(w))_{i \in \mathcal{N}} \downarrow \zeta(w). \qquad (6.10.19)$$

In addition, $\forall j \in \mathcal{N} : \zeta_{j+1} \sqsubseteq \zeta_j$. From (6.3.12) we have $\forall j \in \mathcal{N}$ :

$$\gamma_U(\zeta_{j+1}) \sqsubseteq \gamma_U(\zeta_j). \qquad (6.10.20)$$

By (6.10.19) we have $\forall j \in \mathcal{N} : \zeta \sqsubseteq \zeta_j$. Therefore we obtain the analogous property for the images under the operation $\gamma_U$. Namely, by (6.3.12) $\forall j \in \mathcal{N}$ :

$$\gamma_U(\zeta) \sqsubseteq \gamma_U(\zeta_j). \qquad (6.10.21)$$

Fix $w_* \in U$. Then by (6.10.20) we have $\forall j \in \mathcal{N}: \gamma_U(\zeta_{j+1})(w_*) \subset \gamma_U(\zeta_j)(w_*)$. From (6.10.21) we obtain that $\forall j \in \mathcal{N}: \gamma_U(\zeta)(w_*) \subset \gamma_U(\zeta_j)(w_*)$. As a corollary,

$$\gamma_U(\zeta)(w_*) \subset \bigcap_{i\in\mathcal{N}} \gamma_U(\zeta_i)(w_*). \tag{6.10.22}$$

Choose arbitrarily $f_* \in \bigcap_{i\in\mathcal{N}} \gamma_U(\zeta_i)(w_*)$. From (6.3.12) and (6.10.19) we have $f_* \in \zeta(w_*)$. Choose arbitrarily $A^* \in \mathcal{X}$ and $w^* \in (\mathrm{Ge})[U; w_* \mid A^*]$. Then $w^* \in U$. Using the axiom of choice, for some sequence

$$(f_i^*)_{i\in\mathcal{N}} \in \prod_{i\in\mathcal{N}} \zeta_i(w^*) \tag{6.10.23}$$

we have the property

$$(f_i^*)_{i\in\mathcal{N}} : \mathcal{N} \longrightarrow Z_0(f_* \mid A^*). \tag{6.10.24}$$

By (6.3.12) such a choice (see (6.10.23), (6.10.24)) is possible. From (6.10.19) we have the convergence

$$(\zeta_i(w^*))_{i\in\mathcal{N}} \downarrow \zeta(w^*). \tag{6.10.25}$$

From (6.10.25) we have the property: $\zeta_i(w^*) \subset \zeta_1(w^*)$ for $i \in \mathcal{N}$. Therefore by (6.10.23) we get

$$(f_i^*)_{i\in\mathcal{N}} : \mathcal{N} \longrightarrow \zeta_1(w^*). \tag{6.10.26}$$

Recall that $\zeta_1(w^*) \in \mathbb{T}$. Then $\zeta_1(w^*) \in \mathbb{C}$ (see (6.10.5)). We use (6.10.4). By (2.3.22), (6.10.4), and (6.10.26) we obtain that $\exists_{D_0}\, \mathbf{S}[D_0 \neq \varnothing]\ \exists \preceq \in (\mathrm{DIR})[D_0]\ \exists l \in (\mathrm{isot})[D_0; \preceq]\ \exists \hat{f} \in \zeta_1(w^*)$ :

$$(D_0, \preceq, (f^*_{l(d)})_{d\in D_0}) \xrightarrow{\theta} \hat{f}. \tag{6.10.27}$$

Fix $\hat{f} \in \zeta_1(w^*)$ and some net $(D_0, \preceq, l), l \in (\mathrm{isot})[D_0; \preceq]$, in $\mathcal{N}$ with the property (6.10.27). Then by (2.3.9), (6.10.2), and (6.10.27) we have the convergence

$$(D_0, \preceq, (f^*_{l(d)})_{d\in D_0}) \overset{\otimes^X(\tau)}{\longrightarrow} \hat{f}. \tag{6.10.28}$$

From (2.6.22) and (6.10.28) we obtain (see [35, p. 35]) that $\forall x \in X$ :

$$(D_0, \preceq, (f^*_{l(d)}(x))_{d\in D_0}) \xrightarrow{\tau} \hat{f}(x). \tag{6.10.29}$$

Of course, (6.10.29) can be used in the case $x \in A^*$. By (6.3.8) and (6.10.24) we have $\forall x \in A^*\ \forall k \in \mathcal{N} : f_k^*(x) = f_*(x)$. Consider this property in the totality with (6.10.29). Fix $\bar{x} \in A^*$. Then by (6.10.29) we have

$$(D_0, \preceq, (f^*_{l(d)}(\bar{x}))_{d \in D_0}) \xrightarrow{\tau} \hat{f}(\bar{x}). \tag{6.10.30}$$

By (2.2.48) we obtain that

$$N_\tau(\hat{f}(\bar{x})) \subset (X - \text{ass})[D_0; \preceq; (f^*_{l(d)}(\bar{x}))_{d \in D_0}]. \tag{6.10.31}$$

Here we use (2.2.47). Then $\hat{f}(\bar{x}) = f_*(\bar{x})$. Indeed, let by contradiction $\hat{f}(\bar{x}) \neq f_*(\bar{x})$. In addition, $\hat{f}(\bar{x}) \in Y$ and $f_*(\bar{x}) \in Y \setminus \{\hat{f}(\bar{x})\}$. By (2.2.15), for some $\hat{H} \in N_\tau(\hat{f}(\bar{x}))$ we have $f_*(\bar{x}) \notin \hat{H}$. But by (6.10.31) we obtain that

$$\hat{H} \in (X - \text{ass})[D_0; \preceq; (f^*_{l(d)}(\bar{x}))_{d \in D_0}].$$

By (2.2.47) we have the property that for some $\bar{d} \in D_0$ and for any $d \in D_0$

$$(\bar{d} \prec d) \Rightarrow (f^*_{l(d)}(\bar{x}) \in \hat{H}). \tag{6.10.32}$$

From (6.10.32) we have, in particular, the property

$$\bar{z} \triangleq f^*_{l(\bar{d})}(\bar{x}) \in \hat{H}.$$

But by the choice of $\bar{x}$ we have $f^*_k(\bar{x}) = f_*(\bar{x}), k \in \mathcal{N}$. Then we obtain that $f_*(\bar{x}) = \bar{z} \in \hat{H}$. The last inclusion is impossible owing to the choice of $\hat{H}$. The obtained contradiction means that $\hat{f}(\bar{x}) = f_*(\bar{x})$. Since the choice of $\bar{x}$ was arbitrary, we obtain the property

$$(\hat{f} \mid A^*) = (f_* \mid A^*). \tag{6.10.33}$$

So $\hat{f} \in Z$ has the property (6.10.33). Therefore we have

$$\hat{f} \in Z_0(f_* \mid A^*). \tag{6.10.34}$$

Fix $n \in \mathcal{N}$. Using (2.2.50) we choose $\delta_1 \in D_0$ for which $\forall d \in D_0$ :

$$(\delta_1 \preceq d) \Rightarrow (n \leq l(d)). \tag{6.10.35}$$

We recall that $\zeta_n(w^*) \in \mathbb{T}$. In particular, $\zeta_n(w^*) \in \mathbb{F}$; see (6.10.5). From (6.10.25) and (6.10.35) we have $\forall d \in D_0$ :

$$(\delta_1 \preceq d) \Rightarrow (\zeta_{l(d)}(w^*) \subset \zeta_n(w^*)).$$

From (6.10.23) we obtain for $d \in D_0, \delta_1 \preceq d$, the property

$$f^*_{l(d)} \in \zeta_n(w^*). \tag{6.10.36}$$

From (2.2.4), (2.2.47), (6.10.27), and (6.10.36) we have

$$\forall H \in N_\theta(\hat{f}) : H \cap \zeta_n(w^*) \neq \varnothing.$$

So by (2.3.10) we obtain that $\hat{f} \in \mathrm{cl}(\zeta_n(w^*), \theta)$ and hence $\hat{f} \in \zeta_n(w^*)$, since $\zeta_n(w^*)$ is a closed set. We have established that $\forall k \in \mathcal{N} : \hat{f} \in \zeta_k(w^*)$. From (6.10.25) we have $\hat{f} \in \zeta(w^*)$. From (6.10.34) it follows that

$$Z_0(f_* \mid A^*) \cap \zeta(w^*) \neq \varnothing.$$

Since the choice of $A^*$ and $w^*$ was arbitrary, we have $\forall A \in \mathcal{X}$ $\forall s \in (\mathrm{Ge})[U; w_* \mid A] : Z_0(f_* \mid A) \bigcap \zeta(s) \neq \varnothing$. From (6.3.12) we obtain $f_* \in \gamma_U(\zeta)(w^*)$. The inclusion

$$\bigcap_{i \in \mathcal{N}} \gamma_U(\zeta_i)(w_*) \subset \gamma_U(\zeta)(w_*) \tag{6.10.37}$$

is established. By (6.10.22) and (6.10.37) we have the equality

$$\gamma_U(\zeta)(w_*) = \bigcap_{i \in \mathcal{N}} \gamma_U(\zeta_i)(w_*).$$

As a corollary, the convergence

$$(\gamma_U(\zeta_i)(w_*))_{i \in \mathcal{N}} \downarrow \gamma_U(\zeta)(w_*)$$

is established. Since the choice of $w_*$ was arbitrary, we now have the convergence

$$(\gamma_U(\zeta_i))_{i \in \mathcal{N}} \Downarrow \gamma_U(\zeta); \tag{6.10.38}$$

see definitions of Section 6.3. So (6.10.18) $\Rightarrow$ (6.10.38). Since the choice of $(\zeta_i)_{i \in \mathcal{N}}$ and $\zeta$ was arbitrary, by (6.7.8) we obtain that $\mathbb{T}^U \in \mathfrak{C}_U$. From (6.10.7) the required property follows. □

COROLLARY 6.10.1 *If* $\tau \in (\mathrm{top})_0[Y]$, *then* $\forall U \in \Sigma_0$:

$$(\mathbb{K}^U \in \mathfrak{Z}_U \cap \mathbf{Z}_U \cap \mathfrak{C}_U) \& (\mathcal{K}^U \in \mathfrak{Z}_U \cap \mathbf{Z}_U \cap \mathfrak{C}_U). \tag{6.10.39}$$

We omit the proof of the second part of (6.10.39); see [50, pp. 70–72]. Moreover, in (6.10.39) we use the properties (6.10.6).

COROLLARY 6.10.2 *If* $\tau \in (\mathrm{top})_0[Y]$, *then*

$$(\mathbb{K}^\Omega \in \mathfrak{Z} \cap \mathbf{Z} \cap \mathfrak{C}) \;\&\; (\mathcal{K}^\Omega \in \mathfrak{Z} \cap \mathbf{Z} \cap \mathfrak{C}).$$

COROLLARY 6.10.3 *If* $\tau \in (\mathcal{D} - \mathrm{top})[Y]$, *then* $\mathbb{T}^\Omega \in \mathbf{Z} \bigcap \mathfrak{C}$.

THEOREM 6.10.1 *Let* $(Y, \tau)$ *be a Hausdorff space:* $\tau \in (\mathrm{top})_0[Y]$. *Then* $\forall U \in \Sigma_0$ :

$$(\forall \zeta \in \mathbb{K}^U : (U - \mathrm{Na})[\zeta] = \gamma_U^\infty(\zeta) \in \mathbb{K}^U)$$
$$\&(\forall \mathcal{D} \in \mathcal{K}^U : (U - \mathrm{Na})[\mathcal{D}] = \gamma_U^\infty(\mathcal{D}) \in \mathcal{K}^{\mathcal{U}}).$$

PROOF. Fix $U \in \Sigma_0$. By Corollary 6.10.1 we have $\mathbb{K}^U \in \mathfrak{Z}_U \cap \mathbf{Z}_U \cap \mathfrak{C}_U$ and $\mathcal{K}^U \in \mathfrak{Z}_U \cap \mathbf{Z}_U \cap \mathfrak{C}_U$; see (6.10.39). Then by Proposition 6.7.1 we have

$$(\forall \zeta \in \mathbb{K}^U : (U - \mathrm{Na})[\zeta] = \gamma_U^\infty(\zeta))\&(\forall \mathcal{D} \in \mathcal{K}^U : (U - \mathrm{Na})[\zeta] = \gamma_U^\infty(\zeta)). \tag{6.10.40}$$

We supplement (6.10.40). Namely, recall that $\mathbb{K}^U$ and $\mathcal{K}^U$ are elements of $\mathfrak{Z}_U$. Therefore for $\zeta \in \mathbb{K}^U$ we have $(\gamma_U^k(\zeta))_{k \in \mathcal{N}_0} : \mathcal{N}_0 \longrightarrow \mathbb{K}^U$ (see the clarification after (6.7.9)). Since $\mathbb{K}^U \in \mathbf{Z}_U$, we have

$$\gamma_U^\infty : \mathbb{K}^U \longrightarrow \mathbb{K}^U$$

(see (6.7.8)). From (6.10.40) we have

$$\forall \zeta \in \mathbb{K}^U : (U - \mathrm{Na})[\zeta] = \gamma_U^\infty(\zeta) \in \mathbb{K}^U.$$

The proof of the second part of the basic statement is analogous. □

COROLLARY 6.10.4 *If* $\tau \in (\mathrm{top})_0[Y]$, *then*

$$(\forall \alpha \in \mathbb{K}^\Omega : (\mathrm{na})[\alpha] = \Gamma^\infty(\alpha) \in \mathbb{K}^\Omega)\&(\forall \beta \in \mathcal{K}^\Omega : (\mathrm{na})[\beta] = \Gamma^\infty(\beta) \in \mathcal{K}^\Omega).$$

For the proof it is sufficient to use Theorem 6.10.1, (6.3.15), and (6.4.8).

THEOREM 6.10.2 *Let* $\tau \in (\mathrm{top})_0[Y], U \in \mathbb{H}_0 \setminus \{\Omega\}$ *and* $\alpha \in \mathbb{M}(\Omega, Z)$. *Moreover, let* $(\forall w \in U : \alpha(w) \in \mathbb{K})\&(\forall w \in \Omega \setminus U : \alpha(w) \in \mathcal{K})$. *Then* $(\mathrm{na})[\alpha] = \Gamma^\infty(\alpha)$ *and the following properties hold:*

$$(\forall w \in U : (\mathrm{na})[\alpha](w) \in \mathbb{K}) \ \& \ (\forall w \in \Omega \setminus U : (\mathrm{na})[\alpha](w) \in \mathcal{K}).$$

PROOF. Recall Corollary 6.8.4. By Corollary 6.10.1 all conditions of Corollary 6.8.4 are satisfied. Hence we have the equality $(\mathrm{na})[\alpha] = \Gamma^\infty(\alpha)$. By Theorem 6.10.1 we obtain that $(U - \mathrm{Na})[(\alpha \mid U)] = \gamma_U^\infty((\alpha \mid U)) \in \mathbb{K}^U$. Let $V \triangleq \Omega \setminus U$. Then $V \in \mathbb{H}$ (see properties of $\mathbb{H}$ in Section 6.5). Moreover, $U \neq \Omega$ and $U \subset \Omega$. Therefore $V \neq \varnothing$. So $V \in \mathbb{H}_0$. In addition, $(\alpha \mid V) \in \mathcal{K}^V$. By Theorem 6.10.1 we have

$$(V - \mathrm{Na})[(\alpha \mid V)] = \gamma_V^\infty((\alpha \mid V)) \in \mathcal{K}^V.$$

By (6.5.9) we get

$$(\mathrm{na})[\alpha] = (U - \mathrm{Na})[(\alpha \mid U)]\square(V - \mathrm{Na})[(\alpha \mid V)].$$

Then we obviously have

$$(\forall w \in U : (\text{na})[\alpha](w) = (U - \text{Na})[(\alpha \mid U)](w) \in \mathbb{K})$$
$$\& \ (\forall w \in V : (\text{na})[\alpha](w) = (V - \text{Na})[(\alpha \mid V)](w) \in \mathcal{K}).$$

The proof is completed. □

In Theorems 6.10.1 and 6.10.2 we have the very general sufficient conditions of convergence of the iterated sequences to the extremal hereditary multi-selector of the initial set-valued mapping (see Corollary 6.10.4 too). In this connection we recall [36, 42, 43, 48, 50]. The above mentioned conditions are connected with compactness and sequential compactness of the initial set-valued mappings. Of course, this requirement admits the natural analogies with the corresponding properties of the spaces of generalized elements in Chapter 3. We extend these analogies to constructions connected with investigation of the effective domains of the set-valued mappings realized by procedures of the type (6.8.3).

THEOREM 6.10.3 *Let* $\tau \in (\text{top})_0[Y], U \in \Sigma_0$, *and* $\alpha \in \mathbb{K}^U \cup \mathcal{K}^U$. *Then*

$$((\text{DOM})[\gamma_U^k(\alpha)])_{k \in \mathcal{N}} \downarrow (\text{DOM})[(U - \text{Na})[\alpha]].$$

The proof is similar to that of [50, p.74]. Now we consider only the case $\alpha \in \mathbb{K}^U$. In addition, it is sufficient to establish the inclusion

$$\bigcap_{k \in \mathcal{N}} (\text{DOM})[\gamma_U^k(\alpha)] \subset (\text{DOM})[\gamma_U^\infty(\alpha)] \tag{6.10.41}$$

(see Theorem 6.10.1). Choose an arbitrary element $w_0$ of the set on the left hand side of (6.10.41). Then we have $\forall k \in \mathcal{N}$ :

$$\gamma_U^k(\alpha)(w_0) \neq \varnothing. \tag{6.10.42}$$

Moreover, we have the obvious compactness of $\gamma_U^k(\alpha)(w_0), k \in \mathcal{N}$, in $(Z; \theta)$; see the discussion after (6.7.9), and Proposition 6.10.1. Since $(Y, \tau)$ is a Hausdorff space, the topology $\theta$ (6.10.2) defines a Hausdorff space too. Therefore

$$(\gamma_U^k(\alpha)(w_0))_{k \in \mathcal{N}_0} : \mathcal{N}_0 \longrightarrow \mathbb{F}.$$

As a corollary, by (6.10.42) the family $\mathcal{M} \triangleq \{\gamma_U^k(\alpha)(w_0) : k \in \mathcal{N}\}$ is a centered system of sets closed in the compact TS

$$(\alpha(w_0), \theta \mid_{\alpha(w_0)});$$

in this connection see (6.3.12) and (6.8.2). By (2.2.21) the intersection of all sets of $\mathcal{M}$ is not empty. But by the definition of Section 6.3 we have

$$\gamma_U^\infty(\alpha)(w_0) = \bigcap_{M \in \mathcal{M}} M \neq \varnothing.$$

Then $w_0 \in (\mathrm{DOM})[\gamma_U^{\infty}(\alpha)]$. Since the choice of $w_0$ was arbitrary, the inclusion (6.10.41) is established. This inclusion realizes the basic property of convergence in the case $\alpha \in \mathbb{K}^U$. For the case $\alpha \in \mathcal{K}^U$ the corresponding proof is analogous (see [50, p.74]). □

COROLLARY 6.10.5 *Let $\tau \in (\mathrm{top})_0[Y]$ and $\alpha \in \mathbb{K}^{\Omega} \bigcup \mathcal{K}^{\Omega}$. Then*

$$((\mathrm{DOM})[\Gamma^k(\alpha)])_{k\in\mathcal{N}} \downarrow (\mathrm{DOM})[(\mathrm{na})[\alpha]]. \tag{6.10.43}$$

THEOREM 6.10.4 *Let $\tau \in (\mathrm{top})_0[Y]$, $\alpha \in \mathbb{M}(\Omega, Z)$, and $U \in \mathbb{H}_0 \setminus \{\Omega\}$. Moreover, let*

$$(\forall w \in U : \alpha(w) \in \mathbb{K})\&(\forall w \in \Omega \setminus U : \alpha(w) \in \mathcal{K}).$$

*Then (6.10.43) is true.*

PROOF. From Theorem 6.10.3 we have the convergence

$$((\mathrm{DOM})[\gamma_U^k((\alpha \mid U))])_{k\in\mathcal{N}} \downarrow (\mathrm{DOM})[(U-\mathrm{Na})[(\alpha \mid U)]]. \tag{6.10.44}$$

Introduce $V \triangleq \Omega \setminus U \in \mathbb{H}_0$ (see properties of $\mathbb{H}$ in Section 6.5). Since $(\alpha \mid V) \in \mathcal{K}^V$, we have the convergence

$$((\mathrm{DOM})[\gamma_V^k((\alpha \mid V))])_{k\in\mathcal{N}} \downarrow (\mathrm{DOM})[(V-\mathrm{Na})[(\alpha \mid V)]]. \tag{6.10.45}$$

In addition, by (6.5.9) we have the equality

$$(\mathrm{na})[\alpha] = (U-\mathrm{Na})[(\alpha \mid U)]\Box(V-\mathrm{Na})[(\alpha \mid V)]. \tag{6.10.46}$$

Analogously, by Theorem 6.8.2 we have $\forall k \in \mathcal{N}$ :

$$\Gamma^k(\alpha) = \gamma_U^k((\alpha \mid U))\Box\gamma_V^k((\alpha \mid V)). \tag{6.10.47}$$

In addition, by Theorem 6.10.2 we obtain the equality $(\mathrm{na})[\alpha] = \Gamma^{\infty}(\alpha)$. For the proof it is sufficient to establish the inclusion

$$\bigcap_{k\in\mathcal{N}} (\mathrm{DOM})[\Gamma^k(\alpha)] \subset (\mathrm{DOM})[\Gamma^{\infty}(\alpha)]. \tag{6.10.48}$$

Let $w_0$ be an arbitrary element of the intersection on the left hand side of (6.10.48). In addition, $(w_0 \in U) \vee (w_0 \in V)$.

Let $w_0 \in U$. Then by (6.10.46) we have

$$(\mathrm{na})[\alpha](w_0) = \Gamma^{\infty}(\alpha)(w_0) = (U-\mathrm{Na})[(\alpha \mid U)](w_0). \tag{6.10.49}$$

From (6.10.47) it follows that for $k \in \mathcal{N}$

$$\Gamma^k(\alpha)(w_0) = \gamma_U^k((\alpha \mid U))(w_0) \neq \varnothing.$$

So $w_0 \in (\mathrm{DOM})[\gamma_U^k((\alpha \mid U))], k \in \mathcal{N}$. Therefore by (6.10.44)

$$w_0 \in (\mathrm{DOM})[(U - \mathrm{Na})[(\alpha \mid U)]]$$

and as a corollary we have $(U - \mathrm{Na})[(\alpha \mid U)](w_0) \neq \varnothing$. Then by (6.10.49) we get $\Gamma^\infty(\alpha)(w_0) \neq \varnothing$ and as a result $w_0 \in (\mathrm{DOM})[\Gamma^\infty(\alpha)]$.

Let $w_0 \in V$. By (6.10.46) we have

$$(\mathrm{na})[\alpha](w_0) = \Gamma^\infty(\alpha)(w_0) = (V - \mathrm{Na})[(\alpha \mid V)](w_0). \tag{6.10.50}$$

By (6.10.47) we get $\forall k \in \mathcal{N}$ :

$$\Gamma^k(\alpha)(w_0) = \gamma_V^k((\alpha \mid V))(w_0) \neq \varnothing.$$

So $\forall k \in \mathcal{N} : w_0 \in (\mathrm{DOM})[\gamma_V^k((\alpha \mid V))]$. From (6.10.45) we have the inclusion

$$w_0 \in (\mathrm{DOM})[(V - \mathrm{Na})[(\alpha \mid V)]].$$

By (6.10.50) we obtain $\Gamma^\infty(\alpha)(w_0) \neq \varnothing$. In this case $w_0 \in (\mathrm{DOM})[\Gamma^\infty(\alpha)]$ too. We have established (6.10.48) and hence the convergence (6.10.43). □

THEOREM 6.10.5 *Let $\tau \in (\mathrm{top})_0[Y], \alpha \in \mathbb{M}(\Omega, Z)$, and let the following property be valid:*

$$(\alpha \in \mathbb{K}^\Omega) \vee (\alpha \in \mathcal{K}^\Omega)$$

$$\vee(\exists U \in \mathbb{H}_0 \setminus \{\Omega\} : (\forall w \in U : \alpha(w) \in \mathbb{K}) \& (\forall w \in \Omega \setminus U : \alpha(w) \in \mathcal{K})). \tag{6.10.51}$$

*Then* $\forall w \in \Omega : ((\mathrm{na})[\alpha](w) = \varnothing) \Rightarrow (\exists n \in \mathcal{N} : \Gamma^n(\alpha)(w) = \varnothing)$.

PROOF. From (6.10.51), Corollary 6.10.5, and Theorem 6.10.4 we have the convergence (6.10.43). Let $w \in \Omega$ be such that $(\mathrm{na})[\alpha](w) = \varnothing$. Then

$$w \notin (\mathrm{DOM})[(\mathrm{na})[\alpha]]. \tag{6.10.52}$$

From (6.10.43) we have the equality

$$(\mathrm{DOM})[(\mathrm{na})[\alpha]] = \bigcap_{k \in \mathcal{N}} (\mathrm{DOM})[\Gamma^k(\alpha)].$$

From (6.10.52) it follows that $w \notin (\mathrm{DOM})[\Gamma^n(\alpha)]$ for some $n \in \mathcal{N}$. As a consequence $\Gamma^n(\alpha)(w) = \varnothing$. □

In the sequel we shall supplement Theorem 6.10.5 by some useful properties connected with the factorization based on (6.9.1), (6.9.11).

## 6.11 COMPACT-VALUED MAPPINGS

In this section we consider some non-sequential analogues of Proposition 6.7.3. But we touch upon only compact-valued goal mappings. It is possible to easily verify the validity of the following implication:

$$(\tau \in (\mathcal{D} - \mathrm{top})[Y]) \Rightarrow (\forall A \in \mathcal{X}\ \forall z \in Z : Z_0(z \mid A) \in \mathbb{F}). \qquad (6.11.1)$$

We use an analogue of (6.4.10). Namely, if $T \in \Sigma_0$, $\Lambda \in \mathcal{Z}_T$ and $\mathbb{M} \in 2^{\mathbb{M}(T,Z)}$, then for $\mathbf{A} \triangleq \{\Lambda(\mathcal{E}) : \mathcal{E} \in \mathbb{M}\}$ we use the notation

$$\bigwedge_{\mathcal{E}\in\mathbb{M}}^{[T]} \Lambda(\mathcal{E}) \qquad (6.11.2)$$

instead of (6.4.10); of course, (6.11.2) is the mapping from $T$ into $\mathcal{P}(Z)$ for which $\forall \omega \in T$:

$$\left(\bigwedge_{\mathcal{E}\in\mathbb{M}}^{[T]} \Lambda(\mathcal{E})\right)(\omega) = \bigcap_{\mathcal{E}\in\mathbb{M}} \Lambda(\mathcal{E})(\omega). \qquad (6.11.3)$$

We use (6.4.11) in (6.11.3). Suppose that $\forall \Lambda \in \mathcal{Z}\ \forall \mathbb{M} \in 2^{\mathbb{M}(\Omega,Z)}$:

$$\bigwedge_{\mathcal{E}\in\mathbb{M}} \Lambda(\mathcal{E}) \triangleq \bigwedge_{\mathcal{E}\in\mathbb{M}}^{[\Omega]} \Lambda(\mathcal{E}).$$

The following condition is supposed to be true up to the end of this section.

CONDITION 6.11.1 $\tau \in (\mathrm{top})_0[Y]$.

As a consequence, from (6.10.6) we have $\mathbb{K} \in \mathbb{F}$ and $\mathcal{K} \in \mathcal{F}$. From Corollary 6.10.1 it follows that $\forall U \in \Sigma_0$:

$$(\mathbb{K}^U \in \mathfrak{Z}_U \cap \mathbf{Z}_U \cap \mathfrak{C}_U)\ \&\ (\mathcal{K}^U \in \mathfrak{Z}_U \cap \mathbf{Z}_U \cap \mathfrak{C}_U).$$

Of course, $\mathbb{K}^\Omega \in \mathfrak{Z} \cap \mathbb{Z} \cap \mathfrak{C}$ and $\mathcal{K}^\Omega \in \mathfrak{Z} \cap \mathbb{Z} \cap \mathfrak{C}$. As a consequence of this properties we have (see Theorem 6.10.1) $\forall U \in \Sigma_0$:

$$(\forall \mathcal{E} \in \mathbb{K}^U : (U - \mathrm{Na})[\mathcal{E}] = \gamma_U^\infty(\mathcal{E}) \in \mathbb{K}^U)$$

$$\&\ (\forall \mathcal{D} \in \mathcal{K}^U : (U - \mathrm{Na})[\mathcal{D}] = \gamma_U^\infty(\mathcal{D}) \in \mathcal{K}^U).$$

By Corollary 6.10.3 we get

$$(\forall \alpha \in \mathbb{K}^\Omega : (\mathrm{na})[\alpha] = \Gamma^\infty(\alpha) \in \mathbb{K}^\Omega) \& (\forall \beta \in \mathcal{K}^\Omega : (\mathrm{na})[\beta] = \Gamma^\infty(\beta) \in \mathcal{K}^\Omega).$$

From Theorem 6.10.2 we obtain $\forall U \in \mathbb{H}_0 \setminus \{\Omega\}\ \forall \alpha \in \mathbb{M}(\Omega, Z)$:

$$\Big((\forall \omega \in U : \alpha(\omega) \in \mathbb{K})\ \&\ (\forall \omega \in \Omega \setminus U : \alpha(\omega) \in \mathcal{K})\Big)$$

$$\Rightarrow \Big(((\text{na})[\alpha] = \Gamma^{\infty}(\alpha))\ \&\ (\forall \tilde{\omega} \in U : (\text{na})[\alpha](\tilde{\omega}) \in \mathbb{K})$$

$$\&\ (\forall \hat{\omega} \in \Omega \setminus U : (\text{na})[\alpha](\hat{\omega}) \in \mathcal{K})\Big).$$

From Theorem 6.10.3 (and Condition 6.11.1) we obtain $\forall U \in \Sigma_0\ \forall \alpha \in \mathbb{K}^U \cup \mathcal{K}^U$:

$$((\text{DOM})[\gamma_U^k(\alpha)])_{k \in \mathcal{N}} \downarrow (\text{DOM})[(U - \text{Na})[\alpha]].$$

From Corollary 6.10.4 it follows that $\forall \alpha \in \mathbb{K}^{\Omega} \cup \mathcal{K}^{\Omega}$:

$$((\text{DOM})[\Gamma^k(\alpha)])_{k \in \mathcal{N}} \downarrow (\text{DOM})[(\text{na})[\alpha]].$$

By Theorem 6.10.4 we have $\forall \alpha \in \mathbb{M}(\Omega, Z)\ \forall U \in \mathbb{H}_0 \setminus \{\Omega\}$:

$$\Big((\forall \omega \in U : \alpha(\omega) \in \mathbb{K})\&(\forall \omega \in \Omega \setminus U : \alpha(\omega) \in \mathcal{K})\Big)$$

$$\Rightarrow \Big(((\text{DOM})[\Gamma^k(\alpha)])_{k \in \mathcal{N}} \downarrow (\text{DOM})[(\text{na})[\alpha]]\Big).$$

Finally, by Theorem 6.10.5 we have $\forall \alpha \in \mathbb{M}(\Omega, Z)$:

$$\Big((\alpha \in \mathbb{K}^{\Omega}) \vee (\alpha \in \mathcal{K}^{\Omega})$$

$$\vee(\exists U \in \mathbb{H}_0 \setminus \{\Omega\} : (\forall \omega \in U : \alpha(\omega) \in \mathbb{K})\&(\forall \omega \in \Omega \setminus U : \alpha(\omega) \in \mathcal{K}))\Big)$$

$$\Rightarrow \Big(\forall \tilde{\omega} \in \Omega : ((\text{na})(\alpha)[\tilde{\omega}] = \varnothing) \Rightarrow (\exists n \in \mathcal{N} : \Gamma^n(\alpha)(\tilde{\omega}) = \varnothing)\Big).$$

The above summary is supplemented by some statements having the sense of an ordered continuity.

THEOREM 6.11.1 *Let $U \in \Sigma_0$ and let $\mathbb{M}$ be a nonempty subset of $\mathbb{K}^U$ for which*

$$\forall \alpha \in \mathbb{M}\ \forall \beta \in \mathbb{M}\ \exists \lambda \in \mathbb{M} : (\lambda \sqsubseteq \alpha)\&(\lambda \sqsubseteq \beta). \tag{6.11.4}$$

*Then the following equality holds:*

$$\gamma_U\Big(\bigwedge_{\mathcal{E}\in\mathbb{M}}^{[U]}\mathcal{E}\Big)=\bigwedge_{\mathcal{E}\in\mathbb{M}}^{[U]}\gamma_U(\mathcal{E}). \tag{6.11.5}$$

PROOF. By (6.3.12) we have $\forall\mathcal{D}\in\mathbb{M}$:

$$\gamma_U\Big(\bigwedge_{\mathcal{E}\in\mathbb{M}}^{[U]}\mathcal{E}\Big)\sqsubseteq\gamma_U(\mathcal{D}).$$

As a consequence, by (6.11.3) we get

$$\gamma_U\Big(\bigwedge_{\mathcal{E}\in\mathbb{M}}^{[U]}\mathcal{E}\Big)\sqsubseteq\bigwedge_{\mathcal{E}\in\mathbb{M}}^{[U]}\gamma_U(\mathcal{E}). \tag{6.11.6}$$

Fix $\omega\in U$ and consider (6.11.3) for $T=U$ and $\Lambda=\gamma_U$. Choose arbitrarily

$$\varphi\in\Big(\bigwedge_{\mathcal{E}\in\mathbb{M}}^{[U]}\gamma_U(\mathcal{E})\Big)(\omega).$$

Then $\forall\mathcal{E}\in\mathbb{M}:\varphi\in\gamma_U(\mathcal{E})(\omega)$. Hence $\varphi$ is an element of the intersection of all sets $\mathcal{E}(\omega)$, $\mathcal{E}\in\mathbb{M}$; see (6.3.13). By (6.4.11) we have

$$\varphi\in\Big(\bigwedge_{\mathcal{E}\in\mathbb{M}}^{[U]}\mathcal{E}\Big)(\omega). \tag{6.11.7}$$

From (6.3.12) we get $\forall\mathcal{E}\in\mathbb{M}\ \forall A\in\mathcal{X}$: $\varphi\in\tilde{\gamma}_U[A](\mathcal{E})(\omega)$. Fix $A\in\mathcal{X}$ and $s\in(\mathrm{Ge})[U;\omega\mid A]$. By (6.3.10) we obtain

$$\forall\mathcal{E}\in\mathbb{M}:Z_0(\varphi\mid A)\cap\mathcal{E}(s)\neq\varnothing. \tag{6.11.8}$$

Choose arbitrarily $\tilde{\mathcal{M}}\in\mathbb{M}$ and consider the set $\tilde{\mathcal{M}}(s)\in\mathbb{K}$. Then

$$\vartheta\triangleq\theta\mid_{\tilde{\mathcal{M}}(s)}\in(\mathbf{c}-\mathrm{top})[\tilde{\mathcal{M}}(s)]; \tag{6.11.9}$$

we use (2.3.6) and (6.10.4). Let $\mathbf{F}\triangleq\mathcal{F}_\vartheta$ (see (2.2.17)). By (2.3.12) and (6.11.9) we have

$$\mathbf{F}=\mathcal{F}_\vartheta\mid_{\tilde{\mathcal{M}}(s)}=\mathbb{F}\mid_{\tilde{\mathcal{M}}(s)}.$$

Let $\mathbb{Q}\in\mathrm{Fin}(\mathbb{M})$. Then $\tilde{\mathbb{Q}}\triangleq\mathbb{Q}\cup\{\tilde{\mathcal{M}}\}\in\mathrm{Fin}(\mathbb{M})$ and by (6.11.4) we get for some $\mathcal{E}_{\mathbb{Q}}\in\mathbb{M}$ the relation

$$\mathcal{E}_{\mathbb{Q}}\sqsubseteq\bigwedge_{Q\in\tilde{\mathbb{Q}}}^{[U]}Q. \tag{6.11.10}$$

From (6.11.10) we obviously have

$$Z_0(\varphi \mid A) \cap \mathcal{E}_{\mathbb{Q}}(s) \subset \bigcap_{Q \in \mathbb{Q}} (\tilde{\mathcal{M}}(s) \cap Z_0(\varphi \mid A) \cap Q(s)).$$

Using (6.11.8) we get

$$\bigcap_{Q \in \mathbb{Q}} (\tilde{\mathcal{M}}(s) \cap Z_0(\varphi \mid A) \cap Q(s)) \neq \varnothing.$$

Since the choice of $\mathbb{Q}$ was arbitrary, we obtain that

$$\hat{\mathfrak{F}} \triangleq \{\tilde{\mathcal{M}}(s) \cap Z_0(\varphi \mid A) \cap Q(s) : Q \in \mathbb{M}\} \in \mathbf{Z}[\mathbf{F}]. \tag{6.11.11}$$

In (6.11.11) we use the following properties: 1) by (6.11.1) and Condition 9.11.1, $Z_0(\varphi \mid A) \in \mathbb{F}$; 2) if $Q \in \mathbb{M}$, then $Q(s) \in \mathbb{K}$ and hence $Q(s) \in \mathbb{F}$; 3) for $F \in \mathbb{F}$ the statement $\tilde{\mathcal{M}}(s) \cap F \in \mathbf{F}$ holds.

From (2.2.21), (6.11.9), and (6.11.11) we have

$$\bigcap_{H \in \hat{\mathfrak{F}}} H \neq \varnothing. \tag{6.11.12}$$

Let $\psi$ be an element of the set on the left hand side of (6.11.12). By (6.11.11) $\psi \in \tilde{\mathcal{M}}(s)$ has the two following properties:

$$(\psi \in Z_0(\varphi \mid A)) \;\&\; (\psi \in \bigcap_{Q \in \mathbb{M}} Q(s)). \tag{6.11.13}$$

From (6.4.11) and (6.11.13) it follows that

$$\psi \in Z_0(\varphi \mid A) \cap \Big(\bigwedge_{\mathcal{E} \in \mathbb{M}}^{[U]} \mathcal{E}\Big)(s).$$

Now we have the following statement:

$$Z_0(\varphi \mid A) \cap \Big(\bigwedge_{\mathcal{E} \in \mathbb{M}}^{[U]} \mathcal{E}\Big)(s) \neq \varnothing.$$

Since the choice of $s$ was arbitrary, by (6.3.10) and (6.11.7) we have the inclusion

$$\varphi \in \tilde{\gamma}_U[A]\Big(\bigwedge_{\mathcal{E} \in \mathbb{M}}^{[U]} \mathcal{E}\Big)(\omega).$$

But the choice of $A$ was arbitrary too. Hence we have

$$\varphi \in \gamma_U\Big(\bigwedge_{\mathcal{E}\in\mathbb{M}}^{[U]} \mathcal{E}\Big)(\omega).$$

So we have established the inclusion

$$\Big(\bigwedge_{\mathcal{E}\in\mathbb{M}}^{[U]} \gamma_U(\mathcal{E})\Big)(\omega) \subset \gamma_U\Big(\bigwedge_{\mathcal{E}\in\mathbb{M}}^{[U]} \mathcal{E}\Big)(\omega).$$

Using (6.11.6) we obtain

$$\gamma_U\Big(\bigwedge_{\mathcal{E}\in\mathbb{M}}^{[U]} \mathcal{E}\Big)(\omega) = \Big(\bigwedge_{\mathcal{E}\in\mathbb{M}}^{[U]} \gamma_U(\mathcal{E})\Big)(\omega). \tag{6.11.14}$$

Since the choice of $\omega$ was arbitrary, we obtain the equality (6.11.5) from (6.11.14). □

COROLLARY 6.11.1 *Let* $\mathbb{M}$ *be a nonempty subset of* $\mathbb{K}^{\Omega}$. *Moreover, let (6.11.4) be valid. Then*

$$\Gamma\big(\bigwedge_{\mathcal{E}\in\mathbb{M}} \mathcal{E}\big) = \bigwedge_{\mathcal{E}\in\mathbb{M}} \Gamma(\mathcal{E}).$$

The proof is obvious (see (6.3.15)).

THEOREM 6.11.2 *Let* $U \in \Sigma_0$ *and* $\mathbb{M}$ *be a nonempty subset of* $\mathbb{K}^{\Omega} \cap \mathfrak{N}[U]$. *Moreover, let the relation (6.11.4) be valid. Then*

$$\bigwedge_{\mathcal{E}\in\mathbb{M}}^{[U]} \mathcal{E} \in \mathbb{K}^{\Omega} \cap \mathfrak{N}[U].$$

PROOF. We use (6.3.20). Then by Theorem 6.11.1 we have (see (6.11.5))

$$\gamma_U\Big(\bigwedge_{\mathcal{E}\in\mathbb{M}}^{[U]} \mathcal{E}\Big) = \bigwedge_{\mathcal{E}\in\mathbb{M}}^{[U]} \mathcal{E}.$$

Fix $\omega \in U$. Then by (6.4.11) we obtain the equality

$$F \triangleq \Big(\bigwedge_{\mathcal{E}\in\mathbb{M}}^{[U]} \mathcal{E}\Big)(\omega) = \bigcap_{\mathcal{E}\in\mathbb{M}} \mathcal{E}(\omega). \tag{6.11.15}$$

Recall that $\mathcal{E}(\omega) \in \mathbb{K}$ for $\mathcal{E} \in \mathbb{M}$. Fix $Q \in \mathbb{M}$ (recall that $\mathbb{M} \neq \varnothing$). Then $Q(\omega) \in \mathbb{K}$. In addition, $\mathcal{E}(\omega) \in \mathbb{F}$, $\mathcal{E} \in \mathbb{M}$. By (6.11.15) we have $F \in \mathbb{F}$ and $F \subset Q(\omega)$. But

$$\mathcal{Q} \triangleq \theta \mid_{Q(\omega)} \in (\mathbf{c} - \mathrm{top})[Q(\omega)], \tag{6.11.16}$$

and by (2.3.12) we have the property

$$F = F \cap Q(\omega) \in \mathcal{F}_{\mathcal{Q}}. \tag{6.11.17}$$

From (6.11.16) and (6.11.17) we get $F \in (\mathcal{Q} - \mathrm{comp})[Q(\omega)]$ (see the respective properties in Section 2.3). So

$$\theta \mid_F = \mathcal{Q} \mid_F \in (\mathbf{c} - \mathrm{top})[F]$$

(see (2.3.6)). Then by (2.3.6) and (6.10.4), for $F \in \mathcal{P}(Z)$ we have $F \in \mathbb{K}$. From (6.11.15) it follows that

$$\bigwedge_{\mathcal{E} \in \mathbb{M}}^{[U]} \mathcal{E} \in \mathbb{K}^U.$$

The proof of Theorem 6.11.2 is completed. □

COROLLARY 6.11.2 *Let $\mathbb{M}$ be a nonempty subset of $\mathbb{K}^\Omega \cap \mathbb{N}$. Moreover, let (6.11.4) be fulfilled. Then* $\bigwedge_{\mathcal{E} \in \mathbb{M}} \mathcal{E} \in \mathbb{K}^\Omega \cap \mathbb{N}$.

THEOREM 6.11.3 *Let $U \in \Sigma_0$ and $\mathbb{M}$ be a nonempty subset of $\mathbb{K}^U$. Moreover, let (6.11.4) be fulfilled. Then*

$$(U - \mathrm{Na})\left[\bigwedge_{\mathcal{E} \in \mathbb{M}}^{[U]} \mathcal{E}\right] = \bigwedge_{\mathcal{E} \in \mathbb{M}}^{[U]} (U - \mathrm{Na})[\mathcal{E}].$$

PROOF. Introduce the nonempty set

$$\tilde{\mathbb{M}} \triangleq \{(U - \mathrm{Na})[\mathcal{E}] : \mathcal{E} \in \mathbb{M}\}. \tag{6.11.18}$$

From Theorem 6.10.1 it follows that $\tilde{\mathbb{M}}$ is a nonempty subset of $\mathbb{K}^U$. Using the properties of set-valued mappings like (6.4.7), we conclude (see (6.4.3)) that

$$\mathcal{H} \mapsto (U - \mathrm{Na})[\mathcal{H}] : \mathbb{M}(U, Z) \to \mathbb{M}(U, Z) \tag{6.11.19}$$

has the $\sqsubseteq$-monotonicity property. By (6.11.4) we obtain that

$$\forall \alpha \in \tilde{\mathbb{M}}\ \forall \beta \in \tilde{\mathbb{M}}\ \exists \lambda \in \tilde{\mathbb{M}} : (\lambda \sqsubseteq \alpha) \& (\lambda \sqsubseteq \beta).$$

By Theorem 6.11.2 we have

$$\bigwedge_{\mathcal{E}\in\mathbb{M}}^{[U]} (U-\mathrm{Na})[\mathcal{E}] \in \mathfrak{N}[U]. \tag{6.11.20}$$

Of course, we use the definition of (6.11.2). In addition, by the monotonicity of (6.11.19) we have

$$(U-\mathrm{Na})\left[\bigwedge_{\mathcal{E}\in\mathbb{M}}^{[U]} \mathcal{E}\right] \sqsubseteq \bigwedge_{\mathcal{E}\in\mathbb{M}}^{[U]} (U-\mathrm{Na})[\mathcal{E}]. \tag{6.11.21}$$

But for $\mathcal{E}\in\mathbb{M}$ we have $(U-\mathrm{Na})[\mathcal{E}]\sqsubseteq\mathcal{E}$. As a consequence

$$\bigwedge_{\mathcal{E}\in\mathbb{M}}^{[U]} (U-\mathrm{Na})[\mathcal{E}] \sqsubseteq \bigwedge_{\mathcal{E}\in\mathbb{M}}^{[U]} \mathcal{E}. \tag{6.11.22}$$

From (6.4.2), (6.4.3), (6.11.20), and (6.11.22) we get the property

$$\bigwedge_{\mathcal{E}\in\mathbb{M}}^{[U]} (U-\mathrm{Na})[\mathcal{E}] \in \mathfrak{N}_0[U;\ \bigwedge_{\mathcal{E}\in\mathbb{M}}^{[U]} \mathcal{E}].$$

From (6.4.7) it follows that

$$\bigwedge_{\mathcal{E}\in\mathbb{M}}^{[U]} (U-\mathrm{Na})[\mathcal{E}] \sqsubseteq (U-\mathrm{Na})\left[\bigwedge_{\mathcal{E}\in\mathbb{M}}^{[U]} \mathcal{E}\right].$$

Using (6.11.21) we obtain the equality

$$(U-\mathrm{Na})\left[\bigwedge_{\mathcal{E}\in\mathbb{M}}^{[U]} \mathcal{E}\right] = \bigwedge_{\mathcal{E}\in\mathbb{M}}^{[U]} (U-\mathrm{Na})[\mathcal{E}].$$

COROLLARY 6.11.3 *Let* $\mathbb{M}$ *be a nonempty subset of* $\mathbb{K}^{\Omega}$. *Moreover, let (6.11.4) be valid. Then*

$$(\mathrm{na})\left[\bigwedge_{\mathcal{E}\in\mathbb{M}} \mathcal{E}\right] = \bigwedge_{\mathcal{E}\in\mathbb{M}} (\mathrm{na})[\mathcal{E}].$$

By analogy with Proposition 6.10.1 one can establish that $\forall T\in\Sigma_0$ $\forall A\in\mathcal{X}$:

$$(\forall\mathcal{E}\in\mathbb{K}^T : \tilde{\gamma}_T[A](\mathcal{E})\in\mathbb{K}^T)\ \&\ (\forall\mathcal{D}\in\mathcal{K}^T : \tilde{\gamma}_T[A](\mathcal{D})\in\mathcal{K}^T).$$

We consider simple corollaries of these properties. First, $\forall A \in \mathcal{X}$:

$$(\{\Gamma_A(\mathcal{E}) : \mathcal{E} \in \mathbb{K}^\Omega\} \subset \mathbb{K}^\Omega) \,\&\, (\{\Gamma_A(\mathcal{D}) : \mathcal{D} \in \mathcal{K}^\Omega\} \subset \mathcal{K}^\Omega);$$

here we use (6.3.13). Moreover, returning to definitions of Section 6.6, we get $\forall T \in \Sigma_0\ \forall \alpha \in \mathbb{M}(T, Z)\ \forall A \in \mathcal{X}$:

$$(\mathbb{K}^T \in \mathfrak{M}(T, \alpha, A)) \,\&\, (\mathcal{K}^T \in \mathfrak{M}(T, \alpha, A)).$$

Note the important particular case: if $\alpha \in \mathbb{M}(\Omega, Z)$ and $A \in \mathcal{X}$, then

$$(\mathbb{K}^\Omega \in \mathfrak{M}[\alpha; A])\&(\mathcal{K}^\Omega \in \mathfrak{M}[\alpha; A]).$$

Using (6.6.8) we obtain, in particular, that $\forall T \in \Sigma_0\ \forall \alpha \in \mathbb{M}(T, Z)$ $\forall A \in \mathcal{X}$:

$$\begin{aligned} &(\mathbb{K}^T \cap (\text{f.p.}-\alpha)[A \mid T] = \{\tilde{\gamma}_T[A](\mathcal{H}) : \mathcal{H} \in \mathbb{K}^T \cap S_T(\alpha)\}) \\ &\& \ (\mathcal{K}^T \cap (\text{f.p.}-\alpha)[A \mid T] = \{\tilde{\gamma}_T[A](\mathcal{H}) : \mathcal{H} \in \mathcal{K}^T \cap S_T(\alpha)\}). \end{aligned} \tag{6.11.23}$$

We now use Proposition 6.6.1 and (6.11.23). As a consequence we get $\forall T \in \Sigma_0\ \forall \alpha \in \mathbb{M}(T, Z)$:

$$\begin{aligned} &(\mathbb{K}^T \cap \mathfrak{N}_0[T; \alpha] = \bigcap_{A \in \mathcal{X}} \{\tilde{\gamma}_T[A](\mathcal{H}) : \mathcal{H} \in \mathbb{K}^T \cap S_T(\alpha)\}) \\ &\& \ (\mathcal{K}^T \cap \mathfrak{N}_0[T; \alpha] = \bigcap_{A \in \mathcal{X}} \{\tilde{\gamma}_T[A](\mathcal{H}) : \mathcal{H} \in \mathcal{K}^T \cap S_T(\alpha)\}). \end{aligned} \tag{6.11.24}$$

It is useful to consider (6.11.24) in the totality with Theorem 6.10.1 (note also constructions of Section 6.5 concerning the parallel realization of the hereditary set-valued mappings on $\Omega$: for example, see (5.6.10)). Consider $\mathcal{E} \in \mathbb{M}(\Omega, Z)$ and $\mathcal{U} \in \mathbb{H}_0$. Then for constructing $(\text{na})[\mathcal{E}]$ we can use (6.5.8). In addition, we can realize the parallel scheme of Section 6.8; then the corresponding basic iterated procedure admits the natural sewing with local iterated procedures. But in reality, realization of these local procedures is connected with computational difficulties. Moreover, for some $T \in \mathbb{H}_0$ (6.10.35) can be used under $\alpha = (\mathcal{E} \mid T)$. Suppose that this $\alpha$ is a compact-valued mapping: $\alpha \in \mathbb{K}^T$. Then by Theorem 6.10.1 we have $(U - \text{Na})[\alpha] \in \mathbb{K}^T \cap \mathfrak{N}_0[T; \alpha]$ and what is more,

$$\forall \beta \in \mathbb{K}^T \cap \mathfrak{N}_0[T; \alpha] : \beta \sqsubseteq (U - \text{Na})[\alpha];$$

of course, we use (6.4.7). As a consequence we get the following method. Taking (6.11.24) into account, we construct $\{\tilde{\gamma}_T[A](\mathcal{H}) : \mathcal{H} \in \mathbb{K}^T \cap S_T(\alpha)\}$, $A \in \mathcal{X}$. This construction is realized without iterations. Then we must find the greatest element of the intersection (see (6.11.24)) of these sets. Finally, we find the $\sqsubseteq$-greatest element of this intersection.

Returning to (6.11.24), we note that $\forall \alpha \in \mathbb{M}(\Omega, Z)$:

$$(\mathbb{N}_0[\alpha] \cap \mathbb{K}^\Omega = \bigcap_{A \in \mathcal{X}} \{\Gamma_A(\mathcal{H}) : \mathcal{H} \in \mathbb{K}^\Omega \cap \mathbf{S}[\alpha]\})$$
$$\& \ (\mathbb{N}_0[\alpha] \cap \mathcal{K}^\Omega = \bigcap_{A \in \mathcal{X}} \{\Gamma_A(\mathcal{H}) : \mathcal{H} \in \mathcal{K}^\Omega \cap \mathbf{S}[\alpha]\}).$$

THEOREM 6.11.4 *Let (along with Condition 6.11.1) Condition 6.9.1 be valid. Moreover, let $\mathcal{E} \in \mathbb{M}(\Omega, Z)$, $G \in \mathfrak{G}$ and $(\mathcal{E} \mid G) \in \mathbb{K}^G \cup \mathcal{K}^G$. Then the following four conditions are equivalent:*

1) $(\mathrm{DOM})[(\mathrm{na})[\mathcal{E}]] \cap G = \varnothing$;
2) $G \setminus (\mathrm{DOM})[(\mathrm{na})[\mathcal{E}]] \neq \varnothing$;
3) $\exists k \in \mathcal{N} : (\mathrm{DOM})[\Gamma^k(\mathcal{E})] \cap G = \varnothing$;
4) $\exists k \in \mathcal{N}_0 : G \setminus (\mathrm{DOM})[\Gamma^k(\mathcal{E})] \neq \varnothing$.

PROOF. From Corollary 6.9.2 we have the implication 4) $\Rightarrow$ 3). Owing to Proposition 6.9.5 we get the equivalence 1) $\Leftrightarrow$ 2). Obvious properties following from definitions of Section 6.3 (for example, see (6.3.2)) and Corollary 6.7.1 give the implication 3) $\Rightarrow$ 1). Thus we obtain that

$$4) \Rightarrow 3) \Rightarrow [\ 1) \Leftrightarrow 2)\ ].$$

Let 2) be satisfied, and

$$\omega_0 \in G \setminus (\mathrm{DOM})[(\mathrm{na})[\mathcal{E}]]. \tag{6.11.25}$$

Introduce $\alpha \triangleq (\mathcal{E} \mid G) \in \mathbb{M}(G, Z)$. From conditions of the theorem we have $\alpha \in \mathbb{K}^G \cup \mathcal{K}^G$. In addition, $\mathbf{S}[G \neq \varnothing]$ (see (6.9.1)). Therefore $G \in \Sigma_0$. By Theorem 6.10.3 we have the convergence

$$((\mathrm{DOM})[\gamma_G^k(\alpha)])_{k \in \mathcal{N}} \downarrow (\mathrm{DOM})[(G - \mathrm{Na})[\alpha]\ ]. \tag{6.11.26}$$

But $(\mathrm{DOM})[(\mathrm{na})[\mathcal{E}]\ ] = \{\omega \in \Omega \mid (\mathrm{na})[\mathcal{E}](\omega) \neq \varnothing\}$ and $(\mathrm{DOM})[(G - \mathrm{Na})[\alpha]\ ] = \{\omega \in G \mid (G - \mathrm{Na})[\alpha](\omega) \neq \varnothing\}$. Recall that (6.9.1) implies the property $G \in \mathbb{H}_0$. Hence by (6.5.7) we have

$$(G - \mathrm{Na})[\alpha] = ((\mathrm{na})[\mathcal{E}] \mid G).$$

Then we have the equality

$$\begin{aligned}(\mathrm{DOM})[(G - \mathrm{Na})[\alpha]\ ] &= \{\omega \in G \mid (\mathrm{na})[\mathcal{E}](\omega) \neq \varnothing\} \\ &= G \cap (\mathrm{DOM})[(\mathrm{na})[\mathcal{E}]\ ].\end{aligned}$$

From (6.11.25) and (6.11.26), for some $n \in \mathcal{N}$ we get the inclusion

$$\omega_0 \in G \setminus (\mathrm{DOM})[\gamma_G^n(\alpha)]. \tag{6.11.27}$$

By Corollary 6.8.1 we have

$$(\Gamma^n(\mathcal{E}) \mid G) = \gamma_G^n(\alpha).$$

As a consequence we have the equality

$$\begin{aligned}(\mathrm{DOM})[\gamma_G^n(\alpha)] &= \{\omega \in G \mid \gamma_G^n(\alpha)(\omega) \neq \varnothing\} \\ &= \{\omega \in G \mid \Gamma^n(\mathcal{E})(\omega) \neq \varnothing\} = G \cap (\mathrm{DOM})[\Gamma^n(\mathcal{E})].\end{aligned}$$

Therefore from (6.11.27) we get $\omega_0 \notin (\mathrm{DOM})[\Gamma^n(\mathcal{E})]$. As a consequence

$$G \setminus (\mathrm{DOM})[\Gamma^n(\mathcal{E})] \neq \varnothing.$$

Thus 4) holds. Consequently 2) $\Rightarrow$ 4). □

## 6.12 QUASI-STRATEGIES IN A PROBLEM OF CONTROL WITH INCOMPLETE INFORMATION

We consider an applied problem of control. Namely, we deal with the process of pursuit under incomplete phase information. For simplicity, a particular case is considered. Moreover, in this section we use the informative way of the account. So we admit some nonstrict arguments (for reason of lack of the volume). The singularity of this setting is connected with employment of quasi-strategies as control procedures.

Fix $t_0 \in \mathbb{R}$ and $\vartheta_0 \in ]t_0, \infty[$; let $I_0 \triangleq [t_0, \vartheta_0]$. Fix $n \in \mathcal{N}$; we regard $\mathbb{R}^n$ as the phase space. Two systems $\Sigma_1$ and $\Sigma_2$ function in this phase space on the time interval $I_0$. Suppose that $\Sigma_1$ is controlled by the player I with the control $u(t)$, $t_0 \leq t \leq \vartheta_0$. The player II acts on the system $\Sigma_2$ by the control $v(t)$, $t_0 \leq t \leq \vartheta_0$. The player I strives to seize the system $\Sigma_2$ in some domain of influence of $\Sigma_1$. The aim of the player II is contrary. Suppose that $\Sigma_1$ and $\Sigma_2$ are described by the equations

$$\dot{y} = f(t, y) + B(t)u, \quad u \in P, \tag{6.12.1}$$

$$\dot{z} = g(t, z) + C(t)v, \quad v \in Q, \tag{6.12.2}$$

respectively. We postulate that for some $p \in \mathcal{N}$ and $q \in \mathcal{N}$ the sets $P$ and $Q$ are nonempty convex compacta in the spaces $\mathbb{R}^p$ and $\mathbb{R}^q$, respectively. The functions $f$ and $g$ are required to be continuous on $I_0 \times \mathbb{R}^n$. Moreover, $B(\cdot)$ and $C(\cdot)$ are component-wise continuous matriciants on $I_0$. Finally, $f$ and $g$ are supposed to satisfy the 'local' Lipshitz condition with respect to the second variable; such condition was used, for example, in [87, p.38] and [88]. We supplement this condition by the known condition of sublinear growth. Of course, more general conditions [117] can be used, but we do not consider them now. Denote by $\mathcal{U}$ and $\mathcal{V}$ the sets of all Borel functions from $I_0$ into $P$ and $Q$, respectively; in this connection see Chapter 2. Suppose that $y_0 \in \mathbb{R}^n$ is a given initial state of $\Sigma_1$: $y(t_0) = y_0$. Denote by $\mathbb{Y}_0$ the compactum of all Caratheodory $U$-trajectories of $\Sigma_1$ with the initial condition $y(t_0) = y_0$, $U \in \mathcal{U}$; of course, we use the compactness in the sense of the topology of uniform convergence in the space $C_n(I_0)$ of all continuous $n$-vector functions on $I_0$. In addition, $\mathbb{R}^n$ is equipped with the Euclidean norm.

Fix some nonempty compact set $Z^0$, $Z^0 \subset \mathbb{R}^n$, as the set of all possible initial conditions of the system $\Sigma_2$ (6.12.2). As a result we obtain the compactum $\mathbb{Z}^0$ of Caratheodory $V$-trajectories of $\Sigma_2$ under enumeration of $z_0 \in Z^0$ and $V \in \mathcal{V}$. Certainly, we have the compactum in the sense of the topology of uniform convergence in $C_n(I_0)$. The process of control can be treated as a non-anticipating reaction $y(\cdot) \in \mathbb{Y}_0$ to realization of $z(\cdot) \in \mathbb{Z}^0$. Of course, the character of information about $z(\cdot)$ is very important here.

We define the goal of the first player that constructs a control in $\Sigma_1$. Here and below in this section the symbol $\|\cdot\|$ denotes the Euclidian norm in $\mathbb{R}^n$. We suppose that the goal of the player I consists in realization of the property

$$\{\xi \in I_0 \mid \|y(\xi) - z(\xi)\| \leq \varepsilon\} \neq \varnothing, \tag{6.12.3}$$

where $\varepsilon \in [0, \infty[$ is a given number, $y(\cdot)$ is a trajectory of $\Sigma_1$, and $z(\cdot)$ is a trajectory of $\Sigma_2$. We consider the question of constructing the control procedure guaranteeing (6.12.3) for own trajectories $y(\cdot)$ and arbitrary admissible trajectories $z(\cdot)$ of $\Sigma_2$. We suppose that the domain of influence of $\Sigma_1$ is defined in terms of (6.12.3). In addition, we investigate the case of the following informational situation. Namely, the player I receives a signal depending on the state of $\Sigma_2$. Using the obtained signal, the player I strives to realize (6.12.3).

We will make the above information more precise. Namely, fix $\delta \in [0, \infty[$. Suppose that the possible signal is realized as the sum of the true trajectory $z(\cdot)$ and some $\delta$-bounded continuous hindrance. Determining the set of all possible signals is of important for us. Denote by $\mathbb{B}^0$ the set of all functions

$$\hat{z}(\cdot) = (\hat{z}(t),\ t_0 \leq t \leq \vartheta_0) \in C_n(I_0) \tag{6.12.4}$$

for each of which

$$\exists z(\cdot) \in \mathbb{Z}^0\ \forall t \in I_0 : \|z(t) - \hat{z}(t)\| \leq \delta. \tag{6.12.5}$$

The function (6.12.4) is a signal. The property (6.12.5) characterizes the corresponding informational conditions. Then $\mathbb{B}^0$ is the set of all possible signals. We can use $\mathbb{B}^0$ in place of $\Omega$. Each signal $\hat{z}(\cdot) \in \mathbb{B}^0$ is realized by some choice of $z_0 \in Z^0$, $V \in \mathcal{V}$ and some $\delta$-bounded hindrance of the observation channel. To each $\hat{z} = \hat{z}(\cdot) \in \mathbb{B}^0$ we put into correspondence the set

$$\Lambda[\hat{z}] \triangleq \{z \in \mathbb{Z}^0 \mid \forall t \in I_0 : \|z(t) - \hat{z}(t)\| \leq \delta\}. \qquad (6.12.6)$$

From (6.12.5) we get that (6.12.6) is a nonempty set always. And what is more, from the compactness of $\mathbb{Z}^0$ it follows that for $\hat{z} \in \mathbb{B}^0$ the set $\Lambda[\hat{z}]$ is a nonempty compactum in $C_n(I_0)$ with the topology of uniform convergence. Indeed, from the triangle inequality (see (2.7.2)) we infer that (6.12.6) is a closed set in the above topology. Since $\mathbb{Z}^0$ is a compactum, we have the required property of (6.12.6).

Now we can consider a convenient concrete variant of the general definitions of Section 6.3. Suppose that $X = I_0$, $Y = \Upsilon = \mathbb{R}^n$, $\mathcal{X} = \{[t_0, t] : t \in I_0\}$, $\Omega = \mathbb{B}^0$, $Z = \mathbb{Y}_0$. Introduce the goal set-valued mapping $\alpha^0$. So we consider a concrete variant of the mapping from $\mathbb{M}(\Omega, Z)$, setting that $\alpha^0 \in \mathbb{M}(\Omega, Z)$ is defined by the following natural rule:

$$\alpha^0(\omega) \triangleq \{y \in \mathbb{Y}_0 \mid \forall z \in \Lambda[\omega]\ \exists t \in I_0 : \|y(t) - z(t)\| \leq \varepsilon\}\ (\omega \in \mathbb{B}^0). \qquad (6.12.7)$$

**Remark.** It is useful to discuss the sense of (6.12.7). Let $\omega_0 \in \Omega$ have the property $\alpha^0(\omega_0) = \varnothing$. Then

$$\forall y \in \mathbb{Y}_0\ \exists z \in \Lambda[\omega_0]\ \forall t \in I_0 : \varepsilon < \|y(t) - z(t)\|. \qquad (6.12.8)$$

Let $\beta$ be a non-anticipating mapping (see (6.3.3)) such that for each $z \in \mathbb{Z}^0$, $\delta$-bounded continuous hindrance

$$\xi : I_0 \to \mathbb{R}^n$$

and $y \in \beta(z + \xi)$, where $z + \xi \triangleq (z(t) + \xi(t))_{t \in I_0}$, the property

$$\{t \in I_0 : \|y(t) - z(t_0)\| \leq \varepsilon\} \neq \varnothing \qquad (6.12.9)$$

holds. Recall that $\beta$ is a function from $\Omega = \mathbb{B}^0$ into $\mathcal{P}(Z) = \mathcal{P}(\mathbb{Y}_0)$. Then $\beta(\omega_0) = \varnothing$. Indeed, by contradiction, let $\beta(\omega_0) \neq \varnothing$. Choose arbitrarily $y_0 \in \beta(\omega_0)$. Then $y_0 \in \mathbb{Y}$, and by (6.12.8) one can choose $z_0 \in \Lambda[\omega_0]$ for which

$$\forall t \in I_0 : \varepsilon < \|y_0(t) - z_0(t)\|. \qquad (6.12.10)$$

From (6.12.6) we get that $z_0 \in \mathbb{Z}^0$ and the function $\zeta_0$ from $I_0$ into $\mathbb{R}^n$ with the values

$$\zeta_0(t) = \omega_0(t) - z_0(t) \quad (t \in I_0)$$

has the property that $\forall t \in I_0$: $\|\zeta_0(t)\| \leq \delta$. Then

$$\omega_0 = z_0 + \zeta_0 = (\omega_0(t) + \zeta_0(t))_{t \in I_0},$$

and by (6.12.9) we have a contradiction with (6.12.10) for $y_0$. This contradiction shows that $\beta(\omega_0) = \varnothing$. So under

$$\{\omega \in \Omega \mid \alpha^0(\omega) = \varnothing\} \neq \varnothing$$

it is impossible to find the non-degenerate quasi-strategy solving the problem with the goal defined by (6.12.3). □

Returning to (6.12.7), we note that $\alpha^0$ is a compact-valued mapping; of course, we mean the compactness in the sense of the topology of uniform convergence. This compactness is equivalent to the corresponding sequential compactness (see(2.7.43)). Moreover, in connection with the uniform convergence we note Proposition 2.7.1; besides, it is useful to keep in mind (2.7.52).

Now we use the usual topology of coordinate-wise convergence in $\mathbb{R}^n$ in place of $\tau \in (\text{top})[Y]$ (or $\tau \in (\text{top})[\mathbb{R}^n]$). Then such version of the topology $\theta$ (6.10.2) is weaker than the above topology of uniform convergence of $\mathbb{Y}_0$, $\mathbb{Y}_0 \subset C_n(I_0)$. But for $\omega \in \Omega$ the set $\alpha^0(\omega)$ (6.12.7) is compact in the sense of the uniform topology of $\mathbb{Y}_0$; we use the known transitivity property for the operation of passing to subspace of TS (see Section 2.3). As a result, $\alpha^0 \in \mathbb{K}^\Omega$ in the sense of Section 6.10. Therefore by Corollary 6.10.3 we have

$$(\text{na})[\alpha^0] = \Gamma^\infty(\alpha^0) \in \mathbb{K}^\Omega. \tag{6.12.11}$$

Of course, in (6.12.11) we use the natural definition of hereditary mapping corresponding to (6.3.3). But in this concrete case it is useful to keep in mind Proposition 6.3.1; see (6.3.19) and (6.4.4) in this connection. We note only that, in this case, for $\alpha \in \mathbb{M}(\Omega, Z)$ the property $\alpha \in \mathbb{N}$ holds iff $\forall \omega_1 \in \mathbb{B}^0\ \forall \omega_2 \in \mathbb{B}^0\ \forall t \in I_0$:

$$((\omega_1 \mid [t_0, t]) = (\omega_2 \mid [t_0, t]))$$

$$\Rightarrow (\{(y \mid [t_0, t]) : y \in \alpha(\omega_1)\} = \{(y \mid [t_0, t]) : y \in \alpha(\omega_2)\}).$$

In connection with (6.12.11) we note that the properties considered in Theorem 6.11.4 are very important. The property $(\text{DOM})[(\text{na})[\alpha^0]] = \Omega$ is very essential for us, because only this condition provides some 'real' set-valued quasi-strategy. We do not consider other formalizations of the control problems with incomplete phase information (see [83, 88, 91] and many other publications). The construction of this section can be considered only as an example of application of the general scheme of Sections 6.3–6.11.

## 6.13 CONCLUSION

Different formalizations of the game control are used in the theory of differential games [77, 84, 87, 88, 112]. One of them is the Krasovskii formalization, which corresponds to the idea of feedback control. In the framework of this formalization Krasovskii and Subbotin established an important theorem on the alternative in a positional differential pursuit-evasion game. Another formalization (see [70, 108] and other) is based on the representation of control procedures as non-anticipating responses. If some natural conditions of information consistency are satisfied, these two formalizations became equivalent with respect to the result. In this case multi-valued analogues [6, 7, 8, 14, 85, 86, 112] are used along with 'conventional' (point to point) quasi-strategies. These multi-valued quasi-strategies admit the possibility of choosing a non-anticipating selector [13]. The MPI, which are known in the theory of differential games [5]–[8], [62], allows us not only to construct the solution of a positional differential game, but gives (by a known rule) multi-valued quasi-strategies. However, we need first to construct the function of the value of the game or, in another version of the MPI, the stable bridge in the Krasovskii sense. Constructing, by means of MPI, multi-valued quasi-strategies that solve the control problem with noise was realized quite recently [36, 37, 42, 48, 50, 60, 61]. A systematic presentation of this version of MPI is given in Chapter 6. Conditions of the compactness-valued type for multi-valued mappings, which describe the goal of a problem, play an important role here.

# Chapter 7

# AN EXTENSION CONSTRUCTION FOR SET-VALUED QUASI-STRATEGIES

## 7.1 INTRODUCTION

In this chapter we consider a concrete question of extension. Roughly speaking, this question consists in the following. We have a non-anticipating procedure acting in spaces of 'usual' controls. Moreover, some spaces of generalized 'controls' are given. It is required to construct a new non-anticipating procedure acting in the spaces of generalized controls and 'closing', in some natural sense, the initial non-anticipating procedure. In addition, the space of usual controls admits dense imbedding in the space of generalized elements.

When constructing the above 'extended' non-anticipating mapping we strive to realize some 'correctness' in the sense similar to that of the constructions of Chapter 3. We consider the case of the simplest 'usual' controls defined as graduated non-negative functions. In addition, we use the traditional resource constraint on the choice of these 'usual' controls (see Section 4.9). Generalized elements are defined as non-negative FAM with some special property (weak absolute continuity). Of course, for such FAM we introduce an analogue of the resource constraint too. We realize some effects of the compactification type. Recall (see Sections 6.10 and 6.11) that compact-valued mappings play the important role in the constructions of non-anticipating set-valued mappings. In this sense the subsequent constructions are connected with those of Chapter 6 too.

## 7.2 A VARIANT OF EXTENSION OF SET-VALUED MAPPINGS

We consider the following hypothetical scheme of extension of set-valued mappings. We do not strive to the greatest generality. Conversely, we use only a natural analogue of the corresponding procedures of Chapter 3.

Suppose in this section that $\mathbf{S}[X \neq \varnothing]$, $\mathbf{S}[Y \neq \varnothing]$ and $\varphi \in \mathbb{M}(X, Y)$, i.e.,

$$\varphi : X \to \mathcal{P}(Y). \tag{7.2.1}$$

So we have some 'usual' set-valued mapping. Let $\tau \in (\text{top})[Y]$ and $\mathcal{X} \in \mathcal{B}[X]$. Then the following equality holds:

$$\begin{aligned} \bigcap_{U \in \mathcal{X}} \text{cl}\Big(\bigcup_{x \in U} \varphi(x), \tau\Big) = \{y \in Y \mid \exists_T \, \mathbf{S}[T \neq \varnothing] \; \exists \preceq \in (\text{DIR})[T] \; \exists f \in X^T : \\ (\mathcal{X} \subset (X - \text{ass})[T; \preceq; f]) \\ \& \; (\exists g \in \prod_{t \in T} (\varphi \circ f)(t) : (T, \preceq, g) \overset{\tau}{\to} y)\}. \end{aligned} \tag{7.2.2}$$

**Proof of (7.2.2).** Denote by $\mathbf{Y}_1$ and $\mathbf{Y}_2$ the sets on the left hand and right hand sides of (7.2.2), respectively. Let $y_* \in \mathbf{Y}_1$. Then

$$\forall A \in \mathcal{X} \; \forall B \in N_\tau(y_*) : \left(\bigcup_{x \in A} \varphi(x)\right) \cap B \neq \varnothing. \tag{7.2.3}$$

Consider the nonempty set $\mathcal{X} \times N_\tau(y_*)$ with the natural direction $\ll$ of the product of $\mathcal{X}$ and $N_\tau(y_*)$. Namely, $\ll \in (\text{DIR})[\mathcal{X} \times N_\tau(y_*)]$ for which $\forall A_1 \in \mathcal{X} \; \forall B_1 \in N_\tau(y_*) \; \forall A_2 \in \mathcal{X} \; \forall B_2 \in N_\tau(y_*)$:

$$((A_1, B_1) \ll (A_2, B_2)) \Leftrightarrow ((A_2 \subset A_1) \& (B_2 \subset B_1)). \tag{7.2.4}$$

In (7.2.4) we have the usual product of two directions. From (7.2.3) we have $\forall A \in \mathcal{X} \; \forall B \in N_\tau(y_*)$:

$$\Pi(A, B) \triangleq \{x \in A \mid \varphi(x) \cap B \neq \varnothing\} \in 2^A.$$

Consider the following set-valued mapping:

$$(A, B) \mapsto \Pi(A, B) : \mathcal{X} \times N_\tau(y_*) \to 2^X.$$

By the axiom of choice we find that the product of all sets $\Pi(A, B)$, $(A, B) \in \mathcal{X} \times N_\tau(y_*)$, is a nonempty subset of $X^{\mathcal{X} \times N_\tau(y_*)}$. Let

$$\tilde{f} \in \prod_{(A,B) \in \mathcal{X} \times N_\tau(y_*)} \Pi(A, B).$$

Then $(\mathcal{X} \times N_\tau(y_*), \ll, \tilde{f})$ is a net in $X$. In addition, we have $\forall A \in \mathcal{X}$ $\forall B \in N_\tau(y_*)$:

$$(\varphi \circ \tilde{f})(A, B) \cap B \neq \varnothing. \tag{7.2.5}$$

Moreover, from the choice of $\tilde{f}$ we get $\forall A \in \mathcal{X}$ $\forall B \in N_\tau(y_*)$: $\tilde{f}(A, B) \in A$. From (7.2.4) we infer that $\forall A \in \mathcal{X}$ $\exists z_1 \in \mathcal{X} \times N_\tau(y_*)$ $\forall z_2 \in \mathcal{X} \times N_\tau(y_*)$:

$$(z_1 \ll z_2) \Rightarrow (\tilde{f}(z_2) \in A). \tag{7.2.6}$$

From (2.2.47) and (7.2.6) we conclude that

$$\mathcal{X} \subset (X - \text{ass})[\mathcal{X} \times N_\tau(y_*); \ll; \tilde{f}]. \tag{7.2.7}$$

Returning to (7.2.5) and using the axiom of choice we get the property

$$\prod_{(A,B) \in \mathcal{X} \times N_\tau(y_*)} ((\varphi \circ \tilde{f})(A, B) \cap B) \neq \varnothing. \tag{7.2.8}$$

Choose an element $g$ of the set on the left hand side of (7.2.8). By (7.2.8), $g$ is a selector of the set-valued mappings

$$(A, B) \mapsto (\varphi \circ \tilde{f})(A, B) \cap B : \mathcal{X} \times N_\tau(y_*) \to 2^Y.$$

Therefore $g \in Y^{\mathcal{X} \times N_\tau(y_*)}$ has the following property: for $A \in \mathcal{X}$ and $B \in N_\tau(y_*)$,

$$g(A, B) = g((A, B)) \in \varphi(\tilde{f}(A, B)) \cap B.$$

In particular, for $A \in \mathcal{X}$ and $B \in N_\tau(y_*)$ we have $g(A, B) \in B$. As a result we get the convergence

$$(\mathcal{X} \times N_\tau(y_*), \ll, g) \xrightarrow{\tau} y_*. \tag{7.2.9}$$

By the choice of $g$ we have

$$g \in \prod_{(A,B) \in \mathcal{X} \times N_\tau(y_*)} (\varphi \circ \tilde{f})(A, B). \tag{7.2.10}$$

From (7.2.7) and (7.2.9) we conclude that $y_* \in \mathbf{Y}_2$. So $\mathbf{Y}_1 \subset \mathbf{Y}_2$. Choose arbitrarily $y^* \in \mathbf{Y}_2$. Then $y^* \in Y$ and for some net $(T \preceq, f)$ in $X$ we have

$$(\mathcal{X} \subset (X - \text{ass})[T; \preceq; f]) \& (\exists g \in \prod_{t \in T} (\varphi \circ f)(t) : (T, \preceq, g) \xrightarrow{\tau} y^*)\}. \tag{7.2.11}$$

Using (7.2.11) we choose $g \in \prod_{t \in T} (\varphi \circ f)(t)$ with the property

$$(T, \preceq, g) \xrightarrow{\tau} y^*. \tag{7.2.12}$$

Moreover, from (7.2.11) we get that if $\tilde{U} \in \mathcal{X}$, then for some $t^* \in T$

$$\forall t \in T : (t^* \preceq t) \Rightarrow (f(t) \in \tilde{U}). \tag{7.2.13}$$

Here we use (2.2.47) again. As a result, from (7.2.13) we have

$$\forall t \in T : (t^* \preceq t) \Rightarrow \left(g(t) \in \bigcup_{x \in \tilde{U}} \varphi(x)\right). \tag{7.2.14}$$

From (2.3.11), (7.2.12), and (7.2.14) we get

$$y^* \in \mathrm{cl}\left(\bigcup_{x \in \tilde{U}} \varphi(x), \tau\right).$$

Since the choice of $\tilde{U}$ was arbitrary, the inclusion $y^* \in \mathbf{Y}_1$ holds. So we infer that $\mathbf{Y}_2 \subset \mathbf{Y}_1$. Hence $\mathbf{Y}_1 = \mathbf{Y}_2$, and (7.2.2) is established. □

We use (7.2.2) in extension constructions. But for these constructions we realize a very concrete form oriented towards our problem of extension of the set-valued quasi-strategy under resource constraints. In this connection we exploit some notions of Chapter 4, supposing that (in Chapter 4) $E$ is an interval of the real line $\mathbb{R}$.

In the sequel we fix $t_0 \in \mathbb{R}$ and $\vartheta_0 \in ]t_0, \infty[$. We suppose that $I \triangleq [t_0, \vartheta_0[$ and $T_0 \triangleq ]t_0, \vartheta_0[$ and get two nonempty subsets of $\mathbb{R}$. We use $I$ in place of $E$ from Chapter 4 and the interval $T_0$ to construct a concrete variant of the family $\mathcal{X}$ from the previous chapter.

Fix $\mathcal{L} \in \Pi[I]$ with the following property:

$$\forall t \in I \ \forall \tau \in T_0 : [t, \tau[ \in \mathcal{L}. \tag{7.2.15}$$

From (7.2.15) it follows that the measurable space $(I, \mathcal{L})$ with the semialgebra of sets has the property that $\mathcal{L}$ includes the 'pointer' corresponding to the interval $I$. Fix

$$\eta \in (\mathrm{add})_+[\mathcal{L}]. \tag{7.2.16}$$

**Remark 7.2.1.** Of course, the most natural case of (7.2.16) is connected with the supposition that $\eta$ is the restriction (on $\mathcal{L}$) of the Lebesgue measure. But we admit a more general case. We postulate only (7.2.16), although this very general case is not natural from the point of view of the theory of impulse control.

So $(I, \mathcal{L}, \eta)$ is the finitely additive space with measure. Introduce the cone $(\mathrm{add})^+[\mathcal{L}; \eta]$ (see Section 4.9) of all FAM $\mu \in (\mathrm{add})_+[\mathcal{L}]$ such that $\mu(L) = 0$ for $L \in \mathcal{L}$ with $\eta(L) = 0$. We use the imbedding $J$ of the type

$$f \mapsto f * \eta : B(I, \mathcal{L}) \to \mathbf{A}_\eta[\mathcal{L}]. \tag{7.2.17}$$

But for our goals some restriction of (7.2.17) is sufficient. We exploit $\tau_*(\mathcal{L}) \in (\text{top})_0[\mathbf{A}(\mathcal{L})]$ and $\tau_0(\mathcal{L}) \in (\text{top})_0[\mathbf{A}(\mathcal{L})]$ (see Chapter 4).

If $c \in [0, \infty[$, then we suppose

$$(M_c^+(\mathcal{L}) \triangleq \{f \in B_0^+(I, \mathcal{L}) \mid \int_I f\, d\eta \le c\})$$
$$\& \ (\Xi_c^+(\mathcal{L}) \triangleq \{\mu \in (\text{add})^+[\mathcal{L}; \eta] \mid \mu(I) \le c\}), \tag{7.2.18}$$

obtaining two nonempty sets. In Chapter 4 we have established that

$$\Xi_c^+(\mathcal{L}) = \text{cl}(\{f{*}\eta : f \in M_c^+(\mathcal{L})\}, \tau_*(\mathcal{L})) = \text{cl}(\{f{*}\eta : f \in M_c^+(\mathcal{L})\}, \tau_0(\mathcal{L})). \tag{7.2.19}$$

From (7.2.18) and (7.2.19) we see that for our goals it is sufficient to use the mappings of the type $(J \mid M_c^+(\mathcal{L}))$.

In the sequel we fix $a_0 \in ]0, \infty[$ and $b_0 \in ]0, \infty[$. We consider the cases when, in (7.2.18) and (7.2.19), $c = a_0$ and $c = b_0$. Then we have the two concrete imbedding $(J \mid M_{a_0}^+(\mathcal{L}))$ and $(J \mid M_{b_0}^+(\mathcal{L}))$. In addition, $M_{a_0}^+(\mathcal{L})$ and $M_{b_0}^+(\mathcal{L})$ play the role of two spaces of usual controls. Respectively, $\Xi_{a_0}^+(\mathcal{L})$ and $\Xi_{b_0}^+(\mathcal{L})$ play the role of two spaces of generalized controls. Suppose that $I_0 \triangleq [t_0, \vartheta_0]$.

**Remark 7.2.2.** Note the following particular case. Fix $n \in \mathcal{N}$ and consider $\mathbb{R}^n$ as the phase space. Let the linear system

$$\dot{x}(t) = A(t)x(t) + f(t)b_1(t) + g(t)b_2(t) \tag{7.2.20}$$

function on the time interval $I_0$. The initial condition $x(t_0) = x_0 \in \mathbb{R}^n$ is given. Let $A(\cdot)$ be a continous $n \times n$ matriciant on $I_0$. Suppose that $b_1 = b_1(\cdot)$ and $b_2 = b_2(\cdot)$ are bounded $n$-vector functions on $I$. In addition, let $b_1(\cdot)$ and $b_2(\cdot)$ be uniform limits of some sequences of piece-wise constant and continuous from the right $n$-vector functions on $I$. The functions $f = f(\cdot)$ and $g = g(\cdot)$ are supposed to be non-negative piece-wise constant and continuous from the right and to satisfy the conditions

$$\left(\int_{t_0}^{\vartheta_0} f(t)dt \le a_0\right) \& \left(\int_{t_0}^{\vartheta_0} g(t)dt \le b_0\right). \tag{7.2.21}$$

We consider $f$ and $g$ as controls of players I and II, respectively. Introduce $\mathcal{L}$ in the form $\{[p, q[ : p \in I, q \in I_0\} \in \Pi[I]$. Let $\Phi = \Phi(\cdot, \cdot)$ be the matriciant of the homogeneous system $\dot{x}(t) = A(t)x(t)$. In this example we define $\eta$ as the length: $\eta([p, q[) = q - p$ under $t_0 \le p \le q \le \vartheta_0$. By analogy with [32, p.136], (7.2.20) is supplemented by the following generalized 'system':

$$\tilde{x}(t,\mu,\nu) \triangleq \Phi(t,t_0)x_0 + \int_{[t_0,t[} \Phi(t,\xi)b_1(\xi)\mu(d\xi) + \int_{[t_0,t[} \Phi(t,\xi)b_2(\xi)\nu(d\xi), \tag{7.2.22}$$

where $t \in I_0$, $\mu \in \Xi^+_{a_0}(\mathcal{L})$ and $\nu \in \Xi^+_{b_0}(\mathcal{L})$. In addition, the solution of (7.2.20) (for which $x(t_0) = x_0$) is a particular case of (7.2.22): it is required to use the indefinite $\eta$-integrals of $f$ and $g$ in place of $\mu$ and $\nu$. We consider the possibility of the choice $f = \alpha(g)$ in (7.2.20). In addition, it is logical to require $\alpha$ to be a non-anticipating mapping. The corresponding reaction $\mu = \tilde{\alpha}(\nu)$ is used in (7.2.22). We strive to construct $\tilde{\alpha}$ in the class of non-anticipating mappings too. Moreover, we want to realize $\tilde{\alpha}$ compatible with $\alpha$. So we consider the 'usual' and 'generalized' quasi-strategies. □

Fix the mapping

$$\alpha_0 : M^+_{b_0}(\mathcal{L}) \to 2^{M^+_{a_0}(\mathcal{L})}. \tag{7.2.23}$$

We consider (7.2.23) as some controlling reaction to the hindrance realizations. In the theory of differential games such reactions were called pseudo-strategies [70]. In addition (see Remark 7.2.2), a function $f \in M^+_{a_0}(\mathcal{L})$ is regarded as the usual control. If $f \in \alpha_0(g)$ with $g \in M^+_{b_0}(\mathcal{L})$, then we exploit the mapping (7.2.23) for constructing useful controls. Henceforth we use the approach of [31]. With (7.2.23) we connect some set-valued mapping operating on the space of FAM.

First we introduce the mapping

$$\Lambda : \mathcal{P}(\mathbf{A}(\mathcal{L})) \to \mathcal{P}(M^+_{b_0}(\mathcal{L})) \tag{7.2.24}$$

defined by the following rule: for $H \in \mathcal{P}(\mathbf{A}(\mathcal{L}))$,

$$\Lambda(H) \triangleq M^+_{b_0}(\mathcal{L}) \cap J^{-1}(H). \tag{7.2.25}$$

In (7.2.24), (7.2.25) we have the set to set mapping, which is used for constructing the generalized reaction. Introduce the following new mapping

$$\Lambda^*_0 : \mathcal{P}(\mathbf{A}(\mathcal{L})) \to \mathcal{P}(M^+_{a_0}(\mathcal{L})), \tag{7.2.26}$$

setting for $H \in \mathcal{P}(\mathbf{A}(\mathcal{L}))$

$$\Lambda^*_0(H) \triangleq \bigcup_{f \in \Lambda(H)} \alpha_0(f). \tag{7.2.27}$$

In (7.2.26), (7.2.27) we use $\alpha_0$ for some fuzzy arguments processing. But in reality we exploit (7.2.26), (7.2.27) under the condition when a neighborhood of some generalized element is used in place of $H$. Now introduce the generalized point to set mapping

$$A : \Xi^+_{b_0}(\mathcal{L}) \to \mathcal{P}(\Xi^+_{a_0}(\mathcal{L})), \tag{7.2.28}$$

assuming that for each $\nu \in \Xi^+_{b_0}(\mathcal{L})$ the set $A(\nu)$ is defined as follows:

$$A(\nu) \triangleq \bigcap_{H \in N_{\tau_0(\mathcal{L})}(\nu)} \mathrm{cl}(J^1(\Lambda^*_0(H)), \tau_*(\mathcal{L})). \tag{7.2.29}$$

In (7.2.28), (7.2.29) we use the two topologies of $\mathbf{A}(\mathcal{L})$: $\tau_0(\mathcal{L})$ and $\tau_*(\mathcal{L})$. But in fact, from (7.2.29) we have in (7.2.28) action of the corresponding closure operator for comparable subspaces. In this connection see Chapter 4. We now note only the following aspect. In fact, in (7.2.29) we use the $\alpha_0$-reactions for usual solutions near to the corresponding generalized elements in the sense of $\tau_0(\mathcal{L})$. This conclusion is connected with (7.2.27). It is worth noting the following proposition.

PROPOSITION 7.2.1 $\forall \nu \in \Xi^+_{b_0}(\mathcal{L}) : A(\nu) \neq \varnothing$.

PROOF. Fix $\nu \in \Xi^+_{b_0}(\mathcal{L})$. Then by (7.2.17) and (7.2.19) we have

$$\forall H \in N_{\tau_0(\mathcal{L})}(\nu) : J^1(M^+_{b_0}(\mathcal{L})) \cap H \neq \varnothing.$$

From this we have the statement that $\forall H \in N_{\tau_0(\mathcal{L})}(\nu)$:

$$M^+_{b_0}(\mathcal{L}) \cap J^{-1}(H) \neq \varnothing.$$

Then from (7.2.25) we get $\forall H \in N_{\tau_0(\mathcal{L})}(\nu) : \Lambda(H) \neq \varnothing$. Thus

$$\forall H \in N_{\tau_0(\mathcal{L})}(\nu) : \Lambda(H) \in 2^{M^+_{b_0}(\mathcal{L})}.$$

As a consequence, from (7.2.27) we have the property that for $H \in N_{\tau_0(\mathcal{L})}(\nu)$ :

$$\Lambda^*_0(H) \in 2^{M^+_{a_0}(\mathcal{L})}.$$

We get

$$\forall H \in N_{\tau_0(\mathcal{L})}(\nu) : \mathrm{cl}(J^1(\Lambda^*_0(H)), \tau_*(\mathcal{L})) \in \mathcal{F}_{\tau_*(\mathcal{L})} \setminus \{\varnothing\}. \tag{7.2.30}$$

Using properties of the image and closure operations we obtain that

$$\mathcal{F}^* \triangleq \{\mathrm{cl}(J^1(\Lambda^*_0(H)), \tau_*(\mathcal{L})) : H \in N_{\tau_0(\mathcal{L})}(\nu)\} \tag{7.2.31}$$

is the filter base of the compactum $\Xi^+_{a_0}(\mathcal{L})$ with the induced $*$-weak topology. In particular, in (7.2.31) we have a centered system of closed sets in a compact space:

$$\mathcal{F}^* \in \mathbf{Z}[\mathcal{F}_\tau],$$

where $\tau = \tau_*(\mathcal{L})\,|_{\Xi^+_{a_0}(\mathcal{L})}$. This property is an obvious corollary of (2.2.20) and (7.2.30). From (2.2.21) we obtain that

$$A(\nu) = \bigcap_{F \in \mathcal{F}^*} F \neq \varnothing;$$

we use (7.2.29). The proof is completed. □

If $t \in T_0$, then $(\mathbf{S}[[t_0, t[\neq \varnothing])\&(\mathbf{S}[[t, \vartheta_0[\neq \varnothing])$. Moreover, $\forall t \in T_0$:

$$\begin{aligned}&(\mathcal{L}_t^{\leftarrow} \triangleq \{\Lambda \in \mathcal{L} \mid \Lambda \subset [t_0, t[\} = \{L \cap [t_0, t[: L \in \mathcal{L}\} = \mathcal{L}\,|_{[t_0,t[})\\ &\&\ (\mathcal{L}_t^{\rightarrow} \triangleq \{\Lambda \in \mathcal{L} \mid \Lambda \subset [t, \vartheta_0[\} = \{L \cap [t, \vartheta_0[: L \in \mathcal{L}\} = \mathcal{L}\,|_{[t,\vartheta_0[}).\end{aligned} \tag{7.2.32}$$

We use (2.3.2) in (7.2.32); for $t \in T_0$ we have the two MS $([t_0, t[, \mathcal{L}_t^{\leftarrow})$ and $([t, \vartheta_0[, \mathcal{L}_t^{\rightarrow})$. It is obvious that $\forall \mu \in (\mathrm{add})_+[\mathcal{L}]\ \forall t \in T_0$:

$$((\mu \mid \mathcal{L}_t^{\leftarrow}) \in (\mathrm{add})_+[\mathcal{L}_t^{\leftarrow}])\&((\mu \mid \mathcal{L}_t^{\rightarrow}) \in (\mathrm{add})_+[\mathcal{L}_t^{\rightarrow}]). \tag{7.2.33}$$

We use (7.2.33) in the subsequent constructions connected with investigation of non-anticipating mappings.

Return to the mapping $A$. By (2.2.6), $\forall \nu \in \Xi^+_{b_0}(\mathcal{L})$:

$$\begin{aligned}\mathcal{X}_\nu &\triangleq \{J^{-1}(H) \cap M^+_{b_0}(\mathcal{L}) : H \in N_{\tau_0(\mathcal{L})}(\nu)\}\\ &= \{\Lambda(H) : H \in N_{\tau_0(\mathcal{L})}(\nu)\} = \Lambda^1(N_{\tau_0(\mathcal{L})}(\nu)).\end{aligned} \tag{7.2.34}$$

Now we consider a representation of $A$ in terms of (7.2.2).

PROPOSITION 7.2.2 $\forall \nu \in \Xi^+_{b_0}(\mathcal{L})$:

$$\begin{aligned}A(\nu) = \{\mu \in \Xi^+_{a_0}(\mathcal{L}) \mid \exists_T\, \mathbf{S}[T \neq \varnothing]\ \exists \preceq \in (\mathrm{DIR})[T]\ \exists l \in M^+_{b_0}(\mathcal{L})^T :\\ (\mathcal{X}_\nu \subset (M^+_{b_0}(\mathcal{L}) - \mathrm{ass})[T; \preceq; l])\\ \&\ (\exists g \in \prod_{t \in T}\{h * \eta : h \in (\alpha_0 \circ l)(t)\} : (T, \preceq, g) \overset{\tau_*(\mathcal{L})}{\to} \mu)\}.\end{aligned}$$

PROOF. Suppose that in (7.2.2) $X = M^+_{b_0}(\mathcal{L})$, $Y = \Xi^+_{a_0}(\mathcal{L})$, and $\varphi$ is the mapping

$$f \mapsto \{g * \eta : g \in \alpha_0(f)\} : M^+_{b_0}(\mathcal{L}) \to 2^{\Xi^+_{a_0}(\mathcal{L})};$$

we use (7.2.23). In other words, for $f \in M^+_{b_0}(\mathcal{L})$ we have $\varphi(f) = J^1(\alpha_0(f))$. Note that $\forall \nu \in \Xi^+_{b_0}(\mathcal{L}) : \mathcal{X}_\nu \in \mathcal{B}[M^+_{b_0}(\mathcal{L})]$. In this statement we use (2.2.6) and (7.2.34). For simplicity we fix $\nu \in \Xi^+_{b_0}(\mathcal{L})$. By

(7.2.27), for $H \in N_{\tau_0(\mathcal{L})}(\nu)$ we have

$$J^1(\Lambda_0^*(H)) = \{g * \eta : g \in \Lambda_0^*(H)\} = \{g * \eta : g \in \bigcup_{f \in \Lambda(H)} \alpha_0(f)\}$$
$$= \bigcup_{f \in \Lambda(H)} \{g * \eta : g \in \alpha_0(f)\} = \bigcup_{f \in \Lambda(H)} \varphi(f).$$

We now use (7.2.25), (7.2.29), and (7.2.34). By (7.2.29) we have

$$\begin{aligned} A(\nu) &= \bigcap_{H \in N_{\tau_0(\mathcal{L})}(\nu)} \mathrm{cl}\Big(\bigcup_{f \in \Lambda(H)} \varphi(f), \tau_*(\mathcal{L})\Big) \\ &= \bigcap_{H \in N_{\tau_0(\mathcal{L})}(\nu)} \mathrm{cl}\Big(\bigcup_{f \in M_{b_0}^+(\mathcal{L}) \cap J^{-1}(H)} \varphi(f), \tau_*(\mathcal{L})\Big) \qquad (7.2.35) \\ &= \bigcap_{U \in \mathcal{X}_\nu} \mathrm{cl}\Big(\bigcup_{f \in U} \varphi(f), \tau_*(\mathcal{L})\Big). \end{aligned}$$

In (7.2.2) we suppose that $\mathcal{X} = \mathcal{X}_\nu$. Moreover, we note the following property: if $\mathbf{S}[T \neq \varnothing]$ and $l \in M_{b_0}^+(\mathcal{L})^T$, then

$$\prod_{t \in T} (\varphi \circ l)(t) = \prod_{t \in T} \{h * \eta : h \in (\alpha_0 \circ l)(t)\}.$$

Then from (7.2.2) and (7.2.35) we get the required representation for $A(\nu)$. □

Thus for values of $A$ we obtain the representation in terms of AS from Chapter 3. We have the natural application of the extension constructions for investigation of non-anticipating mappings considered below. We begin the investigation of some questions connected with non-anticipating mappings.

If $f \in \mathbb{R}^I$, $\theta \in T_0$ and $g \in \mathbb{R}^{[\theta, \vartheta_0[}$, then in this chapter we define $f \Box g \in \mathbb{R}^I$ by the following rule:

$$f \Box g \triangleq \begin{cases} f(t), & t \in [t_0, \theta[, \\ g(t), & t \in [\theta, \vartheta_0[. \end{cases} \qquad (7.2.36)$$

Of course, we can use graduated functions in (7.2.36). For our goals it is sufficient to consider only the following case. Namely, we use in (7.2.36) $f \in B_0(I, \mathcal{L})$ and $g = (h \mid [\theta, \vartheta_0[)$, where $h \in B_0(I, \mathcal{L})$. Under these conditions (including the requirement $\theta \in T_0$) we have $g \in B_0([\theta, \vartheta_0[, \mathcal{L}_\theta^{\rightarrow})$ and $f \Box h \in B_0(I, \mathcal{L})$. Thus $\forall f \in B_0(I, \mathcal{L})\ \forall g \in B_0(I, \mathcal{L})\ \forall \theta \in T_0$

$$f \Box (g \mid [\theta, \vartheta_0[) \in B_0(I, \mathcal{L}). \qquad (7.2.37)$$

In (7.2.37) we get the usual sewing of $\mathcal{L}$-graduated functions. Recall that by properties established in Section 4.9, for $\mu \in (\text{add})^+[\mathcal{L};\eta]$ and $\mathcal{H} \in \mathbf{D}(I,\mathcal{L})$ the function $\Theta_\eta[\mu;\mathcal{H}] \in B_0^+(I,\mathcal{L})$ satisfies the condition

$$\int_I \Theta_\eta[\mu;\mathcal{H}]d\eta = \mu(I). \tag{7.2.38}$$

In (7.2.38) we use the variant of the construction of Section 4.9, for which $E = I$. From (7.2.18) and (7.2.38) we have $\forall c \in [0,\infty[\ \forall \mu \in \Xi_c^+(\mathcal{L})$ $\forall \mathcal{H} \in \mathbf{D}(I,\mathcal{L})$:

$$\Theta_\eta[\mu;\mathcal{H}] \in M_c^+(\mathcal{L}). \tag{7.2.39}$$

Of course, (7.2.39) is supplemented by Theorems 4.9.1 and 4.9.5. From (7.2.36) and (7.2.37) we get $\forall \mu \in (\text{add})^+[\mathcal{L};\eta]\ \forall \nu \in \Xi_c^+(\mathcal{L})\ \forall \mathcal{H} \in \mathbf{D}(I,\mathcal{L})$ $\forall \theta \in T_0$:

$$\Theta_\eta[\mu;\mathcal{H}]\square(\Theta_\eta[\nu;\mathcal{H}] \mid [\theta,\vartheta_0[) \in B_0^+(I,\mathcal{L}).$$

PROPOSITION 7.2.3 *Let $\mu \in (\text{add})^+[\mathcal{L};\eta]$, $\nu \in (\text{add})^+[\mathcal{L};\eta]$, and $\vartheta \in T_0$ satisfy the condition $(\mu \mid \mathcal{L}_\vartheta^{\leftarrow}) = (\nu \mid \mathcal{L}_\vartheta^{\leftarrow})$. Then the operator $\psi \in B_0^+(I,\mathcal{L})^{\mathbf{D}(I,\mathcal{L})}$ of the type*

$$\mathcal{K} \mapsto \Theta_\eta[\mu;\mathcal{K}]\square(\Theta_\eta[\nu;\mathcal{K}] \mid [\vartheta,\vartheta_0[) : \mathbf{D}(I,\mathcal{L}) \to B_0^+(I,\mathcal{L})$$

*has the following property:*

$$\left((\mathbf{D}(I,\mathcal{L}),\prec,J\circ\psi) \overset{\tau_0(\mathcal{L})}{\to} \nu\right)$$

$$\& \left(\exists \mathcal{K}_0 \in \mathbf{D}(I,\mathcal{L}) \forall \mathcal{K}^0 \in \mathbf{D}(I,\mathcal{L}) \colon (\mathcal{K}_0 \prec \mathcal{K}^0) \Rightarrow \left(\int_I \psi(\mathcal{K}^0)d\eta = \nu(I)\right)\right).$$

PROOF. We use Theorem 4.9.1. By (4.9.10) $\forall L \in \mathcal{L}$:

$$(\mu(L) = \theta_\eta[\mu](L)\eta(L))\&(\nu(L) = \theta_\eta[\nu](L)\eta(L)).$$

Recall that (see(4.9.8)) for $\mathcal{K} \in \mathbf{D}(I,\mathcal{L})$ the function $\Theta_\eta[\mu;\mathcal{K}] \in B_0^+(I,\mathcal{L})$ satisfies the following condition: if $L \in \mathcal{K}$, then $\forall x \in L : \Theta_\eta[\mu;\mathcal{K}](x) = \theta_\eta[\mu](L)$. Similarly, for $\mathcal{K} \in \mathbf{D}(I,\mathcal{L})$ the function $\Theta_\eta[\nu;\mathcal{K}] \in B_0^+(I,\mathcal{L})$ satisfies the following condition: if $L \in \mathcal{K}$, then $\forall x \in L : \Theta_\eta[\nu;\mathcal{K}](x) = \theta_\eta[\nu](L)$. We take into account (4.6.18) and (2.6.36). From Theorem 4.9.1 we have the properties

$$\left((\mathbf{D}(I,\mathcal{L}),\prec,\Theta_\eta[\mu;\cdot]*\eta) \overset{\tau_0(\mathcal{L})}{\to} \mu\right) \& \left((\mathbf{D}(I,\mathcal{L}),\prec,\Theta_\eta[\nu;\cdot]*\eta) \overset{\tau_0(\mathcal{L})}{\to} \nu\right). \tag{7.2.40}$$

From (7.2.40) we obtain the property that if $L \in \mathcal{L}$, then $\exists \mathcal{K}_L \in \mathbf{D}(I, \mathcal{L})\ \forall \mathcal{K} \in \mathbf{D}(I, \mathcal{L})$:

$$(\mathcal{K}_L \prec \mathcal{K}) \Rightarrow \Big( ((\Theta_\eta[\mu; \mathcal{K}] * \eta)(L \cap [t_0, \vartheta[) = \mu(L \cap [t_0, \vartheta[)) \\ \& \ ((\Theta_\eta[\nu; \mathcal{K}] * \eta)(L \cap [\vartheta, \vartheta_0[) = \nu(L \cap [\vartheta, \vartheta_0[)) \Big). \tag{7.2.41}$$

Taking into account properties of $\mu$ and $\nu$ and (7.2.41), we get $\forall L \in \mathcal{L}$ $\exists \mathcal{K}_L \in \mathbf{D}(I, \mathcal{L})\ \forall \mathcal{K} \in \mathbf{D}(I, \mathcal{L})$:

$$(\mathcal{K}_L \prec \mathcal{K}) \Rightarrow \Big( ((\Theta_\eta[\mu; \mathcal{K}] * \eta)(L \cap [t_0, \vartheta[) = \nu(L \cap [t_0, \vartheta[)) \\ \& \ ((\Theta_\eta[\nu; \mathcal{K}] * \eta)(L \cap [\vartheta, \vartheta_0[) = \nu(L \cap [\vartheta, \vartheta_0[)) \Big).$$

As a result, for $L \in \mathcal{L}$ we have the following equality from some moment:

$$\nu(L) = (\Theta_\eta[\mu; \mathcal{K}] * \eta)(L \cap [t_0, \vartheta[) + (\Theta_\eta[\nu; \mathcal{K}] * \eta)(L \cap [\vartheta, \vartheta_0[).$$

From (4.5.8) we obtain the following representation of the last equality:

$$\nu(L) = \int_{L \cap [t_0, \vartheta[} \Theta_\eta[\mu; \mathcal{K}] d\eta + \int_{L \cap [\vartheta, \vartheta_0[} \Theta_\eta[\nu; \mathcal{K}] d\eta.$$

Using the definition of $\psi$, (4.4.7), and (4.4.8) we infer that if $L \in \mathcal{L}$, then from a certain moment

$$\nu(L) = \int_L \psi(\mathcal{K}) d\eta = (\psi(\mathcal{K}) * \eta)(L).$$

Then by (7.2.17) we have the statement: $\forall L \in \mathcal{L}\ \exists \mathcal{K}_1 \in \mathbf{D}(I, \mathcal{L})\ \forall \mathcal{K}_2 \in \mathbf{D}(I, \mathcal{L})$:

$$(\mathcal{K}_1 \prec \mathcal{K}_2) \Rightarrow ((J \circ \psi)(\mathcal{K}_2)(L) = \nu(L)).$$

This statement implies the convergence

$$(\mathbf{D}(I, \mathcal{L}), \prec, J \circ \psi) \overset{\tau_0(\mathcal{L})}{\to} \nu; \tag{7.2.42}$$

we use (4.6.18) and the representations of Section 2.6 (see, for example, (2.6.35)). From (7.2.42) we infer, in particular, that $\exists \mathcal{K}' \in \mathbf{D}(I, \mathcal{L})\ \forall \mathcal{K}'' \in \mathbf{D}(I, \mathcal{L})$:

$$(\mathcal{K}' \prec \mathcal{K}'') \Rightarrow ((J \circ \psi)(\mathcal{K}'')(I) = \nu(I)).$$

The following assertion is a natural generalization of Proposition 7.2.3.

PROPOSITION 7.2.4 *Let* $\mu \in (\mathrm{add})^+[\mathcal{L};\eta]$, $\nu \in (\mathrm{add})^+[\mathcal{L};\eta]$, $\vartheta \in T_0$ *and*

$$(\mu \mid \mathcal{L}_\vartheta^{\leftarrow}) = (\nu \mid \mathcal{L}_\vartheta^{\leftarrow}). \tag{7.2.43}$$

*Moreover, suppose that* $(T, \ll, \psi)$ *is a net in* $B_0^+(I,\mathcal{L})$ *for which*

$$(T, \ll, J \circ \psi) \overset{\tau_0(\mathcal{L})}{\to} \mu, \tag{7.2.44}$$

*and* $\tilde{\psi} : T \times \mathbf{D}(I,\mathcal{L}) \to B_0^+(I,\mathcal{L})$ *is defined by the following rule:* $\forall t \in T$ $\forall \mathcal{K} \in \mathbf{D}(I,\mathcal{L})$,

$$\tilde{\psi}(t,\mathcal{K}) \triangleq \psi(t)\Box(\Theta_\eta[\nu;\mathcal{K}] \mid [\vartheta,\vartheta_0[). \tag{7.2.45}$$

*Finally, let* $\preceq \in (\mathrm{DIR})[T \times \mathbf{D}(I,\mathcal{L})]$ *be the product of* $\ll$ *and* $\prec$*: for* $s_1 \in T$, $\mathcal{K}_1 \in \mathbf{D}(I,\mathcal{L})$, $s_2 \in T$ *and* $\mathcal{K}_2 \in \mathbf{D}(I,\mathcal{L})$,

$$((s_1,\mathcal{K}_1) \preceq (s_2,\mathcal{K}_2)) \Leftrightarrow ((s_1 \ll s_2)\&(\mathcal{K}_1 \prec \mathcal{K}_2)). \tag{7.2.46}$$

*Then the following properties hold:*

$$((T \times \mathbf{D}(I,\mathcal{L}), \preceq, J \circ \tilde{\psi}) \overset{\tau_0(\mathcal{L})}{\to} \nu)$$

$$\&(\exists z_0 \in T \times \mathbf{D}(I,\mathcal{L})\ \forall z \in T \times \mathbf{D}(I,\mathcal{L}) : (z_0 \preceq z) \Rightarrow (\int_I \tilde{\psi}(z)d\eta = \nu(I)).$$

PROOF. The basic scheme is analogous to the reasoning of Proposition 7.2.3. We note that (7.2.46) defines a direction in $T \times \mathbf{D}(I,\mathcal{L})$. This property is well known [81, ch. 2]; in this connection see (2.2.4). From (7.2.44) we have $\forall L \in \mathcal{L}\ \exists t_1 \in T\ \forall t_2 \in T$:

$$(t_1 \ll t_2) \Rightarrow ((J \circ \psi)(t_2)(L) = \mu(L)). \tag{7.2.47}$$

In (7.2.47) we use (4.6.18) and the properties established in Section 2.6. By (7.2.17) we obtain the property that if $L \in \mathcal{L}$, then for some $t_1 \in T$

$$\forall t \in T : (t_1 \ll t) \Rightarrow ((J \circ \psi)(t_2)(L) = (\psi(t_2) * \eta)(L) = \mu(L)).$$

Using (4.4.8) we obtain that $\forall L \in \mathcal{L}\ \exists \alpha \in T\ \forall \beta \in T$:

$$(\alpha \ll \beta) \Rightarrow \left(\mu(L) = \int_L \psi(\beta)d\eta\right). \tag{7.2.48}$$

In particular, we can use $L \in \mathcal{L}_\vartheta^{\leftarrow}$ in (7.2.48); see (7.2.32). Moreover, we use Theorem 4.9.1 and (4.9.5). In particular, we have the convergence of $(\mathbf{D}(I,\mathcal{L}), \prec, \Theta_\eta[\nu;\cdot] * \eta)$ to the point $\nu$ in the sense of $\tau_0(\mathcal{L})$. Therefore $\forall L \in \mathcal{L}\ \exists \tilde{\mathcal{K}} \in \mathbf{D}(I,\mathcal{L})\ \forall \mathcal{K} \in \mathbf{D}(I,\mathcal{L})$:

$$(\tilde{\mathcal{K}} \prec \mathcal{K}) \Rightarrow \left(\int_L \Theta_\eta[\nu;\mathcal{K}]d\eta = \nu(L)\right). \tag{7.2.49}$$

Note that $L \in \mathcal{L}_\vartheta^{\rightarrow}$ can be used in (7.2.49). Fix $\Lambda \in \mathcal{L}$. Then $\Lambda_1 \triangleq \Lambda \cap [t_0, \vartheta[ \in \mathcal{L}_\vartheta^{\leftarrow}$ and $\Lambda_2 \triangleq \Lambda \cap [\vartheta, \vartheta_0[ \in \mathcal{L}_\vartheta^{\rightarrow}$. In particular (see (7.2.32)), we have $\Lambda_1 \in \mathcal{L}$ and $\Lambda_2 \in \mathcal{L}$. Using (7.2.48) we choose $\alpha' \in T$ for which $\forall \beta \in T$: $(\alpha' \ll \beta) \Rightarrow (\mu(\Lambda_1) = \int_{\Lambda_1} \psi(\beta) d\eta)$. But by (7.2.43) we have $\mu(\Lambda_1) = \nu(\Lambda_1)$ and as a consequence

$$\forall \beta \in T : (\alpha' \ll \beta) \Rightarrow (\nu(\Lambda_1) = \int_{\Lambda_1} \psi(\beta) d\eta). \tag{7.2.50}$$

Using (7.2.49) we choose $\mathcal{K}' \in \mathbf{D}(I, \mathcal{L})$ for which $\forall \mathcal{K} \in \mathbf{D}(I, \mathcal{L})$:

$$(\mathcal{K}' \prec \mathcal{K}) \Rightarrow (\nu(\Lambda_2) = \int_{\Lambda_2} \Theta_\eta[\nu; \mathcal{K}] d\eta). \tag{7.2.51}$$

Since $\Lambda \subset I$, we get $\Lambda_1 \cup \Lambda_2 = \Lambda$ and $\Lambda_1 \cap \Lambda_2 = \varnothing$. Then by (4.2.2) we have $(\Lambda_i)_{i \in \overline{1,2}} \in \Delta_2(\Lambda, \mathcal{L})$. From (4.2.8) it follows that $\forall \omega \in (\text{add})[\mathcal{L}]$:

$$\omega(\Lambda) = \omega(\Lambda_1) + \omega(\Lambda_2). \tag{7.2.52}$$

We use (7.2.52) in the case when $\omega = (J \circ \tilde{\psi})(z)$ for some $z \in T \times \mathbf{D}(I, \mathcal{L})$. Note that $z' \triangleq (\alpha', \mathcal{K}') \in T \times \mathbf{D}(I, \mathcal{L})$. Choose arbitrarily $z'' \in T \times \mathbf{D}(I, \mathcal{L})$ for which $z' \preceq z''$. Then (see (4.10.15) and (4.10.16)) $\alpha'' \triangleq \text{pr}_1(z'') \in T$ and $\mathcal{K}'' \triangleq \text{pr}_2(z'') \in \mathbf{D}(I, \mathcal{L})$. By (7.2.46) we have

$$(\alpha' \ll \alpha'') \& (\mathcal{K}' \prec \mathcal{K}''). \tag{7.2.53}$$

From (7.2.50) and (7.2.53) we get the equality

$$\nu(\Lambda_1) = \int_{\Lambda_1} \psi(\alpha'') d\eta. \tag{7.2.54}$$

From (7.2.51) and (7.2.53) it follows that

$$\nu(\Lambda_2) = \int_{\Lambda_2} \Theta_\eta[\nu; \mathcal{K}''] d\eta. \tag{7.2.55}$$

Taking into account (7.2.52), (7.2.54), and (7.2.55) we obtain

$$\nu(\Lambda) = \int_{\Lambda_1} \psi(\alpha'') d\eta + \int_{\Lambda_2} \Theta_\eta[\nu; \mathcal{K}''] d\eta. \tag{7.2.56}$$

Note that by (7.2.45) we have the equality

$$\tilde{\psi}(z'') = \tilde{\psi}(\alpha'', \mathcal{K}'') = \psi(\alpha'') \square (\Theta_\eta[\nu; \mathcal{K}''] \mid [\vartheta, \vartheta_0[).$$

Therefore, using (7.2.36) and (7.2.56), we obtain the equality

$$\nu(\Lambda) = \int_\Lambda \tilde{\psi}(z'') d\eta = (\tilde{\psi}(z'') * \eta)(\Lambda) = (J \circ \tilde{\psi})(z'')(\Lambda). \tag{7.2.57}$$

Since the choice of $\Lambda$ and $z''$ was arbitrary, we get that $\forall L \in \mathcal{L}\ \exists z_1 \in T \times \mathbf{D}(I, \mathcal{L})\ \forall z_2 \in T \times \mathbf{D}(I, \mathcal{L})$:

$$(z_1 \preceq z_2) \Rightarrow (\nu(L) = (J \circ \tilde{\psi})(z_2)(L)).$$

As a consequence we have the convergence

$$(T \times \mathbf{D}(I, \mathcal{L}), \preceq, J \circ \tilde{\psi}) \overset{\tau_0(\mathcal{L})}{\to} \nu. \tag{7.2.58}$$

Owing to (7.2.58) we have, in particular, $\int_I \tilde{\psi}(z) d\eta = (J \circ \tilde{\psi})(z)(I) = \nu(I)$ from some moment. □

Note that by (7.2.18) $\forall c \in [0, \infty[$:

$$\Xi_c^+(\mathcal{L}) = \{\mu \in (\mathrm{add})^+[\mathcal{L}; \eta] \mid V_\mu \leq c\} = U_c(\mathcal{L}) \cap (\mathrm{add})^+[\mathcal{L}; \eta].$$

Then by Theorem 4.9.5 we obtain $\forall c \in [0, \infty[$:

$$\Xi_c^+(\mathcal{L}) \in \mathcal{F}_{\tau_*(\mathcal{L})}. \tag{7.2.59}$$

In particular, from (2.3.4) and (7.2.59) we have

$$\mathcal{F}^*(\mathcal{L}) \triangleq \mathcal{F}_{\tau_*(\mathcal{L})} \cap 2^{\Xi_{a_0}^+(\mathcal{L})} = \mathcal{F}_{\tau_*(\mathcal{L})} \mid_{\Xi_{a_0}^+(\mathcal{L})} \setminus \{\varnothing\} = \mathcal{F}_\tau \setminus \{\varnothing\} \tag{7.2.60}$$

for $\tau = \tau_*(\mathcal{L}) \mid_{\Xi_{a_0}^+(\mathcal{L})}$. Taking into account (7.2.60) we consider the set $\mathcal{Q}$ of all operators $A'$ from $\Xi_{b_0}^+(\mathcal{L})$ into $\mathcal{F}^*(\mathcal{L})$ such that $\forall \nu \in \Xi_{b_0}^+(\mathcal{L})$ $\forall H_1 \in \mathrm{N}_{\tau_*(\mathcal{L})}[A'(\nu)]\ \exists H_2 \in N_{\tau_0(\mathcal{L})}(\nu)$:

$$\mathrm{cl}\Bigg( \bigcup_{f \in \Lambda(H_2)} \{g * \eta : g \in \alpha_0(f)\}, \tau_*(\mathcal{L}) \Bigg) \subset H_1. \tag{7.2.61}$$

From (2.3.13), (7.2.23), and (7.2.59) we infer that the closure on the left hand side of (7.2.61) is analogous to that in $\Xi_{a_0}^+(\mathcal{L})$ with the induced $*$-weak topology

$$\tau_*(\mathcal{L}) \mid_{\Xi_{a_0}^+(\mathcal{L})} .$$

We consider now the construction of the smallest element of $\mathcal{Q}$ in some natural sense. We use the order defined by $\sqsubseteq$ in the previous chapter. But first we remark that by (7.2.27) $\forall \nu \in \Xi_{b_0}^+(\mathcal{L})\ \forall H \in N_{\tau_0(\mathcal{L})}(\nu)$:

$$\{g * \eta : g \in \Lambda_0^*(H)\} = \bigcup_{f \in \Lambda(H)} \{g * \eta : g \in \alpha_0(f)\}. \tag{7.2.62}$$

Recall that $\forall A_1 \in \mathcal{Q}\ \forall A_2 \in \mathcal{Q}$:

$$(A_1 \sqsubseteq A_2) \Leftrightarrow (\forall \nu \in \Xi_{b_0}^+(\mathcal{L}) : A_1(\nu) \subset A_2(\nu)). \tag{7.2.63}$$

Some ordered space is defined in terms of (7.2.63). We connect the definition of $A$ with (7.2.63).

LEMMA 7.2.1 *The property $A \in Q$ holds.*

PROOF. Fix $\nu \in \Xi^+_{b_0}(\mathcal{L})$. Then by Proposition 7.2.1 and (7.2.28) we have $A(\nu) \in 2^{\Xi^+_{a_0}(\mathcal{L})}$. Fix $H_1 \in \mathbb{N}_{\tau_*(\mathcal{L})}[A(\nu)]$. Choose $G_1 \in \mathbb{N}^0_{\tau_*(\mathcal{L})}[A(\nu)]$ with the property $G_1 \subset H_1$ (see (2.2.13)). Introduce the mapping

$$\Omega : N_{\tau_0(\mathcal{L})}(\nu) \to \mathcal{P}(\mathbf{A}(\mathcal{L}))$$

for which $\forall H \in N_{\tau_0(\mathcal{L})}(\nu)$:

$$\Omega(H) \triangleq \mathrm{cl}(\{g * \eta : g \in \Lambda_0^*(H)\}, \tau_*(\mathcal{L})) = \mathrm{cl}(J^1(\Lambda_0^*(H)), \tau_*(\mathcal{L})). \quad (7.2.64)$$

Using (7.2.23), (7.2.27), and (7.2.59) we obtain that

$$\Omega : N_{\tau_0(\mathcal{L})}(\nu) \to \mathcal{F}^*(\mathcal{L}).$$

And what is more, $\Omega$ is a monotone compact-valued mapping. From (2.2.46) we infer that

$$N_{\tau_0(\mathcal{L})}(\nu) \in \mathbb{F}^*(\mathcal{P}(\mathbf{A}(\mathcal{L}))).$$

As a consequence we have the property:

$$\mathfrak{J} \triangleq \Omega^1(N_{\tau_0(\mathcal{L})}(\nu)) \in \mathcal{B}_0[\Xi^+_{a_0}(\mathcal{L})].$$

Moreover, from (7.2.29) and (7.2.64) we obtain that $A(\nu) \in 2^{\Omega(H)}$, $H \in N_{\tau_0(\mathcal{L})}(\nu)$. Taking into account the properties of filter bases we get $\forall \mathcal{K} \in \mathrm{Fin}(\mathfrak{J})\ \exists \Gamma \in \mathfrak{J}$:

$$\Gamma \subset \bigcap_{H \in \mathcal{K}} H. \quad (7.2.65)$$

Recall that by (7.2.29) and (7.2.64) we have the property

$$A(\nu) = \bigcap_{H \in N_{\tau_0(\mathcal{L})}(\nu)} \Omega(H). \quad (7.2.66)$$

So $A(\nu)$ is the intersection of all sets of $\mathfrak{J}$. Note that from the choice of $G_1$ we have (see (7.2.65))

$$\bigcap_{H \in \mathfrak{J}} H \subset G_1. \quad (7.2.67)$$

In addition, $\mathfrak{J} \subset (\tau_*(\mathcal{L}) - \mathrm{comp})[\mathbf{A}(\mathcal{L})]$. Taking into account (7.2.65) and (7.2.67), we get the inclusion $\Gamma \subset G_1$ for a certain $\Gamma \in \mathfrak{J}$. We used such properties in Chapter 3 (see [88, Sect. 3.1]). By definition of $\mathfrak{J}$ (see (7.2.64)) we get that

$$\exists \hat{H} \in N_{\tau_0(\mathcal{L})}(\nu) : \mathrm{cl}\Big(\{g * h : g \in \Lambda_0^*(\hat{H})\}, \tau_*(\mathcal{L})\Big) \subset G_1 \subset H_1.$$

So by (7.2.62) we obtain

$$\exists \hat{H} \in N_{\tau_0(\mathcal{L})}(\nu) : \mathrm{cl}\Big(\bigcup_{f \in \Lambda(\hat{H})} \{g * h : g \in \alpha_0(f)\}, \tau_*(\mathcal{L})\Big) \subset H_1.$$

Since the choice of $\nu$ and $H_1$ was arbitrary, the property $A \in \mathcal{Q}$ is established. □

LEMMA 7.2.2 $\forall \tilde{A} \in \mathcal{Q} : A \sqsubseteq \tilde{A}$.

PROOF. Fix $\tilde{A} \in \mathcal{Q}$. Then we have

$$\tilde{A} : \Xi_{b_0}^+(\mathcal{L}) \to \mathcal{F}^*(\mathcal{L}).$$

In addition, (7.2.61) is valid for $A' = \tilde{A}$.

Fix $\nu \in \Xi_{b_0}^+(\mathcal{L})$. Recall that TS (4.6.9) has the regularity property:

$$\tau_*(\mathcal{L}) \in (\mathrm{top})^0[\mathbf{A}(\mathcal{L})]$$

(see (2.2.18), (4.6.1), (4.6.6), and (4.6.7)). Therefore by (7.2.60) we have

$$\forall \mu \in \mathbf{A}(\mathcal{L}) \setminus \tilde{A}(\nu)\ \exists \hat{H}_1 \in N_{\tau_*(\mathcal{L})}(\mu)\ \forall \hat{H}_2 \in \mathbb{N}_{\tau_*(\mathcal{L})}[\tilde{A}(\nu)] : \hat{H}_1 \cap \hat{H}_2 = \varnothing. \tag{7.2.68}$$

Let us show that $A(\nu) \subset \tilde{A}(\nu)$. Indeed, let by contradiction $A(\nu) \setminus \tilde{A}(\nu) \neq \varnothing$. Choose $\mu_* \in A(\nu) \setminus \tilde{A}(\nu)$. By (7.2.68), for some neighborhoods

$$(\hat{H}_1^* \in N_{\tau_*(\mathcal{L})}(\mu_*))\&(\hat{H}_2^* \in \mathbb{N}_{\tau_*(\mathcal{L})}[\tilde{A}(\nu)])$$

we have the property $\hat{H}_1^* \cap \hat{H}_2^* = \varnothing$. We use the property $\tilde{A} \in \mathcal{Q}$. In this connection we recall that $\nu \in \Xi_{b_0}^+(\mathcal{L})$ and $\hat{H}_2^* \in \mathbb{N}_{\tau_*(\mathcal{L})}[\tilde{A}(\nu)]$. Then for some $H_2^* \in N_{\tau_0(\mathcal{L})}(\nu)$

$$\mathrm{cl}\Big(\bigcup_{f \in \Lambda(H_2^*)} \{g * h : g \in \alpha_0(f)\}, \tau_*(\mathcal{L})\Big) \subset \hat{H}_2^*. \tag{7.2.69}$$

From (7.2.29) we get the inclusion

$$A(\nu) \subset \mathrm{cl}(J^1(\Lambda_0^*(H_2^*)), \tau_*(\mathcal{L})). \tag{7.2.70}$$

Using (7.2.17) and (7.2.27) we infer that

$$J^1(\Lambda_0^*(H_2^*)) = J^1\Big(\bigcup_{f\in\Lambda(H_2^*)} \alpha_0(f)\Big) = \bigcup_{f\in\Lambda(H_2^*)} J^1(\alpha_0(f))$$
$$= \bigcup_{f\in\Lambda(H_2^*)} \{g * \eta : g \in \alpha_0(f)\}.$$

Note that by (7.2.70) the following inclusion holds:

$$A(\nu) \subset \mathrm{cl}\Big(\bigcup_{f\in\Lambda(H_2^*)} \{g * \eta : g \in \alpha_0(f)\}, \tau_*(\mathcal{L})\Big).$$

From (7.2.69) we obtain the inclusion $A(\nu) \subset \hat{H}_2^*$. Then $\mu_* \in \hat{H}_2^*$. But by the choice of $\hat{H}_1^*$ we have $\mu_* \in \hat{H}_1^*$; see definitions of Section 2.2. Hence $\hat{H}_1^* \cap \hat{H}_2^* \neq \varnothing$, which is impossible. Thus $A(\nu) \subset \tilde{A}(\nu)$. Since the choice of $\nu$ was arbitrary, we have $A \sqsubseteq \tilde{A}$. □

THEOREM 7.2.1 *The set-valued mapping $A$ is the $\sqsubseteq$-smallest element of $\mathcal{Q}$:*

$$(A \in \mathcal{Q})\ \&\ (\forall \tilde{A} \in \mathcal{Q} : A \sqsubseteq \tilde{A}).$$

The proof consists in combination of Lemmas 7.4.1 and 7.4.2. So we have the natural construction of extension of $\alpha_0$; this extension is 'correct' in the sense of inexact realization of an argument. Theorem 7.4.1 gives the very economical variant of the extension above. Remark 7.2.2 indicates the natural application of Theorem 7.4.1 to control problems with impulse constraints. Certainly, we have a game setting. This setting is more interesting in the case when employing non-anticipating set-valued mappings.

## 7.3 EXTENSION OF A NON-ANTICIPATING PROCEDURE OF CONTROL

Return to (7.2.23). We have a variant of the reaction to hindrance. Suppose that $\alpha_0$ (7.2.23) is a non-anticipating mapping. Namely, in this section we assume that $\forall f_1 \in M_{b_0}^+(\mathcal{L})\ \forall f_2 \in M_{b_0}^+(\mathcal{L})\ \forall t \in T_0$:

$$((f_1 \mid [t_0, t[) = (f_2 \mid [t_0, t[))$$
$$\Rightarrow (\{(g \mid [t_0, t[) : g \in \alpha_0(f_1)\} = (\{(g \mid [t_0, t[) : g \in \alpha_0(f_2)\}). \quad (7.3.1)$$

Similar procedures were considered in Chapter 6. Note that, in particular, the representation in terms of fixed point (see constructions based on the operator Γ) can be applied to this procedure. But we now consider only the question connected with preserving (7.3.1) for $A$. In addition, the generalized 'control' $\nu \in \Xi^+_{b_0}(\mathcal{L})$ plays the role of the hindrance control $f \in M^+_{b_0}(\mathcal{L})$.

LEMMA 7.3.1 *If* $\nu_1 \in \Xi^+_{b_0}(\mathcal{L})$, $\nu_2 \in \Xi^+_{b_0}(\mathcal{L})$, *and* $\theta \in T_0$, *then*

$$((\nu_1 \mid \mathcal{L}^{\leftarrow}_\theta) = (\nu_2 \mid \mathcal{L}^{\leftarrow}_\theta))$$

$$\Rightarrow (\forall \mu_1 \in A(\nu_1)\ \exists \mu_2 \in A(\nu_2) : (\mu_1 \mid \mathcal{L}^{\leftarrow}_\theta) = (\mu_2 \mid \mathcal{L}^{\leftarrow}_\theta)).$$

PROOF. Fix $\nu_1 \in \Xi^+_{b_0}(\mathcal{L})$, $\nu_2 \in \Xi^+_{b_0}(\mathcal{L})$, and $\theta \in T_0$ with the property

$$(\nu_1 \mid \mathcal{L}^{\leftarrow}_\theta) = (\nu_2 \mid \mathcal{L}^{\leftarrow}_\theta). \tag{7.3.2}$$

The sense of (7.3.2) corresponds to constructions of Chapter 6. Fix $\mu_1 \in A(\nu_1)$ and use Proposition 7.2.2. Choose a net $(T, \ll, r)$ in $M^+_{b_0}(\mathcal{L})$ with the property

$$(\mathcal{X}_{\nu_1} \subset (M^+_{b_0}(\mathcal{L}) - \text{ass})[T; \ll; r])$$

$$\& \left( \exists g \in \prod_{s \in T} \{h * \eta : h \in (\alpha_0 \circ r)(s)\} : (T, \ll, g) \overset{\tau_*(\mathcal{L})}{\to} \mu_1 \right). \tag{7.3.3}$$

Recall that by (7.2.34) the equality $\mathcal{X}_{\nu_1} = \Lambda^1(N_{\tau_0(\mathcal{L})}(\nu_1))$ holds. So $\mathcal{X}_{\nu_1}$ is the family of all sets $\Lambda(H)$, $H \in N_{\tau_0(\mathcal{L})}(\nu_1)$. Choose (see (7.3.3))

$$g \in \prod_{s \in T} \{h * \eta : h \in (\alpha_0 \circ r)(s)\}$$

with the following property of convergence:

$$(T, \ll, g) \overset{\tau_*(\mathcal{L})}{\to} \mu_1. \tag{7.3.4}$$

Note that for $s \in T$ we have

$$M_s \triangleq \{f \in (\alpha_0 \circ r)(s) \mid g(s) = f * \eta\} \in 2^{M^+_{a_0}(\mathcal{L})}.$$

We obtain the set-valued mapping

$$s \mapsto M_s : T \to 2^{M^+_{a_0}(\mathcal{L})}.$$

Using the axiom of choice we can state that $\prod_{s \in T} M_s \neq \varnothing$. Choose

$$\tilde{r} \in \prod_{s \in T} M_s.$$

Then $(T, \ll, \tilde{r})$ is a net in $M^+_{a0}(\mathcal{L})$. In addition, $\forall s \in T : \tilde{r}(s) \in M_s$. In particular, $\forall s \in T : \tilde{r}(s) \in (\alpha_0 \circ r)(s)$. So we have the inclusion $\tilde{r}(s) \in \alpha_0(r(s))$ for each $s \in T$. We can consider the net $(T, \ll, \tilde{r})$ as an 'approximate reaction' to the generalized 'control' $\nu_1$ (see (7.3.3)). We now construct an analogous 'reaction' to $\nu_2$. In this connection we use Proposition 7.2.4 with $\vartheta = \theta$, $\mu = \nu_1$, $\nu = \nu_2$, and $\psi = r$. It is useful to return to the first part of (7.3.3). From (7.2.34) we infer that $N_{\tau_0(\mathcal{L})}(\nu_1) \subset (\mathbf{A}(\mathcal{L}) - \text{ass})[T; \ll; J \circ r]$. As a consequence, from (2.2.48) we get the convergence

$$(T, \ll, J \circ r) \overset{\tau_0(\mathcal{L})}{\to} \nu_1.$$

So we have a variant of (7.2.49). We note that (7.3.2) is the concrete variant of (7.2.43). In (7.2.45) we use the concrete variant $\psi = r$ and $\nu = \nu_2$. Of course, for $s \in T$ and $\mathcal{K} \in \mathbf{D}(I, \mathcal{L})$ we have

$$\tilde{\psi}(s, \mathcal{K}) = r(s) \square (\Theta_\eta[\nu_2; \mathcal{K}] \mid [\theta, \vartheta_0[). \tag{7.3.5}$$

In the form of (7.3.5) we obtain some version of (7.2.45). Introduce $\preceq \in (\text{DIR})[T \times \mathbf{D}(I, \mathcal{L})]$ by the rule (7.2.46). As a result, $(T \times \mathbf{D}(I, \mathcal{L}), \preceq, \tilde{\psi})$ is a net in $B^+_0(E, \mathcal{L})$. In addition, we have two important properties. Firstly, we have the convergence

$$(T \times \mathbf{D}(I, \mathcal{L}), \preceq, J \circ \tilde{\psi}) \overset{\tau_0(\mathcal{L})}{\to} \nu_2. \tag{7.3.6}$$

Secondly, $\exists z_0 \in T \times \mathbf{D}(I, \mathcal{L})\ \forall z \in T \times \mathbf{D}(I, \mathcal{L})$:

$$(z_0 \preceq z) \Rightarrow \left( \int_I \tilde{\psi}(z) d\eta = \nu_2(I) \right). \tag{7.3.7}$$

Consider the natural combination of (7.3.6) and (7.3.7). By (7.3.5) we get $\forall s \in T\ \forall \mathcal{K} \in T \times \mathbf{D}(I, \mathcal{L})$:

$$(\tilde{\psi}(s, \mathcal{K}) \mid [\theta, \vartheta_0[) = (r(s) \mid [\theta, \vartheta_0[). \tag{7.3.8}$$

We note that $\mathbf{Z}_0 \triangleq \{z \in T \times \mathbf{D}(I, \mathcal{L}) \mid z_0 \preceq z\} \in 2^{T \times \mathbf{D}(I, \mathcal{L})}$. By (2.2.4) and (2.2.7) we have $\mathbf{Z}_0 \in (\preceq -\text{cof})[T \times \mathbf{D}(I, \mathcal{L})]$. Therefore by (2.2.8)

$$\lhd \triangleq \preceq \cap (\mathbf{Z}_0 \times \mathbf{Z}_0) \in (\text{DIR})[\mathbf{Z}_0].$$

Of course, $(\mathbf{Z}_0, \lhd)$ is a directed set and $\forall z_1 \in \mathbf{Z}_0\ \forall z_2 \in \mathbf{Z}_0$:

$$(z_1 \lhd z_2) \Leftrightarrow (z_1 \preceq z_2). \tag{7.3.9}$$

Introduce $\tilde{\psi}_0 \triangleq (\tilde{\psi} \mid \mathbf{Z}_0)$. We get in the form of $(\mathbf{Z}_0, \lhd, \tilde{\psi}_0)$ a net in $B_0^+(I, \mathcal{L})$. For $z \in \mathbf{Z}_0$ we have $\tilde{\psi}_0(z) = \tilde{\psi}(z)$. From (7.3.7) it follows that $\forall z \in \mathbf{Z}_0$:

$$(J \circ \tilde{\psi}_0)(z)(I) = J(\tilde{\psi}_0(z))(I) = (\tilde{\psi}_0(z) * \eta)(I) = \int_I \tilde{\psi}_0(z)\, d\eta = \nu_2(I).$$

From (7.2.18) we obtain the inclusion $\tilde{\psi}_0(z) \in M_{b_0}^+(\mathcal{L})$ for $z \in \mathbf{Z}_0$. So $(\mathbf{Z}_0, \lhd, \tilde{\psi}_0)$ is a net in $M_{b_0}^+(\mathcal{L})$. Taking into account the cofinality of $\mathbf{Z}_0$ and using (7.3.6) and (7.3.9), we obtain the convergence

$$(\mathbf{Z}_0, \lhd, J \circ \tilde{\psi}_0) \stackrel{\tau_0(\mathcal{L})}{\rightarrow} \nu_2. \tag{7.3.10}$$

In (7.3.10) we use the equality

$$(J \circ \tilde{\psi}_0)(z) = (J \circ \tilde{\psi})(z)$$

which is valid for $z \in \mathbf{Z}_0$. Note that in accordance with (7.2.23) we have the mapping

$$z \mapsto (\alpha_0 \circ \tilde{\psi}_0)(z) :. \mathbf{Z}_0 \to 2^{M_{a_0}^+(\mathcal{L})}.$$

For $z \in \mathbf{Z}_0$ we have $(\alpha_0 \circ \tilde{\psi}_0)(z) = \alpha_0(\tilde{\psi}_0(z))$. As usual, for $z \in T \times \mathbf{D}(I, \mathcal{L})$ we have elements $\mathrm{pr}_1(z) \in T$ and $\mathrm{pr}_2(z) \in \mathbf{D}(I, \mathcal{L})$ such that $z = (\mathrm{pr}_1(z), \mathrm{pr}_2(z))$. From (7.3.5) we get $\forall z \in T \times \mathbf{D}(I, \mathcal{L})$:

$$(\tilde{\psi}(z) \mid [t_0, \theta[) = (r(\mathrm{pr}_1(z)) \mid [t_0, \theta[).$$

In particular, the last property holds for $z \in \mathbf{Z}_0$. By (7.3.1) we have $\forall z \in \mathbf{Z}_0$:

$$\{(\tilde{f} \mid [t_0, \theta[) : \tilde{f} \in \alpha_0(r(\mathrm{pr}_1(z)))\} = \{(\tilde{g} \mid [t_0, \theta[) : \tilde{g} \in \alpha_0(\tilde{\psi}_0(z))\}. \tag{7.3.11}$$

But for any $z \in \mathbf{Z}_0$ we have $\tilde{r}(\mathrm{pr}_1(z)) \in \alpha_0(r(\mathrm{pr}_1(z)))$. From (7.3.11) it follows that $\forall z \in \mathbf{Z}_0$

$$\mho_z \triangleq \{\tilde{g} \in \alpha_0(\tilde{\psi}_0(z)) \mid (\tilde{g} \mid [t_0, \theta[) = (\tilde{r}(\mathrm{pr}_1(z)) \mid [t_0, \theta[)\} \neq \varnothing.$$

Of course, we have the property

$$(\mho_z)_{z \in \mathbf{Z}_0} \in (2^{M_{a_0}^+(\mathcal{L})})^{\mathbf{Z}_0}.$$

Using the axiom of choice we choose

$$\rho \in \prod_{z \in \mathbf{Z}_0} \mho_z. \tag{7.3.12}$$

From (7.3.12) it follows that $(\mathbf{Z}_0, \lhd, \rho)$ is a net in $M^+_{a_0}(\mathcal{L})$. So we have

$$\rho : \mathbf{Z}_0 \to M^+_{a_0}(\mathcal{L}).$$

In addition, $\rho(z) \in \mho_z$ for any $z \in \mathbf{Z}_0$. Hence $\rho(z) \in \alpha_0(\tilde{\psi}_0(z))$ for any $z \in \mathbf{Z}_0$. As a consequence we have

$$\rho \in \prod_{z \in \mathbf{Z}_0} (\alpha_0 \circ \tilde{\psi}_0)(z).$$

In addition, $\forall z \in \mathbf{Z}_0$:

$$(\rho(z) \mid [t_0, \theta[) = (\tilde{r}(\mathrm{pr}_1(z)) \mid [t_0, \theta[). \tag{7.3.13}$$

Introduce $\tilde{\rho} \triangleq J \circ \rho$. Then by (7.2.19), $(\mathbf{Z}_0, \lhd, \tilde{\rho})$ is a net in $\Xi^+_{a_0}(\mathcal{L})$. Recall that by (4.6.10) and (7.2.18)

$$\Xi^+_{a_0}(\mathcal{L}) \in (\tau_*(\mathcal{L}) - \mathrm{comp})[\mathbf{A}(\mathcal{L})]. \tag{7.3.14}$$

By combining (2.3.23) and (7.3.14) we get that $\exists_\Sigma \mathbf{S}[\Sigma \neq \varnothing]\ \exists \sqsubseteq \in (\mathrm{DIR})[\Sigma]\ \exists \sigma \in (\mathrm{Isot})[\Sigma; \sqsubseteq; \mathbf{Z}_0; \lhd]\ \exists \mu^\natural \in \Xi^+_{a_0}(\mathcal{L})$ :

$$(\Sigma, \sqsubseteq, \tilde{\rho} \circ \sigma) \overset{\tau_*(\mathcal{L})}{\to} \mu^\natural. \tag{7.3.15}$$

We fix the directed set $(\Sigma, \sqsubseteq)$ with $\Sigma \neq \varnothing$, $\sigma \in (\mathrm{Isot})[\Sigma; \sqsubseteq; \mathbf{Z}_0; \lhd]$, and $\mu^\natural \in \Xi^+_{a_0}(\mathcal{L})$ with the property (7.3.15). From (7.3.15) we get the convergence

$$(\Sigma, \sqsubseteq, J \circ \rho \circ \sigma) \overset{\tau_*(\mathcal{L})}{\to} \mu^\natural. \tag{7.3.16}$$

We supplement (7.3.16). Namely, (7.3.10) implies the convergence

$$(\Sigma, \sqsubseteq, J \circ \tilde{\psi}_0 \circ \sigma) \overset{\tau_0(\mathcal{L})}{\to} \nu_2. \tag{7.3.17}$$

Recall that $\tilde{\rho}(z) = (J \circ \rho)(z) \in \Xi^+_{a_0}(\mathcal{L})$ for $z \in \mathbf{Z}_0$. From the definition of $\rho$ it follows that $\rho(z) \in (\alpha_0 \circ \tilde{\psi}_0)(z)$ for $z \in \mathbf{Z}_0$. Hence for $z \in \mathbf{Z}_0$ we get $\tilde{\rho}(z) \in J^1((\alpha_0 \circ \tilde{\psi}_0)(z))$ or

$$\tilde{\rho}(z) \in \{h * \eta : h \in (\alpha_0 \circ \tilde{\psi}_0)(z)\}.$$

As a consequence, $\forall \xi \in \Sigma$

$$(\tilde{\rho} \circ \sigma)(\xi) = \tilde{\rho}(\sigma(\xi)) \in \{h * \eta : h \in (\alpha_0 \circ \tilde{\psi}_0 \circ \sigma)(\xi)\}.$$

Therefore we obtain the property

$$J \circ \rho \circ \sigma = \tilde{\rho} \circ \sigma \in \prod_{\xi \in \Sigma} \{h * \eta : h \in (\alpha_0 \circ (\tilde{\psi}_0 \circ \sigma))(\xi)\}.$$

Returning to (7.3.16) we get the following property:

$$\exists \hat{g} \in \prod_{\xi \in \Sigma} \{h * \eta : h \in (\alpha_0 \circ (\tilde{\psi}_0 \circ \sigma))(\xi)\} : (\Sigma, \sqsubseteq, \hat{g}) \overset{\tau_*(\mathcal{L})}{\to} \mu^{\natural}. \quad (7.3.18)$$

Choose arbitrarily $U^{\natural} \in \mathcal{X}_{\nu_2}$. Then $U^{\natural} \in \mathcal{P}(M_{b_0}^+(\mathcal{L}))$. Using (7.2.34) we choose

$$H^{\natural} \in N_{\tau_0(\mathcal{L})}(\nu_2) \quad (7.3.19)$$

for which $U^{\natural} = J^{-1}(H^{\natural}) \cap M_{b_0}^+(\mathcal{L})$. From (2.2.48), (7.3.17), and (7.3.19) it follows that

$$H^{\natural} \in (\mathbf{A}(\mathcal{L}) - \text{ass})[\Sigma; \sqsubseteq; J \circ \tilde{\psi}_0 \circ \sigma].$$

From the choice of $H^{\natural}$ (7.3.19) we have the property

$$J^{-1}(H^{\natural}) \cap M_{b_0}^+(\mathcal{L}) \in (M_{b_0}^+(\mathcal{L}) - \text{ass})[\Sigma; \sqsubseteq; \tilde{\psi}_0 \circ \sigma]. \quad (7.3.20)$$

In (7.3.20) we use the property that $(\Sigma, \sqsubseteq, \tilde{\psi}_0 \circ \sigma)$ is a net in $M_{b_0}^+(\mathcal{L})$. As a consequence we obtain

$$U^{\natural} \in (M_{b_0}^+(\mathcal{L}) - \text{ass})[\Sigma; \sqsubseteq; \tilde{\psi}_0 \circ \sigma].$$

Since the choice of $U^{\natural}$ was arbitrary, the inclusion

$$\mathcal{X}_{\nu_2} \subset (M_{b_0}^+(\mathcal{L}) - \text{ass})[\Sigma; \sqsubseteq; \tilde{\psi} \circ \sigma] \quad (7.3.21)$$

is established. By Proposition 7.2.2, (7.3.18), and (7.3.21) we have

$$\mu^{\natural} \in A(\nu_2). \quad (7.3.22)$$

From (7.3.22) we have the property $(\mu_1, \mu^{\natural}) \in A(\nu_1) \times A(\nu_2)$. Let $\Lambda \in \mathcal{L}_{\theta}^{\leftarrow}$. From (7.2.32) we obtain that $\Lambda \in \mathcal{L}$ and $\Lambda \subset [t_0, \theta[$. Then by (7.3.13) we have $\forall z \in \mathbf{Z}_0\ \forall t \in \Lambda$:

$$\rho(z)(t) = \tilde{r}(\text{pr}_1(z))(t).$$

From (4.5.8), for $z \in \mathbf{Z}_0$ we get the equalities

$$\begin{aligned}\tilde{\rho}(z)(\Lambda) &= (J \circ \rho)(z)(\Lambda) = \int_{\Lambda} \rho(z) d\eta = \int_{\Lambda} \tilde{r}(\text{pr}_1(z)) d\eta \\ &= (J \circ \tilde{r})(\text{pr}_1(z))(\Lambda).\end{aligned}$$

As a consequence we obtain $\forall \xi \in \Sigma$:

$$(\tilde{\rho} \circ \sigma)(\xi)(\Lambda) = \tilde{\rho}(\sigma(\xi))(\Lambda) = (J \circ \tilde{r})(\text{pr}_1(\sigma(\xi)))(\Lambda) = J(\tilde{r}(\text{pr}_1(\sigma(\xi))))(\Lambda). \quad (7.3.23)$$

We use the topology $\tau_{\otimes}(\mathcal{L})$ (4.6.18). From (4.6.21) and (7.3.16) we have

$$(\Sigma, \sqsubseteq, \tilde{\rho} \circ \sigma) \stackrel{\tau_{\otimes}(\mathcal{L})}{\rightarrow} \mu^{\natural}.$$

This convergence implies the property

$$(\Sigma, \sqsubseteq, (\tilde{\rho} \circ \sigma)(\cdot)(\Lambda)) \stackrel{\tau_{\mathbb{R}}}{\rightarrow} \mu^{\natural}(\Lambda). \tag{7.3.24}$$

Note that $g(s) = \tilde{r}(s) * \eta = (J \circ \tilde{r})(s)$ for $s \in T$. So $g = J \circ \tilde{r}$, and by (7.3.4)

$$(T, \ll, J \circ \tilde{r}) \stackrel{\tau_{*}(\mathcal{L})}{\rightarrow} \mu_1.$$

From (4.6.21) we have the convergence

$$(T, \ll, J \circ \tilde{r}) \stackrel{\tau_{\otimes}(\mathcal{L})}{\rightarrow} \mu_1.$$

As a result we get

$$(T, \ll, (J \circ \tilde{r})(\cdot)(\Lambda)) \stackrel{\tau_{\mathbb{R}}}{\rightarrow} \mu_1(\Lambda). \tag{7.3.25}$$

Fix $\varepsilon \in ]0, \infty[$ and choose (see (7.3.24)) $\xi_1^* \in \Sigma$ such that $\forall \xi \in \Sigma$:

$$(\xi_1^* \sqsubseteq \xi) \Rightarrow (\mid (\tilde{\rho} \circ \sigma)(\xi)(\Lambda) - \mu^{\natural}(\Lambda) \mid < \varepsilon/2). \tag{7.3.26}$$

Taking into account (7.3.25) we choose $s_1 \in T$ for which $\forall s \in T$:

$$(s_1 \ll s) \Rightarrow (\mid (J \circ \tilde{r})(s)(\Lambda) - \mu_1(\Lambda) \mid < \varepsilon/2). \tag{7.3.27}$$

To employ (7.3.27) we use $\varsigma \in T^{\Sigma}$ for which $\forall \xi \in \Sigma : \varsigma(\xi) \triangleq \mathrm{pr}_1(\sigma(\xi))$. In other words, we have

$$\varsigma = (\mathrm{pr}_1(\sigma(\xi)))_{\xi \in \Sigma} \in T^{\Sigma}.$$

From (7.3.23) we get $\forall \xi \in \Sigma$:

$$(\tilde{\rho} \circ \sigma)(\xi)(\Lambda) = J(\tilde{r}(\varsigma(\xi)))(\Lambda) = (J \circ \tilde{r} \circ \varsigma)(\xi)(\Lambda).$$

Consider $z_0$. We have $s_0 \triangleq \mathrm{pr}_1(z_0) \in T$ and $\mathcal{K}_0 \triangleq \mathrm{pr}_2(z_0) \in \mathbf{D}(I, \mathcal{L})$. Let $s^* \in T$ be such that $s_1 \ll s^*$ and $s_0 \ll s^*$. Then

$$z^* \triangleq (s^*, \mathcal{K}_0) \in T \times \mathbf{D}(I, \mathcal{L}).$$

In addition, by the definition of $\preceq$ we obtain $z_0 \preceq z^*$. So $z^* \in \mathbf{Z}_0$. We now use properties of $\sigma$. By (2.2.7) and (2.2.9) we have $z^* \lhd \sigma(\xi_2^*)$ for some $\xi_2^* \in \Sigma$. In addition, $\sigma(\xi_2^*) \in \mathbf{Z}_0$. In particular, $\sigma(\xi_2^*) \in T \times \mathbf{D}(I, \mathcal{L})$, and $z^* \preceq \sigma(\xi_2^*)$ by (7.3.9). Choose $\xi^* \in \Sigma$ such that $(\xi_1^* \sqsubseteq \xi^*)\&(\xi_2^* \sqsubseteq \xi^*)$. From (2.2.9) we have

$$(\sigma(\xi_1^*) \lhd \sigma(\xi^*))\&(\sigma(\xi_2^*) \lhd \sigma(\xi^*)).$$

By (7.3.9) we get that $\sigma(\xi_1^*) \preceq \sigma(\xi^*)$ and $\sigma(\xi_2^*) \preceq \sigma(\xi^*)$. Of course, $z^* \lhd \sigma(\xi^*)$ and hence $z^* \preceq \sigma(\xi^*)$. Then we have

$$(s^* \ll \mathrm{pr}_1(\sigma(\xi^*)))\&(\mathcal{K}_0 \prec \mathrm{pr}_2(\sigma(\xi^*))).$$

By the choice of $s^*$ we have $s_1 \ll \mathrm{pr}_1(\sigma(\xi^*))$ and $s_0 \ll \mathrm{pr}_1(\sigma(\xi^*))$. Recall that $\varsigma(\xi^*) = \mathrm{pr}_1(\sigma(\xi^*))$. Then $s_1 \ll \varsigma(\xi^*)$ and $s_0 \ll \varsigma(\xi^*)$. We have

$$(\tilde{\rho} \circ \sigma)(\xi^*)(\Lambda) = J(\tilde{r}(\varsigma(\xi^*)))(\Lambda) = (J \circ \tilde{r} \circ \varsigma)(\xi^*)(\Lambda).$$

From (7.3.27), for $\varsigma(\xi^*)$ we get the inequality

$$\mid (J \circ \tilde{r})(\varsigma(\xi^*))(\Lambda) - \mu_1(\Lambda) \mid = \mid (J \circ \tilde{r} \circ \varsigma)(\xi^*)(\Lambda) - \mu_1(\Lambda) \mid < \varepsilon/2.$$

As a consequence we deduce that

$$\mid (\tilde{\rho} \circ \sigma)(\xi^*)(\Lambda) - \mu_1(\Lambda) \mid < \varepsilon/2. \tag{7.3.28}$$

But from (7.3.26) we have the inequality

$$\mid (\tilde{\rho} \circ \sigma)(\xi^*)(\Lambda) - \mu^\natural(\Lambda) \mid < \varepsilon/2.$$

Using (7.3.28) we conclude that $\mid \mu_1(\Lambda) - \mu^\natural(\Lambda) \mid < \varepsilon$. Since the choice of $\varepsilon$ was arbitrary, we have $\mu_1(\Lambda) = \mu^\natural(\Lambda)$. Moreover, the choice of $\Lambda$ was arbitrary too. Therefore $\forall L \in \mathcal{L}_\theta^{\leftarrow} : \mu_1(L) = \mu^\natural(L)$. As a result $(\mu_1 \mid \mathcal{L}_\theta^{\leftarrow}) = (\mu^\natural \mid \mathcal{L}_\theta^{\leftarrow})$. From (7.3.22) it follows that

$$\mu^\natural \in A(\nu_2) : (\mu_1 \mid \mathcal{L}_\theta^{\leftarrow}) = (\mu^\natural \mid \mathcal{L}_\theta^{\leftarrow}).$$

But the choice of $\mu_1$ was arbitrary. Taking into account (7.3.2) we obtain that $\forall \mu_1 \in A(\nu_1)\ \exists \mu_2 \in A(\nu_2)$: $(\mu_1 \mid \mathcal{L}_\theta^{\leftarrow}) = (\mu_2 \mid \mathcal{L}_\theta^{\leftarrow})$. The lemma is proved. □

THEOREM 7.3.1 *The operator $A$ is a non-anticipating set-valued mapping:*

(1) *$A$ is an operator from $\Xi_{b_0}^+(\mathcal{L})$ into $2^{\Xi_{a_0}^+(\mathcal{L})}$;*

(2) *If $\nu_1 \in \Xi_{b_0}^+(\mathcal{L})$, $\nu_2 \in \Xi_{b_0}^+(\mathcal{L})$ and $\theta \in T_0$, then $((\nu_1 \mid \mathcal{L}_\theta^{\leftarrow}) = (\nu_2 \mid \mathcal{L}_\theta^{\leftarrow})) \Rightarrow (\{(\mu_1 \mid \mathcal{L}_\theta^{\leftarrow}) : \mu \in A(\nu_1)\} = \{(\mu_2 \mid \mathcal{L}_\theta^{\leftarrow}) : \mu \in A(\nu_2)\})$.*

The proof is a direct combination of (7.2.28), Proposition 7.2.1, and Lemma 7.3.1. Of course, we can consider $A$ in terms of fixed point for the operator of the $\Gamma$-type from the previous chapter. But the fact that $A$ is a non-anticipating extension of $\alpha_0$ is most important in this chapter. In reality, we here return to constructions of Chapter 3.

## 7.4 SOME EXAMPLES

The examples considered below are illustrative and very simple. We use the informative variant of presentation. First, we deal with the point to point reaction (7.2.23) defined by a 'counter-control'. This notion was used in differential games [88] with geometric constraints on the choice of controls. We consider the case of impulse constraints.

Suppose that $\kappa \triangleq a_0/b_0$; then $\kappa$ defines the energetic advantage of the useful control with respect to a 'hindrance'. Introduce the mapping (see (7.2.18))

$$\alpha_0' \triangleq (\kappa f)_{f\in M_{b_0}^+(\mathcal{L})} \in M_{a_0}^+(\mathcal{L})^{M_{b_0}^+(\mathcal{L})}. \tag{7.4.1}$$

So $\alpha_0' : M_{b_0}^+(\mathcal{L}) \to M_{a_0}^+(\mathcal{L})$; in addition, $\alpha_0'(f) = \kappa f$ for any $f \in M_{b_0}^+(\mathcal{L})$. Hence $\forall f \in M_{b_0}^+(\mathcal{L})\ \forall t \in I$:

$$\alpha_0'(f)(t) = \kappa f(t).$$

Then $\forall f_1 \in M_{b_0}^+(\mathcal{L})\ \forall f_2 \in M_{b_0}^+(\mathcal{L})\ \forall \theta \in T_0$:

$$((f_1 \mid [t_0,\theta[) = (f_2 \mid [t_0,\theta[)) \Rightarrow ((\alpha_0'(f_1) \mid [t_0,\theta[) = (\alpha_0'(f_2) \mid [t_0,\theta[)).$$

Suppose that $\alpha_0$ is defined by the equality $\alpha_0(f) = \{\alpha_0'(f)\}$ for $f \in M_{b_0}^+(\mathcal{L})$. Consider a concrete version of set-valued mapping $A$ (7.2.28). By Theorem 7.3.1, $A$ is a non-anticipating set-valued mapping. Fix $\nu \in \Xi_{b_0}^+(\mathcal{L})$ and $\mu \in A(\nu)$. Using Proposition 7.2.2 choose a net $(\mathbb{T}, \preceq, r)$ in $M_{b_0}^+(\mathcal{L})$ for which

$$(\mathcal{X}_\nu \subset (M_{b_0}^+(\mathcal{L}) - \mathrm{ass})[\mathbb{T}; \preceq; r])$$

$$\& \left(\exists g \in \prod_{s\in\mathbb{T}} \{h * \eta : h \in (\alpha_0 \circ r)(s)\} : (\mathbb{T}, \preceq, g) \overset{\tau_*(\mathcal{L})}{\to} \mu\right). \tag{7.4.2}$$

Note that (7.4.2) is similar to representations from Chapter 3. By (7.2.34) and (7.4.2) we have

$$N_{\tau_0(\mathcal{L})}(\nu) \subset (\mathbf{A}(\mathcal{L}) - \mathrm{ass})[\mathbb{T}; \preceq; J \circ r].$$

In other words, we obtain the convergence

$$(\mathbb{T}, \preceq, J \circ r) \overset{\tau_0(\mathcal{L})}{\to} \nu. \tag{7.4.3}$$

Moreover, $(\mathbb{T}, \preceq, J \circ r)$ is a net in $\Xi_{b_0}^+(\mathcal{L})$. In particular, we have

$$J \circ r : \mathbb{T} \to (\mathrm{add})_+[\mathcal{L}].$$

Therefore from (2.3.9), (4.6.35), and (7.4.3) we get the convergence

$$(\mathbb{T}, \preceq, J \circ r) \overset{\tau_0^+(\mathcal{L})}{\to} \nu. \tag{7.4.4}$$

Taking into account (4.6.35) and (7.4.4) we conclude that

$$(\mathbb{T}, \preceq, J \circ r) \overset{\tau_*^+(\mathcal{L})}{\to} \nu.$$

By (2.3.9) we get the obvious convergence in the TS (4.6.9)

$$(\mathbb{T}, \preceq, J \circ r) \overset{\tau_*(\mathcal{L})}{\to} \nu. \tag{7.4.5}$$

Note that always $\kappa r(s) \in B_0^+(I, \mathcal{L})$ for $s \in \mathbb{T}$. From (7.2.18) and the definition of $\kappa$ we have $\forall s \in \mathbb{T} : \kappa r(s) \in M_{a0}^+(\mathcal{L})$. As a consequence, for $s \in \mathbb{T}$ we get

$$J(\kappa r(s)) = (\kappa r(s)) * \eta = \kappa(r(s) * \eta) = \kappa J(r(s)) = \kappa \cdot (J \circ r)(s) \in \Xi_{a0}^+(\mathcal{L}).$$

Therefore $(\mathbb{T}, \preceq, (J(\kappa r(s)))_{s \in \mathbb{T}}) = (\mathbb{T}, \preceq, (\kappa \cdot (J \circ r)(s))_{s \in \mathbb{T}})$ is a net in $\Xi_{a0}^+(\mathcal{L})$. From (7.4.5) we have the convergence

$$(\mathbb{T}, \preceq, (\kappa \cdot (J \circ r)(s))_{s \in \mathbb{T}}) \overset{\tau_*(\mathcal{L})}{\to} \kappa\nu. \tag{7.4.6}$$

In (7.4.6) we use the properties of the TS (4.6.9). Namely (see (4.6.12)), this TS is a topological vector space. Therefore (7.4.6) follows from (7.4.5). Using the second part of (7.4.2) we choose the mapping

$$g \in \prod_{s \in \mathbb{T}} \{h * \eta : h \in (\alpha_0 \circ r)(s)\}$$

for which

$$(\mathbb{T}, \preceq, g) \overset{\tau_*(\mathcal{L})}{\to} \mu.$$

From (7.2.18) and (7.2.23) it follows that $g \in \Xi_{a0}^+(\mathcal{L})^{\mathbb{T}}$. Of course, $\forall s \in \mathbb{T}\ \exists h \in (\alpha_0 \circ r)(s) : g(s) = h * \eta$. Thus $\forall s \in \mathbb{T}$:

$$H_s \triangleq \{h \in (\alpha_0 \circ r)(s) \mid g(s) = h * \eta\} \in 2^{M_{a0}^+(\mathcal{L})}.$$

As a result we obtain the set-valued mapping

$$s \mapsto H_s : \mathbb{T} \to 2^{M_{a0}^+(\mathcal{L})}.$$

Fix $\tilde{s} \in \prod_{s \in \mathbb{T}} H_s$. Then $\tilde{s} \in M_{a0}^+(\mathcal{L})^{\mathbb{T}}$ and $\tilde{s}(s) \in H_s$ for $s \in \mathbb{T}$. Hence $\tilde{s} \in \prod_{s \in \mathbb{T}} (\alpha_0 \circ r)(s)$. In addition, $g(\tau) = \tilde{s}(\tau) * \eta$ for $\tau \in \mathbb{T}$. But

$\forall \tau \in \mathbb{T} : \tilde{s}(\tau) \in \alpha_0(r(\tau))$. Recall that $\alpha_0$ is defined by the point to point mapping $\alpha'_0$. Then for $\tau \in \mathbb{T}$ we have $\tilde{s}(\tau) = \alpha'_0(r(\tau)) = \kappa r(\tau)$. We get

$$\tilde{s} = (\kappa r(\tau))_{\tau \in \mathbb{T}} \in M^+_{a_0}(\mathcal{L})^{\mathbb{T}}.$$

As a result, for $\tau \in \mathbb{T}$ we have $g(\tau) = (\kappa r(\tau)) * \eta = \kappa \cdot (r(\tau) * \eta)$. Then

$$g = J \circ (\kappa r(\tau))_{\tau \in \mathbb{T}} = (\kappa \cdot (J \circ r)(s))_{s \in \mathbb{T}}.$$

From (7.4.6) we get the convergence

$$(\mathbb{T}, \preceq, g) \overset{\tau_*(\mathcal{L})}{\to} \kappa\nu. \tag{7.4.7}$$

By (7.4.7) and the choice of $g$ we have $\mu = \kappa\nu$ (recall that (4.6.9) is a Hausdorff TS). So $A(\nu) \subset \{\kappa\nu\}$ and $A(\nu) \neq \varnothing$ (see Proposition 7.2.1); as a consequence $A(\nu) = \{\kappa\nu\}$. Since the choice of $\nu$ was arbitrary, we conclude that $A$ is generated by the mapping $A'$ of the type

$$\nu \mapsto \kappa\nu : \Xi^+_{b_0}(\mathcal{L}) \to \Xi^+_{a_0}(\mathcal{L});$$

i.e., for $\nu \in \Xi^+_{b_0}(\mathcal{L})$ the equality $A(\nu) = \{A'(\nu)\}$ holds. In fact, $\alpha'_0$ generates $A'$; the latter is an extension of $\alpha'_0$.

Let us consider the second example. Here we suppose $\forall f \in M^+_{b_0}(\mathcal{L})$:

$$\alpha_0(f) \triangleq \left\{ g \in B^+_0(I, \mathcal{L}) \mid \forall \tau \in ]t_0, \vartheta_0] : \int_{[t_0,\tau[} g \, d\eta \leq \kappa \int_{[t_0,\tau[} f \, d\eta \right\}. \tag{7.4.8}$$

Of course, $\forall f \in M^+_{b_0}(\mathcal{L}) : \kappa f \in \alpha_0(f)$. Therefore for $f \in M^+_{b_0}(\mathcal{L})$ we have the property $\alpha_0(f) \neq \varnothing$, and by (7.2.18) $\forall g \in \alpha_0(f)$:

$$\int_I g \, d\eta \leq \kappa \int_I f \, d\eta \leq \kappa b_0 = a_0.$$

We again use (7.2.18). Then, in our concrete variant, (7.2.23) is valid. We note that for $f \in M^+_{b_0}(\mathcal{L})$, $g \in \alpha_0(f)$, and $\tau \in ]t_0, \vartheta_0]$ the inclusion $g\chi_{[t_0,\tau[}[I] \in \alpha_0(f)$ holds (in reality, $g\chi_{[t_0,\tau[}[I] \leqq g$).

Let us show that (7.4.8) defines a non-anticipating mapping. Fix $f_1 \in M^+_{b_0}(\mathcal{L})$, $f_2 \in M^+_{b_0}(\mathcal{L})$, and $\theta \in T_0$ for which

$$(f_1 \mid [t_0, \theta[) = (f_2 \mid [t_0, \theta[).$$

Consider the sets $G_1 \triangleq \{(g \mid [t_0, \theta[) : g \in \alpha_0(f_1)\}$ and $G_2 \triangleq \{(g|[t_0, \theta[) : g \in \alpha_0(f_2)\}$. Let $g_1 \in G_1$ and $\hat{g}_1 \in \alpha_0(f_1)$ be such that $g_1 = (\hat{g}_1|[t_0, \theta[)$. Then $\hat{g}_1 \in M^+_{a_0}(\mathcal{L})$ and

$$\forall \tau \in ]t_0, \vartheta_0] : \int_{[t_0,\tau[} \hat{g}_1 \, d\eta \leq \kappa \int_{[t_0,\tau[} f_1 \, d\eta. \tag{7.4.9}$$

Recall that $\hat{g}_1\chi_{[t_0,\theta[} \leqq \hat{g}_1$ and

$$\hat{g}_1\chi_{[t_0,\theta[} \in \alpha_0(f_1).$$

As a consequence, by (7.4.9) we have $\forall \tau \in ]t_0, \theta]$:

$$\int_{[t_0,\tau[} \hat{g}_1\chi_{[t_0,\theta[}\, d\eta \leq \kappa \int_{[t_0,\tau[} f_2\, d\eta. \tag{7.4.10}$$

On the other hand, from (7.4.10) we have $\forall \tau \in ]\theta, \vartheta_0]$:

$$\int_{[t_0,\tau[} \hat{g}_1\chi_{[t_0,\theta[}\, d\eta = \int_{[t_0,\theta[} \hat{g}_1\, d\eta \leq \kappa \int_{[t_0,\theta[} f_2\, d\eta \leq \kappa \int_{[t_0,\tau[} f_2\, d\eta.$$

Here we take into account the negativity of $f_2$. Using (7.4.10) again, we get $\forall \tau \in ]t_0, \vartheta_0]$:

$$\int_{[t_0,\tau[} \hat{g}_1\chi_{[t_0,\theta[}\, d\eta \leq \kappa \int_{[t_0,\tau[} f_2\, d\eta.$$

From (7.4.8) we have $\hat{g}_1\chi_{[t_0,\theta[} \in \alpha_0(f_2)$. But in this case $g_1 = (\hat{g}_1 \mid [t_0, \theta[) = (\hat{g}_1\chi_{[t_0,\theta[} \mid [t_0, \theta[) \in G_2$. Thus $G_1 \subset G_2$. We have established that for any $f_1 \in M_{b_0}^+(\mathcal{L})$, $f_2 \in M_{b_0}^+(\mathcal{L})$, and $\theta \in T_0$ the following implication is valid:

$$((f_1 \mid [t_0, \theta[) = (f_2 \mid [t_0, \theta[))$$

$$\Rightarrow (\{(g \mid [t_0, \theta[) : g \in \alpha_0(f_1)\} = (\{(g \mid [t_0, \theta[) : g \in \alpha_0(f_2)\}).$$

Thus (7.4.8) can be regarded as a variant of the non-anticipating expenditure of energy.

## 7.5 CONCLUSION

The main contents of the chapter is connected with the problem of 'extension' of the multi-valued quasi-strategy that is defined on spaces of integrally bounded functions. The generalized quasi-strategy obtained acts as a multi-valued mapping on spaces of finitely additive measures. Conceptually, the constructions of generalized multi-valued quasi-strategies go back to the works on MPI in the theory of differential games [5, 6, 8, 85, 86, 112]. The construction considered in Chapter 7 corresponds to [31] and, in the sense of the approach to constructing extensions, is logically connected with Chapters 3 and 4.

# Conclusion

This monograph continues a series of the authors' works devoted to the problem of constructing extensions and relaxations. It is also connected with questions related to both applied and 'pure' mathematics. In the first case it is a matter mainly of problems of control theory, including control under uncertainty. In the second case, questions of asymptotic (in essence) realization of FAM are investigated; i.e., questions of universal integrability and measurability. The idea of extension of either space of solutions in an appropriate class of FAM turned out to be useful in these questions. Sometimes, more general extension schemes, in which only a topological equipment of the space of generalized elements is essential, are discussed. The authors think that quite general constructions of Chapter 3 were rendered concrete either in a direct form or in the form of some natural analogues. In this monograph questions of extremum and related regularizations are not considered (corresponding modifications of basic constructions for this case are given in [32, 35] and many journal articles). The point is that it is not difficult to change the basic constructions for variants of perturbation of conditions of the problem which are examined in this book. Such a change allows us to obtain representations of corresponding analogues of the 'usual' extremum and to investigate the constructions of optimal generalized elements. The authors think that the nature of correct extensions is revealed, in the best way, in problems of attainability.

There are many others of the author's investigations that are adjacent to the subject of this monograph. However, they are not included here by reason of volume. For example, in this work the problem of extension with the use of $(0,1)$-measures or ultrafilters of the corresponding measurable space is not considered in detail. This problem is given as a particular case of application of the general extension scheme. Nevertheless, there are many important questions in the investigations on $(0,1)$-measures that are of interest themselves. In particular, they are concerned with the problem of the existence and the structure of non-Dirac countably additive $(0,1)$-measures. The famous measure problem (i.e., the problem of the existence of measurable cardinal numbers [71, Ch. 3], [90, Ch. IX]) and a series of constructions of the theory of Boolean algebras [111, Ch. II] have a bearing on the questions above. In this connection we note that in the works [55, 57, 59] it is established that measurable spaces admitting countably additive $(0,1)$-measures different from Dirac measures have an essential pathology of the space of bounded measurable functions. In addition, the set of non-Dirac countably additive $(0,1)$-measures itself is the difference of the closure and sequential closure of the set of Dirac measures in some zero-dimensional topologies of the space $\mathbf{A}(\mathcal{L})$ (see Chapter 4). Moreover, the set of Dirac measures turned out to be sequential closed in these topologies. It is of interest that the proof of the above mentioned representation for the set of non-Dirac countably additive (0, 1)-measures is given in [59] with the use of attraction sets (AS), which are similar to those in Chapter 3. These AS are constructed without applying compactifications. This shows that the extension methods which are given in Chapters 3 and 4 can be applied to various problems, where the relaxation of constraints can be natural or can be constructed intentionally.

Note that purely finitely additive 'uniform distributions' on the family of all subsets of the interval $[0, 1]$ which are equal to 1 on the set of all rational numbers from $[0, 1]$ form the set of power not less than continuum (see Chapter 5 and [52, 53]). In these works the connection of these purely finitely additive 'uniform distributions' on the family of all subsets of $[0, 1]$ and purely finitely additive measures on the family of all subsets of the positive integers is outlined in the terms of homomorphisms of measurable spaces (see [110, Ch. V] and [54]).

Finally, we should mention the constructions of MPI (see Chapter 6) and multi-valued quasi-strategies (Chapters 6,7). The scheme of MPI in Chapter 6 can be regarded as 'direct' in the sense of constructing control procedures—multi-valued quasi-strategies. In fact, the origin dynamical system is not used when constructing iterations in Chapter 6 (we say about problems similar to those considered in the theory of differential games); only an initial element is defined in terms of this system. Conversely, in early constructions of MPI, i.e., in 'indirect' versions [5, 6, 8], a concrete dynamical system is actively used in constructing iterations. The natural question arises of how these two very different versions of MPI are related in the case of problems that are similar to those considered in the theory of differential games. Such a comparison is given in [56, 58], where the duality of the 'direct' and 'indirect' versions of MPI is established. This duality allows us to describe one iterative sequence (direct) in terms of another simpler one (indirect). Both of these versions of MPI are based on the idea of searching for a fixed point of 'programmed' (in some sense) operator. In this connection we note another variant of the employment of MPI in [114], where generalized solutions of the Hamilton Jacobi equation, which is widely applicable in the theory of differential games and control theory, are constructed with the use of MPI.

This work was supported by the Russian Foundation for Basic Research, projects no. 00-01-00348 and no. 01–01–96450, and the International Science and Engineering Centre, project no. 1293.

# References


[1] Alexandryan, R.A. and Mirzahanyan, E.A. (1979) *General topology*. Vys. Shkola, Moscow (Russian).

[2] Belov, E.G. (1987) Some operations on the space of discontinuous functions and their representation in the class of finitely additive measures, *Cand. Sc. (Math.) Dissertation*. Ural State University, Sverdlovsk, p. 110.

[3] Belov, E.G. and Chentsov, A.G. (1987) Some properties of two-valued measures and universal integrability condition, *Mat. Zametki*, **42**, pp. 288-297 (Russian).

[4] Bourbaki, N. (1968) *General topology*. Nauka, Moscow (Russian).

[5] Chentsov, A.G.(1975) On the structure of a game problem of convergence, *Soviet Math. Dokl.*, **16**, pp. 1404 - 1408.

[6] Chentsov, A.G. (1976) On a game problem of guidance, *Soviet Math. Dokl.*, **17**, pp. 73–77.

[7] Chentsov, A.G. (1976) On a game problem of guidance with information memory, *Soviet Math. Dokl.*, **17**, pp. 411–414.

[8] Chentsov, A.G. (1976) On a game problem of converging at a given instant of time, *Math. USSR Sbornik*, **28**, pp. 353–376.

[9] Chentsov, A.G. (1976) An example of an irregular differential game, *J. Appl. Math. Mech.*, **40**, no. 6.

[10] Chentsov, A.G. (1976) On formalization of a differential game, *Trudy IMM UNTS AN SSSR*, **19**, Sverdlovsk(Russian).

[11] Chentsov, A.G. (1978) On a game problem of converging at a given instant of time, *Izv. Acad. Nauk USSR, ser. Matematika*, **42**, no. 2 (Russian).

[12] Chentsov, A.G. (1978) Iterative program construction for a differential game with a fixed terminal time, *Dokl. Acad. Nauk SSSR* , **240**, no. 1 (Russian).

[13] Chentsov, A.G. (1978) *Selectors of multi-valued quasi-strategies in differential games*, Dep. in VINITI, no. 3101-78, Sverdlovsk (Russian).

[14] Chentsov, A.G. (1979) *The programmed iteration method for a differential pursuit evasion game,* Dep. in VINITI, no. 1933-79, Sverdlovsk, (Russian).

[15] Chentsov, A.G. (1980) *On differential games with restriction to the number of corrections, I.* Dep. in VINITI, no. 5272-80, Sverdlovsk (Russian).

[16] Chentsov, A.G. (1980) *On differential games with restriction to the number of corrections, II.* Dep. in VINITI, no. 5406-80, Sverdlovsk (Russian).

[17] Chentsov, A.G. (1981) *The structure of differential games with restriction to the number of program corrections, I.* Dep. in VINITI, no. 2772-81, Sverdlovsk (Russian).

[18] Chentsov, A.G. (1981) *On differential games with restriction to the number of program corrections, II.* Dep. in VINITI, no. 2898-81, Sverdlovsk (Russian).

[19] Chentsov, A.G. (1983) *An order structure of scalar finitely additive measures.* Dep. in VINITI, no. 5690-83, Sverdlovsk (Russian).

[20] Chentsov, A.G. (1985) *Finitely additive measures and integrals (theory and applications).* Dep. in VINITI, no. 1143-85, Sverdlovsk (Russian).

[21] Chentsov, A.G. (1985) *Applications of measure theory to control problems.* Sredne-Ural. kn. izd., Sverdlovsk (Russian).

[22] Chentsov, A.G. (1986) On the question of universal integrability of bounded functions, *Mat. sbornik*, **131**, pp. 73–93 (Russian); transl. in *Math. USSR Sb.* (1988) **59**, no. 1.

[23] Chentsov, A.G. (1987) *Infinitive products of additive set functions.* Dep. in VINITI, no. 3728-87, Sverdlovsk, 1987 (Russian).

[24] Chentsov, A.G. (1988) Finitely additive measures and problems on the minimum, *Cybernetics*, no. 3, pp. 353–757.

[25] Chentsov, A.G. (1988) Two-valued measures on a semialgebra of sets and some of their applications to infinite-dimensional problems in mathematical programming, *Cybernetics*, **24**, no. 6, pp. 767–773.

[26] Chentsov, A.G. (1990) The Darboux integral with respect to a finitely additive measure of bounded variation. In: *Control problems*

*and some questions of non-smooth optimization*, UrO AN SSSR, Sverdlovsk (Russian).

[27] Chentsov, A.G. and Morina, S.I. (1994) A problem of asymptotic optimization. *Vestnik Chelyabinsk. Univ.*, **1**, pp. 80–86 (Russian).

[28] Chentsov, A. and Morina, S. (1994) Asymptotically attainable elements under the perturbation of functional constraints and conditions of their bounded realization. *Functional differential equations*, **2**, pp. 23–37.

[29] Chentsov, A.G. (1995) Asymptotic attainability under a perturbation of integral restrictions in an abstract control problem, 1. *Izvestiya Vuzov. Matematika.*, **2**, pp. 60–71 (Russian).

[30] Chentsov, A.G. (1995) Asymptotic attainability under a perturbation of integral restrictions in an abstract control problem, 2. *Izvestiya Vuzov. Matematika.*, **3**, pp. 62–73 (Russian).

[31] Chentsov, A. and Savinova, L.(1995) To the question of realizability of some set-valued mappings, *Functional Differential Equations*, **3**, no. 1–2, pp. 45–67.

[32] Chentsov, A.G. (1996) *Finitely additive measures and relaxations of extremal problems.* Plenum, New York.

[33] Chentsov, A.G. and Pak, V.E. (1996) On the extension of the nonlinear problem of optimal control with non-stationary phase restrictions, *NATMA*, **26**, no. 2, pp. 383–394.

[34] Chentsov, A.G. and Serov, V.P. (1996) Representation for some set-theoretic limits in the class of two-valued finitely additive measures, *Functional Differential Equations*, **3**, no. 3–4, pp. 265–278.

[35] Chentsov, A.G. (1997) *Asymptotic attainability.* Kluwer, Dordrecht.

[36] Chentsov, A.G. (1997) The iterative realization of non-anticipating set-valued mappings, *Dokl. Akad. Nauk,* **357**, pp. 595–598 (Russian).

[37] Chentsov, A.G. (1997) The program iteration method in the class of finitely additive control measures, *Differential Equations*, no. 11, pp. 1528-1536 (Russian).

[38] Chentsov, A.G. and Podorozhnyi, D. (1997) On the question of representation for attraction sets in the class of two-valued

measures, *Vestnik Permsk. Univ. Functional differential equetions.*, no. 4, pp.132–138 (Russian).

[39] Chentsov, A.G. (1998) Universal asymptotic realization of integral constraints and extension constructions in the class of finitely additive measures. *Trudy IMM UrO RAN*, **5**, pp. 328-356 (Russian).

[40] Chentsov, A.G. (1998) Universal properties of generalized integral constraints in the class of finitely additive measures, *Functional Differential Equations*, **5**, no. 1–2, pp. 69–105.

[41] Chentsov, A.G. (1998) Relaxation of integral constraints in the class of vector finitely-additive maesures, *Dokl. Akad. Nauk*, **358**, no. 5, pp.609–613 (Russian).

[42] Chentsov, A.G. (1998) On the question of a parallel version of the abstract analogue of the programmed iteration method, *Dokl. Akad. Nauk*, **362**, no. 5, pp.602–605 (Russian).

[43] Chentsov, A.G. (1998) Non-anticipating selectors of set-valued mappings, *Electronic Journal: Differential Equations and Control Processes*, http://www.neva.ru/journal, no. 2 (Russian).

[44] Chentsov, A.G. (1999) The representation of attraction sets arising when subsequent relaxing constraints, *Dokl. Akad. Nauk*, **365**, no. 2, pp.174–177 (Russian).

[45] Chentsov, A.G. (1999) Well posed extensions of unstable control problems with integral constraints, *Izvestiya: Mathematics*, **63**:3, pp. 599–630.

[46] Chentsov, A.G. (1999) On the question of correct extension of a problem on the choice of the probability density under restrictions on a system of mathematical expectation, *Usp. Mat. Nauk*, **50**, no. 5, pp. 232–242 (Russian).

[47] Chentsov, A.G. (1999) Finitely additive measures and problems of asymptotic analysis, *Non-smooth and discontinuos problems of control and optimization. Proc. volume from the IFAC Workshop, Chelyabinsk, Russia, 17-20 June 1998*, pp.1–12, Pergamon.

[48] Chentsov, A.G. (1999) Non-anticipating selectors of set-valued mapping and iterated procedures, *Functional Differential Equations*, **6**, no. 3–4, pp. 249–274.

[49] Chentsov A.G., Morina S.I., and Zobnin B.B. (1999) On some constructions of control by systems with a varying structure, *Mathematics and Computers in Simulation*, **49**, pp. 319-334.

[50] Chentsov, A.G. (2000) On the question of iterated realization of non-anticipating selectors of set-valued mappings, *Izvestiya Vuzov. Matematika*, **454**, pp. 66–76 (Russian).

[51] Chentsov, A.G. (2000) Topological constructions of extensions and representations of attraction sets, *Proceedings of the Steklov Institute of Mathematics, Suppl. Issue 1*, pp. 35-60.

[52] Chentsov, A.G. and Morina, S.I. (2000) On analogues of uniform distribution in the class of finitely additive probabilities, *Dokl. Akad. Nauk*, **370**, no. 6, pp.741–744 (Russian).

[53] Chentsov, A.G. and Morina, S.I. (2000) On the extension of the length function in the class of finitely additive probabilities, *Functional Differential Equations*, **7**, no. 1–2, pp. 83–98.

[54] Chentsov, A.G. (2000) Some properties of finitely additive measures and their transformation on the base of homomorphisms of measurable structures, *Dokl. Akad. Nauk*, **370**, no. 4, pp.449-452 (Russian).

[55] Chentsov, A.G. (2000) Some properties of two-valued measures and representations of limits with respect to filter, *Dokl. Akad. Nauk*, **370**, no. 5, pp.595-598 (Russian).

[56] Chentsov, A.G. (2000) On consistency of different versions of the method of programmed iterations, *Dokl. Akad. Nauk*, **372**, no. 5, pp.600-603 (Russian).

[57] Chentsov, A.G. (2000) Two-valued measures as generalized elements: the problem of extension for a system of conditions, *Dokl. Akad. Nauk*, **374**, no. 5, pp.611-614 (Russian).

[58] Chentsov A.G. (2001) Topological variant of the program iteration method in an abstract control problem, *Dokl. Akad. Nauk*, **376**, no. 3, pp.311–314 (Russian).

[59] Chentsov A.G. (2001) On the representation of the set of countably additive (0,1)-measures different from Dirac measures. *Dokl. Akad. Nauk*, **377**, no. 3, pp.313–316 (Russian).

[60] Chentsov, A.G. (2001) Non-anticipating multi-valued mappings and constructing them by means of the method of programmed iterations, I. *Differential Equations*, no. 4, pp. 470–480 (Russian).

[61] Chentsov, A.G. (2001) Non-anticipating multi-valued mappings and constructing them by means of the method of programmed iterations, II. *Differential Equations*, no. 5, pp. 679–688 (Russian).

[62] Chistyakov, S.V. (1977) To the solution of game problem of pursuit. *Priklad. Mat. i Mech.*, **41** (Russian).

[63] Devis, (1993) *Applied nonstandard analysis*. John Wiley & Sons, New York.

[64] Dubins, L.E and Savage, L. (1965) *How to gamble if you must*. McGraw-Hill, New York.

[65] Duffin, R.J. (1956) Linear inequalities and related systems, *Ann. of Math. Studies*, **38**, pp. 157–170.

[66] Dunford, N and Schwartz, J.T. (1958) *Linear operators. Vol. 1*. Interscience, New York.

[67] Dyatlov, V.P. and Chentsov, A.G. (1987) Monotone iterations of sets and their applications to game-theoretic control problems, *Kibernetika*, no. 2, pp. 92–99 (Russian).

[68] Dyatlov, V.P. and Chentsov, A.G. (1987) Control with flexible corrections under resriction to the number of switchings, *Gagarinskie nauchnye chteniya po kosmonavtike i aviatsii*, pp. 70–74, Moscow: Nauka (Russian).

[69] Dyatlov, V.P. (1984) Determining the iterations in a differential game with resriction to the number of switchings, In: *Algorithms and programs for solving linear differential games*, pp. 81–102, UNTS AN SSSR, Sverdlovsk (Russian).

[70] Elliot, R.J. and Kalton, N.J. (1972) The existence of value in differential games, *Memoirs of the Amer. Math. Soc.*, no. 126.

[71] Engelking, R. (1977) *General topology*. PWN, Warszawa.

[72] Fikhtengol'ts, G.M. and Kantorovich, L.V. (1934) Sur les opérations linéaires dans l'espace des fonctions bornées, *Studia Math.*, **5**, pp. 69–98.

[73] de Finetti, B. (1937) La prévision: ses lois logigues, ses sources subjectives, *Annales de l'Institut Henri Poincaré*, **7**, pp. 1–68.

[74] Fishburn, P.(1970) *Utility theory for decision making.* Wiley, New York.

[75] Halmos, P. (1950) *Measure theory.* New-York.

[76] Hildebrandt, T.N. (1934) On bounded functional operations, *Trans. Amer. Math. Soc.*, **36**, pp. 868–875.

[77] Isaacs, R. (1965)*Differential Games.* Wiley, New York.

[78] Gamkrelidze, R.V. (1977) *Foundations of optimal control theory.* Izdat. Tbil. Univ., Tbilissi (Russian).

[79] Gol'stein, E.G. (1971) *Duality theory in mathematical programming and its applications.* Nauka, Moscow (Russian).

[80] Kantorovich, L.V. and Akilov, G.P. (1977) *Functional analysis.* Nauka, Moscow (Russian).

[81] Kelley, J.L. (1957) *General Topology.* Van Nostrand, Prnceton, NJ.

[82] Kolmogorov, A.N. and Fomin, S.V. (1981) *Elements of the function theory and of functionalal analisys.* Nauka, Moscow (Russian).

[83] Krasovskii, N.N (1968) *The theory of the control of motion.* Nauka, Moscow (Russian).

[84] Krasovskii, N.N (1970) *Game problems on motion encounter.* Nauka, Moscow (Russian).

[85] Krasovskii, N.N. and Chentsov, A.G. (1977) On the design of differential games, I, *Problems of Control and Information Theory,* **6**, pp. 381–395.

[86] Krasovskii, N.N. and Chentsov, A.G. (1977) On the design of differential games, II, *Problems of Control and Information Theory,* **8**, pp. 3–11.

[87] Krasovskii, N.N. (1985) *Dynamic system control. Problem of the minimum of guaranteed result.* Nauka, Moscow (Russian).

[88] Krasovskii, N.N. and Subbotin, A.I (1988) *Game-theoretical control problems.* Springer Verlag, Berlin.

[89] Kuratowski, K. (1966) *Topology, vol. 1.* Academic Press, New-York.

[90] Kuratowski, K. and Mostowski, A. (1967) *Set theory.* North-Holland, Amsterdam.

[91] Kurzhanskii, A.B. (1987) *Control and observation under conditions of difiniteness*. Nauka, Moscow (Russian).

[92] Leader, S. (1955) On universally integrable functions. *Proc. Amer. Math. Soc.*, **6**, pp. 232–234.

[93] Lipecki, Z. (1998) Quasi-measures with finitely or countably many extreme extensions. *Manuscripta Math.*, **97**, pp. 469–481.

[94] Maharam, D. (1976) Finitely additive measures on the integers. *Sankhyā Ser. A.*, **38**, pp. 44–59.

[95] Maharam, D. (1988) Jordan fields and improper integrals. *J. Math. Anal. Appl.*, **133**, pp. 163–194.

[96] Maynard, H. (1979) A Radon Nikodym theorem for finitely additive bounded measures, *Pacific J. Math.*, **83**, pp. 401–413.

[97] Melentsov, A.A., Baidosov, V.A. and Zmeev, G.M. (1980) *Elements of measure theory and integral (handbook)*. Ural State Univ., Sverdlovsk, p. 100 (Russian).

[98] Morina, S.I. and Chentsov, A.G. (1995) Bounded realization of asymptotically attainable elements. *J. Appl. Maths. Mechs.*, **59**, no. 6, pp. 951–966.

[99] Morina, S.I. and Chentsov, A.G. (1997) The extension of control problems in the class of vector finitely additive measures, *Avtomatika i Telemekhanika*, no. 7, pp. 207–216.

[100] Morina, S.I. (1999) On approximation of asymptotic attainability domains, *Nonsmooth and discontinuos problems of control and optimization. Proc. volume from the IFAC Workshop, Chelyabinsk, Russia, 17-20 June 1998*, pp.165–170, Pergamon.

[101] Morina, S.I. (2000) The procedure of chosing correction instants in a contol problem with restriction on the number of switchings, *Avtomatika i Telemekhanika*, no. 10, pp. 35–47.

[102] Natanson, I.P. (1974) *Theory of Real Variable Functions*. Nauka, Moscow (Russian).

[103] Neveu, J. (1964) *Bases mathématiques du calcul des probabilités*. Masson, Paris.

[104] Nikaido, H. (1968) *Convex structures and economic theory*. Academic Press, New York.

[105] Pak, V.E. and Chentsov, A.G. (1994) Regularization of the function of asymptotic value for a control problem under perturbation of the set of initial positions, *Differential Equations*, **30**, no. 11 (Russian).

[106] Rao, K.P.S.B. and Rao, M.B. (1983) *Theory of charges. A study of finitely additive measures*. Academic Press, London.

[107] Reed, M. and Simon, B. (1972) *Methods of modern mathematical phisics. Vol. 1. Functional analysis.* Academic Press, New York.

[108] Roxin, E. (1969) Axiomatic approach in differential games. *J.Optim. Theory and Appl.*, **3**, pp.153–163.

[109] Schaefer, H. (1966) *Topological vector spaces*. Macmillan, New York.

[110] Semadeni, Z. (1971) *Banach spaces of continuous functions.* PWN, Warszawa.

[111] Sikorski, P. (1964) *Boolean algebras*. Springer Verlag, Berlin.

[112] Subbotin, A.I. and Chentsov, A.G. (1981) *Optimization of guarantee in control problems*. Nauka, Moscow (Russian).

[113] Subbotin, A.I. and Chentsov, A.G. (1996) Iteration procedure for construction of minimax and viscosity solutions of Hamilton Jacobi equations, *Dokl. Acad. Nauk*, **348**, pp. 736–739 (Russian).

[114] Subbotin, A.I. and Chentsov, A.G. (1999) Iteration procedure for construction of minimax and viscosity solutions of Hamilton Jacobi and its generalization, *Proceedings of Steklov's institute*, **224**, pp. 311–334.

[115] Varaiya, P. and Lin, J. (1969) Existence of saddle points in differential games, *SIAM J. Control*, **7**, pp. 141–157.

[116] Vladimirov, D.A. (1969) *Boolean algebras*. Nauka, Moscow (Russian).

[117] Warga, J. (1972) *Optimal control of differential and functional equations*. Academic Press, New York.

[118] Yosida, K. and Hewitt, E.H. (1952) Finitely additive measures, *Trans. Amer. Soc.*, **72**, pp. 44-66.

[119] Yosida, K. (1965) *Functional analysis.* Springer Verlag, Berlin.

[120] Young, L.C. (1969) *Lectures on the calculus of variations and optimal control theory*. Saunders, Philadelphia, Pa.

[121] Zhdanok, A. (1983 ) The Gel'fand compactification and two-valued measures. In: *Topological spaces and their mappings.* Riga.

# Notation

| | |
|---|---|
| $\mathcal{P}(H)$ | –family of all subsets of the set $H$; |
| $2^H$ | –family of all nonempty subsets of the set $H$; |
| $\mathrm{Fin}(H)$ | –family of all nonempty finite subsets of $H$; |
| $\mathrm{Fin}[H \mid P]$ | –family of all sets $Q \in \mathrm{Fin}(H)$ with the property $P \subset Q$, where $P \in \mathrm{Fin}(H)$; |
| $\mathcal{N}$ | –positive integers ($\mathcal{N} \triangleq \{1; 2; \ldots\}$); |
| $\overline{1, m}$ | –set of all numbers $k \in \mathcal{N}$ such that $k \leq m$; |
| $\overrightarrow{m, \infty}$ | –set of all numbers $k \in \mathcal{N}$ such that $m \leq k$; |
| $(\mathrm{Ord})[T]$ | –set of all pre-orders on $T$ (i.e., reflexive and transitive binary relations on $T$); |
| $(\mathrm{DIR})[T]$ | –set of all directions on the set $T$; |
| $(\mathrm{Isot})[A; \ll; B; \preceq]$ | –set of all isotone operators from $(A, \ll)$ into $(B, \preceq)$ with the property of cofinality of the image ($(A, \ll)$ and $(B, \preceq)$ are directed sets); |
| $(\mathrm{top})[X]$ | –set of all topologies of the set $X$; |
| $(\mathrm{BAS})[X]$ | –set of all topological bases of $X$; |
| $(\mathbf{s} - \mathrm{BAS})[X]$ | –set of all topological subbases of $X$; |
| $(\tau - \mathrm{BAS})_0[X]$ | –set of all bases of the topology $\tau$; |
| $N^0_\tau(x), N_\tau(x)$ | –family of all open and arbitrary neighborhoods of a point $x$, resp.; |
| $\mathbb{N}^0_\tau[M], \mathbb{N}_\tau[M]$ | –family of all open and arbitrary neighborhoods of a set $M$, resp.; |
| $\mathcal{F}_\tau$ | –family of all closed (relative to the topology $\tau$) sets; |
| $\mathrm{cl}(\cdot, \tau)$ | –closure operator relative to the topology $\tau$; |
| $(\tau - \mathrm{LIM})[\mathcal{U} \mid f]$ | $= \bigcap_{U \in \mathcal{U}} \mathrm{cl}(f^1(U), \tau)$ – set-theoretic limit of sets of the family $\mathcal{U}$ under imbedding via the mapping $f$ into a space with the topology $\tau$; |
| $(\mathcal{D} - \mathrm{top})[X]$ | –set of all attainable topologies of $X$; |
| $(\mathrm{top})_0[X]$ | –set of all Hausdorff topologies of $X$; |
| $(\mathrm{top})^0[X]$ | –set of all regular topologies of $X$; |
| $(\mathbf{n} - \mathrm{top})[X]$ | –set of all topologies turning $(X, \tau)$ into the normal space; |
| $(\mathbf{c} - \mathrm{top})[X]$ | –set of all 'compact' topologies of $X$; |
| $(\mathbf{c}_{\mathcal{N}} - \mathrm{top})[X]$ | –set of all countably compact topologies of $X$; |
| $(\mathbf{c}_{\mathrm{seq}} - \mathrm{top})[X]$ | –set of all sequentially compact topologies of $X$; |
| $(\mathrm{top})_I[X]$ | –set of all topologies of $X$ with the first axiom of countability; |

| | |
|---|---|
| $(0-\text{top})[X]$ | –set of all zero-dimensional topologies of $X$; |
| $\otimes^X(\tau)$ | –topology of the Tichonoff product of $X$ samples of a space with the topology $\tau$; |
| $\otimes^m[\tau]$ | –topology of the product of $m$ samples of a space with the topology $\tau$ (here $m \in \mathcal{N}$); |
| $(\tau-\text{comp})[X]$ | –family of all compact subsets of $X$ in the sense of the topology $\tau \in (\text{top})[X]$; |
| $(\mathcal{N}-\tau-\text{comp})[X]$ | –family of all countably compact subsets of $X$ in the sense of the topology $\tau \in (\text{top})[X]$; |
| $(\tau-\text{comp}_{\text{seq}})[X]$ | –family of all sequentially compact subsets of $X$ in the sense of the topology $\tau \in (\text{top})[X]$; |
| $(\text{seqcl})[A;\tau]$ | –sequential closure of a set $A$ relative to the topology $\tau$; |
| $\mathcal{F}_{\text{seq}}[\tau]$ | –family of all sequentially closed sets relative to the topology $\tau$; |
| $\mathbf{Z}[\mathcal{H}]$ | –set of all centered subfamilies of a family $\mathcal{H}$; |
| $\mathcal{B}[X]$ | –set of all families $\mathcal{X}$ of subsets of $X$ such that $\forall A \in \mathcal{X}\ \forall B \in \mathcal{X}\ \exists C \in \mathcal{X} : C \subset A \cap B$; |
| $\mathcal{B}_0[X]$ | –set of all filter bases on $X$ (i.e., $\mathcal{B}_0[X] = \{\mathcal{X} \in \mathcal{B}[X] \mid \emptyset \notin \mathcal{X}\}$); |
| $\mathcal{B}_{\mathcal{N}}[X]$ | –set of all families from $\mathcal{B}[X]$ with a countable determining subsystem; |
| $\mathbb{F}^*(\mathcal{X})$ | –family of all $\mathcal{X}$-filters of $X$; |
| $\mathbb{F}_0^*(\mathcal{X})$ | –family of all ultrafilters of the space $(X,\mathcal{X})$; |
| $(X-\text{ass})[\mathbf{D};\preceq;h]$ | – the filter associated with the net $(\mathbf{D},\preceq,h)$; |
| $\pi[X]$ | –set of all multiplicative families of subsets of $X$; |
| $\Pi[X]$ | –set of all semi-algebras of subsets of $X$; |
| $(\text{alg})[X]$ | –set of all algebras of subsets of $X$; |
| $(\sigma-\text{alg})[X]$ | –set of all $\sigma$-algebras of subsets of $X$; |
| $a_X^0(\mathcal{X})$ | –the algebra generated by a family $\mathcal{X}$; |
| $\sigma_X^0(\mathcal{X})$ | –the $\sigma$-algebra generated by a family $\mathcal{X}$; |
| $C(X,\tau,Y,\vartheta)$ | –set of all continuous mappings from $(X,\tau)$ into $(Y,\vartheta)$; $\tau$ and $\vartheta$ are topologies of $X$ and $Y$, resp.; |
| $C_{\text{cl}}(X,\tau,Y,\vartheta)$ | –set of all closed mappings from $(X,\tau)$ into $(Y,\vartheta)$; |
| $C_{\mathbf{ap}}(X,\tau,Y,\vartheta)$ | –set of all almost perfect mappings from $(X,\tau)$ into $(Y,\vartheta)$; |
| $C_{\mathbf{qp}}(X,\tau,Y,\vartheta)$ | –set of all quasi-perfect mappings from $(X,\tau)$ into $(Y,\vartheta)$; |
| $((\text{Meas}))[X;\mathcal{X};Y;\mathcal{Y}]$ | –set of all measurable mappings from the measure space $(X,\mathcal{X})$ into the measure space $(Y,\mathcal{Y})$; |
| $(\mathbf{p}-\text{Dist})[X]$ | –set of all pseudo-metrics of the set $X$; |
| $(\text{Dist})[X]$ | –set of all metrics of the set $X$; |

| | |
|---|---|
| $\mathbf{B}_\rho(x,\varepsilon), \mathbf{B}^0_\rho(x,\varepsilon)$ | –closed and open ball in a pseudo-metric space; |
| $\mathbf{B}_\rho[A,\varepsilon], \mathbf{B}^0_\rho[A,\varepsilon]$ | –closed and open $\varepsilon$-neighborhood of a set $A$ in a pseudo-metric space; |
| $(\text{add})[\mathcal{L}], (\sigma-\text{add})[\mathcal{L}]$ | –set of all finitely and countably additive measures (FAM and CAM) on $\mathcal{L}$, resp.; |
| $(\text{add})_+[\mathcal{L}], (\sigma-\text{add})_+[\mathcal{L}]$ | –set of all non-negative FAM and CAM, resp.; |
| $\mathbf{A}(\mathcal{L})$ | –set of all FAM of bounded variation |
| $v_\mu$ | –variation of $\mu \in \mathbf{A}(\mathcal{L})$ (as function of sets); |
| $(\mathbf{p}-\text{add})_+[\mathcal{L}]$ | –set of all nonnegative purely FAM on $\mathcal{L}$; |
| $\mathbb{P}(\mathcal{L}), \mathbb{P}_\sigma(\mathcal{L})$ | –set of all finitely and countably additive probabilities on $\mathcal{L}$, resp.; |
| $\mathbb{P}^0_\sigma(\mathcal{L})$ | –set of all finitely distributed probabilities $\mu = \sum_{i=1}^n \alpha_i(\delta_{x_i} \mid \mathcal{L})$; |
| $\mathbb{T}(\mathcal{L}), \mathbb{T}_\sigma(\mathcal{L}), \mathbb{T}_\mathbf{p}(\mathcal{L})$ | –set of all two-valued finitely, countably, and purely finitely additive probabilities on $\mathcal{L}$, resp.; |
| $\mathbb{B}(X)$ | –set of all bounded functionals on $X$; |
| $f^1(\cdot), f^{-1}(\cdot)$ | –image and inverse image operations, resp.; |
| $B_0(E,\mathcal{L})$ | –linear manifold of $\mathcal{L}$-step functionals on $E$; |
| $B(E,\mathcal{L})$ | –closure of $B_0(E,\mathcal{L})$ in the sense of sup-norm $\|\cdot\|$ of the space $\mathbb{B}(E)$; |
| $U_b(\mathcal{L})$ | $= \{\mu \in \mathbf{A}(\mathcal{L}) \mid v_\mu(E) \le b\}$, where $b \ge 0$; |
| $\tau_*(\mathcal{L})$ | –$*$-weak topology of the set $\mathbf{A}(\mathcal{L})$ considered as the space topologically conjugate to $B(E,\mathcal{L})$; |
| $\tau_\otimes(\mathcal{L}) = \otimes^\mathcal{L}(\tau_\mathbb{R}) \mid_{\mathbf{A}(\mathcal{L})}$ | –where $\tau_\mathbb{R}$ is the ordinary $\lvert\cdot\rvert$-topology of $\mathbb{R}$; |
| $\tau_0(\mathcal{L}) = \otimes^\mathcal{L}(\tau_d) \mid_{\mathbf{A}(\mathcal{L})}$ | –where $\tau_d$ is the discrete topology of $\mathbb{R}$; |
| $\tau^+_*(\mathcal{L}), \tau^+_0(\mathcal{L}), \tau^+_\otimes(\mathcal{L})$ | –relative topologies of $(\text{add})_+[\mathcal{L}]$; |
| $\tau^*_\mathbb{B}(\mathcal{L})$ | –bounded $*$-weak topology of $\mathbf{A}(\mathcal{L})$; |
| $(\text{add})^+[\mathcal{L};\eta]$ | –set of all $\mu \in (\text{add})_+[\mathcal{L}]$ such that $(\eta(L)=0) \Rightarrow (\mu(L)=0)$, where $\eta \in (\text{add})_+[\mathcal{L}]$; |
| $\mathbf{A}_\eta[\mathcal{L}]$ | $= \{\mu \in \mathbf{A}(\mathcal{L}) \mid v_\mu \in (\text{add})^+[\mathcal{L};\eta]\}$— linear subspace of $\mathbf{A}(\mathcal{L})$ generated by the cone $(\text{add})^+[\mathcal{L};\eta]$; |
| $\mathcal{M}[T;Z]$ | –set of all multi-valued mappings from $T$ into $Z$; |
| $\Omega_0(\omega \mid A)$ | –set of all $\tilde{\omega} \in \Omega$ with the property $(\omega \mid A) = (\tilde{\omega} \mid A)$; |
| $Z_0(z \mid A)$ | –set of all $\tilde{z} \in Z$ with the property $(z \mid A) = (\tilde{z} \mid A)$; |
| $\gamma_T$ | –operator from $\mathcal{M}[T;Z]$, where $T \subset \Omega$; |
| $\Gamma$ | $= \gamma_\Omega$; |
| $\mathbb{N}$ | –set of all fixed points of $\Gamma$; |
| $S_T(\alpha)$ | –set of all multi-valued mappings $\beta$ from $T$ into $Z$ such that $\forall t \in T : \beta(t) \subset \alpha(t)$; |

| | |
|---|---|
| $S[\alpha]$ | $= S_\Omega(\alpha)$; |
| $\mathbb{N}_0[\alpha]$ | $= \mathbb{N} \cap S[\alpha]$; |
| $(\mathrm{na})[\alpha]$ | –greatest hereditary multi-selector of $\alpha$; |
| $\square_{U \in \mathcal{U}} \alpha_U$ | –multi-valued mapping which is agglutinated from the fragments $\alpha_U, U \in \mathcal{U}$; |
| $\alpha_0$ | –multi-valued mapping on spaces of integrally bounded controls defined on the interval $I = [t_0, \vartheta_0[,\ t_0 < \vartheta_0$; |
| $A$ | –multi-valued mapping on spaces of non-negative FAM; |
| $\mathcal{L}_t^{\leftarrow}$ | $= \{\Lambda \in \mathcal{L} \mid \Lambda \subset [t_0, t[\}$; |
| $\mathcal{L}_t^{\rightarrow}$ | $= \{\Lambda \in \mathcal{L} \mid \Lambda \subset [t, \vartheta_0[\}$; |
| $\mathcal{Q}$ | –set of all closed multi-valued mappings from $\Sigma_{b_0}^{+}(\mathcal{L})$ into $\Sigma_{a_0}^{+}(\mathcal{L})$ with the property (7.2.61). |

# Index